Sediments, Diagenesis, and Sedimentary Rocks

Citations

Please use the following example for citations:

Li Y.-H. and Schoonmaker J.E. (2003) Chemical composition and mineralogy of marine sediments, pp. 1–35. In *Sediments, Diagenesis, and Sedimentary Rocks* (ed. F.T. Mackenzie) Vol. 7 *Treatise on Geochemistry* (eds. H.D. Holland and K.K. Turekian), Elsevier–Pergamon, Oxford.

Cover photo: Paleozoic section overlying Precambrian at the Grand Canyon: looking eastward toward Hermit's Rest, Grand Canyon, USA. (Photograph provided by William Graustein)

Sediments, Diagenesis, and Sedimentary Rocks

Edited by

F. T. Mackenzie
University of Hawaii, HI, USA

TREATISE ON GEOCHEMISTRY
Volume 7

Executive Editors

H. D. Holland
Harvard University, Cambridge, MA, USA

and

K. K. Turekian
Yale University, New Haven, CT, USA

ELSEVIER

2005

AMSTERDAM – BOSTON – HEIDELBERG – LONDON – NEW YORK – OXFORD
PARIS – SAN DIEGO – SAN FRANCISCO – SINGAPORE – SYDNEY – TOKYO

ELSEVIER B.V.
Radarweg 29
P.O. Box 211, 1000 AE Amsterdam
The Netherlands

ELSEVIER Inc.
525 B Street, Suite 1900
San Diego, CA 92101-4495
USA

ELSEVIER Ltd
The Boulevard, Langford Lane
Kidlington, Oxford OX5 1GB
UK

ELSEVIER Ltd
84 Theobalds Road
London WC1X 8RR
UK

First edition 2005

Library of Congress Cataloging in Publication Data
A catalog record is available from the Library of Congress.

British Library Cataloguing in Publication Data
A catalogue record is available from the British Library.

ISBN: 0-08-044849-6 (Paperback)

♾ The paper used in this publication meets the requirements of ANSI/NISO Z39.48-1992 (Permanence of Paper).
Printed in Italy

DEDICATED
TO

ROBERT GARRELS
(1916–1988)

Photograph provided by Cynthia Garrels

Contents

Executive Editors' Foreword

H. D. Holland

Harvard University, Cambridge, MA, USA

and

K. K. Turekian

Yale University, New Haven, CT, USA

Geochemistry has deep roots. Its beginnings can be traced back to antiquity, but many of the discoveries that are basic to the science were made between 1800 and 1910. The periodic table of elements was assembled, radioactivity was discovered, and the thermodynamics of heterogeneous systems was developed. The solar spectrum was used to determine the composition of the Sun. This information, together with chemical analyses of meteorites, provided an entry to a larger view of the universe.

During the first half of the twentieth century, a large number of scientists used a variety of methods to determine the major-element composition of the Earth's crust, and the geochemistries of many of the minor elements were defined by V. M. Goldschmidt and his associates using the then new technique of emission spectrography. V. I. Vernadsky founded biogeochemistry. The crystal structures of most minerals were determined by X-ray diffraction techniques. Isotope geochemistry was born, and age determinations based on radiometric techniques began to define the absolute geologic timescale. The intense scientific efforts during World War II yielded new analytical tools and a group of people who trained a new generation of geochemists at a number of universities. But the field grew slowly. In the 1950s, a few journals were able to report all of the important developments in trace-element geochemistry, isotopic geochronometry, the exploration of paleoclimatology and biogeochemistry with light stable isotopes, and studies of phase equilibria. At the meetings of the American Geophysical Union, geochemical sessions were few, none were concurrent, and they all ranged across the entire field.

Since then the developments in instrumentation and the increases in computing power have been spectacular. The education of geochemists has been broadened beyond the old, rather narrowly defined areas. Atmospheric and marine geochemistry have become integrated into solid Earth geochemistry; cosmochemistry and biogeochemistry have contributed greatly to our understanding of the history of our planet. The study of Earth has evolved into "Earth System Science," whose progress since the 1940s has been truly dramatic.

Major ocean expeditions have shown how and how fast the oceans mix; they have demonstrated the connections between the biologic pump, marine biology, physical oceanography, and marine sedimentation. The discovery of hydrothermal vents has shown how oceanography is related to economic geology. It has revealed formerly unknown oceanic biotas, and has clarified the factors that today control, and in the past have controlled the composition of seawater.

Seafloor spreading, continental drift and plate tectonics have permeated geochemistry. We finally understand the fate of sediments and oceanic crust in subduction zones, their burial and their exhumation. New experimental techniques at temperatures and pressures of the deep Earth interior have clarified the three-dimensional structure of the mantle and the generation of magmas.

Moon rocks, the treasure trove of photographs of the planets and their moons, and the successful search for planets in other solar systems have all revolutionized our understanding of Earth and the universe in which we are embedded.

Geochemistry has also been propelled into the arena of local, regional, and global anthropogenic problems. The discovery of the ozone hole came as a great, unpleasant surprise, an object lesson for optimists and a source of major new insights into the photochemistry and dynamics of the atmosphere. The rise of the CO_2 content of the atmosphere due to the burning of fossil fuels and deforestation has been and will continue to be at the center of the global change controversy, and will yield new insights into the coupling of atmospheric chemistry to the biosphere, the crust, and the oceans.

The rush of scientific progress in geochemistry since World War II has been matched by organizational innovations. The first issue of *Geochimica et Cosmochimica Acta* appeared in June 1950. The Geochemical Society was founded in 1955 and adopted *Geochimica et Cosmochimica Acta* as its official publication in 1957. The International Association of Geochemistry and Cosmochemistry was founded in 1966, and its journal, *Applied Geochemistry*, began publication in 1986. *Chemical Geology* became the journal of the European Association for Geochemistry.

The Goldschmidt Conferences were inaugurated in 1991 and have become large international meetings. Geochemistry has become a major force in the Geological Society of America and in the American Geophysical Union. Needless to say, medals and other awards now recognize outstanding achievements in geochemistry in a number of scientific societies.

During the phenomenal growth of the science since the end of World War II an admirable number of books on various aspects of geochemistry were published. Of these only three attempted to cover the whole field. The excellent *Geochemistry* by K. Rankama and Th.G. Sahama was published in 1950. V. M. Goldschmidt's book with the same title was started by the author in the 1940s. Sadly, his health suffered during the German occupation of his native Norway, and he died in England before the book was completed. Alex Muir and several of Goldschmidt's friends wrote the missing chapters of this classic volume, which was finally published in 1954.

Between 1969 and 1978 K. H. Wedepohl together with a board of editors (C. W. Correns, D. M. Shaw, K. K. Turekian and J. Zeman) and a large number of individual authors assembled the *Handbook of Geochemistry*. This and the other two major works on geochemistry begin with integrating chapters followed by chapters devoted to the geochemistry of one or a small group of elements. All three are now out of date, because major innovations in instrumentation and the expansion of the number of practitioners in the field have produced valuable sets of high-quality data, which have led to many new insights into fundamental geochemical problems.

At the Goldschmidt Conference at Harvard in 1999, Elsevier proposed to the Executive Editors that it was time to prepare a new, reasonably comprehensive, integrated summary of geochemistry. We decided to approach our task somewhat differently from our predecessors. We divided geochemistry into nine parts. As shown below, each part was assigned a volume, and a distinguished editor was chosen for each volume. A tenth volume was reserved for a comprehensive index:

(i) *Meteorites, Comets, and Planets*: Andrew M. Davis

(ii) *Geochemistry of the Mantle and Core*: Richard Carlson

(iii) *The Earth's Crust*: Roberta L. Rudnick

(iv) *Atmospheric Geochemistry*: Ralph F. Keeling

(v) *Freshwater Geochemistry, Weathering, and Soils*: James I. Drever

(vi) *The Oceans and Marine Geochemistry*: Harry Elderfield

(vii) *Sediments, Diagenesis, and Sedimentary Rocks*: Fred T. Mackenzie

(viii) *Biogeochemistry*: William H. Schlesinger

(ix) *Environmental Geochemistry*: Barbara Sherwood Lollar

(x) *Indexes*

The editor of each volume was asked to assemble a group of authors to write a series of chapters that together summarize the part of the field covered by the volume. The volume editors and chapter authors joined the team enthusiastically. Altogether there are 155 chapters and 9 introductory essays in the Treatise. Naming the work proved to be somewhat problematic. It is clearly not meant to be an encyclopedia. The titles *Comprehensive Geochemistry* and *Handbook of Geochemistry* were finally abandoned in favor of *Treatise on Geochemistry*.

The major features of the Treatise were shaped at a meeting in Edinburgh during a conference on Earth System Processes sponsored by the Geological Society of America and the Geological Society of London in June 2001. The fact that the Treatise is being published in 2003 is due to a great deal of hard work on the part of the editors, the authors, Mabel Peterson (the Managing Editor), Angela Greenwell (the former Head of Major Reference Works), Diana Calvert (Developmental Editor, Major Reference Works),

Bob Donaldson (Developmental Manager), Jerome Michalczyk and Rob Webb (Production Editors), and Friso Veenstra (Senior Publishing Editor). We extend our warm thanks to all of them. May their efforts be rewarded by a distinguished journey for the Treatise.

Finally, we would like to express our thanks to J. Laurence Kulp, our advisor as graduate students at Columbia University. He introduced us to the excitement of doing science and convinced us that all of the sciences are really subdivisions of geochemistry.

Contributors to Volume 7

W. B. N. Berry
University of California, Berkeley, CA, USA

D. J. DeMaster
North Carolina State University, Raleigh, NC, USA

R. B. Finkelman
US Geological Survey, Reston, VA, USA

M. B. Goldhaber
US Geological Survey, Denver, CO, USA

L. R. Kump
The Pennsylvania State University, PA, USA

L. Lefticariu
Northern Illinois University, DeKalb, IL, USA

Y.-H. Li
University of Hawaii, Honolulu, HI, USA

T. W. Lyons
University of Missouri, Columbia, MO, USA

F. T. Mackenzie
University of Hawaii, Honolulu, HI, USA

W. R. Martin
Woods Hole Oceanographic Institution, MA, USA

J. B. Maynard
University of Cincinnati, OH, USA

K. L. Milliken
The University of Texas at Austin, TX, USA

J. W. Morse
Texas A&M University, College Station, TX, USA

W. H. Orem
US Geological Survey, Reston, VA, USA

E. C. Perry Jr.
Northern Illinois University, DeKalb, IL, USA

R. P. Philp
University of Oklahoma, Norman, OK, USA

B. B. Sageman
Northwestern University, Evanston, IL, USA

F. L. Sayles
Woods Hole Oceanographic Institution, MA, USA

J. E. Schoonmaker
University of Hawaii, Honolulu, HI, USA

J. Veizer
Ruhr University, Bochum, Germany and University of Ottawa, ON, Canada

B. Velde
Ecole Normale Supérieure, Paris, France

Volume Editor's Introduction

F. T. Mackenzie
University of Hawaii, HI, USA

This volume is dedicated to Robert M. Garrels in recognition of his contributions to our understanding of the fundamental geochemical processes governing the formation and transformation of sedimentary rocks through geologic time and their use in deducing the evolution of the Earth's exogenic system. In retrospect, Bob's research career can probably be divided into two main phases. The first dealt with theoretical and experimental aspects of geochemistry, which led to the publication of his book *Mineral Equilibria at Low Temperature and Pressure* in 1960 and later to the book's successor *Solutions, Minerals, and Equilibria* co-authored with Charles Christ in 1965. These books have become classics in the field of aqueous and sedimentary geochemistry, as they were one of the first efforts to apply chemical thermodynamics to geological processes with an emphasis on the construction of stability diagrams from thermodynamic data. The second phase of Bob's career led him to consider the importance of the chemical cycling of sediments and the processes involved: weathering, erosion, transport, sedimentation, burial, diagenesis, and uplift restarting the sedimentary cycle. This recycling concept had lain dormant for 150 years before the publication of *Evolution of Sedimentary Rocks* with Fred Mackenzie in 1971. The papers in this volume largely address this phase of Bob's career.

Biogeochemical processes and the products of those processes, the sedimentary rocks, are discussed in this volume as is the use of various geochemical and isotopic proxies found in sedimentary rocks to constrain the chemical evolution of the Earth surface environment. Not all sedimentary rock facies are given a separate chapter because of space constraints, but the geochemistry of most major facies is considered, particularly in terms of the evolution of sedimentary rocks and their usefulness in deciphering Earth history. The volume begins with four chapters (Chapters 7.01, 7.02, 7.03, and 7.04) dedicated primarily to the geochemistry of major, modern marine, sediment types and their diagenesis. Sediments at their origin have measurable properties such as mineralogy, chemistry, isotopic composition of organic and inorganic phases, biological makeup, and texture. The diagenetic processes of compaction, cementation, and solution and reprecipitation leading to mineral stabilization can modify all of these properties. Thus, as sediments pass through the "diagenetic fence" (Chave, 1960), they lose information necessary for environmental interpretation. Chapter 7.05 continues the theme of Chapter 7.04 on the diagenesis of silica and discusses the product of that diagenesis, chert, in the context of Earth's early temperature history. This is followed by chapters dealing with geochemistry and mass transfer in siliciclastic rocks and the geochemistry of the organic-rich materials in coal, oil, and gas and the important element sulfur. Two minor but environmentally important sediment components are then considered in the chapters on manganiferous sediments and the "green" clay minerals. The final three chapters (Chapters 7.13–7.15) are devoted to the chronological and historical aspects of sediments and sedimentary rocks and the use of biogeochemical and isotopic proxies within them in interpreting the history of Earth surface environments.

In Chapter 7.01 Telu Li and Jane Schoonmaker review the chemical composition of modern marine sediments, building on the classical works of Tizzard *et al.* (1885), Murray and Menard (1891) and of more recent authors (e.g., Arrhenius, 1963). Li and Schoonmaker address the mineralogy and geochemistry of pelagic sediments in general and consider several specific sediment types, particularly in terms of their hydrogenous components. Sediment elemental enrichment factors and factor analysis are used to determine the major components and mineralogical phases in which the elements of marine sediments reside. The major components of marine pelagic sediments are shown to be (i) aluminosilicates, derived mainly from the weathering of shales and eolian and riverine transport of the weathered products to the ocean; and (ii) ferromanganese oxides, carbonate fluorapatite, zeolites, biogenic carbonate, silica, and barite, derived from the products of organic productivity and chemical reactions involving elements directly derived from the general background of elements in seawater or resulting from diagenetic reactions in the oxic, suboxic, and anaerobic realms of early diagenesis (Chester, 1990).

In Chapter 7.02, Bill Martin and Fred Sayles consider the early diagenetic reactions occurring in sediments, including those involving organic matter degradation, calcium carbonate dissolution, and silica cycling, and the role of bioturbation in mixing particles downward from the sediment–water interface. The authors primarily use the wealth of solute concentration data from pore-water vertical profiles that they themselves (e.g., Sayles, 1979, 1981; Martin *et al.*, 2000), as well as other investigators (e.g., Archer *et al.*, 1989; Bender *et al.*, 1989), have obtained during the past several decades to interpret diagenetic processes. Mass balances provided by the authors show that only a small fraction of the organic matter and the opal that falls to the seafloor survives early diagenesis; thus, only a small portion accumulates in the sediment and is available for downcore interpretations of accumulation rates. The incorporation of aluminum into opal and the hypothesized reprecipitation of H_4SiO_4 from dissolving opal into neoformed minerals highlight the significant influence of the silica cycle on the cycling behavior of the major constituents of pore water and the likelihood that "reverse weathering" (e.g., Mackenzie and Garrels, 1966; Michalopoulos and Aller, 1995; Mackenzie and Kump, 1995) is an important process affecting ocean chemistry. For calcium carbonate, the role of dissolution in sediments above the calcite lysocline driven by acids generated by oxic respiration is highlighted, which also brings into question present interpretations of the temporal record of $CaCO_3$ accumulation rates. About half of the O_2 consumption in the ocean below 1,000 m occurs in deep-sea sediments, and sedimentary denitrification is shown by Martin and Sayles to be a dominant term in the marine fixed nitrogen budget. As an aside, these authors point out that pore-water studies probably began with Murray and Irvine's (1885) work.

In Chapter 7.03, John Morse discusses the sources and early diagenesis of deep-sea and shoal-water ("shallow-water") carbonate sediments. One of the long-standing controversial areas of carbonate geochemistry has been the relationship between calcium carbonate accumulation in deep-sea sediments and the saturation state of the overlying waters. Hypotheses considered have ranged from a nearly "thermodynamic" ocean where the calcite and aragonite compensation depths are at the calcite and aragonite saturation horizons (e.g., Li *et al.*, 1969) to a strongly kinetically controlled ocean system where major differences exist between the carbonate compensation depths, lysoclines, and saturation depths (e.g., Morse and Berner, 1972). Morse points out that if recent calculations related to these features are correct, the long-cherished idea of a tight coupling between seawater chemistry and carbonate depositional facies requires re-examination. That such a coupling may not exist has been suggested by Milliman *et al.* (1999), who evaluated several lines of evidence and concluded that considerable dissolution of calcium carbonate, perhaps as much as 60–80%, occurs in the upper 500–1,000 m of the ocean, well above the chemical lysocline. The role of organic matter in this water-column dissolution process and in microbially driven organic oxidation processes in both shoal-water and deep-sea sediments (as discussed by Morse) is now well recognized. Morse also considers what is known about early dolomite formation. He points out that most modern dolomite is forming from relatively high-ionic-strength seawater solutions in which the Mg^{2+} to Ca^{2+} solution ratios are considerably higher than in normal seawater. However, he also recognizes that factors other than the Mg^{2+}/Ca^{2+} ratio may be involved in dolomite formation. Sulfate depletion and alkalinity increase (increase in saturation state) in pore waters, resulting from anaerobic oxidation of organic matter by bacteria using SO_4^{2-} as a substrate, as well as direct participation by bacteria, may play a role in the dolomite formation process. Dolomite is not found in abundance in modern sediments but is an important component of older sedimentary rocks, and its origin has important implications for interpreting the history of seawater and the atmosphere through geologic time (see Chapters 6.21 and 7.15).

Except for the sponges of the latest Precambrian Ediacaran fauna, most of the organisms that make their skeletons of amorphous silica are of Cambrian age or younger. The appearance of radiolarians and diatoms in the Phanerozoic changed the manner of deposition of amorphous silica on the seafloor from a primarily abiotic process to one in which biochemical processes play a strong role, and led to a shift in the locus of silica deposition from the shallow shelf to the deep sea. Prior to the evolution and spread of siliceous biota in the Cambrian, the oceans contained significant concentrations of dissolved silica (Garrels and Perry, Jr., 1974; Siever, 1992), reaching saturation levels with respect to amorphous silica in Precambrian aquatic systems. As siliceous organisms became more prevalent in the Phanerozoic, the surface waters of the ocean became depleted in dissolved silica, and the average concentration of dissolved silica in seawater fell from a level of about 1,000 μM to 100 μM. In the context of these significant events in biotic evolution and seawater chemistry, Dave DeMaster in Chapter 7.04 and Ed Perry and Liliana Lefticariu in Chapter 7.05 discuss the diagenesis of biogenic silica (chemical transformations occurring in the water column, seabed, and crust), and the formation and geochemistry of Precambrian cherts, respectively. DeMaster examines the chemical and structural transformations that take place following the death of siliceous organisms in the oceans and that lead to changes in the solubility and dissolution kinetics of the original biogenic amorphous silica phase. He documents the stabilization pathways of the transformation of biogenic silica from amorphous opal-A to opal-CT and finally in many cases to microcrystalline quartz (the rock chert) as the biogenic silica is buried beneath meters to kilometers of sediment and is exposed to higher temperatures and pressures (e.g., Williams and Crerar, 1985). DeMaster also recognizes the importance of the tie between the aluminum and silicon cycles and "reverse weathering" diagenetic processes as a sink for elements in the ocean.

Perry and Lefticariu highlight the fact that cherts precipitated from aqueous solution have the potential to preserve a record of the oxygen isotopic composition of contemporaneous seawater or lake water and hence the temperature of the environmental medium. The problems, of course, are that the oxygen isotopic signal of cherts is a function of both temperature and solution composition and that the siliceous precipitates undergo diagenesis and may encounter temperatures and pressures of metamorphism in their passage through the sedimentary cycle leading to isotopic exchange. Recognizing these problems, Perry and Lefticariu, nevertheless, suggest that Earth surface temperatures derived from the δ^{18}O signatures of 1,900–2,400-Myr-old cherts could have been as high as 64 °C and those obtained from Archean, 3,400–3,700-Myr-old cherts suggest temperatures as high as 90 °C. Such evidence of high early Earth surface temperatures, although controversial, can be used in conjunction with other evidence to suggest that one of the earliest progenitors of life was an anaerobic thermophilic microorganism evolving in a hydrothermal or volcanic environment (e.g., Kasting and Chang, 1992).

Chapters 7.06 and 7.07 focus mainly on the geochemistry and sedimentology of detrital sediments and sedimentary rocks (siliciclastics), These sedimentary materials preserve records of multiple biogeochemical proxies that reveal the processes and conditions of sediment formation, transport, deposition, burial, and diagenesis and provide information regarding the evolution of Earth's surface environment. The proxies include sedimentary mineralogical suites and their distribution in time and space, major-, minor-, and trace-element concentration data, the stable isotopic composition of carbon, oxygen, sulfur, and nitrogen, radiogenic isotopes such as ^{40}K, and the abundance and isotopic composition of biomarker compounds. Brad Sageman and Tim Lyons deal mainly with the fine-grained, mixed siliciclastic–biogenic sedimentary facies, commonly referred to as hemipelagic facies. Based on the results of case studies of modern Black Sea and Cariaco basin sediments, Cretaceous hemipelagic sediments, Devonian black-to-gray shale transitional zones, and Precambrian shales and argillites, these authors propose a unified "Earth-systems" model for the biogeochemical analysis of hemipelagic rocks. Important linkages in the model include those between detrital fluxes and bulk sedimentation, productivity and nutrients, productivity and δ^{13}C records, molybdenum and iron, and controls on redox conditions, and elemental ratios and accumulation rates. This work demonstrates an abundance of biogeochemical information and proxies is present in the often-overlooked sedimentary hemipelagic facies that can be used to interpret Earth surface conditions through geologic time.

Chapter 7.07 discusses what is known about the chemical and physical processes that lead to the transformation of detrital siliciclastic sediments into rocks. These processes take place during burial diagenesis and low-grade metamophism and certainly involve basin-scale mass transfer of helium, water, and petroleum, and significant transport of carbon out of shales into the ocean–atmosphere system. Large-scale silica transfer from shales to sandstones is also a persistent feature of burial diagenetic sequences, as the processes of smectite dissolution and illite precipitation and quartz pressure dissolution in shales and quartz precipitation in sandstones act in concert, and silica solubility increases with increasing burial and temperature. Calcium is lost from shales, as is carbonate, due to the dissolution of $CaCO_3$ and, although controversial, the probable loss of calcium from smectite as it is transformed to illite (Garrels and Mackenzie, 1974). Potassium progressively increases in some shale sequences undergoing burial diagenesis but decreases in others. However, the global effect appears to be the transport of potassium from other lithologies into shales (Garrels and Mackenzie, 1972). The fate of magnesium and sodium is not well constrained, although Garrels and Mackenzie (1971, 1974) argue that both elements are lost from shales with diagenesis and time, and that some magnesium is transferred via late dolomitization processes to the carbonate reservoir with increasing age of the rock mass (see Chapters 6.21 and 7.15). Milliken concludes her chapter with emphasis on the conclusion that most precipitation reactions occurring during the diagenesis of sandstones and shales are acid-releasing "reverse weathering" reactions and can influence the crustal CO_2 cycle (e.g., Kastner, 1974; Kerrick *et al.*, 1995).

Chapters 7.08–7.10 deal with the formation and fate of materials derived from living organic matter, its degradation products, and elements generally associated with organic-rich sediments deposited in aquatic environments. In Chapter 7.08 Bill Orem and Bob Finkelman point out that because coal is an organic "rock" largely composed of an assemblage of amorphous, degraded plant material, an admixture of syngenetic, diagenetic, epigenetic, and detrital mineral grains, and contains within its structure water, oils, and gases, it is one of the most complex and challenging sedimentary rocks to analyze and understand. Coal is the most abundant of the fossil fuels and the second-largest source of energy in the world. In 2002 it supplied ~30% of the world's commercial energy (Mackenzie, 2003). In the USA more than half of the electricity generated is by coal-fired power plants.

Coal formation begins during the process of peatification in which microorganisms preferentially degrade plant biomolecules such as carbohydrates and leave other materials such as lignin and waxes selectively preserved. In the early stages of coal formation (coalification), numerous chemical changes to biomolecules occur such as loss of oxygen functionality and changes in lignin composition but the overall structure of plant materials of low-rank coals remains relatively intact. Later-stage coalification producing bituminous and higher-rank coals involves more substantial changes in the structure of the coal. These include the transformation of coal to a hard, black, lustrous organic rock depleted in both oxygen and hydrogen relative to the precursor material and with aromatic structures that are more condensed.

In contrast to coal formation, in which the most important starting materials are biopolymers found in vascular plants with some admixture of resistant algae polymers as well as bacterial polymers, kerogens are the insoluble organic materials that are the immediate precursors to the formation of most oil and gas. Kerogen is a very heterogeneous and complex aggregate of macerals, discrete particles of insoluble organic material identifiable under the microscope and representing residual detritus from various sources of organic matter (e.g., Tissot *et al.*, 1974). The types and maturity levels of kerogen play an important role in determining the type of oil or gas generated. In 1992, oil provided 40%, and gas provided 20% of the commercial global energy consumption (Mackenzie, 2003).

In Chapter 7.09, Paul Philp describes the early steps of oil and gas formation, starting with the process of photosynthesis, ultimately responsible for most oil, gas, and coal formation. Today only ~0.1% of global net primary production of organic matter accumulates in sediments (Tissot and Welte, 1978; Mackenzie *et al.*, 1993) and is available for the formation of fossil fuels; thus, these commercial deposits of the stored energy of photosynthesis actually represent but a very small portion of the gross production of organic matter and of total organic matter present in the sedimentary mass. Philp discusses the nature of the insoluble organic materials, kerogen, and the source, depositional environment, maturity, biodegradation, age dating, and migration of the soluble components of kerogen. It is now generally accepted that the major mechanism for the formation of most oil (and gas) is through thermal degradation of kerogen: kerogen → bitumen → oil + gas + residue.

In conclusion, Philp argues that the integration of geochemical parameters, such as biomarkers, with sequence stratigraphy models will become more important in the future in petroleum exploration in terms of predicting how the processes of oil and gas generation, expulsion from source rocks, migration to reservoir rocks and subsequent trapping, and preservation control the volume, quality, and distribution of petroleum.

In Chapter 7.10, Marty Goldhaber considers in detail the geochemistry of sulfur, one of the important elements commonly associated with organic-rich sediments and a principal element found in living organisms. He points out that nearly all marine sediments with more than a few tenths of a percent of organic carbon and those nonmarine sediments with significant concentrations of SO_4^{2-} in depositional waters contain the mineral pyrite (FeS_2). Pyrite, along with sulfur-bearing organic compounds and other mineral phases of sulfur such as mackinawite, greigite, and elemental sulfur, forms at Earth surface temperatures and pressures primarily through the metabolic activities of sulfate-reducing microorganisms. Goldhaber discusses the biochemistry of bacterial sulfate reduction along with the ecology of the bacteria responsible for the overall process, emphasizing the fact that in general the ultimate products of sulfate reduction are depleted in ^{34}S. He then considers the forms, mechanisms, diagenesis, abundance, and isotopic systematics of sulfur in marine sediments. In the global context, the evolution of the cycling behavior of sulfur through geologic time is closely tied to the carbon cycle and the evolution of atmospheric oxygen (e.g., Petsch and Berner, 1998). In addition, the variations in the sulfur isotopic composition of sedimentary sulfate minerals and pyrite provide constraints on the isotopic composition of seawater through geologic time and suggest that both atmospheric O_2 and oceanic SO_4^{2-} were very low prior to 2.4 Ga. The presence of very significant mass-independent sulfur fractionations in sulfate and sulfide minerals older than 2.45 Ga (Farquhar *et al.*, 2000) supports this conclusion.

Chapters 7.11 and 7.12 are concerned with the formation of manganiferous phases in sediments, rocks, and ores and the sedimentary "green" clay minerals. Manganese is the tenth most abundant element in the crust and is very similar to iron in its chemical properties; however, the fact that it has higher valence states distinguishes its behavior from iron and gives rise to a plethora of Mn-oxide minerals. In Chapter 7.11 Barry Maynard discusses the basic geochemical properties of manganese and its common minerals, its distribution in rocks and natural waters, and the composition of its significant accumulations. He addresses the behavior of manganese especially in mid-ocean ridge vent systems and during sedimentation. Throughout the chapter, Maynard emphasizes the difference between the geochemistry of manganese and iron, and points out that under reducing and mildly oxidizing conditions, manganese is exported from low-oxygen environments, such as basalt–hydrothermal systems or euxinic basins, and it accumulates in oxidizing environments of the shallow ocean or in low-productivity areas of the deep sea. Thus the Mn/Fe ratios of sediments and waters provide a clue to the oxidation structure of ocean basins, soils, and groundwater systems. The strong affinity of manganese for certain transition elements and the rare earth elements is an important property. In the case of cerium and europium, these elements when preserved in managnese accumulations can define the relative importance of hydrothermal and diagenetic processes in the sedimentary record.

In Chapter 7.12, Bruce Velde demonstrates that the basic green color of glauconite, celadonite, berthierine, verdine, chamosite, nontronite, and talc is due to the presence of iron in the structures of these minerals. This is not a startling revelation to mineralogists, but it is the key to understanding the origin and stability of the phases in nature. Velde concludes that green clay mineral names can reflect compositional differences, mineral structural differences, or differences in the geological occurrence of these phases. This in itself leads to much confusion in the literature concerning the origin and diagenesis of the green clay minerals. Regardless, the mineralogy and chemistry of the green clay minerals provide valuable clues to the temperature and solution composition of the environment in which they formed. For example, both glauconite and berthierine–verdine minerals are low-temperature phases that form in shallow-water sediments at low sedimentation rates. The formation of glauconite probably requires the more oxidizing conditions of organic-poor sediments rather than the lower redox conditions for berthierine formation in organic-rich sediments. The original glauconite precursor evolves toward a potassic, ferric clay mineral, whereas that of berthierine evolves toward an alkali-free, ferrous mineral. Thus, the two minerals can be used to infer their environmental conditions of deposition and diagenesis. Many of Velde's conclusions are based on work started in the 1980s (e.g., Velde, 1985).

The final three chapters (Chapter 7.13–7.15) of the volume concentrate on the longer-term historical aspects of sediments and sedimentary rocks, their cycling behavior, and the biogeochemical and isotopic proxies found within them that permit the interpretation of the evolution of Earth's surface environment through geologic time. Interpretations of this nature first require a temporal framework in which to place information and events, and this is the subject of Bill Berry's chapter (Chapter 7.13). Berry points out that it was economically imperative to develop a chronometry of sedimentary rocks in order to find and recover larger volumes of nature's resources of coal, ores, building stone, sand, gravel, and eventually petroleum. The basic principle of chronometry based on the fossil record is embodied in the Table of Strata that William Smith dictated to his ecclesiastic associates in June 1799 (see Berry, 1987) and still holds true today, that of the succession of fauna and flora found in the rock record. Berry's description of the development of chronometry based on the fossil record is essentially that of the early development of the science of geology itself. Chorology, the science that deals with the geographical distribution of living organisms, is discussed in terms of the constraints it imposes on chronometry. As Berry points out, chronometry cannot provide the absolute ages of rocks. Fossil-based chronometry has been enhanced through studies of radiochronometry and magnetobiochronometry, as well as orbital chronometry. The chronometry of sedimentary rocks supplies the necessary framework for interpretation of events in Earth history and the longer-term secular evolution and cycling of the sediment-ocean–atmosphere–biota system, the subject matter of the next two chapters.

In Chapter 7.14, Lee Kump points out that the course of biological evolution is inextricably linked to that of the environment in a complicated network of feedback at all time and space scales. Perturbations of the environment and disruptions within have biological consequences and vice versa. Kump considers mainly the elemental and carbon, sulfur, strontium, and oxygen isotopic evidence for the environmental nature of the "big five" extinctions of the Phanerozoic (Sepkoski, 1993): the Late Ordovician (~440 Ma), the Late Devonian (~367 Ma), the Permian–Triassic (~251 Ma), the Triassic–Jurassic (~200 Ma), and the Cretaceous–Tertiary (~65 Ma). His conclusion is that there are no universal geochemical precursors or responses to extinction events in the Phanerozoic. However, the development of widespread ocean anoxia appears to be associated with all three Paleozoic mass extinctions. Such anoxic conditions can certainly lead to mass mortality of extant organisms, and potential rapid turnover of anoxic ocean waters could produce elevated concentrations of CO_2 and H_2S in surface waters resulting in the death of aerobic organisms. Rapid ocean turnover could be the result of an oceanic impactor during the Late Devonian and Permian–Triassic events. However, a purely terrestrial origin of the Late Ordovician extinction event is probable due to a glacio-eustatic sea-level fall causing a loss of shallow-water marine habitats. In contrast, it appears that both the Triassic–Jurassic and Cretaceous–Tertiary mass extinctions occurred during periods of well-oxygenated oceans. An asteroid or cometary impact appears to be the proximal cause of these events. At least for the K–T event, the consequent release of huge quantities of CO_2 and SO_2 to the atmosphere may have changed the climate and caused mass mortality of organisms.

In the final chapter (Chapter 7.15), Veizer and Mackenzie show that the progressive increase with decreasing age of the observed extant thickness and areal extent of Phanerozoic sedimentary strata and the secular variations in the relative proportions of lithological types and their chemistry and mineralogy are a function of both evolution and recycling of the sedimentary mass. They also conclude that the global sedimentary mass of 2.7×10^{24} g is largely "cannibalistic" and has a half-mass age of about 600 Myr, resulting in total sediment deposition and destruction equivalent to about five sedimentary masses over the course of geologic time (e.g., Garrels and Mackenzie, 1971). The differential recycling of various components of the sedimentary rock mass is primarily controlled by the probability of preservation of different tectonic settings in which the sediments are found and the recycling rates of the tectonic realms (e.g., Veizer, 1988). Because of differential recycling, the preservation of the sedimentary mass is inherently biased. For example, the half-life of the post-Devonian sedimentary mass is about 130 Ma, whereas the half-life of post-Permian carbonates is only ~86 Ma; thus, the carbonates are recycled at a rate ~1.5 times faster than the total sedimentary mass. This difference is due to the fact that during post-Permian time, the locus of carbonate deposition has progressively shifted to the deep sea, where post-Permian carbonates are more susceptible to destruction because of subduction. The absence of pre-Permian deep-sea carbonates implies that continental carbonates are the only record of carbonate deposition and chemical, mineralogical, and isotopic data for most of geologic time. Using the full range of mineralogical, chemical, and strontium, osmium, sulfur, carbon, and oxygen isotopic evidence, Veizer and Mackenzie also consider the chemical evolution of the ocean–atmosphere system from the anoxic conditions of the early Precambrian to the modern oxic global environment (see also Chapter 6.21).

REFERENCES

Archer D., Emerson S., and Smith C. R. (1989) Dissolution of calcite in deep-sea sediments: pH and O_2 microelectrode results. *Geochim. Cosmochim. Acta* **53**, 2831–2845.

Arrhenius G. (1963) Pelagic sediments. In *The Sea, Ideas and Observations on Progress in the Study of the Seas*, The Earth Beneath the Sea, History (ed. M. N. Hill). Interscience Publishers, New York, vol. 3, pp. 655–727.

Bender M., Jahnke R., Weiss R., Martin W., Heggie D., Orchardo J., and Sowers T. (1989) Organic carbon oxidation and benthic nitrogen and silica dynamics in San Clemente Basin, a continental borderland site. *Geochim. Cosmochim. Acta* **53**, 685–697.

Berry W. B. N. (1987). *Growth of the Prehistoric Time Scale* (revised edn.). Blackwell, Oxford.

Chave K. E. (1960) Carbonate skeletons to limestones: problems. *Trans. NY Acad. Sci.* **23**, 14–24.

Chester R. (1990) *Marine Geochemistry*. Unwin Hyman, London, 698p.

Farquhar J., Huiming B., and Thiemens M. H. (2000) Atmospheric influence of Earth's earliest sulfur cycle. *Science* **289**, 756–758.

Garrels F. T. and Mackenzie F. T. (1971) *Evolution of Sedimentary Rocks*. WW Norton, New York, 397p.

Garrels R. M. and Mackenzie F. T. (1972) A quantitative model for the sedimentary rock cycle. *Mar. Chem.* **1**, 27–41.

Garrels R. M. and Mackenzie F. T. (1974) Chemical history of the oceans deduced from post-depositional changes in sedimentary rocks. In *Studies in Paleo-Oceanography*. Society of Economic Paleontologists and Mineralogists Special Publication No. 20 (ed. W. H. Hay). Society of Economic Paleontologists and Mineralogists, Tulsa, OK, pp. 193–204.

Garrels R. M. and Perry E. A., Jr. (1974) Cycling of carbon, sulfur, and oxygen through geologic time. In *The Sea* (ed. E. D. Goldberg). Wiley, New York, vol. 5, pp. 303–357.

Kasting J. F. and Chang S. (1992) Formation of the earth and the origin of life. In *The Proterozoic Biosphere: An Interdisciplinary Study* (eds. J. W. Schopf and C. Klein). Cambridge University Press, New York, pp. 9–12.

Kastner M. (1974) The contribution of authigenic feldspars to the geochemical balance of the alkali metals. *Geochim. Cosmochim. Acta* **38**, 650–653.

Kerrick D. M., McKibben M. A., Seward T. M., and Caldeira K. (1995) Convective hydrothermal CO_2 emission from high heat flow regions. *Chem. Geol.* **121**, 17–27.

Li Y.-H., Takahashi T., and Broecker W. S. (1969) Degree of saturation of $CaCO_3$ in the oceans. *J. Geophys. Res.* **74**, 5507–5525.

Mackenzie F. T. (2003) *Our Changing Planet: An Introduction to Earth System Science and Global Environmental Change*. Prentice Hall, New York.

Mackenzie F. T. and Garrels R. M. (1966) Chemical mass balance between rivers and oceans. *Am. J. Sci.* **284**, 507–525.

Mackenzie F. T. and Kump L. R. (1995) Reverse weathering, clay mineral formation, and oceanic element cycles. *Science* **270**, 586–587.

Mackenzie F. T., Ver L. M., Sabine C., and Lane M. (1993) C, N, P, S global biogeochemical cycles and modeling of global change. In *Interactions of C, N, P and S Biogeochemical Cycles and Global Change* (eds. R. Wollast, F. T. Mackenzie, and L. Chou). Springer, Berlin, pp. 1–61.

Martin W. R., McNichol A. P., and McCorkle D. C. (2000) The radiocarbon age of calcite dissolving at the sea floor: estimates from pore water data. *Geochim. Cosmochim. Acta* **64**, 1391–1404.

Michalopoulos P. and Aller R. C. (1995) Rapid clay mineral formation in Amazon Delta sediments: reverse weathering and oceanic elemental cycles. *Science* **270**, 614–617.

Milliman J. D., Troy P. J., Balch W. M., Adams A. K., Li Y.-H., and Mackenzie F. T. (1999) Biologically mediated dissolution of calcium carbonate above the chemical lysocline? *Deep-Sea Res.* **46**, 1653–1669.

Morse J. W. and Berner R. A. (1972) Dissolution kinetics of calcium carbonate in seawater: II. A kinetic origin for the lysocline. *Am. J. Sci.* **274**, 638–647.

Murray J. and Irvine R. (1885) On the chemical changes which take place in the composition of seawater associated with blue muds on the floor of the ocean. *Trans. Roy. Soc. Edinburgh* **37**, 481–507.

Murray J. and Menard A. (1891) Deep sea deposits. In *Challenger Reports*. Longmans, London, 525p.

Petsch S. T. and Berner R. A. (1998) Coupling the geochemical cycles of C, P, Fe, and S: the effect on atmospheric O_2 and the isotopic records of carbon and sulfur. *Am. J. Sci.* **298**, 246–262.

Sayles F. L. (1979) The composition and diagenesis of interstitial solutions: I. Fluxes across the seawater-sediment interface in the Atlantic Ocean. *Geochim. Cosmochim. Acta* **43**, 527–545.

Sayles F. L. (1981) The composition and diagenesis of interstitial solutions: II. Fluxes and diagenesis at the sediment-water interface in the high latitude North and South Atlantic. *Geochim. Cosmochim. Acta* **45**, 1061–1086.

Sepkoski J., Jr. (1993) Ten years in the library: new data confirm paleontological patterns. *Paleobiology* **19**, 43–51.

Siever R. (1992) The silica cycle in the Precambrian. *Geochim Cosmochim. Acta* **56**, 3265–3272.

Tissot B. and Welte D. H. (1978) *Petroleum Formation and Occurrence*. Springer, Berlin.

Tissot B., Durand B., Espitalie J., and Combaz A. (1974) Influence of nature and diagenesis of organic matter in formation of petroleum. *Am. Assoc. Petrol. Geol. Bull.* **58**, 499–506.

Tizzard T. H., Moseley H. N., Buchanan M. A., and Murray J. (1885) *Report of the Scientific Results of the Voyage of H. M. S. Challenger during the Years 1873–1876*, vol. 1, pt. 1. Her Majesty's Stationery Office, London, 509p.

Veizer J. (1988) The evolving exogenic cycle. In *Chemical Cycles in the Evolution of the Earth* (eds. C. B. Gregor, R. M. Garrels, F. T. Mackenzie, and J. B. Maynard). Wiley, New York, pp. 175–220.

Velde B. (1985) *Clay Minerals: A Physio-chemical Explanation of Their Occurrence*. Elsevier, Amsterdam, 427p.

Williams L. A. and Crerar D. A. (1985) Silica diagenesis: II. General mechanisms. *J. Sedim. Petrol.* **55**, 312–321.

7.01
Chemical Composition and Mineralogy of Marine Sediments

Y.-H. Li and J. E. Schoonmaker
University of Hawaii, Honolulu, HI, USA

7.01.1 INTRODUCTION

The earliest reports on the composition of deep-sea sediments resulted from the *Challenger Expedition* (1873–1876) (e.g., Tizzard *et al.*, 1885; Murray and Renard, 1891). Many review papers on marine sediment composition have subsequently been published, including the ones by Revelle (1944), El Wakeel and Riley (1961), Arrhenius (1963), Goldberg (1963), Chester and Aston (1976), Glasby (1977), Bischoff and Piper (1979), Baturin (1982, 1988), Notholt and Jarvis (1990), Nicholson *et al.* (1997), Glenn *et al.* (2000), and Li (2000). The constituents of a marine sediment are often classified according to their origin (Table 1; after Goldberg, 1963). The detrital component is made up of cosmogenous and lithogenous materials. Cosmic spherules contain particles of FeNi that are formed by ablation of iron meteorites as they pass through Earth's atmosphere, as well as fragments of silicate chondrules (Arrhenius, 1963). Lithogenous constituents of marine sediments are the minerals derived from weathering of rock on land or on the seafloor, or from the volcanic eruptions (Goldberg, 1963; see review in Windom (1976)). The biogenous component is made up of the tests of planktic and benthic organisms, as well as biogenic apatite (see review in Berger (1976)). The hydrogenous fraction of marine sediment encompasses phases formed by inorganic precipitation from seawater. Elderfield (1976) and Piper and Heath (1989) provide comprehensive reviews of hydrogenous material in marine sediments.

In this chapter, we present a review of the composition of the marine sediments, first addressing pelagic sediments in general, and then considering several specific types of sediment, each dominated by hydrogenous components. In each section, a review of mineralogy is followed by an examination of the geochemistry using enrichment factors and factor analysis.

Table 2 summarizes the average compositions of the Earth's upper crust, shale, and marine sediments, including pelagic sediments, ferromanganese nodules and crusts, metalliferous

Table 1 Mineral constituents of marine sediments classified by origin.

Cosmogenous		*Biogenous*	
Spherules		Calcite	$CaCO_3$; $(Ca_{1-x}Mg_x)CO_3$
Iron	FeNi	Aragonite	$CaCO_3$
Olivine	$(Mg,Fe)_2SiO_4$	Opal	$SiO_2{\cdot}nH_2O$
Pyroxene	$(Mg,Fe)_2Si_2O_6$	Francolite	$Ca_{10-x-y}Na_xMg_y(PO_4)_{6-z}(CO_3)_zF_{0.4z}F_2$
		Barite	$BaSO_4$
		Celestite	$SrSO_4$
Lithogenous		*Hydrogenous*	
Quartz	SiO_2	FeMn oxides/ oxyhydroxides	See Table 3
Plagioclase	$(Ca,Na)(Al,Si)AlSi_2O_8$		
Clay minerals		Francolite	$Ca_{10-x-y}Na_xMg_y(PO_4)_{6-z}(CO_3)_zF_{0.4z}F_2$
Illite	$K_xAl_2(Si_{4-x}Al_x)O_{10}(OH)_2$	Barite	$BaSO_4$
Chlorite	$(Mg,Fe)_5(Al,Fe)_2Si_3O_{10}(OH)_8$	Celestite	$SrSO_4$
Kaolinite	$Al_2Si_2O_5(OH)_4$	Montmorillonite	$(Na,K)_x(Al_{2-x}R_x)Si_4O_{10}(OH)_2$
Smectite	$(M)_{x-y}(R^{3+}_{2-y}R^{2+}_y)(Si_{4-x}Al_x)O_{10}(OH)_2$	Nontronite	$(Na,K)_xFe_2(Al_xSi_{4-x})O_{10}(OH)_2$
		Glauconite	$K_{0.85}(Fe,Al)_{1.34}(Mg,Fe)_{0.66}(Si_{3.76}Al_{0.24})O_{10}(OH)_2$
Volcanic glass			
Amphiboles	$Ca_2(Mg,Fe)_5Si_8O_{22}(OH)_2$	Zeolites	
Pyroxene	$(Mg,Fe)_2Si_2O_6$	phillipsite	$K_{2.8}Na_{1.6}Al_{4.4}Si_{11.6}O_{32}{\cdot}10H_2O$
Olivine	$(Mg,Fe)_2SiO_4$	clinoptilolite	$K_{2.3}Na_{0.8}Al_{3.1}Si_{14.9}O_{36}{\cdot}12H_2O$
		Geothite	FeOOH
		Palygorskite	$(OH_2)_4Mg_5Si_8O_{20}(OH)_2{\cdot}4H_2O$
		Sepiolite	$(OH_2)_4Mg_8Si_{12}O_{30}(OH)_2{\cdot}4H_2O$

After Goldberg (1986).

ridge and basal sediments, and marine phosphorites. Also given in Table 2 are the compositions of seawater and the hydrothermal vent solution from the Hanging Garden vent on the mid-Pacific rise at 21° N (Li (2000) and references therein; additional data sources are given in the footnote to Table 2).

One convenient way to compare a given sample to a chosen reference material is the so-called enrichment factor. The enrichment factor (E^i_j) is defined as the concentration ratio of a given element i and the normalizing element $j(X_i/X_j)$ in the given sample divided by the same ratio in the reference material, i.e.,

$$E^i_j = \frac{(X_i/X_j)_{\text{sample}}}{(X_i/X_j)_{\text{reference}}}$$

An E^i_j value of greater than one represents enrichment of element i in the sample as compared to the reference; whereas a value less than one means depletion. In order to avoid possible confusion, sample and reference names can be added after E^i_j. As shown in Figure 1(a), the average shale composition is very similar to that of the upper continental crust ($E^i_{\text{Al}} = 1.0 \pm 0.3$). The obvious exceptions are Li, and volatile elements B, C, N, S, Se, Te, Br, I, As, Cd, In, Sb, Hg, and Bi, which are enriched in the shale in comparison to the upper continental crust when aluminum is chosen as the normalizing element (Li, 2000). These excess volatile elements came from the interior of the Earth by magmatic degassing processes during the Earth's early history (Li, 1972). The depletion of Ca, Sr, and Na in the shale relative to the Earth's upper crust is balanced by the increased presence of those elements found in carbonate rocks, evaporates, and seawater. The average compositions of shale and marine pelagic clay are similar (Figure 1(b); within a factor of 2). The obvious exceptions are B, Na, P, Mn, Co, Ni, Cu, Mo, Pd, Te, Ba, W, Os, Ir, Pt, Tl, and Pb, which are enriched in the pelagic clay. As will be shown later, many of these elements are associated with manganese oxide phases in pelagic clay. Shale materials are continuously transported to the oceans via rivers as suspended particles and via air as aerosols.

One disadvantage in dealing with an average composition of any rock type is that important information on the variability of original data, interrelationships among measured elements, and chemical kinship or uniqueness of individual samples is lost during the process of averaging. Therefore, in the following sections, the statistical technique of factor analysis (Davis, 1973) is applied to original data in order to find the underlying interrelationship among elements (old variables), and kinship or variability among a set of given samples. Useful outputs from the factor analysis are the means, the standard deviations of variables, correlation coefficient matrices, eigenvalues, factor loadings, and factor scores. The extracted new factor 1 (F1), or new variable 1, from a factor analysis is the best linear

Table 2 Average compositions of the Earth's upper continental crust, shale, marine sediments (all in units of ppm, noted otherwise), along with seawater and hydrothermal vent solution from the East Pacific Rise (both in units of 10^{-9} g L^{-1}).

Element	*Z*	*Upper crust (1)*	*Shale (1)*	*Pelagic clay (1)*	*Fe–Mn nodule (1)*	*Fe–Mn crust (1)*	*Basal sed. (2)*	*Ridge sed. (3)*	*Phosphorite (1)*	*Seawater (1)*	*Vent solution (1)*
Ag	47	0.06	0.07	0.11	0.09		0.18	6.2h	2	2.5	4,000
Al (%)	13	7.83	8.8	8.4	2.7	0.41	2.73	0.5	0.91	300	120,000
As	33	1.6	13	20	140	230		145	23	1,700	35,000
Au (ppb)	79	2.3	2.5	2	2	250w		16c	1.4	0.03	
B	5	12	100	230	300		123	500	16	4.5E+6	6.0E+6
Ba	56	570	580	2,300	2,300	1,000	6,230	6,000hd	350	1.5E+4	1.5E+6
Be	4	3.2	3	2.6	2.5		6.7		2.6	0.21	120
Bi	83	0.054	0.43	0.53	7	29w	0.17		0.06	0.004	
Br	35	2.1	20		21		58			7E+7	6.9E+7
C (%)	6	0.023	1.2	0.45	0.1	0.12			2.1	3E+7	7.1E+7
Ca (%)	20	3.15	1.6	1	2.3	2.2	1.47		31.4	5E+8	4.8E+8
Cd	48	0.1	0.3	0.42	10	3	0.4	4	18	76	20,000
Ce	58	58	82	101	530	900	34	8.4	104	1.6	1,640
Cl	17	150	180						300	1.9E+10	1.8E+10
Co	27	17	19	74	2,700	8,400	82	105	7	1.2	13,000
Cr	24	69	90	90	35	9.1	15 d	55	125	252.6	
Cs	55	3.7	5	6	1					310	28,000
Cu	29	39	45	250	4,500	380	790	730	75	210	2.8E+6
Dy	66	3.5	4.7	7.4	31		20.7	7.3	19.2	1.5	69
Er	68	2	3	4.1	18	24	12.9	5.6	23.3	1.3	35
Eu	63	1.1	1.2	1.85	9	8.1	5.4	1.5	6.5	0.21	275
F	9	700	740	1,300	200		466		31,000	1.3E+6	0.14E+6
Fe (%)	26	4.17	4.72	6.5	12.5	12.3	20	18	0.77	250	1.39E+8
Ga	31	18	19	20	10		6.8		4	1.7	
Gd	64	3.9	5.1	8.3	32	39	22.6	6	12.8	1.3	92
Ge	32	1.5	1.6	1.6	0.8		3.3			4.3	
Hf	72	4	5	4.1	8		1.6			3.4	
Hg	80	0.08	0.18	0.1	0.15			0.4bf	0.06	0.42	
Ho	67	0.74	1.1	1.5	7	8.2	4.7		4.2	0.45	
I	53	0.5	19	28	400				24	58,000	
In	49	0.05	0.1	0.08	0.25					0.1	
Ir (ppb)	77	0.05	0.05	0.4	7	10v		0.8c		0.0015	
K (%)	19	2.56	2.66	2.5	0.7	0.38	1.15		0.42	3.9E+8	9.5E+8
La	57	30	43	42	157	190	98	29	133	5.6	
Li	3	23	66	57	80		125		5	1.8E+5	94E+5
Lu	71	0.32	0.42	0.55	1.8	3.9	2.2	0.88	2.7	0.32	
Mg (%)	12	1.64	1.5	2.1	1.6	0.88	2.08		0.18	1.3E+9	~0
Mn (%)	25	0.077	0.085	0.67	18.6	20.4	6.1	6	0.12	72	4.9E+7
Mo	42	1.6	2.6	27	400	370		30	9	10,300	
N	7	20	1,000	600	200				100	4.20E+5	
Na (%)	11	2.54	0.59	2.8	1.7	1.5	2.56		0.45	1.8E+10	1.0E+10
Nb	41	15	11	14	50		5.1		10	10	
Nd	60	26	33	43	158	150	87	23	98	4.2	500
Ni	28	55	50	230	6,600	3,900	460	430	53	530	
Os (ppb)	76	0.05	0.05	0.14	2	0.8v				0.0017	
P	15	860	700	1,500	2,500	3,900		9,000b	138,000	65,000	18,000
Pb	82	17	20	80	900	1,400	100	152h	50	2.7	75,000
Pd (ppb)	46	1	(1)	6	6	1.1		21c		0.07	
Pr	59	6.6	9.8	10	36	34	19.3		21	0.87	
Pt (ppb)	78	1	(1)	5	200	350				0.1	
Rb	37	110	140	110	17		16			0.12E+6	2.8E+6
Re (ppb)	75	0.4	0.4	0.3	1					8	
Rh (ppb)	45			0.4	13	14					
Ru (ppb)	44			0.2	8					0.005	
S	16	530	2,400	2,000	4,700				7,200	8.98E+8	1.3E+7
Sb	51	0.2	1.5	1	40		17d		7	150	
Sc	21	14	13	19	10				11	0.86	

(continued)

Table 2 (continued).

Element	*Z*	*Upper crust (1)*	*Shale (1)*	*Pelagic clay (1)*	*Fe–Mn nodule (1)*	*Fe–Mn crust (1)*	*Basal sed. (2)*	*Ridge sed. (3)*	*Phosphorite (1)*	*Seawater (1)*	*Vent solution (1)*
Se	34	0.14	0.6	0.2	0.6		2.6		4.6	145	4,800
Si (%)	14	30	27.5	25	7.7	2.2	10.8	6.1	5.6	2.5E+6	4.5E+8
Sm	62	4.5	6.2	8.35	35	30	18.6	5	20	0.84	137
Sn	50	3.3	3	4	2		0.6		3	0.6	
Sr	38	350	170	180	830	1,200	351		750	7.8E+6	5.8E+6
Ta	73	1.5	1.3	1	10		2.1			2.5	
Tb	65	0.6	0.84	1.42	5.4	5			3.2	0.21	
Te	52	0.003	0.07	1	10					0.07	
Th	90	11	12	13	30		2.4		6.5	0.05	0.3
Ti	22	3,300	4,600	4,600	6,700	7,700		240	640	10	
Tl	81	0.53	0.7	1.8	150		4.8	34h		14	
Tm	69	0.32	0.44	0.57	2.3	3.6			1.2	0.25	
U	92	2.8	2.7	2.6	5		4.2	22f	120	3,200	~0
V	23	140	130	120	500	500		450	100	2,150	
W	74	1.3	1.8	4	100					100	
Y	39	22	26	40	150	190	128		260	13	
Yb	70	2	2.8	3.82	20	24	13	5.7	13	1.5	33
Zn	30	67	95	170	1,200	540	470	380	200	320	6.9E+6
Zr	40	170	160	150	560		225		70	17	

Data sources: (1) Li (2000) and references therein; (2) Cronan (1976); (3) Bostrom and Peterson (1969); b = Bostrom (1973); bf = Bostrom and Fisher (1969); c = Crocket *et al.* (1973); d = Dymond *et al.* (1973); f = Fisher and Bostrom (1969); h = Horowitz (1970); hd = Heath and Dymond (1977); v = Vonderhaar *et al.* (2000); w = Wen *et al.* (1997); REE = Jarvis (1985) for basal sediments, and Bender *et al.* (1971) for ridge sediments. Pd and Pt values for shale are an educated guess based on the upper crust values.

combination of the old variables to account for the largest fraction of the total variance in the whole data set. The factor 2 (F2) is the next best linear combination to account for the residual variance, and so on. Only those factors with eigenvalues greater than one are extracted here. Therefore, the number of extracted factors (new variables) is much smaller than the number of original variables (elements). One may visualize the extracted factor loadings as the correlation coefficients between old variables and new factors, and the extracted factor scores as the new "concentrations" of new factors in each sample. Factor loading ranges between −1 and +1, and factor scores are allowed to have negative "concentrations." Any correlation coefficient (γ) or factor loading value between 0.7 and 1 represents in this context a strong correlation, between 0.5 and 0.69 a moderate correlation, between 0.3 and 0.49 a weak correlation, and between 0 and 0.29, no correlation at all. Similarly, a negative correlation coefficient represents anti- or inverse correlation.

7.01.2 PELAGIC SEDIMENTS

Pelagic sediments are generally defined as those deposited from dilute suspensions of detrital material that are distributed throughout deep-ocean water (Arrhenius, 1963). They are characterized by low accumulation rates of terrigenous material, and presence of relatively high percentages of authigenic minerals, cosmogenic material, and biogenic planktonic debris. Because of their slow depositional rates (~mm Kyr^{-1}), pelagic sediments tend to undergo high degrees of oxidation. The two major types of pelagic sediment are pelagic clay and biogenous oozes. Pelagic clays accumulate at abyssal depths (below the compensation depths for carbonate minerals), in oligotrophic regions of the ocean, far from sources of terrigenous turbidites (e.g., Leinen, 1989). They consist of fine-grained (generally <3 μm; Horn *et al.* (1970)) terrigenous material, largely eolian in origin, and generally contain a few percent of authigenic minerals. In shallower regions of the oceans, or under areas of high productivity, calcareous and/or siliceous tests accumulate to form biogenous oozes (Figure 2).

The dominant lithogenous components of pelagic sediments are quartz, plagioclase, and clay minerals (Table 1). The primary clay minerals are illite, chlorite, kaolinite, and smectite. Relative abundance of these constituents depends on proximity to sources; on an average, the clay-size fraction of North Pacific pelagic clay contains 30–40% illite, 10–15% chlorite, 10–15% quartz, 10–15% plagioclase, 10–15% kaolinite, and 0–5% smectite (Leinen (1989) and references therein). Submarine weathering of basaltic rock, and marine volcanic activity, are sources of plagioclase, amphiboles, pyroxenes, and olivines (Windom, 1976). Windom (1976) notes a number of other lithogenous phases that are generally present in marine sediments in minor amounts,

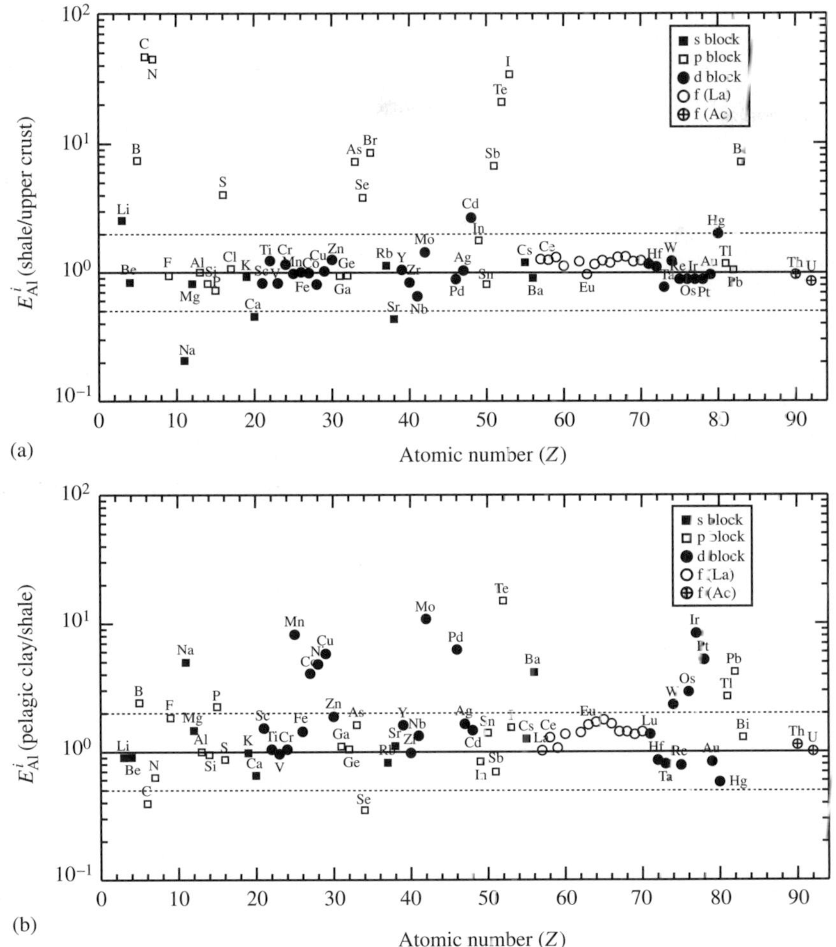

Figure 1 Enrichment factors of elements (E^i_{Al}) in: (a) average shale as compared with the Earth's upper continental crust, and in (b) marine pelagic clay as compared with average shale. Data are from Table 2.

particularly in coastal regions near their sources. Most lithogenous constituents undergo little or no transformation during deposition in the ocean basins (Windom, 1976). Elderfield (1976) discusses the cation exchange and sorption reactions that clay minerals undergo in seawater. In addition to its detrital source, the clay mineral smectite forms authigenically in the ocean basins. The formation of smectite at low temperature via the hydrogenous and hydrothermal means is discussed below.

Biogenous oozes are either calcareous or siliceous. Calcareous oozes are predominantly the calcitic tests of coccolithophores and/or foraminifera, or the aragonitic tests of pteropods. The solubility of $CaCO_3$ increases with decreasing temperature and increasing pressure, and thus with increasing depth in the oceans. Aragonite is ~1.45 times more soluble than calcite (Morse and Mackenzie, 1990), so aragonitic oozes are confined to shallower depths than the calcitic oozes. The compensation depth for each mineral is defined as the depth at which the rates of deposition and dissolution of that mineral are equal, so the content of that mineral drops to 0 wt.% (e.g., Morse and Mackenzie, 1990; Pinet, 2000). Compensation depths are dependent on a number of factors and vary between, and across, ocean basins (e.g., Berger *et al.*, 1976; Biscaye *et al.*, 1976; Kolla *et al.*, 1976; see discussion in Morse and Mackenzie, 1990). The aragonite compensation depth (ACD) generally averages around 3,000 m (Berger, 1978), whereas the calcite compensation depth (CCD) ranges from 5,000–6,000 m in the Atlantic (Biscaye *et al.*, 1976) to 4,000–5,000 m in the Pacific (Berger *et al.*, 1976). Calcareous oozes therefore commonly occur on ridge crests and other topographic highs.

Although calcite and aragonite are both made up of calcium carbonate, the different structures of the two minerals accommodate different elemental substitutions (see review in Mucci and Morse (1990)). Strontium and, to a lesser extent, magnesium and sodium can substitute for calcium in skeletal aragonite (Speer, 1983). Magnesian

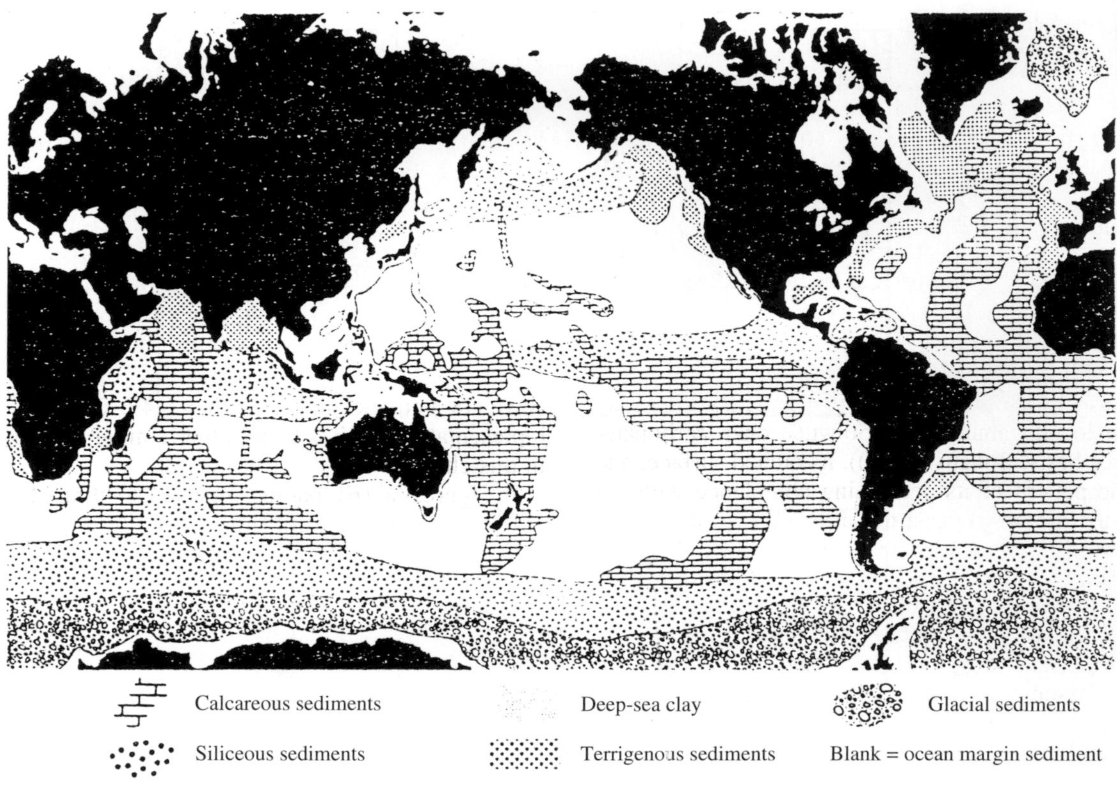

Figure 2 Distribution of the principal types of marine sediment on the seafloor. Calcareous oozes are found in relatively shallow oceanic regions, such as along ridge crests. Siliceous oozes are found predominantly under the equatorial regions of high productivity and in the high-latitude oceans. Deep-sea or pelagic clay occupies the abyssal depths of the ocean basins, far from terrigenous sources of river-borne or glacial sediment (source Davies and Gorsline, 1976) (reproduced by permission of Elsevier from *Chemical Oceanography*, **1976**, *5*).

calcites contain up to 30 mol.% $MgCO_3$ (e.g., Tribble *et al.*, 1995). Magnesian calcites are precipitated by a wide variety of shallow water marine organisms, but the increase in solubility accompanying magnesium substitution limits accumulation of these phases to shallow depths in the oceans (e.g., Bischoff *et al.*, 1987; Mackenzie *et al.*, 1983). Calcite also may contain minor substituents such as Sr^{2+}, Na^{+}, Mn^{2+}, and SO_4^{2-} ions.

Siliceous oozes are accumulations of opaline silica (opal-A, an amorphous phase of high water content and porosity) in the tests of diatoms, radiolarians, and/or silicoflagellates. Opal-A solubility at 25 °C is 60–130 ppm SiO_2(aq) (e.g., Williams *et al.*, 1985), and solubility increases with increasing temperature and pressure (Walther and Helgeson, 1977). Adsorption of aluminum and iron on the surfaces of siliceous tests decreases their solubility (Iler, 1955; Lewin, 1961). Opal-A is a metastable phase that with burial eventually recrystallizes to quartz, often with another metastable intermediary phase, opal-CT (e.g., Hein *et al.*, 1978; Williams *et al.*, 1985; Williams and Crerar, 1985). Opal-CT structurally resembles an inter-layering of the two silica phases, cristobalite and tridymite. Siliceous oozes are usually found on the floor of the Southern Ocean around Antarctica, along the equatorial oceans (especially in the Pacific), and around the northern edge of the Pacific Ocean (Broecker and Peng, 1982; Pinet, 2000).

The hydrogenous fraction of pelagic sediment may contain a wide variety of authigenic minerals (Table 1). The discussion here will be limited to the more abundant phases not considered in subsequent sections, or in separate chapters in this volume (e.g., glauconite, see Chapter 7.12). Barite is the only sulfate mineral present in abundance in pelagic sediment (e.g., Piper and Heath, 1989). Barite of hydrothermal origin is associated with mid-ocean ridge deposits (e.g., Edmond *et al.*, 1980). Crystals of nonhydrothermal barite have been recovered by seawater filtration (Dehairs *et al.*, 1980), and barite is commonly found at concentrations of 1–5% in sediments that are also rich in organic matter (Piper and Heath, 1989). Church (1979) suggested that the microenvironment of pore water might be suitable for precipitation of barite. Most of the world's oceans are undersaturated with respect to barite. The solution model for barite, thanks to Monnin *et al.* (1999), shows that an equilibrium with pure barite is reached or slightly exceeded only in the cold surface waters of the Southern Ocean, at intermediate depths in the Pacific, and in the deep waters of the Bay of Bengal. It is possible that the solubility of barite decreases upon substitution of strontium for barium stabilizing the phase (Church, 1979). Barite is secreted by certain marine plankton, and it is possible that much of marine barite is biogenic in origin (Dehairs *et al.*, 1980, 1990; Bishop, 1988; Bertram and Cowen, 1997).

Phillipsite is the most abundant zeolite in the surface sediments of the Pacific (Boles, 1977; Kastner and Stonecipher, 1978). Although it may be locally abundant (>50 wt.% on a carbonate-free basis, Bonatti (1963)), its etched surface, and the pattern of its decreasing abundance with the burial depth, suggest that it is a metastable phase under deep-sea conditions (Kastner, 1979). The primary mechanism of formation is thought to be alteration of basaltic glass, but it may also form by reaction of biogenic silica and dissolved Al^{3+} (Arrhenius, 1963). Phillipsite is commonly found in association with authigenic smectite, and the combined formation of the two minerals may be represented as

$$\text{basalt} + K^+ + Na^+ + SiO_2 + H_2O + H^+ \rightarrow \text{phillipsite} + \text{smectite} + Ca^{2+} + Fe_2O_3$$

(Piper and Heath, 1989).

The present-day sediments of the Atlantic do not contain a substantial amount of zeolite (Elderfield, 1976). Clinoptilolite has been found, however, in association with sepiolite, quartz, and montmorillonite (Hathaway and Sachs, 1965; Bonatti and Joensuu, 1968). The clay minerals palygorskite and sepiolite are usually minor constituents of marine sediments (Hathaway, 1979), and may be detrital (Weaver and Beck, 1977) or authigenic. Their hydrogenous occurrences are usually in basal sediment sections exposed to fluids of elevated temperatures (Bonatti and Joensuu, 1968; Church and Velde, 1979; see below).

Geochemically, pelagic sediments can be considered as mixtures of the major constituents: pelagic clay, calcium carbonate, opaline silica, and apatite. The following subsections provide some examples to show how the compositions of pelagic sediments change in different environments and to elucidate the factors controlling observed compositional changes.

7.01.2.1 Equatorial Pacific

Results of factor analysis (varimax; used in this chapter throughout) on chemical composition data for the surface siliceous oozes from the Wahine survey area (8°20′ N, 153° W; 34 samples; Calvert *et al.* (1978)) are shown in Figures 3(a) (factor loadings) and (b) (factor scores). The results for the surface pelagic sediments from the wider equatorial Pacific (28° N–13° S, 117° W–175° E; 17 samples; Calvert and Price (1977)) are shown in Figures 3(c) and (d). One unusual sample AMPH-85PG from the equatorial Pacific, which contains micronodules, is excluded from the factor analysis. A group of elements in Figures 3(a) and (c) is enclosed by dotted ellipse, whenever the correlation coefficients are greater than 0.49 among all pairs within a dotted ellipse (based on the correlation coefficient matrix). Any two elements connected by a solid line also have a correlation coefficient higher than 0.49. The same representation applies to all other similar figures in this chapter.

In Figure 3(a) (F1 loading versus F2 loading), F1 is characterized by Al, Ti, K, and Rb, which are associated with terrigenous shale in siliceous oozes of the Wahine survey area. Silicon is only partly associated with this group due to additional independent input of siliceous radiolarian shells. F2 is represented by Mn, Ni, Cu, Zn, Mo, Ba, and Sr. These elements are preferentially incorporated into MnO_2 phases. Factor 3 (F3) (not plotted here) represents a carbonate-fluorapatite (CFA) phase (Ca, P, and Y). Mixing of these four end-members (F1 to F3 and biogenic silica) in different proportions forms the observed sediments in the Wahine survey area. Rare earth elements (REEs), except cerium, are shown to be closely associated with CFA (Ca, P, and Y; Elderfield *et al.* (1981)) in the sediments from the Wahine survey area. As shown in Figure 3(b) (factor scores of F1 versus F2), most samples are chemically similar, except that two samples have high F2 group elements and one sample has high F1 group elements. The elements Fe, Co, As, Pb, Ce, and Mg are not included in the final factor analysis, because the concentrations of these elements correlate only weakly, or not at all, with one another and with any other element. The implication is that these elements are more or less evenly distributed among different phases of samples, and do not concentrate in any particular phase.

Factor analysis of the equatorial Pacific sediments (Figure 3(c)) indicates four major components: F1 (aluminosilicates: Al, Si, Ti, Th, Zr, K, Rb, Fe, As, and Mg), negative F1 (carbonates: Ca and Sr), F2 (manganese oxides: Mn, Ni, Co, Mo, Pd, Cu, and Zn), and F3 (CFA: P and Y). Most samples are chemically similar (Figure 3(d)), except that four samples from topographic highs are carbonate rich and one pelagic clay sample is manganese rich.

In the equatorial Pacific (8°40′ N–10°10′ N, 173°50′ W–175° W), several 6–7 m long cores and many surface sediment samples at water depths of ~6,000 m (below the carbonate compensation depth) were collected during the

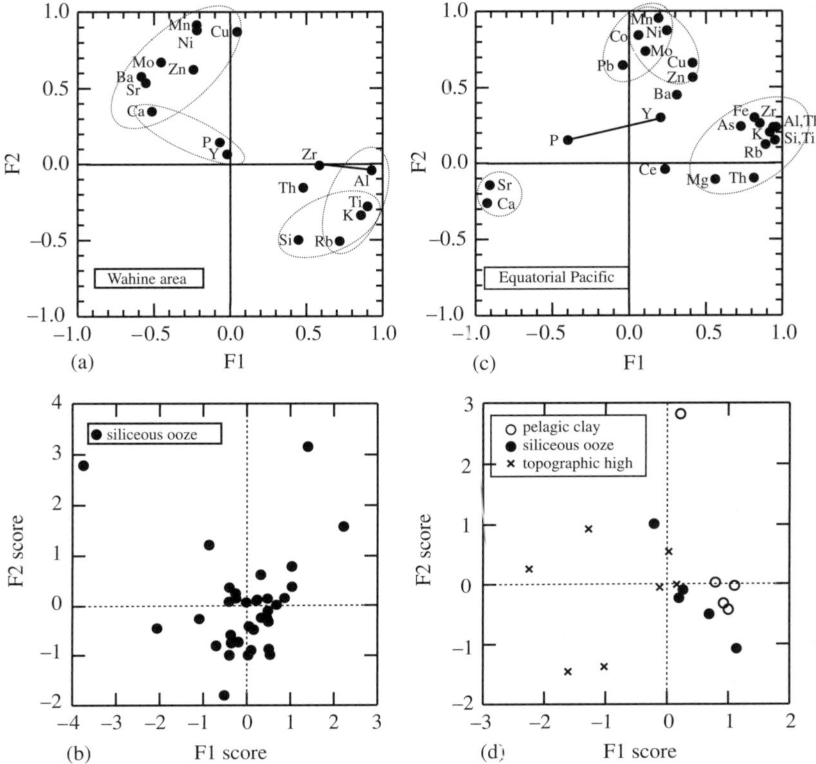

Figure 3 Plots of: (a) factor loadings and (b) factor scores of factors F1 and F2 for the surface siliceous ooze samples from Wahine survey area near Hawaii (Calvert *et al*., 1978). (c) and (d) are similar plots for surface pelagic sediment samples from the equatorial Pacific (Calvert and Price, 1977).

GH80-5 cruise by the Geological Survey of Japan. Samples were analyzed for major elements (Sugisaki and Yamamoto, 1984) and trace elements (Mita and Nakao, 1984). They classified the samples into three major groups: siliceous clay (radiolarian-rich) on the core top, pelagic clay in mid-core, and zeolitic clay (zeolite >5%) below. Factor analysis of 101 carbonate-free samples (excluding the four samples containing Al_2O_3 less than 6.1%, caused by a large dilution effect of biogenic silica) shows four major components (Figures 4(a) and (b)): F1 (aluminosilicates: Al, Fe, and Ti), negative F1 (CFA: Ca and P), F2 (manganese oxides: Mn, Ni, Co, Cu, Zn, and Mo), and F3 (zeolite + biogenic silica: K, Na, and Si). According to the correlation coefficient matrix, K, Na, and Si are also moderately correlated to Al. According to Figures 4(c) and (d), pelagic clay samples are mostly characterized by the presence of high aluminosilicate (F1), siliceous clay by low MnO_2 (F2) and moderately high silica (F3), and zeolitic clay by high MnO_2 (F2) and high CFA (negative F1). The *xy*-plots of the original concentration data of some selected key elements for F1–F4 (Figures 5(a)–(f)) confirm these characterizations. Notice the similarity between Figures 4(c) (F1 versus F2) and 5(a) (Al versus Mn), and Figures 4(d) (F1 versus F3) and 5(b) (Al versus Si). Al and Si are highly correlated for pelagic clay (Figure 5(b); open circles with a dotted dilution line). In contrast, silica concentration is high in siliceous and zeolitic clays relative to pelagic clay at a given aluminum concentration. Similarly, K and Na are strongly correlated with Al for pelagic and siliceous clays (Figure 5(d); open circles and crosses). However, K and Na concentrations can be very high in zeolitic clay. Data points in Figure 5(e) fall on the dotted line with a CaO/P_2O_5 slope of 1.32 (in weight) for fluorapatite, instead of a higher ratio for CFA, probably indicating a systematic error for estimation of phosphate.

7.01.2.2 South Pacific

Zhou and Kyte (1992) provided the concentration data of major and trace elements in the top 24.5 m of Deep Sea Drilling Project hole #596 (located at 23°51′S, 169°39′W with a water depth of 5,700 m). This core contains the K/T (Cretaceous/Tertiary) boundary with characteristic high Ir anomaly, and major and trace element data are available for 65 down-core samples. The results of factor analysis on those samples are summarized in Figures 6(a)–(d) (elements with both factor loadings of less than

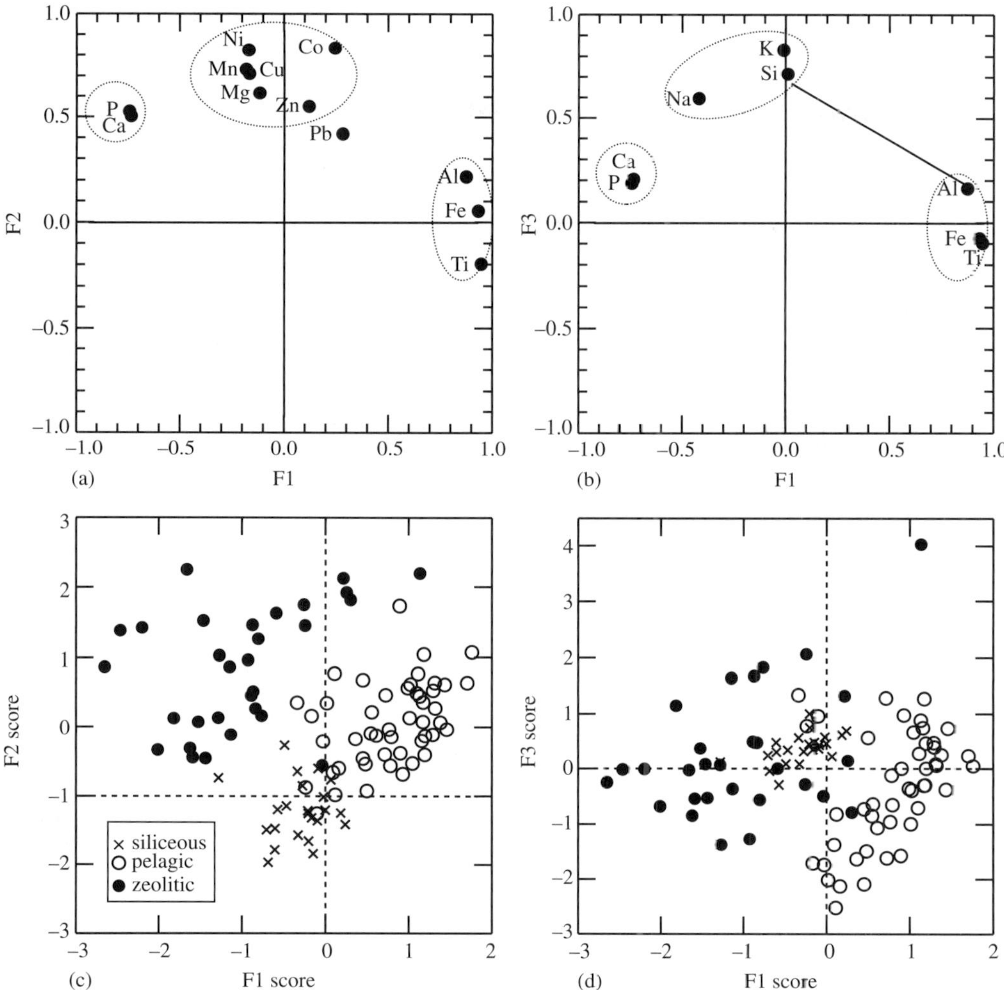

Figure 4 Plots of (a, b) factor loadings and (c, d) factor scores for the pelagic sediment samples from the equatorial Pacific (Sugisaki and Yamamoto, 1984; Mita and Nakao, 1984). Sediments are classified into siliceous clay, pelagic clay, and zeolitic clay.

0.49 are not shown here, and hereafter, for clarity). Iridium data are not included in the final factor analysis due to the iridium anomaly present at the K/T boundary. Six distinct components can be identified in Figures 6(a) and (b). They are F1 (hydrogenous component: Mn and Co; P, Th, and Ta are also partially related to this factor), negative F1 (shale: Al and Ti), F2 (volcanic ash input: Mg, Cr, Rb, and Cs), F3 (hydrothermal input: Fe, As, and Sb), F4 (CFA: Ca, P, and Th), and negative F3 plus negative F4 (zeolite and biogenic silica: K, Al, and Si). Sodium is quite diffused in these samples, and thus does not show any obvious correlation with the other elements. Samples with high F2 factor scores (open circles in Figure 6(c); high in volcanic ash input) are essentially present in the core top section (0.1–3.6 m) and a short section between 5.5 m and 6.1 m. The sections between 3.7 m and 5.2 m, and between 6.3 m and 10.3 m (solid circles in Figure 6(c)) are characterized by a high negative F1 factor score (shale) and low F2 (volcanic ash), whereas a high F1 factor score (hydrogenous input) along with a high F3 factor score (hydrothermal input) is observed in the section between 11.7 m and 16.4 m (crosses in Figures 6(c) and (d)). As one would have expected, Figure 7(a) (Mn versus Cr) appears similar to Figure 6(c) (F1 versus F2). In Figure 7(b), Mn and Fe are inversely correlated except for the hydrothermal samples (crosses), which are high in both Fe and Mn. The high phosphate content in hydrothermal and some hydrogenous samples (Figure 7(e)) is related to the observed high concentrations of ichthyoliths (fish teeth and bone debris; Zhou and Kyte (1992)). Ichthyoliths might serve as nuclei for precipitation of manganese oxide. The dotted line in Figure 7(c) represents a dilution line passing through pelagic samples. Hydrothermal and hydrogenous samples in all cases show excess biogenic silica as compared to the dilution line of

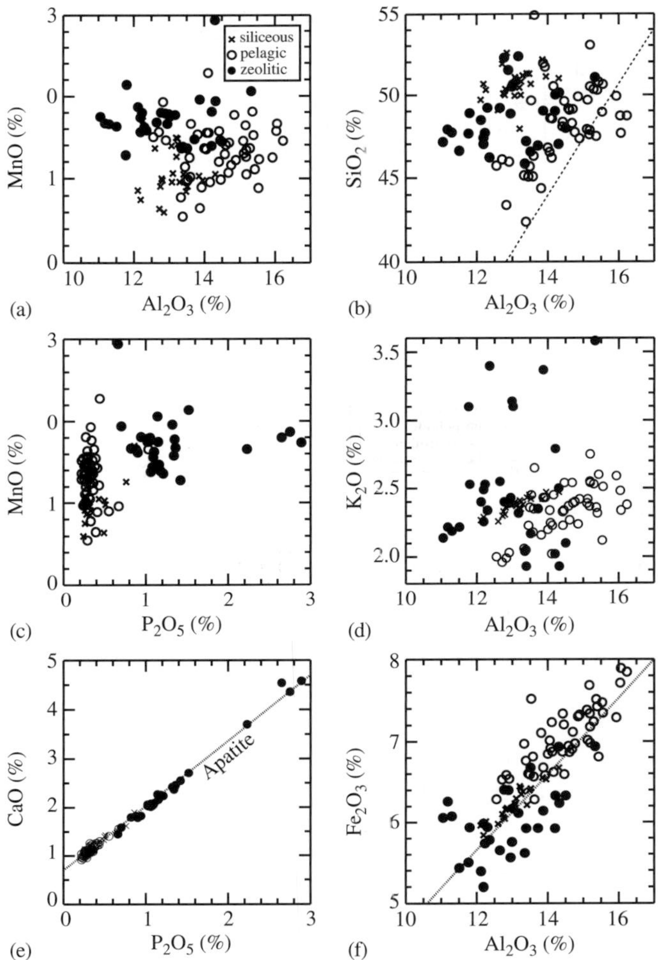

Figure 5 Scatter diagrams of selected elements for the same samples from Figure 3. The dotted lines in (b) and (f) are the dilution lines roughly passing through the pelagic clay samples. The dotted line in (e) has a nonzero intercept and the slope expected from fluorapatite ($Ca_5[PO_4]_3F$).

pelagic samples. This excess biogenic silica is probably important for the formation of zeolite in such samples. When one ignores the Ir anomaly samples in Figure 7(d), Ir is correlated with manganese as well as P. In Figure 7(f), the hydrogenous and hydrothermal samples lie on the dotted line that represents the CaO/P_2O_5 slope of CFA with a possible chemical formula of $Na_{0.5}Ca_5(PO_4)_{2.5}(CO_3)(F,OH)$. The original ichthyoliths no longer exist as hydroxyapatite.

7.01.2.3 Central Indian Basin

Fagel *et al.* (1997, personal communication) provided major and trace element data for five piston cores (20–30 m long) taken from the central Indian Basin at water depths ~4,800–5,400 m (below the carbonate compensation depth). Additional $^{87}Sr/^{86}Sr$ and $^{143}Nd/^{144}Nd$ data for some samples are given by Fagel *et al.* (1994). Biogenic silica (mainly radiolarian tests with minor diatoms and sponge spicules) and barite crystals are ubiquitous in the samples. Chemical data for clay-sized fractions of 81 samples were used in the factor analysis. Unusually high uranium contents of some samples were replaced with interpolated values. The elements Mn, Co, Ni, Pb, Ca, and W do not significantly correlate to one another or to any other analyzed elements, and thus are not included in the final factor analysis. The implication is that manganese oxide phases are not well developed (especially in reducing environments). The results are summarized in Figures 8(a)–(f).

There are seven distinct components: F1 (REE other than Ce), F2 (continental detritus: Rb, Th, U, Ta, Nb, and illite and chlorite clays), F3 (biogenic barite input: Ba, Sr, Ce, Hf, and Zr), F4 (aluminosilicates: Ti, Fe, Al, and Mg), negative F4 (biogenic silica), F5 (chromium minerals: Cr and V), and F6 (zeolite: K, Na, and Al). For convenience, samples are separated into three

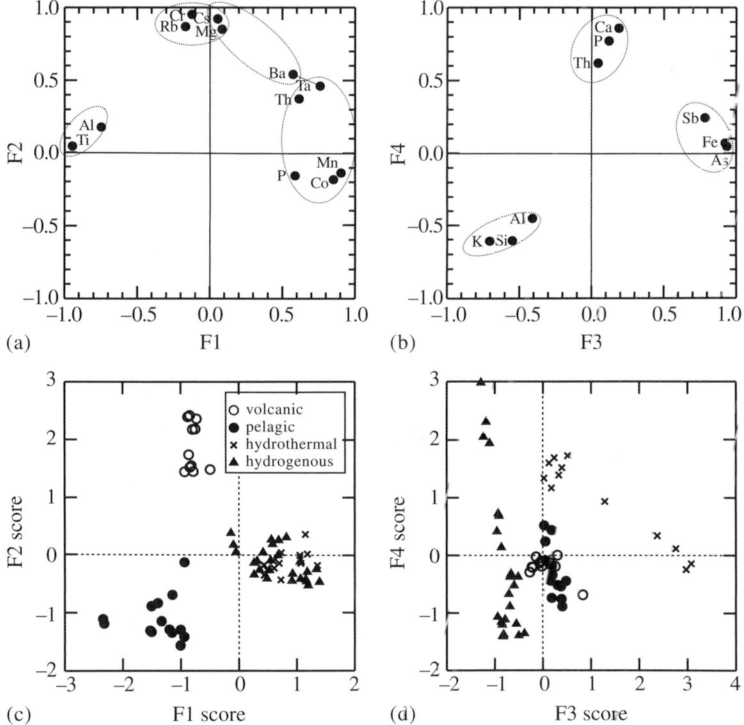

Figure 6 (a, b) Factor loadings and (c, d) factor scores for the pelagic sediment samples from Deep Sea Drilling Project hole #596 in the south Pacific. Samples are characterized by some inputs from volcanic ash, pelagic clay, hydrothermal solutions, and hydrogeneous precipitation (source Zhou and Kyte, 1992).

major groups (Figures 9(a)–(d)): continental detritus, authigenic, and biogenic groups. The continental detritus group is characterized by presence of low Nd and Ba (Figure 9(a)), and high $^{87}Sr/^{86}Sr$ ratio (Figure 9(b)). The authigenic group has high Ba and REEs including Ce (Figure 9(a)) and no excess Ce (Ce/Ce* ratio of ~1, where $Ce^* = 0.5[La + Pr]$; Figure 9(d)). The biogenic group is high in Ba, Ce and low in other REE (Figures 9(a) and (c)) and has high excess Ce (Ce/Ce* ratio of greater than one; Figure 9(d)). According to the factor score plots (Figures 8(d)–(f)), most of the continental detritus group samples have high F2 factor scores (i.e., high in Rb, Th, U, Ta, Nb, and illite and chlorite clay contents) and F5 factor scores (Cr and V). Most of the authigenic group samples have high F1 factor scores (i.e., high in REE other than Ce), and some also have high F3 factor scores (i.e., high in Zr, Hf, Ba, Sr, and Ce). Biogenic group samples have high F3 factor scores but low F1 and F2 factor scores. Since no phosphate analysis was performed, it is hard to pinpoint the host phases for the REE.

Continental detritus group samples contain minerals such as illite, chlorite, and probably polycrase [(Y,Ca,Ce,U,Th)$(Ti,Nb,Ta)_2O_6$] and chromite [$FeO{\cdot}(Cr,V)_2O_3$], supplied from the Himalayas region via the Ganges River (Fagel *et al.*, 1997). Th and U are highly correlated, but several continental detritus group samples and one authigenic group sample are unusually high in U content (Figure 9(e)). In sediments rich in organics, seawater UO_2^{2+} is often reduced to insoluble UO_2 and accumulates in sediments (Cochran, 1992). The high $^{87}Sr/^{86}Sr$ ratio (Figure 9(b)) confirms their continental origin. A linear correlation for all samples in the plot of 1/Sr versus $^{87}Sr/^{86}Sr$ (Fagel *et al.*, 1997) suggests mixing of two end-members (or sources) for Sr, namely, continental weathering products and seawater. Biogenic and authigenic group samples contain more seawater Sr ($^{87}Sr/^{86}Sr = 0.709$; Faure, 1986) as shown in Figure 9(b). Even though both biogenic barite ($BaSO_4$) and biogenic silica reflect high biological productivity along the equatorial upwelling zone, they are not correlated in the sediments (Figure 9(f)). The association of Zr and Hf with barite is puzzling. One possible explanation is that barite and zircon ($ZrSiO_4$) were sorted together in aqueous environment based on their similar specific gravities (4.5 versus 4.6).

7.01.3 FERROMANGANESE NODULES AND CRUSTS

Ferromanganese nodules are primarily centimeter-sized concretions of iron and manganese oxide. They occur on the deep seafloor where pelagic sediment accumulation rates are slow

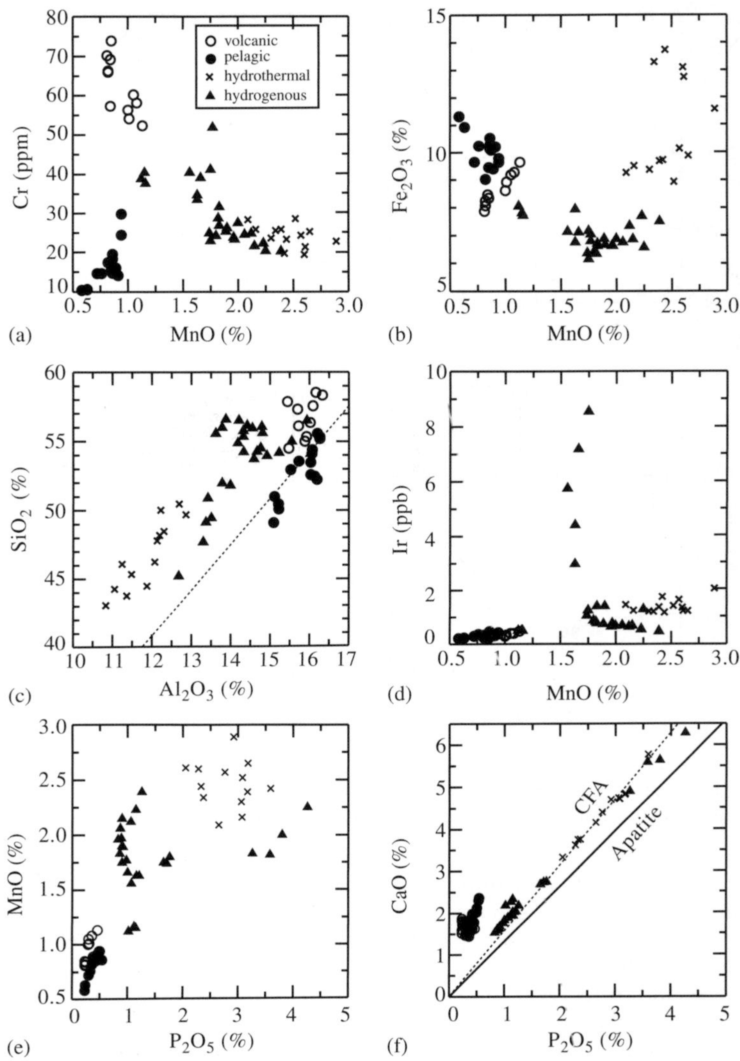

Figure 7 Scatter diagrams of selected elements for the same samples from Figure 5 plus K/T boundary samples. The dotted line in (c) is the dilution line roughly passing through the pelagic clay samples. The dotted and solid lines in (f) have slopes expected from CFA with a possible formula of $Na_{0.5}Ca_5(PO_4)_{2.5}(CO_3)(F,OH)$ and fluorapatite, respectively.

(Baturin, 1988; Pinet, 2000). Some nodules are also found buried in deep-sea sediment cores. Most nodules contain nuclei of different materials (e.g., shark teeth, shells, fish bones, and fragments of hard igneous and sedimentary rocks). The growth rates of nodules as well as crusts are extremely slow (millimeters per million years; Broecker and Peng, 1982; Cowen *et al.*, 1993; Vonderhaar *et al.*, 2000).

Ferromanganese crusts consist of iron and manganese oxide layers deposited on hard-rock substrates of seamounts, ridges, and plateaus, in areas where the ocean currents have been strong enough to keep the hard-rock substrates free of other sediment deposits for at least millions of years. Hein *et al.* (2000) summarized the distribution and geochemistry of ferromanganese crusts in the Pacific Ocean (The average compositions of ferromanganese nodules and crusts are summarized in Table 2).

More than 20 manganese oxide phases have been described from continental manganese deposits, but very few of these minerals have been positively identified from marine environments (Table 3). Useful summaries of the nature and nomenclature of the marine manganese minerals include those of Cronan (1976b), R. G. Burns and V. M. Burns (1979), and Post (1999). Marine manganese oxides are hydrous phases usually with poor crystallinity. They are often intimately intergrown with other phases, particularly iron-oxy-hydroxides and hydrous iron oxides. Marine manganese-oxide minerals host a variety of minor elements of economic

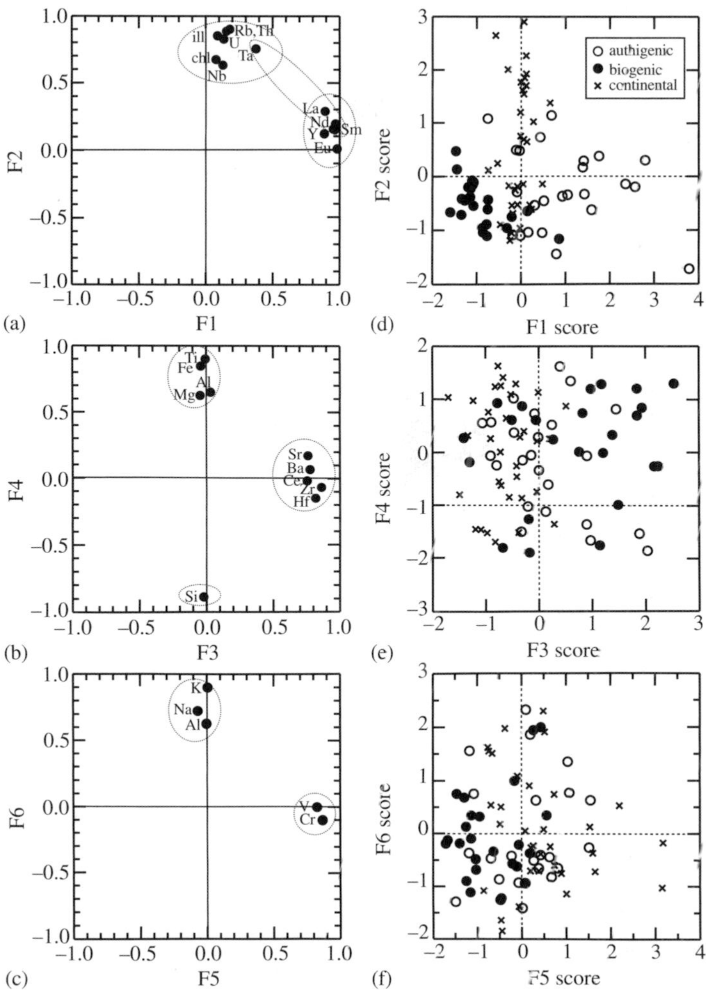

Figure 8 (a, b) Factor loadings and (c, d) factor scores for the clay-sized fractions of core samples taken from the central Indian Basin. Samples are classified into authigenic, biogenic, and continental detritus (source Fagel *et al.*, 1997).

significance, particularly Ni, Cu, and Co. Tetravalent manganese predominates, but minor amounts of Mn^{3+} and Mn^{2+} may also be present. These characteristics, in addition to numerous structural defects, interfere with long-range ordering and make identification by X-ray diffraction difficult (e.g., R. G. Burns and V. M. Burns, 1979).

The nature of marine manganese-oxide phases also makes acquisition of basic thermodynamic data virtually impossible. Gibbs free energy data are available for well-crystallized, anhydrous manganese oxides (Bricker, 1965; Crerar and Barnes, 1974; Crerar *et al.*, 1980; Hem, 1978), but for marine phases, ΔG and E^0 values are reported only for vernadite (Bricker, 1965) and buserite (Jeffries and Stumm, 1976). Using these data, calculated solubilities of vernadite (δ-MnO_2) and buserite at typical seawater Eh values exceed the measured concentrations of 18–38 nM Mn by many orders of magnitude (Crerar *et al.*, 1980). Kinetic factors obviously play a role in the formation and preservation of manganese-oxide minerals in the marine environment. Stumm and Morgan (1970) and Crerar *et al.* (1980) suggest that marine manganese-oxide phases form by catalytic oxidation and adsorption of Mn^{2+} on substrates such as fine-grained MnO_2 and $Fe(OH)_3$, or mineral or rock fragments. Crerar and Barnes (1974) have proposed that bacteria catalyze manganese-oxide precipitation in some environments and Chukhrov *et al.* (1978) have identified vernadite in relict bacterial forms in manganese nodules. Recent studies (see summary in Tebo *et al.* (1997)) have found that the rate of manganese oxidation can be accelerated by five orders of magnitude in the presence of microorganisms.

The major manganese-oxide phases found in ferromanganese crusts and nodules are todorokite, birnessite, and vernadite (e.g., R. G. Burns and

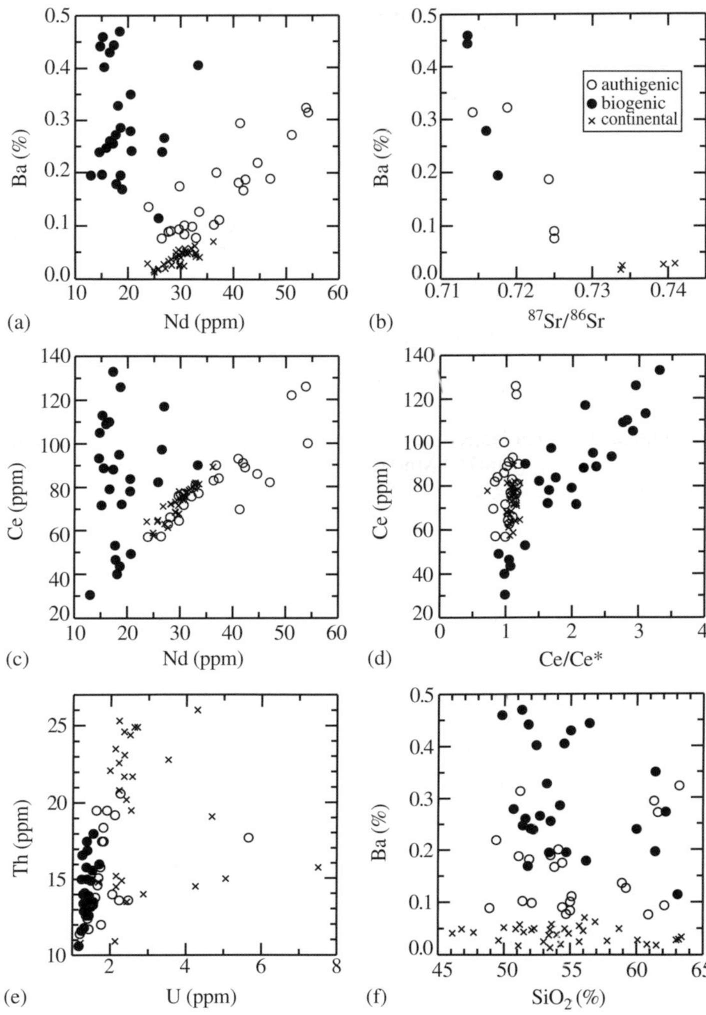

Figure 9 Scatter diagrams of selected elements, $^{87}Sr/^{86}Sr$, and Ce/Ce* for the same samples from Figure 8.

V. M. Burns, 1979; Bolton *et al.*, 1992; DeCarlo and Fraley, 1992; Banerjee *et al.*, 1999). Todorokite and vernadite are the two major manganese-bearing phases present in nodules (e.g., Bolton *et al.*, 1992; Usui and Moritani, 1992; Banerjee *et al.*, 1999; Mukhopadhyay *et al.*, 2002). Vernadite is the primary manganese-bearing phase found in ferromanganese crusts on seamounts and slopes throughout the Pacific (DeCarlo *et al.*, 1987; Alvarez *et al.*, 1990; DeCarlo, 1991; DeCarlo and Fraley, 1992; Hein *et al.*, 1992). Birnessite is reported in hydrothermal deposits and in low-metal nodules (e.g., Glover, 1977; Corliss *et al.*, 1978; Lonsdale *et al.*, 1980; Stoffer *et al.*, 1984). Reports of birnessite in normal hydrogenous and high-grade manganese deposits may be an artifact of analysis of dehydrated or heated samples (Usui and Moritani, 1992).

Todorokite, an iron-free hydrous manganese oxide, is often referred to as 10Å manganite or as buserite. Transmission electron microscopy, X-ray diffraction with Reitveld refinement, and neutron diffraction studies have revealed a tunnel structure made up of triple chains of manganese octahedra; cations and water are held inside the tunnels (Post and Bish, 1988; Mandernack *et al.*, 1995; Post, 1999). Substantial cation substitutions are indicated by the general formula given by R. G. Burns and V. M. Burns (1979): (Ca, Na, K, Ba, Ag)(Mg, Mn^{2+}, Zn)Mn_5^{4+} $O_{12}{\cdot}xH_2O$, and the phase also often contains significant amounts of nickel and copper. Todorokite is thought to form by precipitation of dissolved metals in moderately reducing sediments during early diagenesis (Usui and Moritani, 1992). It has been suggested that todorokite survives in oxidizing environments, despite its reduced manganese content, because of metastability resulting from the substitution of cations (e.g., nickel, copper, zinc, and magnesium) that are less susceptible to oxidation (R. G. Burns and V. M. Burns, 1979). Manganese nodules whose primary manganese-bearing

Table 3 Manganese minerals in marine Mn nodules and crusts.

Mineral	*Other names*	*Approximate formula*	*References*	*Comments*
Birnessite	7Å manganite	$(Na,Ca)Mn_7O_{14}\cdot 2.8H_2O)$	Post (1999)	Found in hydrothermal deposits and low-grade nodules.
		$(Na,Ca,K)(Mg,Mn)$-$Mn_6O_{14}\cdot 5H_2O)$	Burns and Burns (1979)	
Buserite	10Å manganite	NaMn oxide hydrate	Burns and Burns (1979)	Collapses to 7Å birnessite upon heating. Not an approved mineral name.
Chalcophanite		$ZnMn_3O_7\cdot 3H_2O$	Post (1999)	Rare in marine deposits. Weathering product.
Manjiroite		$Na_x[Mn(IV),Mn(III)]_8O_{16}$	Post (1999)	Rare component in ferromanganese crusts.
Todorokite	10Å manganate	$(Ca,Na,K)_{3-5}$ $[Mn(IV),Mn(III),$ $Mg]_6O_{12}\cdot 3\text{-}4.5H_2O$	Post (1999)	Common phase in Mn nodules.
		$(Ca,Na,K)[Mg,$ $Mn(II)]Mn_5O_{12}\cdot x\,H_2O$	Burns and Burns (1979)	
Vernadite	δ-MnO_2;	$MnO_2\cdot n\,H_2O$	Post (1999)	Natural specimens appear to be analogous to synthetic δ-MnO_2 thought to be disordered birnessite. Most common phase in ferromanganese crusts, also found in Mn nodules.
		$MnO_2\cdot n\,H_2O\cdot m(R_2O,RO,R_2O_3)$ where R = Na, Ca, Co, Fe, Mn	Burns and Burns (1979)	

phase is todorokite typically have a rough surface texture (e.g., Bolton *et al.*, 1992; Usui and Moritani, 1992). These nodules tend to have high Mn/Fe ratios (average 3.37) and high concentrations of Cu + Ni + Co (average 2.25%; Bolton *et al.* (1992)).

Vernadite is thought to be analogous to the synthetic phase δ-MnO_2. Although the crystal structure of vernadite is not known, it has been suggested that it is a disordered form of birnessite, a group of manganese oxide minerals with a layered structure (Giovanoli and Arrhenius, 1988; Post, 1999). R. G. Burns and V. M. Burns (1979) report a general formula for vernadite of $MnO_2\cdot n\,H_2O\cdot m(R_2O, RO, R_2O_3)$, where R = Na, Ca, Co, Fe, or Mn. Vernadite is thought to be hydrogenous and to precipitate as nodules and crusts under highly oxidizing conditions (Bolton *et al.*, 1992; Usui and Moritani, 1992). Manganese nodules containing vernadite as the dominant manganese phase, with secondary amounts of todorokite, tend to have smooth surface textures (e.g., Bolton *et al.*, 1992; Usui and Moritani, 1992). These vernadite-rich nodules have relatively low Mn/Fe ratios (average 1.37) and low concentrations of Cu + Ni + Co (average 1.02%; Bolton *et al.*, 1992). There is evidence of post-depositional recrystallization of vernadite to todorokite inside nodules and in metalliferous sediments (V. M. Burns and R. G. Burns, 1978a,b,c).

7.01.3.1 Equatorial Pacific Nodules

Results of factor analysis (varimax) on the composition data for 33 ferromanganese nodules obtained from the general equatorial Pacific region (28° N–13° S, 117° W–175° E; 18 samples by Calvert and Price (1977)) and the specific Wahine survey area (8° 20′ N, 153° W; 15 samples by Calvert *et al.* (1978)) are shown in Figures 10(a)–(d). There are four distinctive components in nodules, i.e., F1 (iron oxides: Fe, Co, As, Pb, Ti, Zr, Ce; and CFA: P, Y, and Sr); negative F1 (manganese oxides: Mn, Cu, Ni, Zn, Mg, Ba, and Mo); F2 (carbonates: Ca and CO_2); and negative F2 (aluminosilicates: Al, Si, K, and Rb). Thorium is related to both iron oxides and aluminosilicates. As shown in Figure 10(c), Mo and Mn are positively correlated, but Mo contents from Wahine area nodules are much lower than those from the equatorial Pacific region for a given Mn content. In general, nodules from the Wahine area are high in Mn group and low in Fe group elements as compared to those from the equatorial Pacific region (Figures 10(b) and (d)). The close association of iron oxides and CFA as F1 (Figure 10(a)) may imply that phosphate was first pre-concentrated by iron oxides, and then transformed into CFA.

The major difference between the ferromanganese nodules and their associated pelagic sediments (Section 7.01.2.1) is the introduction of

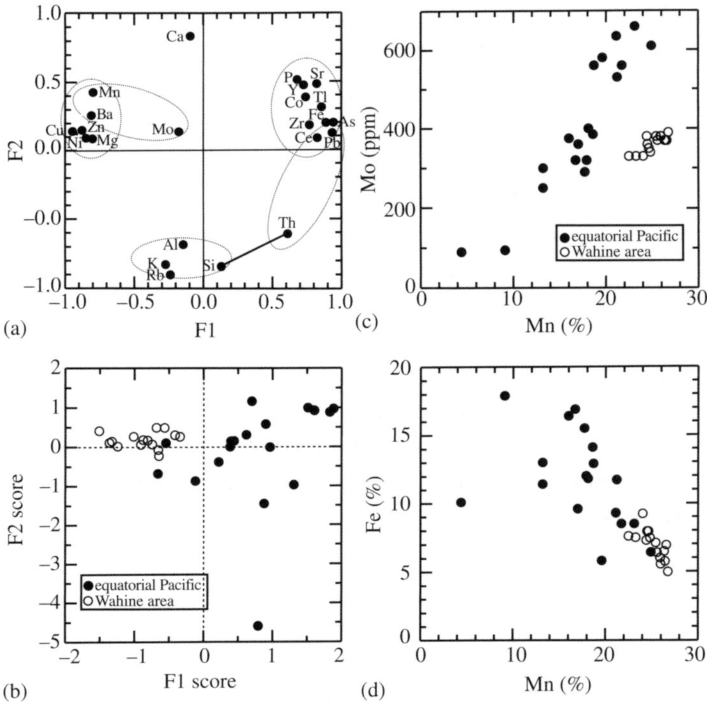

Figure 10 Plots of: (a) factor loadings of F1 and F2, (b) factor scores of F1 and F2, (c) Mn versus Mo, and (d) Mn versus Fe for ferromanganese nodule samples from Wahine survey area (Calvert *et al.*, 1978) and the equatorial Pacific (Calvert and Price, 1977).

a distinct iron oxide phase in the nodules. As summarized by Li (1982, 1991), the elements that associate strongly with the iron-oxide phase of nodules exist in normal seawater mainly as hydroxyl and carbonate complexes of tri- to pentavalent cations (e.g., As, B, Bi, In, Ir, Rh, REEs, Y, Ti, Th, U, Zr, Hf, Nb, and Ta), oxyanions (e.g., P, Re, Ru, Os, S, Se, Te, V, and W), and divalent cations with high first hydrolysis constants (Hg, Be, Pd, Co, and Pb). As shown in Figure 11(a) (based on the average data in Table 2), the E^i_{Al} [nodule/shale] values for most of those elements are similar to $E^{\mathrm{Fe}}_{\mathrm{Al}}$, indicating a similar uptake mechanism (adsorption–desorption equilibrium). The obvious exceptions are Co, Pb, Te, Pt, Rh, Ru, Ir, and Os, whose E^i_{Al} are much higher than $E^{\mathrm{Fe}}_{\mathrm{Al}}$, suggesting additional uptake mechanisms. For example, oxidation of Co^{+2}, Pb^{+2}, and Pt^{+2} into Co^{+3}, Pb^{+4}, and Pt^{+4} on the nodule surface, and the oxidation of Rh, Ir, Ru, and Os from di- or trivalent to tetravalent were suggested by Goldberg *et al.* (1986). In contrast, cobalt, lead and noble-metal cations are mainly associated with MnO_2 phases in pelagic sediments, where iron oxide phases are not well developed. CFA is often closely related to the iron-oxide phase in manganese nodules (Figure 10(a)) and contributes significant amounts of Ca, Sr, P, C, REEs, and Y in manganese nodules. The E^i_{Al} values of near one for Si, K, Na, F, Cr, Rb, Cs, Sc, Ga, Sn, and Ge (Figure 11(a)) suggest that these elements are contributed mainly by aluminosilicate detritus phases (shale) in nodules.

More than one hundred ferromanganese nodules were collected in the equatorial Pacific area (8° 40′ N–10°10′ N, 173°50′ W–175° W) during the GH80-5 cruise to study the relationship among the chemical composition, surface texture (rough, less rough, smooth), and the nodular part analyzed (whole nodule, outer layer, inside, top, and bottom) (Usui and Mochizuki, 1984). Factor analysis results of these samples are shown in Figures 12(a)–(d). There are three distinct components: F1 (manganeses oxides: Mn, Ni, and Cu), negative F1 (iron oxides: Fe and Co), and F2 (aluminosilicates: Al and Si). The result verifies the association of Co with Fe (Figures 12(a) and (c)) in ferromanganese nodules. Top parts of nodules are usually high in iron-oxide group elements and low in manganese oxide group elements as compared with the bottom parts (Figures 12(c) and (d)). By comparing Figures 12(d) and (b), it is evident that the top parts have smooth surface texture, and the bottom parts rough to less-rough textures. The rough samples mostly consist of the manganese oxide mineral of todorokite, which is thought to form at the nodule-sediment interface through the oxidation of Mn^{2+} supplied by the interstitial water of the underlying sediments.

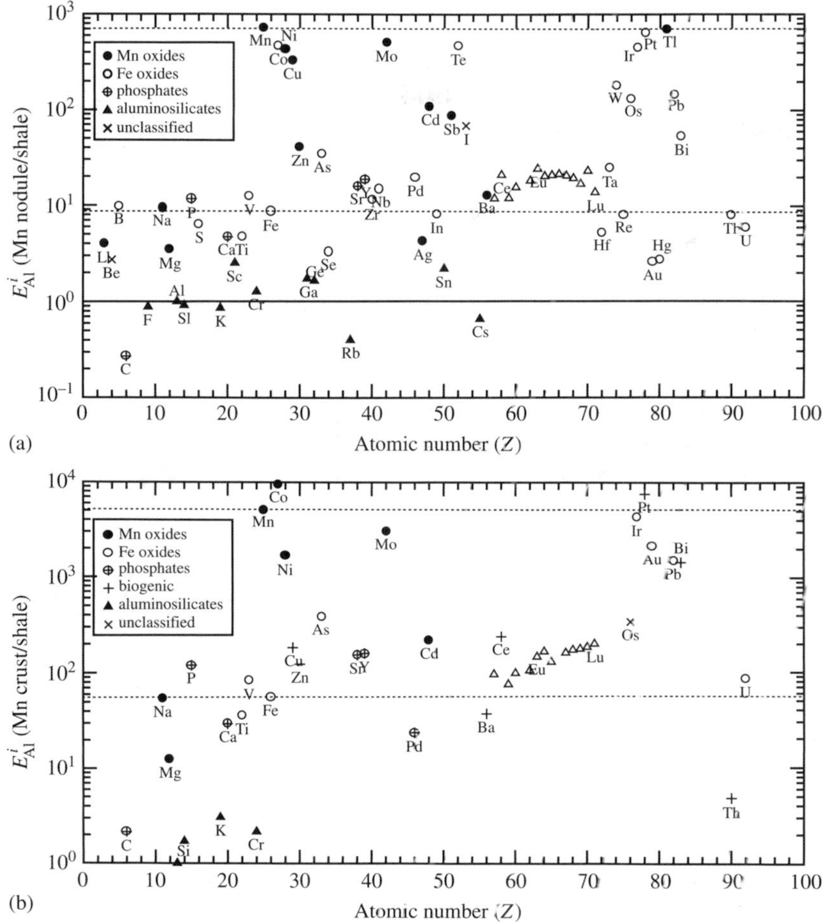

Figure 11 Enrichment factor of elements in (a) average ferromanganese nodule, and (b) average ferromanganese crust, as normalized to Al and relative to shale (Table 2). The specific associations of elements with various phases are shown by different symbols.

7.01.3.2 Seamount Ferromanganese Crusts from the Central Pacific

Seamount crusts near Kiribati (4° S, 170° W; 80 mm thick) and the Hawaiian archipelago (26° N, 160° W; 65 mm thick) were analyzed for minor and trace elements at 2–3 mm depth intervals (Wen *et al.*, 1997). Factor analysis results (Figures 13(a)–(d)) indicate five distinct components: F1 (aluminosilicates and iron oxides: Si, Al, Mg, Y, REE, Fe, Th, and partially V), negative F1 (manganese oxides: Mn, Ni, Co, and partially Zn), F2 (biogenic barite: Ba, Cu, Ce, Pt, Bi, and partially Zn, Sr, and V), F3 (CFA: Ca, P, and Y), and F4 (U and Pb). The Hawaiian crusts are high in aluminosilicate plus iron oxide phases and biogenic barite group elements and are low in manganese oxide group elements as compared to Kiribati crusts (Figures 13(c), 14(a)–(c), and (e)). Only a small fraction of iron is contributed by the shale component (Figure 14(b)), while a large fraction of silicon is contributed by the shale component (Figure 14(a)). REEs are probably contributed equally by iron oxide as well as shale components. Cobalt is strongly correlated with manganese (Figure 13(a)) instead of with iron as in the case of ferromanganese nodules (Figures 10(a) and 12(a)). The difference can be attributed to the fact that the manganese oxide mineral appears mainly as todorokite in nodules, and as vernadite in crusts (Li, 2000). CFA (e.g., $Na_{0.5}Ca_5(PO_4)_{2.5}(CO_3)F$; the dotted line in Figure 14(d)) also contributes some fraction of REE (Figures 13(a) and (b)). The cerium anomaly (Ce/Ce*) increases with cerium concentration in Hawaii and Kiribati crusts separately (Figure 14(f)). The meaning of the Pb and U association (Figure 13(b)) is not clear at first glance, but factor analysis of Kiribati crusts alone showed that lead and uranium belong to the iron oxide component (not shown here).

Based on the present factor analysis results and additional results from Marshall Island crusts (Li, 2000), the associations of elements to specific phases or components in ferromanganese crusts are summarized in Figure 11(b) (data from Table 2).

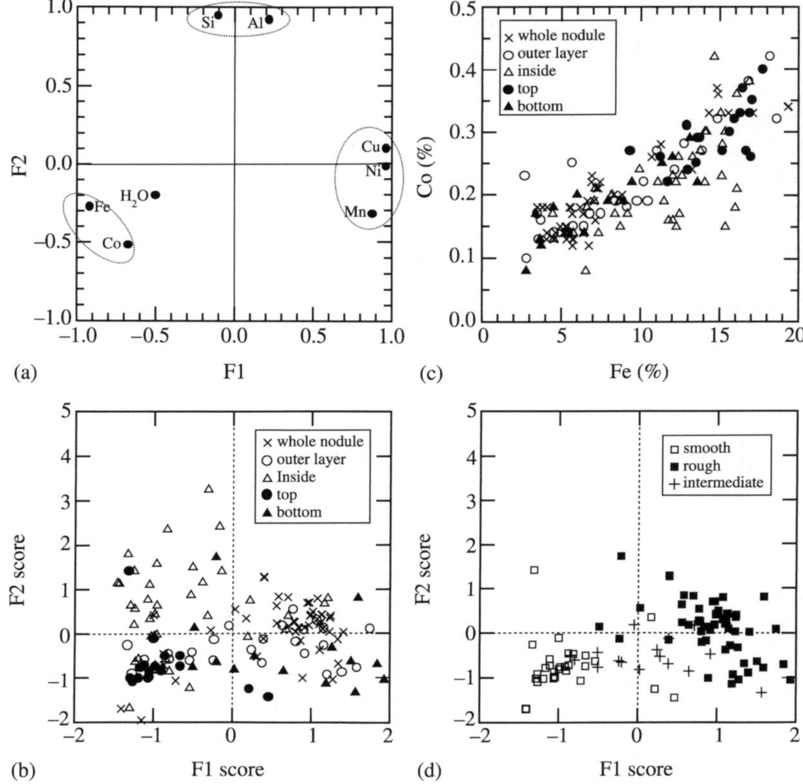

Figure 12 Plots of factor loadings (a), factor scores (c and d), and Fe versus Co (b) for ferromanganese nodules from the equatorial Pacific. See the details in the text (source Usui and Mochizuki, 1984).

7.01.4 METALLIFEROUS RIDGE AND BASAL SEDIMENTS

Hydrothermal vent solutions from the mid-ocean ridges are enriched in many elements as compared with seawater (Table 2). Notable exceptions are Mg, S, P, F, and U, which are depleted in vent solutions. These enriched elements are mostly deposited on the crest and flank of the ridges and form the so-called metalliferous ridge sediments (Figure 15). When metalliferous ridge sediments are eventually covered by regular pelagic sediments during ocean floor spreading, they are referred to as metalliferous basal sediments.

Oceanic hydrothermal metal deposits have been classified into four categories by Bonatti (1983). Pre-discharge deposits form by reaction of the hydrothermal solutions with igneous crust prior to discharge of the solutions at the seafloor. These deposits include massive and disseminated metal sulfides and oxides, and are not considered further here, as they do not form sedimentary deposits. Syn-discharge deposits form at the sites of hydrothermal discharge at the seafloor and include high-temperature smoker-type deposits as well as more disseminated deposits at the locations of diffuse discharge. Because these deposits are confined to regions of active or extinct hydrothermal discharge, their areal extent is limited. Post-discharge deposits form after mixing of the hydrothermal fluids with seawater and may be concentrated near the discharge sites or widely dispersed beneath the hydrothermal plume. In the Pacific, post-discharge deposits are distributed across thousands of kilometers of seafloor. The depth of the median valley of the Mid-Atlantic Ridge, however, limits dispersion of the hydrothermal plume in the Atlantic (Mills and Elderfield, 1995). Bonatti's fourth category of metalliferous sediment is intra-sedimentary deposits formed when hydrothermal solutions are directly discharged into a sedimentary column. Intra-sedimentary deposits are limited to ridge segments in close proximity to terrestrial sources of the sediment.

The minerals produced in these metalliferous deposits reflect the mixing of the two end-member solutions (hydrothermal solution and seawater). The mixing process involves cooling of the hydrothermal fluids and heating of seawater, changes in pH and oxidation state, reaction with previously formed precipitates or sedimentary components, and nonequilibrium kinetic effects

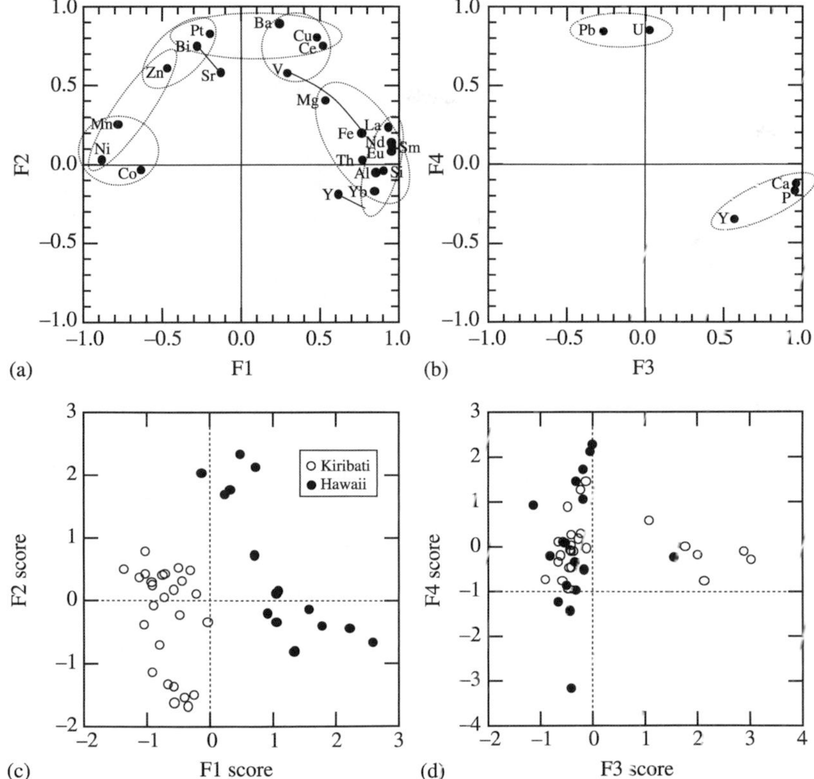

Figure 13 (a, b) Factor loadings and (c, d) factor scores for seamount ferromanganese crusts near Kiribati Island and Hawaiian archipelago (source Wen *et al.*, 1997).

(e.g., Janecky and Seyfried, 1984). Weathering and mass wasting of metalliferous deposits result in new phases formed during replacement reactions and dispersion of sulfide and oxide material into surrounding sediments (Mills and Elderfield, 1995). The wide array of chemical and physical conditions present during the formation of hydrothermal metalliferous deposits leads to very diverse assemblages of minerals. Haymon (1989) compiled mineralogical data from the East Pacific Rise and Galapagos Rift ridge crest and seamount deposits, and reports 37 different minerals. These phases, as well as additional minerals identified in other deposits, are listed in Table 4. Syn-discharge deposits are dominated by sulfide minerals, the most common among them being sphalerite, wurtzite, pyrite, and chalcopyrite (Haymon, 1989). Anhydrite is an abundant component of both black and white smoker chimneys. Barite, goethite, and amorphous silica are also common minerals in syn-discharge deposits (Table 4).

Metal-enriched basal sediments overlying igneous crust are widespread in the Pacific Basin (e.g., Bostrom and Peterson, 1966; Sayles *et al.*, 1975; Heath and Dymond, 1977; Corliss *et al.*, 1978; Edmond *et al.*, 1979; Jarvis, 1985; Mills and Elderfield, 1995). The mineralogy of these deposits reflects the mixing of iron and manganese-bearing phases formed from hydrothermal solutions with normal pelagic sediment. The pelagic component commonly includes calcite, opaline silica, and a variety of clay minerals, with accessory quartz, feldspar, barite, biogenic apatite, and zeolites (Table 4). Metalliferous minerals may be intermixed with the pelagic phases, forming metal-rich carbonates or claystones, or may even be concentrated in the deposits inter-bedded with the pelagic material.

The nature of postdischarge deposits varies with increasing distance from the ridge crest. For example, deposits on the East Pacific Rise ridge crest precipitate from the hydrothermal plume prior to significant dilution of the plume by seawater. These deposits have high accumulation rates and are rich in iron sulfides (e.g., Edmond *et al.*, 1979). Just off the ridge crest, deposits tend to consist of amorphous or poorly crystalline ferromanganese oxyhydroxides with only traces of goethite, vernadite, and iron smectite (Heath and Dymond, 1977; Marchig and Gundlach, 1982). With greater distance from the ridge, the amorphous iron oxyhydroxides recrystallize to goethite, or react with silica to form iron smectite. Accumulation rates are lower, and the metalliferous minerals are intermixed with pelagic sediment to a greater extent. Metal concentrations tend to be higher in distal deposits, because the pelagic minerals provide an additional source of metals,

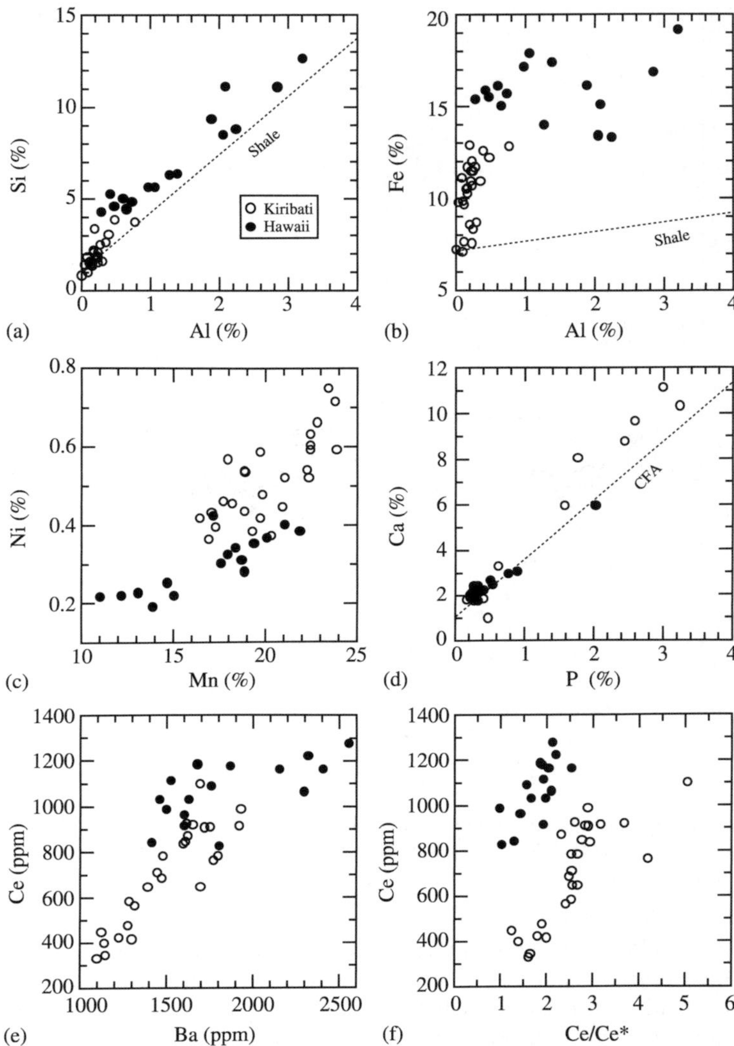

Figure 14 Scatter diagrams of selected elements for the same samples from Figure 13. The dotted lines in (a) and (b) are the slopes for the average shale. The dotted line in (d) is the slope for CFA with a possible formula of $Na_{0.5}Ca_5(PO_4)_{2.5}(CO_3)(F,OH)$.

and the slower accumulation rates allow for more extensive adsorption of metals from seawater (e.g., Heath and Dymond, 1977).

In mound-type deposits such as the Galapagos Mounds (e.g., Corliss *et al.*, 1978; Honnerez *et al.*, 1983), metalliferous deposits are formed by percolation of low-temperature hydrothermal fluids through a sediment pile. Layers of iron smectite are interbedded with pelagic sediment. The sediment column is suboxic, resulting in mobilization of manganese and formation of manganese-oxide crusts of birnessite, todorokite, and vernadite (δ-MnO_2) near the sediment–water interface (e.g., Sayles *et al.*, 1975; Corliss *et al.*, 1978; Edmond *et al.*, 1979). Deposits similar to the Galapagos Mounds have been identified in the Gulf of Aden (Cann *et al.*, 1977) and at the Famous area of the Mid-Atlantic Ridge (Hoffert *et al.*, 1978).

Iron-rich smectite, frequently identified as either iron montmorillonite or nontronite, is a very common component of postdischarge deposits (Table 4). In the Galapagos Mound deposits, nontronite is the major component of the hydrothermal mud, whereas iron montmorillonite occurs in the noncarbonate fraction of the interbedded pelagic ooze (McMurtry *et al.*, 1983). Both of these iron-rich smectites are thought to have been formed by reaction of hydrothermal iron-oxyhydroxides with biogenic silica (e.g., Heath and Dymond, 1977; Corliss *et al.*, 1978; Dymond and Eklund, 1978), or hydrothermal silica (Sayles *et al.*, 1975). Oxygen isotopic analyses on nontronites and iron montmorillonites from the East Pacific Rise, Bauer Basin, and the Galapagos Mounds indicate low-temperature hydrothermal origin (25–50 °C) (McMurtry and Yeh, 1981; McMurtry *et al.*, 1983).

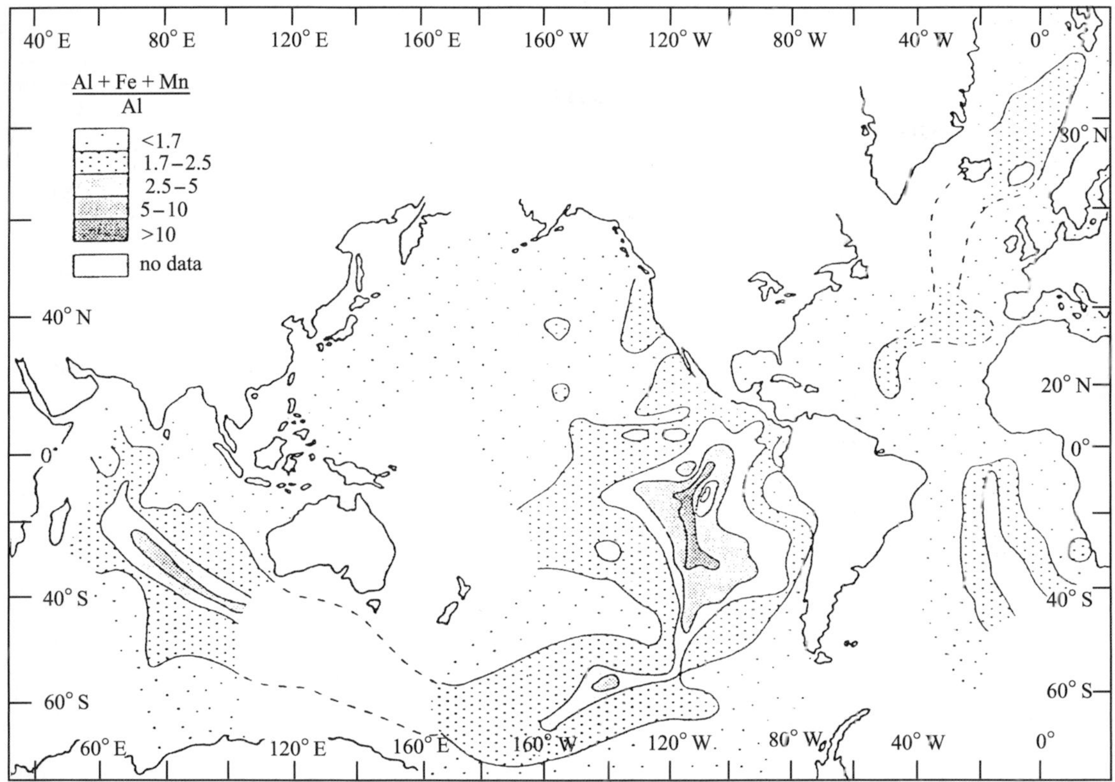

Figure 15 (Fe + Mn + Al)/Al ratio in marine sediment. Metalliferous sediments have elevated ratios compared to background pelagic sedimentation. The hydrothermal source of Fe and Mn to metalliferous sediments is reflected in the elevated ratios along mid-ocean ridges (source Mills and Elderfield, 1995; after Bostrom *et al.*, 1969).

Low-temperature (3–20 °C) experimental synthesis of nontronite has been accomplished by precipitation and aging of iron hydroxides with co-precipitated silica under reducing conditions and at near neutral pH values of 7–9 (Harder, 1976). Iron smectites sometimes coexist with mixed-layer nontronite/celadonite (e.g., Sayles *et al.*, 1975; Honnerez *et al.*, 1983).

Palygorskite and sepiolite are magnesium-rich fibrous aluminosilicates that have been identified in basal deep-sea sediments (Table 4; e.g., Hathaway and Sachs, 1965; Bowles *et al.*, 1971; Bonatti and Joensuu, 1968; Church and Velde, 1979; Jones and Galan, 1988; Velde, 1985). These phases are commonly associated with smectite, and it has been suggested that they originate by alteration of montmorillonite by low-temperature, magnesium-rich, hydrothermal solutions (Bonatti and Joensuu, 1968), e.g.,

$$\text{montmorillonite} + \text{silica} + Ca^{2+} + Mg^{2+} \rightarrow \text{palygorskite} + \text{clinoptilolite}$$

There is not uniform agreement, however, that a montmorillonite precursor is required. Wollast *et al.* (1968) synthesized sepiolite at low temperature by addition of dissolved silica to seawater at pH 8. Bowles *et al.* (1971) suggested that sepiolite should precipitate from seawater to which magnesium-rich solutions have been added.

Intra-sedimentary metalliferous deposits are formed at sites of active venting on sedimented ridge segments. The study of Koski *et al.* (1985) of the Guaymas Basin contrasts the mineralogy of sediment-hosted massive sulfide deposits to those of sediment-starved ridge systems. Reaction of the hydrothermal fluid with the organic-rich sediment pile results in a highly reducing, O_2-depleted solution. Precipitation of minerals within the sediment pile and at the sediment-water interface leads to coexistence of disequilibrium assemblages such as pyrrhotite-barite. The metalliferous deposits of the Guaymas Basin consist of pyrrhotite-rich massive sulfide, barite, barite + calcite, talc, and opaline silica with a substrate of fossiliferous, clay-rich ooze (Table 4). Compared to other massive sulfides the Guaymas deposits have a higher pyrrhotite/pyrite ratio and lower zinc sulfide and total ore metal content, and greater abundances of sulfates, carbonates, and silicates.

Average chemical compositions of metalliferous ridge and basal sediments and their data sources are summarized in Table 2. The calculated enrichment factors E^i_{Al} [sample/shale] and E^i_{Mn} [sample/hydrothermal vent] are plotted in

Table 4 Minerals found in marine metalliferous deposits.

Mineral/compound	*Chemical formula*	*Syn-discharge*		*Post-discharge*	*Intra-sed.*
		Ridge crest[a]	*Seamount*[b]	*Basal sections and mounds*[c]	*Sedimented ridge*[d]
Sulfides/sulfosalts					
Sphalerite	Zn(Fe)S	R-A	R-T		M
Wurzite	Zn(Fe)S	R-A			
Pyrite	FeS_2	R-A	A		m
Chalcopyrite	$CuFeS_2$	R-A	C-A		m
Iss-isocubanite	$CuFe_2S_3$	R-A	T		
Marcasite	Fe_2S	T-A	C-A		
Melnicovite	FeS_{2-x}	C-A	C-A		
Pyrrhotite	$Fe_{1-x}S$	T-A	R-T		M
Bornite-chalcocite	$Cu_5FeS_4{-}Cu_2S$	R-C	R-T		
Covellite	CuS	R	R-T		
Digenite	Cu_9S_5	T-R	R-T		
Idaite	$Cu_{5.5}FeS_{6.5}$	T			
Galena	PbS	T	R		m
Jordanite	$Pb_9As_4S_{15}$	T			
Tennantite	$(Cu,Ag)_{10}(Fe,Zn,Cu)_2As_4S_{23}$	T			
Vallerite	$2(Cu,Fe)_2S_2{\cdot}3(Mg,Al)(OH)_2$	T			
Sulfates					
Anhydrite	$CaSO_4$	A			
Gypsum	$CaSO_4{\cdot}2H_2O$	T-R			m
Caminite	$MgSO_4{\cdot}xMg(OH)_2{\cdot}(1-2x)H_2O$	R			
Barite	$BaSO_4$	T-C	T-A	m	M
Jarosite-natrojarosite	$(K,Na)Fe_3(SO_4)_2(OH)_6$	R	R		m
Chalcanthite	$CuSO_4{\cdot}5H_2O$		R		
Carbonates					
Magnesite	$MgCO_3$	T			
Calcite	$CaCO_3$			M	M
Aragonite	$CaCO_3$				M
Elements					
Sulfur	S	R-A	T		m
Oxides/oxyhydroxides					
Goethite	FeO(OH)	R-C	C	M	M
Limonite	FeO(OH)		C		
Lepidocrocite	FeO(OH)	T-R			m
Hematite	Fe_2O_3	T-R			
Magnetite	Fe_3O_4	T-R			
Psilomelane	$(Ba,H_2O)_2Mn_5O_{10}$		R-T	m	
Birnessite		C	C	M	
Todorokite		T	R	M	
Vernadite	δ-MnO_2	T	R	M	
Silicates					
Amorphous silica (opal)	$SiO_2{\cdot}n\,H_2O$	R-A	R-A	M	M
Quartz	SiO_2		A-T	m	M
Plagioclase	variable	R-C		m	m
Talc	$Mg_3Si_4O_{10}(OH)_2$	R-C	T		M
Chrysotile	$Mg_3Si_2O_5(OH)_4$	A			
Nontronite	$(Fe,Al,Mg)_2(Si_{3.66}Al_{0.34})O_{10}(OH)_2$	T-A	T-A	M	
Smectite/illite-smectite	variable		R-T	M	M
Saponite	$Na_{0.33}Mg_3(Si_{3.67}Al_{0.33})O_{10}(OH)_2$	C		M	
Palygorskite	$(OH_2)_4Mg_5Si_8O_{20}(OH)_2{\cdot}4H_2O$			M	
Sepiolite	$(OH_2)_4Mg_8Si_{12}O_{30}(OH)_2{\cdot}4H_2O$			M	
Phillipsite	$K_{2.8}Na_{1.6}Al_{4.4}Si_{11.6}O_{32}{\cdot}10H_2O$			m	
Clinoptilolite	$K_{2.3}Na_{0.8}Al_{3.1}Si_{14.9}O_{36}{\cdot}12H_2O$			m	
Hydroxychlorides					
Atacamite	$Cu_2Cl(OH)_3$		T-C		
Phosphates					
Carbonate fluorapatite	$Ca_{10-x-y}Na_xMg_y$-$(PO_4)_{6-z}(CO_3)_zF_{0.4z}F_2$			m	

Key: A = abundant; C = common; R = rare; T = trace; M = major; m = minor.
[a] Data for ridge crest deposits from Alt (1995), Dymond and Roth (1988), Haymon(1982, 1989), Haymon and Kastner (1981), Oudin (1983), Styrt *et al.* (1981), Thompson *et al.* (1988), Zierenberg *et al.* (1984). [b] Data for seamount deposits from Alt *et al.* (1987), Alt (1988), Hekinian and Fouquet (1985). [c] Data for basal sections and mounds from Bonatti and Joensuu (1968), Church and Velde (1979), Corliss *et al.* (1978), Elderfield (1976), Heath and Dymond (1977), Honnerez *et al.* (1983), Jarvis (1985), Murchig and Gundlach (1982), Piper and Heath (1989), Sayles *et al.* (1975), and Schoonmaker (1986). [d] Data for intra-sedimentary deposits from Koski *et al.* (1985).

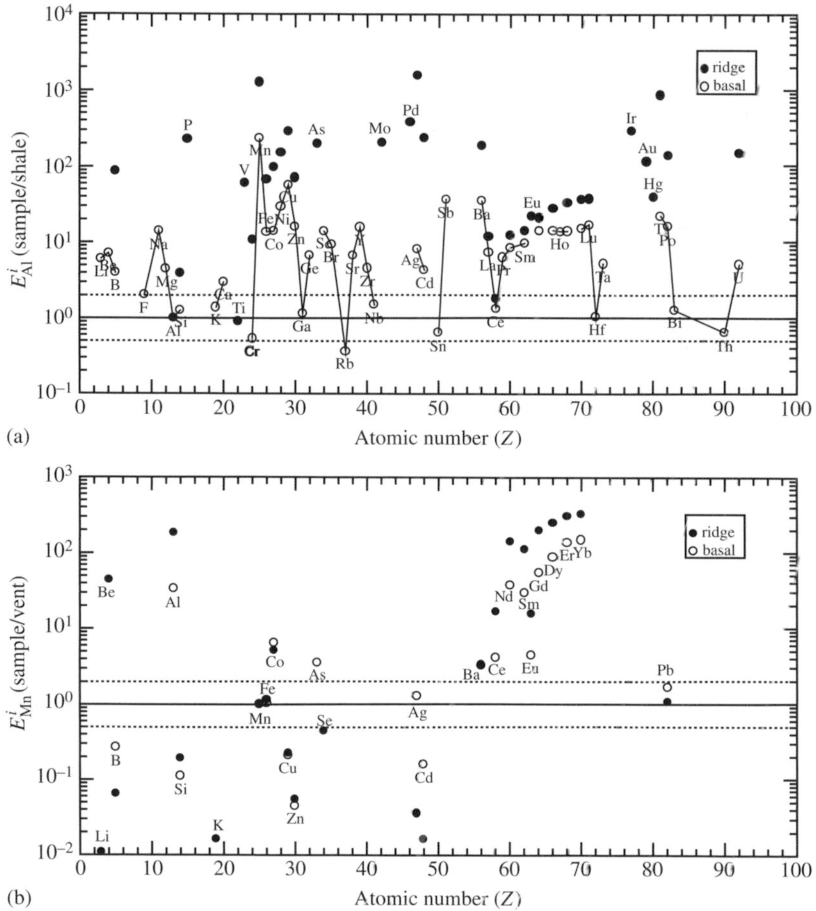

Figure 16 Enrichment factor of elements in metalliferous ridge and basal sediments as normalized to (a) Al and shale, and to (b) Mn and hydrothermal vent solution (Table 2).

Figures 16(a) and (b), respectively. Most of elements are enriched in metalliferrous ridge and basal sediments relative to shale, with the exception of some elements with E^i_{Al} values around one that are mostly contributed by shale. For example, Al and Ti in metalliferous ridge sediments and Al, Si, K, Cr, Ga, Rb, Nb, Sn, Ce, Hf, Bi, and Th in metalliferous basal sediments are mostly contributed by shale (Figure 16(a)). The enriched elements are either contributed from nearby hydrothermal vent solutions or by seawater or both. If one assumes that manganese in metalliferrous ridge and basal sediments is all of vent origin, then the elements with E^i_{Mn} [sample/hydrothermal vent] values of one or less in metalliferous sediments can be fully supplied by hydrothermal vent solution. These elements include Li, B, Si, K, Mn, Fe, Cu, Zn, Se, Ag, Cd, and Pb in Figure 16(b). Elements with E^i_{Mn} greater than 2 are mostly supplied by seawater, such as Be, Co, As, Ba, and REEs (Figure 16(b)). Indeed, Bender *et al.* (1971) showed that the REE distribution pattern in the metalliferrous sediments from the East Pacific Rise is similar to that of seawater, and lead isotopes are similar to those of oceanic basalts. The following sections show some examples.

7.01.4.1 Metalliferous Ridge Sediments

Compositional data from metalliferous ridge sediments across the East Pacific Rise (12–14° S and 90–140° W) as given by Bostrom and Peterson (1969), Horowitz (1970), and Crocket *et al.* (1973) were combined on a carbonate-free basis to perform the factor analysis. The results are summarized in Figures 17(a)–(d). There are three major components: F1 (ferromanganese hydroxide co-precipitates: Fe, Mn, As, V, Pd, B, Zn, and partially Cd), F2 (Mo, Ni, Cu, and partially Co), and F3 (aluminosilicates: Si, Al, Ti and partially Co). Most of the crest samples are enriched in F1 group elements, the flank samples being rich in F2 and F3 group elements (Figures 17(c) and (d)). Fe and Mn correlate nicely (Figure 18(a)), but the apparent partial correlation

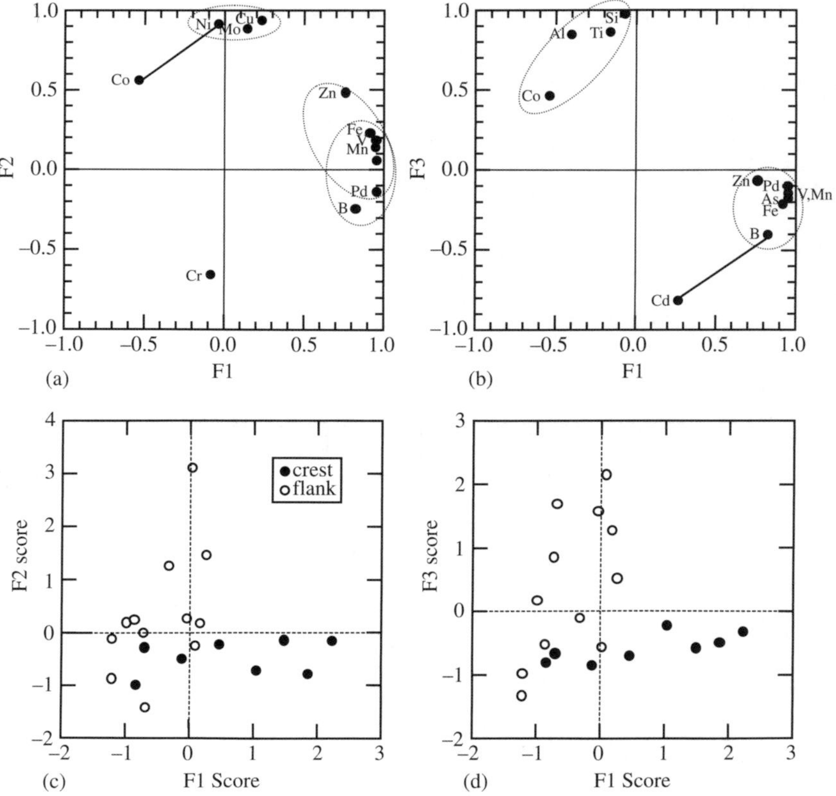

Figure 17 (a, b) Factor loadings and (c, d) factor scores for metalliferous ridge sediments (crest and flank) across the East Pacific Rise (sources Bostrom and Peterson, 1969; Horowitz, 1970; Crocket *et al.*, 1973).

relating Si and Co is probably not real (Figure 18(d)). The crest samples tend to have excess Si over the aluminosilicate contribution (Figure 18(b)). As shown in Figure 18(c), the F2 group element Mo is highly correlated with Mn in the crest and flank samples separately. The implication is that F2 group elements (Mo, Ni, Cu, and partially Co) still precipitate along with F1 group elements (Fe, Mn, etc.), but they tended to precipitate more on the ridge flank than on the crest. The delay mechanism is not clear yet.

7.01.4.2 Metalliferous Basal Sediments

Compositional data from metalliferous sediments of Deep Sea Drilling Project sites 573 (semibasal) and 574 (basal) from the central equatorial Pacific were given by Jarvis (1985). The sediments contain abundant carbonates and biogenic silica. The factor analysis results of data on a carbonate-free basis are summarized in Figures 19(a)–(d). Three samples with unusually low silica content (SiO_2 less than 2.7%) from site 573 were excluded. There are four distinct components: F1 (ferromanganese hydroxide co-precipitates: Fe, Mn, Ni, Cu, Zn, V, B, P, and partially Co), F2 (aluminosilicates: Al, Ti, Mg, Na, K, Sc, and partially Co), negative F1 (biogenic and hydrothermal silica and barite: Si, Ba, and Li), and F3 (REE and Y, and partially scandium). As shown in Figures 19(c), and 20(c)–(f), most samples from site 573 are rich in ferromanganese hydroxide group elements as compared with site 574 samples. Site 574 samples are relatively rich in biogenic silica and barite. REEs and Y are partially related to aluminosilicates through scandium (Figures 19(a) and (b), and 20(a) and (b)). However, the distribution pattern of REEs and Y is similar to that of seawater (Jarvis, 1985) and YREE are probably evenly distributed among major components including carbonates. Phosphate was co-precipitated with ferromanganese hydroxides as F1,as seen in Figure 19(a), and is not transformed into CFA.

7.01.5 MARINE PHOSPHORITES

Precipitation of marine phosphate minerals is a significant exogenic sink for phosphorus (e.g., Froelich *et al.*, 1982; Ruttenberg, 1993; Ruttenberg and Berner, 1993; Compton *et al.*, 2000; Guidry *et al.*, 2000). Marine phosphorites

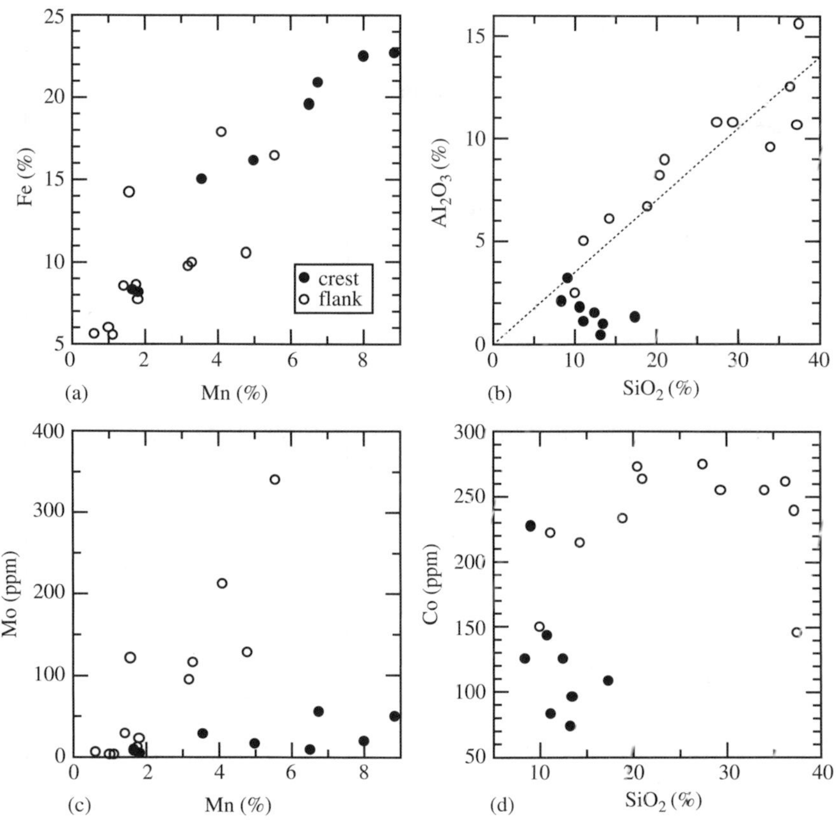

Figure 18 Scatter diagrams of selected elements for the same samples from Figure 17. The dotted line in (b) is the dilution line for the flank samples.

occur mainly on continental shelves and slopes, beneath highly productive upwelling currents, for example, Namibia and South Africa (Parker, 1975; Parker and Siesser, 1972; Watkins *et al.*, 1995; Baturin, 2000) and Peru and Chile (Burnett, 1977; Manheim *et al.*, 1975; Glenn and Arthur, 1988; Glenn *et al.*, 1988). Recent work has also demonstrated the formation of sedimentary apatite in marine nonupwelling environments such as East Australia (O'Brien and Veeh, 1980; O'Brien *et al.*, 1990; Heggie *et al.*, 1990) and Long Island Sound and the Mississippi Delta (Ruttenberg and Berner, 1993). Other important occurrences are on the mid-plate seamounts, submarine plateaus, ridges, banks, atolls, and atoll lagoons (Hein *et al.*, 1993). Supply of phosphate by oxidation of organic matter and pre-concentration of phosphate by iron oxyhydroxides, supply of calcium by dissolution of carbonates or carbonates acting as substrates, and supply of fluorine from seawater are all important factors in controlling the formation of marine phosphorites (Heggie *et al.*, 1990).

The primary marine sedimentary apatite mineral is CFA (or francolite; McClellan, 1980; Kolodny, 1981; Nathan, 1984; McClellan and Van Kauwenbergh, 1990; Follmi, 1996). CFA is a solid solution series in which CO_3^{2-} substitutes for PO_4^{3-} in the apatite [$Ca_5(PO_4)_3F$] structure. Substitution generally ranges from 4 wt.% CO_2 to 7 wt.% CO_2 (Knudsen and Gunter, 2002), but values up to 8.9 wt.% (McArthur, 1978), and perhaps even 16.1 wt.% (Elderfield *et al.*, 1972), have been reported. The charge imbalance created by substitution of CO_3^{2-} for PO_4^{3-} is balanced by the substitution of Na^+ for Ca^{2+}, and by addition of F^- (McClellan and Lehr, 1969) and OH^- (McArthur, 1990). The following idealized formulas for CFA have been suggested: $Ca_{10-x-y}Na_xMg_y(PO_4)_{6-z}(CO_3)_zF_{0.4z}F_2$ (McClellan, 1980), and $Ca_{10-x-y}Na_xMg_y(PO_4)_{6-z}(CO_3)_zF_zFOH$ (McArthur, 1990). Additionally, CFA may contain trace amounts of a wide spectrum of elements, depending on the temperature and solution composition to which the mineral is exposed during formation and subsequent to deposition (Glenn *et al.*, 1994; Jarvis *et al.*, 1994).

Until recently, most experimental work on the thermodynamic and kinetic properties of apatite minerals focused on synthetic fluorapatite and hydroxyapatite, the compositional analogues for dental and skeletal apatite (Budz and Nancollas, 1988; Chin and Nancollas, 1991; Chin *et al.*, 1993, Christoffersen *et al.*, 1998). Recent research,

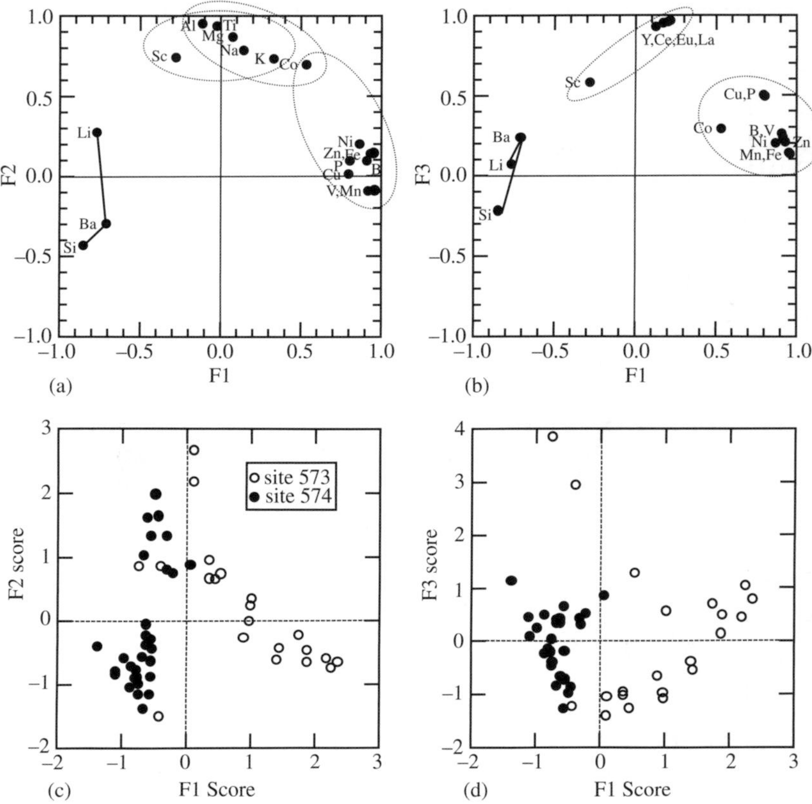

Figure 19 (a, b) Factor loadings and (c, d) factor scores for metalliferous basal sediment cores from Deep Sea Drilling Project sites 573 and 574 in the central equatorial Pacific (source Jarvis, 1985).

however, has investigated how factors such as temperature, pH, saturation state, and mineral composition influence the dissolution and precipitation reactions of the geologically common compositions of apatite, igneous fluorapatite, and sedimentary CFA. Jahnke (1984) performed experiments to determine the solubilities of FAP and a range of CFA compositions from supersaturation as well as undersaturation conditions. FAP was found to have the lowest solubility. K_{sp} values for the CFA phases differed by $\sim 10^{2.3}$, depending on the approach to equilibrium, indicating that equilibrium was not reached in both sets of experiments. The results, however, show that the solubility of CFA increases dramatically with increased CO_3^{2-} substitution (Figure 21; Jahnke, 1984).

Studies of FAP and CFA dissolution indicate initial incongruence, with preferential release of Ca over P and F; dissolution then becomes congruent as equilibrium is approached (Jahnke, 1984; Tribble *et al.*, 1995; Guidry and Mackenzie, 2003). Simultaneous increases in solution pH indicate the formation of a hydrogen-rich, cation-depleted surface layer during dissolution (Guidry and Mackenzie, 2003). Rates of dissolution of FAP and CFA have been determined from experiments in a batch reactor (Valsami-Jones *et al.*, 1998; Welch *et al.*, 2002) and in a fluidized bed reactor (Guidry and Mackenzie, 2003). Dissolution rates decrease sharply in the pH range 2–6 (Figure 22; Guidry and Mackenzie, 2003). Above a pH of 6, FAP dissolution rates level off and become independent of pH. Measurement of CFA dissolution rates at pHs slightly above 7 show a continued decline with increased pH. Dissolution rates of FAP also vary as a function of the degree of undersaturation (Guidry and Mackenzie, 2003). Calculation of the activation energy for FAP dissolution from experiments conducted in a stirred tank reactor yields a value of $8.3 + 0.3\ \text{kcal mol}^{-1}$, a value indicative of a surface controlled reaction (Guidry and Mackenzie, 2003).

Van Cappellen and Berner (1991) studied the growth kinetics of FAP in seeded precipitation experiments using carbonate-free solutions. The growth of FAP was inhibited by Mg^{2+} at concentrations typical of marine pore waters and enhanced by H^+ ions in the pH range of 7–8.5. At moderate degrees of supersaturation, a precursor phase similar in composition to octacalcium phosphate formed on the FAP

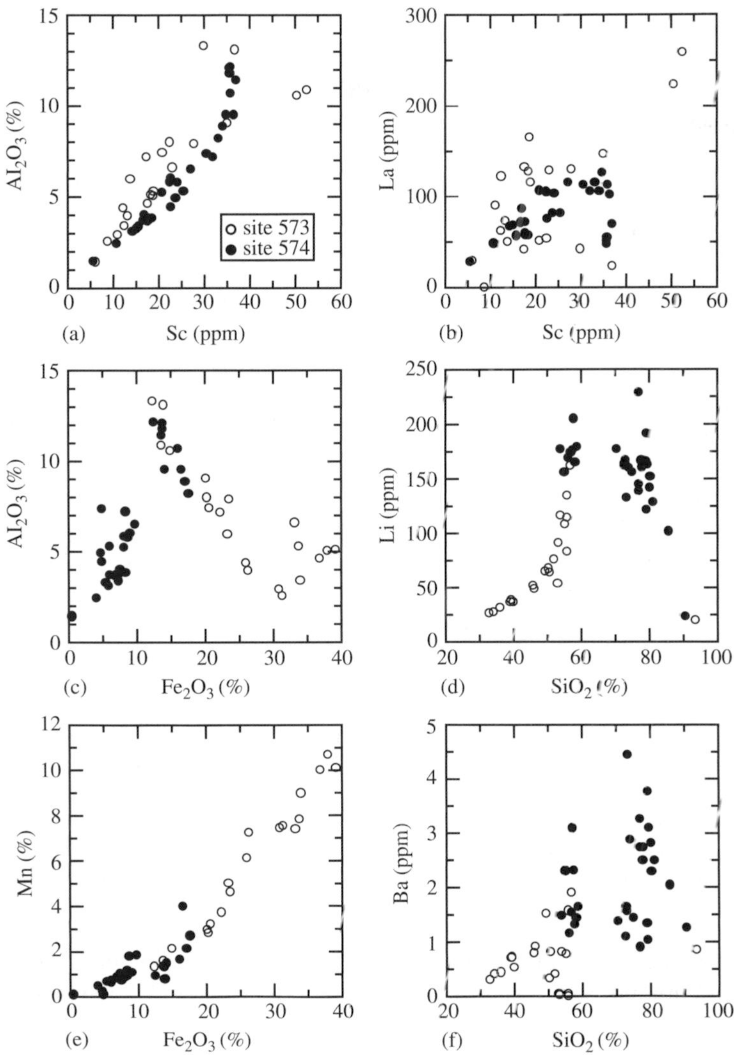

Figure 20 Scatter diagrams of selected elements for the same samples from Figure 19.

seed crystals and acted as a template for FAP precipitation. The growth kinetics determined by Van Cappellen and Berner (1991) indicates that FAP precipitation is a surface-controlled reaction and is consistent with the observed small crystal sizes of authigenic marine apatite.

Marine phosphorites on continental margins are mainly composed of microcrystalline CFA (15–20 wt.% P_2O_5), with associated glauconite ($X_2Y_4Z_8O_{20}(OH, F)$; where $X_2 = (K, Na)_{1.2\text{-}2}$, $Y_4 = (Fe(III), Al, Fe(II), Mg, Mn)_4$, and $Z_8 = Si_{7\text{-}7.6}Al_{1\text{-}0.4}$; Deer *et al.*, 1966), carbonates, organic matter, sulfides, and some aluminosilicate detritus minerals. CFA aggregates are often in the forms of nodules, pellets, oolites, concretions, laminae, etc. Phosphorites on the mid-plate seamounts are thought to result from substitution of carbonates by CFA with an ample supply of phosphate (Hein *et al.*, 1993).

The average composition of marine phosphorites on continental margins is summarized in Table 2, and the enrichment factors E^i_{Al} (phosphorite/shale) for various elements are plotted in Figure 23. E^i_{Al} (phosphorite/shale) values of ~1 (within a factor of 2) for Li, B, Mg, Al, Si, K, Ti, Fe, Ga, and Bi indicate their aluminosilicate origin. The moderate enrichment of Sc, V, Cr, Y, Zr, Nb, REEs, and Th, and high enrichment of Ag, Cd, and U probably result from their replacement of Ca in CFA, facilitated by their similar ionic radii. However, the enrichment of biophile elements S, Se, Cu, Zn, Mo, Sn, Sb, Au, and Hg is also related to the occurrence of sulfides and organic matter in the samples (Li, 2000).

Factor analysis results of marine phosphorite data from the South African continental margin (Parker, 1975; Parker and Siesser, 1972) show four components (Figure 24(a)): F1 (glauconite and shale: Fe, Al, Si, K, and Ti), negative F1 (carbonates: Ca and CO_2), F2 (CFA: P, S, F, and Na), and negative F2 (substitution among divalent Fe, Mg, and Mn cations in the Y position of glauconite). In Figures 24(b)–(f), additional data from other areas (Namibia: Watkins *et al.*, 1995; East Australia: O'Brien *et al.*, 1990; Peru and Chile: Burnett, 1977) as well as seamount phosphorites from the equatorial Pacific (Hein *et al.*, 1993) are also plotted. As shown in Figures 24(b) and (c), excess SiO_2 and Fe_2O_3 over shale contributions (above the dotted lines for shale) for continental margin phosphorites (including South Africa, Namibia, and East Australia) indicate the constant occurrence of glauconite along with apatite. The solid lines in Figures 24(d) and (e) are the slopes expected from CFA with formulas like $Na_{0.5}Ca_5(PO_4)_{2.5}(CO_3)F$ and $Na_{0.7}Ca_5(PO_4)_{2.3}(CO_3)_{1.4}F$. A small fraction of CO_3 in CFA is easily replaced by SO_4 (Figures 24(a) and (f)). The excess calcium over CFA contribution for the continental margin phosphorites (Figure 24(d)) represents carbonate contribution. Chemically, phosphorites from the Peru and Chile continental margins and from the equatorial Pacific seamounts are quite similar (Figures 24(b)–(f)).

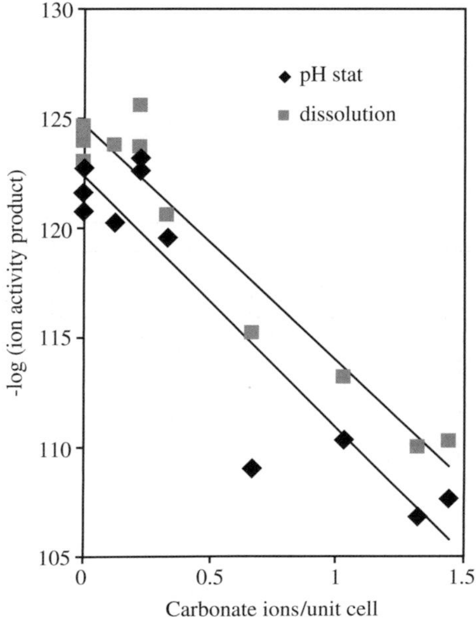

Figure 21 Ion activity product of carbonate fluorapatite as a function of carbonate content as determined from oversaturation (pH stat experiments) and undersaturation (dissolution experiments) by Jahnke (1984). Despite divergence of the two data sets, the trend of increasing solubility with increasing carbonate content is apparent.

7.01.6 CONCLUSIONS

Major components of marine pelagic sediments are aluminosilicates (shale), manganese oxides, CFA, zeolites, biogenic carbonates, silica, and barite. Volcanic ash, hydrothermal precipitates, and rare continental detritus minerals at times become additional important components. The elements shown to associate with manganese

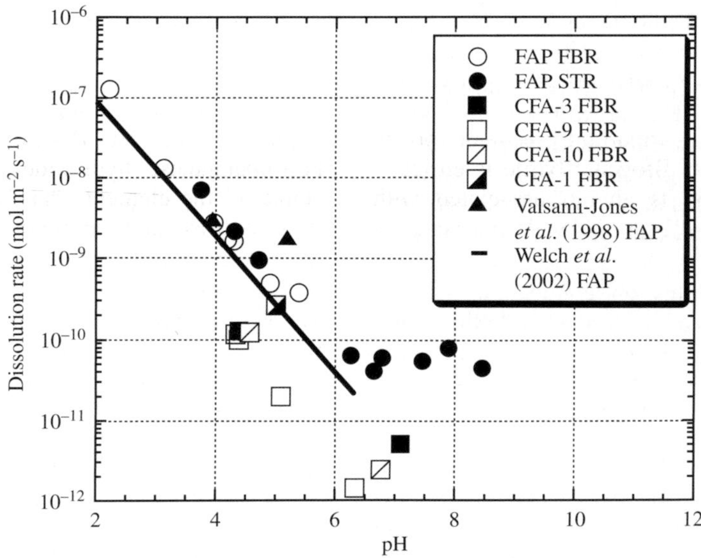

Figure 22 Dissolution rates of FAP and a range of CFA compositions as a function of pH as determined from experiments in a stirred tank reactor (STR) and a fluidized bed reactor (FBR) (reproduced by permission of Elsevier from *Geochim. Cosmochim. Acta*, **2003**, *67*, 2949–2963).

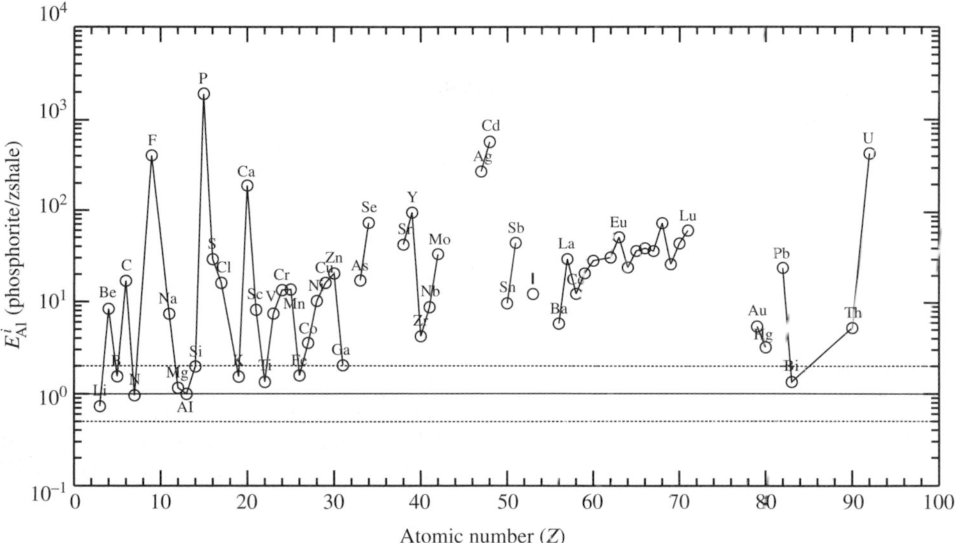

Figure 23 Enrichment factor of elements in average marine phosphorite as normalized to Al and shale (Table 2).

oxide phases of pelagic sediments are usually mono- and divalent cations with low to moderate first hydrolysis constants in seawater (Mn, Co, Ni, Cu, Zn, Pb, Mg, Ca, Sr, Ba, Tl, Ag, etc.) and Mo (Li, 2000). Molybdenum is an oxyanion (MoO_4^{-2}) in normal oxygenated seawater. Its association with manganese may result from the formation of $MnMoO_4$ along with wolframite ($[Mn,Fe]WO_4$). Enrichment of noble metals (Pd, Os, Ir, and Pt) in pelagic clays (Figure 1(b)) is also probably due to their enrichment in manganese oxide phases. The release of Mn^{+2} from reducing sediments of continental margins and its lateral offshore transport provide the necessary sources for manganese oxides in pelagic sediments (Martin and Knauer, 1984; Martin *et al.*, 1985). REEs (except cerium) are often closely associated with CFA (Ca, P, and Y). Biogenic silica is probably an important precursor for the formation of zeolite. Biogenic barite is enriched in Sr and Ce. Iron is always associated with aluminosilicates, and iron oxides do not occur as major phases in the pelagic sediments unless there are hydrothermal inputs.

Ferromanganese nodules and seamount crusts are characterized by distinct alternating layers of iron oxides and manganese oxides in addition to other components such as aluminosilicates (shale), CFA, and sometimes biogenic barite and carbonates in seamount crusts. Iron-oxide phases are enriched in elements that exist in normal seawater mainly as hydroxyl and carbonate complexes of tri- to pentavalent cations (e.g., As, B, Bi, In, Ir, Rh, REEs, Y, Ti, Th, U, Zr, Hf, Nb, and Ta), oxyanions (e.g., P, Re, Ru, Os, S, Se, Te, V, and W), and divalent cations with high first hydrolysis constant (Hg, Be, Pd). The observed specific association of elements with manganese-oxide, iron-oxide, and aluminosilicate phases can be explained by the fundamental differences in physicochemical properties of these phases, such as the pH of zero point of charge, intrinsic acidity constants, and dielectric constants (Li, 2000). REEs are enriched in CFA as well as in iron oxide phases. Biogenic barites are also rich in Bi, Cu, Zn, and Pt in addition to Ba, Ce, and Sr.

Metalliferous ridge and basal sediments mainly consist of ferromanganese hydroxide co-precipitates from hydrothermal vent solutions and aluminosilicates (shale). Biogenic carbonates, silica, and barite can be additional components. Ubiquity of biogenic barite particles in the water column and seafloor is well documented (Dehairs *et al.*, 1980; Bertram and Cowen, 1997). Ferromanganese hydroxide co-precipitates scavenge all the elements that have high affinity to iron oxides and manganese oxides. These elements are supplied from both hydrothermal vent solutions and seawater. Phosphate in seawater is scavenged by ferromanganese hydroxides but does not evolve into apatite.

CFA is omnipresent in pelagic sediments, ferromanganese nodules, and seamount crusts, and marine phosphorites on continental margins and seamounts. Supply of phosphate by oxidation of organic matter and pre-concentration of phosphate by iron oxyhydroxides, supply of calcium by dissolution of carbonates or carbonates acting as substrates, and supply of fluorine from seawater are all important factors in the formation of marine phosphorites.

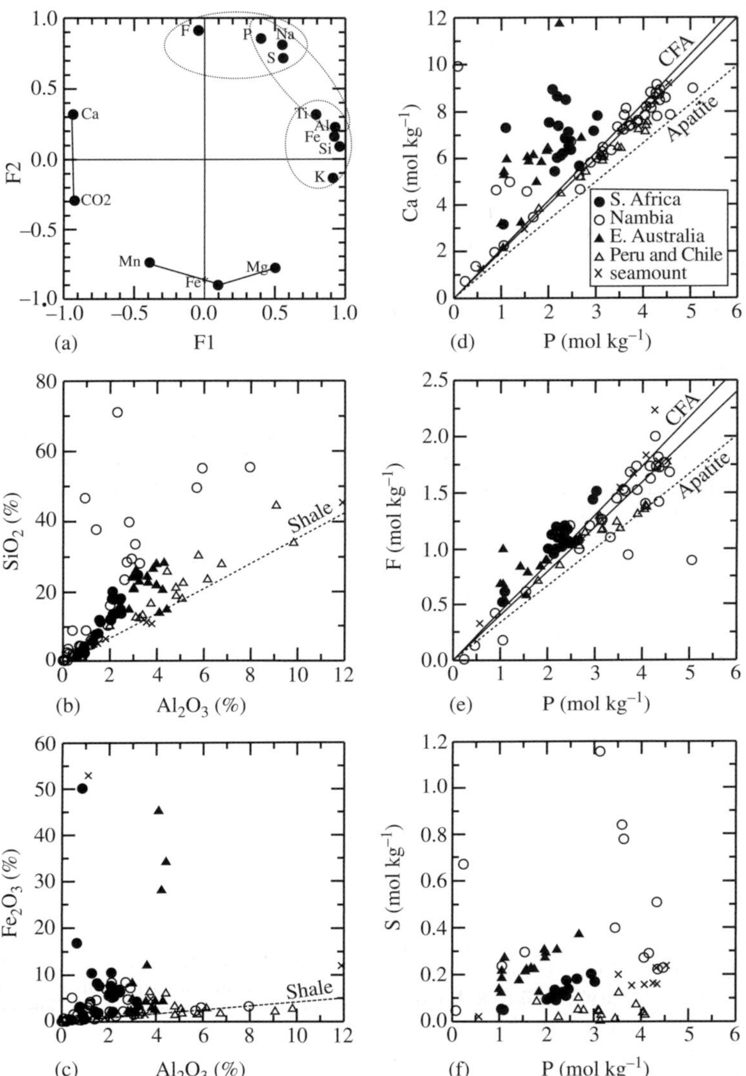

Figure 24 (a) Factor loadings for marine phosphorite samples from the South African continental margin (Parker, 1975; Parker and Siesser, 1972). (b)–(f) Scatter diagrams of selected elements for additional samples from Namibia (Watkins *et al.*, 1995), East Australia (O'Brien *et al.*, 1990), Peru and Chili (Burnett, 1977), as well as seamount phosphorites from the equatorial Pacific (Hein *et al.*, 1993). The dotted lines in (b) and (c) are the slopes for shale. The dotted lines in (d) and (e) are slopes for carbonate fluorapatites with possible formulas of $Na_{0.5}Ca_5(PO_4)_{2.5}(CO_3)F$ and $Na_{0.7}Ca_5(PO_4)_{2.3}(CO_3)_{1.4}F$.

REFERENCES

Alt J. C. (1988) Hydrothermal oxide and nontronite deposits on seamounts in the eastern Pacific. *Mar. Geol.* **81**, 227–239.

Alt J. C. (1995) Subseafloor processes in mid-ocean ridge hydrothermal systems. In *Seafloor Hydrothermal Systems, Physical, Chemical, Biological, and Geological Interactions*, Geophysical Monograph 91 (eds. S. E. Humphris, R. A. Zierenberg, L. S. Mullineaux, and R. E. Thomson). American Geophysical Union, Washington, DC, pp. 85–114.

Alt J. C., Lonsdale P., Haymon R., and Muelenbachs K. (1987) Hydrothermal sulfide and oxide deposits on seamounts near 21 N, East Pacific Rise. *Bull. Geol. Soc. Am.* **98**, 157–168.

Alvarez R., DeCarlo E. H., Cowen J., and Andermann G. (1990) Micromorphological characteristics of a marine ferromanganese crust. *Mar. Geol.* **94**, 239–249.

Arrhenius G. (1963) Pelagic sediments. In *The Sea, Ideas and Observations on Progress in the Study of the Seas: Vol. 3. The Earth Beneath the Sea, History* (ed. M. N. Hill). Interscience, New York, pp. 655–727.

Banerjee R., Roy S., Dasgupta S., Mukhopadhyay S., and Miura H. (1999) Petrogenesis of ferromanganese nodules from east of the Chagos Archipelago, central Indian basin, Indian ocean. *Mar. Geol.* **157**, 145–158.

Baturin G. N. (1982) *Phosphorites on the Sea Floor: Origin, Composition and Distribution.* Elsevier, New York.

Baturin G. N. (1988) *The Geochemistry of Manganese and Manganese Nodules in the Ocean.* Kluwer, Dordrecht, Berlin.

Baturin G. N. (2000) Formation and evolution of phosphorite grains and nodules on the Namibian shelf, from recent to Pleistocene. In *Marine Authigenesis: from Global to Microbial*. Spec. Publ. no. 66 (eds. C. R. Glenn,

L. Prevot-Lucas, and J. Lucas). Society for Seidmentary Geology, Tulsa, pp. 185–199.

Bender M., Broecker W., Gornity V., Middel U., Key R., Sun S. S., and Biscaye P. (1971) Geochemistry of three cores from the East Pacific Rise. *Earth Planet. Sci. Lett.* **12**, 425–433.

Berger W. H. (1976) Biogenous deep-sea sediments: production, preservation and interpretation. In *Chemical Oceanography*, 2nd edn. (eds. J. P. Riley and R. Chester). Academic Press, London, vol. 5, pp. 265–388.

Berger W. H. (1978) Deep-sea carbonate: pteropod distribution and the aragonite compensation depth. *Deep-Sea Res.* **25**, 447–452.

Berger W. H., Adelseck C. G., Jr., and Mayer L. A. (1976) Distribution of carbonate in the surface sediments of the Pacific Ocean. *J. Geophys. Res.* **81**, 2617–2627.

Bertram M. A. and Cowen J. P. (1997) Morphological and compositional evidence for biotic precipitation of marine barite. *J. Mar. Res.* **55**, 577–593.

Biscaye P. E., Kolla V., and Turekian K. K. (1976) Distribution of calcium carbonate in surface sediments of the Atlantic Ocean. *J. Geophys. Res.* **81**, 2595–2603.

Bischoff J. L. and Piper D. Z. (eds.) (1979) *Marine Geology and Oceanography of the Pacific Manganese Nodule Province*. Plenum, New York, 855pp.

Bischoff W. D., Mackenzie F. T., and Bishop F. C. (1987) Stabilities of synthetic magnesian calcites in aqueous solution: comparison with biogenic materials. *Geochim. Cosmochim. Acta* **51**, 1413–1423.

Bishop J. K. B. (1988) The barite-opal-organic carbon association in oceanic particulate matter. *Nature* **331**, 341–343.

Boles J. R. (1977) Zeolites in deep-sea sediments. In *Mineralogy and Geology of Natural Zeolites*, Mineral. Soc. Am. Short Course Notes (ed. F. A. Mumpton). Washington, DC, vol. 4, pp. 137–163.

Bolton B. R., Bogi J., and Cronan D. S. (1992) Geochemistry and mineralogy of ferromanganese nodules from the Kiribati region of the eastern central Pacific Basin. In *Geology and Offshore Mineral Resources of the Central Pacific Basin*, Circum-Pacific Council for Energy and Mineral Resources Earth Science Series (eds. B. H. Keating and B. R. Bolton). Springer, NY, vol. 14, pp. 247–260.

Bonatti E. (1963) *Zeolites in Pacific Pelagic Sediments*. New York Acad. Sci. Series, vol. 25, pp. 933–948.

Bonatti E. (1983) Hydrothermal metal deposits from the oceanic rifts: a classification. In *Hydrothermal Processes at Seafloor Spreading Centers* (eds. P. A. Rona, K. Bostrom, L. Laubier, and K. L. Smith, Jr.). Plenum, New York, pp. 491–502.

Bonatti E. and Joensuu O. (1968) Paygorskite from Atlantic deep sea sediments. *Am. Min.* **53**, 925–983.

Bostrom K. (1973) The origin and fate of ferromanganese active ridge sediments. 27, 149–243.

Bostrom K. and D. E. Fisher (1969) Distribution of mercury in East Pacific sediments. *Geochim. Cosmochim. Acta* **33**, 743–745.

Bostrom K. and Peterson M. N. A. (1966) Precipitates from the hydrothermal exhalations on the East Pacific Rise. *Econ. Geol.* **61**, 1258–1265.

Bostrom K. and Peterson M. N. A. (1969) The origin of aluminum-poor ferromanganoan sediments in areas of high heat flow on the east Pacific rise. *Mar. Geol.* **7**, 427–447.

Bostrom K., Peterson M. N. A., Joensuu O., and Fisher D. E. (1969) Aluminum-poor ferromanganese sediments on active oceanic ridges. *J. Geophys. Res.* **74**, 3261–3270.

Bowles F. G., Angino E. A., Hosterman J. W., and Galles O. K. (1971) Precipitation of deep-sea palygorskite and sepiolite. *Earth Planet. Sci. Lett.* **11**, 324–334.

Bricker O. (1965) Some stability relations in the system Mn–O_2–H_2O at 25° and one atmosphere total pressure. *Am. Mineral.* **50**, 1296–1354.

Broecker W. S. and Peng T. H. (1982) *Tracers in the Sea*. Columbia University, New York.

Budz J. A. and Nancollas G. H. (1988) The mechanism of dissolution of hydroxyapatite and carbonated apatite in acidic solutions. *J. Cryst. Growth* **91**, 490–496.

Burnett W. C. (1977) Geochemistry and origin of phosphorite deposits from off Peru and Chile. *Geol. Soc. Am. Bull.* **88**, 813–823.

Burns R. G. and Burns V. M. (1979) Manganese oxides. In *Reviews in Mineralogy: Vol. 6. Marine Minerals* (ed. R. G. Burns). Mineralogical Society of America, Washington, DC, pp. 1–46.

Burns V. M. and Burns R. G. (1978a) Post-depositional metal enrichment processes inside manganese nodules from the north equatorial Pacific. *Earth Planet. Sci. Lett.* **39**, 341–348.

Burns V. M. and Burns R. G. (1978b) Diagenetic features observed inside deep-sea manganese nodules from the north equatorial Pacific. *Scan. Electr. Micros.* **1978**, 245–252.

Burns V. M. and Burns R. G. (1978c) Authigenic todorokite and phillipsite inside deep-sea manganese nodules. *Am. Mineral.* **63**, 827–831.

Calvert S. E. and Price N. B. (1977) Geochemical variation on ferromanganese nodules and associated sediments from the Pacific Ocean. *Mar. Chem.* **5**, 43–74.

Calvert S. E., Price N. B., Heath G. R., and Moore T. C., Jr. (1978) Relationship between ferromanganese nodule compositions and sedimentation in a small survey area of the equatorial Pacific. *J. Mar. Res.* **36**, 161–183.

Cann J. R., Winter D. K., and Pritchard R. G. (1977) A hydrothermal deposit from the floor of the Gulf of Aden. *Min. Mag.* **41**, 193–199.

Chester R. and Aston S. R. (1976) The geochemistry of deep-sea sediments. In *Chemical Oceanography* (eds. J. P. Riley and R. Chester). Academic Press, London, vol. 6, pp. 281–390.

Chin K. O. A. and Nancollas G. H. (1991) Dissolution of fluorapatite. A constant-composition kinetics study. *Langmuir* **7**, 2175–2179.

Chin K. O. A., Johnsson M., Bergey E. J., Levine M. J., and Nancollas G. H. (1993) A constant composition kinetics study of the influence of salivary cystatins, statherin, amylase, and human serum albumin on hydroxyapatite dissolution. *Coll. Surf. A: Physicochem. Eng. Aspects* **78**, 229–234.

Christoffersen M. R., Dohrup J., and Christoffersen J. (1998) Kinetics of growth and dissolution of calcium hydroxyapatite in suspensions with variable calcium to phosphate ratio. *J. Cryst. Growth* **186**, 283–290.

Chukhrov F. V., Gorshkov A. I., Rudnitskaya E. S., Berezovskaya V. V., and Sivtsov A. V. (1978) On vernadite. *Izvest. Akad. Nauk SSSR, Ser. Geol.* **6**, 5–19.

Church T. M. (1979) Marine barite. In *Reviews in Mineralogy: Vol. 6. Marine Minerals* (ed. R. G. Burns). Mineralogical Society of America, Washington, DC, pp. 175–210.

Church T. M. and Velde B. (1979) Geochemistry and origin of a deep-sea Pacific palygorskite deposit. *Chem. Geol.* **25**, 31–39.

Cochran J. K. (1992) The oceanic chemistry of the uranium- and thorium-series nuclides. In *Uranium-series Disequilibrium*, 2nd edn. (eds. M. Ivanovich and R. S. Harmon). Clarendon Press, Oxford.

Compton J., Mallinson D., Glenn C., Filippelli G., Follmi K., Shields G., and Zanin Y. (2000) Variations in the global phosphorus cycle. In *Marine Authigenesis: from Global to Microbial*, Spec. Publ. 66 (eds. C. R. Glenn, L. Prevot-Lucas, and J. Lucas). Society for Sedimentary Geology, Tulsa, pp. 21–33.

Corliss J. B., Lyle M., Dymond J., and Crane K. (1978) The geochemistry of hydrothermal sediment mounds near the Galapagos rift. *Earth Planet. Sci. Lett.* **40**, 12–24.

Cowen J. P., DeCarlo E. H., and McGee D. L. (1993) Calcareous nannofossil biostratigraphic dating of a ferromanganese crust from Schumann Seamount. *Mar. Geol.* **115**, 289–306.

Crerar D. A. and Barnes H. L. (1974) Deposition of deep-sea manganese nodules. *Geochim. Cosmochim. Acta* **38**, 279–300.

Crerar D. A., Cormick R. K., and Barnes H. L. (1980) Geochemistry of manganese: an overview. In *Geology and Geochemistry of Manganese* (eds. I. M. Varentsov and Gy. Grasselly). E. Schweizerbart'sche Verlagsbuchhandlung Publ., Stuttgart, vol. 1, pp. 293–334.

Crocket J. H., MacDougall J. D., and Harriss R. C. (1973) Gold, palladium and iridium in marine sediments. *Geochim. Cosmochim. Acta* **37**, 2547–2556.

Cronan D. S. (1976a) Basal metalliferous sediments from the eastern Pacific. *Geol. Soc. Am. Bull.* **87**, 928–934.

Cronan D. S. (1976b) Manganese nodules and other ferromanganese oxide deposits, chap. 28. In *Chemical Oceanography*, 2nd edn. (eds. J. P. Riley and R. Chester). Academic Press, London, vol. 5, pp. 217–263.

Davies T. A. and Gorsline D. S. (1976) Oceanic sediments and sedimentary processes. In *Chemical Oceanography*, 2nd edn. (eds. J. P. Riley and R. Chester). Academic Press, London, vol. 5, pp. 1–80.

Davis J. C. (1973) *Statistics and Data Analysis in Geology*. Wiley, New York.

DeCarlo E. H. (1991) Paleoceanographic implications of rare earth element variability within a Fe–Mn crust from the central Pacific Ocean. *Mar. Geol.* **98**, 449–467.

DeCarlo E. H. and Fraley C. M. (1992) Chemistry and mineralogy of ferromanganese deposits from the equatorial Pacific Ocean. In *Geology and Offshore Mineral Resources of the Central Pacific Basin*, Circum-Pacific Council for Energy and Mineral Resources Earth Science Series (eds. B. H. Keating and B. R. Bolton). Springer, NY, vol. 14, pp. 225–245.

DeCarlo E. H., McMurtry G. M., and Kim K. H. (1987) Geochemistry of ferromanganese crusts from the Hawaiian Archipelago: I. Northern survey areas. *Deep-Sea Res.* **34**, 441–467.

Deer W. A., Howie R. A., and Zussman J. (1966) *An Introduction to the Rock Forming Minerals*. Wiley, New York.

Dehairs F., Chesselet R., and Jedwab J. (1980) Discrete suspended particles of barite and the barium cycle in the open ocean. *Earth Planet. Sci. Lett.* **49**, 528–550.

Dehairs F., Shopova D., Ober S., and Goyens L. (1990) Particulate barium stocks and oxygen consumption in the Southern Ocean mesopelagic water column during spring and early summer: relationship with export production. *Deep-Sea Res. II* **44**, 497–516.

Dymond J. and Eklund W. (1978) A microprobe study of metalliferous sediment components. *Earth Planet. Sci. Lett.* **40**, 243–251.

Dymond J. and Roth S. (1988) Plume dispersed hydrothermal particles: a time-series record of settling flux from the Endeavor Ridge using moored sensors. *Geochim. Cosmochim. Acta* **52**, 2525–2536.

Dymond J., Corliss J. B., Heath G. R., Field C. W., Dasch E. J., and Veeh H. H. (1973) Origin of metalliferous sediments from the Pacific Ocean. *Geol. Soc. Am. Bull.* **84**, 3355–3372.

Edmond J. M., Measures C., Mangum B., Grant B., Sclater F. R., Collier R., and Hudson A. (1979) On the formation of metal-rich deposits at ridge crests. *Earth Plant. Sci. Lett.* **46**, 19–30.

Elderfield H. (1976) Hydrogenous material in marine sediments: excluding manganese nodules. In *Chemical Oceanography*, 2nd edn. (eds. J. P. Riley and R. Chester). Academic Press, London, vol. 5, pp. 137–215.

Elderfield H., Holmefjord T., and Summerhayes C. P. (1972) Enhanced CO_2 substitution in carbonate-apatite from the Moroccan continental margin. *Inst. African Geol. Res. 16th Ann. Report* (Univ. Leeds.).

Elderfield H., Hawkesword C. J., Greaves M. J., and Calvert S. E. (1981) Rare earth element geochemistry of oceanic ferromanganese nodules and associated sediments. *Geochim. Cosmochim. Acta* **45**, 513–528.

El Wakeel S. K. and Riley J. P. (1961) Chemical and mineralogical studies of deep sea sediments. *Geochim. Cosmochim. Acta* **25**, 110–146.

Fagel N., Andre L., and Debrabant P. (1997) Multiple seawater-derived geochemical signatures in Indian oceanic pelagic clays. *Geochim. Cosmochim. Acta* **61**, 989–1008.

Fagel N., Debrabant P., and Andre L. (1994) Clay supplies in the Central Indian Basin since the Late Miocene: climate or tectonic control? *Mar. Geol.* **122**, 151–172.

Faure G. (1986) *Principles of Isotope Geology*, 2nd edn. Wiley, New York.

Fisher D. E. and Bostrom K. (1969) Uranium rich sediments on the East Pacific Rise. *Nature* **224**, 64–65.

Follmi K. B. (1996) The phosphorus cycle, phosphogenesis and marine phosphate-rich deposits. *Earth-Sci. Rev.* **40**, 55–124.

Froelich P. N., Bender M. L., Luedtke N. A., Heath G. R., and DeVries T. (1982) The marine phosphorus cycle. *Am. J. Sci.* **282**, 474–511.

Giovanoli R. and Arrhenius G. (1988) Structural chemistry of marine manganese and iron minerals and synthetic model compounds. In *The Manganese Nodule Belt of the Pacific Ocean: Geological Environment, Nodule Formation, and Mining Aspects* (ed. P. Halbach). Ferdinand Enke Verlag, pp. 20–37.

Glasby G. P. (ed.) (1977) *Marine Manganese Deposits*. Elsevier, New York.

Glenn C. R. and Arthur M. A. (1988) Petrology and major element geochemistry of Peru margin phosphorites and associated diagenetic minerals: authigenesis in modern organic-rich sediments. *Mar. Geol.* **80**, 231–267.

Glenn C. R., Arthur M. A., Yeh H.-W., and Burnett W. C. (1988) Carbon isotopic composition and lattice-bound carbonate of Peru–Chile margin phosphorites. *Mar. Geol.* **80**, 287–307.

Glenn C. R., Follmi K. B., Riggs S. R., Baturin G. N., Grimm K. A., Trappe J., Abed A. M., Galli-Olivier C., Garrison R. E., Ilyin A. V., Jehl C., Rohrlich V., Sadaqah R. M. Y., Schidlowski M., Sheldon R. E., and Seigmund H. (1994) Phosphorus and phosphorites: sedimentology and environments of formation. *Eclogae Geologicae Helvetiae* **87**, 747–788.

Glenn C. R., Prevot-Lucas L., and Lucas J. (eds.) (2000) *Marine Authigenesis: from Global to Microbial*. Spec. Publ. no. 66. Society for Sedimentary Geology, Tulsa, 536pp.

Glover E. D. (1977) Characterization of marine birnessite. *Am. Mineral.* **62**, 278–285.

Goldberg E. D. (1963) Mineralogy and chemistry of marine sedimentation. In *SubMar. Geol.* (ed. F. P. Shephard). Harper and Row, New York, pp. 436–466.

Goldberg E. D., Hodge V. F., Kay P., Stallard M., and Koide M. (1986) Some comparative marine chemistries of platinum and iridium. *Appl. Geochem.* **1**, 227–232.

Guidry M. W. and Mackenzie F. T. (2003) Igneous and sedimentary apatite dissolution and the long-term phosphorus cycle. *Geochim. Cosmochim. Acta* **67**(16), 2949–2963.

Guidry M. W., Mackenzie F. T., and Arvidson R. S. (2000) Role of tectonics in phosphorus distribution and cycling. In *Marine Authigenesis: from Global to Microbial*, Spec. Publ. 66 (eds. C. R. Glenn, L. Prevot-Lucas, and J. Lucas). Society for Sedimentary Geology, Tulsa, pp. 35–51.

Harder H. (1976) Nontronite synthesis at low temperatures. *Chem. Geol.* **18**, 169–180.

Hathaway J. C. (1979) Clay minerals. In *Reviews in Mineralogy: Vol. 6*, Marine Minerals (ed. R. G. Burns). Mineralogical Society of America, Washington, DC, pp. 123–150.

Hathaway J. C. and Sachs P. L. (1965) Sepiolite and clinoptilolite from the Mid-Altantic Ridge. *Am. Mineral.* **50**, 852–867.

Haymon R. (1982) Hydrothermal deposition on the East Pacific Rise at 21 N. PhD Thesis, University of California, San Diego.

Haymon R. M. (1989) Hydrothermal processes and products on the Galapagos Rift and East Pacific Rise, chap. 8. In *The Eastern Pacific Ocean and Hawaii*, The Geology of North America (eds. E. L. Winterer, D. M. Hussong, and R. W. Decker). Geol. Soc. Am., Boulder, CO, vol. N, pp. 125–144.

Haymon R. M. and Kastner M. (1981) Hot springs deposits on the East Pacific Rise at 21 N: preliminary description of mineralogy and genesis. *Earth Planet. Sci. Lett.* **53**, 363–381.

Heath G. R. and Dymond J. (1977) Genesis and transformation of metalliferous sediments from the East Pacific Rise, Bauer Deep, and Central Basin, northwest Nazea Plate. *Geol. Soc. Am. Bull.* **88**, 723–733.

Heggie D. T., Skyring G. W., O'Brien G. W., Reimers C., Herczeg A., Moriarty D. J. W., Burnett W. C., and Milnes A. R. (1990) Organic carbon cycling and modern phosphorite formation on the East Australian continental margin: an overview. In *Phosphorite Research and Development*, Geol. Soc. Spec. Publ. #52 (eds. A. J. G. Notholt and I. Jarvis). Geological Society of London, pp. 87–117.

Hein J. R., Scholl D. W., Barron J. A., Jones M. G., and Miller J. (1978) Diagenesis of Late Cenozoic diatomaceous deposits and formation of the bottom simulating reflector in the southern Bering Sea. *Sedimentology* **25**, 155–181.

Hein J. R., Schulz M. S., and Gein L. M. (1992) Central Pacific cobalt-rich ferromanganese crusts: historical perspective and regional variability. In *Geology and Offshore Mineral Resources of the Central Pacific Basin*, Circum-Pacific Council for Energy and Mineral Resources Earth Science Series (eds. B. H. Keating and B. R. Bolton). Springer, NY, vol. 14, pp. 261–283.

Hein J. R., Yeh H. W., Gunn S. H., Sliter W. V., Benninger L. M., and Wang C. H. (1993) Two major Cenozoic episodes of phosphogenesis recorded in equatorial Pacific seamount deposits. *Paleoceanography* **8**, 293–311.

Hein J. R., Koschinsky A., Bau M., Manheim F. T., Kang J. K., and Roberts L. (2000) Cobalt-rich ferromanganese crusts in the Pacific. In *Hand Book of Marine Mineral Deposits* (ed. D. S. Cronan). CRC Press, New York.

Hekinian R. and Fouquet Y. (1985) Volcanism and metallogenesis of axis and off-axis structures on the East Pacific Rise near 13 N. *Econ. Geol.* **80**, 221–249.

Hem J. D. (1978) Redox processes at surfaces of manganese oxide and their effects on aqueous metal ions. *Chem. Geol.* **21**, 199–218.

Hoffert M. A., Perseil A., Hekinian R., Choukroune P., Needham H. D., Francheteau J., and LePichon X. (1978) Hydrothermal deposits sampled by diving saucer in transform fault "A" near 37°N on the Mid-Atlantic Ridge, Famous area. *Oceanol. Acta* **1**, 72–86.

Honnerez J., Karpoff A.-M., and Trauth-Badaut D. (1983) Sedimentology, mineralogy, and geochemistry of green clay samples from the Galapagos hydrothermal mounds, Holes 506, 506C, and 507D. In *Init. Repts. DSDP, 70* (eds. J. Honnerez, R. P. Von Herzen, *et al.*). US Government. Printing Office, Washington, DC, pp. 211–224.

Horn D. R., Horn B. M., and Delach M. N. (1970) Sedimentary provinces in the North Pacific. In *Geological Investigations of the North Pacific*, Geol. Soc. Am. Mem. 126 (ed. J. D. Hayes), Geological society of America, Boulder, CO, pp. 1–22.

Horowitz A. (1970) The distribution of Pb, Ag, Sn, Tl, and Zn in sediments on active oceanic ridge. *Mar. Geol.* **9**, 241–259.

Iler R. K. (1955) *The Colloid Chemistry of Silica and Silicates*. Cornell University Press, Ithaca NY, 324pp.

Jahnke R. A. (1984) The synthesis and solubility of carbonate fluorapatite. *Am. J. Sci.* **284**, 58–78.

Jarvis I. (1985) Geochemistry and origin of Eocene–Oligocene metalliferous sediments from the central equatorial Pacific: Deep Sea Drilling Project sites 573 and 574. In *Initial Reports of the Deep Sea Drilling Project*, vol. 85, pp. 781–804.

Jarvis I., Burnett W. C., Nathan Y., Almbaydin F. S. M., Attia A. K. M., Castro L. N., Flicoteaux R., Ezzeldim Hilmy M., Husain V., Qutawnah A. A., Serjani A., and Zanin Y. N. (1994) Phosphorite geochemistry: state-of-the-art and environmental concerns. *Eclogae Geologicae Helvetiae* **87**, 643–700.

Jeffries D. A. and Stumm W. (1976) The metal-adsorption chemistry of buserite. *Can. Mineral.* **14**, 16–22.

Janecky D. R. and Seyfried W. E. (1984) Formation of massive sulfide deposits on oceanic ridge crests: incremental reaction models for mixing between hydrothermal solutions and seawater. *Geochim. Cosmochim. Acta* **48**, 2723–2738.

Jones B. F. and Galan E. (1988) Sepiolite and palygorskite. In *Hydrous Phyllosilicates (Exclusive of Micas)*, Rev. Mineral. (ed. S. W. Bailey). Mineralogical Society of America, Washington, DC, vol. 19, pp. 631–674.

Kastner M. (1979) Zeolites. In *Reviews in Mineralogy: Vol. 6*, Marine Minerals (ed. R. G. Burns). Mineralogical Society of America, Washington, DC, pp. 111–122.

Kastner M. and Stonecipher S. A. (1978) Zeolites in pelagic sediments of the Atlantic, Pacific, and Indian oceans. In *Natural Zeolites: Occurrence, Properties, Use* (eds. L. B. Sand and F. A. Mumpton). Pergamon, New York, pp. 199–220.

Knudsen A. C. and Gunter M. E. (2002) Sedimentary phosphorites—an example: Phosphoria formation, southeastern Idaho USA. In *Reviews in Mineralogy: Vol. 48. Phosphates—Geochemical, Geobiological, and Materials Importance* (eds. M. J. Kohn, J. Rakovan, and J. M. Hughes). Mineralogical Society of America, Washington, DC, pp. 361–389.

Kolla V., Be A. W. H., and Biscaye P. E. (1976) Calcium carbonate distribution in the surface sediments of the Indian Ocean. *J. Geophys. Res.* **81**, 2605–2626.

Kolodny Y. (1981) Phosphorites. In *The Sea* (ed. C. Emiliani). Wiley, New York, vol. 7, pp. 981–1023.

Koski R. A., Lonsdale P. F., Shanks W. C., Berndt M. F., and Howe S. S. (1985) Mineralogy and geochemistry of a sediment-hosted hydrothermal sulfide deposit from the southern trough of Guaymas Basin, *Gulf of California. J. Geophys. Res.* **90**, 6695–6707.

Leinen M. (1989) The pelagic clay province of the North Pacific Ocean. In *The Eastern Pacific Ocean and Hawaii*, The Geology of North America (eds. E. L. Winterer, D. M. Hussong, and R. W. Decker). Geological Society of America, Boulder, CO, vol. N, pp. 323–335.

Lewin J. (1961) The dissolution of silica from diatom walls. *Geochim. Cosmochim. Acta* **21**, 182–198.

Li Y. H. (1972) Geochemical mass balance among lithosphere, hydrosphere, and atmosphere. *Am. J. Sci.* **272**, 119–137.

Li Y. H. (1982) Interelement relationship in abyssal Pacific ferromanganese nodules and associated pelagic sediments. *Geochim. Cosmochim. Acta* **46**, 1053–1060.

Li Y. H. (1991) Distribution patterns of the elements in the ocean: a synthesis. *Geochim. Cosmochim. Acta* **55**, 3223–3240.

Li Y. H. (2000) *A Compendium of Geochemistry: from Solar Nebula to the Human Brain*. Princeton University Press, Princeton, NJ.

Lonsdale P., Burns V. M., and Fisk M. (1980) Nodules of hydrothermal birnessite in the caldera of a young seamount. *J. Geol.* **88**, 611–618.

Mackenzie F. T., Bischoff W. D., Bishop F. C., Loijens M., Schoonmaker J. E., and Wollast R. (1983) Magnesian calcites: low temperature occurrences, solubility, and solid solution behavior. In *Carbonates: Mineralogy and*

Chemistry, Rev. Mineral. (ed. R. J. Reeder). Mineralogical Society of America, Washington, DC, vol. 11, pp. 97–144.

Mandernack K. W., Post J. E., and Tebo B. M. (1995) Manganese mineral formation by bacterial spores of the marine *Bacillis*, strain SG-1: evidence for the direct oxidation of Mn(II) to Mn(IV). *Geochim. Cosmochim. Acta* **59**, 4393–4408.

Manheim F., Rowe G., and Jipa D. (1975) Marine phosphorite formation off Peru. *J. Sedimen. Petrol.* **45**, 243–251.

Marchig V. and Gundlach H. (1982) Iron-rich metalliferous sediments on the East Pacific Rise: prototype of undifferentiated metalliferous sediments on divergent plate boundaries. *Earth Planet. Sci. Lett.* **58**, 361–382.

Martin J. H. and Knauer G. A. (1984) VERTEX: manganese transport through oxygen minima. *Earth Planet. Sci. Lett.* **67**, 35–47.

Martin J. H., Knauer G. A., and Broenkow W. W. (1985) VERTEX: the lateral transport of manganese in the northeast Pacific. *Deep-Sea Res.* **32**, 1405–1427.

McArthur J. M. (1978) Systematic variations in the contents of Na, Sr, CO_2, and SO_4 in marine carbonate-fluorapatite and their relation to weathering. *Chem. Geol.* **21**, 41–52.

McArthur J. M. (1990) Fluorine-deficient apatite. *Min. Mag.* **54**, 508–510.

McClellan G. H. (1980) Mineralogy of carbonate fluorapatites. *J. Geol. Soc. London* **137**, 675–681.

McClellan G. H. and Lehr J. R. (1969) Crystal chemical investigation of natural apatites. *Am. Mineral.* **54**, 1374–1391.

McClellan G. H. and Van Kauwenbergh S. J. (1990) Mineralogy of sedimentary apatites. In *Phosphorite Research and Development*, Geol. Soc. Spec. Publ. 52 (eds. A. J. G. Notholt and I. Jarvis). Geological Society of London, pp. 23–31.

McMurtry G. M., Wang C.-H., and Yeh H.-W. (1983) Chemical and isotopic investigations into the origin of clay minerals from the Galapagos hydrothermal mounds field. *Geochim. Cosmochim. Acta* **47**, 475–489.

McMurtry G. M. and Yeh H.-W. (1981) Hydrothermal clay mineral formation of East Pacific Rise and Bauer Basin sediments. *Chem. Geol.* **32**, 189–205.

Mills R. A. and Elderfield H. (1995) Hydrothermal activity and the geochemistry of metalliferous sediment. In *Seafloor Hydrothermal Systems, Physical, Chemical, Biological, and Geological Interactions*, Geophysical Monograph 91 (eds. S. E. Humphris, R. A. Zierenberg, L. S. Mullineaux, and R. E. Thomson). American Geophysical Union, Washington, DC, pp. 392–407.

Mita N. and S. Nakao (1984) Minor chemical composition of deep-sea sediments from the GH80-5 area in the central Pacific Basin In *Mar. Geol., Geophysics, and Manganese Nodules in the Northern Vicinity of the Magellan Trough*, Cruise Report no. 20, 215–226, Geol. Surv. Japan.

Monnin C., Jeandel C., Cattaldo T., and Dehairs F. (1999) The marine barite saturation state of the world's oceans. *Mar. Chem.* **65**, 253–261.

Morse J. W. and Mackenzie F.T. (1990) *Geochemistry of Sedimentary Carbonates*. Elsevier, Amsterdam.

Mucci A. and Morse J. W. (1990) Chemistry of low temperature abiotic calcites: experimental studies on coprecipitation, stability, and fractionation. *Rev. Aquat. Sci.* **3**, 217–254.

Mukhopadhyay R., Iyer S., and Ghosh A. (2002) The Indian Ocean nodule field: petrotectonic evolution and ferromanganese deposits. *Earth-Sci. Rev.* **60**, 67–130.

Murray J. and Renard A. (1891) Deep sea deposits. In *Challenger Reports*. Longman, London, 525p.

Nathan Y. (1984) The mineralogy and geochemistry of phosphorites. In *Phosphate Minerals* (eds. J. O. Nriagu and P. B. Moore). Springer, New York, pp. 275–291.

Nicholson K., Hein J. R., Buehn B., and Dasgupta S. (eds.) (1997) *Manganese Mineralization: Geochemistry and Mineralogy of Terrestrial and Marine Deposits*, Geol. Soc. Spec. Publ. 119. Geological Society of London, 370pp.

Notholt A. J. G. and Jarvis I. (eds.) (1990) *Phosphorite Research and Development*. Geol. Soc. Spec. Publ. # 52. Geological Society of London.

O'Brien G. W. and Veeh H. H. (1980) Holocene phosphorite on the east Australian continental margin. *Nature* **288**, 690–692.

O'Brien G. W., Milnes A. R., Veeh H. H., Heggie D. T., Riggs S. R., Cullen D. J., Marshall J. F., and Cook P. J. (1990) Sedimentation dynamics and redox iron-cycling: controlling factors for the apatite-glauconite association on the East Australian continental margin. In *Phosphorite Research and Development*, Geol. Soc. Spec. Publ. #52 (eds. A. J. G. Notholt and I. Jarvis), London, pp. 61–86.

Oudin E. (1983) Hydrothermal sulfide deposits on the East Pacific Rise (21 N): Pt. I, descriptive mineralogy. *Mar. Mining* **4**, 39–72.

Parker R. J. (1975) The petrology and origin of some glauconitic and glauco-conglomeratic phosphrites from the South African continental margin. *J. Sedim. Petrol.* **45**, 230–242.

Parker R. J. and Siesser W. G. (1972) Petrology and origin of some phosphorites from the South African continental margin. *J. Sedim. Petrol.* **42**, 434–440.

Pinet P. R. (2000) *Invitation to Oceanography*, 2nd edn. Jones and Bartlett Publ., Boston.

Piper D. Z. and Heath G. R. (1989) Hydrogenous sediment. In *The Eastern Pacific Ocean and Hawaii, The Geology of North America* (eds. E. L. Winterer, D. M. Hussong, and R. W. Decker). Geological Society of America, Boulder, CO, vol. N, pp. 337–345.

Post J. E. (1999) Manganese oxide minerals: crystal structures and economic and environmental significance. *Proc. Natl. Acad. Sci.* **96**, 3447–3454.

Post J. E. and Bish D. L. (1988) Rietveld refinement of the todorokite structure. *Am. Mineral.* **73**, 861–869.

Revelle R. R. (1944) *Marine Bottom Samples Collected in the Pacific Ocean by the Carnegie on its Seventh Cruise*. Carnegie Inst. Wash. Publ., Washington, DC, 556p.

Ruttenberg K. C. (1993) Reassessment of the oceanic residence time of phosphorus. *Chem. Geol.* **107**, 405–409.

Ruttenberg K. C. and Berner R. A. (1993) Authigenic apatite formation and burial in sediments from non-upwelling, continental margin environments. *Geochim. Cosmochim. Acta* **57**, 991–1007.

Sayles F. L., Ku T.-L., and Bowker P. C. (1975) Chemistry of ferromanganese sediment of the Bauer Deep. *Bull. Geol. Soc. Am.* **86**, 1423–1431.

Schoonmaker J. E. (1986) Clay mineralogy and diagenesis of sediments from deformation zones in the Barbados accretionary wedge (Deep Sea Drilling Project Leg 78A). In *Synthesis of Structural Fabrics of Deep Sea Drilling Project Cores from Forearcs*, Geol. Soc. Am. Mem. 166 (ed. J. C. Moore), pp. 105–116.

Speer J. A. (1983) Crystal chemistry and phase relations of orthorhombic carbonates. In *Carbonates: Mineralogy and Chemistry*, Rev. Mineral. (ed. R. J. Reeder). Mineral. Soc. Am., Washington, DC, vol. 11, pp. 145–190.

Stoffer P., Glasby G. P., and Frenzel G. (1984) Comparison of the characteristics of manganese micronodules from the equatorial and southwest Pacific. *Tschermaks Mineral Petrogr. Mitt.* **33**, 1–23.

Stumm W. and Morgan J. J. (1970) *Aquatic Chemistry*. Wiley-Interscience, New York, 583pp.

Styrt M. M., Brackman A. J., Holland H. D., Clark B. C., Pisutha-Arnoud, Eldridge C. S., and Ohmoto H. (1981) The mineralogy and the isotopic composition of sulfur in hydrothermal sulfide/sulfate deposits on the East Pacific Rise, 21°N latitude. *Earth Planet. Sci. Lett.* **53**, 382–390.

Sugisaki R. and Yamamoto K. (1984) Major element chemistry of Pacific marine sediments around 10°N and 170°W. In *Geophysics, and Manganese Nodules in the Northern Vicinity of the Magellan Trough*, Mar. Geol., Cruise Report no. 20. Geological Survey of Japan, pp. 198–214.

Tebo B. M., Ghiorse W. C., van Waasbergen L. G., Siering P. L., and Caspi R. (1997) Bacterially-mediated mineral formation: insights into manganese(II) oxidation from molecular genetic and biochemical studies. In *Reviews in Mineralogy: Vol. 35. Geomicrobiology: Interactions between Microbes and Minerals* (eds. J. F. Banfield and K. H. Nealson). Mineralogical Society of America, Washington, DC, pp. 225–266.

Thompson G., Mottl M. J., and Rona P. A. (1988) Morphology, mineralogy and chemistry of hydrothermal deposits from the TAG area, 26°N Mid-Atlantic Ridge. *Chem. Geol.* **49**, 243–257.

Tizzard T. H., Moseley H. N., Buchanan M. A., and Murray J. (1885) *Report of the Scientific Results of the Voyage of HMS Challenger during the Years 1873–1876*, vol. 1, part. 1. Her Majesty's Stationery Office, London, 509p.

Tribble J. S., Arvidson R. S., Lane M., III, and Mackenzie F. T. (1995) Crystal chemistry, and thermodynamic and kinetic properties of calcite, dolomite, apatite, and biogenic silica: applications to petrologic problems. *Sedim. Geol.* **95**, 11–37.

Usui A. and Mochizuki T. (1984) Local variability of manganese nodule chemistry and its relationship to mineralogy in the GH80-5 area. In *Geophysics, and Manganese Nodules in the Northern Vicinity of the Magellan Trough*, Mar. Geol., Cruise Report no. 20. Geological Survey of Japan, pp. 242–262.

Usui A. and Moritani T. (1992) Manganese nodule deposits in the central Pacific Basin: distribution, geochemistry, mineralogy, and genesis. In *Geology and Offshore Mineral Resources of the Central Pacific Basin*. Circum-Pacific Council for Energy and Mineral Resources Earth Science Series (eds. B. H. Keating and B. R. Bolton). Springer, NY, vol. 14, pp. 205–223.

Valsami-Jones E., Ragnarsdottir K. V., Putnis A., Bosbach B., Kemp A. J., and Cressey G. (1998) The dissolution of apatite in the presence of aqueous metal cations at pH 2–7. *Chem. Geol.* **151**, 215–233.

Van Cappellen P. and Berner R. A. (1991) Fluorapatite crystal growth from modified seawater solutions. *Geochim. Cosmochim. Acta* **55**, 1219–1234.

Velde B. (1985) *Clay Minerals, a Physico-chemical Explanation of their Occurrence*. Developments in Sedimentology 40,. Elsevier, New York.

Vonderhaar D. L., McMurtry G. M., Garge-Schoenberg D., Stueben D., and Esser B. K. (2000) Platinum and other related element enrichments in Pacific ferromanganese crust deposits. In *Marine Authigenesis: from Global to Microbial*, Society for Sedimentary Geology, Tulsa, Spec. Publ. no. 66 (eds. C. R. Glenn, L. Prevot-Lucas, and J. Lucas), pp. 287–309.

Walther J. V. and Helgeson H. C. (1977) Calculation of the thermodynamic properties of aqueous silica and the solubility of quartz and its polymorphs at high pressures and temperatures. *Am. J. Sci.* **184**, 1315–1351.

Watkins R., Nathan Y., and Bremner J. M. (1995) Rare earth elements in phosphorite and associated sediment from the Namibian and South African continental shelves. *Mar. Geol.* **129**, 111–128.

Weaver C. E. and Beck R. C. (1977) *Miocene of the SE United States: a Model for Chemical Sedimentation in a Peri-marine Environment*, *Developments in Sedimentology*, Elsevier, Amsterdam, vol. 22, 234p.

Welch S. A., Taunton A. E., and Banfield J. F. (2002) Effect of microorganisms and microbial metabolites on apatite dissolution. *Geomicrobiol. J.* **19**, 343–367.

Wen X., DeCarlo E. H., and Li Y. H. (1997) Inter-element relationships in ferromanganese crusts from the central Pacific Ocean: their implications for crust genesis. *Mar. Geol.* **136**, 277–297.

Williams L. A. and Crerar D. A. (1985) Silica diagenesis: II. General mechanisms. *J. Sedim. Petrol.* **55**, 312–321.

Williams L. A., Parks G. A., and Crerar D. A. (1985) Silica diagenesis: I. Solubility controls. *J. Sedim. Petrol.* **55**, 301–311.

Windom H. L. (1976) Lithogenous material in marine sediments. In *Chemical Oceanography*, 2nd edn. (eds. J. P. Riley and R. Chester). Academic Press, New York, vol. 5, pp. 103–136.

Wollast R., Mackenzie F. T., and Bricker O. P. (1968) Experimental precipitation and genesis of sepiolite at earth-surface conditions. *Am. Mineral.* **53**, 1645–1662.

Zierenberg R. A., Shanks W. C., and Bischoff J. L. (1984) Massive sulfide deposits at 21 N, East Pacific Rise: chemical composition, stable isotopes, and phase equilibria. *Bull. Geol. Soc. Am.* **95**, 922–929.

Zhou L. and Kyte F. T. (1992) Sedimentation history of the south Pacific pelagic clay province over the last 85 million years inferred from the geochemistry of Deep Sea Drilling Project hole 596. *Paleoceanography* **7**, 441–465.

7.02
The Recycling of Biogenic Material at the Seafloor

W. R. Martin and F. L. Sayles

Woods Hole Oceanographic Institution, MA, USA

7.02.1 INTRODUCTION

The upper 10–50 cm of the marine sediment column is a site of active processing of particles that are deposited onto the sediments from the water column. Upon their arrival at the sediment surface, particles are simultaneously mixed by physical and biological processes through a layer that is ~10 cm thick, and altered by chemical and microbial processes. These primary particles react with pore-water solutes to produce altered particles and new solutes. In turn, the products of these initial reactions can undergo further alteration. The overall effect of these reactions is to produce solute exchanges across the sediment–water interface and alter the primary particles, so that those that accumulate in the sediments may be substantially different from the rain to the seafloor (Figure 1). The solute exchanges resulting from reactions in surface sediments can be important for the chemistry of oceanic deep water, and to the overall cycles of several elements in the oceans. In addition, the alteration of particles by these reactions must be taken into account when downcore records of the accumulation of sedimentary components are interpreted in terms of past oceanic and atmospheric chemistry. Understanding early diagenetic reactions in the upper few centimeters of the marine sediment column is important both to the study of geochemical cycles in the contemporary ocean and to the reconstruction of past oceanic conditions.

The particles that arrive at the seafloor consist, in varying proportions, of lithogenic and biogenic material; the latter consists of organic matter, $CaCO_3$, and biogenic opal. Biogenic

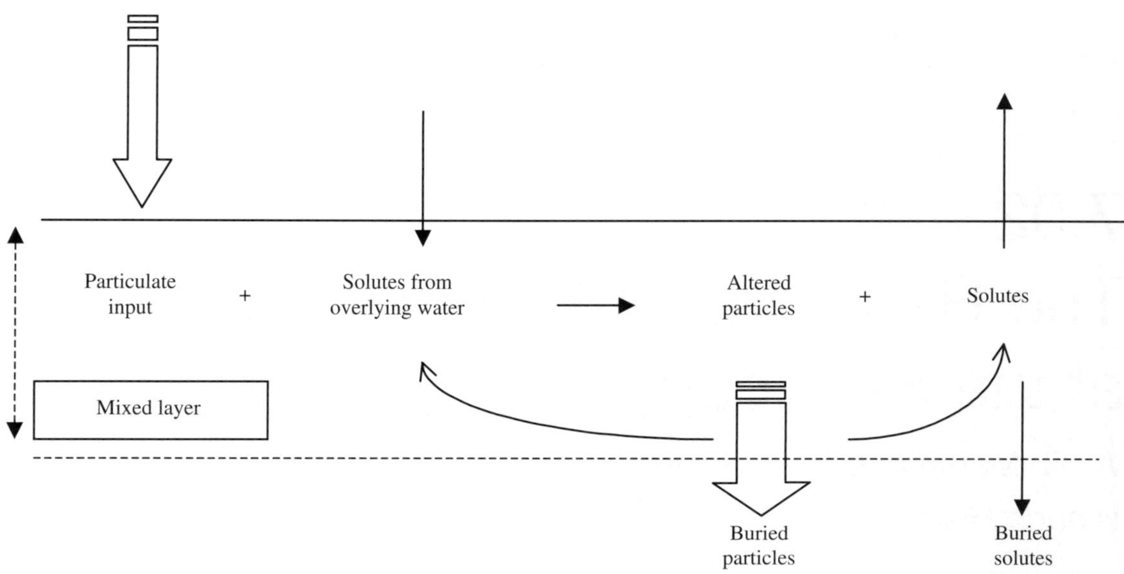

Figure 1 Particle cycling in the surface mixed layer of marine sediments. The processing of the rain of particles to the seafloor results in exchanges of solutes across the sediment–water interface and alteration of the particulate reactants, so that the composition of accumulating sediment is significantly different from that of the particulate rain to the seafloor.

material—especially organic matter—typically undergoes extensive remineralization in the water column, so that the material that arrives at the seafloor is a mixture of relatively "fresh" material that has escaped degradation in the water column and extensively altered, "aged" components. Nonetheless, sedimentary decomposition of biogenic material—in particular, of organic matter—is nearly complete in many locations. The efficient remineralization of biogenic particles depends on the high concentration of particles, active benthic ecosystems, and time. Particles that survive watercolumn decomposition will have resided there for weeks to a few years before being deposited on the sediment surface. In contrast, particles that are unreactive in sediments may reside in the mixed layer at the sediment surface for hundreds to thousands of years. The mixed layer at the surface of deep-sea sediments is typically ~10 cm thick, and sediment burial rates are ~ 1 cm kyr^{-1}. Therefore, the average particle can remain in the mixed layer for $\sim 10^4$ yr. Continental margin sediments experience faster sediment accumulation rates so that, despite the presence of a thicker mixed layer, unreactive particles typically reside in the mixed layer for hundreds of years. In both the cases, there is ample time for degradation of biogenic material from which the most reactive components have been removed by watercolumn processes.

In addition to the relatively long residence times of particles in the sedimentary mixed layer, the layered nature of reaction zones in the sediments is key to their significance in geochemical cycles. As shown in Figure 1, the primary reactants for remineralization reactions are supplied from above in the form of the shower of particles to the sediment surface and diffusive fluxes of solutes from bottom water to sediments. The supply of reactants from above, coupled with the relatively slow resupply of solutes consumed by sedimentary reactions by diffusion from bottom water, results in the observed layered structure, with a different dominant electron acceptor for organic matter oxidation in each layer. Oxygen, whose reaction with organic matter yields the most energy, is used first, followed by the nitrogen bound in NO_3^-, manganese oxides, iron oxides, and the sulfur in SO_4^{2-}. The consequence of this sequence of reactions is that dissolved oxygen is completely removed from pore waters in the upper few millimeters to centimeters of sediments on continental margins throughout the oceans. Therefore, processes, which are rare and localized in the marine water column—most importantly, denitrification and sulfate reduction—are widespread and quantitatively significant in continental margin sediments. These sedimentary processes have important consequences for geochemical cycles at the Earth's surface.

The reactions affecting the balance between preservation and dissolution of biogenic materials in sediments have been studied using measurements of solid-phase and pore-water constituents as well as direct measurements of solute fluxes across the sediment–seawater interface. Each type of measurement has advantages and disadvantages. In this review, we focus on the use of pore-water solute profiles for the study of the fate of biogenic materials in sediments. Pore-water solute concentrations are the most sensitive indicator of sedimentary reactions. Solid-phase concentrations are often quite high, so that reactions, which leave clear signals in pore waters, are not always readily observed in the solid phase. The duration of benthic flux chamber deployments is limited by practical considerations, so that, in low-flux environments, concentration changes in the water in flux chambers can be near the limits of detection. In addition, flux chambers yield only depth-integrated reaction rates. Although, especially in the presence of nondiffusive solute exchange across the sediment–seawater interface, they may give better flux estimates than can be calculated from pore-water profiles, they do not allow the determination of parameters that describe the mechanisms of sedimentary reactions. These parameters can be quantified only with pore-water solute studies. They are essential for the generalization of results from specific locations to ocean-wide models that are useful for the quantification of the role of sedimentary reactions in global marine geochemical cycles. After a review of methods of determining pore-water profiles, we will discuss the use of pore-water data for the study of benthic cycles of organic matter, $CaCO_3$, and biogenic silica.

7.02.2 PORE-WATER SAMPLING AND PROFILING

The vast majority of pore-water studies carried out over the past century have relied on extracting pore water from sediments retrieved from the seafloor in cores. Throughout most of this period, the methods of extraction have relied on expressing pore fluid through a filter that does not pass the sediment particles by the application of pressure to the wet sediment sample. Murray and Irvine (1885) applied this technique in what we believe to be the first pore-water study. They placed "blue terrigenous muds" from the seafloor off the Scottish coast in a canvas bag and caught and analyzed the drippings. Since that time "squeezing" has been a mainstay for the collection of pore waters. The approaches developed since Murray and Irvine may be more "sophisticated," but the principle is the same: forcing pore fluid through a membrane by applying pressure to the wet sediment. In its many forms, the "squeezer" is essentially a cylinder with a supported membrane at its base into which the sediment sample is placed. A piston (or diaphragm) is then inserted into the cylinder and force is applied to the piston, forcing pore water from the base of the cylinder (cf. Manheim, 1966). A variant of forcing of pore fluid across a filter under pressure is the "whole core squeezer" (wcs) (Bender *et al.*, 1987). In this method pressure is applied from the bottom of an intact core section and fluid is forced through a filter at the top. As the sediment collapses against the filter, pore fluid is expressed with essentially no mixing, thus preserving the depth–concentration relationships in the effluent. The advantage of this method is the recovery of a very high resolution pore-water profile. For example, for a 10 cm diameter core and sediment porosity of 0.87, the yield of solution for 1 mm of depth is 3.3 cc. Depending upon analytical requirements, it is quite reasonable to achieve a depth resolution 1 mm or 2 mm for the analyzed constituents. As a result of increasing loss of permeability with increasing compaction (depth), wcs sampling is limited to a depth of 2–4 cm, depending on the type of sediment. An alternative to all of the above pressure/filtration approaches that has been applied to unconsolidated sediments with high water content is the centrifugation of sediment samples to obtain pore waters. This method has been widely applied to most types of surficial sediments in the oceans and is applicable to sediment depths of many tens of centimeters in most marine sediment types. This is the most common method used to obtain pore waters.

Methods for sampling pore waters *in situ* at ambient temperature and pressure were developed in response to the discovery of irreversible artifacts affecting some pore-water constituents in recovered cores (see discussion of artifacts below). *In situ* probe sampling was introduced in the 1970s (Sayles *et al.*, 1973, 1976; Barnes, 1973; Sayles and Curry, 1988). The principle employed is that of expression of pore fluid from sediment across a filter under applied pressure. As implemented *in situ*, ambient pressure provides the driving force through dropping the pressure on the inside of the filter. In most studies reduction in the internal pressure has been accomplished with the use of a spring-loaded syringe assembly (Sayles *et al.*, 1973, 1976), but evacuated cylinders have also been used as the means of reducing internal pressure (Barnes, 1973). The wcs (above) has been implemented in an *in situ* version for a number of studies (Martin and Sayles, 1996; Sayles *et al.*, 2001).

Perhaps the most desirable method of obtaining concentration data for pore-water profiles is that of an *in situ* measurement since all sampling methods appear to suffer from some artifacts. The number of pore-water constituents that has been measured *in situ* in deep-sea sediments is quite limited, but the options are rapidly expanding. The most commonly measured pore-water constituent is oxygen (Revsbeck *et al.*, 1980; Reimers, 1987). pH has also been measured *in situ* and used in conjunction with *in situ* O_2 profiles to characterize benthic metabolic rates (Archer *et al.*, 1989). Extensive discussions of chemical sensors and their application to *in situ* measurement can be found in Varney (2000) and Buffle and Horvai (2000).

As noted above, artifacts of one sort or another affect samples obtained by extraction of pore waters from sediments by each of the methods discussed. Temperature changes experienced by cores retrieved from the deep sea have been shown to alter the concentrations of cations and SiO_2, as well as the pH in pore waters (Mangelsdorf *et al.*, 1969; Bischoff *et al.*, 1970; Fanning and Pilson, 1971). Some of these changes are found to be largely or completely reversible on re-equilibration at seafloor temperatures. Thus, procedures for obtaining pore waters from the retrieved cores routinely minimize warming, and carry out extractions at the water temperatures at the sea bottom. Pressure artifacts are also well documented for components of the carbonate system (Murray *et al.*, 1980; Emerson *et al.*, 1982). These changes are not reversible. *In situ* sampling or measurements are the only viable approaches to obtaining carbonate system data for pore waters in samples retrieved from the deep sea. A quite different set of artifacts is introduced to pore-water chemistry when suboxic and anoxic sediments are exposed to air during extraction. Artifacts include oxidation of reduced species such as Fe^{2+}, Mn^{2+} and dissolved sulfides. In addition, the products of oxidation influence other species (pH, CO_2 system, PO_4^{3-}). These artifacts are believed to be successfully controlled and avoided by extracting pore waters in an inert atmosphere, typically in N_2 gas. In addition to artifacts which generally affect extraction of pore waters from sediments, there are also artifacts specific to particular sampling methods. NO_3^- in pore water obtained by centrifugation is believed to be somewhat enriched above true values in the upper 0.5–3 cm of sediments (Berelson *et al.*, 1990a; Martin and McCorkle, 1993); the mechanism for this enrichment is unknown. SiO_2 concentrations in at least some sediments are modified during wcs processing onboard ship (Berelson *et al.*, 1990a; Martin *et al.*, 1991). The same effect was observed for *in situ* wcs samples from sediments with high content of terrigenous material (>50%), but was not observed in sediments dominated by biogenic silica (>60%) (Sayles *et al.*, 2001). The artifact presumably reflects adsorption of dissolved SiO_2 from solutions originating at depth (high concentrations) as these solutions from deeper beds pass through terrigenous material previously equilibrated with low concentrations characteristic of shallow sediment depth (Berelson *et al.*, 1990a; Martin *et al.*, 1991). That the effect is not observed in biogenic silica-rich sediments where shallow versus deep concentration differences are the greatest implies very limited surface reactivity of the particles. So far as we know, *in situ* probe pore-water sampling (excluding wcs) has not been found to introduce artifacts. However, there have been few, if any, truly independent tests of this hypothesis and making these tests is currently very difficult. It is important to note that *in situ* probe sampling has a severe drawback for some studies. Depth resolution for *in situ* probe samples is coarse, ≥2 cm. In addition, the sample ports near the interface are most prone to contamination by bottom water flowing down the probe inserted into the sediment. Thus for studies requiring high depth resolution near the interface, the *in situ* method is deficient.

In summary, no method for obtaining pore waters meets all needs. Rather, a combination of methods is required to meet the goals of most diagenetic studies. *In situ* measurement is applicable for a few components that are important in the study of sediment diagenesis and holds promise of providing high-resolution, artifact-free profiles for many, but by no means all dissolved components that are important in diagenetic studies. A summary of the anticipated artifacts and methods of choice for characterizing a variety of pore-water components is given in Table 1.

7.02.3 ORGANIC MATTER DECOMPOSITION IN SEDIMENTS

Heterotrophic respiration fueled by the rain of organic matter from the surface ocean is ubiquitous in marine sediments. Its rate determines one of the important characteristics of the sedimentary environment: the depth of redox horizons below the sediment–water interface. Heterotrophic respiration is the process by which carbon and nutrients are returned to the water column; it is important in the marine fixed nitrogen and sulfur cycles; and the accumulation of metabolic products sets the conditions for the removal of phosphorus from the oceans in authigenic minerals. A great deal of effort has been directed toward quantifying the rates, pathways, and effects of metabolism in sediments.

We will not undertake an exhaustive review of the reactions that affect and are driven by the input of organic matter to marine sediments. Rather, we will focus on the use of pore-water data to quantify the rate of decomposition of organic matter in sediments. One simplification of this treatment is that the exact biochemical

Table 1 Pore-water sampling methods.

Constituent	Methods						Comments	References
	Shipboard			In situ				
	sec/cent	*wcs*	*Electrode*	*Harpoon*	*wcs*	*Microelectrode profile*		
T_{CO_2}				~2–60 cm	~0–3 cm		The carbon isotopic composition of pore-water T_{CO_2} has been measured using shipboard pore-water separation methods. The basis for this use is the presumed absence of isotopic fractionation upon $CaCO_3$ precipitation.	Sayles (1981), Emerson *et al.* (1982), McCorkle *et al.* (1985), and Martin *et al.* (2000)
	Precipitation of $CaCO_3$ upon core retrievable makes use of shipboard methods for carbonate system components in deep-sea sediments impossible							
Alkalinity				~2–60 cm	~0–3 cm			
pH						~0–6 cm		Archer *et al.* (1989), Hales *et al.* (1994), and Cai *et al.* (1995)
P_{CO_2}						~0–6 cm		Cai *et al.* (1995)
O_2		~0–3 cm	~0–6 cm			~0–6 cm		Revsbech *et al.* (1980) and Reimers (1987)
NO_3^-	Near-interface artifacts	Near-interface artifacts, less pronounced than for sec/cent		~2–60 cm	~0–3 cm Pore-water gradients obtained to date appear to be low. The reasons for the low gradients are not yet known			Berelson *et al.* (1990a), Martin and McCorkle (1993), and Martin and Sayles (1996)
N_2							Concentration changes are too small to be measurable given the high concentration of N_2 in seawater.	

(continued)

Table 1 (continued).

Constituent	*Methods*						*Comments*	*References*
	Shipboard			*In situ*				
	sec/cent	*wcs*	*Electrode*	*Harpoon*	*wcs*	*Microelectrode profile*		
NH_4^+	Near-interface artifacts	Not applicable: artifacts		$\sim$2–60 cm			Near-interface samples yield concentrations that are somewhat to very much greater than concentrations from *in situ* harpoon sampling.	Jahnke *et al.* (1989)
TH_3PO_4	Artifacts	Artifacts		$\sim$2–60 cm			Shipboard and *in situ* harpoon methods yield different results; however, differences are not consistent from site to site and the reasons for the differences are unclear.	Jahnke *et al.* (1982, 1989)
Mn^{2+}	OK	Artifacts		$\sim$2–60 cm				
Fe^{2+}	OK	Artifacts		$\sim$2–60 cm				
SO_4^{2-}	OK	?		$\sim$2–60 cm			Concentration changes very small except in shallow-water sediments and anoxic basins.	
TH_2S	OK	?	OK				Concentration changes very small except in shallow-water sediments and anoxic basins.	
$Si(OH)_4$	Must be sampled at *in situ* temperature	Artifacts, except in sediments that are nearly pure biogenic opal		$\sim$2–60 cm	Artifacts, except in sediments that are nearly pure biogenic opal			Fanning and Pilson (1971), Berelson *et al.* (1990a), and Sayles *et al.* (2001)

composition of sedimentary organic matter is not taken into consideration, but it is considered to be composed of organic carbon, nitrogen, and phosphorus that requires empirically determined quantities of available electron acceptors for its oxidation to inorganic products. This simplification implies that we will neglect the fact that, in general, the particulate organic matter that reaches the seafloor is not immediately available for microbial respiration. First, it must undergo extracellular degradation (e.g., by hydrolysis or oxidation) that converts it to substrates useful for microbes (e.g., Henrichs, 1992). In effect, we assume that these extracellular reactions are rapid relative to the microbial oxidation of organic substrates to CO_2. Although data substantiating this assumption are limited, recent work by Arnosti and Holmer (1999) on the extracellular decomposition of polysaccharides did suggest that the extracellular processes were rapid relative to oxidation to CO_2 for the compounds studied.

Our focus on the quantification of organic matter oxidation rates in sediments also means that we will not discuss some reactions which, while geochemically important for other reasons, are generally of minor importance in this context. For instance, methanogenesis occurs only in rare circumstances, making it quantitatively unimportant to the total oxidation of organic matter in near-surface marine sediments. Nonetheless, it is an important pathway for the creation of large pools of methane hydrates, whose fate may have significant climatic effects (e.g., Kvenvolden, 1988). In addition, we will not describe the sedimentary cycling of phosphorus. Reactive phosphorus is lost from the oceans by burial in sediments (Ruttenberg, 1993); understanding the sedimentary chemistry of reactive phosphorus is essential to evaluating its potential long-term influence on marine productivity (van Cappellen and Ingall, 1994). Nonetheless, because of its complex sedimentary chemistry (e.g., Froelich *et al.*, 1988; Jahnke *et al.*, 1983), it is not used to evaluate the rate of decomposition of organic matter.

This review will describe our view of the current state of knowledge of the rate of oxidation of organic matter in marine sediments. First, we will discuss the electron acceptors that are used for organic matter oxidation. Then, we will turn to the carbon, nitrogen, oxygen stoichiometry of oxic organic matter decomposition, which dominates organic carbon cycling on most of the seafloor. Following that discussion, we will turn to the efforts describing the distribution of oxidation relative to the sediment–water interface and the timescale of the oxidation reactions.

7.02.3.1 Electron Acceptors for Sedimentary Organic Matter Oxidation

The pathways for sedimentary microbial metabolism are outlined in Table 2. They are presented in order of decreasing free energy yield for reaction of each oxidant (shown in bold type in the table) with sedimentary organic matter (Froelich *et al.*, 1979). Pore-water data support the assertion that the electron acceptors are used in this order of decreasing free energy yield. The order of the $NO_3^- \rightarrow N_2$ and $MnO_2 \rightarrow Mn(II)$ reactions is uncertain, however, and examples exist in the literature for which Mn(IV) appears to be used before NO_3^- (Froelich *et al.*, 1979; Klinkhammer, 1980) or for which NO_3^- appears to be used first (e.g., Shaw *et al.*, 1990; Lohse *et al.*, 1998). Thus, the order of electron acceptor use is: O_2, NO_3^- or Mn(IV), Fe(III), SO_4^{2-}.

Table 2 Oxidation/reduction reactions in marine sediments.

	Pelagic sediments		*Continental margin sediments*	
	Reactants	*Products*	*Reactants*	*Products*
Increasing depth below the sediment–water interface ↓	Organic C, N, P Mn(II) $\mathbf{O_2}$	CO_2, NO_3^-, $H_2PO_4^-$ Mn(IV)	Organic C, N, P NH_4^+, Mn(II), Fe(II), S(−II) $\mathbf{O_2}$	CO_2, NO_3^-, $H_2PO_4^-$ Mn(IV), Fe(III), SO_4^{2-}
	Organic C, N, P ? Fe(II) $\mathbf{NO_3^-}$	CO_2, N_2, $H_2PO_4^-$? Fe(III)	Organic C, N, P NH_4^+ Fe(II), S(−II) $\mathbf{NO_3^-}$	CO_2, N_2, $H_2PO_4^-$? Fe(III), SO_4^{2-}
	Organic C, N, P ? Fe(II) **Mn(IV)**	CO_2, NH_4^+, $H_2PO_4^-$? Fe(III) Mn(II)	Organic C, N, P NH_4^+, Fe(II), S(−II) **Mn(IV)**	CO_2, NH_4^+, $H_2PO_4^-$ Fe(III), SO_4^{2-} Mn(II)
	Organic C, N, P **Fe(III)**	CO_2, NH_4^+, $H_2PO_4^-$ Fe(II)	Organic C, N, P **Fe(III)**	CO_2, NH_4^+, $H_2PO_4^-$ Fe(II)
	Organic C, N, P $\mathbf{SO_4^{2-}}$	CO_2, NH_4^+, $H_2PO_4^-$ S(−II)	Organic C, N, P $\mathbf{SO_4^{2-}}$	CO_2, NH_4^+, $H_2PO_4^-$ S(−II)

Methanogenesis occurs after sulfate reduction, but it is important in early diagenesis only in rare circumstances, and it will not be discussed any further in this chapter.

Table 2 emphasizes the differences between organic matter oxidation pathways in pelagic versus continental margin sediments. For example, the first row in the table shows that only organic matter and Mn(II) are oxidized by O_2 in pelagic sediments. In contintental margin sediments, O_2 is also an oxidant for reduced Fe, S, and NH_4^+. In general, redox cycles occurring within the sediment column contribute more to organic matter oxidation in continental margin than in pelagic sediments. These internal cycles occur when reduced products of organic matter decomposition processes with relatively low free energy yields are transported toward the sediment–water interface by solute diffusion or by solid-phase mixing. As a result of this upward transport, they are re-oxidized. In this way manganese and iron oxides, as well as sulfate-sulfur, can be repeatedly used as electron acceptors many times before burial. While, in the presence of these redox cycles, O_2 may still be the ultimate electron acceptor, other organic matter decomposition pathways take on added importance. Bender and Heggie (1984) have argued that, because the oxic layer in most deep-sea sediments is over 10 cm thick and the supply of reactive iron and manganese oxides is relatively small, the metal oxides do not contribute significantly to oxygen consumption or organic matter decomposition in the deep sea despite the potential for cycling between oxidized and reduced forms. The oxic layer is much thinner in continental margin sediments than in pelagic environments (Reimers *et al.*, 1992; Lohse *et al.*, 1998; Martin and Sayles, submitted; Figure 2). In continental margin sediments, several factors make internal cycles between oxidized and reduced forms of manganese, iron, and sulfur more important: large supplies of organic matter to the seafloor deplete pore-water O_2 and enhance rates of anoxic organic matter decomposition; compressed redox zones lead to steeper gradients of reduced solutes and greater solute fluxes; and more rapid particle mixing enhances the transport of solid-phase products of early diagenetic reactions toward the sediment–water interface. The degree to which this cycling enhances the importance of manganese, iron, and sulfur as electron acceptors, as well as the importance of their oxidation to overall sedimentary oxygen demand, is not clear. Pore-water studies generally show that they are a significant, but secondary factor (e.g., Lohse *et al.*, 1998; Martin and Sayles, submitted). Studies that have relied on sediment incubations to determine reaction rates (summarized in Thamdrup, 2000) have concluded that

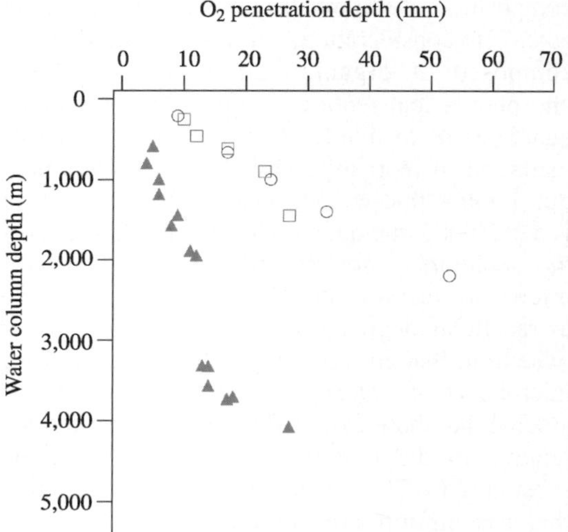

Figure 2 The depth below the sediment–water interface at which pore-water dissolved O_2 concentrations fall to zero on three continental margins: one in the northeast Atlanic (circles; Lohse *et al.*, 1998); one in the northwest Atlantic (squares; Martin and Sayles, submitted); and one in the northeast Pacific (triangles; Reimers *et al.*, 1992). In all cases, the thickness of the oxic layer is less than that in pelagic sediments. It is thicker in Atlantic margin sediments than in the northeast Pacific because of the intense oxygen minimum in the northeast Pacific.

iron and sulfur can be dominant electron acceptors in some cases. However, these studies must be considered with caution, as they require extremely rapid, and perhaps unreasonable, solid-phase mixing rates to support the observed reaction rates (Van Cappellen and Wang, 1996). What is not controversial is the significance of upward diffusion of NH_4^+, released by oxidation of organic matter by sulfate reduction, and its subsequent oxidation either to N_2 by reaction with NO_3^- or to NO_3^- by reaction with oxygen, to denitrification in continental margin sediments (Bender *et al.*, 1989; Devol and Christensen, 1993). These processes contribute significantly to sedimentary denitrification.

On most of the seafloor, organic matter is efficiently recycled by early diagenesis. Therefore, the supply of reactive organic matter is exhausted before all available electron acceptors are used up. As a result, the primary factor influencing the quantitative contribution of reactions involving different electron acceptors to sedimentary organic matter oxidation is the order in which they are used. In pelagic sediments, where the supply of organic matter is generally smaller than that in the ocean margins, oxygen is by far the most important electron acceptor, accounting for well over 90% of organic matter oxidation (Table 3). Denitrification accounts for an additional 1–5%; SO_4^{2-} reduction accounts for

Table 3 Electron acceptors in pelagic sediments.[a]

Site	*Region*	*C_{org} ox. rate* (μmol cm^{-2} yr^{-1})	*% of organic C oxidation by different electron acceptors*				
			O_2	*NO_3^-*	*Mn(IV)*	*Fe(III)*	*SO_4^{2-}*
MANOP H	E. eq. Pacific	12.0	99.2	0.8	0.4		
MANOP C	Central eq. Pacific	20.4	98.1	1.6	0.4		
E. eq. Atlantic	0–3° N, 6–16° W	12.4	93.8	4.4	0.1		1.8

[a] After Bender and Heggie (1984).

Table 4 Electron acceptors in continental margin sediments.

Location	*Water depths*	*Total C_{org} ox.* (μmol cm^{-2} yr^{-1})	*% of organic C oxidation by different electron acceptors*				
			O_2	*NO_3^-*	*Mn(IV)*	*Fe(III)*	*SO_4^{2-}*
NE Atlantic[a]	208–4,500	36–158	67–97	1–8.5	0–2.1	0–1.7	1–20
NW Atlantic[b]	260–2,510	36–52	74–90	1.8–6.0		8–20	
NE Pac.: $O_2 < 50$ μM[c]	780–1,440	66–75	5.0–46	41–69	0.1	0.7–1.3	5.7–25
NE Pac.: $O_2 = 73–145$[c]	1,900–4,070	36–74	69–75	11–18	0.1–6.9	0.3–0.7	5.6–18

[a] Lohse *et al.* (1998). [b] Martin and Sayles (submitted). [c] Reimers *et al.* (1992).

most of the remainder (Canfield, 1989). Iron and manganese are less important than SO_4^{2-} because of the limited supply of reactive metal oxides in most deep-sea sediments. Electron acceptors other than O_2 take on added importance in continental margin sediments. Table 4 summarizes results from studies in the northeast Atlantic (Lohse *et al.*, 1998), northwest Atlantic (Martin and Sayles, submitted), and northeast Pacific (Reimers *et al.*, 1992). The primary reason for this added importance is the greater supply of organic matter to the seafloor in the margin sites, resulting in the exhaustion of pore-water O_2 before all reactive organic matter is consumed. Table 4 shows organic matter oxidation rates that are 2–10 times the rates for pelagic sediments shown in Table 3. The studies in the North Atlantic show that O_2 consumption accounts for 67–97% of organic matter oxidation (after accounting for possible reoxidation of reduced products of anoxic decomposition). NO_3^- reduction accounts for 1–8% of the total. As in deep-sea sediments, iron and manganese appear to be of minor importance, while sulfate reduction can account for as much as 20% of the total. The study in the northeastern Pacific illustrates the importance of bottom water O_2 and NO_3^- concentrations to electron acceptor use. Within the O_2 minimum (bottom water $[O_2] < 50$ μM), O_2 accounted for only 5–46% of sedimentary organic matter oxidation. NO_3^- took on great significance, both because of the low bottom water $[O_2]$ and elevated bottom water $[NO_3^-]$: denitrification was responsible for 41–69% of organic matter oxidation within the O_2 minimum, 11–18% at deeper sites with more elevated bottom water $[O_2]$. As in the North Atlantic, manganese and iron were relatively small contributors to overall oxidation, while sulfate reduction accounted for up to 25% of organic matter oxidation. Sulfate reduction takes on added importance in shallow-water, estuarine sediments, which experience rain rates of organic matter to the seafloor 5–10 times those found on deeper continental margins. In these locations, contribution from sulfate reduction can be equal to, or greater than, oxic organic matter decomposition (Jorgensen, 1982; Canfield, 1989).

7.02.3.2 The Stoichiometry of Oxic Sedimentary Organic Matter Decomposition

An understanding of the stoichiometry of organic matter decomposition is important to the interpretation of pore-water profiles for two reasons. First, the most widely made measurements for the determination of organic matter decomposition rates are pore-water profiles or fluxes of O_2 and NO_3^-. In order to convert these measurements to carbon oxidation rates, the C_{ox}/O_2 or C_{ox}/NO_3^- ratios must be known. Second, even if T_{CO_2} is measured, it is produced both by organic carbon oxidation and by the dissolution of $CaCO_3$. As we will discuss below, it is difficult to separate the contributions of these two processes, and knowledge of the rate of production of T_{CO_2} by organic matter oxidation is useful for determining the rate of $CaCO_3$ dissolution. In this section, we will discuss the state of knowledge of the stoichiometry of oxic organic matter decompostion in pelagic sediments.

For many years, the Redfield–Ketchum–Richards (RKR) model (Redfield *et al.*, 1963) of planktonic organic matter composition has been applied to calculations of sedimentary organic matter oxidation rates. According to this model, the C : N ratio in organic matter is 6.6; all organic carbon is assumed to be in the form of CH_2O, and all organic nitrogen in the form of NH_3. It has been asserted that this model results in unrealistically high oxygen and hydrogen contents in marine organic matter and an unrealistically low O_2 requirement for respiration (Anderson, 1995). Hedges *et al.* (2002) confirmed Anderson's analysis by using ^{13}C NMR to evaluate the biochemical composition of planktonic organic matter. The results of these two studies are in quite good agreement with the composition of the organic matter that is degraded in the ocean, based on an analysis of oxygen consumption and nitrogen, phosphorus, T_{CO_2}, and alkalinity accumulation along isopycnal surfaces (Anderson and Sarmiento, 1994). The O_2/N and O_2/C ratios based on these studies are summarized in Table 5. In applying the results of these studies to sedimentary processes, it is important to recognize that the chemical compositions of the organic matter raining onto the sediment surface and of sedimentary organic matter are largely unkown: while organic geochemists can identify ~80% of the compounds making up planktonic organic matter, they can identify less than 25% of those in deep ocean and sedimentary particulate organic matter (Wakeham *et al.*, 1997). For this reason, the results of the Anderson and Sarmiento (1994) study are perhaps most useful for sediment geochemists, since they yield basin-wide regeneration ratios over the 400–4,000 m depth interval. Still, these results may not apply directly to sedimentary organic matter regeneration at all sites.

We will summarize results from two regions to examine the ratio of oxygen consumed to NO_3^- produced during oxic decomposition of sedimentary organic matter. Results are limited for two reasons. First, relatively few studies include measurements of both O_2 and NO_3^- fluxes or pore-water profiles. Second, the simple analysis used here requires that denitrification be much slower than nitrification. Otherwise, denitrification will draw down pore-water NO_3^-, and the O_2 consumed: NO_3^- produced ratio observed will be greater than that for oxic organic matter decomposition. This criterion is met in the central equatorial Pacific (Bender and Heggie, 1984), where the Eq. Pac. study (2° S–2° N, 140° W) yielded both fluxes across the sediment–water interface (Hammond *et al.*, 1996) and rain rates of particulate organic matter to the seafloor (Honjo *et al.*, 1995). The criterion is also met along the AESOPS transect in the Southern Ocean (we will show results from 64° 11.8′ S, 170° 04.8′ W and 63° 10′ S, 169° 51′ W) (Sayles *et al.*, 2001; Honjo *et al.*, 2000). The results from these studies are shown in Table 5. The two equatorial Pacific data sets include numerous pore-water profiles, from several sites. The results from the Southern Ocean are from measurements made on single cores in two locations. In three of the four sets of measurements, the average O_2/NO_3^- ratio was consistent with the Anderson and Sarmiento (1994) (henceforth A&S) basin-wide water column decompostion ratios; however, in the Martin *et al.* (1991) equatorial Pacific study, the range in measured values also included the RKR value. In the fourth measurement (AESOPS site M3), the measured ratio is consistent with the RKR value. When the A&S O_2/C ratio is used to infer the C : N ratio of the degrading organic matter from the O_2/NO_3^- flux ratios, the inferred

Table 5 Stoichiometric ratios in marine organic matter and for sedimentary oxic organic matter decomposition.

Planktonic organic matter				
Source	$-O_2/NO_3$[a]	$-O_2/C$[b]	C/N[c]	
RKR[d]	8.63	1.3	6.6	
Anderson[e]	9.31	1.41	6.6	
Hedges[f]	9.06	1.45	6.2	
Basin-wide decomposition ratios				
A&S[g]	10.6 ± 0.6	1.45 ± 0.19	7.3 ± 0.8	
Pore waters and sediment traps				
Region	$-O_2/NO_3$	$-O_2/C$	C/N	Trap C/N
Eq. Pac.[h]	9.89 ± 0.60		6.8 ± 1.0	
Eq. Pac. (jgofs)[i]	11.3 ± 1.1	1.45 ± 0.19	7.8 ± 1.3	7.3[j]
Aesops-5/4[k]	10.8 ± 0.4		7.4 ± 1.0	5.8–8.4[l]
Aesops-M3	8.83 ± 0.32		6.1 ± 0.8	7.4

[a] The ratio of O_2 consumed to NO_3 produced during oxidation. [b] The ratio of O_2 consumed to organic C oxidized. [c] The C : N ratio in the decomposing organic matter. [d] Redfield *et al.* (1963). [e] Anderson (1995). [f] Hedges *et al.* (2002). [g] Anderson and Sarmiento (1994). [h] Martin *et al.* (1991). [i] Hammond *et al.* (1996). [j] Honjo *et al.* (1995). [k] Sayles *et al.* (2001). [l] Honjo *et al.* 2000: traps at 1,000 m water depth.

C : N is in agreement, within the rather large uncertainties in the data, with the C : N ratios measured in sediment traps deployed at the same locations. Although the uncertainties in the available data are too large to draw firm conclusions, the data show that the A&S decomposition ratios, determined from water column data, are consistent with O_2 consumption: NO_3^- production ratios in oxic pore waters.

In order to expand the scope of comparison with the A&S ratios, we also consider several sites in the equatorial Atlantic Ocean where denitrification consumes a significant fraction of the NO_3^- released to pore waters by oxic decomposition. In these cases, we use models describing pore-water NO_3^- and O_2 profiles to estimate the $O_2 : NO_3^-$ ratio during oxic decompostion. For example, we consider a site at 3,100 m water depth on the Cape Verde Plateau (18° 27.8′ N, 21° 01.5′ W), where the oxic layer in the sediments was 6 cm thick. We used a fit to pore-water O_2 versus depth data to determine parameters describing O_2 consumption. We then multiplied these parameters by the NO_3^- production (O_2 consumption ratios from (i) the RKR model and (ii) the A&S measurements) and used the result to fit pore-water NO_3^- data. The A&S ratio yielded a significantly better fit to the data than the RKR model (Figure 3). In addition, the application of the model to the data showed that, in this case, with the oxic decompostion/denitrification boundary at 5 cm below the sediment–water interface, a plot of pore-water O_2 versus NO_3^- concentrations departed from linearity at 1.6 cm below the interface: it would be incorrect to derive the $O_2 : NO_3^-$ ratio for oxic decomposition from a linear fit to these data.

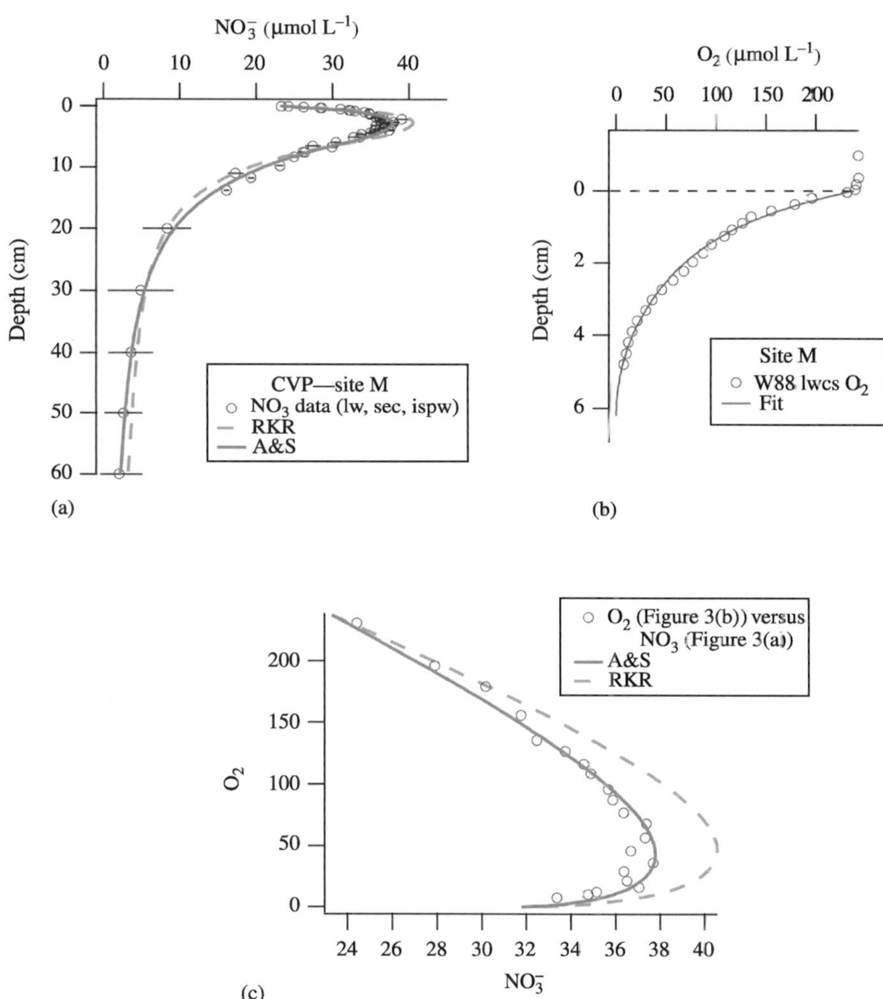

Figure 3 Pore-water data and modeling results from 3,100 m on the Cape Verde Plateau (CVP) at 18° 27.8′ N, 21° 01.5′ W. (a) Shows pore-water NO_3 data, with two model fits for which the oxic portion of the profile is described using parameters based on the fit to pore-water O_2 profile in (b). In one case (dashed line), the O_2 parameters are modified using RKR stoichiometry; in the second (solid line) they are adjusted using Anderson and Sarmiento (1994) stoichiometry. (c) Compares the same model fits to a plot of the pore-water O_2 versus NO_3 concentrations.

We applied a similar procedure to three locations on the Ceara Rise, in the western equatorial Atlantic. At these locations, we used fits to pore-water NO_3^- data (the profiles were the result of combining both *in situ* and shipboard samples, using several profiles at each location (Martin and Sayles, 1996) to obtain parameters describing NO_3^- production during organic matter oxidation). These parameters were multiplied by the $O_2:NO_3^-$ ratio for oxic decomposition of RKR and A&S; and model O_2 profiles using the resulting parameters were compared to the *in situ* microelectrode O_2 profiles of Hales and Emerson (1997). In two of three cases, the A&S ratio was clearly more consistent with the observed data, while the RKR ratio was more consistent with the third (Figure 4). These results confirm the conclusion drawn from the equatorial Pacific and Southern Ocean examples. While there is significant variability in the pore-water results, they are most often consistent with the application of the Anderson and Sarmiento $O_2:NO_3^-$ ratio. Furthermore, when A&S ratios are used to calculate the C : N of the organic matter being oxidized, the resulting value is essentially the same as the C : N ratio measured in the organic matter collected in sediment traps.

7.02.3.3 The Depth Distribution of Organic Matter Oxidation in the Sediment Column

The emergence of methods for the collection of pore-water profiles of O_2 and NO_3^- with millimeter-scale resolution has led to a new examination of the kinetics of sedimentary organic matter oxidation. Armed with high-resolution data, investigators have shown that the rate of organic matter oxidation in pelagic sediments decreases sharply with increasing depth below the sediment–water interface (Martin *et al.*, 1991; Hales *et al.*, 1994; Hales and Emerson, 1996, 1997; Martin and Sayles, 1996; Hammond *et al.*, 1996; Sayles *et al.*, 2001). These results are important for creating basin-wide models of sedimentary processes, for evaluating the timescale of

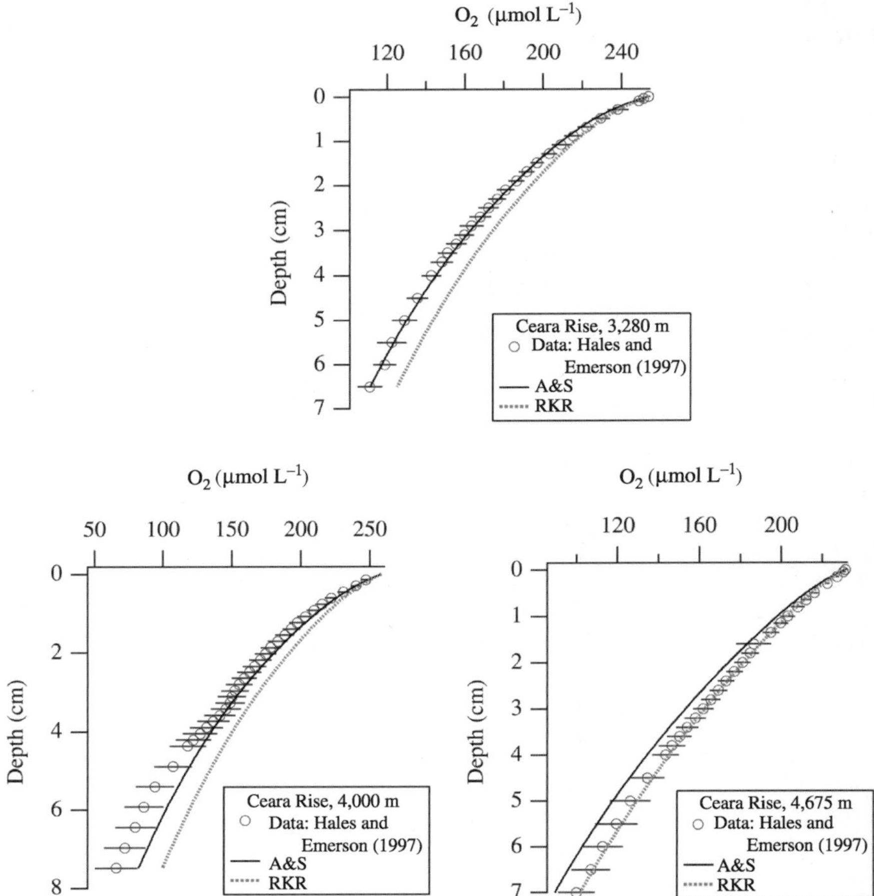

Figure 4 Pore-water oxygen concentrations at three sites on the Ceara Rise, western equatorial Atlantic. In each plot, model pore-water O_2 profiles are shown in which model parameters are calculated from fits to pore-water NO_3 profiles from the same sites, with stoichiometric adjustments based on: (a) RKR stoichiometry (dotted lines) and (b) Anderson and Sarmiento (1994) stoichiometry (solid lines) (source Hales and Emerson, 1997).

organic matter decomposition in sediments, and for determining the role of oxic metabolism in $CaCO_3$ dissolution. Below, we will show a summary of the depth-distribution of oxic organic matter decomposition in pelagic sediments as well as an example of the distribution of oxidation in margin sediments where oxidants other than O_2 are important.

The interpretation of pore-water concentration versus depth profiles of O_2 and NO_3^- in oxic sediments is based on a one-dimensional, steady-state model in which the production or consumption of a solute in a sedimentary layer is balanced by transport into or out of the layer by solute diffusion and burial advection. In mathematical form,

$$0 = \frac{d}{dx}\left\{\phi D_{sed}\frac{dC}{dx}\right\} - \frac{d}{dx}\{\phi wC\} + \phi P(x) \quad (1)$$

In this equation, C is the concentration of the solute in pore water, D_{sed} is the solute diffusion coefficient, corrected for sediment tortuosity, ϕ is porosity, w is the rate of burial of pore water, and $P(x)$ is the production or consumption rate of the solute. For NO_3^-, $P(x)$ has the form

$$P(x) = j_0\,e^{-j_1x} + j_2\,e^{-j_3x} \quad (2)$$

j_0 and j_2 have units of $\mu mol\ cm_{pw}^{-3}\ yr^{-1}$, while j_1 and j_3 are expressed in $1/cm_{sed}$. For O_2, $P(x)$ has the same form and, in principle, j_0 and j_2 for O_2 are related to the similar quantities for NO_3^- through the stoichiometric relationship described above. Martin and Sayles (2003) applied this equation to a set of 60 pore-water O_2 and NO_3^- profiles, obtained by *in situ* and shipboard sampling methods, from three regions in the equatorial Atlantic (Ceara Rise, Cape Verde Plateau, and Sierra Leone Rise: Martin and Sayles, 1996; Hales and Emerson, 1996; Sayles and Martin, unpublished data), the central equatorial Pacific (6° S–10° N, 130° W: Martin *et al.*, 1991), and the AESOPS transect in the Southern Ocean (66° S, 170° W to 56° S, 170° W: Sayles *et al.*, 2001). The sites considered span water column depths from 3,100 m to 5,200 m and total sedimentary organic carbon oxidation rates of 3–50 $\mu mol\ cm^{-2}\ yr^{-1}$. Examples of the fits of the model above to pore-water profiles, used to define values for j_0, j_1, j_2, and j_3, were shown in Figures 3 and 4.

Roughly speaking, the first exponential term in the expression for $P(x)$ (defined by the parameters j_0 and j_1) describes organic matter oxidation occurring closer to the sediment–water interface, while the second term (j_2 and j_3) describes the oxidation that occurs deeper in the sediment column. The first exponential can account for most of the organic matter oxidation (Figure 5(a)).

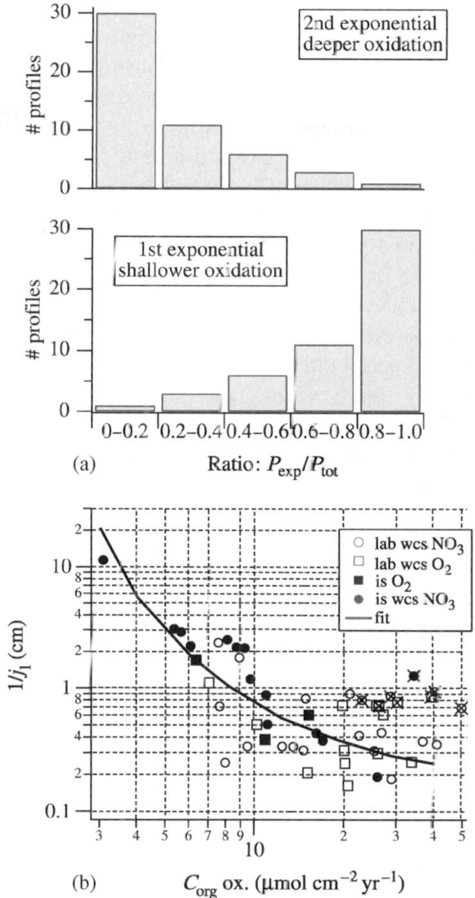

Figure 5 (a) A histogram showing the proportion of total oxic respiration that is accounted for by each of the exponential terms in the pore-water O_2/NO_3 model of Equations (1) and (2). The top part of the figure shows the number of profiles, among the 60 that were analyzed using the model, in which the second (deeper) exponential accounted for 0–20%, 20–40%, etc. of total respiration. The bottom part of the figure shows the same quantities for the shallower exponential. (b) A representation of the e-folding depth for oxic organic matter decomposition as a function of total organic matter oxidation rate (essentially equal to the rain rate of organic C to the seafloor). Lab wcs = shipboard sampling by whole-core squeezing; data are from Martin *et al.* (1991), Martin and Sayles (1996), and Sayles *et al.* (2001). Is O_2 = *in situ* microelectrode profiling (data from Hales and Emerson, 1997). Is wcs = sampling by *in situ* whole-core squeezer; data are from Martin and Sayles (1996), Sayles *et al.* (2001), and Sayles and Martin (unpublished) data from the eastern equatorial Atlantic.

It accounts for over 80% of total oxidation in half the profiles considered, over 60% in two-thirds of the profiles. This result is similar to that obtained by Hammond *et al.* (1996) in the central equatorial Pacific, where they found that 70–90% of organic matter oxidation at the low-latitude sites was accounted for by the first exponential term.

The dominance of the first exponential term in organic matter oxidation means that $1/j_1$ gives a good approximation of the e-folding depth for oxic organic matter decomposition in the sediment column. The variation of $1/j_1$ with total organic matter oxidation rate is shown in Figure 5(b). As might be expected, there is a lot of scatter in the relationship, but there is a steady decrease in the e-folding depth from ~3 cm at organic matter oxidation rates of ~5 μmol cm^{-2} yr^{-1}, to 0.3 cm at 20 μmol cm^{-2} yr^{-1}. There is no clear pattern in e-folding depth above 20 μmol cm^{-2} yr^{-1}. Several profiles show unusually large values of the e-folding depth (symbols overscored by "×" in the figure). These samples are from two regions: three sites on the AESOPS transect in the Southern Ocean where the sediments were covered by a distinct "fluff" layer that was 1–2 cm thick; and two sites at the shallowest depths on the Cape Verde Plateau in the eastern equatorial Atlantic. Nonetheless, the first-order conclusion from this effort is clear. Organic matter oxidation in deep-sea sediments occurs very close to the sediment–water interface. Thus, it is rapid relative to the mixing processes that distribute particles through the upper few centimeters of the sediment column. The rapidity of oxidation implies a short response time of sedimentary oxidation to changes in rain rates to the seafloor. In addition, the proximity of oxic decomposition to the sediment–water interface has important consequences for $CaCO_3$ dissolution, which we discuss below.

The concentration of organic matter oxidation in a layer close to the sediment–water interface is not limited to pelagic sediments. Two examples from a transect of the continental slope in the northwest Atlantic (170° W) illustrate the point (Martin and Sayles, submitted). One of these sites underlies 460 m of water. It has a total organic carbon oxidation rate of 51 μmol cm^{-2} yr^{-1}; 83% of this oxidation used O_2 as an electron acceptor, 5% used NO_3^-, and 12% used manganese or iron oxides or SO_4^{2-}. The second site lies under a 1,467 m water column and has a total organic matter oxidation rate of 43 μmol cm^{-2} yr^{-1}, 88% using O_2, 4% using NO_3^-, and 8% using electron acceptors lower on the free energy yield ladder. At each site, organic matter oxidation rates as a function of depth below the sediment–water interface were determined using fits of Equation (1) to pore-water O_2, NO_3^-, and NH_4^+ profiles. The production/consumption term for NO_3^- allowed production during oxic organic matter decomposition and consumption by denitrification. The NH_4^+ term included production via anoxic organic matter decomposition and oxidation by reaction with either O_2 or NO_3^-. The results are shown in Figure 6, in which the organic matter oxidation rate (in μmol C oxidized per gram of dry sediment per year) is plotted versus depth below the sediment–water interface at each site. The most rapid organic matter oxidation occurs in the oxic zone, which is 1 cm thick at the 460 m site, and 2.2 cm thick at the 1,467 m site. The rate of oxidation at the interface is 2.5 (1,467 m) to over 10 times (460 m) that of the same occurring at 2 cm. Clearly, at these sites, the oxidation of organic matter is concentrated in the upper 1–2 cm of the sediment column.

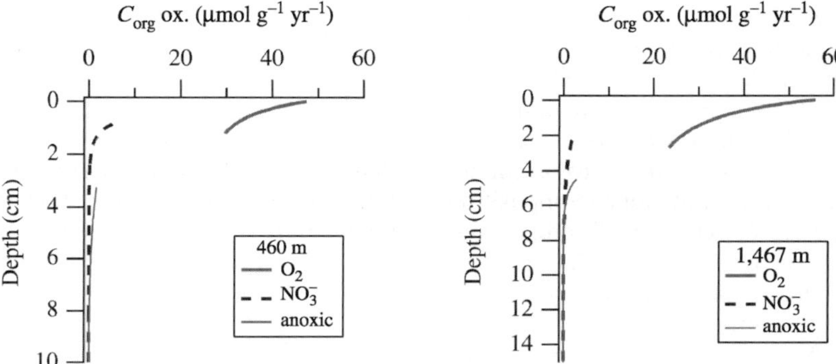

Figure 6 The rate of organic matter oxidation as a function of depth below the sediment–water interface at two locations on the northwest Atlantic margin at 70° W (Martin and Sayles, submitted). The calculated organic C oxidation rates are based on fits to pore-water O_2, NO_3^-, and NH_4^+ profiles. The apparent discontinuity between the rates of oxidation by O_2 and NO_3^- is most likely an artifact of the modeling procedure used to calculate the rates. Fits to dissolved O_2 profiles were entirely independent of fits to NO_3^- profiles; at sites where NO_3^- profiles could be used to calculate both the rate of oxidation by O_2 and that of oxidation by NO_3^-, magnitudes of oxidation rates were similar to those shown here, but no discontinuity was present (Martin and Sayles, submitted). The small overlap in the zones of oxidation by O_2 and NO_3^- is most likely due to the relatively poor resolution (~0.5 cm) of the pore-water NO_3^- concentration profiles.

7.02.3.4 The Response Time for Organic Matter Oxidation

It is important to know the rate at which sedimentary processes respond to changes in the rate of input of particles to the seafloor. The intermediate-term response of ocean chemistry to climate-driven changes in the export of carbon from the surface layer depends in part on the rate at which sediments recycle biogenic material. In addition, knowing the rate of response of sedimentary remineralization to changing inputs is important for interpreting measured fluxes across the sediment–water interface. It is well known that the rain rate of biogenic particles to the seafloor varies throughout the ocean on seasonal and interannual timescales. Given these variable inputs, do benthic remineralization rates reflect brief pulses of inputs, or do they reflect an average response to a longer-term average input? The answer to this question lies in the response time of sedimentary reactions to the changing inputs.

It has been shown experimentally that organic matter decomposition in sediments can be explained as a set of quasi-first-order reactions of the sort

$$\frac{dG_i}{dt} = -k_i G_i \qquad (3)$$

and

$$\frac{dG}{dt} = \sum \frac{dG_i}{dt} \qquad (4)$$

(Westrich and Berner, 1984). G_i is the concentration of a fraction of sedimentary organic matter (e.g., mol cm_{sed}^{-3}) that decays with a rate constant k_i (yr^{-1}). According to this scheme, the organic matter arriving at the seafloor is made up of fractions of differing reactivity, each of which decays following a first-order rate law. Total sedimentary organic matter (G) decomposition is the sum of the decomposition rates of each of these fractions. Two efforts to estimate the response time of organic matter decomposition in sediments used this framework in different ways.

Hammond *et al.* (1996) used fits of the model of Equation (1) to pore-water O_2 and NO_3^- profiles in the central equatorial Pacific to define the rate of organic matter oxidation as a function of depth below the sediment–water interface. Their model was built on a two-exponential description of the O_2 consumption or NO_3^- production rate (Equation (2)). They assumed that the fraction of organic matter decomposition represented by the exponential with a short scale length was the "highly reactive" fraction, the fraction represented by the exponential with the longer scale length was a "less reactive" fraction, and that there was a third fraction that was unreactive on the timescales represented by their O_2 and NO_3^- profiles. These assumptions gave them rates of decomposition of two "reactive" organic matter components. They further assumed that the highly reactive fraction was mixed at a rate described by the short-lived particle tracer, ^{234}Th (24 d half-life) and that the less reactive fraction was mixed at a rate described by ^{210}Pb (22 yr half-life). Their pore-water model gave them scale lengths for decomposition; when coupled with solid-phase mixing rates, they could convert the scale lengths into first-order rate constants for decomposition of two organic matter fractions. Hammond *et al.* found that, at stations between 2° S and 2° N, the response time of the rapidly degrading fraction was days to months, while that of the less labile fraction was decades. Given the quantitative dominance of the most labile fraction, this result implied that, at these sites, sedimentary organic matter decomposition should respond to temporal changes in input rates on a timescale of days to months.

Sayles *et al.* (2001) applied a similar model to organic matter decomposition in sediments along the AESOPS transect in the Southern Ocean. As was done by Hammond *et al.* in the equatorial Pacific, Sayles *et al.* defined the rate of organic matter oxidation versus depth below the sediment–water interface by fits of Equation (1) to pore-water NO_3^- profiles. Since they had measurements of sedimentary organic carbon concentrations at each site, they used Equation (3) to estimate the first-order rate constant for organic matter oxidation. First, observing that sedimentary C_{org} concentrations decreased from the interface to an apparent asymptotic value at depth, they divided the sedimentary organic matter into a "reactive" and an "unreactive" fraction by assuming that the asymptotic concentration was unreactive, and subtracting it from the total C_{org} at each depth to create a reactive C_{org} profile. They then divided this reactive C_{org} concentration profile into the profile of organic matter oxidation rates to yield a continuously varying, depth-dependent rate constant for organic matter oxidation. They obtained a second estimate of the rate constant by applying a solid-phase mixing/organic matter decay model directly to the reactive C_{org} profiles, using fits to the data to define a depth-dependent, quasi-first-order rate constant. They then averaged the depth-dependent rate constants, resulting from the two approaches over the depth interval in which 50% of the total organic matter decomposition occurred, to arrive at an estimate of the response time of organic matter decomposition to short-term changes in the rain rate of organic matter to the seafloor. They estimated response times of 0.3–1.6 yr at sites south of the Polar Front, where organic matter oxidation rates were 17–43 μmol cm^{-2} yr^{-1}, and response times of

4–80 yr at sites north of the Polar Front, where organic matter oxidation rates were 3–11 μmol cm^{-2} yr^{-1}. In the Pacific sector of the Southern Ocean, where organic matter oxidation rates exceeded 15 μmol cm^{-2} yr^{-1}—comparable to the rates observed by Hammond *et al.* (1996) in the central equatorial Pacific—the response time for organic matter oxidation was found to be months.

A very different approach to the estimation of the response time of organic matter oxidation to changing input rates uses contemporaneous measurements of the organic matter rain to the seafloor and the sedimentary O_2 consumption rate. A model, using simple first-order kinetics for sedimentary organic matter oxidation, shows that the more rapid the organic matter oxidation rate, the more closely variations in sedimentary O_2 consumption track variations in the C_{org} rain to the seafloor (Martin and Bender, 1988). The phase relationship between the rain rate and sedimentary O_2 consumption rate, as well as the relative amplitudes of the variations in input and consumption rates, can be used to constrain the rate constant for organic matter oxidation. Sayles *et al.* (1994) applied this model to data from a site in the northeast Pacific (Smith, 1987: 34° 50′ N, 123° W, 4,100 m), with an average C_{org} rain rate of ~35 μmol cm^{-2} yr^{-1} and from the Bermuda time-series site (4400 m water depth), where the average C_{org} rain rate was ~5 μmol cm^{-2} yr^{-1}. They found a rapid response time at the higher flux, northeastern Pacific site, of 0.1–0.2 yr, but a response time no shorter than ~1–5 yr at the low-flux, Atlantic site.

Together, these results indicate that the response time of sedimentary organic carbon oxidation is on the order of weeks when the organic carbon rain to the seafloor is greater than ~15 μmol cm^{-2} yr^{-1}, but is years or more at sites with low rain rates.

7.02.3.5 The Burial of Organic Matter in Marine Sediments

Thus far, we have described the estimation of the rate of oxidation of organic matter in sediments. When this information is coupled with data on the burial rate of organic matter, it allows the calculation of "burial efficiency," or the burial rate divided by the rain rate of organic matter to the seafloor. Sedimentary organic matter oxidation is highly efficient: averaged over all marine sediments, only ~10% of the organic matter that falls to the seafloor is preserved (Hedges and Keil, 1995). A survey of the burial and oxidation rates of organic carbon in the deep sea (>1,000 m water depth) has shown that organic matter oxidation is especially efficient there, as only ~3% of the organic matter that arrives at the seafloor is preserved (Jahnke, 1996). Burial efficiencies are much more variable on continental margins, with values of over 50% in some locations (Henrichs and Reeburgh, 1987).

Organic matter is efficiently oxidized in the oceanic water column as well as in the sediments. Jahnke (1996) estimated that, on an average, 2% of primary productivity reaches the 1,000 m depth level in the ocean, and only 1% reaches the seafloor below 1,000 m. The combination of the high efficiencies of oxidation in the water column and in sediments means that very little organic matter is buried in the deep sea. Berner (1982) estimated that ~90% of the organic carbon burial in the oceans occurs on the continental margins. For that reason, recent efforts to understand the mechanisms of organic matter preservation in the ocean have focused on continental margin sediments. As has been emphasized by Hedges and Keil (1995), the mechanism of organic matter preservation has large-scale geochemical significance: the long-term stability of the atmospheric O_2 level depends on a long-term balance between the rate of kerogen weathering on the continents and the burial rate of organic matter in marine sediments.

The preservation of organic matter may depend on properties of the organic matter itself and on the environment in which the organic matter is deposited. Most work has examined the role of the sedimentary environment. Determining the relative importance of different environmental factors has proven to be very difficult, as it is nearly impossible to find environments that allow the examination of variations in single parameters. It has proven especially difficult to determine whether variations in organic matter accumulation rates are driven primarily by variations in the productivity of overlying waters or by the length of time over which the organic matter is exposed to oxygen. Calvert and Pedersen (1992), for instance, argue that anoxia has little influence on organic carbon preservation. However, recent work on the continental margin in the northeastern Pacific Ocean (Hartnett *et al.*, 1998, 1999) has suggested that the "oxygen exposure time"—roughly speaking, the length of time the average sedimentary particle spends in the oxic layer of sediments—does influence the rate of preservation. Hartnett *et al.* (1998) showed a trend of decreasing organic carbon burial efficiency as oxygen exposure time increased in sediments from two locations (the Washington margin, with high productivity and a moderately intense water column O_2 minimum; and the Mexican margin, with lower productivity and a zone of bottom-water anoxia). Hedges *et al.* (1999) showed, along a transect of the Washington continental slope and rise, that one measure of organic carbon preservation, surface-area normalized organic carbon concentration, was higher in three sediments with O_2 exposure times of less than 100 yr than in three sediments with O_2 exposure times of ~200–1,000 yr. The scatter in the observed trends shows, however, that the simple idea of oxygen exposure time does not fully explain variations in organic carbon preservation. In addition, the mechanism by which oxygen exposure acts is as yet unclear (Hedges *et al.*, 1999). Much work remains to be done to determine the factors that influence organic matter preservation in continental margin sediments.

7.02.3.6 Benthic Organic Matter Cycling and Marine Geochemical Cycles

Jahnke (1996) has examined much of the available data for the consumption of oxygen in sediments at water depths >1,000 m, and has compared total, deep-sea sedimentary respiration to the rain of organic matter to the 1,000 m depth contour. In these deep-sea sediments, 97% of the organic matter raining to the seafloor is oxidized, only 3% being preserved. Forty-five percent of the organic matter that reaches 1,000 m in the ocean is oxidized in the sediments. Thus, sedimentary organic matter oxidation and oxygen consumption are a highly significant part of organic C and O_2 cycles in the deep sea.

Perhaps the best illustration of the importance of benthic processes to marine elemental budgets is the case of denitrification. Denitrification takes place only when ambient O_2 levels are near zero; thus, it occurs only in three open-ocean locations: in the Arabian Sea, in the eastern Pacific north of the equator, and in the eastern Pacific south of the equator. In contrast, sediments on the continental margins become anoxic at depths less than 2 cm below the sediment–water interface throughout the oceans (see Figure 2). Therefore, sedimentary denitrification is ubiquitous in continental margin sediments. Two recent studies, using very different methods, have concluded that sedimentary denitrification is a dominant process in the marine fixed nitrogen cycle. Middelburg *et al.* (1996) used a diagenetic model, with parameters tuned to available data and forcing by observed bottom-water conditions and fluxes to the sediment–water interface, to infer sedimentary denitrification rates in excess of 230 Tg gN yr^{-1}. Brandes and Devol (2002) developed a stable nitrogen isotopic budget for marine fixed nitrogen to arrive at similar rates of sedimentary denitrification. These rates are over twice the rate of water column denitrification. If the modern oceanic fixed nitrogen cycle is in steady state, these values imply that the available estimates of marine nitrogen fixation are underestimates.

7.02.4 PARTICLE MIXING IN SURFACE SEDIMENTS: BIOTURBATION

Sediment accumulation rates in the deep sea are measured in centimeters to as little as millimeters per thousand years. Therefore, if particles were moved below the sediment–water interface by burial alone, the oxidation of organic matter, which occurs on timescales of months to decades, would take place only in the upper millimeter of the sediment column. Although organic matter oxidation is most rapid in the upper few millimeters of the sediments, pore-water profiles of the reactants and products of organic matter oxidation clearly show that the process occurs at measurable rates within a depth interval that is over 10 cm thick. There must be a process other than sediment burial that transports particles downward from the sediment–water interface. This process is the mixing of sediment particles by the feeding, burrowing, and movement of animals living inside the sediment column. These activities are called bioturbation.

The effects of bioturbation are important in shallow-water sediments as well as in the deep sea. In these sediments, a large supply of organic matter leads to the use of both Fe(III) and S(VI) as electron acceptors. Sulfate reduction occurs after iron reduction, so that HS^- is released into pore waters containing Fe^{2+}. These species react to form solid FeS. FeS is not simply buried, but some of it is moved by bioturbation back toward the oxic zone at the sediment surface. There, it is oxidized to SO_4^{2-} and iron oxides (Aller, 1980). This process has important consequences. At most locations, the burial rate of reduced sulfur is much smaller than the sulfate reduction rate, implying that sulfur passes through an S(VI)–S(−II)–S(VI)... cycle several times before being buried (Berner, 1989). Iron undergoes similar redox cycling within the sediments. In addition, the precipitation of authigenic iron oxides at the sediment surface has important consequences, as many other elements are scavenged from pore waters when these oxides precipitate (e.g., phosphorus and many trace metals), and as the oxidation of Fe(II) and S(−II), accompanied by iron oxide precipitation, significantly lowers pore-water pH.

A complete understanding of diagenetic reactions in surface sediments requires an understanding of bioturbation. For many years, bioturbation has been modeled as a quasi-diffusive process (e.g., Goldberg and Koide, 1962). This model implies that particle mixing results from a large number of small, randomly directed particle movements. Given the variety of size and activities of the animals causing bioturbation, this diffusive model of the process is clearly a simplified view. Nonetheless, the simple model has been used along with a wide variety of radiochemical tracers, with half-lives ranging from a few days to thousands of years. We can make some useful generalizations from the results of these studies (Boudreau, 1994). First, the depth of the "mixed layer"—the depth zone over which bioturbation is present—is fairly constant throughout the ocean at 10 ± 5 cm. Second, the bioturbation mixing coefficient varies widely, with values ranging from $\sim 0.1\ cm^2_{sed}\ yr^{-1}$ to $100\ cm^2_{sed}\ yr^{-1}$. Thus, particle mixing is slow relative to solute diffusion (diffusion coefficients in sediments are typically in excess of $150\ cm^2_{sed}\ yr^{-1}$), but rapid relative to sediment burial. Many factors drive the variations in the bioturbation coefficient, but in general they are low in environments with low sedimentation rates (e.g., the deep sea), and high in zones of rapid sedimentation (e.g., coastal sediments) (Boudreau, 1994).

As greater emphasis has been placed on the role of bioturbation in the cycling of organic matter in sediments, greater attention has been paid to the shortcomings of the simple, diffusive mixing model. One key finding is that mixing rates are not the same for all particles. Wheatcroft (1992) spread classes of glass beads of various size, ranging from 8–16 μm to 126–420 μm, onto sediments underlying a 1,240 m water column in the Santa Catalina Basin (southern California borderlands). He returned 997 days later and took sediment cores to recover the beads. He found that the finer beads were present deeper than coarser beads in every sediment core. From this result, he hypothesized that deposit feeders ingest finer particles preferentially over coarser particles. A different type of study examines the particle mixing intensity using tracers of varying half-life. Most often, mixing intensities based on ^{234}Th distributions (half-life = 24 d) are compared to those derived from ^{210}Pb distributions (half-life = 22 yr). In many cases, the mixing intensity derived from the shorter-lived tracer is much greater than that from the longer-lived isotope. For instance, Smith *et al.* (1993) found that, on average, mixing rates in the Santa Catalina Basin based on ^{234}Th profiles were over two orders of magnitude larger than those based on ^{210}Pb profiles. Pope *et al.* (1996), examining bioturbation in equatorial Pacific sediments, concluded that ^{234}Th-based mixing coefficients were up to an order of magnitude greater than ^{210}Pb-based mixing coefficients at 0°, 2°N, and 5° N (140° W), but that the coefficients based on tracers with short and intermediate half-lives were similar at three sites farther from the equator. From these studies, it appears that the tracer used to define the solid-phase mixing intensity in a given study should, ideally, be matched with the timescale of the geochemical process being studied. For instance, mixing intensities based on ^{234}Th (timescale of roughly 100 d) are preferred for studies of the transport of reactive organic matter in the sediments (decay timescale of weeks) over intensities based on ^{210}Pb (timescale of 100 yr).

It has also become clear that not all particle mixing processes can be described by the diffusive model. Wheatcroft *et al.* (1990) considered theoretically the effects of the activities of macrofauna on particle movement in sediments, finding that in many cases, the length scales of macrofaunal activities were too large to be described accurately by a diffusive mixing model. Smith *et al.* (1996) examined bioturbation in the central equatorial Pacific Ocean using bottom photographs, x-radiographs of sediments, and ^{234}Th and ^{210}Pb profiles. They and Pope *et al.* (1996) found that many radiotracer profiles did not match the smooth, exponentially decreasing nature implied by diffusive mixing models. Both ^{234}Th and ^{210}Pb profiles often had subsurface peaks several centimeters below the sediment–water interface, and ^{210}Pb profiles often had irregular features near the interface. They hypothesized that these profiles resulted from a combination of quasi-diffusive mixing, occasional homogenization of the upper 1–2 cm of the sediments, and "subductive transport" resulting from the feeding of animals that ingest particles at the sediment surface while egesting them several centimeters below the sediment–water interface. Subsurface peaks in tracer profiles resulting from subductive transport appear to be widespread features in marine sediments. In addition to the observations of Smith *et al.* (1996) and Pope *et al.* (1996) in the central equatorial Pacific, similar features have been observed in ^{210}Pb profiles by Soetaert *et al.* (1996) in sediments on the continental slope and rise in the northeast Atlantic, by Martin *et al.* (2000) on the Ceara Rise in the tropical northwest Atlantic, and by Sayles *et al.* (2001) in the Pacific sector of the Southern Ocean. Fornes *et al.* (1999) observed similar features in profiles of tracers spread on the surface of sediments at 1,000 m water depth off the coast of N. Carolina in the northwest Atlantic.

These studies of particle mixing in marine sediments make it clear that particle transport in the upper 10 cm of the sediment column has important effects on the distribution of reactive solids in the sediment column. While simple, diffusive mixing models explain many features of particle mixing, it is important to allow for more complex features of infaunal behavior: selective transport of fine-grained, "fresh" sediment particles over coarser, "older" particles and transport over longer depth scales than can be explained by simple, diffusive mixing models. These transport mechanisms have important consequences for the decomposition of biogenic material at the seafloor.

7.02.5 $CaCO_3$ DISSOLUTION IN SEDIMENTS

The solubilities of two polymorphs of $CaCO_3$, aragonite and calcite, follow well-known patterns in the ocean. Aragonite is the more soluble mineral at oceanic T and P, and the dominant carbonate mineral in deep-sea sediments is calcite. Calcite solubility increases with decreasing temperature and increasing pressure, leading to increasing solubility with increasing depth in the water column. The resulting pattern in $CaCO_3$ preservation in marine sediments is well established. Pelagic sediments generally have very high $CaCO_3$ contents (over 80%, with the value depending on the relative rain rates of $CaCO_3$ and lithogenic material) to the water depths at which calcite becomes undersaturated; with increasing depth below the calcite saturation horizon, the percentage of $CaCO_3$ gradually decreases until it reaches zero at the calcite compensation depth. (e.g., Biscaye *et al.*, 1976). Thus, the saturation state of bottom waters has a clear effect on calcite preservation in sediments. In addition to temperature and pressure, the saturation state of seawater is affected by its pH. The higher the pH of seawater, the higher is its CO_3^{2-} concentration, and the higher its degree of saturation with respect to calcite, all other variables being equal. This effect is seen in the difference in saturation horizon between the Atlantic Ocean and Pacific Ocean; since deep-water pH is lower in the Pacific, the calcite saturation horizon is shallower there than in the Atlantic.

The saturation state of bottom water is not the only factor influencing the preservation/dissolution balance for calcite in marine sediments. When oxygen, the dominant electron acceptor for organic matter decomposition in marine sediments, is used to oxidize organic matter, CO_2 is released into pore waters. This CO_2 drives down the pore-water pH until, if there is sufficient oxic metabolism, the pore water becomes undersaturated with respect to calcite. At that point, the CO_2 can be neutralized by the dissolution of sedimentary $CaCO_3$. Through this process, $CaCO_3$ that falls to the seafloor in an area where the bottom water is supersaturated with respect to calcite may be subject to post-depositional dissolution. "Metabolic dissolution" may be an important factor in the marine $CaCO_3$ cycle. Archer (1996) illustrated this point with results from a global model of sedimentary calcite dissolution, finding that the overall marine calcite preservation rate was 40–50% when sedimentary metabolic dissolution was omitted from the model, but only 20–25% when it was included. The model runs also illustrated that sedimentary processes play a key role in the marine $CaCO_3$ cycle: on the order of half the total dissolution of $CaCO_3$ in the oceans occurs in deep-sea sediments.

The marine $CaCO_3$ cycle plays an important role in the atmosphere/ocean CO_2 system. The pool of total dissolved CO_2 in the ocean is ~50 times larger than that in the atmosphere. Several factors can drive changes in oceanic pH through their effects on the flux of alkalinity to the deep sea: changes in terrestrial weathering rates, in the formation rate of shallow-water carbonates, and in marine productivity. The ocean adjusts to these changes in inputs through changes in the saturation state of deep waters with respect to calcite. For instance, if the alkalinity flux to the deep sea were to increase, pH would increase, thereby increasing the area of the seafloor where calcite is preserved until the rate of loss of alkalinity through calcite accumulation balances the new input value, at a higher oceanic pH. The changing oceanic pH, in turn, affects the oceanic P_{CO_2} and atmospheric CO_2 content, with rising pH lowering atmospheric CO_2. The timescale for these adjustments in the marine $CaCO_3$ cycle is thousands of years (Broecker and Peng, 1987). Because of the important role that sedimentary processes play in determining the marine $CaCO_3$ balance, it is important to understand the mechanism of calcite dissolution. Furthermore, the changes in the marine $CaCO_3$ cycle that lead to atmospheric CO_2 changes leave signatures in the carbonate accumulation rate in sediments. It is important to understand sedimentary processes to interpret changes in accumulation rates. The remainder of this section will focus on the mechanism of calcite dissolution in sediments. Since dissolution in sediments underlying undersaturated bottom waters is quite well understood and leaves a clear signal in sedimentary $\%CaCO_3$, we will focus on dissolution in sediments deposited above the calcite saturation horizon. In these sediments, oxic metabolism drives dissolution. Various measurements have shown that metabolic processes may lead to dissolution of up to ~50% of the calcite rain. Because the $CaCO_3$ content of pelagic sediments above the calcite saturation horizon is usually greater than 80%, sedimentary $\%CaCO_3$ is not a sensitive indicator of dissolution. Thus, measurements of pore-water pH, alkalinity, T_{CO_2}, and Ca^{2+}, as well as direct determinations of the benthic flux of these species, have been the key to quantifying metabolic dissolution. We will illustrate the use of pore-water profiles and review the current state of knowledge of the metabolic dissolution process.

7.02.5.1 Metabolic Dissolution in Sediments above the Calcite Lysocline

The occurrence of $CaCO_3$ dissolution in sediments underlying supersaturated bottom water has been recognized for many years, and pore-water data have demonstrated its occurrence since the advent of *in situ* pore-water samplers. Oxic metabolism releases CO_2 into pore waters; in addition, it releases H^+ at a rate of ~16 equivalents per 106 moles of organic C oxidized (when C : N = 106 : 16 and phosphate is neglected). Thus, in the absence of $CaCO_3$ dissolution, the change in alkalinity per mole of T_{CO_2} added to pore waters will be about −0.15 (−16/106). If the CO_2 and H^+ released by oxic metabolism are neutralized by dissolution of $CaCO_3$, then the ratio of the change in alkalinity for a given change in T_{CO_2} is predicted to be ~1.08[2 × (106 + 16)/(2 × 106 + 16)]. Thus, plots of the pore-water alkalinity versus T_{CO_2} concentration should show clearly whether oxic metabolism is accompanied by $CaCO_3$ dissolution. Sayles (1981) showed that such plots have a slope that is essentially identical to the predicted value of 1.08 in sediments both above and below the calcite saturation horizon. Further, he showed that the slope of a $\Delta Ca^{2+}/\Delta Alk$ plot confirmed that the source of the added alkalinity was $CaCO_3$ dissolution.

Plots of alkalinty versus T_{CO_2} and Ca^{2+} versus alkalinity demonstrate conclusively the occurrence of metabolic dissolution. However, they do not show the ratio of the rates of dissolution and carbon oxidation well. The reason is that calcite dissolution is rapid relative to organic matter oxidation. Therefore, pore waters that have become undersaturated due to oxic metabolism re-equilibrate with sedimentary calcite very rapidly. The sedimentary layer in which the release of metabolic acids is not matched by dissolution is expected to be thin, so that the slope of an Alk versus T_{CO_2} plot will be indistinguishable from 1 (e.g., Martin and Sayles, 1996). In order to determine the ratio of $CaCO_3$ dissolution to oxic organic matter decomposition, we must turn to high-resolution, *in situ* pore-water profiles or direct, *in situ* benthic flux measurements. Archer *et al.* (1989) first used *in situ* microelectrode profiling to obtain pore-water O_2 and pH profiles. They interpreted these data using a model of oxic respiration and calcite dissolution to demonstrate the occurrence of metabolic dissolution at several sites in the equatorial Atlantic Ocean. Since then, Hales and co-workers have used an improved version of the *in situ* profiler to examine metabolic dissolution at sites in the northwest Atlantic (Hales *et al.*, 1994), western equatorial Pacific (Hales and Emerson, 1996), and Ceara Rise (western equatorial Atlantic: Hales and Emerson, 1997). Sayles and co-workers have used a different type of sampler, the *in situ* whole-core squeezer, to obtain pore-water alkalinity, T_{CO_2}, Ca^{2+}, and NO_3^- profiles on the Ceara Rise (Martin and Sayles, 1996), in the Southern Ocean (Sayles *et al.*, 2001), and in the eastern equatorial Atlantic (unpublished data: see the example below). The results of these measurements support the occurrence of metabolically driven calcite dissolution above the calcite saturation horizon (Table 6). The pore-water data show that, at each of the sites underlying supersaturated bottom water, the acids released into pore waters by oxic metabolism drive the dissolution of a significant fraction of the rain of $CaCO_3$ to the seafloor (25–68%). However, not all of the metabolic acid is neutralized by dissolution of sedimentary $CaCO_3$: the ratio of $CaCO_3$ dissolved to organic C oxidized (the "efficiency" of metabolic dissolution) is 0.4–0.7. The reason for this is that, when CO_2 and H^+ are released near the sediment–water interface, they give rise to fluxes of CO_2 and H^+ out of the sediments and of CO_3^{2-} and $B(OH)_4^-$ into the

Table 6 $CaCO_3$ dissolution rates at sites above the calcite saturation horizon: (i) *in situ* microelectrode profiling (NW Atlantic and Ceara Rise) and (ii) *in situ* whole-core squeezer (Ceara Rise and Cape Verde Plateau).

Water depth (m)	*Omega-calcite*[a]	C_{org} *ox. rate* (μmol cm^{-2} yr^{-1})	$CaCO_3$ *diss. rate* (μmol cm^{-2} yr^{-1})	*Diss./ox.*	*%*$CaCO_3$ *diss.*
NW Atlantic (Hales et al., 1994)					
2,159	1.7	25–35	13–24	0.62	35–60
4,236	1.1	15–21	10–17	0.75	48–67
4,501	1	12–16	6–12	0.64	37–63
Ceara Rise (Hales and Emerson, 1997)					
3,280	1.23	10–18	3–12	0.54	
4,000	1.0	17	6–9	0.44	
Ceara Rise (Martin and Sayles, 1996)					
4,000	1.0	16	11	0.69	68
Cape Verde Plateau (Sayles and Martin, unpublished data)					
3,100	1.25	28	12	0.43	25

[a] Omega-calcite = the degree of saturation of bottom water with respect to calcite, $([Ca^{2+}][CO_3^{2-}])_{obs.}/([Ca^{2+}][CO_3^{2-}])_{sat.}$.

sediments. In this way, a fraction of the metabolic acids is neutralized by reaction with bottom-water solutes, the primary product of the neutralization being HCO_3^-. The nearer to the interface organic matter oxidation occurs, the smaller becomes the ratio of $CaCO_3$ dissolution to organic carbon oxidation.

In situ benthic chamber deployments have been made in several locations in undersaturated bottom water (e.g., Berelson *et al.*, 1990b; Jahnke *et al.*, 1994). In all cases, the results from these deployments have confirmed the occurrence of calcite dissolution. Jahnke and co-workers have also deployed *in situ* flux chambers at sites with bottom waters that are supersaturated with respect to calcite (Jahnke *et al.*, 1994, 1997; R. A. Jahnke and D. B. Jahnke, in press). In these locations, they have consistently found dissolution when the sedimentary $\%CaCO_3$ was low, but no dissolution when the $\%CaCO_3$ was high. These results are in direct contradiction to those of Hales *et al.* (1997; see Table 6), who found a ratio of dissolution to oxidation of 0.5 at the 3,280 m site on the Ceara Rise, where Jahnke and co-workers found no dissolution; and to the results of Sayles and Martin (see Table 6 and Figure 7) on the Cape Verde Plateau, where they found a ratio of dissolution to oxidation of 0.4 where Jahnke *et al.* (1994) found no dissolution. R. A. Jahnke and D. B. Jahnke (in press) have proposed that reactions very near the sediment–water interface may either buffer pore waters by reaction with $CaCO_3$ surfaces, inhibiting dissolution; or trap the products of metabolic dissolution through precipitation of $CaCO_3$ in slightly supersaturated pore waters at the sediment–water interface. However, to be consistent with the results of Sayles and Martin at the Cape Verde Plateau site, the reactions would have to occur very near the interface and leave no trace in the deeper pore waters. Given the potential importance of metabolic dissolution to the marine $CaCO_3$ cycle, resolving the discrepancy between these different measurements must be a primary goal of further work on calcite dissolution in deep-sea sediments.

7.02.5.2 Metabolic Dissolution and the Kinetics of Organic Matter Oxidation

We noted above that the rate of dissolution of sedimentary calcite resulting from the neutralization of the products of oxic metabolism depends on a competition between neutralization by reaction with bottom water CO_3^{2-} and $B(OH)_4^-$ and by calcite dissolution. The larger the proportion of the metabolic products that is released near the interface, the stronger the bottom-water/pore-water concentration gradients, and the larger the fraction of the products that will be neutralized by bottom-water solutes. In making global models describing the role of sediments in the marine $CaCO_3$ cycle (e.g., Archer, 1996), it will be important to take into account recent findings on the kinetics of organic matter oxidation.

Above, we summarized results showing that the e-folding depth of organic matter oxidation in sediments varies systematically with the organic matter oxidation rate, decreasing from values of ~3 cm at low oxidation rates of 5 μmol cm^{-2} yr^{-1} to ~0.3 cm at and above oxidation rates of 20 μmol cm^{-2} yr^{-1}. A simple modeling experiment illustrates the importance of this result to the evaluation of calcite dissolution rates. To carry out this experiment, we used the respiration/dissolution model of Martin and Sayles (1996), an intermediate organic carbon oxidation rate of 14 μmol cm^{-2} yr^{-1}, and a $C_{org}/CaCO_3$ rain ratio of 0.7 (Archer, 1996). In our model, we described oxic respiration with the two-exponential model of Equations (1) and (2). We varied the e-folding depth of respiration by varying $1/j_1$ from 0.3 cm to 3.5 cm (approximately the range we observed in our survey of pore-water data), covarying j_0 so as to

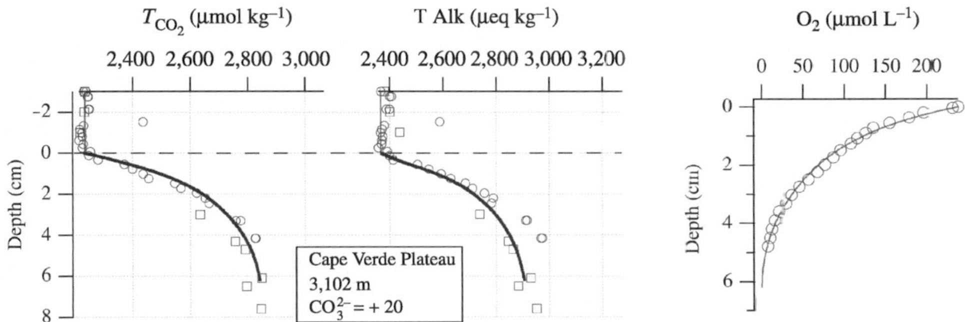

Figure 7 Results from *in situ* whole-core squeezing at a site on the Cape Verde Plateau at 18° 27.8′ N, 21° 01.5′ W (Sayles and Martin). Solid lines are fits of an organic matter oxidation/calcite dissolution model to the data (Martin and Sayles, 1996). Fluxes calculated from the model fits are shown in Table 6. The pore-water O_2 data are from samples collected by shipboard whole-core squeezing. The T_{CO_2} and alkalinity data were collected by *in situ* whole-core squeezing (circles) and *in situ* harpoon sampling (squares).

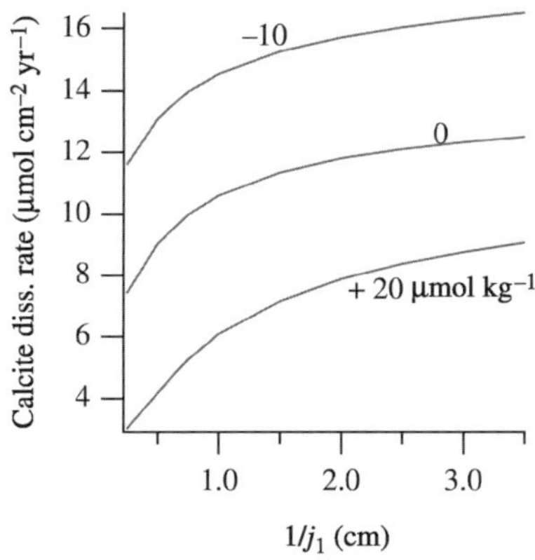

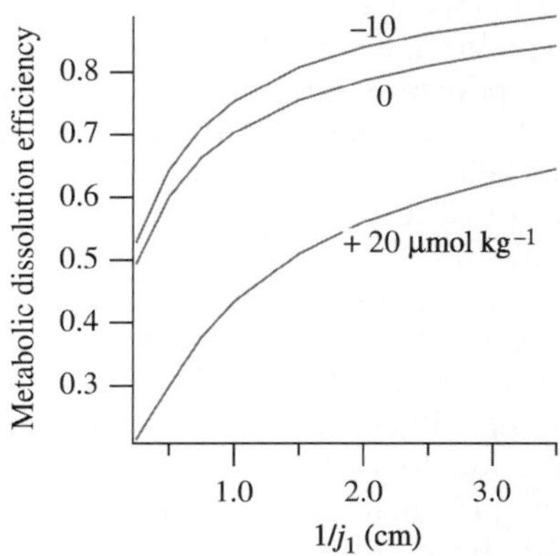

Figure 8 Results of a model study of the influence of the depth distribution of organic matter oxidation below the sediment–water interface on sedimentary calcite dissolution. The e-folding depth for organic C oxidation is approximately equal to $1/j_1$. Results are shown for 3 degrees of saturation with respect to calcite: $+20\ \mu mol\ kg^{-1}$ (supersaturated bottom water), $0\ \mu mol\ kg^{-1}$ (bottom water at saturation), and $-10\ \mu mol\ kg^{-1}$ (undersaturated).

keep the total respiration rate constant. The results are shown in Figure 8. Both the calcite dissolution rate and the ratio of dissolution rate to respiration rate ("metabolic dissolution efficiency") increase markedly, for constant values of supersaturation of bottom water with respect to calcite, as the e-folding depth of respiration increases. The effect is most important in sediments under the most highly supersaturated bottom water. For instance, at a ΔCO_3^{2-} of $+20\ \mu mol\ kg^{-1}$ (quite strongly supersaturated), the metabolic dissolution efficiency increases from ~20% at $1/j_1 = 0.3$ cm (respiration very near the interface) to 60% at $1/j_1 = 3.5$ cm. When the bottom waters are moderately undersaturated ($\Delta CO_3^{2-} = -10\ \mu mol\ kg^{-1}$), metabolic dissolution efficiency increases from ~55% at $1/j_1 = 0.3$ cm to 85% at $1/j_1 = 3.5$ cm. These results show that the kinetics of organic matter oxidation have an important effect on the rate and extent of calcite dissolution in sediments. The use of empirical relationships between oxidation kinetics and the rain rate of organic matter to the seafloor (e.g., Figure 5) should significantly improve the accuracy of global models of calcite dissolution.

7.02.6 SILICA CYCLING IN SEDIMENTS

Biogenic silica, or opal, is the second biogenic material that dominates many marine sediments. Like $CaCO_3$, opal undergoes reactions during early diagenesis, the most important being dissolution. Unlike $CaCO_3$, the reactions affecting opal are little influenced by organic matter breakdown.

Two aspects of the cycling of silica in marine sediments make pore-water H_4SiO_4 interesting and important. There is increasing evidence in support of the breakdown and reprecipitation of siliceous minerals in a variety of sedimentary environments, including coastal/deltaic (Michalopoulos and Aller, 1995) and open ocean biogenic silica sediments (Dixit, 2001). Numerous studies have suggested that aluminum adsorption and incorporation into biogenic silica (opal) influences opal solubility, opal dissolution rates, and hence dissolved silica profiles in marine pore waters (Hurd, 1973; van Bennekom *et al.*, 1988; van Cappellen and Qui, 1997; Dixit *et al.*, 2001). Processes influencing the preservation and dissolution of opal are central to the geochemical mass balance of silicon. In addition, if authigenic silicate or aluminum-silicate formation in marine sediments are significant, then these reactions may be important in the geochemical cycling of many major elements that are likely to be incorporated into these authigenic phases (Mackenzie and Garrels, 1966; Sayles, 1979; Michalopoulos and Aller, 1995). No less important is the potential for biogenic silica to yield information regarding the past productivity of the oceans and export of organic carbon to the deep sea (Archer *et al.*, 1993; Nelson *et al.*, 1995; Ragueneau *et al.*, 2000). Understanding the processes controlling the preservation of opal and the relationship of opal burial to opal productivity in sediments is essential to any attempts to interpret the sedimentary opal record in terms of paleoproductivity and carbon export.

7.02.6.1 Dissolved Silica Profiles in Pore Waters

Dissolved H_4SiO_4 profiles in marine sediments are characteristically quasi-exponential, reaching approximately constant values at sediment depths of, typically, 5–15 cm. These profiles dictate a flux of H_4SiO_4 from the sediments to the oceans everywhere (Figure 9). Opal is presumed to be the

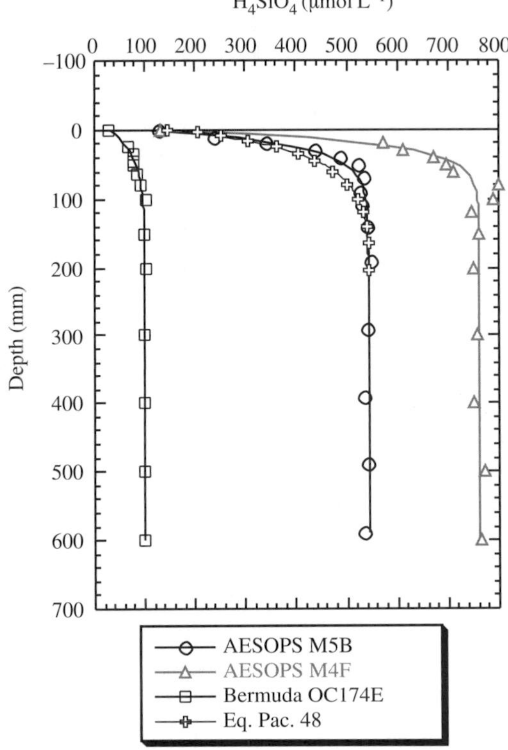

Figure 9 Examples of dissolved silica profiles in a variety of marine sediments. The lines through the data are simple analytical fits of the data to the relation $C_x = a1 * (1 - e^{-a2*x}) + a3$ (see text). AESOPS profiles are from the USJGOFS database (Sayles): M5B is mooring 5, %opal = 33%; M4 is mooring 4, %opal = 90%; Bermuda data are from Sayles *et al.* (1996), %opal = 1.5%; Eq. Pac. 48 data are from the USJGOFS database (Berelson), %opal = 7.5 (source McManus *et al.*, 1995).

most soluble solid silicate in the sediments and is ubiquitous. It has been recognized for many years that the asymptotic concentrations observed rarely reflect equilibrium solubility of opal (Hurd, 1973). H_4SiO_4 profiles often can be fit with a simple analytical expression (Figure 9). While this facilitates the estimation of fluxes across the sediment–water interface, it does not provide any mechanistic understanding of the factors controlling the concentrations profiles. A first-order problem in understanding the controls on H_4SiO_4 in pore waters is explaining the large range in asymptotic concentration observed, of order 100–850 μmol L^{-1} (Figure 9). Three types of hypotheses have been put forward (McManus *et al.*, 1995). Equilibrium models, which consider the asymptotic concentration to represent solubility, explain the variations in this value as reflecting opal phases of differing solubility (Archer *et al.*, 1993). The second approach calls upon two (or more) types of opal one of which is unreactive and will be buried, and a second which has a solubility greater than the asymptotic concentration. The second phase dissolves completely and the asymptotic concentration is determined by the interplay of the flux of reactive opal, the dissolution rate of the opal, and sediment mixing rates (Schink *et al.*, 1975). The third type of model employs a depth-dependent rate constant of dissolution that may reflect decreasing surface area, the buildup of a surface layer on opal that has a lower solubility and restricts the rate of dissolution, and a competition between dissolution and precipitation of authigenic silicates (McManus *et al.*, 1995; Rabouille *et al.*, 1997; Dixit *et al.*, 2001).

Recent experiments with opal-rich sediments provide increasing evidence that aluminum plays a central role in determining pore-water H_4SiO_4 concentrations and consequently in opal preservation. Van Cappellen and Qiu (1997) found that the asymptotic H_4SiO_4 concentrations in sediments varied systematically and nonlinearly with the ratio of detrital to opal material in the sediments. They hypothesized that the asymptotic concentrations are controlled, to a large extent, by the reprecipitation of H_4SiO_4 from opal and aluminum from detrital material and conclude that the main environmental factor controlling pore-water H_4SiO_4 concentration is the ratio of the fluxes of detrital components and opal to the sediment (Figure 10(a)). Dixit *et al.* provide additional evidence for decreases in opal reactivity with aging and increased aluminum content. These experiments also support the hypothesis that precipitation of authigenic aluminum silicates may prevent concentrations of H_4SiO_4 from reaching opal equilibrium values.

The data from the AESOPS program provide an independent opportunity to assess the relationship between opal solubility and the relative abundance of detrital material and opal. Sediment compositions are roughly similar for the AESOPS and Van Cappellen and Qiu (1997) studies, except that the detrital content of the AESOPS sediments reaches somewhat lower values in the opal-rich sediments of that region. The most striking feature is the similarity in the relationship of the asymptotic H_4SiO_4 to %detrital/%opal for the two data sets (Figure 10(a)). They are, however, quantitatively different. Concentrations at a given %detrital/%opal ratio are systematically lower for the AESOPS data set. This implies that the response of opal/detrital mixtures differs in different areas, which is not surprising, given that the "detrital" component varies from region to region with a wide range of minerals, of differing weathering histories, and reactivities. It is also of interest that the influence of the detrital component on asymptotic H_4SiO_4 occurs at very low ratios. When the scale is expanded as a log scale, the influence of the detrital component is apparent

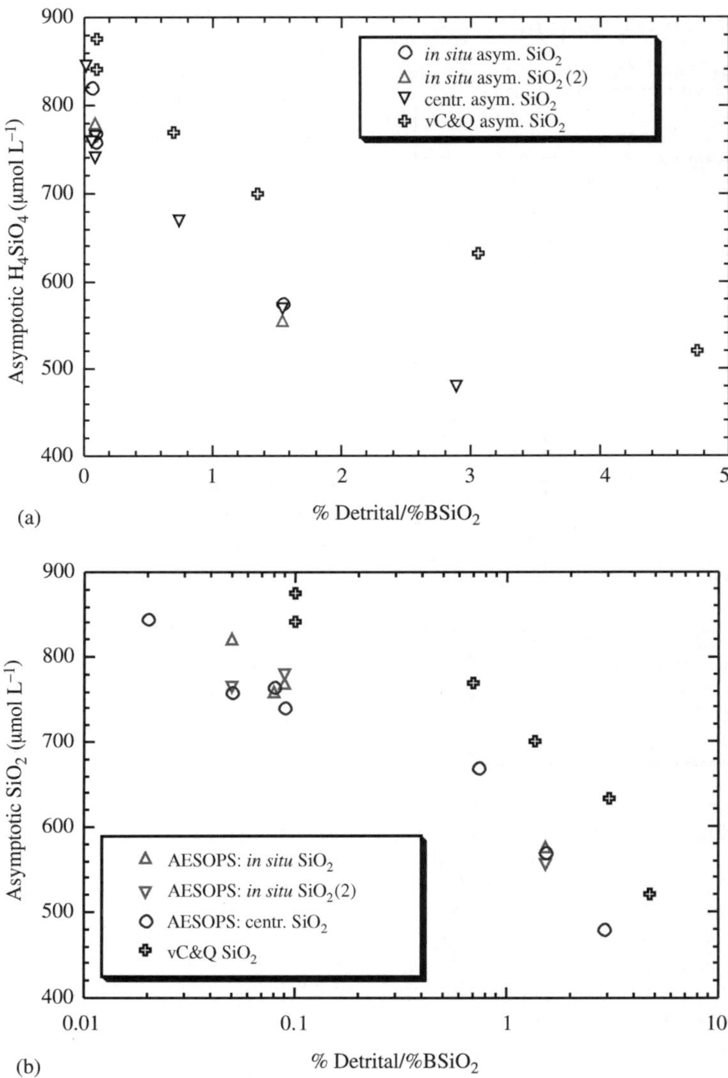

Figure 10 (a) Relationship between pore-water asymptotic H_4SiO_4 concentration and the ratio of detrital % to opal% in the sediments of the AESOPS study (above). *In situ* asym. are concentrations determined on samples collected *in situ*, (2) is just the designation of a replicate data set collected at three of the locations studied. Centr. asym. are data from samples extracted from cores collected at the *in situ* sample sites by centrifugation at ~0 °C on the ship. The three are shown to provide an idea of reproducibility. vC&Q asym. are data redrawn from van Capellen and Qiu (1997). (b) The same data as (a) presented on a log scale to expand the low ratio points.

at ratios above ~0.02 in the AESOPS data, indicating high sensitivity to small additions of aluminum silicates.

7.02.6.2 Preservation Efficiency

Interpretation of sedimentary opal records in terms of past opal productivity requires that the fraction of the opal preserved in the sediments be predictable globally. Estimates of preservation efficiency have been made for many regions of the world ocean. In the Indian Ocean sector of the Southern Ocean, Rabouille *et al.* (1997) report preservation efficiencies of only 3–6% at low opal rain rates but efficiencies of 30–38% at the highest opal rain rates. In the Atlantic sector of the Southern Ocean, Schluter *et al.* (1998) report values ranging from 1% to 39%. In the Pacific sector, Sayles *et al.* (2001) report values of 5–19% (uncorrected for sediment focusing). Dymond and Lyle (1994), summarizing data from a number of measurements in the tropical and subtropical Atlantic and Pacific, report values in the range 1–18% with a single value of 75%. Berelson *et al.* (1997) report opal burial and rain data for a number of locations between 5° S and 5° N in the Pacific from which we calculate an average preservation efficiency of 9%. These data

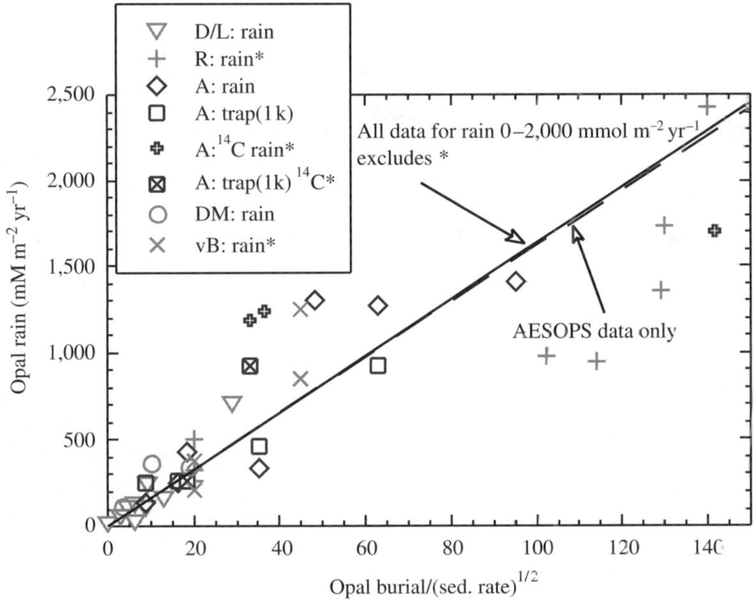

Figure 11 Relationship between the rain of opal and the burial flux of opal normalized to (sedimentation rate (cm kyr^{-1}))$^{1/2}$. The figure is taken from Sayles *et al.* (2001, figure 14(b)). The rain of opal is measured either with sediment traps (direct) or estimated from the SiO_2 remineralization flux plus opal burial flux, as indicated below. The linear regressions shown use different data sets as indicated. Where the data are designated A, they are from Sayles *et al.* (2001). (D/L: rain—Dymond and Lyle (1994); R: rain—Rabouille *et al.* (1997); A: rain—uses benthic SiO_2 remineralization and ^{230}Th-normalized sedimentation rates and opal burial rates (Sayles *et al.*, 2001.) A: trap(1 k) uses particle flux measured in sediment traps at ~1 km depth (Honjo *et al.*, 2000) together with ^{230}Th-normalized sedimentation rates and opal burial rates. AESOPS: ^{14}C—uses SiO_2 remineralization together with ^{14}C-derived sedimentation rates and opal burial rates. A: trap(1 k) ^{14}C particle flux measured in sediment traps at ~1 km depth (Honjo *et al.*, 2000) used together with ^{14}C-derived sedimentation rates and opal burial rates. DM: rain—DeMaster *et al.* (1996); vB: rain—van Bennekom *et al.* (1988). Opal burial is in g m^{-2} yr^{-1}; sedimentation rate is in cm kyr^{-1}. The linear regressions have been forced through the origin. The Rabuoille data were omitted from the regression because sedimentation rates were assumed, not measured. The ^{14}C corrected data were not used for consistency with other studies.

make it clear that variation in preservation efficiency within and between areas is large; this is true of the highly siliceous sediments of the Southern Ocean and sediments elsewhere with an opal content of ≤10%. Further, with few exceptions, preservation is quite limited; the median preservation from the above referenced reports is 10%, while in 80% of the areas preservation is <15%.

Many relationships have been put forward to explain the large observed variations in the efficiency of opal preservation. Broecker and Peng (1982) suggested a correlation between opal rain rate and preservation. While correlations in limited areas may be good, data from a variety of areas do not correlate well (Sayles *et al.*, 2001). Rabouille *et al.* (1997) noted a correlation between percent opal and rain rate for data from the Indian sector of the Southern Ocean. Sayles *et al.* (2001) found a similar correlation in the Pacific sector, but the correlations are quite different, precluding the use of correlation based on experiments in one area to characterize preservation characteristics in the other. An empirical relationship that relates opal burial/$S^{1/2}$ (S = linear sedimentation rate: cm kyr^{-1}) to opal rain was found to correlate fairly well for data from a large number of open ocean studies and environments (Figure 11) (Sayles *et al.*, 2001). The average percent difference between the measured rain rate and rain rate estimated from burial/$S^{1/2}$ was ~25%, an improvement over most relationships but still entailing substantial uncertainty.

7.02.6.3 Summary: Silica in Pore Waters and Sediments

Recent studies have very clearly delineated the importance of aluminum in sediments for both the opal dissolution rate and the observed asymptotic H_4SiO_4 concentration. The incorporation of aluminum into opal and the hypothesized reprecipitation of H_4SiO_4 from dissolving opal suggest that the silica cycle may exert a significant influence on the cycling of major constituents of pore water. Reverse weathering (Mackenzie and Garrells, 1966) may well need to be revived as an important process affecting ocean chemistry.

Understanding the mechanisms controlling the relationship between opal rain and opal burial and quantifying these relationships sufficiently to extract opal rain through time from sedimentary data remains elusive. Taking the further steps of relating opal in sediments to carbon rain and surface ocean productivity poses even more problems.

7.02.7 CONCLUSIONS

Studies of solute concentrations in the pore waters contained in the upper tens of centimeters of marine sediments have yielded a wealth of information about the cycling of biogenic material at the seafloor. They have contributed both to the estimation of solute exchange across the sediment–water interface and to the estimation of reaction parameters that are needed to make diagnostic and predictive models of the effects of sedimentary diagenesis on marine geochemical cycles.

Mass balances for biogenic materials in sediments show that only small fractions of the organic matter and opal that are deposited on the seafloor survive early diagenesis to become part of the sediment record. This observation has important implications for the use of these materials as indicators of past oceanic conditions. When over 90% of a biogenic component arriving at the sediment–water interface is recycled, then small changes in the extent of recycling lead to significant changes in accumulation rates. Under these conditions, it is essential to take into account possible changes in diagenetic reaction rates while interpreting the down-core record of accumulation of organic matter and opal.

The balance between dissolution and preservation of $CaCO_3$ in the oceans has important implications for the marine alkalinity balance and therefore for the atmospheric CO_2 concentration. On the order of half the $CaCO_3$ dissolution in the oceans occurs in sediments. Studies of the sedimentary dissolution process are leading to a quantitative understanding of the role of sediments in the marine $CaCO_3$ cycle. Most importantly, they have highlighted the potential importance of the role of dissolution in sediments above the calcite lysocline, which is driven by neutralization of acids produced by oxic metabolism. Dissolution above the saturation horizon cannot be ignored in the marine $CaCO_3$ cycle, and possible temporal variations in its extent must be considered in interpreting the temporal record of sedimentary $CaCO_3$ accumulation.

The breakdown of organic matter in sediments leads to significant exchanges of solutes across the sediment–water interface. Oxygen consumption in deep-sea sediments accounts for about half the total O_2 consumption in the ocean below 1,000 m. Sedimentary denitrification, once thought to be relatively minor, has been shown by recent studies to be a dominant term in the marine fixed nitrogen budget and, through that dominance, to be a potentially important determinant of marine productivity.

ACKNOWLEDGMENTS

The authors would like to thank the editors, F. Mackenzie and H. Holland, for critical comments. They would also like to thank Lucinda Gathercole for help with manuscript preparation. This work was supported by the National Science Foundation under grants OCE-9617430 and OCE-9876533.

REFERENCES

Aller R. C. (1980) Diagenetic processes near the sediment–water interface of Long Island Sound: II. Fe and Mn. In *Estuarine Physics and Chemistry: Studies in Long Island Sound* (ed. B. Saltzman). Academic Press, New York, vol. 22, pp. 351–415.

Anderson L. A. (1995) On the hydrogen and oxygen content of marine phytoplankton. *Deep-Sea Res. I* **42**, 1675–1680.

Anderson L. A. and Sarmiento J. L. (1994) Redfield ratios of remineralization determined by nutrient data analysis. *Global Biogeochem. Cycles* **8**(1), 65–80.

Archer D. (1996) A data-driven model of the global calcite lysocline. *Biogeochem. Cycles* **10**, 511–526.

Archer D., Emerson S., and Smith C. R. (1989) Dissolution of calcite in deep-sea sediments: pH and O_2 microelectrode results. *Geochim. Cosmochim. Acta* **53**(11), 2831.

Archer D., Lyle M., Rodgers K., and Froelich P. (1993) What controls opal preservation in tropical deep-sea sediments. *Paleoceanography* **8**(1), 7–21.

Arnosti C. and Holmer M. (1999) Carbohydrate dynamics and contributions to the carbon budget of an organic-rich coastal sediment. *Geochim. Cosmochim. Acta* **63**, 393–403.

Barnes R. O. (1973) An *in situ* interstitial water sampler for use in unconsolidated sediments. *Deep-Sea Res.* **20**, 1125–1128.

Bender M., Martin W., Hess J., Sayles F., Ball L., and Lambert C. (1987) A whole-core squeezer for interstitial pore-water sampling. *Limnol. Oceanogr.* **32**(6), 1214–1225.

Bender M., Jahnke R., Weiss R., Martin W., Heggie D., Orchardo J., and Sowers T. (1989) Organic carbon oxidation and benthic nitrogen and silica dynamics in San Clemente Basin, a continental borderland site. *Geochim. Cosmochim. Acta* **53**, 685–697.

Bender M. L. and Heggie D. T. (1984) Fate of organic carbon reaching the deep seafloor: a status report. *Geochim. Cosmochim. Acta* **48**, 977–986.

Berelson W. M., Hammond D. E., O'Nseill D., Xu X.-M., Chin C., and Zukin J. (1990a) Benthic fluxes and pore water studies from the central equatorial North Pacific: nutrient diagenesis. *Geochim. Cosmochim. Acta* **54**, 3001–3012.

Berelson W. M., Hammond D. E., and Cutter G. A. (1990b) *In situ* measurements of calcium carbonate dissolution rates in deep-sea sediments. *Geochim. Cosmochim. Acta* **54**, 3013–3020.

Berelson W. M., Anderson R. F., Dymond J., DeMaster D., Hammond D. E., Collier R., Honjo S., Leinen M.,

McManus J., Pope R., Smith C., and Stephens M. (1997) Biogenic budgets of particle rain, benthic remineralization and sediment accumulation in the equatorial Pacific. *Deep-Sea Res. II* **44**(9–10), 2251–2282.

Berner R. A. (1982) Burial of organic carbon and pyrite sulfur in the modern ocean: its geochemical and environmental significance. *Am. J. Sci.* **282**, 451–473.

Berner R. A. (1989) Biogeochemical cycles of carbon and sulfur and their effect on atmospheric oxygen over phanerozoic time. *Palaeogeogr. Palaeoclimatol. Palaeoecol.* **75**, 97–122.

Biscaye P. E., Kolla V., and Turekian K. K. (1976) Distribution of calcium carbonate in surface sediments of the Atlantic Ocean. *J. Geophys. Res.* **81**(15), 2595–2603.

Bishoff J. L., Greer R. E., and Luistro A. O. (1970) Composition of interstitial waters of marine sediments: temperature of squeezing effect. *Science* **167**, 1245–1246.

Boudreau B. P. (1994) Is burial velocity a master parameter for bioturbation? *Geochim. Cosmochim. Acta* **58**(4), 1243–1249.

Brandes J. A. and Devol A. H. (2002) A global marine-fixed nitrogen isotopic budget: implications for Holocene nitrogen cycling. *Global Biogeochem. Cycles* **16**, 1120–1134.

Broecker W. S. and Peng T. H. (1982) *Tracers in the Sea*. Lament–Doherty Geological Observatory, Columbia University, Palisades, NY.

Broecker W. S. and Peng T. H. (1987) The role of $CaCO_3$ compensation in the glacial to interglacial atmospheric CO_2 change. *Global Biogeochem. Cycles* **1**, 15–29.

Buffle J. and Horvai G. (2000) *In situ* monitoring of aquatic systems: chemical analysis and speciation. In *IUPAC Series on Analytical and Physical Chemistry of Environmenetal Systems* (eds. J. Buffle and H. P. van Leeuwen). Wiley, vol. 6, 623pp.

Cai, Wei-Jun, Clare Reimers, and Timothy Shaw (1995) Microelectrode studies of organic carbon degradation and calcite dissolution at a California continental rise site. *Geochim. Cosmochim. Acta* **59**, 497–511.

Calvert S. E. and Pedersen T. F. (1992) Organic carbon accumulation and preservation in marine sediments: how important is anoxia? In *Organic Matter: Productivity, Accumulation, and Preservation in Recent and Ancient Sediments* (eds. J. K. Whelan and J. W. Farrington). Columbia University Press, New York, pp. 121–138.

Canfield D. E. (1989) Sulfate reduction and oxic respiration in marine sediments: implications for organic carbon preservation in euxinic environments. *Deep-Sea Res.* **36**(1), 121–138.

DeMaster D. J., Ragueneau O., and Nittrouer C. A. (1996) Preservation efficiencies and accumulation rates for biogenic silica and organic C, N, and P in high-latitude sediments: in the Ross Sea. *J. Geophys. Res.* **101**(8), 18501–18518.

Devol A. H. and Christensen J. P. (1993) Benthic fluxes and nitrogen cycling in sediments of the continental margin of the eastern North Pacific. *J. Mar. Res.* **51**, 345–372.

Dixit S. (2001) Dissolution of biogenic silica: solubility, reactivity and the role of aluminum. PhD Dissertation, Georgia Institute of Technology.

Dixit S., Van Cappellen P., and van Bennekom A. J. (2001) Processes controlling solubility of biogenic silica and pore water build-up of silicic acid in marine sediment. *Mar. Chem.* **73**, 333–352.

Dymond J. and Lyle M. (1994) Particle fluxes in the ocean and implications for sources and preservation of ocean sediments. In *Studies in Geophysics-Material Fluxes on the Surface of the Earth* (eds. National Research Council (US), Board on Earth Sciences and Resources). National Academy Press, Washington, DC, pp. 125–142.

Emerson S., Grundmanis V., and Graham D. (1982) Carbonate chemistry in marine pore waters: MANOP sites C and S. *Earth Planet. Sci. Lett.* **61**, 220–232.

Fanning K. A. and Pilson M. E. Q. (1971) Interstitial silica and pH in marine sediments: some effects of sampling procedures. *Science* **173**, 1228–1231.

Fornes W. L., DeMaster D. J., Levin L A., and Blair N. E. (1999) Bioturbation and particle transport in Carolina slope sediments: a radiochemical approach. *J. Mar. Res.* **57**, 335–355.

Froelich P. N., Klinkhammer G. P., Bender M. L., Luedtke N. A., Heath G. R., Cullen D., Dauphin P., Hammond D., Hartman B., and Maynard V. (1979) Early oxidation of organic matter in pelagic sediments of the eastern equatorial Atlantic: suboxic diagenesis. *Geochim. Cosmochim. Acta* **43**, 1075–1090.

Froelich P. N., Arthur M. A., Burnett W. C., Deakin M., Hensley V., Jahnke R., Kaul L., Dim K.-H., Roe K., Soutar A., and Vathakanon C. (1988) Early diagenesis of organic matter in Peru continental margin sediments: phosphorite precipitation. *Mar. Geol.* **80**, 309–343.

Goldberg E. D. and Koide M. (1962) Geochronological studies of deep-sea sediments with the ionium/thorium method. *Geochim. Cosmochim. Acta* **26**, 417–750.

Hales B. and Emerson S. (1996) Calacite dissolution in sediments of the Ontong-Java Plateau: *in situ* measurements of pore water O_2 and pH. *Global Biogeochem. Cycles* **10**(3), 527–541.

Hales B. and Emerson S. (1997) Calcite dissolution in sediments of the Ceara Rise: *in situ* measurements of pore water O_2, pH, and $CO_{2(aq)}$. *Geochim. Cosmochim. Acta* **61**, 501–514.

Hales B., Emerson S., and David A. (1994) Respiration and dissolution in the sediments of the western North Atlantic: estimates from models of *in situ* microelectrode measurements of pore water oxygen and pH. *Deep-Sea Res.* **41**, 695–719.

Hammond D. E., McManus J., Berelson W. M., Kilgore T. E., and Pope R. H. (1996) Early diagenesis of organic material in equatorial Pacific sediments: stoichiometry and kinetics. *Deep-Sea Res. II.* **43**, 1365–1412.

Hartnett H. E., Keil R. G., Hedges J. I., and Devol A. H. (1998) Influence of oxygen exposure time on organic carbon preservation in continental margin sediments. *Nature* **391**, 572–574.

Hedges J. I. and Keil R. G. (1995) Sedimentary organic matter preservation: an assessment and speculative synthesis. *Mar. Chem.* **49**, 81–115.

Hedges J. I., Hu F. S., Devol A. H., Hartnett H. E., Tsamakis E., and Keil R. G. (1999) Sedimentary organic matter preservation: a test for selective degradation under oxic conditions. *Am. J. Sci.* **299**, 529–555.

Hedges J. I., Baldock J. A., Gelinas Y., Lee C., Peterson M. L., and Wakeham S. G. (2002) The biochemical and elemental compositions of marine plankton: a NMR perspective. *Mar. Chem.* **78**, 47–63.

Henrichs S. M. (1992) Early diagenesis of organic matter in marine sediments: progress and perplexity. *Mar. Chem.* **39**, 119–149.

Henrichs S. M. and Reeburgh W. S. (1987) Anaerobic mineralization of marine sediment organic matter: rates and the role of anaerobic processes in the oceanic carbon economy. *Geomicrobiol. J.* **5**(3/4), 191–237.

Honjo S., Dymond J., Collier R. W., and Manganini S. (1995) Export production of particles to the interior of the equatorial Pacific Ocean during the 1992 EqPac experiment. *Deep-Sea Res. II* **42**, 831–870.

Honjo S., Francois R., Manganini S., Dymond J., and Collier R. W. (2000) Particle fluxes to the interior of the Southern Ocean in the western Pacific sector along 170° W. *Deep-Sea Res. II* **47**, 3521–3548.

Hurd D. (1973) Interactions of biogenic opal, sediment and seawater in the central equatorial Pacific. *Geochim. Cosmochim. Acta* **37**, 2257–2282.

Jahnke R. A. (1996) The global ocean flux of particulate organic carbon: areal distribution and magnitude. *Global Biogeochem. Cycles* **10**(1), 71–88.

Jahnke R. A. and Jahnke D. B. Calcium carbonate dissolution in deep sea sediments: reconciling microelectrode, pore water, and benthic flux chamber results. *Geochim. Cosmochim. Acta* (in press).

Jahnke R. A., Heggie D., Emerson S., and Grundmanis V. (1982) Pore waters of the central Pacific Ocean: nutrient results. *Earth Planet. Sci. Lett.* **61**, 233–256.

Jahnke R. A., Emerson S. R., Roe K. K., and Burnett W. C. (1983) The present day formation of apatite in Mexican continental margin sediments. *Geochim. Cosmochim. Acta* **47**, 259–266.

Jahnke R. A., Emerson S. R., Reimers C. E., Schuffert J., Ruttenberg K., and Archer D. (1989) Benthic recycling of biogenic debris in the eastern tropical Atlantic Ocean. *Geochim. Cosmochim. Acta* **53**, 2947–2960.

Jahnke R. A., Craven D. B., and Gaillard J.-F. (1994) The influence of organic matter diagenesis on $CaCO_3$ dissolution at the deep-sea floor. *Geochim. Cosmochim. Acta* **58**, 2799–2809.

Jahnke R. A., Craven D. B., McCorkle D. C., and Reimers C. E. (1997) $CaCO_3$ dissolution in California continental margin sediments: the influence of organic matter remineralization. *Geochim. Cosmochim. Acta* **61**, 3587–3604.

Jorgensen B. B. (1982) Mineralization of organic matter in the sea bed—the role of sulphate reduction. *Nature* **296**, 643–645.

Klinkhammer G. P. (1980) Early diagenesis in sediments from the eastern equatorial Pacific: II. Pore water metal results. *Earth Planet. Sci. Lett.* **49**, 81–101.

Kvenvolden K. A. (1988) Methane hydrates and global climate. *Global Biogeochem. Cycles* **2**, 221–229.

Lohse L., Helder W., Epping E. H. G., and Balzer W. (1998) Recycling of organic matter along a shelf-slope transect across the NW European continental margin (Goban Spur). *Prog. Oceanogr.* **42**, 77–110.

Mackenzie F. T. and Garrels R. M. (1966) Chemical mass balance between rivers and oceans. *Am. J. Sci.* **264**, 507–525.

Mangelsdorf P. C., Jr., Wilson R. S., and Daniell E. (1969) Potassium enrichments in interstitial waters of recent marine sediments. *Science* **165**, 171–174.

Manheim F. T. (1966) A hydraulic squeezer for obtaining interstitial water from consolidated and unconsolidated sediments. *Prof. Pap. US Geol. Surv.* **550**, C256–C261.

Martin W. R. and Bender M. L. (1988) The variability of benthic fluxes and sedimentary remineralization rates in response to seasonally variable organic carbon rain rates in the deep sea: a modeling study. *Am. J. Sci.* **288**, 561–574.

Martin W. R. and McCorkle D. C. (1993) Dissolved organic carbon concentrations in marine pore waters determined by high-temperature oxidation. *Limnol. Oceanogr.* **38**(7), 1464–1479.

Martin W. R. and Sayles F. L. (1996) $CaCO_3$ dissolution in sediments of the Ceara Rise, western equatorial Atlantic. *Geochim. Cosmochim. Acta* **60**(2), 243–263.

Martin W. R. and Sayles F. L. (2003) Metabolically driven calcite dissolution in deep-sea sediments. *Abstracts of the Joint Global Ocean Flux Study Open Science Conference, May 4–8, 2003, Washington, DC.*

Martin W. R. and Sayles F. L. Organic matter cycling in sediments of the continental margin in the northwest Atlantic Ocean. *Deep-Sea Res. I* (submitted).

Martin W. R., Bender M., Leinen M., and Orchardo J. (1991) Benthic organic carbon degradation and biogenic silica dissolution in the central equatorial Pacific. *Deep-Sea Res.* **38**, 1481–1516.

Martin W. R., McNichol A. P., and McCorkle D. C. (2000) The radiocarbon age of calcite dissolving at the seafloor: estimates from pore water data. *Geochim. Cosmochim. Acta* **64**, 1391–1404.

McCorkle D. C., Emerson S. R., and Quay P. D. (1985) Stable carbon isotopes in marine pore waters. *Earth Planet. Sci. Lett.* **74**, 13–26.

McManus J., Hammond D. E., Berelson W. M., Kilgore T. E., Demaster D. J., Ragueneau O. G., and Collier R. W. (1995) Early diagenesis of biogenic opal: dissolution rates, kinetics, and paleoceanographic implications. *Deep-Sea Res. II* **42**(2–3), 871–903.

Michalopoulos P. and Aller R. C. (1995) Rapid clay mineral formation in the Amazon delta sediments: reverse weathering and ocean elemental cycles. *Science* **270**, 614–617.

Middelburg J. J., Soettaert K., Herman P. M. J., and Heip C. H. R. (1996) Denitrification in marine sediments: a model study. *Global Biogeochem. Cycles* **10**, 661–673.

Murray J. and Irvine R. (1885) On the chemical changes which take place in the composition of the seawater associated with blue muds on the floor of the ocean. *Trans. Roy. Soc. Edinburgh* **37**, 481–507.

Murray J. W., Emerson S., and Jahnke R. A. (1980) Carbonate saturation and the effect of pressure on the alkalinity of interstitial waters from the Guatemala Basin. *Geochim. Cosmochim. Acta* **44**, 963–972.

Nelson D. M., Treguer P., Brzezinski M. A., Leynaert A., and Queguiner B. (1995) Production and dissolution of biogenic silica in the ocean: revised global estimates, comparison with regional data and relationship to biogenic sedimentation. *Global Biogeochem. Cycles* **9**(3), 359–372.

Pope R. H., DeMaster D. J., Smith C. R., and Seltmann H., Jr. (1996) Rapid bioturbation in equatorial Pacific sediments: evidence from excess ^{234}Th measurements. *Deep-Sea Res. II.* **43**, 1339–1364.

Rabouille C., Gaillard J.-F., and Treguer P. (1997) Biogenic silica recycling in surficial sediments across the polar front of the Southern Ocean (Indian sector). *Deep-Sea Res. II* **44**(5), 1151–1176.

Ragueneau O., Treguer P., Leynaert A., Anderson R. F., Brzezinski M. A., DeMaster D. J., Dugdale R. C., Dymond J., Fisher G., Francois R., Heinze C., Maier-Reimer E., Martin-Jezequel V., Nelson D. M., and Quequiner B. (2000) A review of the Si cycle in the modern ocean: recent progress and missing gaps in the application of biogenic opal as a paleoproductivity proxy. *Global Planet. Change* **26**, 317–365.

Redfield A. C., Ketchum B. H., and Richards F. A. (1963) The influence of organisms on the composition of seawater. In *The Sea* (ed. M. N. Hill). Wiley-Interscience, New York, vol. 2, pp. 26–77.

Reimers C. E. (1987) An *in situ* microprofiling instrument for measuring interfacial pore water gradients: methods and oxygen profiles from the North Pacific Ocean. *Deep-Sea Res.* **34**, 2019–2035.

Reimers C. E., Jahnke R. A., and McCorkle D. C. (1992) Carbon fluxes and burial rates over the continental slope and rise off central California with implications for the global carbon cycle. *Global Biogeochem. Cycles* **6**, 199–224.

Revsbeck N. P., Sorensen J., Blackburn T. H., and Lomholt J. P. (1980) Distribution of oxygen in marine sediments measured with microelectrodes. *Limnol. Oceanogr.* **25**(3), 403–411.

Ruttenberg K. C. (1993) Reassessment of the oceanic residence time of phosphorous. *Chem. Geol.* **107**, 405–409.

Sayles F. L. (1979) The composition and diagenesis of interstitial solutions: I. Fluxes across the seawater–sediment interface in the Atlantic Ocean. *Geochim. Cosmochim. Acta* **43**, 527–545.

Sayles F. L. (1981) The composition and diagenesis of interstitial solutions: II. Fluxes and diagenesis at the water–sediment interface in the high latitude North and South Atlantic. *Geochim. Cosmochim. Acta* **45**, 1061–1086.

Sayles F. L. and Curry W. B. (1988) δ^{13}C, TCO_2, and the metabolism of organic carbon in deep sea sediments. *Geochim. Cosmochim. Acta* **52**, 2963–2978.

Sayles F. L., Wilson T. R. S., Hume D. N., and Mangelsdorf P. C., Jr. (1973) *In situ* sampler for marine sedimentary pore waters: evidence for potassium depletion and calcium enrichment. *Science* **181**, 154–156.

Sayles F. L., Mangelsdorf P. C., Jr., Wilson T. R. S., and Hume D. N. (1976) A sampler for the *in situ* collection of marine sedimentary pore waters. *Deep-Sea Res.* **23**, 259–264.

Sayles F. L., Martin W. R., and Deuser W. G. (1994) Response of benthic oxygen demand to particulate organic carbon supply in the deep sea near Bermuda. *Nature* **371**, 686–689.

Sayles F. L., Deuser W. G., Goudreau J. E., Dickinson W. H., Jickells T. D., and King P. (1996) The benthic cycle of biogenic opal at the Bermuda Atlantic Time Series site. *Deep-Sea Res. I* **43**(4), 383–409.

Sayles F. L., Martin W. R., Chase Z., and Anderson R. F. (2001) Benthic remineralization and burtial of biogenic SiO_2, $CaCO_3$, organic carbon, and detrital material in the Southern Ocean along a transect at 170° west. *Deep-Sea Res. II* **48**, 4323–4383.

Schink D. R., Guinasso N. L., Jr., and Fanning K. A. (1975) Processes affecting the concentration of silica at the sediment water interface of the Atlantic Ocean. *J. Geophys. Res.* **80**, 3013–3031.

Schluter M., Rutgers van der Loeff M. M., Holby O., and Huhn G. (1998) Silica cycle in surface sediments of the South Atlantic. *Deep-Sea Res. I* **45**, 1085–1109.

Shaw T. J., Gieskes J. M., and Jahnke R. A. (1990) Early diagenesis in differing depositional environments: the response of transition metals in pore water. *Geochim. Cosmochim. Acta* **54**, 1233–1246.

Smith C. R., Pope R. H., DeMaster D. J., and Magaard L. (1993) Age-dependent mixing of deep-sea sediments. *Geochim. Cosmochim. Acta* **57**, 1473–1488.

Smith C. R., Hoover D. J., Doan S. E., Pope R. H., DeMaster D. J., Dobbs F. C., and Altabet M. A. (1996) Phytodetritus at the abyssal seafloor across 10° of latitude in the central equatorial Pacific. *Deep-Sea Res.* **43**, 1309–1338.

Smith K. L. (1987) Food energy supply and demand: a discrepancy between particulate organic carbon flux and sediment community oxygen consumption in the deep ocean. *Limnol. Oceanogr.* **32**, 201–220.

Soetaert K., Herman P. M. J., Middelburg J. J., Heip C. H. R., deStigter H. S., van Weering T. C. E., Epping E. H. G., and Helder W. (1996) Modeling ^{210}Pb-derived mixing activity in ocean margin sediments: diffusive versus nonlocal mixing. *J. Mar. Res.* **54**, 1207–1227.

Thamdrup B. (2000) Bacterial manganese and iron reduction in aquatic sediments. *Adv. Microb. Ecol.* **16**, 41–84.

van Bennekom A. J., Berger G. W., Van der Gaast S. J., and De Vries R. T. (1988) Primary productivity and the silica cycle in the Southern Ocean (Atlantic sector). *Palaeogeogr. Palaeoclimatol. Palaeoecol.* **67**, 19–30.

Van Cappellen P. and Ingall E. D. (1994) Benthic phosphorous regeneration, net primary production, and ocean anoxia: a model of the coupled marine biogeochemical cycles of carbon and phosphorus. *Paleoceanography* **9**(5), 677–692.

Van Cappellen P. and Qiu L. (1997) Biogenic silica dissolution in sediments of the Southern Ocean: I. Solubility. *Deep-Sea Res. II* **44**, 1109–1128.

Van Cappellen P. and Wang Y. (1996) Cycling of iron and manganese in surface sediments: a general theory for the coupled transport and reaction of carbon, oxygen, nitrigon, sulfur, iron, and manganese. *Am. J. Sci.* **296**, 197–243.

Varney M. S. (2000) *Chemical Sensors in Oceanography*. Gordon and Breach, Amsterdam, vol. 1, 333pp.

Wakeham S. G., Lee C., Hedges J. I., Hernes P. J., and Peterson M. L. (1997) Molecular indicators of diagenetic status in marine organic matter. *Geochim. Cosmochim. Acta* **61**, 5363–5369.

Westrich J. T. and Berner R. A. (1984) The role of sedimentary organic matter in bacterial sulfate reduction: the G model tested. *Limnol. Oceanogr.* **29**(2), 236–249.

Wheatcroft R. A. (1992) Experimental tests for particle size-dependent bioturbation in the deep ocean. *Limnol. Oceanogr.* **37**, 90–104.

Wheatcroft R. A., Jumars P. A., Smith C. R., and Nowell R. M. (1990) A mechanistic view of the particulate biodiffusion coefficient: step lengths, rest periods and transport directions. *J. Mar. Res.* **48**, 177–207.

7.03
Formation and Diagenesis of Carbonate Sediments

J. W. Morse

Texas A&M University, College Station, TX, USA

7.03.1 INTRODUCTION AND BACKGROUND

7.03.1.1 General Overview of Sedimentary Marine Carbonates

Carbonate minerals in modern marine sediments can readily be divided into those found in shoal-to-shallow and deep-water environments. The factors controlling the sources, mineralogy, and diagenesis of carbonates in these environments are very different. Within the shoal-to-shallow water environment, the sources and diagenesis of carbonates differ substantially between carbonate-rich and primarily siliciclastic sediments.

Almost all deep-sea carbonate-rich sediments are composed of calcite low in magnesium ($>99\%$ $CaCO_3$). This material is primarily derived from pelagic skeletal organisms. Coccolithophores are usually the most important quantitatively, followed by foraminifera. In sediments overlain by waters of intermediate depth such as mid-ocean ridge crests, aragonite derived from pelagic pteropods and heteropods can be found. Calcite cements containing abundant magnesium can also occur in deep-water sediments, but they are relatively rare.

Shallow-water carbonate-rich sediments are largely confined today to the subtropic and tropic climatic zones, but they are found even at high latitudes. Their occurrence is strongly influenced by factors such as water temperature and terrigenous input. These sediments are generally dominated by aragonite, followed by calcites rich in magnesium (>4 mol.%). Low-magnesian calcite is usually a minor component. Dolomite occurs only in special environments, and even then it is generally not a major component of the sediment. Shallow-water carbonates are primarily produced by the disintegration of the skeletons of benthic organisms, such as corals, echinoids, mollusks, foraminifera, and coralline algae. In some environments, inorganic precipitates such as cements and oöids are also abundant. As discussed later in this chapter, the source of aragonite needle muds remains controversial.

In siliciclastic sediments, the major source of carbonates is also primarily derived from benthic organisms. These include bivalves, other mollusks, sea urchins, and foraminifera. In these sediments there is often a zone of considerable undersaturation produced near the sediment–water interface, where the oxidation of organic matter and bacterially produced sulfides can result in almost complete dissolution of sediment carbonate. Carbonates from organisms that burrow beneath this zone of intense diagenetic activity are often well preserved. In organic-rich siliciclastic sediments sulfate reduction may be very extensive, with the increase in alkalinity outweighing the decrease in pH, resulting in the precipitation of calcium carbonate.

Carbonates in ancient sedimentary rocks are dominantly calcite and dolomite, as predicted by equilibrium thermodynamics. This indicates that metastable shallow-water sedimentary carbonates have undergone major diagenetic changes. The distribution of calcium carbonate in deep-sea sediments indicates that extensive early diagenesis has led to the loss of most carbonate minerals from these sediments. This loss appears to be linked to the saturation state of the overlying waters. However, kinetic and biogeochemical processes also play a major role in controlling the distribution of carbonate minerals. This chapter examines the geochemistry of the early diagenesis of carbonates in marine sediments.

7.03.1.2 Geochemistry of Major Sedimentary Carbonate Minerals

Before proceeding to the "natural world," a brief commentary on the chemistry of major marine carbonate minerals is offered first. A vast literature exists on this topic: Morse and Mackenzie (1990) can provide an introduction.

Marine carbonate minerals have both biotic (dominant) and abiotic (minor) sources. Their formation is often controlled by kinetic factors or biomediated processes in organisms. Surface seawater is most highly supersaturated (the ion activity product (IAP) is much greater than the solubility product) with respect to dolomite ($\sim 50\times$), followed by pure calcite ($\sim 6\times$), then by aragonite ($\sim 4\times$). It may be close to equilibrium with respect to high ($\sim$13 mol.%) magnesian calcites. However, dolomite is rarely found in modern sediments. Sediments containing "recent" dolomite are generally found associated with exceptionally high salinity environments. Aragonite and high-magnesian calcites dominate shoal-to-shallow water carbonate sediments. Seawater must be $\sim$22 times or more supersaturated with respect to calcite before pseudo-homogeneous precipitation occurs (e.g., Morse *et al.*, 1997). However, metastable aragonite precipitates, not calcite, due largely to the inhibiting effects of Mg^{2+} on calcite precipitation (Pytkowicz, 1965; Berner, 1975).

Dolomite is one of the most abundant sedimentary carbonate minerals. However, after years of intense study its mode of formation remains controversial, and its properties under Earth surface conditions are less well known than for most other carbonate minerals. The primary reason for this is that its formation is kinetically hindered by its well-ordered structure. Another problem in understanding dolomite may be as stated by Land (1985): "there are dolomites and dolomites and dolomites." Most recently formed marine dolomites are classified as "protodolomites." They are far from being perfectly ordered structurally and usually contain a few percent excess calcium. Attempts to measure their solubility have been frustrating, but it does appear that they can be one or two orders of magnitude more soluble than ideal dolomite. Thus, seawater may not be supersaturated with respect to protodolomites.

High-magnesian calcites are common components of shallow marine sediments. They are derived from the skeletons of organisms, such as benthic foraminifera and sea urchins, and by direct precipitation of marine cements. Their solubility is strongly influenced by their magnesium content. The solubilities of biotic and synthetic magnesium calcites differ substantially (Figure 1). In seawater, biogenic magnesian calcite containing $\sim$11 mol.% $MgCO_3$ may have about the same solubility as aragonite.

Berner (1976) reviewed the problems of measuring calcite solubility in seawater, and it is these problems, in part, that have led to the use of stoichiometric solubility constants for calcite and aragonite (see Section 7.03.3). The most difficult problem is that, although the solubility of pure calcite is sought in experiments with seawater solutions, extensive magnesium co-precipitation can produce magnesian calcites. The solubility of these magnesian calcites differs from that of pure calcite. Thus, it is not possible to measure the solubility of pure calcite directly in seawater.

Another problem is the determination of aragonite solubility in seawater. A study of the solubility of calcite and aragonite in seawater

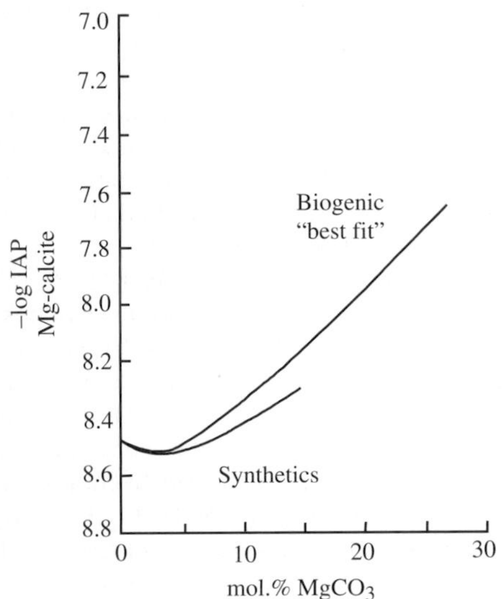

Figure 1 The equilibrium IAP for magnesian calcites of biogenic and synthetic origin versus magnesium concentration (after Morse and Mackenzie, 1990).

was conducted by Morse *et al.* (1980); their results for calcite solubility were in good agreement with earlier work (e.g., Ingle, 1975). However, for aragonite they found a much lower solubility than observed by previous investigators. It was demonstrated that the primary reason for this difference was that earlier investigators had not waited for a sufficiently long period of time to reach equilibrium and that strangely, for equilibration periods of less than about a month, different solubility values can be obtained in a reversible manner. This result points to some type of short-term, solubility controlling surface phase. The value obtained by Morse *et al.* (1980) for aragonite was in good agreement with the value predicted from thermodynamic solubility products and ion activity coefficients in seawater, and the value for the solubility ratio of aragonite to calcite in seawater was 1.51, in good agreement with dilute-solution studies. From these considerations, it appears that the co-precipitation of magnesium (~8 mol.% $MgCO_3$) with calcite forming slowly from seawater does not change its solubility beyond the uncertainty in the measurements, which is estimated to be less than 10%.

In addition to magnesium, at least trace amounts of many components present in seawater can be incorporated into marine carbonates. Concentrations range from trace (e.g., heavy metals), to minor (e.g., strontium), to major (e.g., magnesium, see previous discussion). This means that there is potentially a large amount of information that can be obtained from the study of carbonate mineral composition. This type of information allied with stable isotope ratio data, which are influenced by many of the same environmental factors, has become a major area of study in carbonate geochemistry. Much of the research on co-precipitation reactions with calcite and aragonite has been reviewed by Mucci and Morse (1990).

7.03.1.3 The CO_2 System in Oceanic Waters

The chemistry of the carbonic acid system in seawater has been one of the more intensely studied areas of carbonate geochemistry. This is because a very precise and detailed knowledge of this system is necessary to understand carbon dioxide cycling and the deposition of carbonate sediments in the marine environment. A major concept applicable to problems dealing with the behavior of carbonic acid and carbonate minerals in seawater is the idea of a "constant ionic medium." This concept is based on the observation that the salt in seawater has almost constant composition, i.e., the ratios of the major ions are the same from place to place in the ocean (Marcet's principle). Possible exceptions can include seawater in evaporative lagoons, pores of marine sediments, and near river mouths. Consequently, the major ion composition of seawater can generally be determined from its salinity. It has been possible, therefore, to develop equations in which the influence of seawater composition on carbonate equilibria is described simply in terms of salinity.

In theory, it should be possible to deal with all carbonate geochemistry in seawater by knowing the appropriate activity coefficients and their response to changes in how salinity, temperature, and pressure affect them. In practice, we are only now beginning to approach the treatment of activity coefficients with sufficient accuracy to be useful for most problems of interest. That is why "apparent" and stoichiometric equilibrium constants, which do not involve the use of activity coefficients, have been in widespread use in the study of marine carbonate chemistry since the early 1970s. The stoichiometric constants, usually designated as K^*, involve only the use of concentrations (m), whereas expressions for apparent equilibrium constants (K') contain both concentrations and apparent hydrogen activities (a_{H^+}) or concentrations. Examples of these different types of constants are

$$\text{stoichiometric constant}: K^*_{\text{calcite}} = m_{Ca^{2+}} m_{CO_3^{2-}} \tag{1}$$

$$\text{apparent constant}: K'_2 = \frac{m_{CO_3^{2-}} a_{H^-}}{m_{HCO_3^-}} \tag{2}$$

It should be noted that in seawater the molinity concentration scale (mol kg^{-1} of seawater) is often used, and care must be taken to make certain that stoichiometric and apparent constants are on the same concentration scale as the measured values.

pH values are "apparent" because electrodes do not measure hydrogen ion activity. The surface chemistry of glass electrodes and liquid junction potentials between the reference electrode filling solution and seawater contribute to this complexity. NBS buffer standards have a much lower ionic strength than seawater; this further complicates the problem. One way in which this last problem has been attacked is to make up buffered artificial seawater solutions and very carefully determine the relation between measurements and actual hydrogen ion activities or concentrations (e.g., see review of Millero, 2001). As a practical matter, pH values of seawater are generally measured at ~25 °C and atmospheric pressure. These pH measurements must be corrected for pressure and temperature changes for application to *in situ* conditions.

Another practical consideration when dealing with the seawater carbonic acid system is that in addition to carbonate alkalinity, H^+ and OH^-, a number of other components can contribute to the total alkalinity (TA). The seawater constituent that is usually most important is boric acid. Under most conditions, boric acid contributes ~0.1 mmol L^{-1} alkalinity; it is usually taken into consideration when making calculations. Nutrient compounds, such as ammonium, phosphate, and silica, whose concentrations in seawater are highly variable, can also influence alkalinity. They must be taken into account for very precise work. In anoxic pore waters a number of compounds, such as hydrogen sulfide and dissolved organic matter, can be significant contributors to alkalinity (e.g., see Berner *et al.*, 1970).

One of the primary aims in the study of the geochemistry of carbonates in marine waters is the calculation of the saturation state of the seawater with respect to carbonate minerals. The saturation state of a solution with respect to a given mineral is simply the ratio of the ion activity or concentration product to the thermodynamic or stoichiometric solubility product (Equation (3)). In seawater the latter is generally used and Ω_{mineral} is the symbol used to represent the ratio. If $\Omega = 1$, the solid and solution are in equilibrium; if $\Omega < 1$, the solution is undersaturated and mineral dissolution can occur, and if $\Omega > 1$, the solution is supersaturated and precipitation should occur:

$$\Omega_{\text{calcite}} = \frac{a_{Ca^{2+}} a_{CO_3^{2-}}}{K_{\text{calcite}}} \quad \text{or} \quad \frac{m_{Ca^{2+}} m_{CO_3^{2-}}}{K^*_{\text{calcite}}} \qquad (3)$$

The calcium concentration in normal seawater can be calculated from the salinity for most purposes (accuracy is almost always better than within 1%). As previously noted, care must be taken in studies of pore waters and in unusual areas such as coastal waters, carbonate banks, and lagoons where significant deviations from normal seawater concentrations can occur.

The apparent constants are not those for standard seawater where seawater composition has been significantly altered. Composition changes can be especially important in anoxic environments where extensive sulfate concentration changes are produced. The carbonate ion concentration can be calculated from any two of the four parameters: pH, total CO_2 (TCO_2), the partial pressure of CO_2 (p_{CO_2}), or TA (e.g., Morse and Mackenzie, 1990).

7.03.2 SOURCES AND DIAGENESIS OF DEEP-SEA CARBONATES

7.03.2.1 Sources and Sedimentation

The primary sources of calcium carbonate in deep-sea sediments are pelagic coccolithophores and foraminifera that grow dominantly in the photic zone. The distribution of calcium carbonate-secreting pelagic organisms is primarily controlled by the fertility and temperature of the near-surface ocean. The fertility of seawater is largely a result of ocean circulation patterns and, in particular, processes leading to upwelling of nutrient-rich waters. In general, coccolithophores are common in temperate waters, but rare in high-latitude cold waters where diatoms dominate. Coccolithophores are numerically much more abundant (usually $\sim 10^4 \times$) than foraminifera. Benthic foraminifera consist either of aragonitic or high-magnesian calcite. Because of their scarcity in deep-sea sediments, aragonitic pelagic organisms have received relatively little attention. Pteropods are the most abundant pelagic aragonitic organisms.

The mechanisms and rates of transfer of biogenic carbonate material from near-surface waters to deep-sea sediments have been investigated intensely. Major studies have dealt with the transition from living organism to carbonate test, rates of sinking, extent of dissolution in the water column and on the sea bottom, and the relation between life and death assemblages. These studies have raised the question of where and when carbonate starts. Formally, diagenesis should probably not be considered to start until the deposition on the seafloor, but pelagic carbonates can be altered considerably between death in the upper water column and deposition on the seafloor several kilometers below this ocean surface (e.g., Milliman *et al.*, 1999).

Because of their small size, individual coccoliths sink slowly (Lerman and Dacey, 1974; Lerman, 1979) and may spend ~100 yr in the water column. This long residence time should lead to substantial or complete dissolution of coccoliths in the undersaturated part of the water column. The origin of coccolith ooze on portions of the seafloor overlain by undersaturated waters, therefore, is difficult to explain in terms of the settling of individual particles (e.g., Honjo, 1975, 1976). Data obtained from sediment traps have shown that most coccoliths reach the seafloor as aggregates produced by zooplankton grazing and encapsulation in fecal matter.

An extensive investigation of planktonic foraminifera in sediment trap samples from the tropical Pacific and central Atlantic oceans was conducted by Thunell and Honjo (1981). At both study sites, the total foraminiferal flux and the carbonate flux tend to decrease with increasing depth. In addition, the flux of individual species of planktonic foraminifera varies significantly with depth. The number of small, solution-susceptible species decreases with increasing depth. These results indicate that the dissolution of small (<150 μm) foraminifera is significant as they settle through the water column. Mineral collected from the sediment–water interface directly below the Pacific sediment trap array contained no planktonic foraminifera, suggesting that the residence time of an individual skeleton on the seafloor before it dissolves is extremely short (Thunell and Honjo, 1981).

About 12% of the $CaCO_3$ flux made by Berner and Honjo (1981) consists of aragonite. A study of sedimentation of pteropods and foraminifera in the North Pacific by Betzer *et al.* (1984) using sediment traps confirmed that considerable dissolution of pteropods takes place in the water column. They estimate that ~90% of the aragonite flux was dissolved in the upper 2.2 km of the water column. It should be noted that the depth for total dissolution of carbonates in the water column is considerably more than the aragonite compensation depth (ACD). This is probably due to the short residence time of pteropods in the water column because of their rapid rates of sinking.

7.03.2.2 Distribution of $CaCO_3$ in Deep-sea Sediments

Diagenesis of carbonates in the deep sea almost exclusively involves the dissolution of calcium carbonate, where only 20–30% of the flux to the seafloor is preserved (Archer, 1996a). As such, it is primarily reflected in large variations in the weight percent calcium carbonate in deep-sea sediments that can range from 0 wt.% to over 90 wt.%. The distribution of calcium carbonate in sediments with ocean depth has wide variations. In open ocean basins, where rates of detrital sedimentation are moderate to low, sediments above 3,000 m water depth are generally high in calcium carbonate, whereas sediments below 6,000 m generally have a very low calcium carbonate content. Between these depths there is a poor correlation between the weight percent of calcium carbonate and depth (Smith *et al.*, 1968; Archer, 1996a).

Calcium carbonate is more abundant in Atlantic Ocean sediments, and generally occurs in significant amounts to deeper depths than in Pacific Ocean sediments. The major reason for this is that at a given depth Pacific Ocean waters are less saturated with respect to calcium carbonate than Atlantic Ocean waters. A secondary factor leading to greater preservation in Atlantic Ocean sediments is that at approximately the same degree of undersaturation in Atlantic and Pacific Ocean waters, the input of terrigenous material is higher in the Atlantic Ocean. This leads to more rapid burial and to better preservation of carbonates (Archer, 1996b). Higher productivity in overlying waters, often in upwelling areas, can also lead to more rapid carbonate deposition and to higher carbonate concentrations in sediments. A final major factor influencing calcium carbonate preservation is the relative calcium carbonate to organic carbon rain rate ratio. This is important, because oxidation of sedimentary organic matter can decrease the saturation of pore waters with respect to $CaCO_3$ by increasing p_{CO_2}. This leads to significant carbonate dissolution (for a review of work in this area and recent research, see Mekik *et al.*, 2002). This diagenetic process is discussed in the next section of this chapter.

With increasing water depth, aragonitic pteropods disappear first, followed by the small and fragile types of calcitic foraminifera, until in the deeper parts of the oceans only badly damaged tests of large, thick-shelled foraminifera remain in the surface sediments.

The carbonate compensation depth (CCD) occurs where the rate of calcium carbonate dissolution is balanced by the rate of infall, and the calcium carbonate content of surface sediments is close to 0 wt.% (e.g., Bramlette, 1961). The CCD has been confused with the calcium carbonate *critical* depth (sometimes used interchangeably with the lysocline discussed next), where the carbonate content of the surface sediment drops below 10 wt.%. A similar marker level in deep-sea sediments is the ACD, below which aragonite is no longer observed to accumulate in sediments.

The name "lysocline" was applied by Berger (1968) to the region of rapid increase in dissolution rate that could be recognized in surface sediments as the level at which small thin-shelled foraminifera disappear or start showing strong signs of dissolution. The foraminiferal lysocline (FL) was defined by Berger (1968) as the depth where the dominant type of foraminifera shifts in surface sediments from "soluble" to "resistant" species (~50% change in ratio). Berger (1970) also defined another useful marker level in pelagic sediments, which is similarly based on the change in the ratio of "soluble" to "resistant" species of foraminifera. This is the R_o level at which the first significant (~10%) change in the ratio of these two types of foraminifera is observed in sediments. Laboratory experiments on sediments to determine how much carbonate must be dissolved to produce assemblages of foraminifera that are characteristic of the FL and R_o levels were conducted by Berger (1968). He found that ~80% dissolution is necessary to produce the FL assemblage and ~50% dissolution is needed to produce the R_o assemblage. Other attempts to quantify the extent of dissolution required to produce these sedimentary carbonate marker horizons and to relate this to the factors that control dissolution rates are discussed by Mekik *et al.* (2002).

The distribution patterns of coccoliths and foraminifera differ significantly in deep-sea sediments (Schneidermann, 1973; Roth *et al.*, 1975). In most sediments the more easily dissolved coccoliths are missing. Even in areas where the overlying water is supersaturated with respect to calcite (e.g., 3,000 m in the Atlantic Ocean; Schneidermann, 1973), significant dissolution of coccoliths has been observed. Solution-resistant coccoliths, however, may survive in sediments in which all foraminiferal tests have been dissolved. Schneidermann (1973) used these characteristics to create additional subdivisions of the region near the CCD, and Roth *et al.* (1975) emphasized that coccolith and foraminifera solution indices can be used in a complementary manner. Coccoliths are good indicators of dissolution above the lysocline, whereas foraminifera are better indicators of dissolution below the lysocline.

The distribution of aragonite and magnesian calcite derived from shallow-water areas and sediments and deposited in surrounding deep-sea sediments has also been investigated. Notable studies include those of Berner *et al.* (1976) on the Bermuda Pedestal, Land (1979) on the north Jamaican island slope, Droxler *et al.* (1988a) on the Bahama Banks region, and Droxler *et al.* (1988b) on the Nicaragua Rise.

7.03.2.3 $CaCO_3$ Diagenesis in Deep-sea Sediments

7.03.2.3.1 General relations

One of the most controversial areas of carbonate geochemistry has been the relation between calcium carbonate accumulation in deep-sea sediments and the saturation state of the overlying water. The CCD, FL, R_o, and ACD have been carefully mapped in many areas. However, with the exception of complete dissolution at the CCD and ACD, the extent of dissolution that has occurred in most sediments is difficult to determine. Consequently, it is generally not possible to make reasonably precise plots of percent dissolution versus depth. In addition, the analytical chemistry of the carbonate system (e.g., GEOSECS data) and constants used to calculate the saturation states of seawater have been a source of contention. One of the reasons for the controversy regarding the relation between the extent of carbonate dissolution occurring in deep-sea sediments and the saturation state of the overlying water is that models for the processes controlling carbonate deposition depend strongly on this relation. Hypotheses have ranged from a nearly "thermodynamic" ocean where the CCD and ACD are close to coincident with calcite and aragonite saturation levels (e.g., Turekian, 1964; Li *et al.*, 1969), to a strongly kinetically controlled system (e.g., Morse and Berner, 1972) where major differences in the CCD, FL, and saturation depth exist.

Calculations such as those by Morse and Mackenzie (1990) indicate that the calcite saturation depth is generally ~1 km greater than proposed by Berger (1977) and that it is much greater than R_o. It appears only loosely related to the FL. In the equatorial eastern Atlantic Ocean, FL is ~600 m shallower than the saturation depth. If these observations are close to correct, the long cherished idea of a "tight" relation between seawater chemistry and carbonate depositional facies must be reconsidered (Mekik *et al.*, 2002). The influence of near interfacial diagenetic processes on these relationships is discussed in the next section.

7.03.2.3.2 Early diagenetic processes in deep-sea sediments

Numerous models have been proposed for the processes occurring near the sediment–water interface in deep-sea sediments which lead to a balance between dissolution and retention of calcium carbonate in these sediments. A major difficulty in studying and modeling these processes is that many of the most important changes take place over distances of only a few millimeters in a highly dynamic environment.

Early models for the process of calcium carbonate dissolution from deep-sea sediments (e.g., Takahashi and Broecker, 1977) were based on simple diagenetic models, in which calcium carbonate dissolved into the pore waters of the sediments. The alteration of concentrations from an initial saturation state of the overlying water to equilibrium with calcite results in diffusion of the reactants and products and exchange with the overlying waters. The process is governed by the difference between equilibrium concentrations and those found in the overlying waters, the rate of dissolution as a function of disequilibrium diffusion rates, and the thickness of the stagnant benthic boundary layer above the interface. Because the dissolution rate of calcium carbonate is relatively rapid, the depth in the sediment calculated for equilibrium has generally been only a few millimeters.

A major process that had only casually been considered for its potential influence on benthic carbonate accumulation in deep-sea sediments is the oxidation of organic matter. The general reaction for this process is

$$\begin{aligned}&(CHOH)_{106}(NH_3)_{16}H_3PO_4 + 138O_2 \\ &\quad + 124CaCO_3 \rightarrow 16H_2O + 16NO_3^- \\ &\quad + HPO_4^{2-} + 124Ca^{2+} + 230HCO_3^- \quad (4)\end{aligned}$$

(Note that this reaction overestimates the oxygen and water content of average organic matter (Hedges *et al.*, 1999).) Although the importance of this reaction in coastal and continental slope sediments that are relatively rich in organic matter was widely recognized, its effect was generally ignored in deep-sea sediments, which usually contain less than 0.2 wt.% organic carbon. With the advent of sediment traps, however, it became apparent that significant amounts of organic matter are reaching the sediment–water interface and that near interfacial oxidation of organic matter is potentially a major diagenetic process even in deep-sea sediments.

A model for the influence of organic matter on carbonate dissolution in deep-sea sediments was introduced by Emerson and Bender (1981). In their model, they emphasized the importance of determining the depth distribution of organic matter oxidation. If organic matter is rapidly oxidized after arrival at the sediment–water interface, the CO_2 generated will have little chance to interact with calcium carbonate, and the influence of the oxidation of organic matter on dissolution will only be of secondary importance. If, alternatively, organic matter is rapidly mixed into the sediment by bioturbation, the oxidation of the organic matter can be very important.

Subsequent to the work of Emerson and Bender (1981), numerous studies have quantified and modeled various aspects of the interaction of organic matter with deep-sea sedimentary carbonates (e.g., Bender and Heggie, 1984; Emerson *et al.*, 1985; Peterson and Prell, 1985; Jahnke, 1988; Jahnke *et al.*, 1985, 1994, 1997; Martin *et al.*, 1986; Sayles and Curry, 1988; Archer *et al.*, 1989, 2002; Berelson *et al.*, 1990; Emerson and Archer, 1990; Archer, 1991; Hales *et al.*, 1994, 1997; Cai *et al.*, 1995; Hales and Emerson, 1996, 1997a,b; Martin and Sayles, 1996; Wenzhöfer *et al.*, 2001; R. A. Jahnke and D. B. Jahnke, in press). It is beyond the scope of this chapter to present a detailed review of this substantial and often complex body of scientific literature. Only the major concepts and points of contention will be presented here for the discussion of the diagenesis of shoal-water carbonate-rich and siliciclastic sediments where similar processes control carbonate accumulation.

The approach that has generally been used is to estimate carbonate dissolution rates via diagenetic models (e.g., Archer *et al.*, 2002), incorporating organic matter oxidation by oxygen (assuming a C:O ratio) that produces undersaturated conditions. However, in deep sediments that contain more substantial concentrations of metabolizable organic matter, suboxic electron acceptors, such as nitrate and MnO_2, must also be considered. A relationship is then established between the calcite dissolution rate and pore-water saturation state. The general kinetic equation for calcite introduced by Morse and Berner (1972) for calcite dissolution in seawater has been most commonly used:

$$R(\%/\mathrm{d}) = k(\%/\mathrm{d})(1 - \Omega)^n \quad (5)$$

where R is the rate, k the rate constant, Ω the saturation state, and n the reaction order. Several variables influence the relationship between the rate and the saturation state, including temperature and dissolved components that act as reaction inhibitors.

The fundamental parameter in all models for calcium carbonate dissolution in the deep sea is the saturation state of pore waters. In order to determine the saturation state, not only must the composition of the pore waters be known, but also the solubility of the calcium carbonate. Therefore, many studies of carbonate chemistry in deep-sea sediment pore waters have focused on the apparent solubility behavior of carbonates in these sediments.

The results of these studies have shown a surprising degree of variability and have further demonstrated the complexity of calcium carbonate accumulation in deep-sea sediments. Several studies by different groups of investigators appeared at about the same time (Emerson *et al.*, 1980; Murray *et al.*, 1980; Sayles, 1980). The results of Emerson *et al.* (1980) and Sayles (1980) indicated that, in spite of considerable variability in the pore-water chemistry of sediments at a number of different sites in the Atlantic and Pacific oceans, the pore waters are generally close to equilibrium with calcite. Higher ion concentration products in pore waters where Mn^{2+} was measurable were observed by Emerson *et al.* (1980), indicating possible control by a carbonate phase other than calcite. Subsequently, Sayles (1985) observed similar changes below ~20 cm in many sediments and Sayles (1980) also found much higher ion concentration products in one sediment containing aragonite. A similar study in the Guatemala Basin was conducted by Murray *et al.* (1980). There the sediment contained only a few tenths of a weight percent $CaCO_3$. A sharp drop in pH below the sediment–water interface was observed as the result of organic carbon oxidation. The pore waters were undersaturated in the top few centimeters, but came close to the predicted value for saturation at depth. Hales and Emerson (1997a) observed different apparent solubilities in sediments from the Ceara Rise and the Ontong-Java Plateau.

Perhaps no aspect of the diagenesis of deep-sea carbonates has been more controversial than the values of rate constant and reaction order in Equation (5) (Figure 2). Most investigators have modeled rates which demand rate constants to be orders of magnitude less than those measured in the laboratory (e.g., Morse, 1978; Keir, 1980; Morse and Arvidson, 2002). The values for the reaction order have ranged from 1 (Hales and Emerson, 1997b) to 4.5 (e.g., Wenzhöfer *et al.*, 2001). Figure 3 illustrates the difficulty of

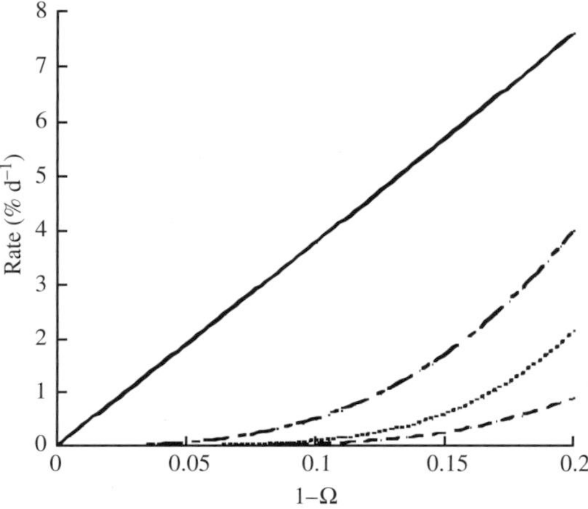

Figure 2 Rate of carbonate dissolution from deep-sea sediment versus $(1 - \Omega)$. Solid line from Hales and Emerson (1996), dotted line from Keir (1980), dashed line from Atlantic Ocean, and dotted and dashed line from Pacific Ocean sediment results of Morse (1978). Note that Hales and Emerson (1996) used a different calcite solubility product.

selecting the proper combination of rate constants and reaction orders for deep-sea sediments. Even within a given study, different values of apparent solubility products and the rate constants have been required (e.g., Hales and Emerson, 1997a; Wenzhöfer *et al.*, 2001). Although there has been a considerable amount of speculation in attempts to explain these large variations in the kinetics of deep-sea carbonate dissolution, our ability to make predictions is still quite limited. A number of parameters may contribute to the variability.

A puzzling observation has recently been made by R. A. Jahnke and D. B. Jahnke (in press). They found that in sediments above the saturation depth that contain high concentrations of calcium carbonate, the ratio of the calcium carbonate dissolution rate to the organic matter remineralization rate was substantially less than at other types of sites. They have suggested that this may be the result of exchange on carbonate particle surfaces coupled with particle mixing, but this process has yet to be clearly substantiated.

In summary, our understanding of the diagenetic factors that control the partial or complete dissolution of deep-sea carbonates can largely be considered to be one of scale. At the scale of the (for differences among) major ocean basins, it is clear that the primary variable is the saturation state of water at a given depth. This is largely controlled by its potential p_{CO_2}. A secondary influence is the deposition rate of siliciclastic sediments. Within ocean basins, variations in the near-surface primary productivity and the ratio of the rain rate of organic carbon to calcium carbonate to sediments can exert a strong influence and, at least in some areas, lead to substantial dissolution above the saturation depth of the overlying water. This process will be explored further in the discussions of carbonate diagenesis in shoal-water carbonate-rich and shallow-to-intermediate depth siliciclastic sediments, where metabolizable organic matter is at higher concentrations and where it exerts even a greater influence on carbonate diagenesis.

Submarine lithification and precipitation of cements in deep-sea carbonate sediments are relatively rare in typical major ocean basin sediments (Milliman, 1974; Milliman and Müller, 1973, 1977). The cements consist of aragonitic and magnesian calcite mineralogies. They are largely restricted to shallow seas such as the Mediterranean and the Red Sea, and to sediments in the shallower parts of major ocean basins in which biogenic aragonite is also present. Mucci (1987) (see also Garrels and Wollast, 1978) summarized much of the data on the composition of magnesian calcite cements in different environments. He found that many of the shallow-water and

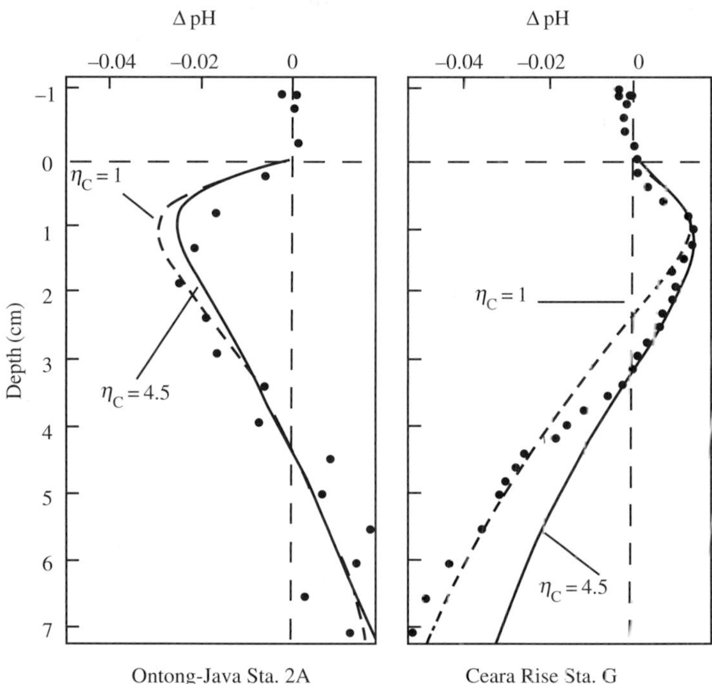

Figure 3 ΔpH versus depth for sediments from the Ontong-Java Plateau and Ceara Rise. *n* is the reaction order used to fit the data (after Hales and Emerson, 1996).

deep-sea carbonate cements contain 10–15 mol.% magnesium; there is a strong maximum in magnesium abundance at ~13 mol.%.

7.03.3 SOURCES AND DIAGENESIS OF SHOAL-WATER CARBONATE-RICH SEDIMENTS

7.03.3.1 Sources of Shoal-water Carbonates

7.03.3.1.1 General considerations

Because shelves in tropical to subtropical environments are capable of producing major quantities of carbonates, they are often referred to as "carbonate factories." A model in which carbonate production decreases with increasing water depth was produced by Bosscher and Schlager (1992). Others (e.g., Bowman and Vail, 1999) have included the influence of distance from the platform edge as an important factor in carbonate production in shallow-to-shoal water environments.

Carbonate sediments can be subdivided into skeletal and nonskeletal components. Nonskeletal carbonate grains have been divided into five major types: mud, pellets, oöids, lithoclasts, and relict. Carbonate muds are common deposits in low-energy environments, such as tidal flats and subtidal areas. Pellets are formed by the ingestion of sediment by marine organisms and excretion of fecal material. Oöids are spherical to ovoid, 0.2–1 mm grains with an internal concentric or radial structure. Lithoclasts are fragments of previously deposited, and usually somewhat lithified, carbonate sediment. Relict grains are of older origin, having formed under previous environmental conditions.

The skeletal components of carbonate sediments represent the complete or partial skeleton, or the decomposed and disaggregated skeletal remains, of organisms extant at the time of deposition of the sediment. Sedimentary particles can be produced from calcareous organisms by disaggregation of their skeletons, by mechanical means related to wave and current energy, or by bioerosion of carbonate substrates such as corals, mollusks, and rocks. The wide diversity of shoal-water organisms and their complex composition and morphology, in combination with the varied processes causing their disaggregation, account for the very complex mixture of biogenic carbonate grains in shoal-water sediments.

Seawater in the upper regions of the ocean is strongly supersaturated with respect to calcite, aragonite, and ideal dolomite. Why these minerals fail to precipitate directly from it has been a problem of major interest. In experimental studies (e.g., Pytkowicz, 1965, 1973; Berner, 1975), the magnesium ion has been observed to be a strong inhibitor of calcite precipitation and largely responsible for its failure to precipitate directly from seawater. The presence of magnesium can also elevate the supersaturation necessary for aragonite precipitation in the absence of nuclei, to a level far beyond that likely to be found in natural waters. Natural organic matter can also strongly inhibit $CaCO_3$ precipitation (e.g., Berner *et al.*, 1978). It is unlikely, based on this experimental work, that conditions appropriate for direct (homogeneous) nucleation of calcium carbonate will be found in normal modern seawater (Morse and He, 1993). However, these experimental results do not preclude the precipitation of carbonate minerals on pre-existing carbonate mineral nuclei.

7.03.3.1.2 Sources of carbonate muds

The origin of carbonate mud is a long-standing topic of considerable debate and controversy. The three major potential sources generally considered are (e.g., Reid *et al.*, 1992): (i) physicochemical or biochemical precipitation, (ii) postmortem disintegration of calcareous algae, and (iii) physical or biological abrasion of skeletal material. Research has generally focused on either the aragonitic or magnesium calcite mud components. Andrews *et al.* (1997) have found that generally within the silt and clay sizes ($\leq$63 μm), there is little mineralogic, elemental, or stable isotopic variation.

Fine-grained carbonate-rich sediments that are dominated by small (a few micrometers) needle-shaped aragonite grains are called aragonite needle muds. The two major sources for the aragonite needles are generally considered to be acalcareous algae and abiotic precipitates. Milliman (1974), Bathurst (1975), Scoffin (1987), and Morse and Mackenzie (1990) have reviewed numerous studies of the source of aragonite in calcareous muds. Most of the interpretations of its origin are based on chemical data, although needle morphology has also been used.

Attempts at making budgets of carbonate production and accumulation have also been used in attempts to define the source of aragonite needle muds. In areas such as British Honduras and Florida Bay, the supply of biogenic material appears to be sufficient to provide the sediment for the carbonate muds, although some abiotic precipitation cannot be ruled out. However, in other areas such as the Great Bahama Bank and the Persian Gulf, the biogenic supply appears to be insufficient, and an abiotic source of aragonite needles is needed (e.g., Milliman *et al.*, 1993).

Unfortunately, the precipitation of aragonite from seawater produces needles of a size and morphology that is very similar to those produced by the breakdown of common codiacean green algae. It is necessary, therefore, to examine the chemistry of the needles to determine their origin. However, even their chemical characteristics are not unambiguously diagnostic (e.g., Lowenstam and Epstein, 1957; Milliman, 1974; Bathurst, 1975; Loreau, 1982).

Not nearly as much attention has been paid to the production of lime mud from calcitic sources, but these muds can be the dominant component in peri-reefal (Debenay, 1985) and lagoonal (Reid *et al.*, 1992) environments where 70–90% of the mud may consist of magnesium calcite. This can form from the micritization and recrystallization of skeletal grains (Reid *et al.*, 1992), and the breakdown of foraminiferal tests (Debenay *et al.*, 1999).

7.03.3.1.3 Formation of carbonate sands

Carbonate-rich sediments that are dominated by sand-sized grains are common in shoal-water sediments. In most cases they are associated with moderate- to high-energy environments such as beaches and behind fringing reefs. They are often composed of carbonates of mixed mineralogy from a variety of biogenic sources produced by physical abrasion, boring, and other biological processes.

Subspherical carbonate grains, generally referred to as oöids because of their resemblance to fish eggs, comprise sandy sediments called oölites. They are common members of modern shoal-water tropic and subtropic carbonate-rich sediments. Oöids are variable in both mineralogy and structure. Aragonite, high-magnesian calcite, and calcite have been observed to occur in oöids (e.g., Land *et al.*, 1979). Radial, concentric-tangential (laminated), and micritic structures are commonly observed around a central nucleus of non-oöid material. The majority of modern, and probably most ancient, oöids are (or originated as) either aragonite with a laminated structure or high-magnesian calcite with a radial structure. The laminated aragonitic oöids are by far the most abundant type of oöids in "recent" sediments and have received the majority of attention.

Oöids are believed to be of nonbiogenic origin and represent one of the most important modes of nonbiogenic removal of $CaCO_3$ from the ocean. In addition, it has frequently been pointed out that their mineralogy and structure reflect those of marine carbonate cements (e.g., Fabricius, 1977; Land *et al.*, 1979). In fact, they grade into each other in the case of Bahamian grapestone. Land *et al.* (1979) stressed the tie between carbonate cements and oöids, pointing out that oöids can be viewed as cements centrifugally deposited on grains, whereas submarine cements can be viewed as centripetally coated pores.

Hypotheses for the processes responsible for the existence of oöids are based on bacterial-mechanical, algal, and chemical mechanisms for oöid formation. Some investigators have suggested models that involve more than one of these mechanisms. An extensive review and discussion of the hypotheses for oöid formation can be found in Morse and Mackenzie (1990).

7.03.3.2 Early Marine Diagenesis of Shoal-water Carbonate-rich Sediments

7.03.3.2.1 Pore-water chemistry

The chemistry of sediment pore waters provides important information regarding chemical reactions in sediments. In many cases, the reactions between pore waters and solids are not obvious from observations of the solids alone, but because the solid-to-solution ratio in sediments is so large, major changes can occur in pore waters during diagenesis. If equilibration alone were to take place in carbonate-rich sediments between the pore waters and metastable solids, precipitation of a stable phase would occur until equilibrium was reached. However, the oxidation of organic matter can also be an important process in these sediments. The carbon dioxide generated by this process can be sufficient to cause undersaturation, resulting in dissolution—not precipitation—as the dominant early reaction.

A major complicating factor in the diagenesis of shoal-water carbonate-rich sediments is that much of the organic matter oxidation can take place via bacterially mediated sulfate reduction. If marine plankton-like organic matter is oxidized via sulfate reduction, the nutrients phosphate and ammonia are also produced (Equation (6)):

$$\frac{1}{53}(CH_2O)_{106}(NH_3)_{16}H_3PO_4 + SO_4^{2-} \rightarrow CO_2 + HCO_3^- + HS^- + \frac{16}{53}NH_3 + \frac{1}{53}H_3PO_4 + H_2O \quad (6)$$

Because the products are generated nearly constant ratios, fixed pH can be established after a moderate degree of sulfate reduction in a closed system (Ben-Yaakov, 1973). During the early stages of reaction in seawater the pH drops to ~6.9 and then remains constant. This can result in undersaturated pore waters and carbonate dissolution. Further sulfate reduction causes the alkalinity to rise while the pH remains constant. At ~35% sulfate reduction, the pore waters regain supersaturation with respect to calcium carbonate. Further sulfate reduction can result in calcium carbonate precipitation (Figure 4). Furukawa *et al.* (2000) have emphasized that, due to the high

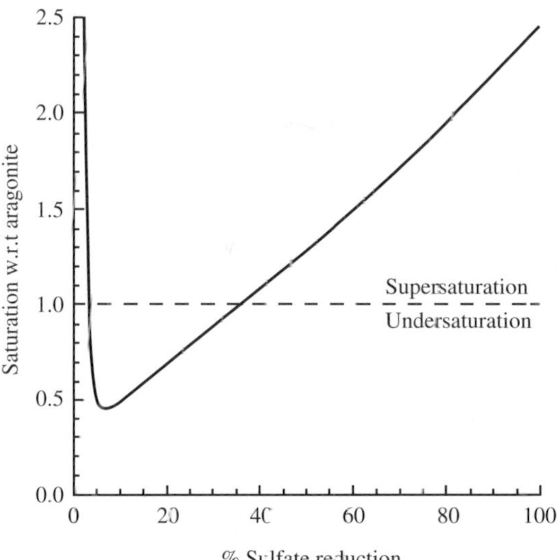

Figure 4 The saturation state with respect to aragonite versus the extent of sulfate reduction for a closed system containing seawater (after Morse and Mackenzie, 1990).

bioirrigation rates in many carbonate sediments, depth profiles of dissolved sulfate may be poor indicators of the importance of sulfate reduction in the oxidation of sedimentary organic matter in shoal-water carbonate sediments.

Several investigations of shallow-water carbonate-rich sediments have been made to determine which processes control the calcium carbonate IAP. Two major studies that investigated this topic were conducted by Berner (1966) in south Florida and Bermuda and by Morse *et al.* (1985) in the Bahamas. Berner's pioneering study was important in demonstrating through pore-water chemistry that, in sediments exposed to normal seawater, there is little diagenetic alteration of sedimentary carbonates. Evidence for the dissolution of magnesian calcites was found in pore waters from sediments in the northern part of Florida Bay where pore waters are fresher. Based on carbonate equilibrium constants available at the time, Berner concluded that most pore waters are close to equilibrium with low-magnesian calcite. Calculations using new constants indicate that this is not generally the case.

A number of pore waters in both coarse- and fine-grained sediments from the northern Great Bahama Bank and Little Bahama Bank were studied by Morse *et al.* (1985). Figure 5 summarizes their findings for the pH and TA in pore waters from sediments in different environments. Lines are also given in this figure for pore water in equilibrium with calcite, aragonite, and an 18 mol.% magnesian calcite, using the solubility constants of Walter and Morse (1984). Clearly, there is great variability

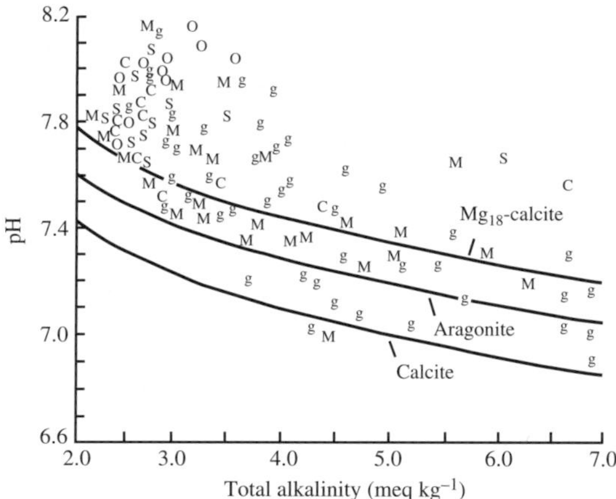

Figure 5 pH versus total alkalinity for pore waters in Bahamian carbonate-rich sediments. M = mud; S = sand; O = oölite; C = coarse-grained grassbed; g = fine-grained grassbed (after Morse *et al.*, 1985).

in pore-water composition, and many sediment pore-waters are substantially supersaturated with respect to all these phases. p_{CO_2} values are high in these sediments, typically at least 10 times the atmospheric value. Even within a given core it was found that calcium carbonate IAPs were highly variable (Morse *et al.*, 1985). Because aragonite is generally the most abundant carbonate phase in these sediments, control of the IAP by the most abundant phase does not explain these observations.

A number of solubility experiments, on sediments from the Bahama sites at which pore waters were collected, were performed by Bernstein and Morse (1985) in an attempt to understand the processes controlling the calcium carbonate IAP in these sediments. Equilibrium was approached from both supersaturation and undersaturation for different time periods up to 50 d. Generally good agreement was found between the field and laboratory observations for fine-grained sediments, indicating at least a dynamic steady state within the sediments between the pore water and some solid phase. Less agreement was found for coarse-grained sediments, where flow of water through sediments (e.g., oölite banks) may result in a residence time too short for dynamic equilibrium to be reached. An interesting observation was that oöids, although composed almost entirely of aragonite, were significantly more soluble than aragonite, in agreement with the Weyl (1965) hypothesis of a magnesian calcite coating.

Field and laboratory observations are consistent with the idea that dissolution in carbonate sediments can proceed faster than precipitation, and that the pore waters reach steady-state IAPs close to those of the most unstable phase (dissolution processes will be discussed later in this chapter). Carbonate ion may be "pumped" down to values at saturation with less soluble phases, as dissolution of the more soluble material eventually causes its removal. However, the persistence of high-magnesian calcites in sediments for long periods of time indicates that this process does not involve a large amount of mass transfer under normal marine conditions.

The influence of seagrass beds on fine-grained sediments near San Salvador Island was investigated by Short *et al.* (1985), and Morse *et al.* (1987) studied their influence on pore-water chemistry in coarser-grained sediments near the Berry Islands. Their influence was less in the coarse-grained sediments, but even in these sediments, elevated p_{CO_2} and alkalinity values in sediments beneath the seagrass beds were apparent. This is probably the result of elevated organic matter concentrations associated with roots and debris from the plants. Jensen *et al.* (1998) observed that the carbonate matrix is dissolved in the seagrass rhizosphere resulting in a release of phosphorus, which is believed to limit seagrass production in carbonate sediments.

7.03.3.2.2 Precipitation of early carbonate cements

The formation of carbonate overgrowths and cements is certainly one of the most important and highly studied aspects of carbonates in sediments and sedimentary rocks. This section discusses the geochemical aspects of the formation of early diagenetic cements and overgrowths. An extensive literature exists on the occurrence of early carbonate precipitates in marine sediments, where they are generally termed cements. Included in this literature are books devoted solely to carbonate cements (e.g., Bricker, 1971; Schneidermann and Harris, 1985) and numerous reviews (e.g., Milliman, 1974; Bathurst, 1974, 1975; Harris *et al.*, 1985; Morse and Mackenzie, 1990). Many investigations have been largely descriptive and have focused primarily on the distribution, mineralogy, and morphology of the cements.

Carbonate cements in calcareous sediments belong to three major groups. The most common are those that occur in voids found in biogenic carbonates. This group is especially important in reefs, which are discussed in a separate section of this chapter. Cements also occur on the exterior of carbonate particles, where their intergrowth can cause formation of hardened pellets, grapestones, crusts, hardgrounds, and beachrock. This type of cementation may also be important in oöids (e.g., Fabricius, 1977). Micritic cements associated with boring algae comprise the third common type of cement.

One of the most interesting questions about carbonate cements in the marine environment is why they are not more abundant. In carbonate-rich sediments carbonate surfaces for overgrowths are abundant, and both overlying seawater and many pore waters are supersaturated with respect to carbonate minerals. Morse and Mucci (1984) demonstrated that precipitation of cements from pore waters of carbonate sediments is severely inhibited by studying Iceland spar calcite crystals that had been buried for several months in a variety of carbonate sediments in the Bahamas. After recovery and analysis using very sensitive depth-profiling Auger electron spectroscopy, overgrowths were detected on only a few crystals. The calculated growth rates of the precipitates were much less than those observed in laboratory experiments using seawater at the same supersaturation state.

At least part of the reason for these observations must be that precipitation is severely inhibited by organics such as humic acids (Berner *et al.*, 1978). Mitterer and co-authors (e.g., Mitterer and Cunningham, 1985) have explored the possible role of organic matter in cement formation. These authors suggested that, whereas some types of organic matter inhibit precipitation, other types, particularly those rich in aspartic acid, favor precipitation by complexing calcium. Inhibition of precipitation, coupled with slow transfer of fresh supersaturated seawater into sediment pores, seems to account for the lack of extensive early cementation.

The factors controlling the mineralogy and chemical composition of carbonate cements in the marine environment have also been investigated extensively. Bathurst (1975, 1987) summarized many of the observations (e.g., Glover and Pray, 1971). In some cases the host carbonate mineral is a factor in determining mineral composition. Usually aragonite grains have aragonite overgrowths, whereas high-magnesian calcite grains have high-magnesian calcite overgrowths of similar magnesium content. In some instances, syntaxial formation of cement is observed. Although this explanation of host control is a "comfortable" one, it does not explain the common occurrence of cements of mixed mineralogy or those that differ in mineralogy from their host grains.

Environmental factors have also been found to correlate loosely with cement formation, mineralogy, and morphology. The energy (waves, currents, etc.) of the environment and the rate of sedimentation are most often cited as important in shallow-water environments. The energy of the environment is important in supplying seawater, from which the carbonate cements can be derived. Formation of hardgrounds was originally thought to necessitate a low-energy environment and low sedimentation rates (e.g., Taft *et al.*, 1968; Shinn, 1969). However, Dravis (1979) found that hardgrounds could form quickly even under high-energy and rapid sedimentation rate conditions. He found that in such environments endolithic algal filaments play an important role in binding the sediment and providing sites of cement formation. Aragonite and high-magnesian calcite cements can form under most environmental conditions, but aragonite is generally favored under higher-energy conditions. Given and Wilkinson (1985) suggested that this is because precipitation rates are faster under higher-energy conditions and that a faster precipitation rate favors aragonitic cements.

A major concern in studies of the chemistry of carbonate cement formation is the source of the calcium and carbonate ions necessary to form the cements. The obvious source is seawater, but large volumes of seawater are necessary if significant amounts of cement are to be produced. Cement formation is consequently favored near the sediment–water interface and in high-energy environments where water can be flushed through porous structures such as reefs. The observation that cements usually form only in thin crusts near the sediment–water interface also demonstrates the importance of normal seawater for cement precipitation. Further evidence for cement formation in normal seawater comes from stable isotopes. The $\delta^{13}C$ values of cements are usually close to those predicted for carbonates precipitating from seawater (e.g., see Given and Wilkinson, 1985). Another possible source of the ions necessary for cement formation is the dissolution of carbonate phases more soluble than the cements. High-magnesian calcites could provide such a source, as has been demonstrated in periplatform oozes in the Bahamas (Mullins *et al.*, 1985) and for coastal carbonate sediments of the Bay of Calvi in Corsica (Moulin *et al.*, 1985). This process is consistent with the previously discussed chemistry of many pore waters.

The precipitation of calcium carbonate (usually, but not exclusively, from petrographic evidence in the form of high-magnesian calcite, e.g., Alexandersson and Milliman, 1981) is generally inferred from a decrease in dissolved pore-water calcium (e.g., Thorstenson and Mackenzie, 1974; Gaillard *et al.*, 1986). Most studies of the impact of chemical diagenesis on the carbonate chemistry of anoxic sediments have focused primarily on the fact that sulfate reduction results in the production of alkalinity, which can cause precipitation of carbonate minerals (e.g., Berner, 1971). Carbonate precipitation can also occur via methane oxidation instead of organic matter oxidation, resulting in distinctly "light" $\delta^{13}C$ values for authigenic carbonates (e.g., Ritger *et al.*, 1987).

Finally, it should be noted that while most investigators have rejected the idea that carbonate cements are directly or even closely linked to biological mechanisms of formation, some investigators hold the position that organisms may be very important for cement formation processes. Marine peloids (aggregates of cemented carbonate grains) have received considerable attention in this regard because of their close association with bacterial clumps (e.g., Chafetz, 1986). Fabricius (1977) has presented a lengthy brief in favor of the formation of grapestone cements and oöids by algae.

7.03.3.2.3 Dissolution of carbonates

Many studies of the impact of chemical diagenesis on the carbonate chemistry of anoxic sediments have focused primarily on the fact that sulfate reduction results in the production of alkalinity, which can cause precipitation of carbonate minerals (see previous discussion). However, during the early stages of sulfate reduction ($\sim$2–35%), this reaction may not cause precipitation, but dissolution of carbonate minerals, because the impact of a lower pH is greater than that of increased alkalinity (Figure 4). Carbonate ion activity decreases rapidly as it is "titrated" by CO_2 from organic matter decomposition leading to a decrease in pore-water saturation state. This process is evident in data for the Fe-poor, shallow-water carbonate sediments of Morse *et al.* (1985) from the Bahamas and has been confirmed in studies by Walter and Burton (1990), Walter *et al.* (1993), and Ku *et al.* (1999) for Florida Bay, Tribble (1990) in Checker Reef, Oahu, and Wollast and Mackenzie (unpublished data) for Bermuda sediments.

Three primary processes can lead to undersaturation, in addition to the one that may result during the early stages of sulfate reduction. These are early post-death microenvironments within organisms, oxidation of organic matter by processes preceding sulfate reduction, and oxidation of sulfides. Commonly, these processes are most important near the sediment–water interface.

Early studies by Hecht (1933) showed that dissolution of carbonate in mollusk shells can begin immediately after death. In laboratory studies he found shell weight losses of as much as 25% in two weeks. Other examples of this type of behavior come from a study by Berner (1969), who found calcium carbonate dissolution during bacterial decomposition of two types of clams, possibly as a result of the production of organic acids in addition to CO_2 during early degradation of organic matter. Birnbaum and Wiremen (1984) have also pointed out that a proton gradient can be established across the cell wall of *Desulfovibrio desulfuricans,* resulting in calcium carbonate dissolution in the nearby microenvironment.

Prior to the onset of sulfate reduction, extensive organic matter degradation can occur by bacterially mediated oxygen use as described for deep-sea sediments. The influence of benthic bacterial activity on carbonate mineral dissolution under aerobic conditions has been demonstrated nicely by Moulin *et al.* (1985) for pore waters from sediments of the Gulf of Calvi in Corsica. Other processes, but probably less important, include nitrate reduction and fermentation (e.g., Aller, 1980).

Walter and co-workers (Walter and Burton, 1990; Walter *et al.*, 1993; Ku *et al.*, 1999) have made extensive efforts to demonstrate the importance of dissolution of calcium carbonate in shallow-water carbonate sediments. Up to $\sim$50% carbonate dissolution can be driven by the sulfate reduction–sulfide oxidation process. In calcium carbonate-rich sediments there is often a lack of reactive iron to produce iron sulfide minerals. The sulfide that is produced by sulfate reduction can only be buried in dissolved form in pore waters, oxidized, or can diffuse out of the sediments. In most carbonate-rich sediments the oxidative process strongly dominates the fate of sulfide. Figure 6 (Walter *et al.*, 1993) shows the strong relationship that generally occurs in the carbonate muds of Florida Bay between total carbon dioxide, excess dissolved calcium (calcium at a concentration above that predicted from salinity), and the amount of sulfate that has been reduced. It is noteworthy that the burrowed banks show much more extensive increase in calcium than the other mud banks. This is in good agreement with the observations of Aller and Rude (1988) that in Long Island Sound siliciclastic sediments an increased bioturbation leads to increased sulfide oxidation and carbonate dissolution.

7.03.3.2.4 Carbonate diagenesis associated with reefs

Because of the difficulties inherent in obtaining pore waters from the interior of reefs, the chemical environment in reef structures went largely unstudied until the 1980s. Most of the studies of early reef diagenesis are those of Hawaiian and Australian reefs by Sansone and co-workers (Sansone, 1985; Sansone *et al.*, 1988a,b; Tribble, 1990, 1993; Tribble *et al.*, 1990, 1992). Fichez *et al.* (1997) and Andrié *et al.* (1998) have conducted studies of chemical changes in pore waters in a deep borehole within a Tahitian fringing reef. These changes occur during time periods substantially longer than those

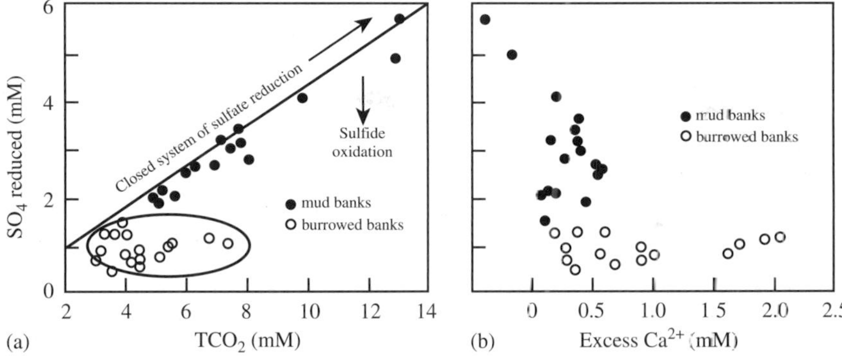

Figure 6 Dissolved sulfate versus: (a) TCO_2 and (b) excess Ca^{2+} in the pore waters of carbonate-rich Florida Bay sediments (after Walter *et al.*, 1993).

encompassed by the "early" diagenesis considered in this chapter.

The studies of Checker Reef, Oahu, Hawaii (a lagoonal patch reef) and Davies Reef, Great Barrier Reef, Australia (a platform reef), show that microbially mediated processes of oxic respiration, sulfate reduction, and, to a lesser extent, nitrate reduction and methanogenesis are important in these reef structures. The interstitial waters of these reefs are highly depleted in oxygen, and exhibit lower pH values and elevated concentrations of dissolved methane, sulfide, ammonium, phosphate, and silica relative to the seawater surrounding the reef. In Checker Reef the degree of alteration of interstitial waters increases from the margin to the center of the reef; in Davies Reef the most altered pore waters are at the margin of the reef.

The differences between these reefs probably reflect differences in their structural framework and variations in wave characteristics and tidal range in the two environments. Anaerobic reactions in reef interstitial waters may not progress far if reef structures are open and well flushed. If, however, the systems are nearly closed, little fresh reactant will enter via seawater exchange and mass transfer will be limited by the reactants trapped in the reef interstitial waters. An important conclusion of Sansone's studies was that thermodynamic disequilibrium among dissolved species such as CH_4 and SO_4^{2-} implies microzonation of chemical reactions. Microzonation resulting in slight differences in reef interstitial water compositions may account for the coexistence of different cement mineralogies in reef structures.

A pattern of initial carbonate dissolution followed by carbonate precipitation as a function of the extent of sulfate reduction occurs within reefs in a manner similar to that previously described for sediments (Tribble, 1993). Throughout this process the interstitial waters maintain close to equilibrium compositions with aragonite. This buffers the pH of the waters. It was found that FeS formation can play an important role in regulating pore-water chemistry. Systems with more intense FeS formation have a tendency to become more supersaturated with respect to aragonite. N-poor organic matter appeared to result in more corrosive conditions.

7.03.3.2.5 Early dolomite formation

The formation of dolomite under conditions encountered in modern marine sediments and sediment burial to moderate depths is strongly controlled by reaction kinetics that are slow even at high supersaturations (e.g., Lippmann, 1973; Morrow, 1982). A plausible explanation for the slow precipitation kinetics of dolomite is the requirement that cation ordering puts a major limit on the rate at which it can form (e.g., the "simplexity" principle, Goldsmith, 1953). It is likely that the dehydration kinetics of Mg^{2+} also play a role (e.g., Lippmann, 1973). Because the precipitation kinetics of dolomite are too slow to be studied in the laboratory at near Earth-surface temperatures (Land, 1998), experiments on dolomite reaction kinetics (e.g., Arvidson and Mackenzie, 1999) have generally been conducted at elevated temperatures (typically between 100 °C and 300 °C). Considerable interesting information has been gathered from these experiments. However, the applicability of the results to processes at lower temperatures and slower reaction kinetics remains highly questionable.

Observations of dolomite formation in natural systems have been used for defining additional factors that may influence the rate of dolomite formation. These include catalysis by certain clay minerals (e.g., Wanless, 1979) and production of organic by-products by bacteria (e.g., Gunatilaka *et al.*, 1985). Mg^{2+} transport to sites of dolomite precipitation can inhibit the reaction in hemipelagic sediments (e.g., Baker and Burns, 1985). However, the true influence of reaction rates is largely speculative, because the kinetic factors are generally deduced from the presence or absence of dolomite in different environments.

Most modern dolomite is forming from high ionic strength solutions that are usually derived from the evaporation of seawater or lakes in arid regions. These environments have been studied extensively, because they provide an opportunity to observe directly systems in which substantial amounts of dolomite are currently forming. The associated hydrology and solution chemistry of the dolomitizing fluids can also be determined. Sites of particular note are the Persian Gulf sabkhas, the Coorong district in South Australia, Bonaire Island in the Caribbean Sea, Sugarloaf Key in Florida, supratidal sediments on Andros Island in the Bahamas, and Deep Springs Lake in California. A general feature of many of these areas is that high Mg^{2+}/Ca^{2+} ratios in solution promote dolomitization. Friedman (1980) discussed the importance of modern high salinity environments for dolomite formation and evidence in the rock record, indicating a strong relation between evaporites and dolomite. He claimed that most sedimentary dolomite should be considered as evaporite mineral, but this opinion is not universally accepted.

The inhibition of dolomite formation by sulfate has been widely debated (e.g., see discussion in Hardie, 1987). While dolomite formation is usually associated with sediments where sulfate reduction is active, the observation that dolomite can form even when dissolved sulfate concentrations are high argues against a dominant control by dissolved sulfate. Products of sulfate reduction such as increased alkalinity may increase the formation rate of dolomite more than it is inhibited by sulfate. This conclusion has been expanded by Compton (1988), who emphasized that highly elevated solution supersaturations with respect to dolomite are usually associated with sediments rich in organic matter. He termed dolomite that forms under these conditions "organogenic dolomite." Mazzullo (2000) has summarized the data supporting the early formation of dolomite in organic-rich sediments where extensive sulfate reduction and methanogenesis occur.

Direct participation by bacteria may also promote dolomite formation during early diagenesis. The literature relating to this manner of low-temperature dolomite formation was reviewed by Vasconcelos and McKenzie (1997) in their study of microbial mediation of dolomite formation in Lagoa Vermelha in Brazil.

REFERENCES

Alexandersson E. T. and Milliman J. D. (1981) Intragranular Mg-calcite cement in *Halimeda* plates from the Brazilian continental shelf. *J. Sediment. Petrol.* **51**, 1309–1314.

Aller R. C. (1980) Diagenetic processes near the sediment–water interface of Long Island Sound: I. Decomposition and nutrient element geochemistry (S, N, P). *Adv. Geophys.* **22**, 237–350.

Aller R. C. and Rude P. D. (1988) Complete oxidation of solid phase sulfides by manganese and bacteria in anoxic marine sediments. *Geochim. Cosmochim. Acta* **52**, 751–765.

Andrews J. E., Christidis S., and Dennis P. F. (1997) Assessing mineralogical and geochemical heterogeneity in the sub 63 micron size fraction of Holocene lime muds. *J. Sediment. Res.* **67**, 531–535.

Andrié C., Jean-Baptiste P., Pierre C., Déjardin P., Fichez R., Poupeau J.-J., and Rougerie F. (1998) Pore water geochemistry and mixing processes within the Tahiti barrier reef. *Geochim. Cosmochim. Acta* **62**, 2809–2822.

Archer D. (1991) Modeling the calcite lysocline. *J. Geophys. Res.* **96**, 17037–17050.

Archer D. (1996a) An atlas of the distribution of calcium carbonate in sediments of the deep sea. *Glob. Biogeochem. Cycles* **10**, 159–174.

Archer D. (1996b) A data driven model for the lysocline. *Glob. Biogeochem. Cycles* **10**, 511–526.

Archer D., Emerson S., and Reimers C. (1989) Dissolution of calcite in deep-sea sediments: pH and O_2 microelectrode results. *Geochim. Cosmochim. Acta* **53**, 2831–2845.

Archer D. E., Morford J. L., and Emerson S. R. (2002) A model of suboxic sedimentary diagenesis suitable for automatic tuning and gridded global domains. *Glob. Biogeochem. Cycles* **16**, 10.1029/2000BG001288.

Arvidson R. S. and Mackenzie F. T. (1999) The dolomite problem: control of precipitation kinetics by temperature and saturation state. *Am. J. Sci.* **299**, 257–288.

Baker P. A. and Burns S. J. (1985) Occurrence and formation of dolomite in organic-rich continental margin sediments. *AAPG Bull.* **69**, 1917–1930.

Bathurst R. G. C. (1974) Marine diagenesis of shallow water calcium carbonate sediments. *Ann. Rev. Earth Planet. Sci.* **2**, 257–274.

Bathurst R. G. C. (1975) *Carbonate Sediments and their Diagenesis*, 2nd edn., Developments in Sedimentology No. 12. Elsevier, Amsterdam, 658p.

Bathurst R. G. C. (1987) Diagenetically enhanced bedding in argillaceous platform limestones: stratified cementation and selective compaction. *Sedimentology* **34**, 749–778.

Bender M. L. and Heggie D. T. (1984) Fate of organic carbon reaching the deep sea floor: a status report. *Geochim. Cosmochim. Acta* **48**, 977–986.

Ben-Yaakov S. (1973) pH buffering of pore water of recent anoxic marine sediments. *Limnol. Oceanogr.* **18**, 86–94.

Berelson W. M., Hammond D. E., and Cutter G. A. (1990) *In situ* measurements of calcium carbonate dissolution rates in deep-sea sediments. *Geochim. Cosmochim. Acta* **54**, 3013–3020.

Berger W. H. (1968) Planktonic foraminifera: selective solution and paleo-climatic interpretation. *Deep-Sea Res.* **15**, 31–43.

Berger W. H. (1970) Planktonic foraminifera: selective solution and the lysocline. *Mar. Geol.* **8**, 111–138.

Berger W. H. (1977) Deep-sea carbonate and the deglaciation preservation spike in pteropods and foraminifera. *Nature* **269**, 301–304.

Berner R. A. (1966) Chemical diagenesis of some modern carbonate sediments. *Am. J. Sci.* **264**, 1–36.

Berner R. A. (1969) Chemical changes affecting dissolved calcium during the bacterial composition of fish and clams in seawater. *Mar. Geol.* **7**, 253–261.

Berner R. A. (1971). *Principles of Chemical Sedimentology.* McGraw-Hill, New York, 240p.

Berner R. A. (1975) The role of magnesium in the crystal growth of calcite and aragonite from seawater. *Geochim. Cosmochim. Acta* **39**, 489–504.

Berner R. A. (1976) The solubility of calcite and aragonite at one atmosphere and 34.5 parts per thousand. *Am. J. Sci.* **276**, 713–730.

Berner R. A., Berner E. K., and Keir R. S. (1976) Aragonite dissolution on the Bermuda Pedestal: its depth and geochemical significance. *Earth Planet. Sci. Lett.* **30**, 169–178.

Berner R. A. and Honjo S. (1981) Pelagic sedimentation of aragonite: its geochemical significance. *Science* **211**, 940–942.

Berner R. A., Scott M. R., and Thomlinson C. (1970) Carbonate alkalinity in the pore waters of anoxic sediments. *Limnol. Oceanogr.* **15**, 544–549.

Berner R. A., Westrich J. T., Graber R., Smith J., and Martens C. S. (1978) Inhibition of aragonite precipitation from supersaturated seawater: a laboratory and field study. *Am. J. Sci.* **278**, 816–837.

Bernstein L. D. and Morse J. W. (1985) The steady-state calcium carbonate ion activity product of recent shallow water carbonate sediments in seawater. *Mar. Chem.* **15**, 311–326.

Betzer P. R., Byrne R. H., Acker J. G., Lewis C. S., Jolley R. R., and Feely R. A. (1984) The oceanic carbonate system: a reassessment of biogenic controls. *Science* **226**, 1074–1077.

Birnbaum S. J. and Wireman J. W. (1984) Bacterial sulfate reduction and pH: implications for early diagenesis. *Chem. Geol.* **43**, 143–149.

Bosscher H. and Schlager W. (1992) Computer simulation of reef growth. *Sedimentology* **39**, 503–512.

Bowman S. A. and Vail P. R. (1999) Interpreting the stratigraphy of the Baltimore Canyon section, offshore New Jersey with PHIL, a stratigraphic simulator. In *Numerical Experiments in Stratigraphy: Recent Advances in Stratigraphic and Sedimentologic Computer Simulations*, SEPM Special Publication 62 (eds. J. W. Harbaugh *et al.*). Society of Economic Paleontologists and Mineralogists, Tulsa, OK, pp. 117–138.

Bramlette M. N. (1961) Pelagic sediments. In *Oceanography*, AAAS Publication 67 (ed. M. Sears). American Association for the Advance of Science, Washington, DC, pp. 345–366.

Bricker O. P. (1971) *Carbonate Cements*, Studies in Geology No. 19, Johns Hopkins University, Baltimore, MD, 376p.

Cai W.-J., Reimers C. E., and Shaw T. (1995) Microelectrode studies of organic carbon degradation and calcite dissolution at a California continental rise site. *Geochim. Cosmochim. Acta* **59**, 497–511.

Chafetz H. S. (1986) Marine peloids: a product of bacterially induced precipitation of calcite. *J. Sediment. Petrol.* **56**, 812–817.

Compton J. S. (1988) Degree of supersaturation and precipitation of organogenic dolomite. *Geology* **16**, 318–321.

Debenay J. P. (1985) Le lagon sud-ouest et la marge insulaire sud de Nouvelle-Calédonie: importance et répartition des foraminifères de grande taille, leur importance dans la sédimentogenèse. *Océanogr. Trop.* **20**, 171–192.

Debenay J. P., Andre J. P., and Lesourd M. (1999) Production of lime mud by breakdown of foraminiferal tests. *Mar. Geol.* **157**, 159–170.

Dravis J. (1979) Rapid and widespread generation of recent oölitic hardgrounds on a high energy Bahamian platform, Eleuthera Bank, Bahamas. *J. Sediment. Petrol.* **49**, 195–208.

Droxler A. W., Morse J. W., and Baker V. (1988a) Good agreement between carbonate mineralogical depth variations of surficial periplatform ooze and carbonate saturation levels of the overlying intermediate waters: new data from the Nicaragua Rise. *EOS* **91**, 1345.

Droxler A. W., Morse J. W., and Kornicker W. A. (1988b) Controls on carbonate mineral accumulation in Bahamian Basins and adjacent Atlantic Ocean sediments. *J. Sediment. Petrol.* **58**, 120–130.

Emerson S. R. and Archer D. E. (1990) Calcium carbonate preservation in the ocean. *Phil. Trans. Roy. Soc. London A* **331**, 29–40.

Emerson S. R. and Bender M. L. (1981) Carbon fluxes at the sediment–water interface in the deep-sea: calcium carbonate preservation. *J. Mar. Res.* **39**, 139–162.

Emerson S. R., Fischer K., Reimers C., and Heggie D. (1985) Organic carbon dynamics and preservation in deep-sea sediments. *Deep-Sea Res.* **32**, 1–21.

Emerson S., Jahnke R., Bender M., Froelich P., Klinkhammer G., Bowser C., and Setlock G. (1980) Early diagenesis in sediments from the eastern equatorial Pacific: I. Pore water nutrient and carbonate results. *Earth Planet. Sci. Lett.* **49**, 57–80.

Fabricius F. H. (1977) *Origin of Marine Oöids*. Contributions to Sedimentology No. 7, E. Schweizerbartshe Verlagsbuchhandlung, Stuttgart, 113p.

Fichez R., Harris P., Cauwet G., and Dejardin P. (1997) Dissolved carbon in pore waters from the carbonate barrier reef of Tahiti (French Polynesia) and its basalt basement. *Aquat. Geochem.* **2**, 255–271.

Friedman G. M. (1980) Dolomite is an evaporite mineral: evidence from the rock record and from sea-marginal ponds of the Red Sea. *Soc. Econ. Paleontologists and Mineralogists*, Spec. Pub. 28, Tulsa, OK, pp. 69–80.

Furukawa Y., Bentley S. J., Shiller A. M., Lavoie D. L., and Van Cappellen P. (2000) The role of biologically-enhanced pore water transport in early diagenesis: an example from carbonate sediments in the vicinity of North Key Harbor, Dry Tortugas National Park, Florida. *J. Mar. Res.* **58**, 493–522.

Gaillard J.-F., Jeandel C., Michard G., Nicolas E., and Renard D. (1986) Interstitial water chemistry of Villefranche Bay sediments: trace metal diagenesis. *Mar. Chem.* **18**, 233–247.

Garrels R. M. and Wollast R. (1978) Discussion of: equilibrium criteria for two-component solids reacting with fixed composition in an aqueous phase—example: the magnesian calcites. *Am. J. Sci.* **278**, 1469–1474.

Given R. K. and Wilkinson B. H. (1985) Kinetic control of morphology composition and mineralogy of abiotic sedimentary carbonates. *J. Sediment. Petrol.* **55**, 109–119.

Glover E. D. and Pray L. C. (1971) High-magnesian calcite and aragonite cementation within modern subtidal carbonate sediment grains. In *Carbonate Cements*, Studies in Geology No. 19 (ed. O. P. Bricker). Johns Hopkins University, Baltimore, MD, pp. 80–87.

Goldsmith J. R. (1953) A "simplexity principle" and its relation to "ease" of crystallization. *J. Geol.* **61**, 439–451.

Gunatilaka A., Saleh A., and Al-Temeemi A. (1985) Sulfate reduction and dolomitization in a Holocene lagoon in Kuwait, Northern Arabian Gulf. SEPM Mid-Year Meeting. Golden, Colorado, Abstracts, p. 38.

Hales B. and Emerson S. (1996) Calcite dissolution in sediments of the Ontong-Java Plateau: *in situ* measurements of pore water O_2 and pH. *Glob. Biogeochem. Cycles* **10**, 527–541.

Hales B. and Emerson S. (1997a) Calcite dissolution in sediments of the Ceara Rise: *in situ* measurements of porewater O_2, pH, and CO_2(aq). *Geochim. Cosmochim. Acta* **61**, 501–514.

Hales B. and Emerson S. (1997b) Evidence in support of first-order dissolution kinetics of calcite in seawater. *Earth Planet. Sci. Lett.* **148**, 317–327.

Hales B., Burgess L., and Emerson S. (1997) An absorbance-based fiber-optic sensor for CO_2(aq) measurement in porewater of seafloor sediments. *Mar. Chem.* **59**, 51–62.

Hales B., Emerson S., and Archer D. (1994) Respiration and dissolution in the sediments of the western North Atlantic: estimates from models of *in situ* microelectrode measurements of pore water oxygen and pH. *Deep-Sea Res.* **41**, 695–719.

Hardie L. A. (1987) Perspectives on dolomitization: a critical view of some current views. *J. Sediment. Petrol.* **57**, 166–183.

Harris P. J., Kendall G., and Lerche I. (1985) Carbonate cementation—a brief review. In *Carbonate Cements* (eds. N. Schneidermann and P. M. Harris). Society of Economic Paleontologists and Mineralogists, Tulsa, OK, pp. 79–95.

Hecht F. (1933) Der Verbleib der organische Substanz der Tiere bei meerischer Einbettung. *Senckenbergiana* **15**, 165–219.

Hedges J. I., Hu F. S., Devol A. H., Hartnett H. E., Tsamakis E., and Keil R. G. (1999) Sedimentary organic matter preservation: a test for selective degradation under oxic conditions. *Am. J. Sci.* **299**, 529–555.

Honjo S. (1975) Dissolution of suspended coccoliths in the deep-sea water column and sedimentation of coccolith ooze. In *Dissolution of Deep-sea Carbonates*, Spec. Publ. No. 13 (eds. W. Sliter, A. W. H. Bé, and W. H. Berger). Cushman Found. Foraminiferal Research, Washington, DC, pp. 115–128.

Honjo S. (1976) Coccoliths: production, transportation, and sedimentation. *Mar. Micropaleon.* **1**, 65–79.

Ingle S. E. (1975) Solubility of calcite in the ocean. *Mar. Chem.* **3**, 301–319.

Jahnke R. A. (1988) A simple, reliable, and inexpensive porewater sampler. *Limnol. Oceanogr.* **3**, 483–486.

Jahnke R. A., Craven D. B., and Gaillard J.-F. (1994) The influence of organic matter diagenesis on $CaCO_3$ dissolution at the deep-sea floor. *Geochim. Cosmochim. Acta* **58**, 2799–2809.

Jahnke R. A., Craven D. B., McCorkle D. C., and Reimers C. E. (1997) $CaCO_3$ dissolution in California continental margin sediments: the influence of organic matter mineralization. *Geochim. Cosmochim. Acta* **61**, 3587–3604.

Jahnke R. A. and Jahnke D. B. Calcium carbonate dissolution in deep-sea sediments: implications of bottom water saturation state and composition. *Geochim. Cosmochim. Acta* (in press).

Jahnke R. A., Emerson S. R., Cochran J. K., and Hirschberg D. J. (1986) Fine scale distribution of porosity and particulate excess ^{210}Pb, organic carbon and $CaCO_3$ in surface sediments of the deep equatorial Pacific. *Earth Planet. Sci. Lett.* **77**, 59–69.

Jensen H. S., McGlathery K. J., Marino R., and Howarth R. W. (1998) Forms and availability of sediment phosphorus in carbonate sand of Bermuda seagrass beds. *Limnol. Oceanogr.* **43**, 799–810.

Keir R. S. (1980) The dissolution kinetics of biogenic calcium carbonates in seawater. *Geochim. Cosmochim. Acta* **44**, 241–252.

Ku T. C. W., Walter L. M., Coleman M. L., Blake R. E., and Martini A. M. (1999) Coupling between sulfur recycling and syndepositional carbonate dissolution: evidence from oxygen and sulfur isotope composition of pore water sulfate, South Florida Platform, USA. *Geochim. Cosmochim. Acta* **63**, 2529–2546.

Land L. S. (1979) The fate of reef-derived sediment on the North Jamaica island slope. *Mar. Geol.* **29**, 55–71.

Land L. S. (1985) The origin of massive dolomite. *J. Geol.* **33**, 112–125.

Land L. S. (1998) Failure to precipitate dolomite at 25 °C from dilute solution despite 1,000-fold oversaturation after 32 years. *Aquat. Geochem.* **4**, 361–368.

Land L. S., Behrens E. W., and Frishman S. A. (1979) The oöids of Baffin Bay, Texas. *J. Sediment. Petrol.* **49**, 1269–1278.

Lerman A. L. (1979). *Geochemical Processes: Water and Sediment Environments*. Wiley Interscience, New York, 481pp.

Lerman A. L. and Dacey M. F. (1974) Stokes' settling and chemical reactivity of suspended particles in natural waters. In *Suspended Solids in Water* (ed. R. J. Gibbs). Plenum, New York, pp. 17–47.

Li Y.-H., Takahashi T., and Broecker W. S. (1969) Degree of saturation of $CaCO_3$ in the oceans. *J. Geophys. Res.* **74**, 5507–5525.

Lippmann F. (1973) *Sedimentary Carbonate Minerals*. Springer, New York, 228pp.

Loreau J. P. (1982) *Sediments Aragonitiques et leur Genese*. Mem. Mus. Nat.d'Histoire Naturelle, Serie C, Tome, vol. XLVII, 300pp.

Lowenstam H. A. and Epstein S. (1957) On the origin of sedimentary aragonite needles of the Great Bahama Bank. *J. Geol.* **65**, 364–375.

Martin W. R. and Sayles F. L. (1996) $CaCO_3$ dissolution in sediments of the Ceara Rise, western equatorial Atlantic. *Geochim. Cosmochim. Acta* **60**, 243–263.

Martin G. D., Wilkinson B. H., and Lohmann K. C. (1986) The role of skeletal porosity in aragonite neomorphism—*Strombus* and *Montastrea* from the Pleistocene Key Largo Limestone, Florida. *J. Sediment. Petrol.* **56**, 194–203.

Mazzullo S. J. (2000) Organogenic dolomitization in peritidal to deep-sea sediments. *J. Sediment. Res.* **70**, 10–23.

Mekik F. A., Loubere P. W., and Archer D. E. (2002) Organic carbon flux and organic carbon to calcite flux ratio recorded in deep sea carbonates: demonstration and a new proxy. *Glob. Biogeochem. Cycles* **16**, 1–15.

Millero F. J. (2001) *Physical Chemistry of Natural Waters*. Wiley-Interscience Series in Geochemistry, New York.

Milliman J. D. (1974) *Recent Sedimentary Carbonates 1, Marine Carbonates*. Springer, New York.

Milliman J. D., Freile D., Steinen R. P., and Wilber R. J. (1993) Great Bahama Bank argonitic muds: mostly inorganically precipitated, mostly exported. *J. Sediment. Petrol.* **63**, 589–595.

Milliman J. D. and Müller J. (1973) Precipitation and lithification of magnesian calcite in the deep-sea sediments of the eastern Mediterranean Sea. *Sedimentology* **20**, 29–45.

Milliman J. D. and Müller J. (1977) Characteristics and genesis of shallow-water and deep-water limestones. In *The Fate of Fossil Fuel CO_2 in the Oceans* (eds. N. R. Anderson and A. Malahoff). Plenum, New York, pp. 655–672.

Milliman J. D., Troy P. J., Balch W. M., Adams A. K., Li Y.-H., and Mackenzie F. T. (1999) Biologically mediated dissolution of calcium carbonate above the chemical lysocline. *Deep-Sea Res. I* **46**, 1653–1669.

Mitterer R. M. and Cunningham R., Jr. (1985) The interaction of natural organic matter with grains surfaces: implications for calcium carbonate precipitation. In *Carbonate Cements* (eds. N. Schneidermann and P. M. Harris), Soc. Econom. Paleontologists and Mineralogists, Tulsa, OK, pp. 17–31.

Morrow D. W. (1982) Diagenesis 1. Dolomite: Part 1. The chemistry of dolomitization and dolomite precipitation. *Geosci. Canada* **9**, 5–13.

Morse J. W. (1978) Dissolution kinetics of calcium carbonate in sea water: VI. The near-equilibrium dissolution kinetics of calcium carbonate-rich deep sea sediments. *Am. J. Sci.* **278**, 344–353.

Morse J. W. and Arvidson R. S. (2002) Dissolution kinetics of major sedimentary carbonate minerals. *Earth Sci. Rev.* **58**, 51–84.

Morse J. W. and Berner R. A. (1972) Dissolution kinetics of calcium carbonate in seawater: II. A kinetic origin for the lysocline. *Am. J. Sci.* **274**, 638–647.

Morse J. W. and He S. (1993) Influences of T, S and P_{CO_2} on the pseudo-homogeneous nucleation of calcium carbonate from seawater: implications for whiting formation. *Mar. Chem.* **41**, 291–298.

Morse J. W. and Mackenzie F. T. (1990) *Geochemistry of Sedimentary Carbonates*. Elsevier, Amsterdam.

Morse J. W. and Mucci A. (1984) Composition of carbonate overgrowths produced on calcite crystals in Bahamian pore waters. *Sediment. Geol.* **40**, 287–291.

Morse J. W., Mucci A., and Millero F. J. (1980) The solubility of calcite and aragonite in seawater of 35‰ salinity and atmospheric pressure. *Geochim. Cosmochim. Acta* **44**, 85–94.

Morse J. W., Wang Q., and Tsio M. Y. (1997) Influences of temperature and Mg : Ca ratio on the mineralogy of $CaCO_3$ precipitated from seawater. *Geology* **25**, 85–87.

Morse J. W., Zullig J. J., Bernstein L. D., Millero F. J., Milne P., Mucci A., and Choppin G. R. (1985) Chemistry of calcium carbonate-rich shallow water sediments in the Bahamas. *Am. J. Sci.* **285**, 147–185.

Morse J. W., Zullig J. J., Iverson R. L., Choppin G. R., Mucci A., and Millero F. J. (1987) The influence of seagrass beds on carbonate sediments in the Bahamas. *Mar. Chem.* **22**, 71–83.

Moulin E., Jordens A., and Wollast R. (1985) Influence of the aerobic bacterial respiration on the early dissolution of carbonates in coastal sediments. *Proc. Progress in Belgium Oceanographic Research, Brussels*, pp. 196–208.

Mucci A. (1987) Influence of temperature on the composition of magnesian calcite overgrowths precipitated from seawater. *Geochim. Cosmochim. Acta* **51**, 1977–1984.

Mucci A. and Morse J. W. (1990) The chemistry of low temperature abiotic calcites: experimental studies on coprecipitation, stability and fractionation. *Rev. Aquat. Sci.* **3**, 217–254.

Mullins H. T., Wise S. H., Jr., Gardulski A. F., Hinchey E. J., Masters P. M., and Siegel D. I. (1985) Shallow subsurface diagenesis of Pleistocene periplatform ooze: northern Bahamas. *Sedimentology* **32**, 473–494.

Murray J. W., Emerson S., and Jahnke R. (1980) Carbonate saturation and the effect of pressure on the alkalinity of interstitial waters from the Guatemala Basin. *Geochim. Cosmochim. Acta* **44**, 963–972.

Peterson L. C. and Prell W. H. (1985) Carbonate dissolution in recent sediments of the eastern equatorial Indian Ocean: preservation patterns and carbonate loss above the lysocline. *Mar. Geol.* **64**, 259–290.

Pytkowicz R. M. (1965) Rates of inorganic calcium carbonate nucleation. *J. Geol.* **73**, 196–199.

Pytkowicz R. M. (1973) Calcium carbonate retention in supersaturated seawater. *Am. J. Sci.* **273**, 515–522.

Reid R. P., Macintyre I. G., and Post J. E. (1992) Micritized skeletal grains in Northern Belize Lagoon: a major source of Mg-calcite mud. *J. Sediment. Petrol.* **62**, 145–156.

Ritger S., Carson B., and Suess E. (1987) Methane-derived authigenic carbonates formed by subduction-induced pore-water expulsion along the Oregon/Washington margin. *Geol. Soc. Am. Bull.* **98**, 147–156.

Roth P. H., Mullin M. M., and Berger W. H. (1975) Coccolith sedimentation by fecal pellets: laboratory experiment and field observations. *Geol. Soc. Am. Bull.* **86**, 1079–1084.

Sansone F. J. (1985) Methane in the reef flat pore waters of Davies Reef, Great Barrier Reef (Australia). *Proc. Fifth Int. Coral Reef Congr Tahiti* **3**, 415–420.

Sansone F. J., Andrews C. G., Buddemeier R. W., and Tribble G. W. (1988a) Well point sampling of reef interstitial water. *Coral Reefs* **7**, 19–22.

Sansone F. J., Tribble G. W., Buddemeier R. W., and Andrews C. C. (1988b) Time and space scales of anaerobic diagenesis within a coral reef framework. *Proc. Sixth Int. Coral Reef Congr. Australia* **3**, 367–372.

Sayles F. L. (1980) The solubility of $CaCO_3$ in seawater at 2 °C based upon *in situ* sampled pore water composition. *Mar. Chem.* **9**, 223–235.

Sayles F. L. (1985) $CaCO_3$ solubility in marine sediments: evidence for equilibrium and non-equilibrium behavior. *Geochim. Cosmochim. Acta* **49**, 877–888.

Sayles F. L. and Curry W. B. (1988) $\delta^{13}C$, TCO_2, and the metabolism of organic carbon in deep sea sediments. *Geochim. Cosmochim. Acta* **52**, 2963–2978.

Schneidermann N. (1973) Deposition of coccoliths in the compensation zone of the Atlantic Ocean. In *Proceedings of Symposium on Calcareous Nannofossils* (eds. L. A. Smith and J. Hardenbol). Gulf Coast Section SEPM, Houston, TX, pp. 140–151.

Schneidermann N. and Harris P. M. (1985) *Carbonate Cements*. Society of Economic Paleontologists and Mineralogists, Tulsa, OK.

Scoffin T. P. (1987) *An Introduction to Carbonate Sediments and Rocks*. Blackie, Glasgow.

Shinn E. A. (1969) Submarine lithification of Holocene carbonate sediments in the Persian Gulf. *Sedimentology* **12**, 109–144.

Short F. T., Davis M. W., Gibson R. A., and Zimmerman C. F. (1985) Evidence for phosphorous limitation in carbonate sediments of the seagrass *Syringodium filiforme*. *Estuar. Coast. Shelf Sci.* **20**, 419–430.

Smith S. V., Dygas J. A., and Chave K. E. (1968) Distribution of calcium carbonate in pelagic sediments. *Mar. Geol.* **6**, 391–400.

Taft W. H., Arrington F., Haimovitz A., MacDonald C., and Woolheater C. (1968) Lithification of modern carbonate sediments at Yellow Bank, Bahamas. *Mar. Sci. Gulf Caribbean Bull.* **18**, 762–828.

Takahashi T. and Broecker W. S. (1977) Mechanisms for calcite dissolution on the seafloor. In *The Fate of Fossil Fuel CO_2 in the Oceans* (eds. N. R. Anderson and A. Malahoff). Plenum, New York, pp. 455–477.

Thorstenson D. C. and Mackenzie F. T. (1974) Time variability of pore water chemistry in recent carbonate sediments, Devil's Hole, Harrington Sound, Bermuda. *Geochim. Cosmochim. Acta* **38**, 1–19.

Thunell R. C. and Honjo S. (1981) Planktonic foraminiferal flux to the deep oceanic sediment trap results from the tropical Atlantic and the central Pacific. *Mar. Geol.* **40**, 237–253.

Tribble G. W. (1990) Early diagenesis in a coral reef framework. PhD Thesis, University of Hawaii.

Tribble G. W. (1993) Organic matter oxidation and aragonite diageneses in a coral reef. *J. Sediment. Petrol.* **63**, 523–527.

Tribble G. W., Sansone F. J., Buddemeier R. W., and Li Y.-H. (1992) Hydraulic exchange between a coral reef and surface seawater. *Geol. Soc. Bull.* **104**, 1280–1291.

Tribble G. W., Sansone F. J., and Smith S. V. (1990) Stoichiometric modeling of carbon diagenesis within a coral reef framework. *Geochim. Cosmochim. Acta* **54**, 2439–2449.

Turekian K. K. (1964) The geochemistry of the Atlantic Ocean basin. *Trans. New York Acad. Sci.* **26**, 312–330.

Vasconcelos C. and McKenzie J. A. (1997) Microbial mediation of modern dolomite precipitation and diagenesis under anoxic conditions (Lagoa Vermelha, Rio de Janeiro, Brazil). *J. Sediment. Res.* **67**, 378–390.

Walter L. M., Bischof S. A., Patterson W. P., and Lyons T. W. (1993) Dissolution and recrystallization in modern shelf carbonates: evidence from pore water and solid phase chemistry. *Phil. Trans. Roy. Soc. London A* **344**, 27–36.

Walter L. M. and Burton E. A. (1990) Dissolution of recent platform carbonate sediments in marine pore fluids. *Am. J. Sci.* **290**, 601–643.

Walter L. M. and Morse J. W. (1984) Magnesian calcite solubilities: a reevaluation. *Geochim. Cosmochim. Acta* **48**, 1059–1069.

Wanless H. R. (1979) Limestone response to stress: pressure solution and dolomitization. *J. Sediment. Petrol.* **49**, 437–462.

Wenzhöfer F., Holby O., and Kohls O. (2001) Deep penetrating oxygen profiles measured *in situ* by oxygen optodes. *Deep-Sea Res.* **48**, 1741–1755.

Weyl P. K. (1965) The solution behavior of carbonate materials in sea water. *Int. Conf. Trop. Oceanogr., Univ. Miami*, pp. 178–228.

7.04
The Diagenesis of Biogenic Silica: Chemical Transformations Occurring in the Water Column, Seabed, and Crust

D. J. DeMaster
North Carolina State University, Raleigh, NC, USA

NOMENCLATURE

B	the concentration of "reactive" biogenic silica per volume of sediment (mol cm^{-3})
C	the pore-water silicate concentration (mol cm^{-3})
C_s	the pore-water asymptotic silicate value observed at depth in the sediment column (mol cm^{-3})
D_B	sediment bioturbation coefficient (cm^2 yr^{-1})
D_s	the molecular diffusion coefficient for silicate in the seabed corrected for tortuosity effects (cm^2 of sediment y^{-1})
k	the dissolution rate constant for biogenic silica (yr^{-1})
z	depth in the sediment column (cm)
μM	micromolar
Φ	the sediment porosity (volume pore-water/volume total sediment)

7.04.1 INTRODUCTION

Biota that make their skeletal material out of amorphous silica have occurred in the geologic record since Cambrian time. Siliceous sponges first appeared in the lower Cambrian, whereas radiolaria and diatoms first appeared in the Ordovician and Jurassic, respectively (Blatt *et al.*, 1980). Silicoflagellates also form their skeletal material out of amorphous silica (commonly called biogenic silica), but their delicate skeletons are rarely preserved to a significant extent in the sedimentary record (Treppke *et al.*, 1996). Siliceous biota have been studied extensively from biological, paleontological, and geochemical perspectives. Diatoms, a type of unicellular algae (Figure 1), are responsible for as much as 30–40% of the primary production occurring in the surface ocean, and even a greater percentage of the organic-carbon flux

Figure 1 Diatoms are one of the most abundant biota containing siliceous skeletal material in the marine environment. Radiolaria, silicoflagellates, and siliceous sponges also deposit this amorphous phase for structural support. The diatoms in the figure (*Corethron* sp.) were collected from the Antarctic continental shelf, occurring as a monotypic assemblage in a large phytoplankton bloom. The siliceous frustules in this figure typically are 100–200 μm in length.

exported from the euphotic zone (Buesseler, 1998). Radiolaria, a unicellular zooplankton, as well as diatoms commonly are used to establish time horizons in paleoceanographic studies of sediment stratigraphy. Geochemists study the burial of biogenic silica in the oceans because it offsets the supply of riverine-dissolved silicate to the ocean that results from continental-rock weathering reactions. Approximately 400 Mt of biogenic silica are buried each year in marine sediments, making this phase second only to calcium carbonate in terms of biogenic sediment burial rate. Prior to the evolution of siliceous biota in the Cambrian, dissolved silicate was likely to have been removed from the oceans via inorganic precipitation of opal-A and opal-CT. Dissolved silicate concentrations in the ocean during the Precambrian may have been as high as several 1,000 μM (Siever, 1992). As siliceous biota became more prevalent in oceanic environments hundreds of millions of years ago, the primary mode of silica precipitation became biogenic and surface-ocean waters (where most of the siliceous biota live) became depleted in dissolved silicate. In the modern ocean, siliceous biota commonly precipitate their skeletal material from surface waters that are 10–100 fold undersaturated with respect to amorphous silica. This situation is very different than that for calcareous biota, which precipitate their skeletons from surface waters that typically are supersaturated by a factor of 2–4.

The focus of this chapter is to examine the chemical and structural transformations that take place in biogenically precipitated amorphous silica following the death of the biota, including chemical reactions that occur as this material sinks through the water column, as it is buried in the seabed, and ultimately as it is exposed to high pressures and temperatures in the Earth's crust. The initial discussion will focus on the physical properties of the skeletal material in the biota and then the chapter will examine how chemical reactions occurring in the water column and seabed change the chemical structure of the material, altering its solubility and dissolution kinetics. The final section of the chapter examines the long-term diagenetic changes of biogenic silica as it is transformed from opal-A to opal-CT and ultimately to chert.

7.04.2 THE PRECIPITATION OF BIOGENIC SILICA

A brief description of the processes involved in biogenic-silica precipitation is useful in understanding many of the physical properties of this amorphous material. There have been significant advances in our understanding of the biochemical reactions involved in biogenic-silica formation since the early 1990s (see Kroger and Sumper (2000), Hildebrand (2000), and Wetherbee *et al.* (2000) for a review). In diatoms, the protoplast of the cell is surrounded by a cell wall made up of amorphous silica. In order to precipitate this siliceous material, the abundance of silicic acid must be concentrated from surface seawater values (typically $<10\ \mu M$ in low and mid latitudes and $<70\ \mu M$ at high latitudes) up to at least saturation values (1,000–2,000 μM, depending primarily on temperature). This concentration of silicate occurs in the silica-deposition vesicle (SDV), where sodium-dependent silicon transporter proteins are used to build silicon concentrations to levels as high as $10^3-10^5\ \mu M$ (Kroger and Sumper, 2000). As concentration in the SDV increases, the monosilicic acid is transformed into polysilicic acid, which condenses to form a colloid with particle sizes ~1–10 nm. The colloidal particles are the nuclei for silica precipitation, which occurs via a flocculation process, enhanced by the presence of a low-molecular polypeptide compound called silaffin. The cell wall ultimately is composed of a network of nearly spherical particles with diameters ~10–100 nm (Kroger and Sumper, 2000). Biogenic silica nucleation in the SDV is regulated by controlling the pH, salt concentration, and metal-ion abundance (especially aluminum and iron). The exact mechanism responsible for the precipitation of the ornate skeletal shapes, however, is uncertain. The two most promising hypotheses are: (i) an organic macromolecule matrix is used as a template for silica precipitation; or (ii) cytoplasmic moulding creates voids that are filled with precipitating siliceous material (Wetherbee *et al.*, 2000). An organic coating occurs on the outside of the siliceous cell wall to protect it from dissolution in the highly undersaturated environment of ambient seawater.

7.04.3 THE PHYSICAL PROPERTIES OF BIOGENIC SILICA

The nanostructural nature of biogenic silica commonly is described as globular spheres constructed of colloidal silica. The spheres can be as large as 100 nm in diameter and in many cases the void space between the spheres is filled with silica, giving the surface a smooth appearance (Kroger and Sumper, 2000). When the biogenic silica undergoes dissolution, the silica deposited between the spheres commonly dissolves first, yielding a highly porous structure. Based on gas adsorption measurements, Hurd (1983) suggests that the mean diameter of the pore spaces in modern and ancient biogenic silica is ~5–10 nm with an overall porosity ranging from 35% to 50%. The density of biogenic silica equals $2.0\ g\ cm^{-3}$ for diatoms and sponges, whereas radiolarian densities range from $1.7\ g\ cm^{-3}$ to $2.0\ g\ cm^{-3}$. The water content of biogenic silica varies from 8 wt.% to 17 wt.%, depending on the type of siliceous biota as well as their age (Hurd and Theyer, 1977). Typical surface area values for siliceous biota range from $25\ m^2\ g^{-1}$ to $150\ m^2\ g^{-1}$.

Mineralogically, biogenic silica commonly is called opal-A, with its highly disordered, nearly amorphous structure, exhibiting only a broad, diffuse X-ray peak centered at ~20–26° two theta (Figure 2). Integrating the area under this peak has been utilized as a technique for quantifying biogenic silica abundance (Eisma and van der Gaast, 1971). Biogenic silica also exhibits several infrared absorption peaks including a strong peak at 9.2 μm (common to many siliceous phases) and a moderate peak at 12.55 μm (Kamatani, 1971), which has been used to quantify biogenic silica abundance (Chester and Elderfield, 1968). Scanning electron microscopy (SEM) and transmission electron microscopy (TEM) have proven to be very useful in revealing the internal structure and surface chemistry of various diatoms, radiolaria, and siliceous sponges (e.g., see Hurd *et al.*, 1981).

The solubility of biogenic silica is highly temperature dependent varying from 800–1,000 μM Si at 3 °C to 1,500–1,700 μM Si at 23 °C (Lawson *et al.*, 1978; Kamatani, 1982). Pressure also affects the solubility of biogenic silica (increasing solubility with increasing depth), but the total effect is only ~200 μM (or 20%) over the depth ranges commonly observed in the ocean (Willey, 1974). Combining these temperature and pressure effects for typical low- and mid-latitude waters yields a maximum solubility in surface waters (1,600 μM), a minimum solubility (~1,000 μM) at ~1,000 m depth, increasing to intermediate values (~1,100 μM) at typical deep-ocean depths of 4,000–5,000 m (Hurd, 1983). The solubility of biogenic silica can be affected by the adsorption of cations on its surface (Lewin, 1961), which is why "acid-cleaned" biogenic silica commonly has higher solubilities (and a more rapid rate of dissolution) than those from natural siliceous assemblages of phytoplankton (Lawson *et al.*, 1978).

The solubility of biogenic silica also is affected by its aluminum content. Van Bennekom *et al.* (1991) reported that the Al/Si atomic ratio in siliceous material collected from various marine environments varied from 0.6×10^{-3} to 7×10^{-3}, with a corresponding change in silica solubility from 1,080 μM to 660 μM. The amount of aluminum incorporated into siliceous skeletons during growth appears to be dependent on the Al/Si ratio in the surrounding water as well as the particular species of siliceous biota.

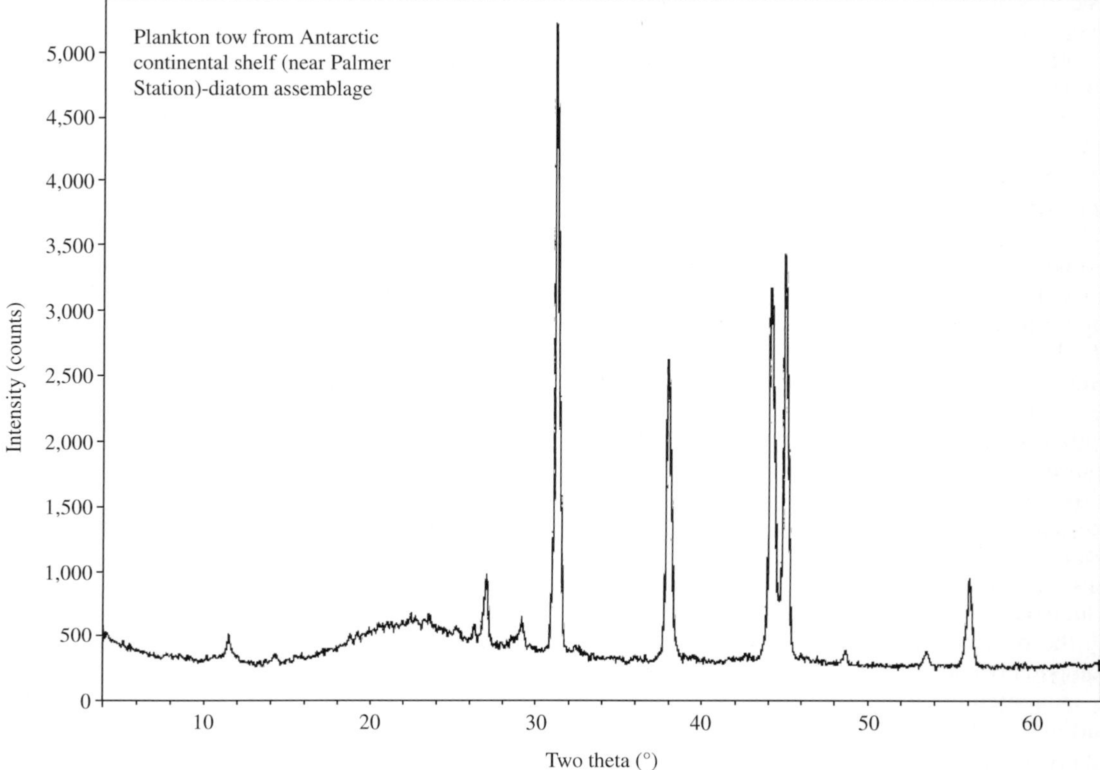

Figure 2 X-ray diffraction pattern of Antarctic diatoms from plankton tow. The amorphous nature of the siliceous material is apparent from the very broad peak, which occurs at an angle of ~20–26° two theta.

Consequently, the solubility of biogenic silica precipitated from coastal waters (where dissolved aluminum concentrations are relatively high) may be somewhat lower than the solubility of similar biota grown in open-ocean waters (where dissolved aluminum concentrations are lower).

7.04.4 CHANGES IN BIOGENIC SILICA CHEMISTRY OCCURRING IN THE WATER COLUMN

Prior to the past several years of research, nearly all of the diagenetic transformations in biogenic silica chemistry were believed to occur after burial in the seabed (Ragueneau *et al.*, 2000). Nelson and Goering (1977) noted that the specific dissolution rate for biogenic silica (expressed in per second) increased from fivefold to sixfold as depth increased throughout the 25–60 m euphotic zone of northwest Africa. However, a likely explanation for the increase in specific dissolution rate with depth is that the phytoplankton are less viable near the base of the euphotic zone and some of the organic matter protecting the siliceous cell walls has been removed by zooplankton or microbial activity (Bidle and Azam, 1999). Grill and Richards (1964) estimate that the removal of the organic membrane protecting the siliceous cell wall is nearly complete within a few days of death of the biota. Temperature changes in the water column could not explain these large changes in the specific dissolution rate. In the Nelson and Goering (1977) study there was no direct evidence that the nature of the biogenic silica chemistry had changed (although additional siliceous surface area may have been exposed). There may, however, be some small differences in the solubility of biogenic silica formed in surface water depending on the biota forming the material and the nature of the biomineralization process (Archer *et al.*, 1993).

Chemical measurements from particle traps have been used to document changes in the solubility of biogenic silica as particles settle through the North Atlantic Ocean water column (Gallinari *et al.*, 2002). Not only were the biogenic silica solubilities of particle-trap materials collected at 1,000 m and 3,000 m water depth measured using flow-through reactors (Van Cappellen and Qiu, 1997a), but also the silicate concentrations of the solutions coexisting with the sediment in the sealed particle-trap cups were used to estimate temporal variations in the biogenic silica solubility. The flow-through reactor data indicated that during periods of high particle flux

(March–June) the biogenic silica solubility of the bulk material was ~900 μM in the 1,000 m trap, whereas it was only 630 μM in the 3,000 m trap. After correcting for a small difference due to temperature, most of the change in solubilities was attributed to "aging" in the water column, probably as a result of surface-silica adsorption of cations or clay mineral formation (Gallinari *et al.*, 2002). In contrast, the solutions in the particle-trap cups showed relatively high values both at 1,000 m (as high as 1,150–1,300 μM Si) and at 3,000 m depth (as high as 1,150 μM Si). The difference in these results was explained by invoking the presence of a high-soluble fraction of siliceous material in the 3,000 m trap in such small quantities that it would affect the particle-trap supernatant solutions, but dissolve prior to or during the flow-through reactor experiments. During low-flux periods, flow-through reactor data indicate that the trap particles at these depths have a solubility of ~500 μM, which the investigators attribute to a longer settling time for the particles in the water column. The Gallinari *et al.* (2002) study is one of the first to be able to document chemical changes in biogenic silica occurring in the water column. The exact mechanism responsible for the diminished solubility is not known. A likely mechanism responsible for the reduction in solubility is adsorption of aluminum on the biogenic silica surface (Iler, 1973); however, the formation of clay minerals cannot be totally excluded (Michalopoulos and Aller, 1995). The changes in silica solubility observed by Gallinari *et al.* (2002) are significantly larger than typically attributed differences in biomineralization processes in the euphotic zone at a single site.

An important factor affecting the chemical changes to biogenic silica in the water column is the amount of settling time (i.e., exposure time to ambient seawater). The primary control on this settling time is the nature of the packaging of siliceous material during its downward descent. If siliceous particles are packaged as fecal pellets (e.g., Schrader, 1971; Smetacek, 1985), settling rates in the open ocean can be several hundred meters per day (corresponding to 15 days of settling to reach 3,000 m depth). In addition, the fecal pellets of many zooplankton (typically 50–250 μm in size) have membranes that provide a barrier for the exchange of dissolution products with ambient seawater, thus further reducing the dissolution of the siliceous material. If siliceous particles settle as marine snow, little or no protective membrane is usually present and the settling velocities typically are less than fecal pellets' (Brzezinski *et al.*, 1997; Alldredge and Gotschalk, 1989). Lastly, if siliceous material settles as discrete particles, the water column residence times (several years) become so long that much of the siliceous material dissolves in the water column (Wollast, 1974). According to the Wollast kinetic model, for example, siliceous particles,10 μm or less in size, totally dissolve before reaching a water depth of 3,000 m. The extent of biogenic silica recycling in the water column is important to recognize in this discussion of silica preservation and transformation. Approximately half of the biogenic silica produced in the euphotic zone redissolves without being exported to depth. Only 10% of the biogenic silica produced in surface waters makes it to the seabed, with silica burial accounting for less than 3% of surface production (Treguer *et al.*, 1995).

Van Cappellen and Qiu (1997a) have shown that the dissolution of biogenic silica in the highly undersaturated water column is inherently different than that in the seabed, where pore waters reach near-saturation levels. In areas of high undersaturation the dissolution rate of biogenic silica increases exponentially with the increasing departure from undersaturation. In contrast, most sedimentary environments exhibit a linear relationship between departure from undersaturation and dissolution rate. The difference in dissolution mechanism is believed to be related to the onset of localized dissolution centered on surface defects (Van Cappellen and Qiu, 1997a).

7.04.5 DIAGENESIS OF BIOGENIC SILICA IN THE UPPER METER OF THE SEABED

7.04.5.1 Modeling Biogenic Silica Distributions and Pore-water Silicate Concentrations

There are numerous models describing the dissolution of biogenic silica in marine sediments. Schink *et al.* (1975) were one of the first quantitative treatments of this subject, but there have been many other articles published since then (e.g., Hurd and Birdwhistell, 1983; Boudreau, 1990; Rabouille and Gaillard, 1990; McManus *et al.*, 1995). The basic modeling approach has been to set up diagenetic equations describing the distribution of "reactive" biogenic silica and dissolved silicate. The equations given below come from the Schink *et al.* (1975) model:

For reactive biogenic silica

$$\frac{dB}{dt} = D_B \frac{\partial^2 B}{\partial z^2} - \frac{kB(C_s - C)}{C_s} = 0 \qquad (1)$$

For dissolved silicate

$$\frac{dC}{dt} = D_s \frac{\partial^2 C}{\partial z^2} - \frac{kB(C_s - C)}{\Phi C_s} = 0 \qquad (2)$$

where D_B is the sediment bioturbation coefficient ($cm^2\ yr^{-1}$), B is the concentration of "reactive" biogenic silica per volume of sediment (mol cm^{-3}), z is depth in the sediment column (cm), k is the dissolution rate constant for biogenic silica (yr^{-1}), C_s is the pore-water asymptotic silicate value observed at depth (mol cm^{-3}), C is the pore-water silicate concentration (mol cm^{-3}), D_s is the molecular diffusion coefficient for silicate in the seabed corrected for tortuosity effects (cm^2 of sediment yr^{-1}), and Φ is the sediment porosity (volume pore-water/volume total sediment). One nonsatisfying aspect of this model is that use of B, the concentration of "reactive" biogenic silica. Essentially, this is the biogenic silica that will dissolve in the sediment column prior to burial at depths where the silicate concentration reaches its asymptotic value. Defining B this way is useful for modeling, but it is impossible to measure B directly at any particular depth in the sediment column. Other assumptions in the Schink *et al.* (1975) model have been summarized by Berner (1980) and they are given as follows.

(i) Reactive-biogenic silica dissolves with first-order kinetics relative to the silicate-concentration deviation from the asymptotic silicate value observed at depth.

(ii) The specific surface area of the biogenic silica does not change with depth.

(iii) Advection of pore water and biogenic silica due to burial is small relative to reaction and diffusive fluxes (model is best suited for deep-sea sediments).

(iv) There is no compaction, bioirrigation, or advective pore-water flow.

(v) Diagenesis is a steady-state process.

(vi) Adsorption of silicate can be ignored because of steady-state conditions and lack of advective processes.

(vii) At great depth the amount of reactive biogenic silica (B) goes to zero.

The analytical solution to this model can be found in Wong and Grosch (1978). Schink and Guinasso (1980) report a more comprehensive model describing silica diagenesis in marine sediments, but it is sufficiently complex that the associated equations can only be solved using advanced numerical techniques. The model assumptions described above have been verified in the field and in the laboratory (using flow-through reactor experiments) making this modeling approach useful in most deep-sea environments.

These diagenetic models have proven effective in understanding the effects of various sedimentary processes on silica dynamics in the seabed (Johnson, 1976; Ragueneau *et al.*, 2000). For example, the effect of bioturbation by benthic fauna on pore-water silicate profiles has been examined (Schinck and Guinasso, 1978). Higher intensities of bioturbation lead to higher asymptotic silicate values at depth because the benthic fauna are mixing down particles from the surface seabed with higher reactivities and higher biogenic silica contents than the sediments at depth (Berner, 1980). Rabouille and Gaillard (1990) examined the silica/silicate steady-state assumptions for a pore-water/sediment system undergoing bioturbation and turbidite deposition. They concluded that following these perturbations the solid phase would return to a steady-state distribution on a timescale of months, whereas the pore-water profile would resume steady-state conditions in a matter of days.

Details concerning the mechanism responsible for the variation in asymptotic silicate values observed in deep-sea sediments have been difficult to establish. Hurd (1973) recognized that the silicate concentrations in marine pore waters reached asymptotic values (200–800 μM) that were significantly less than saturation with respect to amorphous silica (typically ≥1,000 μM at deep-ocean temperatures and pressures) and attributed the differences to the growth of aluminosilicate clay minerals on the surfaces of the biogenic silica. Schink *et al.* (1975) hypothesized that the low-silicate concentrations in marine sediments were the result of a dynamic balance between biogenic silica dissolution and reprecipitation reactions involving silica and probably aluminum. McManus *et al.* (1995) suggest that the dominant process controlling pore-water silicate could be the solubility of the bulk biogenic silica, the solubility of the most-soluble biogenic silica fraction, or surface silica reactions involving aluminum. Flow-through reactor experiments (Van Cappellen, 1996; Van Cappellen and Qiu, 1997a,b; Dixit *et al.*, 2001; Dixit, 2001) indicate that aluminum substitution in the siliceous matrix, aluminum adsorption, and precipitation of aluminum-rich siliceous phases on skeletal surfaces all contribute to the low-asymptotic concentrations of silicate in marine pore waters. As mentioned above, some of this addition of aluminum occurs in the water column, but most of it occurs in the seabed, where the proximity to detrital minerals and higher dissolved aluminum concentrations are greater. Not only is the solubility of biogenic silica decreased with the addition of aluminum to the skeletal structure, but the silica dissolution rate also is diminished (Van Bennekom *et al.*, 1989, 1991, 1997; Van Cappellen and Qiu, 1997b). In general, the Al/Si ratios of skeletal material in the seabed (e.g., $(1-3) \times 10^{-3}$ atomic ratio in Southern Ocean sediments) are many times greater than the Al/Si ratios observed in siliceous biota in the euphotic zone ($(0.1-0.5) \times 10^{-3}$ in the Scotia Sea). The asymptotic silicate concentration observed at depth in marine sediments can be predicted with reasonable certainty from the wt.% detrital/wt.% opal ratio (Van Cappellen and Qiu, 1997a). Based on potentiometric titrations of siliceous material,

Dixit (2001) suggests that the density of reactive-surface silanol groups (which is 4–8 times lower in marine sediments than in fresh diatoms) may be an important factor controlling silica solubility and dissolution kinetics in the marine environment. Greenwood *et al.* (2001) recently reported that Activated Complex Theory may be useful is describing reaction kinetics and the changing solubilities of biogenic silica following burial in the sediment column.

7.04.5.2 Transformation of Biogenic Silica to Authigenic Clay Minerals

In addition to the observation that clay minerals such as sepiolite are forming on siliceous skeletal surfaces in the deep ocean (e.g., Hurd, 1973), evidence has been presented documenting the formation of K–Fe–Mg aluminosilicate clay minerals in the deltaic deposits at the mouth of the Amazon River (Michalopoulos and Aller, 1995; Michalopoulos *et al.*, 2000). In these studies, biogenic silica is found to be dissolving in the seabed releasing silicate, which then is precipitating as newly formed clay minerals. SEM, TEM, and TEM-EDS (energy-dispersive system) have been used to document the chemical transformations occurring in these deposits. Because of the rapid burial of detrital sediment near the mouth of the Amazon River (nearly a billion tons per year), the biogenic silica contents of the Amazon-shelf sediments typically are less than 0.5 wt.% biogenic SiO_2. With these low-silica abundances and the highly reactive lithogenic material discharging from the river, quantifying the amount of authigenic clay mineral precipitation has proven difficult. The authigenic minerals formed on the Amazon shelf are described as potassium-rich, iron-rich mica-type clay minerals (Michalopoulos and Aller, 1995; Michalopoulos *et al.*, 2000). Incubation experiments indicate that these minerals can form on a timescale of months. The precipitation of authigenic clay minerals in these deltaic deposits has been cited as an example of "reverse weathering" reactions in which amorphous aluminosilicates are transformed into cation aluminosilicates (Mackenzie *et al.*, 1981). The precipitation of the authigenic clay minerals may account for as much as 10% of the riverine K^+ and 10–20% of the dissolved silicate coming down this major dispersal system (Michalopoulos and Aller, 1995).

7.04.5.3 Preservation of Biogenic Silica in the Seabed

In addition to the effects of aluminum on biogenic silica chemistry, there are several other factors and processes associated with silica diagenesis in marine sediments that should be considered in the context of this chapter. These include changes in particle surface area during diagenesis as well as the effects of pH variations and sediment accumulation rate. The specific surface area of siliceous material generally decreases between the surface ocean ($25–150\ m^2\ g^{-1}$) and the seafloor and then continues to decrease with depth in the seabed ($<25\ m^2\ g^{-1}$). Much of the decrease in surface area in the water column results from the preferential dissolution of high surface-area biogenic silica (i.e., biota with delicate skeletal structures). This process continues in the seabed, however, surface area also is reduced because of the precipitation of silica and aluminosilica in skeletal pore spaces (Hurd, 1973). Therefore, just the difference in surface area alone can cause siliceous particles in the seabed to dissolve as much as 6 times slower than the siliceous particles collected from the surface ocean (Van Cappellen *et al.*, 2002; Dixit, 2001). Silica-solubility and silica-dissolution rate constants both increase as pH increases between 6 and 9 (Hurd, 1973; Van Cappellen and Qiu, 1997a,b). However, given the dynamic range for pH commonly observed in marine waters and sediments (7.5–8.2), variations in pH produce less than a 5% change in solubility, making this parameter relatively inconsequential. Lastly, sediment accumulation rate can have a substantial effect on silica diagenesis and silica preservation because this parameter controls the time that siliceous particles are exposed to low-saturation levels near the sediment–water interface (Figure 3). For example, DeMaster *et al.* (1996) observed in Ross Sea sediments that seabed biogenic silica preservation efficiencies were near zero when the sediment accumulation rate was $<2\ cm\ kyr^{-1}$, whereas the seabed preservation efficiencies were ~60–80% when the sediment accumulation rate was $\geq 16\ cm\ kyr^{-1}$.

Many of the factors favoring biogenic silica preservation occur in continental margin deposits. Detrital material is abundant (releasing dissolved aluminum in pore waters for solid-phase interactions), the water column is relatively shallow as compared to deep-sea depths (avoiding extensive dissolution during settling), and the sediment accumulation rates generally are an order of magnitude greater than in open-ocean deposits. Until recently, however, marine geochemists believed that nearly all biogenic silica burial in the marine environment occurred in deep-sea deposits, primarily in the siliceous ooze belt surrounding Antarctica (see DeMaster, 1981; Treguer *et al.*, 1995). In these deep-ocean deposits the silica/organic carbon weight ratio commonly reaches several hundred because organic matter is

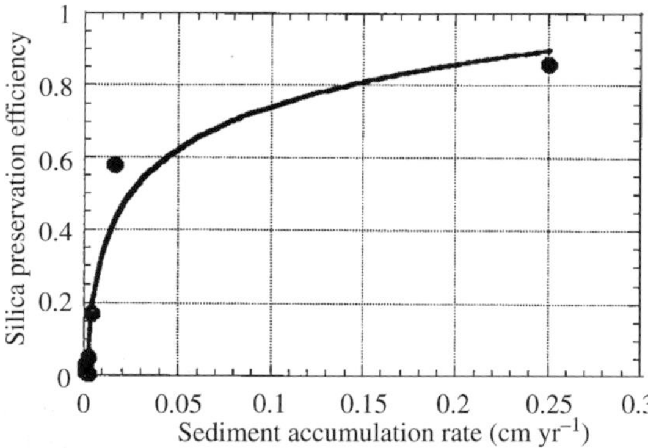

Figure 3 Seabed silica-preservation efficiency plotted as a function of sediment accumulation rate in Antarctic Ross Sea deposits. As sediment accumulation rates increase, the amount of time that siliceous material is exposed to the highly undersaturated bottom waters at the sediment–water interface decreases, which enhances the preservation of biogenic silica in the seabed.

nearly totally removed as the biogenic particles settle through the water column. By incorporating ^{230}Th sediment inventories to correct for lateral transport and focusing over rugged bottom topography, scientists have been able to obtain better estimates for biogenic-silica accumulation rates in the Southern Ocean (see DeMaster (2002) for a review). The new data suggest that previous values for deep-sea, Southern Ocean biogenic silica-accumulation rates were overestimated by as much as 35%. To maintain an approximate balance between silicate supply to the ocean and marine-biogenic silica burial, an additional repository for biogenic silica must exist to compensate for these previous overestimates. Heath (1974) and DeMaster (2002) believe that biogenic-silica accumulation in continental-margin deposits has been underestimated in past marine-silica budgets. Based on the amount of marine-organic matter buried in continental-margin sediments ($(3-4)\times 10^{13}$ g C yr^{-1}; Hedges and Keil, 1995) and typical silica/organic carbon weight ratios in productive continental-margin sediments (~3), as much as half of the world's biogenic silica may be accumulating in continental-margin sediments, where surface-silica production rates are high and the silica-preservation potentials are high (DeMaster, 2001). The biogenic silica content of continental margin sediments is quite low (generally less than 10%) because of the rapid co-accumulation of continentally derived lithogenic material. Burial of organic carbon is enhanced in these continental-margin sediments (silica/organic carbon ratios of ~3 as compared to 100–300 in the deep sea), which means that the new marine-silica budgets have the cycles of biogenic silica and organic carbon much more tightly coupled than the previous oceanic budgets, in which nearly all biogenic-silica accumulation occurred in the deep sea.

7.04.6 SILICA DIAGENESIS ON TIMESCALES OF MILLIONS OF YEARS

7.04.6.1 Formation of Opal-CT and Chert

As biogenic silica is buried beneath meters to kilometers of sediment, exposure to higher temperatures and pressures alters the structure of this biogenic material, transforming it from amorphous opal-A to opal-CT and in some cases to chert. Review articles describing opal-CT and chert formation include: Williams and Crerar (1985), Williams *et al.* (1985), Hesse (1989), and Knauth (1994). The chemical, mineralogical, and structural changes that occur during this process have been documented in *Deep-Sea Drilling Project Studies* (e.g., Hurd and Theyer, 1977; Hurd *et al.*, 1981; Hurd, 1983) as well as in research examining uplifted marine sediments such as the Monterey Formation (e.g., Calvert, 1983) in California. Biogenic silica as precipitated by diatoms, radiolaria, sponges, and silicoflagellates is in the form of opal-A, which is highly disordered and almost completely amorphous. As mentioned earlier the X-ray diffraction pattern for biogenic silica yields only a very broad peak at ~20–26° two theta. As this siliceous material is buried in the seabed and exposed to greater pressure and temperature conditions over a time period of millions of years, the nature of the siliceous test is altered (Figure 4). Research from the *Deep-Sea Drilling Project* indicates that skeletal surface area diminishes from initial values of 50–100 m^2 g^{-1} to values <20 m^2 g^{-1} over a 40 Myr period. In addition, the average pore volume also decreases with increased age, pressure, and temperature from initial values of 0.3–0.5 cm^3 g^{-1} to <0.1 cm^3 g^{-1} (Hurd *et al.*, 1981). With age there is a general trend of decreased solubility, decreased water content, and increased density (Hurd and Theyer, 1977).

The transformation of opal-A to opal-CT generally begins at 35–50 °C, corresponding to burial depths of hundred of meters. In some environments this temperature may be as low as 17–21 °C (Matheney and Knauth, 1993; Monterey Formation) or even 0–4 °C (Botz and Bohrmann, 1991; Antarctic deep sea). The acoustic properties of the sediment are altered during the transformation to opal-CT, typically providing an acoustic reflector of the diagenetic front (Calvert, 1983; Tribble *et al.*, 1992). Opal-CT, also known as porcelanite, exhibits X-ray characteristics of low cristobalite and tridymite (Figure 4). This mineral exists as

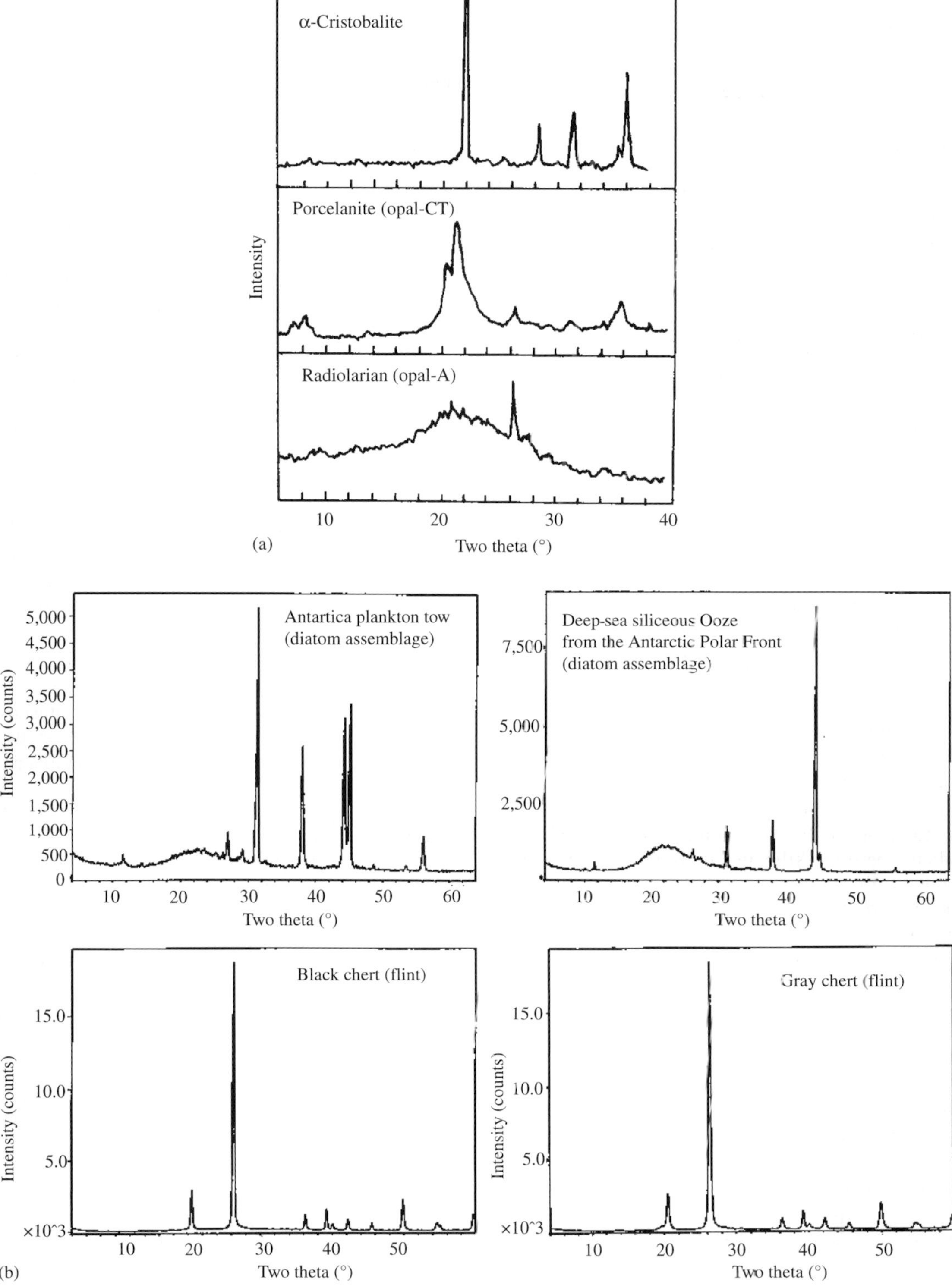

Figure 4 X-ray diffraction patterns contrasting various crystallinities of silica: (a) radiolarian silica, Porcelanite (opal-CT) and α-Cristobalite (made by heating silica gel at 1,350 °C for 4 h) from Calvert (1983); (b) diatom assemblage from Antarctic plankton tow, deep-sea siliceous ooze (Holocene in age) from beneath the Antarctic Polar Front, and two chert deposits from state of New York. The sharpness of the silica peak(s) between 20° and 26° two theta increases as silica undergoes diagenetic transformation from a fresh-diatom assemblage to buried sediment for thousands of years to the eventual formation of chert.

10–15 μm diameter spheres, called lepispheres, which are composed of bladed crystals 30–50 nm thick. Kastner *et al.* (1977) reported that the formation of opal-CT results from dissolution followed by reprecipitation reactions that require a source of magnesium, alkalinity, and hydroxide ion. The presence of aluminosilicate phases commonly retards the formation of opal-CT (Hinman, 1998).

The transformation of opal-CT to chert is a highly temperature-dependent reaction in which a solid-state inversion takes place converting opal-CT to microcrystalline quartz (Calvert, 1983). The quartz crystals in chert typically range in size from 0.1 to 10s of micrometers. The amount of extra-crystalline water in this metamorphosed siliceous phase is very low (<1 wt.%). In the laboratory, synthesized microcrystalline quartz can have elongate euhedral crystals suggesting that the mineral has precipitated from solution (instead of by solid-state inversion). In the Monterey Formation the transformation of opal-CT to chert takes place at a temperature of 80–110 °C. Laboratory experiments suggest that at 200 °C the transformation of opal-CT to microcrystalline quartz occurs on a timescale of decades, whereas at a temperature of 100 °C, the conversion requires only 4×10^4 yr. At 50 °C the transformation occurs in ~4 Myr, and at 20 °C it takes ~200 Myr (Blatt *et al.*, 1980). At extremely high pressures (>300 MPa) or extremely high hydroxide concentrations, opal-A apparently can be converted to microcrystalline quartz (chert) without the opal-CT intermediate phase. If opal-A is heated to 1,100 °C (1 atm pressure), alpha cristobalite (with its distinct X-ray diffraction pattern) is formed, which Goldberg (1958) used to determine the abundance of biogenic silica in marine sediments.

7.04.6.2 Geological Settings for Chert Formation

There are two common stratigraphic occurrences of chert: as bedded cherts associated with shales or iron formations and as nodules in carbonate rocks (Blatt *et al.*, 1980). The bedded cherts are predominant in Precambrian time, reaching a maximum extent 2–3 Ga, when they represented as much as 15% of the sedimentary record. The Precambrian bedded cherts contain microspheres of quartz, suggesting that they may have precipitated inorganically. Commonly, bedded cherts are associated with ophiolite sequences, which may have hydrothermal or metasomatic sources of silicate. The co-occurrence of bedded cherts and shales (typically dark in color) suggests that many cherts form in a hemipelagic or deep-sea, open-ocean setting, far from sources of coarse clastic material. In many of the bedded cherts deposited during post-Precambrian time, there is fossil evidence that some of the silica was provided by alteration of siliceous biota.

Nodular cherts occur primarily during Phanerozoic time. The nodules commonly are tens of centimeters in diameter, embedded in limestone, and elongate in the plane of bedding. The source of the silica for these cherts is believed to be primarily from the solution of siliceous biota (Blatt *et al.*, 1980), as indicated by trace-fossil evidence. The internal structure of the nodules is consistent with silica precipitation as an amorphous silica gel (albeit from the solution of siliceous-skeletal material). The depositional environments for these sediments can be shelf and/or near-shore basins associated with carbonate reefs. In addition, modern sediments north of the Southern Ocean Polar Front contain high abundances of carbonate and biogenic silica (without a high abundance of lithogenic sediment), suggesting the potential for a high-latitude environment of chert formation (Bohrmann *et al.*, 1994).

ACKNOWLEDGMENTS

The X-ray patterns for the various materials kindly were run by Dr. Skip Stoddard of the Department of Marine, Earth, and Atmospheric Sciences, North Carolina State University.

REFERENCES

Alldredge A. L. and Gotschalk C. C. (1989) Direct observations of the mass flocculation of diatom blooms: characteristics, settling velocities and formation of diatom aggregates. *Deep-Sea Res.* **36**, 159–171.

Archer D., Lyle M., Rodgers K., and Froelich P. (1993) What controls opal preservation in tropical deep-sea sediments? *Paleoceanography* **8**, 7–21.

Berner R. A. (1980) *Early Diagenesis: A Theoretical Approach.* Princeton University Press, Princeton, 241pp.

Bidle K. D. and Azam F. (1999) Accelerated dissolution of diatom silica by marine bacterial assemblages. *Nature* **397**, 508–512.

Blatt H., Middleton G., and Murray R. (1980) *Origin of Sedimentary Rocks.* Prentice-Hall, Englewood Cliffs, 782pp.

Bohrmann G., Abelmann A., Gersonde R., Hubberten H., and Kuhn G. (1994) Pure siliceous ooze, a diagenetic environment for early chert formation. *Geology* **22**, 207–210.

Botz R. and Bohrmann G. (1991) Low-temperature opal-CT precipitation in Antarctic deep-sea sediments—evidence from oxygen isotopes. *Earth Planet. Sci. Lett.* **107**, 612–617.

Boudreau B. P. (1990) Asymptotic forms and solutions of the model for silica-opal diagenesis in bioturbated sediments. *J. Geophys. Res.* **95**, 7367–7379.

Brzezinski M. A., Alldredge A. L., and O'Bryan L. M. (1997) Silica cycling within marine snow. *Limnol. Oceanogr.* **42**, 1706–1713.

Buesseler K. O. (1998) The decoupling of production and particulate export in the surface ocean. *Global Biogeochem. Cycles* **12**, 297–310.

Calvert S. E. (1983) Sedimentary geochemistry of silicon. In *Silicon Geochemistry and Biogeochemistry* (ed. S. R. Aston). Academic Press, London, pp. 143–186.

Chester R. and Elderfield H. (1968) The infrared determination of opal in siliceous deep-sea sediments. *Geochim. Cosmochim. Acta* **32**, 1128–1140.

DeMaster D. J. (1981) The supply and removal of silica from the marine environment. *Geochim. Cosmochim. Acta* **45**, 1715–1732.

DeMaster D. J. (2001) Marine silica cycle. In *Encyclopedia of Ocean Sciences* (eds. J. H. Steele, S. A. Thorpe, and K. K. Turekian). Academic Press, San Diego, pp. 1659–1667.

DeMaster D. J. (2002) The accumulation and cycling of biogenic silica in the Southern Ocean: revisiting the marine silica cycle. *Deep-Sea Res. II* **49**, 3155–3167.

DeMaster D. J., Ragueneau O., and Nittrouer C. A. (1996) Preservation efficiencies and accumulation rates for biogenic silica and organic C, N, and P in high-latitude sediments: the Ross Sea. *J. Geophys. Res.* **101**, 18501–18518.

Dixit S. (2001) Dissolution of biogenic silica: solubility, reactivity and the role of aluminum. PhD Dissertation, Georgia Institute of Technology, Atlanta, 197pp.

Dixit S., Van Cappellen P., and Van Bennekom A. J. (2001) Processes controlling solubility of biogenic silica and pore water build-up of silicic acid in marine sediments. *Mar. Chem.* **73**, 333–352.

Eisma D. and van der Gaast S. J. (1971) Determination of opal in marine sediments by X-ray diffraction. *Neth. J. Sea Res.* **5**, 382–389.

Gallinari M., Ragueneau O., Corrin L., DeMaster D., and Treguer P. (2002) The importance of water column processes on the dissolution properties of biogenic silica in deep-sea sediments: I. Solubility. *Geochim. Cosmochim. Acta* **66**, 2701–2717.

Goldberg E. D. (1958) Determination of opal in marine sediments. *J. Mar. Res.* **17**, 178–182.

Greenwood J. E., Truesdale V. W., and Rendell A. R. (2001) Biogenic silica dissolution in seawater *in vitro* chemical kinetics. *Prog. Oceanogr.* **48**, 1–23.

Grill E. V. and Richards F. A. (1964) Nutrient regeneration from phytoplankton decomposing in seawater. *J. Mar. Res.* **22**, 51–56.

Heath G. R. (1974) Dissolved silica in deep-sea sediments. In *Studies in Pale-Oceanography* (ed. W. W. Hay). Soc. Econ. Paleont. Min., Tulsa, OK vol. 20, pp. 77–93.

Hedges J. I. and Keil R. G. (1995) Sedimentary organic matter preservation: an assessment and speculative synthesis. *Mar. Chem.* **49**, 81–115.

Hesse R. (1989) Silica diagenesis: origin of inorganic and replacement cherts. *Earth Sci. Rev.* **26**, 253–284.

Hildebrand M. (2000) Silicic acid transport and its control during cell wall silicification in diatoms. In *Biomineralization* (ed. E. Baeuerlein). Wiley, Weinheim, Germany, pp. 171–188.

Hinman N. W. (1998) Sequences of silica phase transitions: effects of Na, Mg, K, Al, and Fe ions. *Mar. Geol.* **147**, 13–24.

Hurd D. C. (1973) Interactions of biogenic opal, sediment and seawater in the central equatorial Pacific. *Geochim. Cosmochim. Acta* **37**, 2257–2282.

Hurd D. C. (1983) Physical and chemical properties of siliceous skeletons. In *Silicon Geochemistry and Biogeochemistry* (ed. S. R. Aston). Academic Press, London, pp. 187–244.

Hurd D. C. and Birdwhistell S. (1983) On producing a more general model for biogenic silica dissolution. *Am. J. Sci.* **283**, 1–28.

Hurd D. C. and Theyer F. (1977) Changes in the physical and chemical properties of biogenic silica from the central equatorial Pacific: Part II. Refractive index, density, and water content of acid-cleaned samples. *Am. J. Sci.* **277**, 1168–1202.

Hurd D. C., Pankratz H. S., Asper V., Fugate J., and Morrow H. (1981) Changes in the physical and chemical properties of biogenic silica from the central equatorial Pacific: Part III. Specific pore volume, mean pore size, and skeletal ultrastructure of acid-cleaned samples. *Am. J. Sci.* **281**, 833–895.

Iler R. K. (1973) Effect of adsorbed alumina on the solubility of amorphous silica in the water. *J. Colloid Interface Sci.* **43**, 399–408.

Johnson T. C. (1976) Controls on the preservation of biogenic opal in sediments of the eastern tropical Pacific. *Science* **192**, 887–890.

Kamatani A. (1971) Physical and chemical characteristics of biogenous silica. *Mar. Biol.* **8**, 89–95.

Kamatani A. (1982) Dissolution rates of silica from diatoms decomposing at various temperatures. *Mar. Biol.* **68**, 91–96.

Knauth L. P. (1994) Petrogenesis of chert. *Rev. Min.* **29**, 233–258.

Kastner M., Keene J. B., and Gieskes J. M. (1977) Diagenesis of siliceous oozes: 1. Chemical controls on the rate of opal-A to opal-CT transformation—experimental study. *Geochim. Cosmochim. Acta* **41**, 1041–1058.

Kroger N. and Sumper M. (2000) The biogeochemistry of silica formation in diatoms. In *Biomineralization* (ed. E. Baeuerlein). Wiley, Weinheim, Germany, pp. 151–170.

Lawson D. S., Hurd D. C., and Pankratz H. S. (1978) Silica dissolution rates of decomposing phytoplankton assemblages at various temperatures. *Am. J. Sci.* **278**, 1373–1393.

Lewin J. C. (1961) The dissolution of silica from diatom walls. *Geochim. Cosmochim. Acta* **21**, 182–198.

Mackenzie F. T., Ristvet B. L., Thorstenson D. C., Lerman A., and Leeper R. H. (1981) Reverse weathering and chemical mass balance in a coastal environment. In *River Inputs to Ocean Systems* (eds. J. M. Marten, J. D. Burton, and D. Eisma). UNEP and UNESCO, Switzerland, pp. 152–187.

Matheney R. K. and Knauth L. P. (1993) New isotopic temperature estimates for early silica diagenesis in bedded cherts. *Geology* **21**, 519–522.

McManus J., Berelson W. M., Hammond D. E., Kilgore T. E., DeMaster D. J., Ragueneau O., and Collier R. (1995) Early diagenesis of biogenic silica: dissolution rates, kinetics, and paleoceanographic implications. *Deep-Sea Res. II* **42**, 871–903.

Michalopoulos P. and Aller R. C. (1995) Rapid clay mineral formation in Amazon delta sediments: reverse weathering and oceanic elemental cycles. *Science* **270**, 614–617.

Michalopoulos P., Aller R. C., and Reeder R. J. (2000) Conversion of diatoms to clays during early diagenesis in tropical, continental shelf muds. *Geology* **28**, 1095–1098.

Nelson D. A. and Goering J. J. (1977) Near-surface silica dissolution in the upwelling region off northwest Africa. *Deep-Sea Res.* **24**, 65–73.

Rabouille C. and Gaillard J. F. (1990) The validity of steady-state flux calculations in early diagenesis: a computer simulation of deep-sea silica diagenesis. *Deep-Sea Res.* **37**, 625–646.

Ragueneau O., Treguer P., Leynaert A., Anderson R., Brzezinski M., DeMaster D., Dugdale R., Dymond J., Fischer G., Francois R., Heinze C., Maier-Reimer E., Martin-Jezequel V., Nelson D., and Queguiner B. (2000) A review of the Si cycle in the modern ocean: recent progress and missing gaps in the application of biogenic opal as a paleoproductivity proxy. *Global Planet. Change* **26**, 317–365.

Schink D. R. and Guinasso N. L., Jr. (1978) Redistribution of dissolved and adsorbed materials in abyssal marine sediments undergoing biological stirring. *Am. J. Sci.* **278**, 687–702.

Schink D. R. and Guinasso N. L., Jr. (1980) Processes affecting silica at the abyssal sediment-water interface. In *Biocochimie de la Matiere Organique a l'Interface Eau-sediment Marin.* Colloq. Int. du CNRS, vol. 293, pp. 81–92.

Schink D. R., Guinasso N. L., Jr., and Fanning K. A. (1975) Processes affecting the concentration of silica at the sediment-water interface of the Atlantic Ocean. *J. Geophys. Res.* **80**, 3013–3031.

Schrader H. J. (1971) Fecel pellets—role in sedimentation of pelagic diatoms. *Sciences* **174**, 55–58.

Siever R. (1992) The silica cycle in the Precambrian. *Geochim. Cosmochim. Acta* **56**, 3265–3272.

Smetacek V. S. (1985) Role of sinking in diatom life-history cycles: ecological, evolutionary and geological significance. *Mar. Biol.* **84**, 239–251.

Treguer P., Nelson D. M., Van Bennekom A. J., DeMaster D. J., Laynaert A., and Queguiner B. (1995) The silica balance in the world ocean: a re-estimate. *Science* **268**, 375–379.

Treppke U. F., Lange C. B., and Wefer G. (1996) Vertical fluxes of diatoms and silicoflagellates in the eastern equatorial Atlantic, and their contribution to the sedimentary record. *Mar. Micropaleontol.* **28**, 73–96.

Tribble J. S., Mackenzie F. T., Urmos J., O'Brien D. K., and Manghnani M. H. (1992) Effects of biogenic silica on acoustic and physical properties of clay-rich marine sediments. *Am. Assoc. Petrol. Geol. Bull.* **76**, 792–804.

Van Bennekom A. J., Jansen J. H. F., Van der Gaast S. J., Van Ieperen M. M., and Pieters J. (1989) Aluminum-rich opal: an intermediate in the preservation of biogenic silica in the Zaire (Congo) deep-sea fan. *Deep-Sea Res.* **36**, 173–190.

Van Bennekom A. J., Buma A. G. J., and Nolting R. F. (1991) Dissolved aluminum in the Weddell-Scotia confluence and effect of Al on the dissolution kinetics of biogenic silica. *Mar. Chem.* **35**, 423–434.

Van Beusekom J. E. E., Van Bennekom A. J., Treguer P., and Morvan J. (1997) Aluminum and silicic acid in water and sediments of the Enderby and Crozet basins. *Deep-Sea Res. II* **44**, 987–1003.

Van Cappellen P. (1996) Reactive surface area control of the dissolution kinetics of biogenic silica in deep-sea sediments. *Chem. Geol.* **132**, 125–130.

Van Cappellen P. and Qiu L. (1997a) Biogenic silica dissolution in sediments of the Southern Ocean: I. Solubility. *Deep-Sea Res. II* **44**, 1109–1128.

Van Cappellen P. and Qiu L. (1997b) Biogenic silica dissolution in sediments of the Southern Ocean: II. Kinetics. *Deep-Sea Res. II* **44**, 1129–1149.

Van Cappellen P., Dixit S., and Van Beusekom J. E. E. (2002) Biogenic silica dissolution in the oceans: reconciling experimental and field-based dissolution rate. *Global Biogeochem. Cycles* **16**(25), 1–10.

Wetherbee R., Crawford S., and Mulvaney P. (2000) The nanostructure and development of diatom biosilica. In *Biomineralization* (ed. E. Baeuerlein). Wiley, Weinheim, Germany, pp. 189–206.

Willey J. D. (1974) The effect of pressure on the solubility of amorphous silica in seawater at 0 °C. *Mar. Chem.* **2**, 239–250.

Williams L. A. and Crerar D. A. (1985) Silica diagenesis: II. General mechanisms. *J. Sedim. Petrol.* **55**, 312–321.

Williams L. A., Parks G. A., and Crerar D. A. (1985) Silica diagenesis: I. Solubility controls. *J. Sedim. Petrol.* **55**, 301–311.

Wollast R. (1974) The silica problem. In *The Sea* (ed. E. D. Goldberg). Wiley, New York, vol. 5, pp. 359–392.

Wong G. T. F. and Grosch C. E. (1978) A mathematical model for the distribution of dissolved silicon in interstitial waters—an analytical approach. *J. Mar. Res.* **36**, 735–750.

7.05
Formation and Geochemistry of Precambrian Cherts

E. C. Perry, Jr. and L. Lefticariu
Northern Illinois University, DeKalb, IL, USA

7.05.1 INTRODUCTION

Chert is a chemically precipitated sedimentary rock chiefly composed of microcrystalline quartz, chalcedonic quartz, and minor megaquartz. Precambrian chert differs from Phanerozoic chert most importantly in its mode of formation. There is no evidence that silica-secreting organisms were present in the Precambrian in sufficient abundance to have had a significant influence on the silica cycle, although some Neoproterozoic protists had scales that were probably siliceous, and Ediacaran sponges produced siliceous spicules. This contrasts with the Phanerozoic, in which the appearance of radiolaria and diatoms has changed the locus of silica precipitation from peritidal and shallow shelf deposition to the deep ocean. Silica-secreting organisms also control the silica concentration of much of the ocean at values of 1 ppm or less, in marked contrast to estimated concentrations of ~60 ppm or more in the Precambrian. A major part of the interest in Precambrian chert arises from the exquisite preservation of assemblages of microfossils that they sometimes contain. Well-preserved 1,900-Myr-old microfauna from chert have been abundantly documented, and forms that putatively are the remains of microbes have been reported in chert as old as 3,400 Myr. Cherts (and carbonates) precipitated from surface water have the potential to preserve a record of the oxygen isotopic composition of contemporary sea- or lake water. The problem is complicated by the possibility of diagenetic and/or metamorphic isotopic exchange and by the fact that the isotopic composition of unaltered chert is a function of both the original isotopic composition of the precipitating water and temperature. Evidence is presented that the maximum $\delta^{18}O$ of +24.7‰ for chert in the 1,900 Myr Gunflint iron formation and +24.0‰ for chert in the ~2,450–2,500 Myr Kuruman and Hamersley iron formations closely approximates the original sedimentary value. Here, $\delta^{18}O$ is the deviation in parts per thousand of the ratio $^{18}O/^{16}O$ of the material analyzed with respect to the international standard VSMOW (Coplen, 1996). If the isotopic composition of seawater was controlled at near zero per mil, Earth-surface temperature could have been as high as 64 °C at that time. There is less confidence in data obtained from more strongly metamorphosed rocks of the Archean. $\delta^{18}O$ values of +22‰ to +20‰ (VSMOW) are registered for rocks that are 3,400–3,700 Myr old. If these are the primary values inherited from the time of chert precipitation, they could record temperatures as high as 90 °C. In fact, if these data record actual Earth-surface temperatures, the Early Archean surface environment may have approximated that of a hydrothermal system. Late Archean Weld Range (Australia) cherty iron formation has an exceptionally low $\delta^{18}O_{SMOW}$ value of +12‰ and may be a lacustrine deposit.

7.05.1.1 Definition and Importance

Chert is defined here, following Folk (1980, p. 79), as "a chemically precipitated sedimentary rock, essentially monomineralic and composed chiefly of microcrystalline and/or chalcedonic quartz, with subordinate megaquartz and minor amounts of impurities." Chalcedony, which consists of sheath-like bundles of thin, radiating fibers of SiO_2, is rare in most Precambrian chert. Precambrian chert occurs as distinct beds or lenses, particularly in Archean rocks, as nodules or silicified laminae in carbonate rocks, or as a siliceous end-member, as granules or cements in iron formation.

The importance attached to Precambrian chert derives primarily from the fact that it often forms near the Earth's surface either during early diagenesis of a chemically precipitated precursor or as a primary precipitate. Because of its genetic relation to surface water bodies such as the ocean and/or shallow groundwater, chert can retain a chemical/isotopic signature of surface processes. Because of its origin, its manner of formation, and its fine grain size, some chert has preserved the morphological outlines and some of the chemical properties of microorganisms that are our most direct evidence of life forms as old as 2,000 Myr (Barghoorn and Tyler, 1965; House *et al.*, 2000) and perhaps much older (Awramik *et al.*, 1983; Schopf, 1993, and references therein; and Schopf *et al.*, 2002; cf. critics: Buick, 1990; Brasier *et al.*, 2002).

An oxygen isotope record of past Earth-surface conditions is partially preserved in cherts (and carbonate rocks). However, isotopic fractionation of oxygen between quartz and water is a function of temperature. Hence, either the surface temperature or the isotopic composition of the precipitating fluid (such as seawater or formation water) must be known to determine the other. The following semi-empirical equation proposed by Knauth and Epstein (1976) relates the temperature to the isotopic composition of chert and water:

$$1{,}000 \ln \alpha_{\text{chert-water}} = 3.09 \times 10^6/T^2 - 3.29 \approx (\delta_{\text{chert}} - \delta_{\text{water}}) \qquad (1)$$

Furthermore, the oxygen isotopic composition of a given chert depends not only on the oxygen isotopic composition of the initial silica precipitate, but may also depend on the temperature, water isotopic composition, and water–rock ratio attained during intermediate diagenetic steps as was demonstrated by Jones and Knauth (1979) and Matheney and Knauth (1993). Thus, understanding the diagenetic pathways of chert is essential when using chert isotopic composition to interpret paleo-environmental conditions.

The problem of distinguishing between temperature of formation and the isotopic composition of water in Equation (1) is not easily resolved. Perry (1967) and Perry and co-workers (1972, 1983 and references therein) argued that the low $^{18}O/^{16}O$ ratios in Precambrian chemically precipitated minerals of marine sedimentary rocks probably resulted primarily from a lower value of this ratio in contemporary seawater. However, Muehlenbachs and Clayton (1976) presented strong arguments that the oxygen isotopic composition of the modern ocean is closely controlled by crust–mantle–hydrosphere interaction related to plate tectonics, and Holmden and Muehlenbachs (1993) asserted that this buffering process has existed at least as far back in time as the Proterozoic. Nevertheless, even the significance of the isotopic composition of oxygen in Phanerozoic and Neoproterozoic marine precipitates (chert and carbonate rocks) remains to be resolved (Veizer *et al.*, 1999; Jacobsen and Kaufman, 1999; Muehlenbachs *et al.*, 2003). Interpretation of the $\delta^{18}O$ data in the Paleoproterozoic and the Archean rocks is even more daunting.

7.05.1.2 Fundamental Difference between Precambrian and Phanerozoic Chert

The diatoms, radiolaria, and sponges that control the silica concentration of seawater at a very low level (<1 ppm at the surface to 10–15 ppm in some bottom water) are relatively recent evolutionary newcomers. Siliceous sponges and protists with what were probably siliceous scales appeared in the Neoproterozoic (Gehling and Rigby, 1996; Brasier *et al.*, 1997;

Porter and Knoll, 2000); silica-secreting radiolaria and diatoms first appeared in the Phanerozoic. There is no direct evidence that any organisms precipitated silica in the Precambrian in sufficient quantity to affect the silica cycle (Maliva *et al.*, 1989; Porter and Knoll, 2000), although organic matter and possibly bacteria (Birnbaum and Wireman, 1985; Siever, 1992; Knauth, 1994; Simonson and Hassler, 1996; Cady and Farmer, 1996) may have played a role in the nucleation and/or precipitation of the extensive chert-bearing iron formations and other cherty rocks of the Archean and Paleoproterozoic eons. The modern surface environment, therefore, offers no satisfactory analogues for the formation of Precambrian marine chert, and apparent similarities between Phanerozoic cherts, such as the Arkansas novaculite or the Monterey chert, and the major chert occurrences of the Precambrian are very superficial and offer no reliable models for Precambrian silica concentration, precipitation, or diagenesis. Chert represents an extreme example of the nonapplicability of uniformitarian concepts to the study of the Precambrian and of the importance of biological processes in regulating geologic systems that superficially appear to be purely inorganic.

The Phanerozoic organisms that maintain the concentration of silica in seawater at such low levels also control the distribution of siliceous sediments. Since diatoms became major participants in the silica cycle, these sediments have dominantly been deep-sea accumulations of relatively soluble opal-A. This phase is converted diagenetically over a period of thousands (Bohrmann *et al.*, 1994) to tens of millions of years to opal-CT (solubility 25–60 ppm; Kastner *et al.*, 1977; Siever, 1992) and later to crystalline chert. Factors responsible for the rate of this conversion are not well understood, but they include temperature during diagenesis and the presence or absence of clay minerals. There is strong evidence that these transformations occur primarily by solution–reprecipitation reactions (Knauth, 1994). If the conditions of diagenesis are such that the concentration of silica remains below the solubility of opal-CT, then chert can apparently form directly from opal-A. This is the mechanism favored by Knauth (1994) for the formation of the nodular cherts that are common in Phanerozoic limestones.

7.05.1.3 Aspects of Inorganic Geochemistry of Silica

In the absence of silica-secreting organisms, it is not clear what inorganic mechanisms buffered or otherwise regulated the concentration of silica in Precambrian seawater. Temperature and pH (Figure 1) are factors that control silica solubility. Equally important are exchange reactions with silicate minerals. A concentration of 60 ppm silica in Precambrian seawater was suggested by Siever (1992), and this figure is widely accepted (Simonson and Hassler, 1996). Values up to 120 ppm have been considered (Morris, 1993).

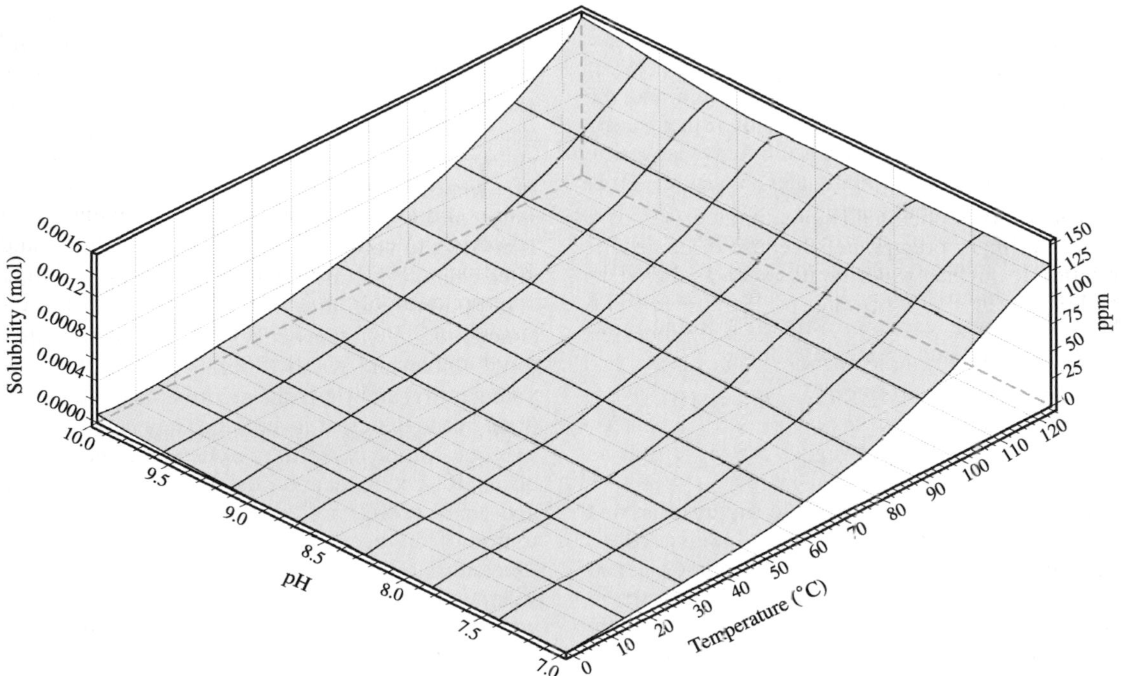

Figure 1 Silica solubility versus temperature and pH (source Fleming and Crerar, 1982).

A relatively constant major ion chemistry of the Precambrian ocean was postulated by Holland (1984), Walker (1977), and Walker *et al.* (1983) but has been challenged by Kempe and Degens (1985) and Grotzinger (1989). The chemistry of Late Neoproterozoic and Phanerozoic primary fluid inclusions in marine halite documents temporal variations in the Mg^{2+}, Ca^{2+}, and SO_4^{2-} concentration, suggestive of long-term cycles in seawater chemistry (Horita *et al.*, 2002). Grotzinger and Kasting (1993), reviewed temporal changes in the nature of chemical sediments during the Precambrian and the possible relation of these changes to seawater composition. Changes in ocean chemistry are to be expected during a time interval of almost 4,000 Myr that included long, quiescent intervals, glacial events, a probable decrease in the partial pressure of CO_2 (Grotzinger and Kasting, 1993), and an increase in the partial pressure of oxygen in the atmosphere (Berkner and Marshall, 1965; Cloud, 1972; Rasmussen *et al.*, 1999; Anbar and Knoll, 2002; cf. Ohmoto, 1999; Beukes *et al.*, 2002), changes in the abundance and importance of other atmospheric gases such as hydrogen (Holland, 2002) and methane (Pavlov *et al.*, 2003), intervals of intense volcanism and/or meteorite bombardment, a progressive increase in solar luminosity (Gough, 1981; Newman and Rood, 1977) and possible changes in the crust–mantle–hydrosphere interactions (Zegers and van Keken, 2001; cf. Komiya *et al.*, 1999). Detailed exploration of these phenomena is well beyond the scope of this review (see the review by Holland, Chapter 6.21).

General information on the geochemistry of silicon can be found in Iler (1979), Astor (1983), and Carozzi (1993). Knauth (1992, 1994) has reviewed chert geochemistry and the uses of chert data. It has been shown that quartz can precipitate directly from seawater (Mackenzie and Gees, 1971) or groundwater (Wray, 1999) at Earth-surface temperatures.

The solubility of quartz is insensitive to pH in acid and intermediate water but quite sensitive to pH at values above 9 (Figure 1) according to the following reactions (Iler, 1979; Robie *et al.*, 1978; Fleming and Crerar, 1982):

$$H_2O + SiO_2 = H_4SiO_4, \quad K_{25} = 10^{-3.95}$$

$$H_4SiO_4 = H^+ + H_3SiO_4^-, \quad K_{25} = 10^{-9.8}$$

Inorganic pH controls include hydrolysis and oxidation of minerals and are sensitive to atmospheric partial pressures of both CO_2 and oxygen. Higher partial pressure of CO_2 results in increased acidity, which can be neutralized by hydrolysis. Oxidation produces acidity, whereas hydrolysis reactions increase pH or alkalinity and release silicic acid.

Kempe and Degens (1985) presented arguments in support of an early soda ocean, with pH values high enough to produce significantly greater silica concentrations. Grotzinger and Kasting (1993) countered that seawater pH is unlikely ever to have exceeded 9. Critical assumptions in their argument are: (i) a significantly greater partial pressure of atmospheric CO_2 to counter low solar luminosity and keep Earth's surface from freezing; (ii) limited total carbon in the hydrosphere–atmosphere system; and (iii) limited seawater concentration of sulfate in the Archean ocean as evidenced by absence of gypsum in evaporite deposits.

Modern abiogenic precipitation of chert or chert precursors is a relatively rare phenomenon, but it has been documented. Wray (1999) reported speleothems of chert in sandstone, showing that direct precipitation of quartz at low temperature is possible, even from water in equilibrium with quartz and hence having a low silica concentration.

7.05.1.4 Association of Chert and Evaporite

Eugster and Jones (1968) observed the association of chert with the evaporite mineral magadiite ($NaSi_7O_{13}(OH)_3 \cdot 4H_2O$) at Lake Magadi, Kenya and postulated that this mineral was a possible chert-forming precursor in the Precambrian. Schubel and Simonson (1990) evaluated this as a mechanism for Precambrian chert precipitation and were not convinced of its viability. Shallow water features such as shrinkage cracks and association with stromatolites occur in Neoproterozoic and Paleoproterozoic cherts (Simonson, 1987; Eriksson and Warren, 1983), but the association of these cherts with evaporite mineral pseudomorphs is relatively rare, and there is no general agreement regarding the necessity of evaporative concentration as a mechanism for Precambrian chert precipitation.

According to Grotzinger and Kasting (1993), it is difficult to verify most reports of pseudomorphs after calcium sulfate mineral in rocks older than ~1,900 Myr and they have argued that sulfate concentration of the ocean was low before that time. Anbar and Knoll (2002) proposed that the younger Precambrian rock record is also indicative of an ocean deficient in sulfate. They attribute this to sequestration of sulfur in a deep H_2S layer in the ocean. If Grotzinger and Kasting (1993) and Anbar and Knoll (2002) are correct that sulfate concentration of the ocean was low during much of the Precambrian, it would not be surprising that even sediments formed in an evaporitic environment would rarely contain sulfate evaporite minerals. This would lessen the importance of gypsum/anhydrite pseudomorphs as a criterion for an evaporitic environment of SiO_2 precipitation.

The time when verifiable pseudomorphs of $CaSO_4$ precipitates appear in marine sediments has assumed importance in the interpretation of paleo-seawater composition. A simple technique exists to verify the origin of at least some of these pseudomorphs. Hemzacek and co-workers (1982, 1986) and Feng (1986) were able to extract enough $CaSO_4$ for isotopic analysis from chert pseudomorphs of well-identified gypsum crystals in the 2,300 Myr Kona Dolomite and other rocks by grinding carefully cleaned chert and soaking the powder in deionized water. The technique is not reliable if oxidizable pyrite is present.

The granular Paleoproterozoic iron formations of the Lake Superior region are undoubtedly of shallow-water origin and hence formed in an environment that may have been at least mildly evaporitic. However, the Paleoproterozoic, evenly banded Hamersley and Kuruman iron formations obviously formed below wave base. This does not exclude the possibility of evaporative concentration in huge restricted basins or on continental shelves with deep barrier bars as proposed by Trendall and Blockley (1970). Diagenetically formed riebeckite is abundant in both the Hamersley and Kuruman iron formations as well as in the Paleoproterozoic Ukrainian iron formation of Krivoy Rog. The prevalence of riebeckite in these relatively deep-water Paleoproterozoic iron formations is not compelling evidence that evaporative concentration played some part in deposition of these rocks, but it is consistent with such a mechanism. However, rare earth and other data mentioned below do not support a restricted basin model.

7.05.2 NEOPROTEROZOIC AND MESOPROTEROZOIC ENVIRONMENTS OF CHERT FORMATION

Cherts within the ~2,000 m, 700–800-Myr-old Academikerbreen Group, Spitzbergen (Maliva *et al.*, 1989) contain microfossils preserved in cellular detail (Knoll, 1982 and references cited therein), which are useful for defining the depositional/diagenetic environment of Neoproterozoic cherts. "Akademikerbreen cherts occur as ellipsoidal to irregular nodules, individually silicified grains in oolites and flake conglomerates, and individually silicified laminae of … probable microbial mat origin" (Maliva *et al.*, 1989, p. 525). Thus, in marked contrast to younger cherts of Silurian to Cretaceous age, these (and other cherts of similar age; Horodyski and Donaldson (1980, 1983); and other references cited in Maliva *et al.* (1989)) formed in peritidal environments, and at least some chert emplacement occurred very early, near the sediment–water interface (Maliva *et al.*, 1989). Similar peritidal or lacustrine chert sequences of Meso- and Neoproterozoic age, some possibly associated with evaporites, are found in the Shaler Group, Victoria Island, Canada and the Bitter Springs Formation, Australia (Southgate, 1986; Maliva *et al.*, 1989 and references therein).

Mesoproterozoic nodular chert of the Belt Supergroup, Glacier National Park, Montana has formed by diagenetic replacement of limestone (Maliva, 2001), but this chert presents something of a contrast to the (mostly younger) Precambrian cherts described above in that most, though not necessarily all, nodular chert formation and partial replacement of oöids by megaquartz in the Belt cherts occurred relatively late in the diagenetic process: before extensive dolomitization but after compaction and burial to a "depth on the order of tens of meters or greater" (Maliva, 2001, p. 894). The Belt microtextures and paragenesis features are similar to those in Phanerozoic nodular cherts. Maliva (2001, p. 894) therefore reported: "no evidence to suggest that the Belt Sea was supersaturated with respect to quartz or opal-CT to a degree adequate for widespread early diagenetic silica precipitation." The water from which the Belt rocks precipitated may have been lacustrine or seawater of a restricted basin that was not typical of contemporary seawater. The source of silica could have been abiogenic but intraformational (e.g., from desorption, breakdown of complexes, or clay mineral alteration).

Maliva *et al.* (2002) have called attention to the unique transitional features of Neo- and Mesoproterozoic cherts. These are distinguished from Phanerozoic cherts in that they "contain evidence of a peritidal locus of early diagenetic chert formation, which strongly contributed to the preservation of paleontologically important microbiotas" (Maliva *et al.*, 2002). They differ from Paleoproterozoic granular chert of some iron formations (described below) in that the Neo- and Mesoproterozoic cherts typically occur as diagenetic replacements of carbonates in nodules or discontinuous beds, whereas the older granular iron-formation cherts contain quartz that appears to be a direct precipitate.

7.05.3 CHERT OF PALEOPROTEROZOIC IRON FORMATION

Iron formation, the sedimentary rock characterized by the association of chert and megaquartz with silicate, carbonate, and iron oxide minerals, is one of the most distinctive Precambrian sedimentary rocks. These rocks, which reached their peak in abundance during the Paleoproterozoic (Figure 2) (James, 1983; Isley, 1995), consist of both banded iron formations (BIFs) and granular (arenitic) iron formations (GIFs) and

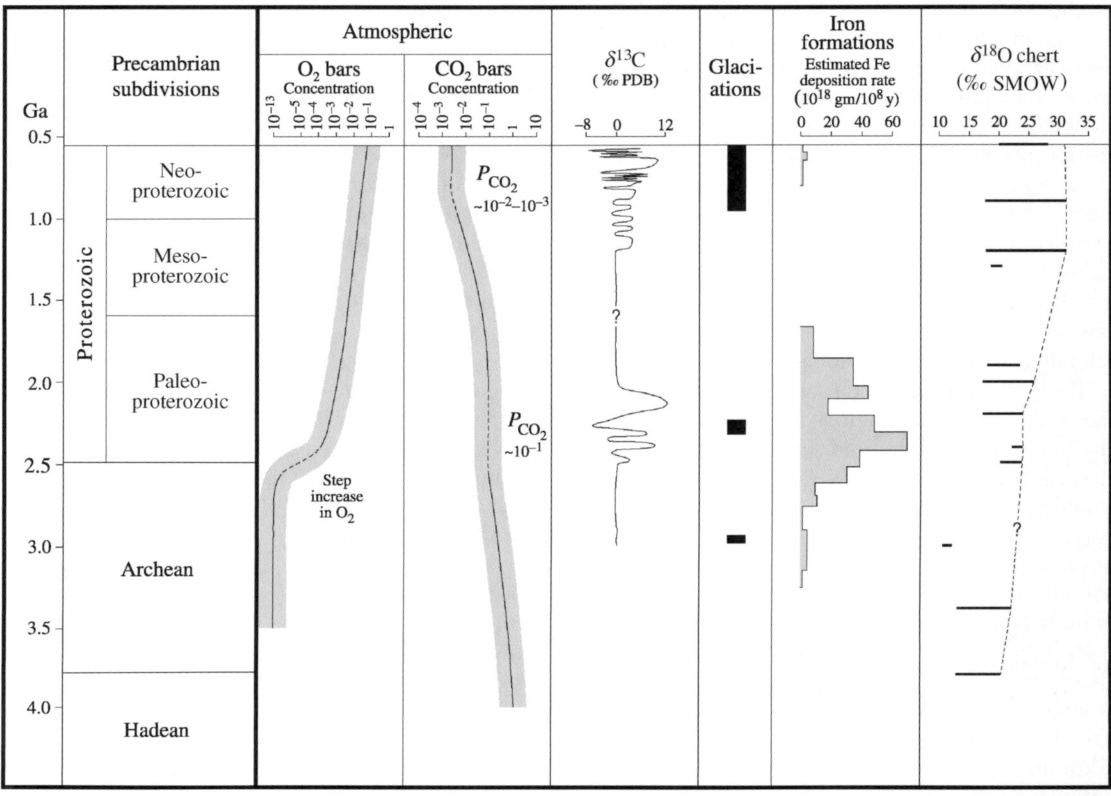

Figure 2 Age relations of: atmospheric P_{O_2} and P_{CO_2} (Kasting, 1987), $\delta^{13}C_{PDB}$ (Anbar and Knoll, 2002), major glaciations (Crowell, 1999), times of major iron formation (James, 1983; Isley, 1995), and $\delta^{18}O_{SMOW}$ of chert (from text).

constitute the world's major industrial source of iron. Although technically the designation "iron formation" applies only to rocks that typically contain 20–35% or more iron (and 40–50% SiO_2) (James, 1983), the iron content of units within rocks labeled iron formations, such as the 1,900-Myr-old Gunflint iron formation of Ontario, Canada, can be very low. Some of the best-preserved Paleoproterozoic iron formations, such as the Biwabik iron formation of Minnesota, contain cherty members that are dominantly shallow-water arenites. Others, like the Dales Gorge Member of the Brockman iron formation of Western Australia, which are generally somewhat older, are deeper-water marine shelf or basinal deposits consisting of: (i) microlaminae (few to tens of micrometers), (ii) ~5 mm scale microbands or "aftvarves," (iii) millimeter to decimeter scale mesobands, and (iv) decimeter to tens of meter scale macrobands, all of which may be continuous for many kilometers (Trendall and Blockley, 1970; Morris, 1993).

Various sources, transport mechanisms, and prediagenetic precursors of iron-formation minerals have been proposed. These are of direct concern here only to the extent that they are related to or can elucidate the formation of their chert component. The newer models of Precambrian iron-formation development assume an atmosphere of negligible or low partial pressure of oxygen that will allow iron to be present and transported in some part of the ocean as Fe^{2+} (Klein and Beukes, 1989; Morris, 1993; Isley, 1995; Anbar and Knoll, 2002). A mechanism is required to oxidize some of this iron to Fe^{3+} so that it will precipitate as a precursor to the hematite and magnetite now associated with chert in oxide facies iron formation. The discovery of Archean molecular fossils of cyanobacteria (Brocks *et al.*, 1999) suggests that these organisms may have oxidized iron by electron transfer before the evolution of free oxygen into the atmosphere. In Paleoproterozoic basins of South Africa and Australia, chert-bearing iron formation and chert-bearing limestone and dolomite precipitated essentially simultaneously in different parts of the water column (Klein and Beukes, 1989; Morris, 1993; Simonson *et al.*, 1993). Most of the newer proposed iron-formation models postulate a Paleoproterozoic ocean vertically stratified into two or more layers (Morris, 1993; Klein and Beukes, 1989; Canfield, 1998; Anbar and Knoll, 2002). Co-precipitation of iron oxide and quartz from silica–iron colloids has been discussed by

Melnik (1973, 1982) and Ewer (1983), but the environment of iron-formation deposition is perhaps too poorly constrained to adequately interpret what is known about iron–silica colloids. Furthermore, the simultaneous precipitation of iron-free chert and ferruginous iron formation demonstrates that an iron-rich colloid is not necessary for chert precipitation.

7.05.3.1 Chert of Lake Superior-type Iron Formation

Simonson (1987) has made a detailed study of the petrography of four North American GIFs: the Gunflint, Biwabik, Gogebic, and Sokoman iron formations. These 1,900 Myr rocks (Ojakangas *et al.*, 2001), which have been labeled Lake Superior type by Gross (1973), are among the Precambrian sedimentary rocks that are least altered by post-depositional processes. The Gunflint contains some of the oldest microfossils whose biological origin has not been seriously questioned (Barghoorn and Tyler, 1965; Cloud, 1965). According to Simonson (1987) these formations contain a high proportion of oölitic and nonoölitic granules consisting primarily of quartz, hematite, magnetite, minnesotaite, and/or greenalite that are cemented by microquartz and megaquartz. Granule and cement composition and textures provide valuable information about the environment in which Proterozoic cherty iron formations were deposited. Simonson makes the following observations. (i) The primary arenite was highly porous. In the Biwabik arenites, for example, he estimates that ~44% (by volume) of the rock is quartz and that 16% of this (at least 7% of the total rock) is interstitial chert and megaquartz. (ii) Most interstitial chert and megaquartz originated as cement rather than as detrital material or as a replacement of nonsiliceous precursors. (iii) Intergranular cementation began during early diagenesis, as evidenced by the presence of chert and megaquartz cements in intraclastic pebbles and by shrinkage cracks containing internal sediment as well as the more normal cement filling. Simonson suggests that these cracks formed beneath but near the sediment interface. Fossils of cavity-dwelling microbiota have been found in some cracks (Simonson and Lanier, 1987). (iv) Quartz is the only silica mineral present in many cavity fillings; some voids were filled with primary quartz, but cement shrinkage cracks indicate that some gelatinous material was originally present. (v) Neomorphism including syntaxial overgrowths is widespread, and Simonson suggests that the precipitation mechanism involves solution–reprecipitation.

In summary, the four GIFs investigated by Simonson (1987) formed as granular, porous, shallow-water sediments with textures so similar to those of shallow-water carbonates that other researchers (Dimroth, 1976; Kimberley, 1979; Loughheed, 1983) actually considered them to be replaced limestones. Simonson has discounted a replacement origin, especially because (as stated above) interstitial chert and megaquartz originated as cement rather than as a replacement of nonsiliceous precursors and also because of the sheer volume of material to be replaced and the absence of mechanisms to accomplish such replacement.

Winter and Knauth (1992) examined granular cherts of the Gunflint iron formation and some deeper-water facies as well. They reported a predominance of chalcedonic, drusy, and microgranular quartz over megaquartz and considered this, together with other evidence, to indicate absence of metamorphic effects (except in the immediate vicinity of contact metamorphism by the 1,100 Myr Duluth Complex). Winter and Knauth's research is broadly consistent with the work of Simonson (1987) in concluding that chert of the Gunflint iron formation is a primary or very early diagenetic precipitate.

In the absence of modern analogues, several hypotheses for the direct precipitation of quartz in granular Lake Superior iron formations are worth consideration. One of these involves a soda ocean such as that proposed by Kempe and Degens (1985) in which considerable silica could dissolve (Figure 1). Objections to an early soda ocean raised by Grotzinger and Kasting (1993) have been discussed above. Simonson rejected the soda ocean hypothesis for a different reason—a lack of sodium carbonate mineral pseudomorphs in Early Precambrian rocks (but see previous discussion of the sodium amphibole riebeckite ($Na_2Fe_3^{+2}Fe_2^{+3}(Si_8O_{22})(OH)_2$), common in some iron formations). Simonson's preferred explanation for near-surface formation of chert is that upwelling silica-rich water was expelled up-dip along a steep thermal gradient from sediments deeper in the sedimentary basin, a process viewed as analogous to the formation of Mississippi Valley ore deposits.

It is quite likely that precipitation of quartz in shallow-water Paleoproterozoic iron formations occurred directly from seawater or from a siliceous gel without complicated intermediate steps. Winter and Knauth's (1992) isotopic evidence for this is presented below. Knauth (1994) pointed out that experiments by Mackenzie and Gees (1971) show that direct precipitation of quartz from seawater is possible at room temperature, and he believes that extraordinary hydrothermal circulation systems are not required to explain the observed characteristics of iron formations. One explanation of the oxygen isotope data for chert, discussed below, is that Earth-surface temperatures during the Paleoproterozoic were considerably higher than modern surface temperatures, and that this would both increase silica solubility (Figure 1) and promote direct quartz precipitation. As mentioned above, Wray (1999) has documented the formation of chert speleothems in quartz sandstone under present-day surface conditions, presumably from low-SiO_2 groundwater.

7.05.3.2 Chert of the Transvaal Supergroup and Hamersley Group

The ~2,500 Myr iron formations of the Transvaal Supergroup of South Africa (Eriksson and Warren, 1983; Klein and Beukes, 1989) and of the Hamersley Group of Western Australia (Ewers, 1983; Morris, 1993; Simonson *et al.*, 1993) contain peritidal-to-shallow subtidal carbonate formations transitional to deeper-water iron formation with microscopically laminated units that are continuous over long distances. These two important chert-bearing groups are discussed together, because they are quite similar and because deposition of iron formation in both groups was essentially contemporaneous (Martin *et al.*, 1998; Altermann and Nelson, 1998; Pickard, 2002).

Chert-bearing, stromatolitic limestone and dolomite of the Campbellrand–Malmani sequence, which conformably underlie Transvaal iron formation (Eriksson and Warren, 1983; Klein and Beukes, 1989), have lithologies which Eriksson and Warren interpreted as ranging from tidal flat to deeper subtidal (below photic zone) conditions. Chert and dolomite both display replacement textures. Eriksson and Warren suggested that this replacement can be explained by the mixing zone model originally proposed for dolomitization by Badiozamani (1973) and applied by Knauth (1979) to Phanerozoic replacement chert formation. In this model, fresh groundwater mixing with seawater in a coastal environment produces mixtures that are undersaturated with respect to calcite but supersaturated with respect to dolomite (and/or silica). For two waters that are saturated with calcite to form an unsaturated mixture, these waters must have previously equilibrated with gas phases containing different partial pressures of CO_2. During the Phanerozoic, groundwater typically equilibrated with soil gas at a high partial pressure of CO_2 produced by the oxidation of organic matter. The implicit assumption is that free oxygen is available to participate in this reaction. It is not obvious whether this mechanism is viable in the Paleoproterozoic, with an atmosphere of low P_{O_2} and high P_{CO_2} (Figure 2; Grotzinger and Kasting, 1993; Anbar and Knoll, 2002 and references therein). The model of Ericksson and Warren (1993) thus warrants careful critical evaluation. Their interpretation, if correct, could lead to insight into the contemporary atmosphere.

The considerable rare earth data available for rocks of the Transvaal Supergroup (Klein and Beukes, 1989) show significant differences between the iron-formation units (with positive europium anomalies and relative depletion in cerium and light rare earths) and the limestones, which have rare earth patterns that suggest an appreciable clastic component. Klein and Beukes (1989) suggested that this is strong evidence for a hydrothermal source for the iron of the iron formation, and they indicated that such a source is consistent with a stratified ocean in which Fe^{2+} is present only within a lower stratum characterized by low oxidation potential. Because chert is abundant in both the limestone and the iron-formation units, it appears that silica was present throughout the water column, perhaps because of multiple sources and/or because silica distribution was not affected by redox reactions within the water column. Morris (1993) reported similar rare earth patterns for the Dales Gorge Member of the Brockman iron formation, but he further distinguished "high iron" bands, with higher rare earth concentrations, from repetitive varve-like laminae (microbands or "aftvarves"), which have lower rare earth concentrations and nearly constant light rare earth concentrations. Morris used this and other evidence to posit a mixed source for iron and silica including hydrothermal solution and chemical weathering of continental sediments into the ocean. The major source of iron would then be from a hydrothermal source limited in time and/or space. Silica, which precipitated as chert in rocks of widely varying iron content including iron-free shallow-water limestone, would have had a variety of sources and would have been present throughout the water column.

7.05.4 ARCHEAN CHERT

A Late Archean (estimated age >2,950 Myr) 15 m thick chert of the Gorge Creek Group, Western Australia, contains four distinct layers with varying contents of iron carbonate, iron oxide, and organic matter (Sugitani *et al.*, 1998). Their high TiO_2/Al_2O_3 ratio was thought to indicate the presence of material weathered by acid meteoric water, whereas the rare earth pattern suggests a hydrothermal input as well. Textures were interpreted to indicate both direct crystallization of quartz from a gel and quartz precipitation in cavity fillings. Chert pseudomorphs were interpreted as replaced gypsum crystals.

Massive cherty iron formation is present in the 3,700 Myr Isua Supercrustal sequence in Greenland, and chert and cherty iron formation is common throughout the Archean. Many of the

Archean iron formations are of the "Algoma type" (Gross, 1973) consisting of lenses, often very large, in greenstone terrain. Dymek and Klein (1988) reported that rare earth patterns in the Isua iron formation have hydrothermal affinities. Simonson and Hassler (1996) considered Archean iron formations to have formed in a marine environment in close proximity to hydrothermal systems. In many Archean occurrences, and particularly in the carefully studied ones including the Onverwacht sequence in South Africa, silicification has taken place not only in the cherty units, but in much of the volcanic sequence as well. Knauth (1994) has remarked on the difficulty of producing such through-going hydrothermal alteration of huge piles of relatively impermeable rock such as the Onverwacht lava sequence, and has interpreted their oxygen isotope evidence (discussed in the next section) to show that the surface temperature of the Earth in the Archean may have been sufficiently high to constitute an essentially open hydrothermal system at the surface of the Earth.

7.05.5 PRESERVATION OF OXYGEN ISOTOPE RATIOS IN PRECAMBRIAN CHERT

Carbon, oxygen, and sulfur isotopes are strongly fractionated during exogenic processes, and there are large temporal variations in the isotopic composition of each of these elements in sedimentary rocks. Whereas temporal trends in sulfur and carbon isotopes continue to provide powerful insights about Earth-surface processes, the interpretation of temporal trends in the oxygen isotopes of low-temperature minerals precipitated from aqueous solution has proved controversial. Difficulty in interpretation arises from: (i) the mobility of water and the likelihood of exchange of oxygen isotopes with meteoric or formation waters during diagenesis and (ii) the dual dependence of the oxygen isotope composition of chert on temperature and the isotopic composition of the surface water (ocean or lake) from which that chert was deposited (Equation (1)). The problem is complicated by uncertainties in the crystallization and early diagenetic history of Precambrian cherts. This introduces questions about whether equilibration of the chert occurred with seawater, evaporative brine, fresh groundwater, pore water, or some mixture of these. A knowledge of the oxygen isotopic composition of seawater is important for understanding crust–mantle–hydrosphere evolution, but it has proved difficult to extract this information unambiguously from the isotopic composition of marine chert.

With the exception of a few Neoproterozoic cherts, including those in the Bass and Mescal Limestones (Knauth and Epstein, 1976), the maximum $\delta^{18}O$ values of all Precambrian cherts are lower than those of Phanerozoic marine cherts (Figure 2). Representative $\delta^{18}O_{SMOW}$ values of Phanerozoic cherts are +35‰ for fibrous quartz from Eocene deep-sea chert (Knauth and Epstein, 1976) and +30‰ to +34‰ for Devonian Arkansas novaculite chert precipitated at shallow depth (Jones and Knauth, 1979). In contrast, the *maximum* $\delta^{18}O$ of chert in Paleoproterozoic iron formation is +24.0‰ to +24.7‰ (Winter and Knauth, 1992; Becker and Clayton, 1976; Perry *et al.*, 1973), and $\delta^{18}O$ values for Archean chert are even lower, +22‰ for 3,400 Myr Onverwacht formation chert (Knauth and Epstein, 1976) and +20‰ for metachert in 3,700 Myr Isua iron formation (Perry *et al.*, 1978).

Before using temporal trends in the oxygen isotopic composition of cherts to answer fundamental questions of paleoenvironmental conditions and/or hydrosphere–lithosphere interaction, it is essential to ascertain whether cherts retain an isotopic signature from the time of initial crystallization or early diagenesis. The simplest explanation for the observed temporal trend in chert data follows from the observation that virtually all processes of weathering, diagenesis, and metamorphism tend to lower the $^{18}O/^{16}O$ ratio of chert and carbonate minerals derived from seawater at low temperature on or near Earth's surface. According to this argument, old cherts (and old carbonate rocks) are depleted in ^{18}O simply because they have had more time to exchange isotopes with their surroundings than young rocks. Jones and Knauth (1979) and Matheney and Knauth (1993) have challenged the validity of this hypothesis with careful studies of the diagenetic history of the Arkansas Novaculite and the Monterey Chert. They demonstrated that, once converted to crystalline quartz, silica is highly resistant to further isotopic exchange. Furthermore, despite their Precambrian age, chert in the Bass and Mescal Limestone are as ^{18}O-enriched as many Phanerozoic cherts (Knauth and Epstein, 1976). The late alteration hypothesis has also been rejected as an explanation of the observation that the $\delta^{18}O$ values of carefully studied Paleoproterozoic limestones are 10‰ lower than that of modern marine carbonates (Burdett *et al.*, 1990).

In this chapter the case of the oxygen isotopic composition of the Dales Gorge Member of the Brockman iron formation will be considered (Figure 3). This formation is underlain by shales of the Marra Mamba Formation that contain extractable hydrocarbon biomarkers generated before or during a peak regional metamorphic event 2,450–2,000 Ma (Brocks *et al.*, 1999). It is unlikely that such labile material could have survived the pervasive fluid migration that would

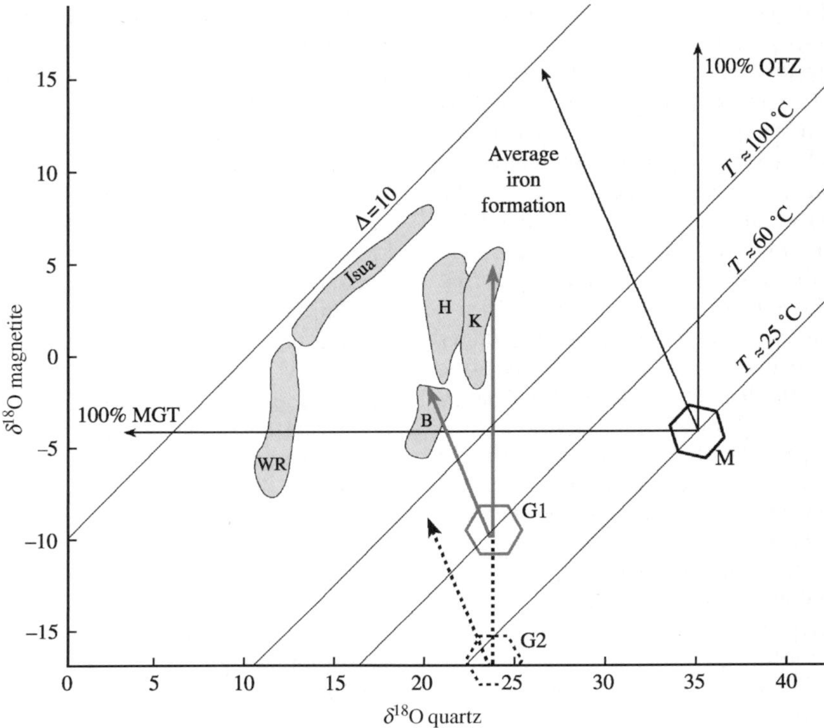

Figure 3 $\delta^{18}O_{quartz}$ versus $\delta^{18}O_{magnetite}$ of Hamersley (H), Kuruman (K), Biwabik (B), and Weld Range (WR) iron formations, all of low metamorphic grade. Primary or early diagenetic quartz from Gunflint iron formation (from samples with no magnetite) are plotted as hexagons in positions G2 (corresponding to 25 °C) and G1 (corresponding to 60 °C). Also plotted: Modern chert (M) and Isua cherty iron formation of amphibolite facies. In addition to quartz and magnetite, all of the low-grade iron formations have carbonate and hydrous silicate phases that can react with magnetite (which is secondary and of several generations) much more than with quartz, which is early diagenetic or primary. Compare with Gregory (1986). There is no evidence to support Gregory's hypothesis that chert of modern isotopic composition at normal Earth-surface temperature (M) has exchanged oxygen with hypothetical fluids so as to occupy the fields indicated.

have been required to significantly alter the isotopic composition of quartz in cherty iron formation.

In both shallow- and deeper-water chert-carbonate facies of the Gunflint iron formation, Winter and Knauth (1992) found a narrow range of +21.3‰ to +24.7‰ in $\delta^{18}O$ of quartz independent of water depth and whether the quartz was a primary precipitate or a diagenetic product of a hydrous silica precursor. From this they concluded that all of the cherts had formed early and at near-surface temperatures. Although the isotopic composition of oxygen of Gunflint quartz showed little variation, the oxygen isotopic composition of coexisting siderite was found to vary over a range of ~8‰, indicating neomorphic recrystallization, perhaps in the presence of isotopically light water. Hydrogen isotope values of Gunflint cherts were found to vary over a range that was relatively narrow compared to cherts altered by meteoric water (−51‰ to −95‰ SMOW) from which Winter and Knauth concluded that meteoric water was not involved in the precipitation process. Their overall conclusion was that Gunflint chert has a primary oxygen isotope signature reflecting that of the water body from which chert formed; the water body need not have been the open ocean.

In the Biwabik iron formation, oxygen isotope values in quartz are ~3‰ lower than the range of +23.5‰ to +24.7‰ in the microfossiliferous chert units of the correlative Gunflint iron formation (Figure 3). The Biwabik iron formations are commercial sources of iron ore; they contain a high percentage magnetite with a $\delta^{18}O$ value lower than that of any other rocks discussed here. It is probable that isotopic exchange has taken place between iron oxides and SiO_2, either during early diagenesis of precursor phases of both or during low-grade metamorphism. Because of other indications that quartz, once formed, resists exchange at low temperature, diagenetic reaction among precursor phases is the preferred explanation.

The resistance of quartz in iron-formation chert to undergo isotopic exchange during diagenesis is illustrated in Figure 3, which shows the exchange paths of quartz and magnetite in several iron formations. In a multiphase closed system, one criterion that a phase has not participated in exchange reactions is that it has maintained its isotopic composition while other phases show isotopic variation (cf. Gregory, 1986 and references therein). Figure 3 shows that the isotopic composition of quartz in the Kuruman and Brockman iron formations is virtually constant, whereas the isotopic composition of magnetite varies by several per mil. Furthermore, the oxygen isotopic composition of quartz in the iron formation (which consists of mixtures of quartz, iron carbonate, low-temperature hydrous iron silicates, and magnetite) is almost the same as the +23.5‰ to +24.0‰ values reported in units of pure chert (Becker and Clayton, 1976; Perry and Ahmad, 1983).

These observations are consistent with the petrographic studies of Simonson (1987) and Simonson and Goode (1989) which showed that quartz, which is formed very early in iron formation, retains its primary oxygen isotope signature whereas magnetite, which is a secondary mineral, has a variable isotopic composition because it usually occurs in several generations during which it interacted with iron-rich carbonate and silicate phases. Gregory (1986), who incorrectly treated these multicomponent systems as if they were composed only of quartz and magnetite, interpreted the same data differently.

7.05.6 ALTERNATIVE EXPLANATIONS FOR THE ISOTOPIC COMPOSITION OF PRECAMBRIAN CHERT

From Equation (1) it can be seen that the oxygen isotopic composition of cherts (as well as carbonates and other primary sedimentary precipitates) has the potential to establish the temperature of the water in which they formed. To apply this technique it is necessary to know the isotopic composition of that water. Thus, secular trends in the oxygen isotope composition of seawater have been of interest ever since isotopic paleotemperature measurements were first proposed (Urey *et al*., 1951). Early researchers assumed that exchange of oxygen isotopes with surface rocks had occurred early in Earth history and that subsequent variations in the isotopic composition of seawater were minimal. Later, Muehlenbachs and Clayton (1976) proposed that the oxygen isotopic composition of the ocean is buffered by interaction between seawater and the oceanic crust. According to Muehlenbachs and co-workers, this mechanism has been operating for as long as the present plate tectonic regime has been active. Holmden and Muehlenbachs (1993) reported that the plate tectonic control of seawater $\delta^{18}O$ was active as long ago as the Paleoproterozoic, and that the isotopic composition of oxygen in seawater has remained near 0‰ for at least 2,000 Myr. Nevertheless, the question of variations in the isotopic composition of seawater remains unresolved, even for the Phanerozoic. Veizer *et al*. (1999) determined the oxygen isotopic composition of carefully screened Paleozoic fossils and found low $^{18}O/^{16}O$ ratios that are difficult to explain by high Earth-surface temperature, whereas Muehlenbachs (2001) and Muehlenbachs *et al*. (2003) reported isotopic evidence that they interpreted to indicate seawater isotopic compositions near 0‰ at precisely the same time.

To summarize the Paleoproterozoic oxygen isotope data, the most ^{18}O-enriched cherts of the Gunflint, Kuruman, and Hamersley iron formations, with ages of 1,900–2,500 Myr, have maximum $\delta^{18}O_{SMOW}$ values of +24.0‰ to +24.7‰. If this value has been retained since crystallization, three important questions remain: (i) what were the precursor phases; (ii) what processes and what fluids were involved in diagenetic conversion of the precursor phase to microcrystalline quartz and/or megaquartz; and (iii) at what temperature did this diagenetic/metamorphic process occur?

It seems likely that microcrystalline quartz and megaquartz precipitated directly from Paleoproterozoic oceans, which probably had a high silica content. Lack of clear-cut evidence for evaporation, especially in iron formations, and the fact that shallow- and intermediate-water Gunflint cherts and that deeper-water Kuruman and Hamersley cherts give essentially identical isotopic values suggests that neither evaporation nor dilution by fresh (^{18}O-depleted) groundwater is likely to have been important during deposition of many Precambrian cherts.

If the Muhlenbachs and Clayton mechanism for control of ocean isotopic composition at 0‰ to −1‰ was operative in the Paleoproterozoic, the temperature of chert deposition according to Equation (1) could have been as high as 65 °C. At least two processes could affect this estimate. First, dilution of seawater by fresh groundwater in the case of permeable near-shore facies of the Gunflint–Biwabik iron formation would result in a lowered $\delta^{18}O$ and a corresponding overestimate of the temperature of deposition. However, groundwater–marine-water mixing is not a convenient explanation for either the narrow range in hydrogen isotope data of Gunflint chert (Winter and Knauth, 1992) or for the essentially identical values of the deeper-water deposits of the Kuruman and Hamersley iron formations. Furthermore, if Precambrian chert normally precipitated from mixtures of seawater and fresh groundwater, the question arises why the

Bass and Mescal Limestone should not have been similarly depleted by this mechanism. Second, an evaporitic depositional environment would result in compensating trends of ^{18}O-enriched water and anomalously high local temperature that might effectively cancel each other. For example, the cherts from Lake Magadi, which have $\delta^{18}O_{SMOW}$ values ranging up to +44‰, are among the most ^{18}O-enriched of all terrestrial sedimentary rocks even though they formed in an evaporitic environment (O'Neil and Hay, 1973).

Perry and Ahmad (1983) used Precambrian glacial events as an argument that Paleoproterozoic temperatures probably were not particularly high and interpreted the chert data to indicate a Paleoproterozoic ocean depleted in ^{18}O compared to the modern ocean. This argument would not apply if surface-temperature fluctuations were relatively rapid, and chert/iron-formation deposition was not contemporaneous with glaciations. Rapid temperature fluctuations can perhaps be accommodated by ice-house–greenhouse climate models (Fischer, 1984). It appears that most iron formations of the Neoproterozoic are separated from glacial events by major erosional intervals (Simonson and Hassler, 1996; Ojakangas *et al.*, 2001) (Figure 2). However, the Boolgeeda iron formation of the Hamersley Province of Western Australia was deposited during a glacial cycle (Martin, 1999). Examination of the $\delta^{18}O$ of its chert may help to clarify the question of seawater ^{18}O composition versus Precambrian surface temperature.

7.05.7 THE OXYGEN ISOTOPE RECORD IN METAMORPHOSED ARCHEAN CHERT

Almost all Archean chert has undergone substantial metamorphism that makes it impossible to evaluate its primary isotopic composition with confidence. Two criteria for choosing the most reasonable data are: (i) in a suite of chemical metasediments from a given unit, the highest $\delta^{18}O$ values are likely to be closest to the primary values and (ii) in a metamorphosed rock composed of minerals that vary greatly in their tendency to concentrate ^{18}O, such as quartz and magnetite, a strong correlation between the isotopic composition of these phases and the bulk chemical composition of the rock implies that isotope exchange occurred but only over limited distances.

For example, with chert of the Isua iron formation (Figure 3) the most ^{18}O-rich chert occurs in thick layers that contain little magnetite (Perry *et al.*, 1978). If the highest $\delta^{18}O_{quartz}$ values of +20‰ to +22‰ for the Isua and Onverwacht cherts, respectively (Perry *et al.*, 1978; Knauth and Lowe, 1978), represent primary values, and if $\delta^{18}O$ of the ocean was maintained at 0‰ at that time, then Earth-surface temperature between 3,400 Myr and 3,800 Myr might have been as high as 84–90 °C. Knauth (1994) has argued from these data and on the basis of the extensive silicification of the Onverwacht lavas that an essentially open hydrothermal system may have existed at the surface of the Earth at that time.

Alternatively, although Holmden and Muehlenbachs (1993) have argued effectively that a tectonic buffering system controlled the oxygen isotope composition of the ocean as long ago as 1,900 Myr, that mechanism may not have operated as effectively in the Early Archean. In that case exogenic processes could have been more important, and the oceanic $\delta^{18}O$ could have been lower.

What can be said about the late Archean Weld Range iron formation (Figure 3)? This cherty iron formation of relatively low metamorphic grade (Gole, 1980) forms a vertical array in Figure 3, which can be interpreted to indicate that quartz did not undergo isotopic exchange with its surroundings, but that the array is displaced by ~10‰ below Paleoproterozoic cherty iron formations in its $\delta^{18}O_{quartz}$ values and has one of the lowest $\delta^{18}O_{magnetite}$ values on record. There seems to be no simple explanation for these data. The Weld Range vertical array is similar to the vertical quartz arrays of the Hamersley and Kuruman iron formations. Taken together with the low $\delta^{18}O_{magnetite}$ of Weld Range samples, the data for quartz suggest that measured oxygen isotopic composition of quartz in Weld Range is near primary. No similar isotopic values are known for bona fide sedimentary marine cherts. Perhaps the chert is lacustrine; its low $\delta^{18}O$ implies both a high surface temperature and a large oxygen isotope fractionation in the hydrosphere (Perry and Ahmad, 1983). If this is so, the study of Precambrian lacustrine cherts could lead to a better understanding of the Precambrian meteoric water cycle.

NOTE ADDED IN PROOF

After review of the manuscript of this chapter, a paper by Knauth and Lowe (2003) has been published in which structure, texture, and oxygen isotope composition of cherts of the Swaziland Supergroup (including the Fig Tree and Onverwacht Formations) are discussed. The maximum recorded $\delta^{18}O$ value of +22‰ reported in this paper is not significantly different from the value reported in this chapter for these formations, but it does reinforce the interpretation that this is a value that has been preserved from sedimentation or very early diagenesis.

ACKNOWLEDGMENTS

Perry acknowledges interesting discussions about chert, spanning many years, with a number of researchers, especially K. Muehlenbachs, P. Knauth, J. Veizer, and H. Taylor. He further appreciates the opportunity provided by the editors of this Treatise to revisit the topic of Precambrian chert in light of recent advances in the understanding of various aspects of Precambrian geology and geochemistry. Both ECP and LL acknowledge constructive comments by H. D. Holland.

REFERENCES

Altermann W. and Nelson D. R. (1998) Sedimentation rates, basin analysis and regional correlations of three Neoarchean and Paleoproterozoic sub-basins of the Kaapvaal craton as inferred from precise U–Pb zircon ages from volcanoclastic sediments. *Sedim. Geol.* **120**, 225–256.

Anbar A. D. and Knoll A. H. (2002) Proterozoic ocean chemistry and evolution: a bioinorganic bridge? *Science* **297**, 1137–1142.

Astor S. R. (ed.) (1983) Natural water and atmospheric chemistry of silicon. In *Silicon Geochemistry and Biogeochemistry.* Academic Press, London, pp. 77–100.

Awramik S. M., Schopf J. W., and Walter M. R. (1983) Filamentous fossil bacteria from the Archean of Western Australia. *Precamb. Res.* **20**, 357–374.

Badiozamani K. (1973) The dorag dolomitization model—application to the Middle Ordovician of Wisconsin. *J. Sedim. Petrol.* **43**, 965–984.

Barghoorn E. S. and Tyler S. A. (1965) Microorganisms of middle Precambrian age from the Animikie series. Current aspects of exobiology, Ontario, Canada, chap. 3, pp. 93–118.

Becker R. H. and Clayton R. N. (1976) Oxygen isotope study of a Precambrian banded iron-formation, Hamersley Range, Western Australia. *Geochim. Cosmochim. Acta* **40**(10), 1153–1165.

Berkner L. V. and Marshall L. C. (1965) History of major atmospheric components: Symposium on the evolution of the Earth's atmosphere. *Proc. Natl. Acad. Sci. USA* **53**(6), 1215–1226.

Beukes N. J., Dorland H., Gutzmer J., Nedachi M., and Ohmoto H. (2002) Tropical laterites, life on land, and the history of atmospheric oxygen in the Paleoproterozoic. *Geology* **30**, 491–494.

Birnbaum S. J. and Wireman J. W. (1985) Sulfate-reducing bacteria and silica solubility: a possible mechanism for evaporite diagenesis and silica precipitation in banded iron formations. In *Role of Organisms and Organic Matter in Ore Deposition–Le role des organisms et de la matiere organique dans la formation des gisements metalliferes* (eds. R. W. Macqueen and J. A. Coope). *Can. J. Earth Sci.* **22**(12), 1904–1909.

Bohrmann G., Abelmann A., Gersonde R., Hubberten H., and Kuhn G. (1994) Pure siliceous ooze, a diagenetic environment for early chert formation. *Geology* **22**(3), 207–210.

Brasier M., Green O., and Shields G. (1997) Ediacarian sponge spicule clusters from southwestern Mongolia and the origins of the Cambrian fauna. *Geology* **28**, 303–306.

Brasier M. D., Green O. R., Jephcoat A. P., Kleppe A. K., Van Kranendonk M. J., Lindsay J. F., Steele A., and Grassineau N. V. (2002) Questioning evidence for Earth's oldest fossils. *Nature* **416**, 76–81.

Brocks J. J., Logan G. A., Buick R., and Summons R. E. (1999) Archean molecular fossils and the early rise of eukaryotes. *Science* **285**, 1033–1036.

Buick R. (1990) Microfossil recognition in Archean rocks: an appraisal of spheroids and filaments from a 3,500 m.y, chert-barite unit at north pole, Western Australia. *Palaios* **5**, 441–459.

Burdett J. W., Grotzinger J. P., and Arthur M. A. (1990) Did major changes in the stable isotope composition of Proterozoic seawater occur? *Geology* **18**, 227–230.

Cady L. and Farmer J. D. (1995) Fossilization processes in siliceous thermal springs: trends in preservation along thermal gradients. In *Evolution of Hydrothermal Ecosystems on Earth (and Mars?).* Wiley, Chichester, pp. 150–173.

Canfield D. E. (1998) A new model for Proterozoic ocean chemistry. *Nature* **396**(6710), 450–453.

Carozzi A. V. (1993) *Sedimentary Petrology.* Prentice-Hall, New Jersey, 263pp.

Cloud P. (1972) A working model of the primitive earth. *Am. J. Sci.* **272**(6), 485–486.

Cloud P. E., Jr. (1965) Significance of the Gunflint (Precambrian) microflora. *Science* **148**, 27–35.

Coplen T. B. (1996) New guidelines for reporting stable hydrogen, carbon, and oxygen isotope-ratio data. *Geochim. Cosmochim. Acta* **60**, 3359–3360.

Crowell J. C. (1999) Pre-Mesozoic ice ages: their bearing on understanding the climate system. *GSA Memoir* **192**, 106pp.

Dimroth E. (1976) Aspects of the sedimentary petrology of cherty iron formation. In *Handbook of Stratabound and Strataform Ore Deposits* (ed. K. H. Wolf). Elsevier, Amsterdam, vol. 7, pp. 203–254.

Dymek R. and Klein C. (1988) Chemistry, petrology, and origin of banded iron-formation lithologies from the 3,800 Ma Isua Supracrustal Belt, West Greenland. *Precamb. Res.* **39**, 247–302.

Eriksson K. A. and Warren J. L. (1983) A paleohydrologic model for Early Proterozoic dolomitization and silicification. *Precamb. Res.* **21**, 299–321.

Eugster H. P. and Jones B. F. (1968) Gels composed of sodium–aluminum silicate, Lake Magadi, Kenya. *Science* **161**, 160–163.

Ewers W. E. (1983) Chemical factors in the deposition and diagenesis of banded iron-formation. In *Iron-formations: Facts and Problems* (eds. A. F. Trendall and R. C. Morris). Elsevier, Amsterdam, pp. 491–512.

Feng J. (1986) Sulfur and oxygen isotope geochemistry of Precambrian marine sulfate and chert. MD Thesis, Northern Illinois University (unpublished).

Fischer A. G. (1984) The two Phanerozoic supercycles. In *Catastrophes and Earth History* (eds. W. A. Berggen and J. A. Van Couvering). Princeton University Press, Princeton, NJ, pp. 129–150.

Fleming B. A. and Crerar D. A. (1982) Silicic acid ionization and calculation of silica solubility at elevated temperature and pH, application to geothermal fluid processing and reinjection. *Geothermics* **11**(1), 15–29.

Folk R. L. (1980) *Petrology of Sedimentary Rocks.* Hemphill Publishing Company, Austin, p. 185.

Gehling J. G. and Rigby J. (1996) Long expected sponges from the Neoproterozoic Ediacara fauna of South Australia. *J. Paleonotl.* **70**, 135–195.

Gole M. J. (1980) Mineralogy and petrology of very-low-metamorphic grade Archean banded iron formations, Weld Range, Western Australia. *Am. Mineral.* **65**, 8–25.

Gough D. O. (1981) Solar interior structure and luminosity variations. *Solar Phys.* **74**, 21–34.

Gregory R. T. (1986) Oxygen isotope systematic of quartz-magnetite pairs from Precambrian iron formations: evidence for fluid–rock interaction during diagenesis and metamorphism. In *Fluid–Rock Interactions during Metamorphism* (eds. J. V. Walther, J. B. Wood, and S. K. Saxena). *Adv. Phys. Geochem.* **5**, 132–153.

Gross G. A. (1973) The depositional environments of the principal types of Precambrian iron-formations. In *Genesis of Precambrian Iron and Manganese Deposits. UNESCO Earth Sci. Ser.* **9**, 15–21.

Grotzinger J. P. (1989) Facies and evolution of Precambrian carbonate depositional systems; emergence of the modern platform archetype. In *Controls on Carbonate Platform and Basin Development* (eds. P. D. Crevello, J. L. Wilson, J. F. Sang, and J. F. Read). Soc. Econ. Paleont. Mineral., pp. 79–106.

Grotzinger J. P. and Kasting J. F. (1993) New constraints on Precambrian ocean composition. *J. Geol.* **101**, 235–243.

Hemzacek J. M. (1986) Replaced evaporites and the sulfur isotope age curve of the Precambrian. MS Thesis, Northern Illinois University (unpublished).

Hemzacek J. M., Perry E. C., Jr., Larue D. K., and Feng J. (1982) Sulfur isotope composition of sulfate in chert horizons of the Proterozoic (Precambrian X) Kona Dolomite, Marquette Region, Michigan, GSA Abstracts with Programs, New Orleans.

Holland H. D. (1984) *The Chemical Evolution of the Atmosphere and Oceans*. Princeton series in geochemistry, p. 582.

Holland H. D. (2002) Volcanic gases, black smokers, and the Great Oxidation Event. *Geochem. Cosmochem. Acta* **66**, 3811–3826.

Holmden C. and Muehlenbachs K. (1993) The $^{18}O/^{16}O$ ratio of 2-billion-year-old seawater inferred from ancient oceanic crust. *Science* **259**, 1733–1736.

Horita J., Zimmermann H., and Holland H. D. (2002) Chemical evolution of seawater during the Phanerozoic: implications from the record of marine evaporates. *Geochim. Cosmochim. Acta* **66**, 3733–3756.

Horodyski R. J. and Donaldson J. A. (1980) Microfossils from the Middle Proterozoic Dismal Lakes Group, Arctic Canada. *Precamb. Res.* **11**, 125–159.

Horodyski R. J. and Donaldson J. A. (1983) Distribution and significance of microfossils in cherts of the Middle Proterozoic Dismal Lakes Group, District of MacKenzie, Northwest Territories, Canada. *J. Paleontol.* **57**, 271–288.

House C. H., Schopf J. W., McKeegan K. D., Coath C. D., Harrison T. M., and Stetter K. O. (2000) Carbon isotopic composition of individual Precambrian microfossils. *Geology* **28**, 707–710.

Iler R. (1979) *The Chemistry of Silica*. Wiley, New York, p. 866.

Isley A. E. (1995) Hydrothermal plumes and the delivery of iron to banded iron formation. *J. Geol.* **103**, 169–185.

Jacobsen S. B. and Kaufman A. J. (1999) The Sr, C and O isotopic evolution of Neoproterozoic seawater. *Chem. Geol.* **161**, 37–57.

James H. L. (1983) Distribution of banded iron-formation in space and time. In *Iron-formation: Facts and Problems* (eds. A. F. Trendall and R. C. Morris). Elsevier, Amsterdam, pp. 471–490.

Jones D. L. and Knauth L. P. (1979) Oxygen isotopic and petrographic evidence relevant to the origin of the Arkansas Novaculite. *J. Sedim. Petrol.* **49**, 581–598.

Kasting J. F. (1987) Theoretical constraints on oxygen and carbon dioxide concentrations in the Precambrian atmosphere. *Precamb. Res.* **34**, 205–229.

Kastner M., Keene J. B., and Gieskes J. M. (1977) Diagenesis of siliceous oozes: 1. Chemical controls on the rate of opal-A to opal-CT transformation—an experimental study. *Geochim. Cosmochim. Acta* **41**, 1041–1059.

Kempe S. and Degens E. T. (1985) An early soda ocean? *Chem. Geol.* **53**, 95–108.

Kimberley M. M. (1979) Origin of oolitic iron formations. *J. Sedim. Petrol.* **49**, 111–132.

Klein C. and Beukes N. J. (1989) Geochemistry and sedimentology of a facies transition from limestone to iron-formation deposition in the Early Proterozoic Transvaal Supergroup, South Africa. *Econ. Geol.* **84**, 1733–1774.

Knauth L. P. (1979) A model for the origin of chert in limestone. *Geology* **7**, 274–277.

Knauth L. P. (1992) Origin and diagenesis of cherts: an isotopic perspective. In *Isotopic Signatures and Sedimentary Records* (eds. N. Clauer and S. Chaudhuri). Springer, Berlin, pp. 123–152.

Knauth L. P. (1994) Petrogenesis of chert. *In Silica: Physical Behavior, Geochemistry and Materials Applications*, Reviews in Mineralogy (eds. P. J. Heaney, C. T. Prewitt, and G. V. Gibbs). Minerological Society of America, vol. 29, pp. 233–258.

Knauth L. P. and Epstein S. (1976) Hydrogen and oxygen isotope ratios in nodular and bedded cherts. *Geochim. Cosmochim. Acta* **40**(9), 1095–1108.

Knauth L. P. and Lowe D. R. (1978) Oxygen isotope geochemistry of cherts from the Onverwacht Group (3.4 billion years), Transvaal, South Africa, with implication for secular variations in the isotopic composition of cherts. *Earth Planet. Sci. Lett.* **41**, 209–222.

Knauth L. P. and Lowe D. R. (2003) High Archean climatic temperature inferred from oxygen isotope geochemistry of cherts in the 3.5 Ga Swaziland Supergroup, South Africa. *GSA Bull.* **115**, 566–580.

Knoll A. H. (1982) Microfossils from the late Precambrian Draken Conglomerate, Ny Friesland, Svalbard. *J. Paleontol.* **56**(3), 755–790.

Komiya T., Maruyama S., Masuda T., Nohda S., Hayashi M., and Okamoto K. (1999) Plate tectonics at 3.8–3.7 Ga: field evidence from the Isua Accretionary Complex, southern West Greenland. *J. Geol.* **107**, 515–554.

Loughheed M. S. (1983) Origin of Precambrian iron formation in the Lake Superior region. *GSA Bull.* **94**, 325–340.

Mackenzie F. T. and Gees R. (1971) Quartz: synthesis at Earth-surface conditions. *Science* **173**, 533–535.

Maliva R. G. (2001) Silicification in the Belt Supergroup (Mesoproterozoic) Glacier National Park, Montana, USA. *Sedimentology* **48**, 887–896.

Maliva R. G., Knoll A. H., and Siever R. (1989) Secular change in chert distribution: a reflection of evolving biological participation in the silica cycle. *Palaios* **4**, 519–532.

Maliva R. G., Simonson B. M., and Knoll A. H. (2002) Chert and the Proterozoic oceanic silica cycle (abstr.). GSA Abstract with Programs 2002, pp. 243–215.

Martin D. McB. (1999) Depositional setting and implications of Paleoproterozoic glaciomarine sedimentation in the Hamersley Province, Western Australia. *GSA Bull.* **111**, 189–203.

Martin D. McB., Clendenin C. V., Krapez B., and McNaughton N. J. (1998) Tectonic and geochronological constraints on Late Archean and Paleoproterozoic statigraphic correlation within and between the Kaapvaal and Pilbara Cratons. *J. Geol. Soc.* **155**, 311–322.

Matheney R. K. and Knauth L. P. (1993) New isotopic temperature estimates for early silica diagenesis in bedded cherts. *Geology* **21**, 519–522.

Melnik, Y. P. (1973) Physiochemical conditions of formation of the Precambrian ferruginous quartzite. Kiev, Akad. Nauk Ukrainskoe SSR Institut geokhimii I fiziki mineralov (Izdatel; stvo "Naukova Dumka") p. 272.

Melnik Y. P. (1982) *Precambrian Banded Iron Formation: Physiochemical Conditions of Formation*. Elsevier, Amsterdam, p. 310.

Morris R. C. (1993) Genetic modeling for banded iron-formation of the Hamersley Group, Pilbara Craton, Western Australia. *Precamb. Res.* **60**, 243–286.

Muehlenbachs K. (2001) Ophiolites as faithful records of the oxygen isotope ratio of ancient seawater. *Abstr. Progr.—Geol. Soc. Am.* **33**(6), 225.

Muehlenbachs K. and Clayton R. N. (1976) Oxygen isotope composition of the oceanic crust and its bearing on seawater. *J. Geophys. Res.* **81**(23), 4365–4369.

Muehlenbachs K., Furnes H., Fonneland H. C., and Hellevang B (2003) Ophiolites as faithful records of the oxygen isotope ratio of ancient seawater: the Solund–Stavfjord Ophiolite Complex as a Late Ordovician example. In *Ophiolites in Earth History* (eds. Y. Dilek and P. T. Robinson). Geol. Soc. of London, Spec. Publ. (in press).

Newman M. J. and Rood R. T. (1977) Implications of solar evolution for the Earth's early atmosphere. *Science* **198**, 1035–1037.

Ohmoto H. (1999) Redox state of the Archean atmosphere: evidence from detrital heavy minerals in ca. 3250–2750 Ma sandstones from the Pilbara Craton, Australia: Comment. *Geology* **27**, 1151–1152.

Ojakangas R. W., Morey G. B., and Southwick D. L. (2001) Paleoproterozoic basin development and sedimentation in the Lake Superior region, North America. *Sedim. Geol.* **141–142**, 319–341.

O'Neil J. R. and Hay R. L. (1973) $^{18}O/^{16}O$ ratios in cherts associated with the saline lake deposits of East Africa. *Earth Planet. Sci. Lett.* **19**, 257–266.

Pavlov A. A., Hurtgen M. T., Kasting J. M., and Arthur M. A. (2003) Methane-rich Proterozoic atmosphere? *Geology* **31**, 87–90.

Perry E. C. (1967) The oxygen isotope chemistry of ancient cherts. *Earth Planet. Sci. Lett.* **3**, 62–66.

Perry E. C. and Ahmad S. N. (1983) Oxygen isotope geochemistry of Proterozoic chemical sediments. *GSA Memoir* **161**, 253–263.

Perry E. C. and Tan F. C. (1972) Significance of oxygen and carbon isotope variations in Early Precambrian cherts and carbonate rocks of southern Africa. *GSA Bull.* **83**, 647–664.

Perry E. C., Tan F. C., and Morey G. B. (1973) Geology and stable isotope geochemistry of the Biwabik Iron Formation, northern Minnesota. *Econ. Geol.* **68**, 1110–1125.

Perry E. C., Ahmad S. N., and Swulius T. M. (1978) The oxygen isotope composition of 3,800 m.y. old metamorphosed chert and iron formation from Isukasia, West Greenland. *J. Geol.* **86**, 223–239.

Pickard A. L. (2002) SHRIMP U–Pb zircon ages of tuffaceous mudrocks in the Brockman Iron Formation on the Hamersley Range, Western Australia. *Austral. J. Earth Sci.* **49**, 491–507.

Porter S. M. and Knoll A. H. (2000) Testate amoebae in the Neoproterozoic era: evidence from vase-shaped microfossils in the Chuar Group, Gran Canyon. *Paleobiology* **26**, 360–385.

Rasmussen B., Buick R., and Holland H. D. (1999) Redox state of the Archean atmosphere: evidence from detrital heavy minerals in ca. 3250–2750 Ma sandstones from the Pilbara Craton, Australia: Reply. *Geology* **27**, 1152.

Robie R. A., Hemingway B. S., and Fisher J. R. (1978) Thermodynamic properties of minerals and related substances at 298.15 K and 1 bar (10^5 Pa) pressure and at higher temperatures. *US Geol. Surv. Bull.* **B1452**, 456.

Schopf J. W. (1993) Microfossils of the Early Archean Apex Chert: new evidence of the antiquity of life. *Science* **260**, 640–646.

Schopf J. W., Kudryavtsev A. B., Agresti D. G., Wdowiak T. J., and Czaja A. D. (2002) Laser-Raman imagery of Earth's earliest fossils. *Nature* **416**, 73–76.

Schubel K. A. and Simonson B. M. (1990) Petrography and diagenesis of cherts from Lake Magadi, Kenya. *J. Sedim. Petrol.* **60**, 761–776.

Siever R. (1992) The silica cycle in the Precambrian. *Geochem. Cosmochem. Acta* **56**, 3265–3272.

Simonson B. M. (1987) Early silica cementation and subsequent diagenesis in arenites from four Early Proterozoic iron formations of North America. *J. Sedim. Petrol.* **57**, 494–511.

Simonson B. M. and Goode A. D. T. (1989) First discovery of ferruginous chert arenites in the Early Precambrian Hamersley Group of Western Australia. *Geology* **17**, 269–272.

Simonson B. M. and Hassler S. W. (1996) Was the deposition of large Precambrian iron formations linked to major marine transgressions? *J. Geol.* **104**, 665–676.

Simonson B. M. and Lanier W. P. (1987) Early silica cementation and microfossil preservation in cavities in iron-formation stromatolites. Early Proterozoic of Canada. In *Precambrian Iron-formations* (eds. A. Uitterdijk, W. Peter, and L. L. Gene). Theophrastus Publ. and Pty. Co., Athens, Greece, pp. 187–213.

Simonson B. M., Schubel K. A., and Hassler S. W. (1993) Carbonate sedimentology of the Early Precambrian Hamersley Group, Western Australia. *Precamb. Res.* **60**, 287–335.

Southgate P. (1986) Depositional environment and preservation of microfossils, upper Proterozoic Bitter Springs Formation, Australia. *Geology* **14**, 683–686.

Sugitani K., Yamamoto K., Adachi M., Kawabe I., and Sugisaki R. (1998) Archean cherts derived from chemical, biogenic and clastic sedimentation in a shallow restricted basin: examples from the Gorge Creek Group in the Pilbara Block. *Sedimentology* **45**(6), 1045–1063.

Trendall A. F. and Blockley J. G. (1970) The iron formation of the Precambrian Hamersley Group, Western Australia; with special reference to the associated crocidolite. *West. Aust. Geol. Surv. Bull.* **119**, 336.

Urey H. C., Epstein S., and McKinney C. R. (1951) Measurement of paleotemperatures and temperatures of the Upper Cretaceous of England, Denmark, and the southeastern United States. *GSA Bull.* **62**(4), 399–416.

Veizer J., Ala D., and Azmy K. (1999) $^{87}Sr/^{86}Sr$, $\delta^{13}C$, $\delta^{18}O$ evolution of Phanerozoic seawater. *Chem. Geol.* **161**, 59–88.

Walker J. C. G. (1977) *Evolution of the Atmosphere*. Macmillan, New York, p. 318.

Walker J. C. G., Klein C., Schidlowski M., Schopf J. W., Stevenson D. J., and Walter M. R. (1983) Environmental evolution of the Archean–Early Proterozoic Earth. In *Earth's Earlier Biosphere: Its Origin and Evolution* (ed. J. W. Schopf). Princeton University Press, Princeton, NJ, pp. 260–290.

Wray R. A. L. (1999) Opal and chalcedony speleothems on quartz sandstones in the Sydney region, southeastern Australia. *Austral. J. Earth Sci.* **46**, 623–632.

Winter B. L. and Knauth L. P. (1992) Stable isotope geochemistry of cherts and carbonates from the 2.0 Ga Gunflint Iron Formation; implications for the depositional setting, and the effects of diagenesis and metamorphism. *Precamb. Res.* **59**, 283–313.

Zegers T. E. and van Keken P. E. (2001) Middle Archean continent formation by crustal delamination. *Geology* **29**, 1083–1086.

7.06
Geochemistry of Fine-grained Sediments and Sedimentary Rocks

B. B. Sageman

Northwestern University, Evanston, IL, USA

and

T. W. Lyons

University of Missouri, Columbia, MO, USA

7.06.1 INTRODUCTION

The nature of detrital sedimentary (siliciclastic) rocks is determined by geological processes that occur in the four main Earth surface environments encountered over the sediment's history from source to final sink: (i) the site of sediment production (provenance), where interactions among bedrock geology, tectonic uplift, and climate control weathering and erosion processes; (ii) the transport path, where the medium of transport, gradient, and distance to the depositional basin may modify the texture and composition of weathered material; (iii) the site of deposition, where a suite of physical, chemical, and biological processes control the nature of sediment accumulation and early burial modification; and (iv) the conditions of later burial, where diagenetic processes may further alter the texture and composition of buried sediments. Many of these geological processes leave characteristic geochemical signatures, making detrital sedimentary rocks one of the most important archives of geochemical data available for reconstructions of ancient Earth surface environments. Although documentation of geochemical data has long been a part of the study of sedimentation (e.g., Twenhofel, 1926, 1950; Pettijohn, 1949; Trask, 1955), the development and application of geochemical methods specific to sedimentary geological problems blossomed in the period following the Second World War (Degens, 1965; Garrels and Mackenzie, 1971) and culminated in recent years, as reflected by the publication of various texts on marine geochemistry (e.g., Chester, 1990, 2000), biogeochemistry (e.g., Schlesinger, 1991; Libes, 1992), and organic geochemistry (e.g., Tissot and Welte, 1984; Engel and Macko, 1993).

Coincident with the growth of these subdisciplines a new focus has emerged in the geological sciences broadly represented under the title of "Earth System Science" (e.g., Kump *et al.*, 1999). Geochemistry has played the central role in this revolution (e.g., Berner, 1980; Garrels and Lerman, 1981; Berner *et al.*, 1983; Kump *et al.*, 2000), with a shifting emphasis toward sophisticated characterization of the linkages among solid Earth, oceans, biosphere, cryosphere, atmosphere, and climate, mediated by short- and long-term biogeochemical cycles. As a result, one of the primary objectives of current geological inquiry is improved understanding of the interconnectedness and associated feedback among the cycles of carbon, nitrogen, phosphorous, oxygen, and sulfur, and their relationship to the history of Earth's climate. This "Earth System" approach involves uniformitarian extrapolations of knowledge gained from modern environments to proxy-based interpretations of environmental change recorded in ancient strata. The strength of modern data lies with direct observations of pathways and products of physical, chemical, and biological processes, but available time-series are short relative to the response times of many of the biogeochemical systems under study. By contrast, stratigraphically constrained geological data offer time-series that encompass a much fuller range of system response. But with the enhanced breadth of temporal resolution and signal amplitude provided by ancient sedimentary records comes a caveat—we must account for the blurring of primary paleoenvironmental signals by preservational artifacts and understand that proxy calibrations are extended from the modern world into a nonsubstantively uniformitarian geological past.

Fortunately, detrital sedimentary rocks preserve records of multiple proxies (dependent and independent) that illuminate the processes and conditions of sediment formation, transport, deposition, and burial. An integrated multiproxy approach offers an effective tool for deconvolving the history of biogeochemical cycling of, among other things, carbon and sulfur, and for understanding the range of associated paleoenvironmental conditions (e.g., levels of atmospheric oxygen and carbon dioxide, oceanic paleoredox, and paleosalinity). Authors of a single chapter can hope, at best, to present a cursory glance at the many biogeochemical proxies currently used and under development in sedimentary studies. Our goal, instead, is to focus on a selected suite of tools of particular value in the reconstruction of paleo-environments preserved in fine-grained siliciclastic sedimentary rocks.

Fine-grained, mixed siliciclastic–biogenic sedimentary facies—commonly termed hemipelagic (mainly calcareous or siliceous mudrocks containing preserved organic matter (OM))—are ideal for unraveling the geological past and are thus the focus of this chapter. These strata accumulate in predominantly low-energy basinal environments where the magnitude (and frequency) of lacunae is diminished, resulting in relatively continuous, though generally condensed sequences. Fortunately, condensation tends to benefit geochemical analysis as it helps to amplify some subtle environmental signals. Because hemipelagic facies include contributions from both terrigenous detrital and pelagic biogenic systems, as well as from authigenic components reflecting the burial environment (Figure 1), they are rich archives of geochemical information. In this chapter we present a conceptual model linking the major processes of detrital, biogenic, and authigenic accumulation in fine-grained hemipelagic settings. This model is intended to be a fresh synthesis of decades of prior research on the geochemistry of modern and ancient mudrocks, including our own work.

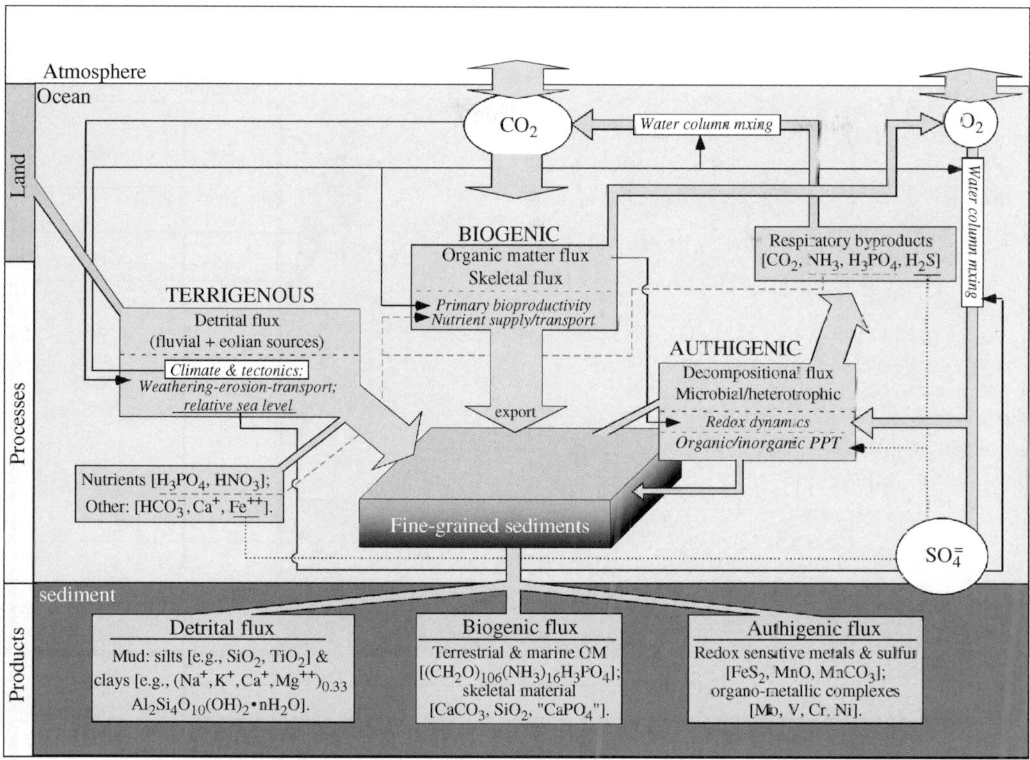

Figure 1 Conceptual model for the origin of mixed detrital–biogenic facies relating the three major inputs to the processes that control them. The major inputs are shown in boxes with bold-type labels. Controlling factors are shown in italics. Large and medium scale arrows represent fluxes of key components involved in sedimentation and the biogeochemical cycles of carbon, sulfur, and oxygen. Thin arrows illustrate relationships between major controlling factors and depositional processes and/or feedback. Dashed thin arrows apply to major nutrient fluxes only. Dotted thin arrows apply to major authigenic fluxes only. See text for further explanation.

In the sections that follow we will first develop and illustrate (Figures 1 and 2) the conceptual model and the proxy methods incorporated within this model—with additional details provided in the cited literature. The remainder of the chapter is devoted to demonstrating the utility of the model through a series of case studies ranging from the modern Black Sea to shales and argillites of the Precambrian. Based on these case studies, we conclude with a summary of similarities and differences in proxy application.

7.06.2 CONCEPTUAL MODEL—PROCESSES

A schematic representation of our model for the major inputs and feedback involved in the formation of fine-grained, mixed siliciclastic–biogenic facies in marine basins is shown in Figure 1. The major inputs include terrigenous detritus, such as material derived from weathering of continental crust or from volcanic sources, biogenic components (both OM and mineralized microskeletal remains) derived from primary photosynthetic production and heterotrophic processes on land and in the sea, and authigenic material precipitated at or near the sediment-water interface as a consequence of Eh-pH-controlled organic and inorganic reactions. Although these major inputs have been recognized for many years (e.g., see reviews by Potter *et al.*, 1980; Gorsline, 1984; Arthur and Sageman, 1994; Wignall, 1994; Hedges and Keil, 1995; Tyson, 1995; Schieber *et al.*, 1998a,b; Chester, 2000, and references therein), the synthesis in Figure 1 is novel in that it integrates physical (sedimentologic and oceanographic) and biogeochemical processes, relates major inputs to proximate and broader paleoenvironmental controls, illustrates important linkages between causes and effects in the model (i.e., feedback), and, lastly, identifies and tracks key components of the major biogeochemical cycles (C, O, S, N, and P) involved in regulating conditions at the Earth's surface.

Climate and plate tectonics are the master controlling factors for the system represented in Figure 1. Climate includes a complex set of phenomena (temperature, evaporation, precipitation, and wind) and interactions among the atmosphere, land surface, ocean surface, biosphere, and cryosphere that are driven largely by variations in the amount and distribution of incoming solar radiation.

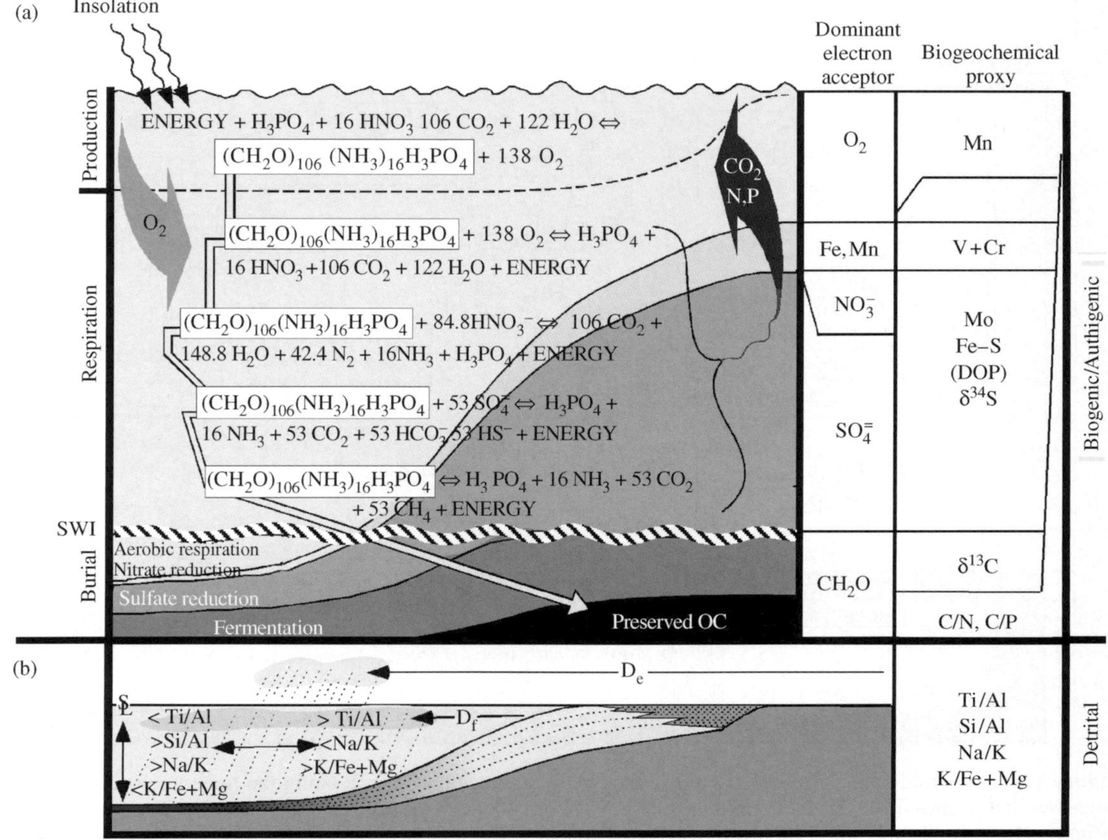

Figure 2 (a) Details of the interrelationship between biogenic and authigenic processes in Figure 1 are shown. Sequential steps in the remineralization of photosynthetically produced OM (represented by gray arrow) are represented by simplified chemical reactions. For each step in which a given electron acceptor is dominant, one or more characteristic biogeochemical proxies may accumulate (in some cases proxies accumulate across a range of conditions; in others the range is narrower than the text font: such cases are represented by slanted borders). The rate at which OM moves through the steps depends on availability of electron acceptors and bulk sedimentation rate, which may be evaluated using detrital proxies. (b) Cartoon illustrating sources of terrigenous flux and proxies for their detection. See text for further discussion.

Tectonic processes, by contrast, can be simplified to two parameters. These are vertical uplift, which creates crustal source areas for weathering and erosion, and subsidence, which together with eustasy acts to control the accommodation space available for accumulation of sediments (e.g., Sloss, 1962).

7.06.2.1 Detrital Flux

Climate and tectonics, particularly the interaction between uplift of continental crust and physical/chemical weathering, controls the generation of terrigenous detritus (sand, silt, and clay), as well as dissolved species (e.g., cations and anions such as Na^+, Ca^{2+}, SO_4^{2-}, Cl^-, and HCO_3^-), many of which play important roles in biogeochemical cycling (Figure 1). The chemical composition of the detrital fraction is largely determined by the mineralogy of the source rocks, the weathering regime, and the reactivity of weathering products during transport. Regional climatic conditions determine weathering regimes (temperature, precipitation, and runoff), and for a given weathering environment, uplift rate and drainage area determine net detrital flux or sediment yield into a basin of deposition (Milliman and Syvitski, 1992; Perlmutter *et al.*, 1998). Tectonic activity drives source area uplift to create relief, and on a global scale also influences long-term eustasy, which together with local subsidence will determine the bathymetric profile of a given sedimentary basin, as well as its relative sea-level history. Geography (latitude and basin orientation) and bathymetric profile combine with climate to determine the oceanographic character of a basin (e.g., circulation), which, in turn, controls delivery of the detrital fine fraction to distal depositional sites and strongly influences biogenic and authigenic processes. The ratio of accommodation (whether controlled by eustasy, local tectonics, or both) to sediment supply is one of the most important regulators of particulate detrital flux to a basin (e.g., Loutit *et al.*, 1988; Pasley *et al.*, 1991) because it controls the extent to which river-transported materials are stored in estuaries or transported beyond the "littoral energy fence" (Allen, 1970; Swift and Thorne, 1991), where they are dispersed by the shelfal transport system.

In general, transport of weathered material is directly mediated by runoff and wind-driven waves and currents. Both of these are controlled by climatic parameters and exhibit broadly predictable latitudinal patterns. Although runoff is the most volumetrically significant agent of detrital and chemical transport to the oceans, under certain circumstances windblown components can have a major influence on sedimentary composition and geochemical interpretations of detrital fluxes (e.g., Bertrand *et al.*, 1996). Once fine-grained material enters a marine basin, for example, from a riverine source, its transport is controlled by surface circulation, and deposition occurs by fall-out from suspension (Gorsline, 1984) often mediated by biologically and chemically induced aggregation. Large sediment plumes associated with major rivers characterize supply-dominated systems (Swift and Thorne, 1991) and are well known from modern observations (e.g., the Amazon).

Although good modern analogues for ancient epicratonic basins do not exist today, it is assumed that turbid plumes of muddy sediment blanketed such ancient settings (e.g., Pratt, 1984). There is, however, relatively little known about the dispersal of fine-sediment in ancient depositional systems beyond what can be implied by trends of increasing thickness, grain size, etc., toward dominant sediment sources (e.g., Elder, 1985). Sediment gravity processes, when bathymetric topography is sufficient (Gorsline, 1984), or impingement of storm wave base during large storms or episodes of decreased relative sea level can redistribute fine-grained sediment into thicker and thinner accumulations on the basin floor. These events can result in winnowing of clays and progressive concentration of coarser grained pelagic and benthic skeletal components (Sageman, 1996) or eolian-derived particles, thus altering the geochemical proportions of the detrital fraction. It is also well known (e.g., in the modern Black Sea; Lyons, 1991) that fine sediment is transported over large lateral distances along isopycnal surfaces within well-stratified water columns, often as a result of internal wave activity along the density interfaces.

The major elemental components of the net particulate flux from modern fluvial sources, in order of decreasing magnitude, are Si, Al, Fe, Ca, K, Mg, Na, Ti, P, Mn, and Ba (Martin and Whitfield, 1983; see also Bewers and Yeats, 1979; Martin and Maybeck, 1979; Chester and Murphy, 1990; Chester *et al.*, 1997). Similar proportions among the most abundant components (e.g., Si, Al, and Fe) generally characterize the mineralogy of the eolian dust flux (e.g., Lantzy and Mackenzie, 1979), but there can be significant regional variations in elemental concentrations depending on the composition of source areas (e.g., Chester, 2000). For example, titanium concentrations can be significantly elevated over average crustal values in eolian fluxes from arid regions (Bertrand *et al.*, 1996; Yarincik *et al.*, 2000b). Deviations from the crustal mean also characterize areas with a high volcanigenic contribution to the atmospheric flux (Weisel *et al.*, 1984; Arthur and Dean, 1991). Silicon is often involved in biogenic processes (see below), but aluminum and titanium are regarded as the most refractory products of crustal weathering (Taylor and McLennan, 1985), except under extreme weathering conditions (Young and Nesbitt, 1998).

7.06.2.2 Biogenic Flux

Photosynthetic production of OM and associated skeletal material defines the second major input to fine-grained, mixed siliciclastic–biogenic facies. Sedimentary organic carbon (C_{org}) has two major sources: terrestrial and marine. About 65% of the total terrestrial carbon flux survives as recalcitrant OM and can be incorporated in marine sediments (Ittekkot, 1988). In general, the concentration of this material varies as a function of proximity of the fluvial source and the dispersion processes. However, some recent studies have shown that much of this material is transported as sorbed phases on mineral surfaces, which can be shed in the nearshore zone and replaced by marine OM (e.g., Keil and Hedges, 1993; Mayer, 1994; Leithold and Blair, 2001). This marine C_{org} is far more labile and thus provides the dominant substrate for microbial decomposers, and through most of the marine realm its export to surface sediments is controlled by surface-water bioproductivity, heterotrophic "repackaging" (as fecal pellets with rapid down-column transit rates; Suess, 1980; Degens and Ittekkot, 1987), and water depth, which regulates the extent of degradation during export through the water column (e.g., Suess, 1980). In ancient epicratonic seaways, water depths probably varied from less than 100 m to not significantly more than 300 m. Consequently, decomposition during down-column transport was appreciably lower than for open-ocean settings (e.g., 61% versus 99% loss; Suess, 1980). Distal hemipelagic deposits in these basins commonly received minimal terrigenous OM, with marine phytoplanktonic and bacterial fractions dominating the sedimentary C_{org}.

The fundamental control on the production of marine OM and associated skeletal material

($CaCO_3$ and SiO_2) is the concentration and temporal continuity of biolimiting nutrients in surface waters (e.g., nitrate, phosphate, and micronutrients such as iron). In general, these are produced by chemical weathering and biogenic activities on land and transported via riverine or eolian processes to the marine realm, or they may be produced by biological processes within the water column (i.e., N-fixation). Although phosphate is transported in both dissolved (orthophosphate) and particulate (organic and inorganic) phases, only the dissolved form is readily bioavailable. However, preferential remineralization of particulate nutrients due to, for example, microbially mediated reactions under oscillating redox conditions (Aller, 1994; Ingall and Jahnke, 1997) may enhance nutrient recycling (Tyson and Pearson, 1991). In general, nitrogen is thought to be effective as a biolimiting nutrient on short (ecological) timescales, while phosphorus availability limits marine bioproductivity on geological timescales (Broecker and Peng, 1982; Berner and Canfield, 1989; Van Cappellen and Ingall, 1994; Falkowski, 1997; Tyrrell, 1999). In modern open-ocean settings the surface-water nutrient inventory is generally low, and upwelling zones represent major sites of enhanced bioproductivity. Nutrients remineralized in the deeper water column are actively recycled to the surface, thus driving new production. Climate exerts control on the biogenic flux by regulating nutrient supply, either through its effect on chemical weathering, soil biology, and rates of riverine transport, or by controlling water-column stability and wind-driven (Ekman-transport-induced) upwelling.

The pelagic biogenic flux includes carbon, nitrogen, and phosphorus in OM, which under conditions of exceptional preservation may be buried in proportions approximating the algal Redfield ratio (Redfield *et al.*, 1963; Murphy *et al.*, 2000a) but more commonly is enriched in carbon relative to nitrogen and phosphorus due to preferential nutrient release (Van Cappellen and Ingall, 1994; Aller, 1994; Ingall and Jahnke, 1997). Similarly, $CaCO_3$ accumulation can directly reflect primary productivity. Because epicratonic settings were sufficiently shallow to allow carbonate deposition with little or no dissolution, relations between $CaCO_3$ accumulation and total organic production in modern oceans (i.e., Broecker, 1982) may provide a means to reconstruct paleoproductivity in ancient hemipelagic deposits (Meyers *et al.*, in review).

7.06.2.3 Authigenic Flux

The third major input of material to fine-grained, mixed siliciclastic–biogenic facies is associated with authigenic processes that occur as a consequence of OM remineralization (bacterial and macrofaunal heterotrophy) in uncompacted sediments (Figure 1). In cases where the supply of O_2 exceeds the demand of microbial and macrofaunal respiration, most OM is efficiently oxidized by aerobes to dissolved inorganic C (ΣCO_2) and other byproducts, such as PO_4^{3-} and ultimately NO_3^-, which can return to surface waters and/or the atmosphere during water-column mixing events (Figures 1 and 2). If such conditions predominate on short to intermediate geological timescales (centuries to 1 Myr), the global carbon cycle remains in relative equilibrium. When O_2 supply fails to meet respiratory demand (e.g., due to diminished O_2 advection in a stratified water column, excess O_2 demand in the water column and sediments from enhanced production, or limitations in diffusional O_2 replenishment), oxygen is depleted and alternate terminal electron acceptors are employed in the remineralization of OM by a series of dysaerobic to anaerobic microbial communities (Figure 2); the dominant metabolic processes are nitrate reduction, manganese and iron reduction, sulfate reduction, and fermentation (methanogenesis), which represent progressively less efficient (i.e., less energy yielding per mole of C_{org} oxidized) forms of respiration (Froelich *et al.*, 1979; Berner, 1980; Canfield and Raiswell, 1991). While the relative net efficiencies of aerobic and anaerobic remineralization are debated (as discussed below), some fraction of the C_{org} is buried, and its removal from the short-term part of the carbon cycle may have direct consequences for climate change (i.e., Arthur *et al.*, 1988).

Nutrient release during OM remineralization appears to depend on the dominant type of bacterial decomposition, as well as its frequency and duration (e.g., Ingall *et al.*, 1993) and is thus related to the redox state of the system. Decoupled elemental release from OM (enhanced phosphorus regeneration relative to carbon) has been shown to occur during decomposition in sediments overlain by O_2-deficient bottom waters (e.g., Ingall and Jahnke, 1997) and may be particularly pronounced for nitrate under oscillating bottom-water redox conditions (Aller, 1994). In modern oceans, preferential nutrient release under normal aerobic conditions results in OM being buried with organic C : P ratios of ~250 : 1, compared to the assimilation ratio of ~106 : 1 (Van Cappellen and Ingall, 1994). The ratios increase dramatically under anoxic depositional conditions (Ingall *et al.*, 1993). Information about the extent to which nitrogen is preferentially regenerated from OM in natural settings is not abundant, as $C_{org} : N_{total}$ ratios are typically interpreted in terms of OM source rather than preservation state (e.g., Meyers, 1994). However, there is evidence that $C_{org} : N_{total}$ ratios increase with depth in sediments independent of source variation (e.g., Stevenson and Cheng, 1972), suggesting that postdepositional release of nitrogen may be enhanced by prolonged exposure to active microbial degradation.

In modern deep oceans, the aerobic zone dominates water columns and often the substrates, and most OM is remineralized by aerobic respiration during export, with less than 1–2% reaching the seafloor (Suess, 1980). This loss reflects the great depth of the oceanic water column, as well as vigorous resupply of O_2 due to thermohaline circulation. By contrast, up to 40% of primary production reaches substrates in shelfal areas where depths are ± 100 m (and by inference shallow epicontinental seaways), even though water columns are fully oxic (Suess, 1980). However, most of this OM is subsequently remineralized by aerobic and anaerobic bacteria and other heterotrophic organisms in the sediments. Consequently, net carbon burial remains very low, but the percent of the flux to the sediment–water interface that becomes buried and preserved beyond early diagenetic remineralization also varies sympathetically with rates of bulk sediment accumulation (Stein, 1986; Henrichs and Reeburgh, 1987; Canfield, 1989, 1994; Betts and Holland, 1991). In open oceanic water columns, nitrate reduction may occur at mid-water depths where the decomposition of descending particulate OM depletes available oxygen (Falkowski, 1997), but thermohaline circulation tends to resupply oxygen to bottom waters, ensuring the near-complete remineralization of pelagic OM. Upwelling zones with high fluxes of recycled nutrients and enhanced production are characterized by benthic anoxia and high burial fluxes of C_{org}, but they tend to be nonsulfidic—suggesting that oxygen and nitrate are resupplied by currents at levels just sufficient to prevent sulfate reduction (Arthur *et al.*, 1998). Well-documented occurrences of water-column sulfate reduction and high sulfide concentrations are known to occur today only in highly restricted (silled) basins with relatively isolated bottom waters and predominantly stratified water columns (Black Sea, Fjords), and/or within restricted settings beneath fertile surface waters (Cariaco Basin, California borderland basins) (e.g., Rhoads and Morse, 1971; Demaison and Moore, 1980; Pedersen and Calvert, 1990; Werne *et al.*, 2000; Lyons *et al.*, 2003).

Ultimately, the redox state of a depositional system represents a dynamic balance between supply of electron acceptors and their consumption during OM remineralization. The key controls include water-column mixing rate, which is influenced by relative sea level and climate, and OM production and export rates (plus feedback expressed in water-column redox and associated effects on OM preservation—see Section 7.06.2.4) controlled mainly by nutrient supply and the biology of the planktic biota. Because the redox state of a depositional system regulates various organic and inorganic reactions by which dissolved constituents are precipitated as geologically preservable minerals (e.g., metal oxyhydroxides versus pyrite) and added to the burial flux, it is possible to reconstruct the history of changes in redox. Examples of these constituents include Fe, Mn, V, Cr, Mo, and U.

7.06.2.4 Carbon Cycle and Climate Feedback

In addition to the intensity of incoming solar radiation, the surface temperature on Earth is regulated by the warming effect of greenhouse gases in the atmosphere, in particular CO_2, and this relationship is influenced by several negative and positive feedback loops (e.g., Berner, 1999). The inorganic part of the carbon cycle (weathering of silicates by carbonic acid to produce Ca^{2+} and HCO_3^-, deposition of $CaCO_3$ in carbonate rocks, and metamorphic recycling of CO_2 from subducted $CaCO_3$) stores and transfers the largest masses of carbon over very long timescales (Berner, 1999). However, the focus of this chapter is fine-grained detrital sediments and sedimentary rocks. Although $CaCO_3$ can be volumetrically important in these fine-grained deposits, these dominantly siliciclastic facies are the primary reservoir for burial of C_{org} in both modern and ancient marine environments. Unlike inorganic (carbonate) carbon, this organic carbon pool has the potential to produce large and rapid perturbations in the carbon cycle, with concomitant climatic effects (e.g., Arthur *et al.*, 1988; Kump and Arthur, 1999, Berner and Kothavala, 2001).

Viewed in the context of Figure 1, when C_{org} fixation by photoautotrophs (biogenic flux) is not matched by sediment and water-column decomposition by bacteria and other heterotrophs, significant masses of C_{org} can be buried in detrital-biogenic and pelagic sediments. If this occurs over sufficiently large areas, geologically rapid and/or prolonged changes in atmospheric p_{CO2} may result (e.g., Arthur *et al.*, 1988; Berner and Canfield, 1989; Kump and Arthur, 1999; Berner *et al.*, 2000; Berner, 2001; Berner and Kothavala, 2001). Interpreted episodes of widespread carbon burial, such as the Miocene Monterey event (Vincent and Berger, 1985), Cretaceous oceanic anoxic events (Schlanger and Jenkyns, 1976; Schlanger *et al.*, 1987), the Permo-Triassic “superanoxia” event (Isozaki, 1997), and the Late Devonian Kellwasser events (Joachimski and Buggisch, 1993), span the geological record and are thought to be responsible for major changes in climate and biotic extinction/evolution. Because of this, the factors controlling C_{org} burial have long been topics of intense interest.

Many past studies have attributed enhanced OM burial to the relative inefficiency of anaerobic bacterial respiration (i.e., Demaison and Moore, 1980). In this model the driving mechanism of carbon burial is pervasive anoxia in paleo-oceans and shallow seas, attributed respectively to sluggish thermohaline circulation during warm climate intervals (e.g., Schlanger and Jenkyns, 1976) or enhanced salinity stratification because of high freshwater flux to shallow epicontinental basins (e.g., Seilacher, 1982; Ettensohn, 1985a,b). More recently, the notion of permanent stratification in shallow epeiric seas associated with the "stagnant basin" model has been challenged by a model invoking dynamic, seasonal (thermal?) stratification (e.g., Oschmann, 1991; Sageman and Bina, 1997; Murphy *et al.*, 2000a; Sageman *et al.*, 2003) based, in part, on observations of modern shallow marine systems (Tyson and Pearson, 1991). At about the same time that this new "dynamic stratification" model was being developed, evidence suggesting that increase in the biogenic flux (i.e., greater productivity) played a significant role in enhanced OM burial was accumulating (e.g., Arthur *et al.*, 1987). Ultimately, Pedersen and Calvert (1990) cited evidence for similar rates of aerobic and anaerobic degradation (e.g., unexceptional C_{org} preservation in the modern Black Sea; Calvert *et al.*, 1991), as well as patterns of accumulation in modern high-productivity zones that argue for enhanced production as the main driver of ancient OM burial. In their view anoxic conditions were a consequence rather than a cause of OM accumulation.

Following another decade of research, neither the "preservation" nor "production" mechanism has emerged as a comprehensive paradigm. Publication of new observational and experimental data has suggested that differential degradation of OM does occur under variable redox conditions (Hartnett *et al.*, 1998; Van Mooy *et al.*, 2002), providing support for the preservation end-member. More recently, secondary C_{org} loss in Mediterranean sapropels associated with downward advancing oxidation fronts—following transitions from anoxic to oxic bottom water conditions—has indicated the role of O_2 exposure in diminished OM preservation (Thomson *et al.*, 1999; Slomp *et al.*, 2002; and reference therein). There are, however, also cases in which increased production without coeval evidence for significant anoxicity appears to account for enhanced OM burial (Meyers *et al.*, 2001), thus providing support for the production end-member. Whereas both production and preservation factors appear to be critical components, depending on circumstances to be explored herein, the role of varying bulk sedimentation rate (i.e., Johnson-Ibach, 1982; Henrichs and Reeburgh, 1987; Mueller and Suess, 1979) has received comparatively less attention. Yet, this process regulates OM concentration and controls the rate at which OM is transported from the oxic zone into underlying suboxic to anoxic zones, where the efficiency of decomposition is reduced, and thus ultimately impacts the extent of OM preservation (Toth and Lerman, 1977; Canfield, 1994).

7.06.3 CONCEPTUAL MODEL: PROXIES

Two fundamental types of geochemical proxies are employed in the reconstruction of ancient depositional environments and geological processes. Broadly defined, these are elemental and compound concentration/accumulation data and stable isotopic data. Depending on the type of data used and the process under investigation, there are specific limitations that characterize each proxy.

7.06.3.1 Limitations of Proxy Data

Ideally, each proxy employed in the study of ancient fine-grained mixed siliciclastic–biogenic facies should be based on a known or inferred relationship between a primary geological process and the corresponding flux of a geochemical component to the sediment—or an isotopic fractionation expressed in elements involved in the process. The most useful proxies are those for which a single or predominant controlling factor can be identified based on direct modern observations and for which preserved signals are particularly sensitive to changes in the primary process. Sources of error in the development of such proxy concepts include: (i) those associated with quantifying the geochemical relationship in modern systems (sampling and instrument error, as well as uncertainties/simplifications about the modern processes—e.g., Popp *et al.*, 1998); (ii) those associated with preservation (diagenetic alteration); (iii) those associated with inferences, assumptions, or extrapolations made when the spatial and temporal scale of an ancient data set exceeds that of modern data; and (iv) those associated with measurement of components in ancient rock samples (also including sampling and instrument error).

Elemental concentration data suffer from the inherent limitations of reciprocal dilution. The risk of spurious suggestions of covariance driven by mutual dilution (and thus false interpretations of coupled delivery) provides a major impediment for unambiguous determinations of elemental fluxes within ancient sediments. Conversion of concentration values to flux terms (mass area^{-1}

time^{-1}) may be accomplished in cases where age–depth relationships are adequately known to allow the calculation of sedimentation rates, and factors like bulk density and porosity can be measured or estimated (e.g., Bralower and Thierstein, 1987; Park and Herbert, 1987; Meyers *et al.*, 2001). However, timescales of sufficient resolution are relatively rare in most pre-Pleistocene sequences, and sedimentary geochemists have devised other means of resolving relative trends in elemental fluxes. These include the calculation of elemental ratios (e.g., Turekian and Wedepohl, 1961; Brumsack, 1989; Arthur *et al.*, 1988; Arthur and Dean, 1991; Calvert and Pedersen, 1993; Piper and Isaacs, 1995; Davis *et al.*, 1999; Hofmann *et al.*, 2003; Lyons *et al.*, 2003), which serve to normalize components to major dilutants and thus allow contributions to flux variability to be better isolated. Absolute concentrations and ratios of elemental constituents can also be evaluated relative to mean values within a stratigraphic section (Sageman *et al.*, 2003) or relative to mean values for a given lithotype (e.g., world average shale or WAS: Turekian and Wedepohl, 1961; North American shale composite or NASC: Gromet *et al.*, 1984; Post-Archean average Australian shales or PAAS: Taylor and McLennan, 1985; mid-continent shales: Cullers, 1994). However, some normalization techniques (e.g., X/Al : Y/Al cross-plots, where X and Y are trace metals) are not without pitfalls and must be applied with caution (Van der Weijden, 2002).

Fractionations among stable isotopes of carbon, nitrogen, and sulfur are by-products of both kinetic and equilibrium effects associated with inorganic and biologically mediated reactions, and many of these have been well characterized in modern experimental and environmental studies (Hayes, 1993; Altabet and Francois, 1993, Canfield, 2001). Although trends in sedimentary bulk isotopic data reflect a net fractionation in the paleo-reservoir being studied, the fact that many elements have multiple sources in the paleo-environment requires a mass balance (or at least species-based) approach to distinguish the major drivers of the fractionation (e.g., marine versus terrestrial C_{org} or pyrite versus organically bound sulfur). Compound-specific isotopic methods (e.g., Hayes *et al.*, 1989, 1990) have made possible the direct assessment of isotopic ratios in molecules that can be linked to specific primary sources.

In the sections below each of the processes introduced in Figure 1 (detrital, biogenic, and authigenic) is related to a corresponding set of proxies. In each case, specific strengths and limitations are reviewed, and working hypotheses for interpretation are evaluated. Given the low ratio of sedimentation rate-to-sample size (duration) that is typical for geochemical analysis in fine-grained facies, all proxy data are subject to a certain range of time averaging. The highest resolution data are commonly derived from 1 cm-thick (or less) samples taken through sequences with effective sedimentation rates (not corrected for compaction) between 0.5 cm kyr^{-1} and 2.0 cm kyr^{-1}, which therefore average, respectively, from 2 kyr to 0.5 kyr of depositional history into a single data point. However, recent undisturbed varved sequences at sites of comparatively rapid sedimentation (e.g., Black Sea and Cariaco Basin) provide a template for the very high level of temporal resolution possible in some settings (Hughen *et al.*, 1996, 1998).

7.06.3.2 Detrital Proxies

The objectives of detrital proxy analyses include: (i) identification of sources (e.g., weathered crust versus extrusive igneous), transport paths (fluvial, eolian, and ice-rafted), and depositional modes (suspension fallout versus gravity flow) of the terrigenous detrital constituents; (ii) determination of bulk and component fluxes of these constituents relative to biogenic and authigenic inputs; and (iii) determination of the controls on terrigenous fluxes (Figures 1 and 2). Determining the overall rate of terrigenous accumulation is important not only because it influences concentrations of biogenic and authigenic components through dilution/condensation (Johnson-Ibach, 1982) but more so because it actively modulates pore- and bottom-water redox conditions and organic carbon preservation by controlling the rate at which labile OM is transported through the successive decompositional zones shown in Figure 2 (Canfield, 1989; Meyers *et al.*, (in review)). While bulk sedimentation reflects the overall behavior of depositional systems tracts (Pasley *et al.*, 1991; Creaney and Passey, 1993), relative changes in the fluxes of individual components provide evidence for variation in dominant transport paths, which can be related to climate history (e.g., Arthur *et al.*, 1985).

Detrital geochemical proxies employed for reconstruction of source rock composition or provenance include rare-earth element suites and neodymium isotopes (e.g., McLennan *et al.*, 1993; McDaniel *et al.*, 1994; Cullers, 1994a,b; Weldeab *et al.*, 2002). Methods employing strontium and osmium isotopes (e.g., DePaolo, 1986; Ravizza, 1993) and Ge/Si ratios (Froelich *et al.*, 1992) have been used to reconstruct source area weathering and uplift history. Although such analyses are important in our understanding of the source-to-sink history of hemipelagic deposits, in this review we focus on the use of major-, minor-, and trace-element data to determine relative fluxes of detrital (weathered riverine or eolian) and/or volcanigenic fractions to bulk sedimentation (e.g., Arthur *et al.*, 1985; Pye and Krinsley, 1986; Arthur and Dean, 1991). These proxies track the relative proportions of detrital and volcanigenic mineral grains through their signature elemental compositions. Variations in these fluxes and the corresponding elemental signatures are conferred by differences in source, mode of transport (sorting), and rate of deposition and thus must be viewed in the appropriate geological context (paleogeography, tectonic framework, climate, etc.).

7.06.3.2.1 Physical methods

Hemipelagic facies generally include finely laminated to bioturbated to massive mudstones with variable concentrations of silt-sized particles (dominantly composed of quartz and authigenic pyrite), carbonate or siliceous content (<10 wt.% to >90 wt.%), and organic carbon (ranging from <1 wt.% to >20 wt.%). Variations in the proportion of a dominant biogenic component, such as calcium carbonate, relative to insoluble residue provide the basis for lithologic terms such as claystone (<10% $CaCO_3$), calcareous shale or mudstone (10–50% $CaCO_3$), marly shale or marlstone (50–75% $CaCO_3$), and limestone (>75% $CaCO_3$). The detrital component of these facies can be described in terms of its specific clay mineral composition, trends in grain size or optical characteristics of grains, and changes in other parameters directly controlled by the concentration of detrital grains relative to pelagic carbonate, such as magnetic susceptibility.

The major clay minerals include discrete illite, kaolinite, and chlorite, which reflect continental weathering and riverine discharge (Pratt, 1984; Leckie *et al.*, 1991), and mixed-layer illite/smectite, which results from postdepositional alteration of volcanic ash (Pollastro, 1980; Arthur *et al.*, 1985) and is an important background constituent of hemipelagic facies deposited near active volcanic belts. Variation in the proportions of different clays may reflect relative changes in dominance of riverine versus volcanigenic inputs (Arthur *et al.*, 1985; Dean and Arthur, 1998). Changes in the relative proportions of clays versus detrital grains larger than clay (>63 μm), including quartz, feldspar, biotite, and heavy mineral grains, are interpreted to reflect changes in bulk detrital flux by either mode of transport. Methods to quantify these changes include analysis of trends in grain size (e.g., Leithold, 1994; Rea and Hovan, 1995; Hassold *et al.*, 2003), quantification of weight percentages of detrital particles (e.g., DeMenocal, 1995), and in cases where sufficient time resolution is available, calculation of accumulation rates for the insoluble residue (Harris *et al.*, 1997) or individual detrital elements such as titanium (Meyers *et al.*, 2001). Petrographic and SEM analysis of grains can also assist in the determination of source and mode of transport (Schieber, 1996; Werne *et al.*, 2002). Finally, magnetic susceptibility—which measures changes in the proportion of magnetizable minerals and shows marked contrast between paramagnetic grains such as clays, ferromagnesian silicates, and iron sulfides versus diamagnetic components such as calcite—has been used as an indicator of changes in detrital flux (e.g., Ellwood *et al.*, 2000).

7.06.3.2.2 Elemental proxies

Elemental proxies of detrital flux are also based on changing proportions of mineral constituents and can be used to track subtle changes in grain size (e.g., Bertrand *et al.*, 1996). In hemipelagic rocks aluminum is generally regarded as the main conservative proxy for clay minerals, which dominate the terrigenous insoluble residue (Arthur *et al.*, 1985; Arthur and Dean, 1991; Calvert *et al.*, 1996). Although aluminum scavenging by sinking biogenic particles documented in the deep equatorial Pacific (Murray *et al.*, 1993; Murray and Leinen, 1996, Dymond *et al.*, 1997) would impair its use as a conservative tracer, this process is unlikely to have been significant in the comparatively shallow water columns of epicontinental basins where the studies described herein are focused. Thus, changes in detrital flux in these settings can be detected via: (i) variations in elements associated with coarser fractions relative to the aluminum proxy, such as changes in silicon related to detrital quartz silt, or in titanium and zirconium related to heavy mineral grains such as zircon, rutile, sphene, titanite, and ilmenite (Arthur *et al.*, 1985; Pye and Krinsley, 1986; Calvert *et al.*, 1996; Bertrand *et al.*, 1996; Davis *et al.*, 1999; Wortmann *et al.*, 1999; Yarincik *et al.*, 2000b; Haug *et al.*, 2003); (ii) changes in elements indicative of detrital clays, such as potassium associated with discrete illite, relative to background aluminum (Pratt, 1984, Arthur *et al.*, 1985; Pye and Krinsley, 1986; Yarincik *et al.*, 2000b; Hofmann *et al.*, 2003); and (iii) changes in elements indicative of altered volcanic ash, such as sodium and iron + magnesium—which reflect a background of eolian delivery—relative to indicators of hemipelagic detrital flux, such as potassium input as discrete illite (Dean and Arthur, 1998).

In each case, changes in the elemental ratios are caused by relative changes in bulk sedimentation—dilution or condensation results from changes in the terrigenous clay flux relative to eolian or other inputs. Correct interpretation of proxies depends on distinguishing elemental sources and transport modes. For example, some increases in Ti/Al are interpreted to reflect increased eolian flux relative to the hemipelagic background (Bertrand *et al.*, 1996; Yarincik *et al.*, 2000b), whereas in other cases the same signal is interpreted to indicate enhanced delivery of riverine detritus (Arthur *et al.*, 1985; Murphy *et al.*, 2000a; Meyers *et al.*, 2001). In situations where biotite is a dominant constituent of volcanic ash, changes in Ti/Al may track bentonite content. Similarly, enrichments in Si/Al that reflect proportions of silicon in excess of the aluminosilicate (mudrock) background may reflect enhanced quartz delivery due to eolian inputs (Pye and Krinsley, 1986; Werne *et al.*, 2002) or enhanced input of biogenic silicon and thus a productivity signal (Davis *et al.*, 1999). Geological context (e.g., proximity to arid source regions or deltaic systems), optical or SEM identification of grain types and surface textures (Schieber *et al.*, 2000; Werne *et al.*, 2002), and the use of multiple complementary proxies can help address these questions of source and transport mode.

7.06.3.3 Biogenic Proxies

The objectives of biogenic proxy analyses include: (i) identification of the sources of OM (terrigenous, marine algal, and bacterial) and biogenic skeletal material, (ii) determination of bulk and component fluxes of biogenic constituents, and (iii) determination of controls on ancient biogenic fluxes (Figures 1 and 2). Due to several factors, bulk organic carbon or pelagic skeletal material in ancient mudrocks cannot necessarily be viewed as reliable quantitative proxies for primary production in overlying surface waters. Firstly, OM and skeletal material such as $CaCO_3$ may derive from multiple sources. Second, as described above, concentrations of C_{org} and other constituents are influenced by bulk sedimentation rate, including relative dilution/condensation, as well as the possible influence of changes in mineral surface area and consequent sorption capacity with changing clay properties (Kennedy *et al.*, 2002). Therefore, identification of relative contributions of marine phytoplankton versus other biogenic components and, ideally, calculation of accumulation rates for individual components is required before primary and export production can be accurately determined. However, because the efficiency of OM remineralization, which depends on factors such as water depth and transport time, redox state of the water column and pore waters, and bulk sedimentation rates, exerts significant control on C_{org} burial flux (e.g., Emerson, 1985; Emerson and Hedges, 1988; Canfield, 1994; Meyers *et al.*, in review), reconstruction of productivity based on OM accumulation may be biased (similar arguments can be made for carbonate if dissolution is significant). Therefore, even if accurate accumulation rates can be determined, they must be calibrated against, among other things, proxies of redox history (see below; note that these arguments are the basis for the complementary multiproxy approach illustrated in Figures 1 and 2).

In most pre-Pleistocene deposits, where timescales typically do not allow high-resolution accumulation rate estimates, assessment of paleoproduction relies mostly on indirect, qualitative, or semiquantitative methods. Nongeochemical approaches include, for example, analyses of changes in planktic bioassemblages (e.g., Watkins, 1989; Burns and Bralower, 1998; Peterson *et al.*, 1991). Among the host of geochemical techniques investigated in the literature we will review a subset that have produced consistent and complementary results for the different time intervals we have studied. These include methods to assess OM and skeletal sources, stable isotopic techniques, analyses of elemental ratios, compound specific approaches, and accumulation rate calculations for biogenic components (e.g., C_{org}, $CaCO_3$, or SiO_2).

7.06.3.3.1 Organic matter sources

Methods employed to determine OM sources in the hemipelagic facies described in our case studies include organic petrography, which allows delineation of relative proportions of terrigenous versus marine algal macerals (Durand, 1980; Pratt, 1984), and calculation of hydrogen and oxygen indices from OM pyrolysis (Rock Eval), which allows characterization of kerogen types and can distinguish between terrestrial and marine, as well as oxidized/thermally mature and well-preserved kerogen OM (Pratt, 1984; Kuhnt *et al.*, 1990). Other methods that contribute to recognition of OM source include C : N ratios (Meyers, 1994) and organic compounds or biomarkers (de Leeuw *et al.*, 1995).

7.06.3.3.2 Stable carbon isotopes of OM ($\delta^{13}C$)

Despite the wide diversity of controls on the carbon isotope composition of preserved OM and biogenic $CaCO_3$ (e.g., Hayes, 1993; Kump and Arthur, 1999), it is possible under certain circumstances to argue that a few variables are dominant and thus to relate changes in $\delta^{13}C_{org}$ to trends in paleoproduction. As argued originally by Scholle and Arthur (1980), Lewan (1986), Arthur *et al.* (1988), and others, changes in the $\delta^{13}C$ of preserved marine algal OM and biogenic $CaCO_3$ reflect changes in the isotopic composition of dissolved inorganic carbon in surface waters, which on short to intermediate timescales (<1 Myr) may be controlled by the balance between net respiration and net burial of OM in sediments. For example, positive shifts in the $\delta^{13}C$ of organic carbon dominantly sourced from marine photoautotrophs have been interpreted to reflect elevated burial fluxes of OM related to global increase in primary productivity (Arthur *et al.*, 1987, 1988), whereas negative shifts have been interpreted to reflect recycling of respired CO_2 in a more localized reservoir (e.g., Saelen *et al.*, 1998; Murphy *et al.*, 2000a; Rohl *et al.*, 2001). For more detailed recent reviews of the controls on C-isotope fractionation see Kump and Arthur (1999), Hayes *et al.* (1999), and papers in Valley and Cole (2001).

7.06.3.3.3 Elemental ratios

A common indirect approach to reconstructing paleoproductivity is the analysis of components that reflect changes in nutrients. Among the numerous methods described in the literature, we have focused on the ratios of carbon, nitrogen, and phosphorous, which have been employed in studies of Paleozoic, Mesozoic, and Cenozoic black shales (Ingall *et al.*, 1993; Murphy *et al.*, 2000a; Slomp *et al.*, 2002; Filippelli *et al.*, 2003; Sageman *et al.*, 2003). Additional work has demonstrated the utility of barium (e.g., Ba/Al or Ba/Ca) as a proxy for paleoproductivity in some settings (Dymond *et al.*, 1992; Francois *et al.*, 1995; Paytan *et al.*, 1996; Van Santvoort *et al.*, 1996; cf. McManus *et al.*, 1999), but has also highlighted the limitations of this method under reducing conditions where low sulfate concentrations can lead to barite undersaturation (McManus *et al.*, 1998). Although promising in oceanic settings, use of the barium proxy in ancient epeiric deposits may be difficult.

Based on the observation that suboxic to anoxic decompositional processes favor the strongly preferential release of nitrogen and phosphorus from OM (Aller, 1994; Van Cappellen and Ingall, 1994; Ingall and Jahnke, 1997), increases in the ratios of C : N : P in preserved OM may imply increasing nitrogen and phosphorus bioavailability. First, preserved OM must be determined to be predominantly marine algal in origin (e.g., by organic petrography, biomarker abundance and isotopic composition, etc.). If so, it is reasonable to assume that its C : N : P content originally approximated the modern Redfield ratio of 106 : 16 : 1 (Redfield *et al.*, 1963). However, remineralized phosphorus is not necessarily released to the overlying water column but may instead become immobilized in sediments in inorganic form (e.g., Filippelli, 1997). Changes in total sedimentary phosphorus can be evaluated as an indication of the amount of phosphorus that was remineralized and not subsequently precipitated and thus potentially bioavailable. Although there may be sources of phosphorus in excess of that associated with sedimented OM (Schenau and DeLange, 2001), such additions would tend to reduce C : P anomalies, suggesting that C_{org} to total phosphorus values provide a minimum estimate of the phosphorus released by the sediment during OM remineralization.

The amount of OM that survives to become sedimentary OC is a function of the rate of production and down-column export, as well as the dominant type of decomposition (metazoan and/or microbial) and its duration relative to bulk sedimentation rate (burial). It should be noted that these generalizations, derived from the study of modern oceans, must be viewed in the context of shallow epicontinental settings where total depth probably did not exceed ±300 m during maximum highstands. Thus, the transit time of OM in the water column was comparatively short, and the relative percentage of production to reach the sediment surface (export production) was high (Suess, 1980). As a consequence, respiratory demand in the bottom waters was likely to have been intense, especially during warm seasons when thermal stratification of the water column prevented downward advection of dissolved O_2. If decompositional release of nutrients occurred during establishment of seasonal (or longer-term) thermoclines in these shallow seas, and such nutrients were recycled to surface waters when the thermocline dissipated, an effective mechanism was available to drive primary production and increase the burial flux of carbon to the sedimentary reservoir (Murphy *et al.*, 2000a). This simplified view omits discussion of many issues, such as sedimentation rate controls on phosphorus burial (Tromp *et al.*, 1995) and dynamics of N-cycling in the water column (e.g., Altabet *et al.*, 1991; Holmes *et al.*, 1997). However, since studies of ancient stratigraphic sequences tend to span large time intervals based on analysis of significantly time-averaged samples, and since phosphorus is regarded as the more significant limiting nutrient on such geological timescales, we argue that the approach described here, in combination with other proxies, provides a reasonable first-order approximation of nutrient–productivity dynamics.

7.06.3.3.4 Biomarkers

Changes in abundance and isotopic composition of biomarker compounds can help delineate relative mass contributions from terrestrial, marine algal, and bacterial OM sources, as well as provide indications of specific processes and environmental conditions in ancient water columns (e.g., Hayes *et al.*, 1989; Sinninghe Damsté *et al.*, 1993; Silliman *et al.*, 1996; Kuypers *et al.*, 2001; Simons and Kenig, 2001; Pancost *et al.*, 2002). In some of our studies discussed below these methods have been used in a limited fashion, mainly to help constrain the relative contributions of different OM sources to observed $\delta^{13}C_{org}$ variations (e.g., Murphy *et al.*, 2000a).

7.06.3.3.5 Accumulation rates

With high-resolution timescales, calculation of accumulation rates (mass area^{-1} time^{-1}) for C_{org}, $CaCO_3$, and SiO_2 (e.g., Bralower and Thierstein, 1984; Archer, 1991; Pedersen *et al.*, 1991; Sancetta *et al.*, 1992; Arthur *et al.*, 1994; Calvert and Karlin, 1998) can provide a direct measure of the net burial flux of biogenic components. However, relating burial fluxes to primary production may be complicated by factors described above. For example, significant remineralization of OM can occur even under the most anoxic conditions (Canfield, 1989). Although the dissolution of $CaCO_3$ provides a similar impediment to its use as a linear proxy for paleoproduction, in shallow seas where surface-water productivity was dominated by calcareous nanoplankton, $CaCO_3$ accumulation rates may provide a reasonable first-order estimate of production. Broecker (1982) estimated a 1 : 4 relationship between $CaCO_3$ and C_{org} fixation rates in modern open-ocean settings. Meyers *et al.* (in review) evaluated this hypothesis using a compilation of modern data representing a range of values for $CaCO_3$ accumulation and C_{org} production and confirmed that the 1 : 4 ratio applied only under the highest rates of $CaCO_3$ accumulation. Although timescales of sufficient resolution to calculate accumulation rates are quite difficult to establish in pre-Pleistocene stratigraphic sequences where datable horizons may be few and disconformities many, in an increasing number of studies orbital timescales developed from analysis of rhythmically bedded hemipelagic facies are being employed for this purpose (Herbert and Fischer, 1986; Herbert *et al.*, 1986; Park and Herbert, 1987; Meyers *et al.*, 2001).

7.06.3.4 Authigenic Proxies

The chemical behavior of various minor and trace elements is relatively well characterized for particular redox conditions, and there has been significant effort directed at the development of geochemical proxies for paleo-oxygenation in black shale sequences (see reviews in Calvert and Pedersen, 1993; Arthur and Sageman, 1994; Jones and Manning, 1994; Wignall, 1994; Schieber *et al.*, 1998a,b). Elements of proven paleoredox utility include Mo (Coveney *et al.*, 1991; Dean *et al.*, 1999; Meyers *et al.*, in review), V–Ni (Lewan and Maynard, 1982; Lewan, 1984; Breit and Wanty, 1991), U (Wignall and Myers, 1988), Mn (Calvert and Pedersen, 1993), Re (Crusius *et al.*, 1996), and rare-earth elements or, more specifically, the Ce anomaly (Wright *et al.*, 1987; Wilde *et al.*, 1996; cf. Bright *et al.*, submitted; see German and Elderfield, 1990, for review). In addition, proxies that directly assess HS^- (ΣH_2S) availability in ancient water columns include: (i) degree of pyritization, which is a measure of the extent to which reactive iron has been transformed to pyrite, and related iron approaches (Berner, 1970; Raiswell *et al.*, 1988; Canfield *et al.*, 1996; Raiswell *et al.*, 2001); (ii) sulfur isotope relationships (Jørgensen, 1979; Goldhaber and Kaplan, 1974, 1980; Anderson *et al.*, 1987; Fisher and Hudson, 1987; Beier and Hayes, 1989; Habicht and Canfield, 1997; Lyons, 1997); (iii) bacterial pigments indicative of anoxygenic photosynthesis in the presence of hydrogen sulfide (e.g., Repeta, 1993; Sinninghe Damsté *et al.*, 1993; Koopmans *et al.*, 1996; Huang *et al.*, 2000); and (iv) pyrite framboid size distributions (Wilkin *et al.*, 1996; Wignall and Newton, 1998).

Studies in modern oxygen-deficient basins have helped to calibrate these methods (Spencer and Brewer, 1971; Jacobs and Emerson, 1982; Jacobs *et al.*, 1985, 1987; de Baar *et al.*, 1988; Anderson *et al.*, 1989a, b; German and Elderfield, 1989; German *et al.*, 1991; Lewis and Landing, 1992; Repeta, 1993; Sinninghe Damsté *et al.*, 1993; Van Cappellen *et al.*, 1998; Morford and Emerson, 1999), including those associated with iron and molybdenum cycling and resulting patterns of (sulfide-driven) mineralization that can preserve deep into the geological record (Francois, 1988; Emerson and Huested, 1991; Lewis and Landing, 1991; Canfield *et al.*, 1996; Calvert *et al.*, 1996; Crusius *et al.*, 1996; Helz *et al.*, 1996; Lyons, 1997; Wilkin *et al.*, 1997; Raiswell and Canfield, 1996, 1998; Dean *et al.*, 1999; Yarincik *et al.*, 2000a; Zheng *et al.*, 2000; Adelson *et al.*, 2001; Wijsman *et al.*, 2001; Wilkin and Arthur, 2001; Lyons *et al.*, 2003). Rather than providing cursory background for each of a large number of proxies, we will emphasize a few approaches of particular value in recent black shale research, while also providing some historical perspective.

7.06.3.4.1 C–S relationships

The C–S (organic carbon–pyrite sulfur) paleoenvironmental method—which has long been applied to ancient shale-bearing sequences—is based on the observation that different factors generally limit sedimentary pyrite formation in normal marine (oxic bottom waters), euxinic marine (anoxic and H_2S-containing), and freshwater to brackish settings (Berner, 1984). In normal marine Phanerozoic environments, the supply of dissolved sulfate for bacterial reduction to H_2S is typically not limiting because of the ample concentrations in seawater (contrasting low marine sulfate availability during the Precambrian is summarized in Lyons *et al.*, in press). As the result of reactions between the bacterially generated H_2S and detritally delivered reactive iron, appreciable pyrite sulfur (Spy) is produced during diagenesis if sufficient C_{org} is also available. The bacteria oxidize C_{org} in the process of reducing sulfate to H_2S. In normal marine settings, the formation of pyrite can be limited by the availability of bacterially metabolizable OM. This linkage is expressed as a positive linear relationship between concentrations of C_{org} and Spy—with a zero sulfur intercept. Normal marine sediments from a wide range of Holocene localities yield a linear trend with a mean C/S weight ratio of 2.8 (Berner, 1982; Morse and Berner, 1995). Expressions of the coupling between C_{org} and Spy and corresponding variations in C/S ratios in normal shales spanning the Phanerozoic are addressed in Raiswell and Berner (1986), although studies are increasingly demonstrating the potential for reactive-iron limitation within organic-rich sediments beneath oxic water columns (Canfield *et al.*, 1992; see Lyons *et al.*, 2003).

Euxinic marine environments are complex in that limitations in iron, rather than C_{org}, are the norm, and the presence of sulfide within the water column can decouple pyrite formation from the local burial flux of C_{org} (Raiswell and Berner, 1985; Lyons and Berner, 1992; Lyons, 1997). Regardless of the complexity, these anoxic marine systems are noted for their relatively high abundances of Spy and C_{org}. By contrast, sediments deposited under the generally sulfate-deficient conditions of natural freshwater to brackish settings display low amounts of Spy despite the potential for high concentrations of C_{org} and thus plot close to the C_{org} axis on a C–S plot (Figure 3; Berner, 1984; Berner and Raiswell, 1984). In practice, specific paleo-environmental variations within the marine realm are often difficult to delineate uniquely using the C–S technique because of the complex interplay among primary and secondary controlling factors (e.g., Lyons and Berner, 1992). As a result, distinctions between fully marine and low salinity freshwater-to-brackish settings are the least ambiguous application of the C–S method. Furthermore, primary C–S relationships can be masked by weathering, metamorphism, and both low-temperature and hydrothermal secondary overprints (Figure 3; Leventhal, 1995; Lyons *et al.*, 2000, 2003; Petsch *et al.*, 2000, 2002).

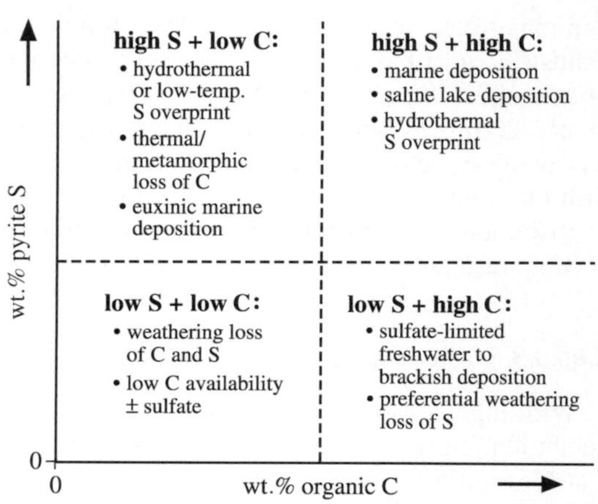

Figure 3 Schematic plot emphasizing the principal environmental controls, both primary and secondary, on C_{org} versus S_{py} distributions in fine-grained siliciclastic sediments and sedimentary rocks (after Lyons *et al.*, 2000). Preferential loss of sulfur during weathering is addressed in Petsch *et al.* (2000, 2002); however, weathering effects are minimized through the use of drill core. Low- and high-temperature sulfur overprints can be distinguished from primary S_{py} signals by their $\delta^{34}S$ characteristics (Lyons *et al.*, 2000, 2003). Background is provided in Section 7.06.3.4.1 and Lyons *et al.* (2000).

7.06.3.4.2 Sulfur isotope relationships

Details of sulfur isotope geochemistry are presented elsewhere in this volume (see Chapter 7.10) and are only highlighted here as related to paleo-environmental interpretations of fine-grained siliciclastic sequences. Formation of sedimentary pyrite initiates with bacterial sulfate reduction (BSR) under conditions of anoxia within the water column or sediment pore fluids. The kinetic isotope effect associated with bacterial sulfate reduction results in hydrogen sulfide (and ultimately pyrite) that is depleted in ^{34}S relative to the $^{34}S/^{32}S$ ratios of residual sulfate (Goldhaber and Kaplan, 1974). The balance between net burial versus oxidative weathering of pyrite controls the $^{34}S/^{32}S$ ratio in the global oceanic sulfate reservoir and, along with the redox cycling of organic carbon, is the principal modulator of PO_2 in the atmosphere over geological time (Claypool *et al.*, 1980; Berner and Petsch, 1998).

Dissimilatory BSR under pure-culture laboratory conditions can produce sulfide depleted in ^{34}S by ~2–46‰ relative to the parent sulfate (Chambers *et al.*, 1975; Canfield, 2001; Detmers *et al.*, 2001). Although this range is generally accepted, controls on the magnitude of this fractionation are subjects of recent debate. For example, contrary to a long-held assumption, the isotopic offset between parent sulfate and HS- produced during BSR ($\Delta^{34}S$) may not vary with a simple inverse relationship to the rate of sulfate reduction (cf. Kaplan and Rittenberg, 1964; Canfield, 2001; Detmers *et al.*, 2001; Habicht and Canfield, 2001). Furthermore, in light of the significantly smaller isotope effects attributable to BSR under pure-culture conditions and by inference in natural settings (cf. Wortmann *et al.*, 2001), recent studies have addressed the fractionations of up to and exceeding 60‰ that abound in the Phanerozoic record. One model invokes bacterial disproportionation of elemental sulfur and other sulfur mediates as a means of exacerbating the ^{34}S depletions observed in HS- and pyrite (Canfield and Thamdrup, 1994; Habicht and Canfield, 2001).

Ultimately, net isotopic fractionations preserved in geological systems reflect both the magnitudes of bacterial fractionations and the properties of the sulfate reservoir—as recorded in the integrated history of pyrite formation (Zaback *et al.*, 1993). Even in the presence of large fractionations during BSR and coupled disproportionation, comparatively high $\delta^{34}S_{sulfide}$ values can occur in environments where renewal of sulfate is restricted relative to the rate of bacterial consumption (e.g., within sediments under conditions of rapid accumulation). Conversely, low $\delta^{34}S$ values are typical of marine systems where sulfate availability does not limit BSR. Because of these multiple controlling factors, bacteriogenic pyrite can display a broad range of $\delta^{34}S$ values that are often very low (^{34}S-depleted) relative to coeval seawater sulfate. In euxinic settings (i.e., those with persistently anoxic and sulfidic bottom waters such as the modern Black Sea and Cariaco Basin), much (often most) of the pyrite forms (syngenetically) within the sulfidic water column and shows the light and uniform $\delta^{34}S$ values expected under conditions of large sulfate and sulfide reservoirs. Conversely, under oxic depositional conditions all of the pyrite forms diagenetically and is thus vulnerable to the ^{34}S enrichments and wide $\delta^{34}S$ variability expected in a restricted pore-water reservoir. At euxinic sites of very rapid sediment accumulation, such as coastal Fjords and marginal locations within larger anoxic basins (Lyons, 1997; Hurtgen *et al.*, 1999), iron sulfides form both diagenetically and syngenetically, which results in ^{34}S-enrichment relative to distal, largely syngenetic pyrite pools. These facies-dependent sulfur isotope trends are well expressed in a number of ancient marine sequences (Figure 4; Anderson *et al.*, 1987; Fisher and Hudson, 1987; Beier and Hayes, 1989).

In the two Mesozoic examples shown in Figure 4, the light and uniform $\delta^{34}S$ values that typify water-column pyrite formation under euxinic conditions, and are further favored by the comparatively slow rates of sediment accumulation, are well expressed. The Cretaceous data of Gautier (1986, 1987) also show the classic broad range of $\delta^{34}S$ values possible for oxic deposition. Under the diffusion-controlled sulfate fluxes of diagenesis, rapid sedimentation can favor ^{34}S enrichment in pyrite by enhancing the rates of BSR and reactive iron availability below surface sediment layers and by facilitating protracted transformations of iron monosulfide precursors to pyrite (Hurtgen *et al.*, 1999). These pore-water reservoir effects occur beneath both oxic and anoxic bottom waters. Conversely, the openness of pore waters beneath oxic bottom waters can be maintained by infaunal mixing and bio-irrigation. These processes of

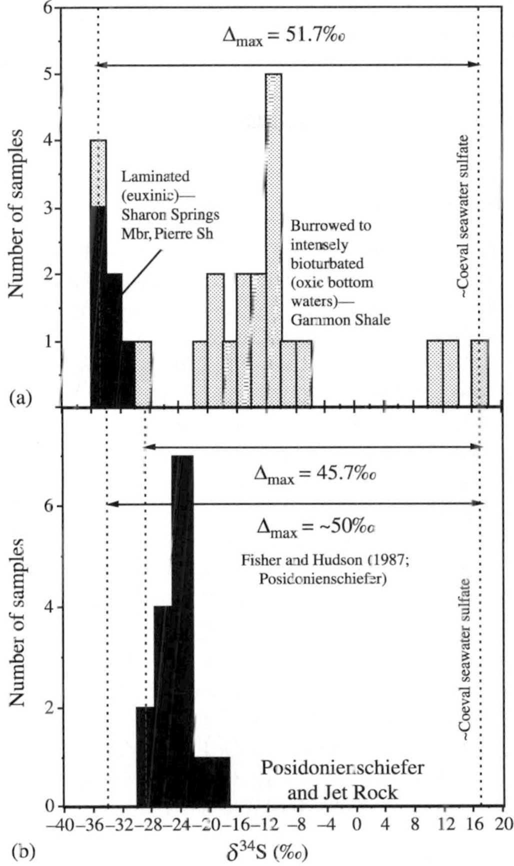

Figure 4 Sulfur isotope summary for black shales from the Pierre Shale of the Cretaceous Western Interior, North America (Gautier, 1986, 1987), and the Jurassic Posidonienschiefer and Jet Rock (Raiswell *et al.*, 1993). For comparison, the maximum fractionation observed in the Posidonienschiefer by Fisher and Hudson (1987) is also shown. The isotopically uniform and strongly ^{34}S-depleted pyrites of the Jurassic shales and the Cretaceous Sharon Springs Member of the Pierre Shale—like the sediments of the modern Black Sea and Cariaco Basin (Figure 7)—are diagnostic of euxinic (water-column) pyrite formation (see Section 7.06.3.4.2). By contrast, the Cretaceous Gammon Shale shows the ^{34}S enrichments and broad range of $\delta^{34}S$ values possible under oxic depositional conditions (Gautier, 1986, 1987).

biologically enhanced sulfate transport, in combination with sulfide reoxidation, can yield comparative ^{34}S deficiencies in pyrite of normal marine sediments.

7.06.3.4.3 Iron

A recent model for iron distributions within marine basins suggests that regional and temporal patterns in iron speciation—as manifested in degrees of pyritization and ratios of reactive (Fe_R)-to-total Fe (Fe_T) and Fe_T-to-Al—are largely controlled by the relative proportions of iron delivered with detrital sediments (e.g., iron oxides and silicates) and as a fraction that is decoupled from the local detrital flux through scavenging of dissolved iron in a euxinic water column during syngenetic pyrite formation (Figure 5; Canfield *et al.*, 1996; Lyons, 1997; Raiswell and Canfield, 1998; Raiswell *et al.*, 2001; Lyons *et al.*, 2003). (Degree of pyritization (DOP) is historically defined as the ratio of pyrite iron (Fe_{py}) to Fe_{py} plus iron extractable with boiling, concentrated HCl—Berner, 1970; Raiswell *et al.*, 1988; Raiswell and Canfield, 1998; Lyons *et al.*, 2003).

By this model, oxic settings (where iron scavenging is precluded) and sites of rapid euxinic siliciclastic accumulation show low to intermediate DOP values in association with Fe_R/Fe_T and Fe_T/Al ratios that are typical of the riverine (continental) flux, while more-condensed euxinic settings can show dramatic augmentation of the iron reservoir (via Fe_{scav})—and thus diagnostic elevation of the three iron-based paleoredox proxies (Raiswell *et al.*, 1988; Raiswell and Canfield, 1998; Raiswell *et al.*, 2001; Lyons *et al.*, 2003). Because the iron paleoredox proxies are a function of (i) the presence or absence of hydrogen sulfide in the water column, (ii) the intensity of water-column pyrite formation and associated iron scavenging, and (iii) the relative siliciclastic flux (Figure 6), iron distributions at euxinic sites near the basin margin can look like those of oxic sediments because the scavenged iron is swamped by the high rates of siliciclastic accumulation (Raiswell *et al.*, 2001; Lyons *et al.*, 2003). As a result, high values of DOP (and Fe_T/Al ratios elevated above the local detrital flux) point uniquely toward strong and likely persistent euxinic conditions in the basin, while very low DOP values are suggestive of oxic deposition under conditions of low OM accumulation. Intermediate DOP values, by contrast, can reflect either oxic deposition associated with appreciable organic accumulation—wherein pyrite formation can be iron limited—or euxinic deposition at sites of rapid siliciclastic influx (Lyons *et al.*, 2003).

While the details are almost certainly more complex than the model presented here and elsewhere (e.g., Raiswell *et al.*, 2001; Lyons *et al.*, 2003), nearshore–offshore gradients in DOP (and overall iron patterns of enrichment as

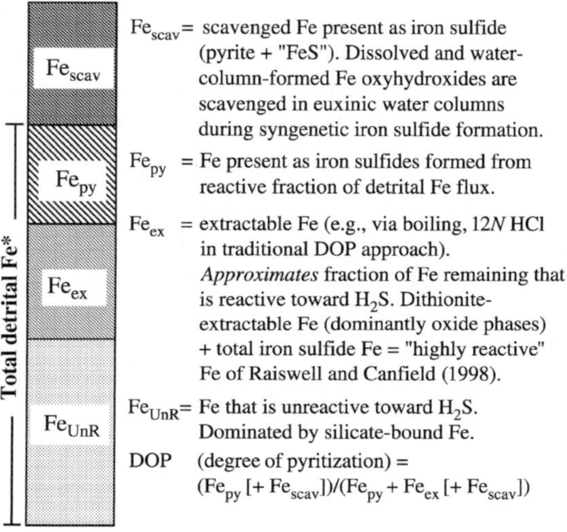

Figure 5 Summary of iron speciation in fine-grained siliciclastic sediments and sedimentary rocks. Total Fe (Fe_T) is equal to the sum of all these fractions. DOP increases in oxic sediments through the conversion of Fe_{ex} to Fe_{py}, although the HCl procedure generally overestimates the readily reactive iron available. In euxinic settings, high DOP values (Fe_T/Al ratios) result from scavenging of dissolved iron during pyrite formation in the water column. See Section 7.06.3.4.3 and Lyons *et al.* (2003) for further discussion and background.

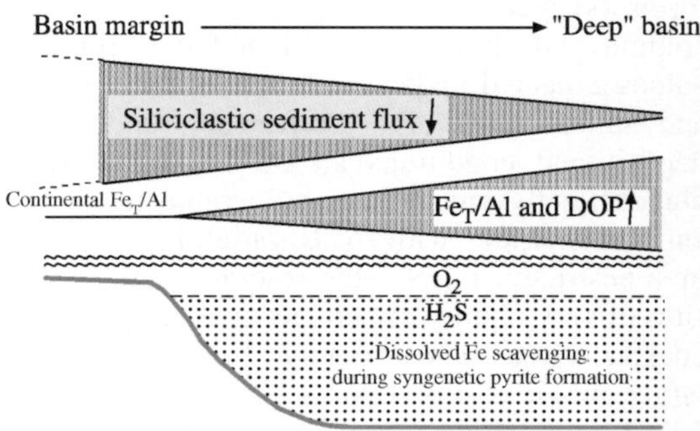

Figure 6 Schematic representation of the spatial gradients in Fe_T/Al ratios and DOP values in the bottom sediments of euxinic basins. Nearshore-to-offshore increases in these parameters derive from corresponding decreases in the siliciclastic flux and thus increases in the proportion of (i) iron scavenged during syngenetic pyrite formation in the euxinic water column to (ii) iron delivered with the local siliciclastic flux. The scavenged iron is initially present as dissolved iron and as oxyhydroxides formed near the chemocline and is therefore decoupled from the local siliciclastic (aluminum) flux. Details and other supporting references are provided in Section 7.06.3.4.3 and Lyons *et al.* (2003).

recorded in Fe_R/Fe_T and Fe_T/Al ratios) cannot simply be attributed to transport-related phenomena independent of depositional redox (e.g., sorting by grain size, mineralogy, or grain type—such as fluxes dominated by riverine versus eolian inputs). For example, the Orca Basin—a small anoxic brine pool on the northern slope of the Gulf of Mexico—shows a transition from Fe_T/Al ratios of 0.40–0.50 for oxic sediments (much like average continental crust; Taylor and McLennan, 1985) to values of ~0.55–0.75 within the anoxic zone. Ratios of ~0.85–1.00 occur in the transition zone where the redox interface (chemocline), with its associated iron cycling and particulate iron maximum (Van Cappellen *et al.*, 1998), impinges on the substrate (Lyons, unpublished data). The essential point here is that the iron pattern is linked to depositional redox, and, from the standpoint of siliciclastic transport and associated fractionation of the iron reservoir, the oxic, transitional, and anoxic sites are all equidistant from the shoreline.

7.06.3.4.4 Trace metals

In our research (e.g., Werne *et al.*, 2002; Lyons *et al.*, 2003), molybdenum has emerged as one of the empirically most useful, but mechanistically least understood proxies for depositional redox (cf. Meyers *et al.*, in review). It is clear to us and others that molybdenum enrichments in organic-rich facies are linked to the presence of appreciable hydrogen sulfide and are coupled, directly or indirectly, to C_{org} accumulation, where the type of OM present may also play a critical role (Coveney *et al.*, 1991; Helz *et al.*, 1996). What is less clear, however, are the specific mineralization mechanisms by which molybdenum accumulates and whether such enrichment can speak specifically to the presence or absence of sulfide in the water column, in contrast to the sediment pore waters. Molybdenum (dominantly as molybdate) is about two orders of magnitude more abundant in oxic seawater than iron. Consequently, the iron enrichments observed in euxinic sediments must reflect scavenging within anoxic water columns where dissolved iron is greatly elevated, and diffusion of dissolved iron into sulfidic sediments beneath an oxic water column is quantitatively negligible. Molybdenum, however, because of its greater abundance in oxic waters can diffuse into sulfide-rich pore waters and become appreciably enriched in the sediments relative to detritally delivered molybdenum.

The absence of elevated molybdenum concentrations in particulate samples from sediment traps deployed in the water columns of modern anoxic basins (e.g., Saanich Inlet, Cariaco Basin, and Black Sea) has suggested to some that molybdenum enrichment occurs only during diagenesis by reaction at or through diffusion across the sediment–water interface (Francois, 1988; Emerson and Huested, 1991; Crusius *et al.*, 1996) and that the link between high molybdenum concentration and accumulation and independently indicated euxinicity is the abundance of organic- and sulfide-rich sediments accumulating in oxygen-deficient settings (Zheng *et al.*, 2000). If true, a euxinic setting is not required for molybdenum enrichment. It should be noted, however, that the hydrogen sulfide concentrations in these anoxic water columns (e.g., <100 μM to 300–400 μM for the Cariaco Basin and Black Sea, respectively) may be below the critical threshold described by Helz *et al.* (1996) for the switch to particle-reactive thiomolybdate and thus may not effectively represent all sites of ancient anoxia. Alternatively, the availability of sediment-trap data remains small and potentially ambiguous, particularly for modern anoxic sites of comparatively high and persistent sulfide concentrations—such as Framvaren Fjord, Norway. It is also interesting to note that molybdenum is enriched significantly within Unit 1 sediments in the Black Sea—even at sites with present-day pore-water concentrations of dissolved sulfide that are not appreciably elevated above bottom-waters values (Lyons and Berner, 1992; Lyons, unpublished data).

In the case studies provided below we see evidence for molybdenum enrichments that are precisely coincident with the onset of euxinic conditions (Werne *et al.*, 2002; Lyons *et al.*, 2003), despite evidence for appreciable diagenetic dissolved sulfide in the underlying sediments deposited under conditions ranging from oxic to perhaps dominantly anoxic/nonsulfidic. In this review we cannot hope to resolve the role water-column sulfide plays in molybdenum accumulation and enrichment other than to suggest that (i) high concentrations of sedimentary molybdenum appear to be diagnostic of depositional environments with euxinic water columns, although high rates of molybdenum accumulation are not necessarily indicative of euxinicity (Meyers *et al.*, in review), and (ii) reactive iron availability may ultimately control molybdenum concentration and accumulation rate by modulating the amount of sulfide buildup in sediments and in the water column through iron sulfide formation (Meyers *et al.*, in review). Also of interest is the commonly observed, strongly positive covariance between the concentrations and accumulation rates of molybdenum and those of C_{org}. We are currently exploring the mechanistic underpinnings of this relationship and how it might constrain paleo-environments over geological time. Currently, we favor two models: (i) reactions between OM and molybdenum in the presence of dissolved sulfide that yield a systematic (stoichiometric?) relationship between the two (e.g., Coveney *et al.*, 1991) and (ii) the possibility that sulfide alone is the link, with increasing OM yielding increasing H_2S via BSR and thus parallel molybdenum sequestration (see review in Lyons *et al.*, 2003). It seems unlikely, however, that there will be a consistent slope for this relationship across time or among basins—as evidenced by the two distinct trends that are present for sediments deposited since the last oxic–anoxic transition in the Cariaco Basin (Lyons *et al.*, 2003).

Recently, mass-dependent isotopic variation for molybdenum has surfaced as a tool of high paleo-environmental potential (Barling *et al.*, 2001; Barling and Anbar, submitted). Specifically, relatively inefficient scavenging of molybdenum from seawater by oxide phases under oxic marine conditions—compared to efficient removal under anoxic settings—yields patterns of molybdenum isotope variability in seawater (recorded in the black shales) that may track the global proportion of oceanic anoxia over geological time.

Other metals that occur as minor and trace elements in marine waters and clastic sediments, but may become concentrated by precipitation from seawater under appropriate redox conditions and thus have redox proxy potential, include Mn, V, Cr, Ni, Co, Cu, U, and Th (Emerson and Huested, 1991; Jones and Manning, 1994; Piper, 1994; Calvert and Pedersen, 1996). Among these we have investigated Mn, V, and Cr relationships in some of our studies, in addition to the proxies described above. The rationale for the V + Cr proxy is as follows: V and Cr, both of which have a variety of common valence states and correspondingly complex chemistries, are precipitated from seawater as hydroxides or hydrated oxides at Eh conditions that correspond to the range in which denitrification occurs (Piper, 1994). Relative increases in the sedimentary concentration of V + Cr may, therefore, be indicative of a proportional increase in the significance of denitrification relative to aerobic respiration. This is an important parameter to constrain for two reasons: (i) denitrification is a dysaerobic to anaerobic metabolism, whose increased prominence suggests the relatively reduced availability of oxygen in the system (Froelich *et al.*, 1979), but unlike sulfate-reduction, does not produce a by-product that is toxic to aerobes; and (ii) denitrification leads to loss of the biolimiting nutrient nitrogen from the system as the chemically reduced N_2, which is insoluble and escapes to the atmosphere.

Below Eh values of ~500 mV, manganese occurs as the soluble Mn^{2+} ion in seawater, whereas Eh values >500 mV favor insoluble Mn^{4+} oxides (Hem, 1981). Thus, under oxidizing conditions insoluble Mn-oxyhydroxides precipitate, while under reducing conditions manganese is maintained as a dissolved species (e.g., Piper, 1994), especially through bacterially mediated MnO_2 reduction (e.g., Stumm and Morgan, 1996; Van Cappellen *et al.*, 1998). Calvert and Pedersen (1993) argued that sediments in oxic depositional settings can show manganese enrichments relative to the continental flux due to diagenetic remobilization under reducing pore-water conditions and corresponding reprecipitation at the redox interface in the uppermost layers of the sediment column. Through a repeated sequence of burial, dissolution, remobilization, and reprecipitation, a manganese pump is established wherein pore waters can locally reach supersaturation with respect to manganese carbonate. Calvert and Pedersen (1993) suggested that in the absence of oxyhydroxide precipitation at the sediment–water interface, such a pump does not develop beneath anoxic waters, and instead manganese can be lost to the water column. They further argued that anoxic water columns generally fail to reach the saturation states necessary for water-column Mn-carbonate precipitation—thus a scavenging mechanism analogous to that for iron (via syngenetic Fe-sulfide formation; Section 7.06.3.4.3) may not operate. By their model, sediment manganese enrichments record bottom-water oxygenation (see also Yarincik *et al.*, 2000a). This pattern of manganese cycling, however, may be most relevant for suboxic to moderately sulfidic settings because it ignores the possible sediment immobilization or water-column precipitation (scavenging) of manganese as a sulfide phase (Lewis and Landing, 1991). Nevertheless, manganese concentrations in ancient hemipelagic sediments can provide a useful redox proxy companion to Mo and V + Cr data and are particularly effective for identifying the oxic–anoxic transition. For example, Lyons *et al.* (1993) reported manganese enrichments in sediments of the Black Sea outer shelf, at sites that are presently under oxic bottom waters, as evidence for the dramatic, short-term vertical (tens of meters) excursions of the chemocline that have been widely reported and debated in the literature (e.g., Anderson *et al.*, 1994).

7.06.4 GEOCHEMICAL CASE STUDIES OF FINE-GRAINED SEDIMENTS AND SEDIMENTARY ROCKS

The data presented for each case study are largely abstracted from prior publications, where they are explained in greater detail (see references below). Rather than repeat these details, discussions here focus on similarities and differences in proxy interpretation among different hemipelagic systems in an effort to develop a more comprehensive understanding of the geochemistry of fine-grained sediments and sedimentary rocks.

7.06.4.1 Modern Anoxic Environments of OM Burial—Black Sea and Cariaco Basin

Voluminous recent work in the Black Sea and Cariaco Basin, the world's first and second largest modern anoxic basins, respectively, has centered on patterns and pathways of carbon–sulfur cycling and sequestration, and the comparative behaviors of redox-sensitive metals and their paleo-environmental implications. At the same time, many workers are challenging the validity of the Black Sea paradigm (e.g., Rhoads and Morse, 1971; Demaison and Moore, 1980)—i.e., deposition under deep, highly stratified water-column conditions—as a universally relevant model for ancient anoxic marine deposits (e.g., Murphy *et al.*, 2000a). Nevertheless, research in these modern anoxic basins continues to illuminate general geochemical pathways (e.g., metal cycling) under oxygen-deficient conditions that are independent of the specifics of basin hydrography, water depths, etc. This approach is allowing us to define, refine, and calibrate proxies of broad paleo-environmental relevance.

Studies since the 1990s have consistently demonstrated that the majority of the pyrite accumulating within the uppermost laminated intervals in the Black Sea and Cariaco Basin formed within the euxinic water column. This prevalence of syngenetic pyrite is supported by (i) sedimentary sulfur isotope data that show uniform $\delta^{34}S$ values matching the isotopic composition of the hydrogen sulfide within the present-day water column (Figure 7; Calvert *et al.*, 1996; Lyons, 1997; Wilkin and Arthur, 2001; Lyons *et al.*, 2003; Werne *et al.*, 2003), (ii) pyrite grains with size distributions controlled by the settling velocities of framboids formed within euxinic waters (Wilkin *et al.*, 1997; Wilkin and Arthur, 2001), and (iii) iron enrichments that are most consistent with iron scavenging within a sulfidic water column during syngenetic pyrite formation (Canfield *et al.*, 1996; Raiswell and Canfield, 1998; Raiswell *et al.*, 2001; Lyons *et al.*, 2003). Ratios of Fe_T/Al and, correspondingly, DOP values in the anoxic portions of the Black Sea increase from nearshore to central basin localities by a factor of roughly two in response to a siliciclastic flux that decreases by two orders of magnitude across the same transect (Lyons *et al.*, 2003). At the nearshore, rapidly accumulating sites in the Black Sea, $\delta^{34}S_{py}$ values are enriched by ~10‰ relative to the deep basin in response the diagenetic effects discussed above (Section 7.06.3.4.2, Figure 7). As expected from their intermediate rates of siliciclastic accumulation relative to the end-members in the Black Sea, microlaminated sediments of the Cariaco Basin show intermediate values for the iron proxies (Figure 8). Each of these expressions of the sulfidic waters of the Black Sea and Cariaco Basin translates into a signature of euxinicity that would readily preserve into the deep geological record—and when viewed collectively, would provide a clear picture of paleoredox.

Despite the Black Sea's long status as the quintessential euxinic basin—the term euxinic derives from the basin's ancient name, *Pontus Euxinus*—it is not presently a site of anomalous C_{org} accumulation or preservation. Calvert *et al.* (1991) demonstrated that rates of C_{org} accumulation in the deep anoxic basin are not meaningfully elevated relative to oxic sites when normalized to primary production, water depth, and sedimentation rate (also Arthur *et al.*, 1994). Furthermore, the transition from the organic-rich (Unit 2) sapropel, with C_{org} concentrations of up to ~20 wt.% (Arthur *et al.*, 1994), to the overlying, carbonate-rich Unit 1 deposit where C_{org} is less abundant and averages ~5.3 wt.% (Lyons and Berner, 1992) is driven by carbonate dilution rather than shifts in production or preservation. Both intervals record euxinic conditions in the water column (Repeta, 1993; Sinninghe Damsté *et al.*, 1993; Huang *et al.*, 2000; Wilkin and Arthur, 2001), which is consistent with the persistence of lamination throughout. The onset of Unit 1 deposition specifically marks the attainment of a threshold salinity for coccolith (*E. huxleyi*) production as the basin evolved from the essentially freshwater conditions of the last glacial to the brackish marine environment of today—but it does not reflect a significant change in the rate of C_{org} accumulation relative to the upper part of Unit 2 (Arthur *et al.*, 1994; Arthur and Dean, 1998; Calvert and Karlin, 1998). Restricted marginal marine basins, such as the Black Sea and Cariaco Basin, are particularly susceptible to secondary sulfur overprints that correspond to a temporal evolution from lacustrine or oxic marine conditions during the sea-level lowstand of the last glacial to the anoxic marine settings now present (Middelburg *et al.*, 1991; Lyons *et al.*, 2003). Such overprints can complicate paleoenvironmental interpretations based solely on C–S relationships (Figure 3), although sulfur isotopes assist in the recognition and characterization of secondary signals (see Figures 7 and 9; Lyons *et al.*, 2003).

The modern Black Sea also highlights the impact of siliciclastic dilution on C_{org} concentrations. At the euxinic sites on the basin margin

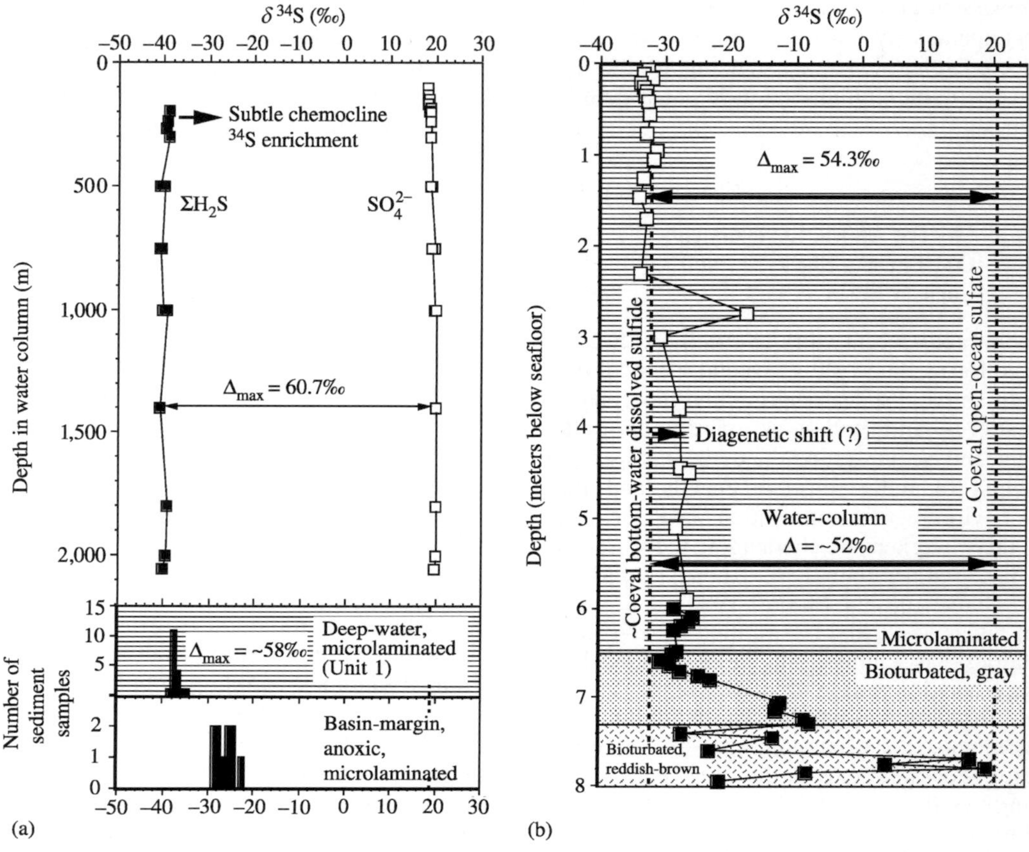

Figure 7 Summary of $\delta^{34}S$ values for pyrite in the sediments of the modern Black Sea (a) and Cariaco Basin (b). Data and discussions for the Black Sea and Cariaco sediments are available in Lyons (1997) and Lyons *et al.* (2003), respectively. $\delta^{34}S$ values from the basin-margin anoxic site in the Black Sea include pyrite and iron monosulfide (Lyons, 1997). Water-column data for the Black Sea and Cariaco, showing net fractions of 50–60‰, are from Sweeney and Kaplan (1980) and Fry *et al.* (1991), respectively. Note the strong correspondence between the data from the water column and the uniform $\delta^{34}S$ values of the bottom sediments—suggesting that most of the pyrite accumulating in the sediments formed (syngenetically) within the water column. Diagenetic effects at the basin margin in the Black Sea and possibly deep within the microlaminated zone of the Cariaco Basin enrich the iron sulfides in ^{34}S relative to the water-column signal. The basal, oxic portion of the Cariaco profile shows the $\delta^{34}S$ signature of the secondary sulfur overprint (see Figure 9 for S_{py} concentrations). Further discussions are available in Sections 7.06.3.4.2 and 7.06.4.1.

characterized by extremely high rates of siliciclastic accumulation, mean C_{org} concentrations are 1.6 wt.% compared to 5.3 wt.% in the central basin despite $CaCO_3$ contents averaging ~52 wt.%, while C_{org} accumulation rates are ~30.9 gC_{org} m^{-2} y^{-1} and 1.9 gC_{org} m^{-2} y^{-1}, respectively (Figure 10; Calvert and Karlin, 1991; Calvert *et al.*, 1991; Lyons and Berner, 1992; Lyons, 1997). Ancient facies analogous to these euxinic upper-slope environments along the Black Sea margin provide a unique challenge for the paleoredox proxies outlined here. However, these complications can also work to our benefit. For example, a multiproxy approach, including benthic ecologies, can speak to the presence or absence of sulfide in the water column, the persistence of such conditions, and the relative clastic fluxes within and among oxygen-deficient basins.

A recurring theme in this paper is the challenge of deducing detailed environmental change from ancient sequences lacking adequate age models (compare Meyers *et al.*, 2001). By contrast, the Black Sea and Cariaco Basin provide high-resolution ^{14}C-, ^{210}Pb- and varve-based chronological control under conditions of comparatively rapid sedimentation (Calvert *et al.*, 1991; Crusius and Anderson, 1992; Anderson *et al.*, 1994; Arthur *et al.*, 1994; Arthur and Dean, 1998; Hughen *et al.*, 1996) and are thus ideally suited to paleoceanographic study—including the principal controls on C_{org} accumulation. Among these important findings, recent work in the Cariaco Basin has yielded a clear low-latitude record of high C_{org} accumulation during the Younger Dryas cold event, which is readily attributable to high productivity during enhanced, trade-wind-driven

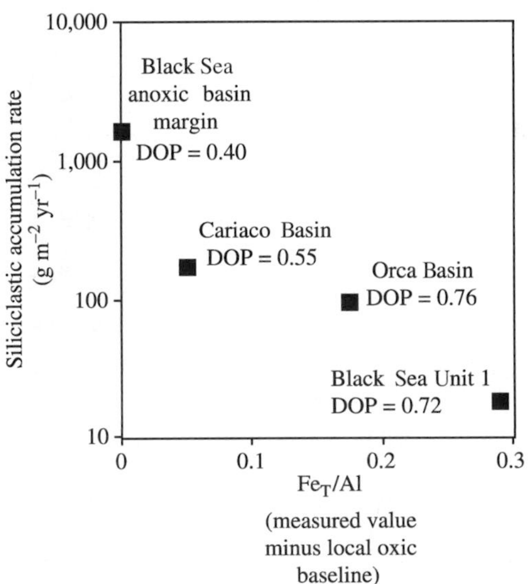

Figure 8 Approximate data showing the general inverse relationship between rate of siliciclastic accumulation and Fe_T/Al ratios as predicted by our model (Sections 7.06.3.4.3 and 7.06.4.1; Lyons *et al.*, 2003). The chronological framework for the Black Sea is from Calvert *et al.* (1991) and Anderson *et al.* (1994) (see also Lyons *et al.*, 2003), and details for the Cariaco Basin are provided in Lyons *et al.* (2003). The plotted Fe_T/Al ratio is the difference between the measured mean at a given euxinic site and a mean value for oxic sediments within the basin (analogous to the "GMS" introduced in Section 7.06.4.3). The Fe_T/Al ratios for the Orca Basin are unpublished, but other details are available in Hurtgen *et al.* (1999). Our unpublished results from Effingham Inlet, British Columbia—an anoxic Fjord where Fe_T/Al ratios are significantly elevated despite very high siliciclastic accumulation rates (see Hurtgen *et al.*, 1999)—have already suggested that this relationship is not universally valid. These complexities are the subjects of ongoing research.

upwelling (Peterson *et al.*, 1991; Hughen *et al.*, 1996; Werne *et al.*, 2000).

Figure 9 shows the relationships among Mo/Al, C_{org}, and pyrite sulfur for the euxinic interval of the last ~14.5 kyr in the Cariaco Basin and the uppermost oxic deposition, which correspond to the last glacial lowstand. Although the gray sediments at the top of the oxic interval bear a strong secondary sulfur overprint marking the onset of euxinic conditions, Mo/Al within the bioturbated gray and reddish-brown clays occurs at roughly average shale (continental) values (Lyons *et al.*, 2003). By contrast, molybdenum is enriched within the laminated euxinic interval, as expressed in Mo/Al ratios, by roughly two orders of magnitude beyond the oxic sediments, and the ratios vary sympathetically with C_{org} concentration. Despite the potential for ambiguity in molybdenum relationships in organic-rich sediments, as outlined in Section 7.06.3.4.4, Mo/Al ratios do sharply delineate the onset of euxinic deposition in the Cariaco Basin and suggest a strong mechanistic linkage between molybdenum and C_{org} accumulation (Figure 9; Lyons *et al.*, 2003; Meyers *et al.*, in review). Our recent work (see Section 7.06.4.2 and Meyers *et al.*, in review) is showing, however, that iron may play a central in controlling patterns of molybdenum enrichment. Specifically, the buildup of HS^-, which seems essential for molybdenum mineralization, is ultimately modulated by the balance between rates of sulfide production through BSR and consumption through iron sulfide formation. Therefore, in the Cariaco Basin, the strong diffusional sulfide overprint linked to the transition from oxic marine to euxinic marine conditions (Figure 9) resulted in extensive pyrite formation but likely little accumulation of dissolved sulfide because of the high availability of reactive iron. This relationship was compounded by the very low levels of OM in the oxic sediments and thus low potential for in situ sulfide production.

7.06.4.2 Cretaceous Western Interior Basin

During the Mesozoic, greenhouse melting of polar ice caps combined with global tectonic processes resulted in eustatic highstands that flooded many continental regions with epeiric seas, significantly expanding the shallow coastal zone of the oceans. These sites not only preserved key records of oceanic and climatic events but also acted as reservoirs of carbon burial and thus played important role in the linkage between biogeochemical cycles and oceanographic/ climatic events. One of the best preserved records of a Cretaceous epeiric sea is the Western Interior seaway, a mid- to high-latitude, meridional retro-arc foreland basin that extended from present day Arctic Canada to the Gulf of Mexico and connected the northern Boreal Sea with the circum-equatorial Tethys Sea (Kauffman, 1984). The basin is bounded on the west by the Sevier Orogenic Belt, which supplied most of its sedimentary fill and was responsible for a significant portion of its subsidence history, especially in the western foredeep (Kauffman, 1984; Kauffman and Caldwell, 1993). This subsidence, in combination with long-term eustatic rise, resulted in accumulation of over 5 km of sediment in the basin spanning Albian through Maastrichtian time (e.g., Dyman *et al.*, 1994). These deposits, which represent mixing of two main sedimentary sources (Sevier-derived siliciclastics and OM, and pelagic-derived carbonate and OM), have been studied in great detail through the years, culminating in what is surely one of the most highly refined and comprehensive chronostratigraphic frameworks and geological

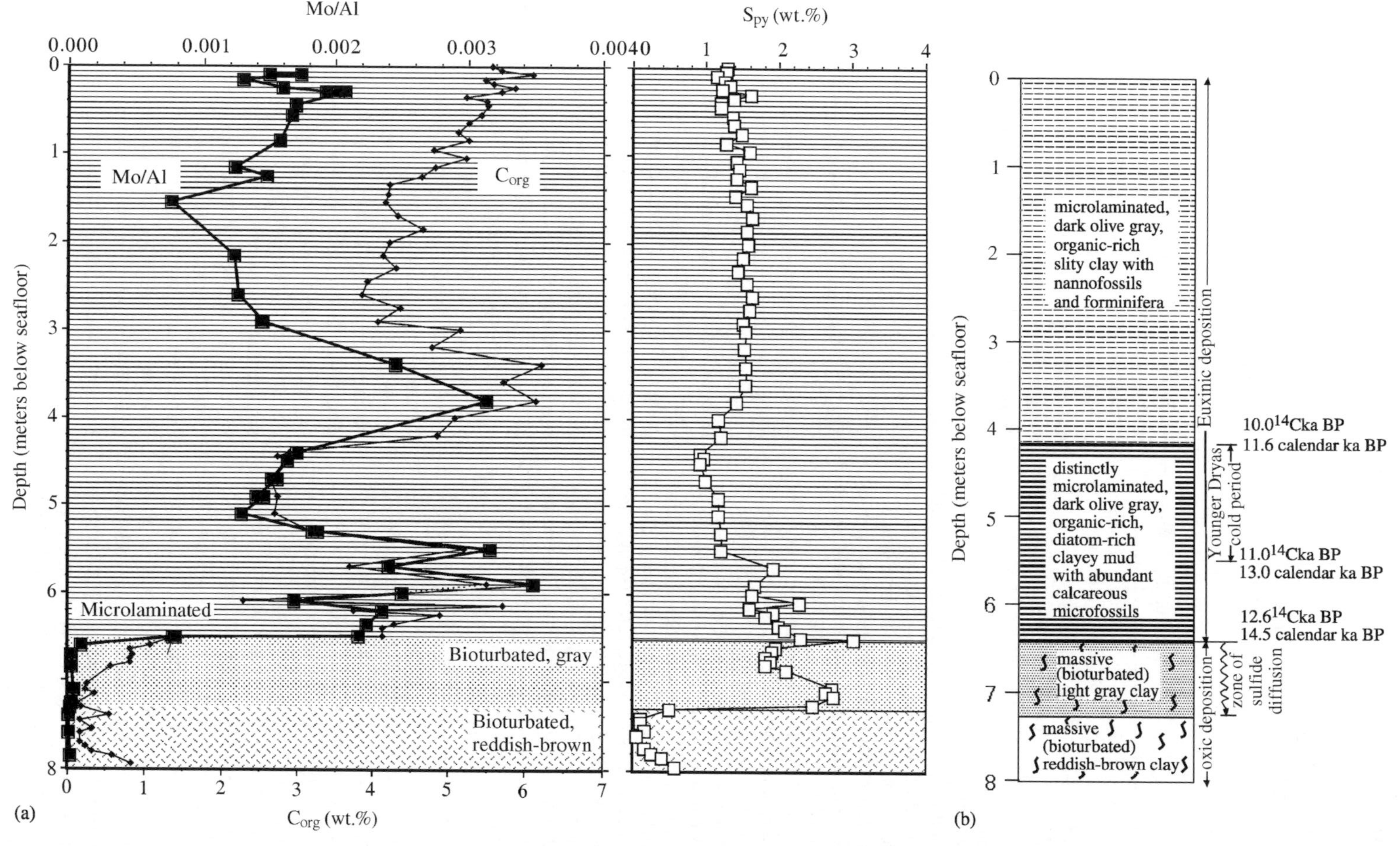

Figure 9 (a) Depth profiles for total Mo/Al ratios and wt.% pyrite sulfur (S_{py}) for both oxic and euxinic sediments spanning the most recent glacial–interglacial transition in the Cariaco Basin, Venezuela (Ocean Drilling Program Site 1002). Corresponding data for C_{org} concentrations are provided for comparison. (b) Generalized lithostratigraphy with corresponding chronology. Note general covariance between C_{org} and Mo/Al within the euxinic interval. See Sections 7.06.3.4.4 and 7.06.4.1 (after Lyons *et al.*, 2003).

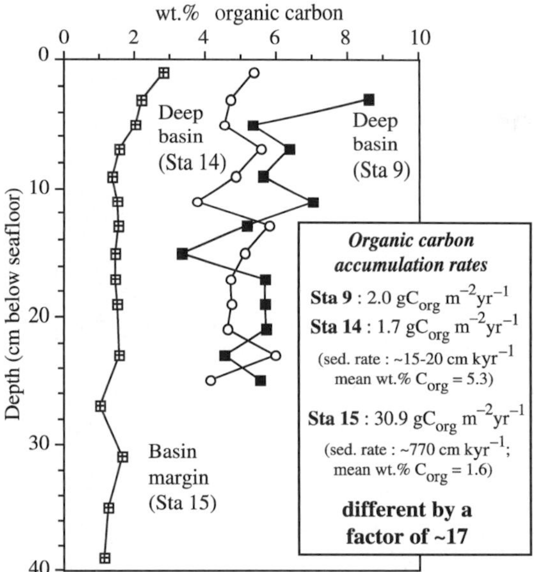

Figure 10 Summary of C_{org} concentration and accumulation rate data for modern euxinic sediments in the Black Sea. Stations 9 and 14 in the deep basin are characterized by microlaminated, carbonate-rich Unit 1 deposits, and Station 15 represents the rapidly accumulating, soupy, black (iron-monosulfide-rich), dominantly siliciclastic sediments on the anoxic upper slope (Calvert and Karlin, 1991; Lyons, 1991, 1997; Calvert *et al.*, 1991; Lyons and Berner, 1992; Anderson *et al.*, 1994; see also Lyons *et al.*, 2003). The sediments at all three sites are laminated. The ^{210}Pb and ^{137}Cs data of Moore and O'Neill (1991) and Anderson *et al.* (1994) for Station 15 confirm that the radiocarbon accumulation rates of Calvert *et al.* (1991) are spurious (see Lyons, 1997) and are related to reworking of older terrestrial and/or marine $CaCO_3$ and OM in the upper slope setting (Lyons, unpublished data). In the deep basin, however, ^{14}C-based chronologies (e.g., Calvert *et al.*, 1991) are consistent with independent measures of sedimentation rate, including varve counts (Arthur *et al.*, 1994). This figure is included to highlight the potential for low C_{org} concentrations as a result of rapid clastic dilution—even under conditions of high C_{org} accumulation (Section 7.06.4.1).

databases of any ancient sedimentary basin. Using this framework, numerous geochemical investigations of Cretaceous hemipelagic sedimentation have been conducted in the Western Interior (e.g., Pratt, 1984, 1985; Arthur *et al.*, 1985; Hayes *et al.*, 1989; Arthur and Dean, 1991; Pratt *et al.*, 1993; Sageman *et al.*, 1997, 1998; Dean and Arthur, 1998; Meyers *et al.*, 2001, submitted). The data shown in Figure 11 represent a distillation of these studies organized using the conceptual model proposed above.

The study interval discussed here includes the Middle to Late Cenomanian and Early Turonian Lincoln Limestone (LLM), Hartland Shale (HSM), and Bridge Creek Limestone Members (BCMs) of the Greenhorn Formation (Figure 11). These units were penetrated in the USGS #1 Portland core as part of a continental drilling study (Dean and Arthur, 1998), sampled at high resolution, and analyzed for many of the proxies described above (Sageman *et al.*, 1997, 1998; Meyers *et al.*, 2001; Sageman, unpublished data). The data series in Figure 11(a) include concentration data (5–10 cm resolution and 2 m moving average), and isotopic data (from Pratt, 1985). In Figure 11(b) the same concentration data are plotted with calculated mass accumulation rates (MAR: 2 m moving average). In the BCM a high-resolution orbital timescale was used for accumulation rate calculations, whereas MARs in the underlying units are based on Ar–Ar dating of bentonite layers that occur within the lowermost Bridge Creek, upper Lincoln, and basal Lincoln Members (Obradovich, 1993); analytical methods used to generate concentration and accumulation rate data are described in detail in Meyers *et al.* (2001). Although parts of some of the plots in Figure 11 are repeated from earlier publications, this compilation of high-resolution data comparing all the members of the Greenhorn Formation has not been presented before. Our main purpose in discussing this data set is to simply highlight the strengths and weaknesses of selected detrital, biogenic, and authigenic proxies.

The data set is excellent for this purpose because paleo-environmental conditions and sea-level history for the units are well known from regional high-resolution chronostratigraphic correlation and detailed lithic and paleobiologic data sets. The lithic character of the Greenhorn Formation reflects a significant part of the transgressive phase of the Greenhorn cyclothem, culminating in peak flooding in the Early Turonian (Kauffman and Caldwell, 1993). The LLM is predominantly characterized by weakly to moderately calcareous shale with intercalated skeletal limestone and bentonite beds in the upper half (Sageman and Johnson, 1985). Based on correlation to progradational units in nearshore western settings, the presence of submarine (wave base) unconformities on the eastern cratonic margin, and evidence of storm influence in the basin center, Sageman (1985, 1996) interpreted upper LLM skeletal limestones to reflect winnowing during a relative sea level fall, followed by condensation during subsequent sea level rise and transgression. The HSM is dominated by well-laminated calcareous shales and a restricted benthic fauna (Sageman, 1985; Sageman and Bina, 1997) and also includes skeletal limestone and bentonite beds in its middle section that correlate to a western prograding clastic wedge, suggesting a relative sea-level oscillation (Sageman, 1985, 1996). However, this sea-level event was likely of lower amplitude than the

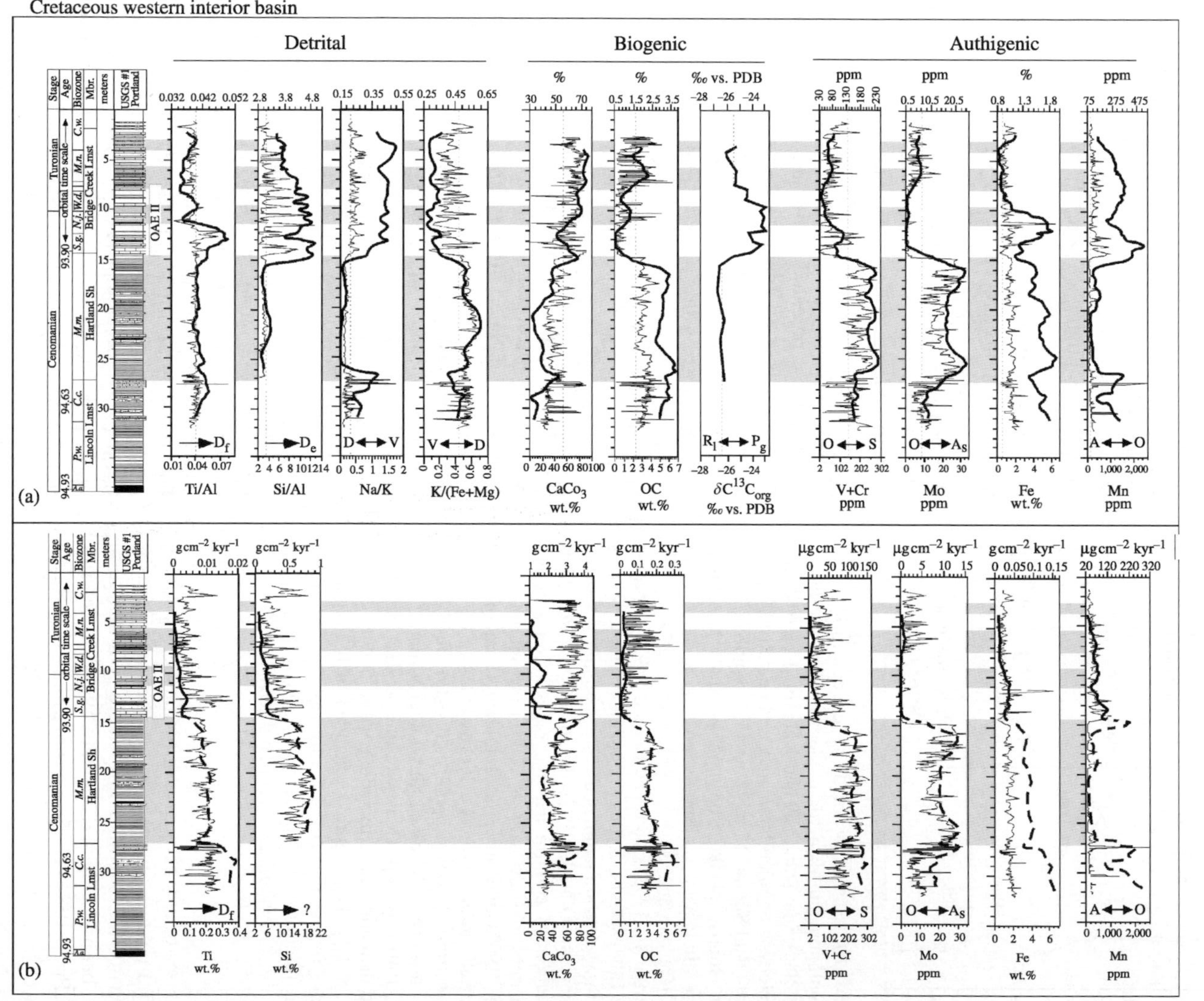

Cretaceous western interior basin
Detrital
Biogenic
Authigenic
Stage
Age
Biozone
Mbr.
meters
USGS #1 Portland
Turonian
Cenomanian
orbital time scale
93.90
94.63
94.93
C.w.
M.n.
W.d.
N.j.
S.g.
M.m.
C.c.
P.w.
Bridge Creek Lmst
Hartland Sh
Lincoln Lmst
OAE II
Ti/Al
Si/Al
Na/K
K/(Fe+Mg)
CaCo3 wt.%
OC wt.%
δC13Corg ‰ vs. PDB
V+Cr ppm
Mo ppm
Fe wt.%
Mn ppm
Ti wt.%
Si wt.%
D ↔ V
V ↔ D
R1 ↔ Pg
O ↔ S
O ↔ As
A ↔ O
(a)
(b)

Lincoln event. Throughout most of HSM deposition the basin was dominated by oxygen-deficient conditions as indicated by faunal data (Sageman and Bina, 1997). The overlying BCM is marked by a conspicuous change to rhythmically interbedded limestone and calcareous shale or marlstone beds with interbedded bentonites and some skeletal limestones. This change is associated with a major increase in faunal diversity and abundance and a shift to extensively burrowed substrates (Elder, 1985, 1991; Sageman *et al.*, 1997). Faunal diversity decreases in a stepwise fashion up to the Cenomanian–Turonian boundary and this biotic decline reflects a major extinction event (Elder, 1989). This event is characterized by global C_{org} burial and oxygen deficient conditions and has been termed Oceanic Anoxic Event II (Schlanger *et al.*, 1987; Arthur *et al.*, 1987); it is marked in Figure 11 by a positive $\delta^{13}C_{org}$ excursion originally documented in Western Interior strata by Pratt (1985). The BCM retains its lithologic character over thousands of square kilometers in the basin (Kansas to western Colorado; South Dakota to New Mexico), and the transition into this facies can be traced to a major flooding surface on the western margin (Elder *et al.*, 1994). This lateral relationship suggests that Bridge Creek deposition was initiated by a rapid pulse of relative sea-level rise and transgression resulting in clastic sediment starvation over a large portion of the basin (Meyers *et al.*, 2001).

The proxies for balance between detrital and volcanigenic input (Na/K, K/(Fe + Mg)), which are interpreted to reflect increase/decrease in riverine siliciclastic flux relative to (± constant) background volcanic ashfall, are consistent with higher detrital flux during HSM deposition, relative condensation during the upper LLM, and a significant decrease in terrigenous flux at the base of the BCM. These data agree with Pratt's (1984) observations for trends in discrete illite. Anomalous increase in Si/Al ratios above the average for hemipelagic clays could reflect enhanced biogenic silicon input (Eicher and Diner, 1989) or greater eolian inputs of quartz silt, suggesting relative condensation (Arthur *et al.*, 1985; Arthur and Dean, 1991). For the Si/Al data shown in Figure 11 we favor the latter interpretation based on the concentration maxima of silicon in BCM limestone beds, which we interpret to reflect times of decreased siliciclastic flux (see also Arthur and Dean, 1991; Dean and Arthur, 1998), as well as the accumulation rate data for biogenic components, which suggest higher overall levels of primary production during HSM deposition. The Ti/Al ratio is weakly indicative of these same trends, showing greater than average values below the lower BCM and less than average values above. The accumulation rate data for titanium are more illustrative, broadly tracking the Greenhorn transgression (Figure 11). Interestingly, the largest pulse in Ti/Al occurs in the lower BCM following the major transition to dominance of carbonate dilution over insoluble residue. This pulse is matched by other indicators of detrital input, and it correlates to a progradational interval in SW Utah (Laurin and Sageman, 2001). However, the progradation is a relatively minor clastic wedge, suggesting that the sensitivity of the Ti/Al proxy may vary as a function of the proportion of insoluble residue relative to carbonate.

The most prominent indicator of biogenic processes in the C–T interval is the positive excursion in $\delta^{13}C_{org}$ (Figure 11), which has been interpreted to reflect a global increase in primary productivity (e.g., Arthur *et al.*, 1988). Interpretations of the isotopic record in the study interval are constrained by the determination that preserved OM in the distal basin is dominated by marine algal input (Pratt, 1984; Pancost *et al.*, 1998). As discussed by Meyers *et al.* (2001), evidence for relative increase in production in the BCM (increased MAR-$CaCO_3$ and MAR-C_{org}) does not coincide with the initial isotopic shift but occurs instead in the latter part of the excursion, suggesting that the onset of environmental conditions characteristic of OAE II in the Western Interior basin lagged behind the global record. In addition, although the $\delta^{13}C_{org}$ excursion interval is marked by indicators of euxinic conditions in the

Figure 11 Geochemical data series and lithologic log for Cretaceous Greenhorn Formation in USGS #1 Portland core, modified from Sageman *et al.* (1997, 1998), Meyers *et al.* (2001; submitted), and Sageman (unpublished data). Data series include: (a) high-resolution concentration data (thin lines) with 2 m moving averages (thick lines) and (b) high-resolution concentration data (thin lines) with 2 m moving averages for accumulation rate. In both (a) and (b) units on the upper scales refer to thick lined plots. Thin vertical dashed lines represent mean values for the data set ($N = 309$ for most proxies). Horizontal shaded areas mark the organic-rich intervals. Data were generated using analytical methods and orbital timescale described in Meyers *et al.* (2001), except MAR values below BCM (dashed), which employ 2 m moving averages of concentration data, bulk density values of 2.6 g cm^{-3}, and linear effective sedimentation rates based on radiometric ages of Obradovich (1993). Also shown are $\delta^{13}C_{org}$ data from Pratt (1985). Location of OAE II, defined by the $\delta^{13}C_{org}$ excursion, is indicated next to the lithologic log. Abbreviations for proxy interpretation terms include: D—detrital source; V—volcanogenic source; D_f—detrital fluvial source; D_e—detrital eolian source; R_l—respiration/local reservoir-dominated; P_g—production/global reservoir-dominant; O—oxic; S—suboxic; A—anoxic; A_s—anoxic/sulfidic.

North Atlantic basin (Brumsack and Thurow, 1986; Sinninghe Damsté and Koster, 1998), trace-metal evidence for pervasive occurrence of such conditions in the BCM occurs only after the C-isotope excursion—during maximum highstand and maximum sediment starvation (see also Simons and Kenig, 2001). Also interesting is the fact that MAR values for $CaCO_3$ and C_{org} a re much higher in the underlying HSM than in the BCM (Figure 11). The higher burial flux of C_{org} in this interval may be explained by enhanced preservation, as authigenic proxies for redox conditions suggest common transitions through the suboxic zone (maximum concentration and MAR of V + Cr) and at least intermittent dominance of anoxic–sulfidic conditions (maximum concentration and MAR of molybdenum). Yet, the HSM water column was never permanently sulfidic, as indicated by the frequency of benthic colonization events (Sageman, 1985, 1989), but probably experienced strong seasonal thermal stratification with only weak mixing by winter storms, perhaps punctuated by 100-year-storm ventilation events (Sageman and Bina, 1997). Consistent with this interpretation is the observation that average values of $\delta^{13}C_{org}$ in the HSM are slightly depleted relative to the post-OAE II background (Figure 11), suggesting that recycled CO_2 may have more strongly influenced the HSM isotopic signal. Manganese data also support these interpretations, showing enrichment (concentration and MAR) during the upper LLM and lower BCM, likely reflecting precipitation of Mn-oxyhydroxides or Mn-carbonates during oxygenation events, and depletion during the HSM, recording time averaged dominance of oxygen deficient conditions and prevalence of soluble Mn^{2+}.

Based on a comparison of modern chemical oceanographic data and the results described above, Meyers *et al.* (submitted) have attempted to link the production, terrigenous dilution, sedimentation rate, and redox processes illustrated graphically in Figure 2 into a quantitative model (Figure 12). This model depends on the fact that OM decomposition in sediments may be approximated using first-order rate equations (see Figure 12(a)), where the degree of OM degradation (ΔG) in each remineralization regime is dependent upon: (i) the amount of time OM resides in each regime (t), (ii) the initial concentration of metabolizable OM at the top of each regime (G), and (iii) the first-order OM degradation rate constants for each regime (k). The rate of organic carbon remineralization at a given time may be expressed as $-kG$, where the concentration of metabolizable organic carbon (G) is dependent upon the balance between export production (OM-export), dilution (MAR-dilutant), and the degree of degradation that has occurred (ΔG). Meyers *et al.* (submitted) demonstrate that the integration of this conceptual framework for OM diagenesis with high-resolution geochemical proxy burial fluxes in ancient organic-rich strata (Figure 12(a)) provides a unique opportunity to (i) evaluate the role of anaerobic remineralization on OM export to the lithosphere during OM burial events, and (ii) to deconvolve the paleo-environmental mechanisms that control OM delivery to these anaerobic microbial zones over geological time.

Application of this framework to evaluate anaerobic remineralization during ancient organic burial events requires a proxy for the rate of anaerobic OM degradation within sediments ($-kG$). The sensitivity of molybdenum accumulation to the rate of hydrogen sulfide production via sulfate reduction (Helz *et al.*, 1996) provides a promising proxy for $-k_sG_s$. Achievement of the critical pore water H_2S levels necessary for molybdenum accumulation depends upon a balance between the processes that source H_2S (*in situ* sulfate reduction rate and the rate of diffusion/advection of hydrogen sulfide into the pore water in the case of euxinic basins), the processes that deplete H_2S (the rate of formation of Fe-sulfides and OM sulfurization and the rate of diffusion/advection of H_2S out of the pore water), and the volume of water in the connected pore space. When the rate of sulfate reduction is high, production of hydrogen sulfide can outpace depletion within pore waters (which is largely controlled by reactive-Fe availability and pyrite formation), and hydrogen sulfide may amass to the levels necessary for molybdenum scavenging. Such high rates of sulfate reduction may be a consequence of shallowing of the sediment sulfate reduction zone ($d_a + d_n$), increased export production (OM-export), changes in inorganic dilutant flux (MAR-dilutant), or changes in bulk sedimentation rate (ω) (Figure 12(a)). Based on these lines of reasoning, Meyers *et al.* (in review) employed molybdenum accumulation as a proxy for the rate of sulfate reduction during and immediately following Oceanic Anoxic Event II (OAE II) and evaluated it in tandem with iron accumulation rates (primarily reflecting pyrite formation) to assess changes in the relative demand for H_2S within the sediment pore water (Figure 12(b)). When compared to C_{org} accumulation rates through the same interval, these data suggest that the highest rates of sulfate reduction and the highest rates of C_{org} accumulation within the sediments occur not during OAE II but following this global event. Iron accumulation rate data from the same interval suggest that lower C_{org} and molybdenum accumulation during OAE II was due to buffering of H_2S levels via reactive-Fe delivery and pyrite formation. Decreased dilution of labile OM and decreased delivery of reactive iron in the post-OAE II interval served to (i) increase the rates of

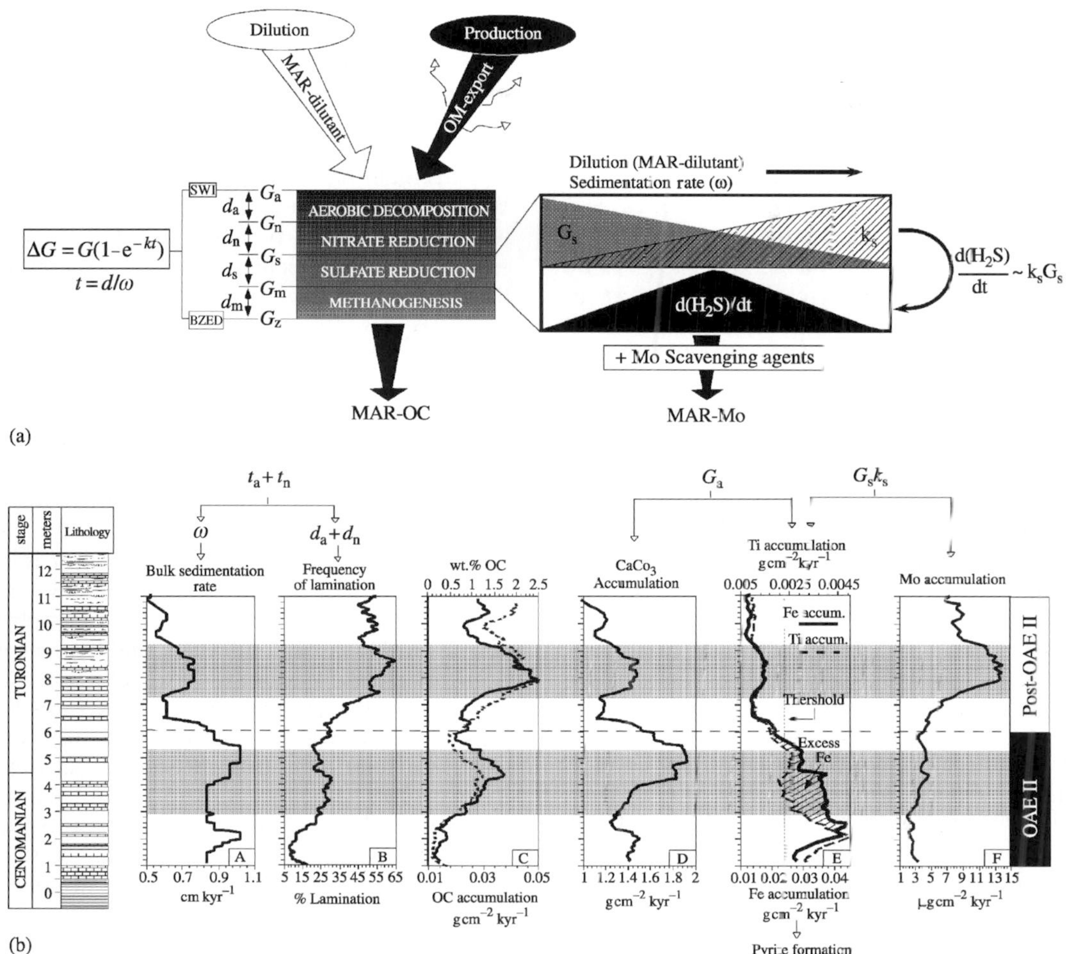

Figure 12 (a) Theoretical model for OM degradation in the sediment, and the linkage between OM burial and molybdenum accumulation. Subscript "a" refers to aerobic decomposition, subscript "n" refers to nitrate reduction, subscript "s" refers to sulfate reduction, and subscript "m" refers to methanogenesis. G = concentration of OC at the top of each remineralization zone; ΔG = wt.% OC change due to decomposition in each remineralization zone; ω = bulk sedimentation rate in cm/kyr; k = decomposition rate constant in kyr^{-1}; t = time spent in decomposition zone in kyr; d = thickness of decomposition zone in cm; (dH_2S/dt) = rate of hydrogen sulfide production via sulfate reduction; MAR-dilutant = dilutant accumulation in $g\,cm^{-2}\,kyr^{-1}$; OM-export = organic carbon export in $g\,cm^{-2}\,kyr^{-1}$; MAR-Mo = authigenic molybdenum accumulation in $g\,cm^{-2}\,kyr^{-1}$; MAR-OC = organic carbon accumulation in $g\,cm^{-2}\,kyr^{-1}$; G_z = final wt.% OC; SWI = sediment–water interface; BZED = base of zone of early diagenesis. (b) Bulk sedimentation rates, 2 m moving-average geochemical proxy accumulation rates, 2 m moving average wt.% OC data, and percent lamination data from the Bridge Creek Limestone Member (#1 Portland core). Percent lamination was calculated using the ORI rank data from Savrda (1998) and a 2 m moving-window, and is employed to assess the average depth of the upper interface of the SRZ. $CaCO_3$ accumulation is employed as a proxy for export production, and titanium accumulation is employed as a proxy for siliciclastic dilution.

hydrogen sulfide generation and (ii) decrease the buffering capacity of the system, driving the sediments to a more sulfidic state. High levels of reactive-Fe delivery during OAE II were a consequence of greater detrital flux, and this iron source was supplemented by an additional dissolved reactive-Fe source. The timing of this excess (dissolved) iron accumulation (Figure 12(b)) correlates with the hypothesized initiation of Tethyan oxygen minimum zone (OMZ) influence within the basin (Leckie *et al.*, 1998; Meyers *et al.*, 2001).

Two additional factors that must be considered when assessing the mechanism for enhanced molybdenum and OC accumulation in the post-OAE II interval are the rate of export production (OM-export) and the amount of time this export production spends prior to entering the sediment SRZ ($t_a + t_n$). Based on calibration with modern oceanic sites, Meyers *et al.* (submitted) employ $CaCO_3$ accumulation as a proxy for export production. The term $t_a + t_n$ is estimated based on the bulk sedimentation rate (ω) and frequency of lamination of the strata, which is believed to reflect changes in the average depth of the upper interface of the SRZ ($d_a + d_n$). Taken together, these data suggest that (i) the highest rates of C_{org} accumulation are decoupled from the highest rates of export production and (ii) although sedimentation rates decrease in the post-OAE II interval (which should otherwise result in an increase in $t_a + t_n$), a decrease in $d_a + d_n$ (increase in the frequency of lamination) exerts the dominant control on labile OM export to the SRZ. Based on these results, Meyers *et al.* (submitted) conclude: (i) OM accumulation in the Bridge Creek Limestone Member is controlled by the rate of export of OM to the SRZ; (ii) the location of the upper boundary of the SRZ is the first-order control (OAE II versus post-OAE II) on OM export into the sulfate reduction zone; and (iii) changes in SRZ location and molybdenum accumulation between the OAE II and post-OAE II interval are attributable to the balance between hydrogen sulfide production via sulfate reduction and hydrogen sulfide depletion through reactive-Fe delivery and pyrite formation. These results suggest that the strong correlation between source rock development and intervals of transgression in the geological record is the biogeochemical consequence of a decrease in the siliciclastic flux, which concentrates labile OM, driving higher rates of hydrogen sulfide production and reduces reactive-Fe flux, permitting hydrogen sulfide levels to escalate and enhancing the preservational state of the system. As the burial flux data for the Bridge Creek demonstrate, the increased rate of export of labile OM into a shallower SRZ resulted in elevated C_{org} accumulation rates, even under lower rates of primary production.

7.06.4.3 Devonian Appalachian Basin

During a significant portion of the Paleozoic, conditions were broadly similar to the Mesozoic greenhouse: elevated p_{CO2} levels, warm temperatures, decreased equator to pole temperature gradients, and widespread marine flooding of continental areas due to reduced ice volumes and possibly tectonoeustatic effects (e.g., Woodrow, 1985; Johnson *et al.*, 1985; Berner and Kothavala, 2001). Within these greenhouse times, the Devonian is of particular biogeochemical interest due to events such as the rise of vascular plants, widespread black shale deposition (Algeo *et al.*, 1995), and the Frasnian–Famennian mass extinction (McGhee, 1982). One of the best-preserved records of Devonian faunal history and environmental conditions is found in deposits of the Appalachian basin. Like the Western Interior, this basin formed in mid-latitudes (southern hemisphere; Witzke and Heckel, 1988) as a retroarc foreland adjacent to an orogenic belt (Acadian Orogen), the uplift of which by terrane collision (Faill, 1985) drove load-induced subsidence to create accommodation space and sourced most of the siliciclastic material to fill it (Ettensohn, 1985a,b). Sedimentation in the distal part of the Appalachian basin was hemipelagic like the Western Interior, but pelagic carbonate production was comparatively limited prior to significant expansion of calcareous nanoplankton and planktic foraminifera later in the Mesozoic (Gartner, 1977; Haynes, 1981). Thus, sources of carbonate were limited to thin-shelled planktic styliolinids (Yochelson and Lindemann, 1986), allodapic carbonate mud transported from shallow areas to the west of the foredeep (Werne *et al.*, 2002), and reworked skeletal material, which is quite common in some intervals due to the highly fossiliferous nature of the strata.

The Devonian stratigraphic succession in the Appalachian basin and adjacent areas (e.g., Illinois and Michigan basins) has been the subject of much study through the years, and has recently been reviewed by Murphy *et al.* (2000b). Geochemical investigations of these Devonian fine-grained facies have contributed significantly to the characterization of petroleum source rocks (e.g., Roen, 1984; Roen and Kepferle, 1993), to development of ideas about carbon isotope systematics (Maynard, 1981), and to understanding the biogeochemical dynamics of ancient oxygen-deficient environments (e.g., Ingall *et al.*, 1993). The data presented in Figure 13 are abstracted from results of a recent study undertaken by the authors and their students (Murphy *et al.*, 2000a,c; Werne *et al.*, 2002; Sageman *et al.*, 2003). The project's main objective was to develop a high-resolution, continuous geochemical database from analysis of core samples spanning Eifelian through Famennian strata in western New York State and to use it to delineate the controls on C_{org} burial based on the conceptual approach illustrated in Figures 1 and 2. The complete data set, with descriptions of analytical methods and detailed interpretations, was summarized recently by Sageman *et al.* (2003). Here we focus on four of the studied intervals that include transitions between black and gray shale facies, but omit data from the thick intervening intervals of relatively homogenous gray shale. Our objective is to highlight the strengths and weaknesses of selected detrital, biogenic, and authigenic proxies.

The four study intervals are, in ascending order, the Marcellus subgroup (including Union Springs Member, Bakoven Formation and Oatka Creek Formation: US and OC), the Geneseo Formation (GS), the Pipe Creek Formation (PC), and a thin section of organic-rich facies in the uppermost Hanover Formation (UH) (Figure 13);

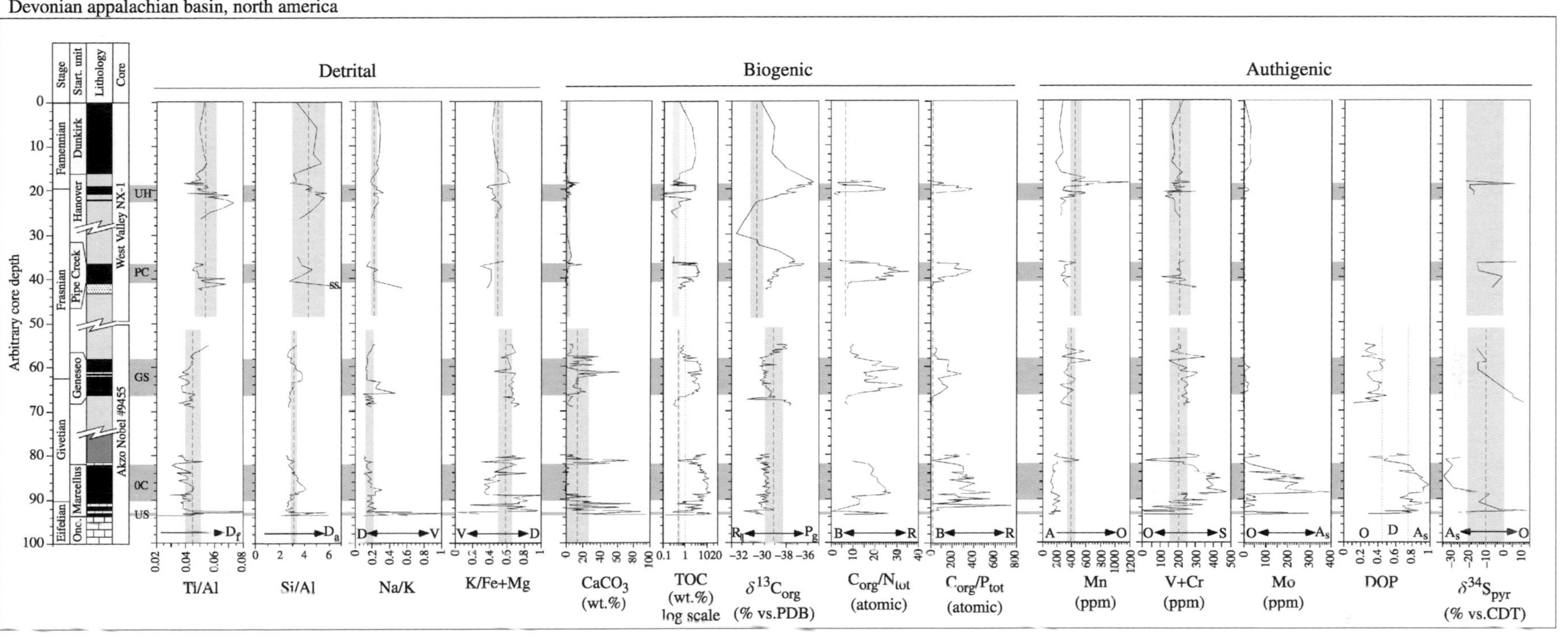

Figure 13 Geochemical data series and lithologic log for Devonian Marcellus subgroup, Geneseo, Pipe Creek, and Hanover–Dunkirk Formations in the Akzo Nobel #9455 and West Valley NX-1 cores, modified from Murphy *et al.* (2000a), Werne *et al.* (2002), and Sageman *et al.* (2003). Data series include concentration data (thin lines) with gray shale mean (GSM) indicated by vertical dashed line and $\pm 1\sigma$ indicated by vertical gray shading. For calculation of GSM in Akzo core, $N = 227$ for most proxies; for West Valley $N = 141$ for most proxies. In some cases vertical dotted lines mark threshold values, such as 1% C_{org} (gray shale cutoff), or 0.45 and 0.75 DOP (divisions for oxic, dysoxic, and anoxic: Raiswell *et al.*, 1988; Canfield *et al.*, 1996). Horizontal shaded areas mark the organic-rich intervals and are labeled with stratigraphic unit abbreviations used in text. Data were generated using analytical methods described in Murphy *et al.* (2000a) and Werne *et al.* (2002). Abbreviations for proxy interpretation terms as in Figure 11, except: B—buried nutrients; R—recycled nutrients; D_x—dysoxic.

the latter two intervals correspond to lower and upper Kellwasser horizons, respectively (Murphy *et al.*, 2000c). Each of these organic-rich shale units has been interpreted to coincide with relative deepening based on biostratigraphic correlation to proposed global eustatic events (i.e., Johnson and Sandberg, 1988), as well as regional sea-level reconstructions based on lithofacies and stratigraphic architecture (Brett and Baird; 1986; House and Kirchgasser, 1993; Brett, 1995; Ver Straeten and Brett, 1995). These deepening events may be related to tectonic evolution of the Appalachian basin as suggested by Ettensohn (1985a,b, and Ettensohn *et al.* (1988), but evidence discussed in Werne *et al.* (2002) suggests that eustasy also played a role. No matter what the mechanism, the bulk of geological evidence indicates that organic carbon burial occurred preferentially during deepening events, which are superimposed upon a long-term secular trend of relative shallowing reflecting the overall progradation of the Catskill Delta complex (Ettensohn, 1985a).

Although geochronology for the Devonian study interval has been refined recently (Tucker *et al.*, 1998), control points are still far too coarse to allow meaningful accumulation rate calculations. Instead, multiple lines of evidence from concentration and isotopic data are employed. Proxies reflecting changes in detrital input or shifts in the proportion of detrital flux relative to eolian flux (siliciclastic or volcanogenic) include Ti/Al, Si/Al, Na/K, and K/(Fe + Mg) (Figure 13). The data are best evaluated relative to deviations from the gray shale mean (GSM), which represents "background" values averaged from shales with $C_{org} < 1\%$. The C_{org}-enriched intervals show, on average, Ti/Al values that are depleted relative to GSM, consistent with sediment starvation during transgression (Figure 13). With the exception of a major excursion at the base of the Marcellus subgroup, which represents the highly condensed distal record of a lowstand event (Cherry Valley/ Halihan Hill bed), Ti/Al values are sub-GSM in the US/OC and GS black shales but increase as C_{org} levels decrease (e.g., late GS, corresponding to progradation and progressive dilution; Murphy *et al.*, 2000a).

In the younger two study intervals, where delta progradation was more pronounced, Ti/Al values are higher overall but still show negative shifts associated with organic-rich units. These observations are generally corroborated by trends in Si/Al and the Na/K and K/(Fe + Mg) proxies, but variations up-section may reflect changes in background volcanogenic fluxes and other processes. For example, increased Si/Al, which clearly characterizes the US, middle OC, and middle GS intervals (Figure 13), suggests that a source of silicon other than that associated with the detrital riverine flux has been enhanced. Although this source could include biogenic silicon and thus reflect enhanced production (i.e., Schieber *et al.*, 2000), petrographic and SEM observations from the OC and GS showed quartz silt with surface features indicating an eolian source. These observations support the interpretation of condensation within the organic-rich units. Similarly, high Na/K and low K/(Fe + Mg) characterize the US and middle OC in the Akzo core (Figure 13). At this location both units are significantly thinned compared to sections to the east (Rickard, 1975), suggesting significant basinward condensation. Widely traceable bentonite beds found in the upper part of the underlying Onondaga Limestone (Brett and Ver Straeten, 1995) provide evidence of volcanogenic activity at this time, which would supply the contrasting elemental flux. In the lower part of the GS, Na/K shows a similar pattern, but the K/(Fe + Mg) proxy remains enriched relative to GSM (suggesting detrital dominance). In the upper two units, fluctuations in these proxies are more subtle. These patterns are consistent with Ettensohn's (1985a,b) reconstruction of Appalachian basin tectonic history, which attributes Marcellus and Geneseo deposition to deepening resulting from load-induced subsidence following collision events (termed tectophases). These events would have likely been associated with enhanced volcanism. However, the volcanogenic signal is diluted by terrigenous flux in the upper two units. Three conclusions can be drawn from analysis of the proxies: (i) significant decreases in bulk sedimentation and a relative increase in eolian over fluvial input accompanies C_{org} burial events, (ii) there is a clear overall trend of increasing siliciclastic dilution through the study interval, and (iii) although the detrital proxies as a group have variable expression over time related to evolving input fluxes, consideration of multiple indicators provides a consistent interpretation.

Potential biogenic proxies in Devonian mudrocks include skeletal carbonate and organic carbon, $\delta^{13}C_{org}$ values, and changes in the ratios of C_{org} to N and P (Figure 13). Unlike the bulk of the carbonate in the Cretaceous strata described above, $CaCO_3$ in the study interval has multiple sources and is controlled by a variety of processes. Over the long term it is negatively correlated with siliciclastic dilution (Figure 13), like carbonate in the Greenhorn Formation. But in many of the individual study intervals there is a positive correlation with detrital proxies. These cases reflect input of $CaCO_3$ dominantly sourced from benthic skeletal material, which is reworked and comminuted along with coarser siliciclastic fractions during sea-level lowstands. However, there is also evidence for ecologic and diagenetic effects on $CaCO_3$ production and dissolution (see discussion in Werne *et al.*, 2002).

Since no clear relationship between $CaCO_3$ concentration and pelagic production can be established, C_{org} is the main potential product of biogenic primary production in surface waters (following confirmation of marine algal source dominance, which was based on organic petrography and compound specific isotopic analysis; Murphy *et al.*, 2000a). As described above, however, the relationship between production and C_{org} concentration may be masked by variations in detrital dilution and OM remineralization. The observation that C_{org} enrichment in the studied units is definitely associated with condensation, and thus probably decreased bulk sedimentation rates, is consistent with evidence for transgression. As for elevated production, both the US/OC and the GS intervals have $\delta^{13}C_{org}$ values that are depleted relative to the GSM, which could reflect dominance of respired CO_2 in a restricted local reservoir (Lewan, 1986; Rohl *et al.*, 2001). Based on the parallel observation in each C_{org}-rich unit of significant deviations in C : N : P ratios from values expected for marine algal sources (Figure 13), Murphy *et al.* (2000a) proposed a model of fluctuating anoxia, nutrient remineralization, and relative eutrophication based on arguments summarized earlier (e.g., Aller, 1994; Ingall and Jahnke, 1997). The driving force for this model is establishment and breakdown of thermal stratification on timescales sufficient to allow nutrient build up and release and to cycle respired CO_2 back to surface waters with regenerated nutrients. In contrast to the US/OC and GS, the upper two organic-rich units (PC and UH) are characterized by major positive $\delta^{13}C_{org}$ excursions (Figure 6) that have been identified in localities around the world (e.g., Kellwasser horizons; Joachimski and Buggisch, 1993). These events are hypothesized to reflect, like the OAE II scenario described above, times when the extent of global net primary production and C_{org} burial exceeded respiration and recycling at a scale sufficient to overprint local reservoir effects.

Redox dynamics play a critical role in the hypothesis described above, and trends in the authigenic proxies support the model. For example, manganese concentrations are generally depleted throughout organic-rich intervals but commonly show enrichments at the top of these intervals, marking transitions from predominantly dysoxic–anoxic to predominantly oxic states. This pattern is particularly pronounced in the UH interval (Figure 13). The other indicators of reducing conditions show enrichments in the organic-rich intervals, but they clearly fall into two groups. The US/OC interval—characterized by the largest enrichments in V + Cr, Mo, and reactive-Fe (expressed in Fe_T/Al and DOP) and the most depleted values of $\delta^{34}S_{py}$ (Figure 13)—is the best candidate for a euxinic system of any Phanerozoic sequence studied by the authors (see detailed discussions in Werne *et al.*, 2002, and the paleoredox proxy details provided above). Although the other organic-rich intervals show depleted manganese values, small enrichments in V + Cr and Mo and shifts toward more depleted $\delta^{34}S_{py}$ values relative to GSM values, these changes are comparatively modest. Collectively, these observations suggest that most of the studied black shales were deposited in a shallow epeiric sea that was dynamically balanced between production, which led to increased oxygen utilization, and water-column mixing, which effectively recycled limiting nutrients to drive production but also bottom waters that were sufficiently oxygenated to ensure that sulfidic conditions could not be maintained for significant periods. The euxinic conditions of the US/OC interval were actually the exception rather than the norm (contrary to prior views—Byers, 1977; Ettensohn, 1985a,b) and resulted from a combination of maximum basin subsidence and eustatic rise, which resulted in water depths sufficient to limit effective mixing of the water column (Werne *et al.*, 2002). Notably, the most euxinic conditions in the studied units, although correlative with evidence for significant oxygen deficiency in other parts of the world (e.g., Truyols-Massoni *et al.*, 1990), do not correlate with a positive C-isotope excursion.

7.06.4.4 Recent Precambrian Advances

Precambrian studies are among the most promising yet least explored frontiers in shale geochemistry. In light of new analytical techniques, an improved understanding of sulfur microbiology, and vast additions to the geochemical database, however, shales are starting to play a central role in the development and validation of important new models for evolution of the early ocean–atmosphere–biosphere (e.g., Kakegawa *et al.*, 1998; Lyons *et al.*, 2000; Shen *et al.*, 2002; Strauss, 2002; see Lyons *et al.*, in press). One model argues for a globally euxinic Proterozoic ocean (Canfield, 1998), reflecting increases in seawater sulfate concentrations (linked to oxic continental weathering) in combination with a deep ocean that remained largely oxygen free. With the buildup of sulfate in an anoxic deep ocean, BSR drove extensive hydrogen sulfide production and accumulation, which may have ultimately led to the disappearance of banded iron formations through the corresponding decrease in iron solubility. (The Archean ocean, although anoxic, would have lacked the sulfate necessary for euxinic conditions (Canfield, 1998; Habicht *et al.*, 2002).

Under these widely sulfidic conditions of the Proterozoic, sequestration of bioessential metals such as molybdenum and iron, which are necessary for the production and utilization of bioavailable nitrogen, may have limited the ecological range and ultimately the evolution of eukaryotic algae (Anbar and Knoll, 2002).

Despite assertions made elsewhere in this report, delineation of an oxygen-deficient, sulfide-rich Proterozoic ocean is particularly elusive. For example, it is generally regarded that sulfate concentrations were substantially lower than those of the present ocean (perhaps only 10% of present levels during the Mesoproterozoic; Canfield, 1998; Hurtgen *et al.*, 2002; Shen *et al.*, 2002; Lyons *et al.*, in press; Kah *et al.*, submitted). Under such conditions, the light and uniform $\delta^{34}S_{py}$ values that are diagnostic of euxinic environments during the Phanerozoic (Section 7.06.3.4.2) are replaced by an abundance of ^{34}S-enriched pyrite that records local and likely global sulfate limitation (Lyons *et al.*, 2000; Luepke and Lyons, 2001; Shen *et al.*, 2002). By contrast, iron-based proxies, such as DOP, show more promise for the recognition of euxinicity on scales of individual basins (Shen *et al.*, 2002), and molybdenum isotope data may speak to the global distribution of such settings (Arnold *et al.*, 2002). Pyritic sulfur in fine-grained sediments spanning the ~2.3 Ga "great oxidation event" of the Precambrian atmosphere (e.g., Canfield, 1998; Holland, 2002; as reviewed in Lyons *et al.*, in press) also records the telltale disappearance of strong mass-independent sulfur isotope fractionation expected with shifting global redox (Farquhar *et al.*, 2000). Similarly, patterns of mass-dependent fraction expressed in these pyrites faithfully track the evolution of oxygen availability in the Precambrian ocean–atmosphere (Canfield and Teske, 1996; Canfield, 1998; Canfield *et al.*, 2000; Habicht *et al.*, 2002; Hurtgen *et al.*, 2002; reviewed in Lyons *et al.*, in press), and trace amounts of sulfate within carbonate rocks are yielding continuous, high-resolution records of the evolving $\delta^{34}S$ of seawater (Burdett *et al.*, 1989; Hurtgen *et al.*, 2002; Gellatly and Lyons, submitted; Kah *et al.*, submitted; Lyons *et al.*, in press).

7.06.5 DISCUSSION: A UNIFIED VIEW OF THE GEOCHEMISTRY OF FINE-GRAINED SEDIMENTS AND SEDIMENTARY ROCKS

In this chapter we first outlined a conceptual model for the relationship between primary geological processes involved in the formation of fine-grained, siliciclastic–biogenic sediments and sedimentary rocks and a selected set of geochemical proxies particularly useful for the reconstruction of these processes in ancient deposits (Figures 1 and 2). We then illustrated applications of this model through four case studies representing Precambrian, Paleozoic, Mesozoic, and modern marine environments. The conceptual model encompasses a formidable array of interrelated processes and proxies and draws on a very broad body of scholarship, which we endeavored to cite as fully as space would allow. Our rationale for pursuing such a broad-based approach is that fine-grained, siliciclastic–biogenic sediments provide perhaps the most fully integrated record of past Earth surface processes of any sedimentary facies and thus offer the opportunity to open a clearer window into the workings of the Earth system *if the interdependent parts of the depositional system can be adequately assessed.*

Although the general aspects of our conceptual model have their roots in the broad base of literature cited throughout the chapter, the specifics were refined in the course of research conducted by ourselves, our students, and others on the intervals described in the case studies. We argue that the model illustrated in Figures 1 and 2 provides the unifying framework for understanding these depositional systems, and analysis of key similarities and differences between the case studies reveals the constraints on ranges of proxy expression across space and time. We conclude with a summary of the key similarities and differences in proxy expression and a prospectus for future work.

- *Applicability of the mode*—A common attribute of the case studies is that all three factors in the tripartite model (Figure 1) play important roles in the origin of fine-grained, siliciclastic–biogenic facies. There are variations in the dominance of one control over another through geological history (e.g., authigenic processes in the anoxic–sulfidic oceans of the Precambrian versus the predominantly oxic waters of Phanerozoic seas) and through the individual histories of each study interval (e.g., changing roles of production and benthic redox in controlling OM accumulation during BCM deposition). End-member models, as exemplified by the "productivity" versus "preservation" dichotomy, are more clearly viewed as process continua to which additional axes must be added (bulk sedimentation rates) to fully assess depositional processes and their feedback on paleo-environmental conditions. It is the changes in process dominance, placed within appropriate geological contexts (paleogeography, paleoclimate, etc.) that provide the most penetrating information about Earth system dynamics.

- *Detrital fluxes and bulk sedimentation*—In hemipelagic systems oscillations in the detrital flux play a critical role in modulating the dynamic interplay between biogenic fluxes and authigenic modification of those fluxes. As a general rule, relative condensation (decreased dilution by detrital or other sediment sources) accompanies most cases of elevated C_{org} concentrations. However, the sources and pathways of dilutants vary widely through space and time. There is no single unambiguous proxy for distinguishing terrigenous dilutants, and even the most conservative tracers may vary in their transport paths (e.g., fluvial versus eolian), resulting in significant differences of interpretation. In each case, proxies must be calibrated to the geological context of a depositional system.
- *Productivity and nutrients*—One of the most interesting linkages represented in the model in Figure 1 is that between OM remineralization and nutrient recycling, a process with significant potential to affect rates of primary production. The key factor for this interpretation is the physical oceanography of the system, which we argue was characterized by strong seasonal thermal stratification and intermittent mixing in ancient shallow epeiric seas. This model provides the best explanation for observations of Cretaceous and Devonian black shales described herein. Only rarely were such basins deep and/or isolated enough to experience near-permanent stratification and euxinicity (Devonian US/OC interval), such as that present in the modern Black Sea. Further, in cases where C_{org} accumulation and paleoproductivity estimates are possible (e.g., Greenhorn Formation), these estimates suggest elevated production during black shale deposition, even at maximum highstands when the distance to terrigenous nutrient sources would be greatest. These observations indirectly support the hypothesis of nutrient recycling.
- *Productivity and C-isotope records*—Attributing trends in bulk $\delta^{13}C_{org}$ only to the trade-off between dominance of respiration-influenced local carbon reservoirs versus dominance of production-influenced global carbon reservoirs is clearly a gross oversimplification, as a number of other factors may influence C-isotope fractionation in the biogenic and aqueous reservoirs of the carbon cycle. However, this hypothesis finds support in the observations of Cretaceous and Devonian basins summarized herein. In each of these case studies, relative depletion of $\delta^{13}C_{org}$ values are observed in association with units interpreted to represent examples of the anoxia–nutrient feedback mechanism, and recycling of respired CO_2 to surface waters along with regenerated nutrients would be a plausible scenario. Similarly, each case study includes examples of major positive shifts in $\delta^{13}C_{org}$ documented from sites on distant continents that have been interpreted to reflect global increases in C_{org} burial fluxes sufficient to well exceed local reservoir effects. Considering the global carbon isotopic record of OAE II, the positive excursion is very well constrained by biostratigraphy and thus precisely correlated. Comparing numerous pelagic and hemipelagic sites where preserved OM is dominated by marine algal material, we observe variation of up to 4‰ in the pre-OAE II mean of $\delta^{13}C_{org}$ among different sites and a similar variation of up to 4‰ in the magnitude of the excursion (e.g., Arthur *et al.*, 1988; Pratt *et al.*, 1993; Kuypers, 2001). There are similar variations in the $\delta^{13}C_{org}$ values from the Frasnian–Famennian Kellwasser horizons at different localities (Murphy *et al.*, 2000c), although the OM sources are less well constrained. We hypothesize that these variations largely reflect the dynamic interplay between a local balance of OM respiration and burial and large-scale reservoir effects forced by global environmental perturbations.
- *Mo, Fe, and redox control*—Oxygen deficiency has been recognized as a key factor influencing the formation of Phanerozoic organic-rich hemipelagic facies and is clearly a critical boundary condition in the Precambrian. Reviewing the literature since the early 1980s on this topic, we note a transition to progressively more quantitative approaches to the chemistry of marine redox dynamics. We believe that one of the most exciting advances is the integration of first-order rate equations for OM degradation, for example, via sulfate reduction, developed based on modern observations, to paleoredox proxies such as molybdenum—the accumulation of which appears to reflect sulfide generation (Meyers *et al.*, in review). As described above, reactive-Fe availability plays a key role in this process, and may in fact be a primary regulator in the OM burial process. This idea needs to be tested in more sequences where accumulation rates for key proxies can be calculated.
- *Elemental ratios versus accumulation rates*—In most sections of the geological record, timescales are insufficiently resolved to calculate sedimentation rates at a scale capable of yielding meaningful accumulation rate data. In recent years, however, orbital timescales have been developed for this purpose in some ancient stratigraphic sequences (Park and Herbert, 1987; Meyers *et al.*, 2001). In this paper, the presentation of accumulation rates with the

more common concentration and elemental ratio data serve to highlight the strengths of each. Because interpretations are commonly built upon recognition of *relative changes* among interrelated components, multiproxy concentration and ratio data can be very useful. However, integration of the two types of information clearly represents the optimal approach. In the case of the Greenhorn Formation, accumulation rate data allow actualistic comparisons to be made—whereas average MAR-C_{org} values in the HSM and BCM ($0.16\ g\ cm^{-2}\ kyr^{-1}$ and $0.03\ g\ cm^{-2}\ kyr^{-1}$, respectively) exceed modern open-ocean oligotrophic sites by at least two orders of magnitude, they are conversely at least one order of magnitude lower than modern high-productivity upwelling sites (Bralower and Thierstein, 1987). Notably, the Late Cenomanian HSM experienced higher MAR C_{org} rates (by a factor of 5) than the overlying BCM, which contains the putative production pulse of OAE II. Although there is certainly a relative change in production that can be linked to this event, these data again underscore the importance of recognizing local records of global events. With the exception of some Cretaceous sites that are analogous to modern upwelling zones (e.g., Tarfaya Basin), it is possible that most areas experienced relatively modest production levels during OAE II (e.g., Thurow *et al.*, 1988; Kuhnt *et al.*, 1990), a conclusion that highlights the importance of changes in bulk sedimentation and redox conditions to explain the increased global C_{org} burial flux at this time. As originally suggested by Arthur *et al.* (1987, 1988), sea-level rise and its effect on bulk terrigenous sediment flux and water depth, as well as reactive-Fe delivery to distal hemipelagic sites (Meyers *et al.*, in review), was likely a master variable. This type of model appears to have similar explanatory power for OM burial in the Devonian Appalachian basin (Sageman *et al.*, 2003).

- *Future prospects*—Some of the topics of future importance not covered in detail in this report include advances in the analysis of OM sources, transport dynamics, and reactivities. These studies center on: (i) identification of terrestrial OM and weathered kerogen in the terrigenous flux and the fate of these components during transport and (re)burial (Leithold and Blair, 2001; Petsch *et al.*, 2000, 2002); (ii) further analysis of the role of mineral surface area in fine-grained, organic-rich facies (Kennedy *et al.*, 2002); and (iii) better quantification of decomposition rates for different components of sedimentary OM (e.g., Van Mooy *et al.*, 2002). Other emerging directions in carbon-cycle research include models that rigorously incorporate coupled nutrient cycles (e.g., Kump and Arthur, 1999), quantitative molecular analysis of bacterial and archeal biomass and the new information yielded about metabolic mechanisms (such as anaerobic methane oxidation) over time and space (e.g., Kuypers *et al.*, 2001), and methane storage in gas hydrates and release as a mechanism for driving abrupt climate change (Froelich *et al.*, 1993; Dickens *et al.*, 1995; Dickens, 1999).

ACKNOWLEDGMENTS

The authors wish to acknowledge financial support of the National Science Foundation (e.g., EAR-9725441, EAR-0001093), as well as Northwestern University and the University of Missouri, which made possible some of the projects upon which this compilation is based. The thesis work of our students Adam Murphy, Josef Werne, Matt Hurtgen, and Steve Meyers contributed greatly to the progression of ideas developed herein, and Steve Meyers is specifically thanked for providing significant scientific and editorial input to the manuscript. Lastly, thanks go to the other members of the NU—Sedimentary Research Group (Joniell Borges, Jason Flaum, Michael Fortwengler, Rob Locklair, Petra Pancoskova, and Ramya Sivaraj), who all helped in compiling parts of the bibliography.

REFERENCES

Adelson J. M., Helz G. R., and Miller C. V. (2001) Reconstructing the rise of recent coastal anoxia: molybdenum in Chesapeake Bay sediments. *Geochim. Cosmochim. Acta* **65**, 237–252.

Algeo T. J., Berner R. A., Maynard J. B., and Scheckler S. E. (1995) Late Devonian oceanic anoxic events and biotic crises: "rooted" in the evolution of vascular land plants? *GSA Today* **5**(45), 64–66.

Allen J. R. L. (1970) Studies in fluviatile sedimentation: a comparison of fining-upwards cyclothems, with special reference to coarse-member composition and interpretation. *J. Sedim. Petrol.* **40**, 298–323.

Aller R. C. (1994) Bioturbation and remineralization of sedimentary organic matter: effects of redox oscillation. *Chem. Geol.* **114**, 331–345.

Altabet M. A. and Francois R. (1993) The use of nitrogen isotopic ratio for reconstruction of past changes in surface ocean nutrient utilization. In *Carbon Cycling in the Glacial Ocean: Constraints on the Ocean's Role in Global Change: Quantitative Approaches in Paleoceanography, Series I, Global Environmental Change*, 17 NATO ASI Series (eds. R. Zahn, T. F. Pedersen, M. A. Kaminski, and L. Labeyrie). Springer, Berlin, pp. 281–306.

Altabet M. A., Deuser W. G., Honjo S., and Stienen C. (1991) Seasonal and depth-related changes in the source of sinking particles in the North Atlantic. *Nature* **354**, 136–139.

Anbar A. D. and Knoll A. H. (2002) Proterozoic ocean chemistry and evolution: a bioinorganic bridge? *Science* **297**, 1137–1142.

Anderson R. F., Fleisher M. Q., and Le Huray A. P. (1989a) Concentration, oxidation state, and particulate flux of uranium in the Black Sea. *Geochim. Cosmochim. Acta* **53**, 2215–2224.

Anderson R. F., Le Huray A. P., Fleisher M. Q., and Murray R. W. (1989b) Uranium depletion in Saanich Inlet sediments, Vancouver Island. *Geochim. Cosmochim. Acta* **53**, 2205–2213.

Anderson R. F., Lyons T. W., and Cowie G. L. (1994) Sedimentary record of a shoaling of the oxic/anoxic interface in the Black Sea. *Mar. Geol.* **116**, 373–384.

Anderson T. F., Kruger J., and Raiswell R. (1987) C–S–Fe relationships and the isotopic composition of pyrite in the New Albany Shales of the Illinois Basin USA. *Geochim. Cosmochim. Acta* **51**, 2795–2805.

Archer D. E. (1991) Equatorial Pacific calcite preservation cycles: production or dissolution? *Paleoceanography* **6**, 561–571.

Arnold G. L., Anbar A. D., and Barling J. (2002) Oxygenation of Proterozoic oceans: insight from molybdenum isotopes. *Geochim. Cosmochim. Acta* **66**, A30 Goldschmidt Conference Abstracts.

Arthur M. A. and Dean W. E. (1991) An holistic geochemical approach to cyclomania: examples from Cretaceous pelagic limestone sequences. In *Cycles and Events in Stratigraphy* (eds. G. Einsele, W. Ricken, and A. Seilacher). Springer, Berlin, pp. 126–166.

Arthur M. A. and Dean W. E. (1998) Organic-matter production and preservation and evolution of anoxia in the Holocene Black Sea. *Paleoceanography* **13**, 395–411.

Arthur M. A. and Sageman B. B. (1994) Marine black shales: a review of depositional mechanisms and environments of ancient deposits. *Ann. Rev. Earth Planet. Sci.* **22**, 499–552.

Arthur M. A., Dean W. E., Pollastro R., Scholle P. A., and Claypool G. E. (1985) A comparative geochemical study of two transgressive pelagic limestone units, Cretaceous Western Interior basin US. In *Fine-Grained Deposits and Biofacies of the Cretaceous Western Interior Seaway: Evidence of Cyclic Sedimentary Processes*. (eds. L. M. Pratt, E. G. Kauffman, and F. B. Zelt). Society Economic Paleontologists and Minerologists, Tulsa, pp. 16–27.

Arthur M. A., Schlanger S. O., and Jenkyns H. C. (1987) The Cenomanian-Turonian Oceanic Anoxic Event: II. Paleoceanographic controls on organic matter production and preservation. In *Marine Petroleum Source Rocks*. Geological Society of London Special Publication 26 (eds. J. Brooks and A. J. Fleet). pp. 401–420.

Arthur M. A., Dean W. E., and Pratt L. M. (1988) Geochemical and climatic effects of increased marine organic carbon burial at the Cenomanian/Turonian boundary. *Nature* **335**, 714–717.

Arthur M. A., Dean W. E., Neff E. D., Hay B. J., Jones G., and King J. (1994) Late Holocene (0–2000 yBP) organic carbon accumulation in the Black Sea. *Global Biogeochem. Cycles* **8**, 195–217.

Arthur M. A., Dean W. E., and Laarkamp K. (1998) Organic carbon accumulation and preservation in surface sediments on the Peru margin. *Chem. Geol.* **152**, 273–283.

Barling J. and Anbar A. D. Molybdenum isotope fractionation during adsorption by manganese oxides. *Earth Planet. Sci. Lett.* (submitted).

Barling J., Arnold G. L., and Anbar A. D. (2001) Natural mass-dependent variations in the isotopic composition of molybdenum. *Earth Planet. Sci. Lett.* **193**, 447–457.

Beier J. A. and Hayes J. M. (1989) Geochemical and isotope evidence for paleoredox conditions during deposition of the Devonian-Mississippian New Albany Shale, southern Indiana. *Geol. Soc. Am. Bull.* **101**, 774–782.

Berner R. A. (1970) Sedimentary pyrite formation. *Am. J. Sci.* **268**, 1–23.

Berner R. A. (1980) *Early Diagenesis—A Theoretical Approach*. Princeton University Press, Princeton, NJ, 241pp.

Berner R. A. (1982) Burial of organic carbon and pyrite sulfur in the modern ocean: its geochemical and environmental significance. *Am. J. Sci.* **282**, 451–473.

Berner R. A. (1984) Sedimentary pyrite formation: an update. *Geochim. Cosmochim. Acta* **48**, 605–615.

Berner R. A. (1999) A new look at the long-term carbon cycle. *GSA Today* **9**, 1–6.

Berner R. A. (2001) Modeling atmospheric O_2 over Phanerozoic time. *Geochim. Cosmochim. Acta* **65**, 685–694.

Berner R. A. and Canfield D. E. (1989) A new model for atmospheric oxygen over Phanerozoic time. *Am. J. Sci.* **289**, 333–360.

Berner R. A. and Kothavala Z. (2001) GEOCARB III: a revised model of atmospheric CO_2 over Phanerozoic time. *Am. J. Sci.* **301**, 182–204.

Berner R. A. and Petsch S. T. (1998) The sulfur cycle and atmospheric oxygen. *Science* **282**, 1426–1427.

Berner R. A. and Raiswell R. (1984) C/S method for distinguishing freshwater from marine sedimentary rocks. *Geology* **12**, 365–368.

Berner R. A., Lasaga A. C., and Garrels R. M. (1983) The carbonate-silicate geochemical cycle and its effect on atmospheric carbon dioxide over the past 100 million years. *Am. J. Sci.* **283**, 641–683.

Berner R. A., Petsch S. T., Lake J. A., Berling D. J., Popp B. N., Lane R. S., Laws E. A., Westley M. B., Cassar N., Woodward F. I., and Quick W. P. (2000) Isotope fractionation and atmospheric oxygen: implications for Phanerozoic O_2 evolution. *Science* **287**, 1630–1633.

Bertrand P., Shimmield G., Martinez P., Grousset F., Jorissen F., Paterne M., Pujol C. J., Bouloubassi I., Buat Menard P., Peypouquet J. P., Beaufort L., Sicre M. A., Lallier-Verges E., Foster J. M., and Ternois Y. (1996) The glacial ocean productivity hypothesis: the importance of regional temporal and spatial studies. *Mar. Geol.* **130**, 1–9.

Betts J. N. and Holland H. D. (1991) The oxygen content of bottom waters, the burial efficiency of organic carbon, and the regulation of atmospheric oxygen. *Global Planet. Change* **5**, 5–18.

Bewers J. M. and Yeats P. A. (1979) The behavior of trace metals in estuaries of the St. Lawrence Basin. *Le Naturaliste Canadien* **106**, 149–160.

Bralower T. J. and Thierstein H. R. (1984) Low productivity and slow deep water circulation in mid-Cretaceous oceans. *Geology* **12**, 614–618.

Bralower T. J. and Thierstein H. R. (1987) Organic-carbon and metal accumulation in Holocene and Mid-Cretaceous marine sediments: palaeoceanographic significance. In *Marine Petroleum Source Rocks*. Geological Society of London Special Publication 26 (eds. J. Brooks and A. J. Fleet), pp. 345–369.

Breit G. N. and Wanty R. B. (1991) Vanadium accumulation in carbonaceous rocks: a review of geochemical controls during deposition and diagenesis. *Chem. Geol.* **91**, 83–97.

Brett C. E. (1995) Sequence stratigraphy, biostratigraphy, and taphonomy in shallow marine environments. *Palaios* **10**, 597–616.

Brett C. E. and Baird G. C. (1986) Symmetrical and upward shallowing cycles in the Middle Devonian of New York State and their implications for the punctuated aggradational cycle hypothesis. *Paleoceanography* **1**, 431–445.

Brett C. E. and Ver Straeten C. A. (1995) Stratigraphy and facies relationships of the Eifelian Onondaga Limestone (Middle Devonian) in western and west central New York State. In *67th Annual Meeting Field Trip Guidebook* (eds. J. I. Garver and J. A. Smith). New York State Geological Association, Albany, pp. 221–269.

Bright C. A., Lyons T. W., MacLeod K. G., Glascock M. D., Rexroad C. B., Brown L. M., and Ethington R. L. Arguments against preservation of primary seawater signals in the rare-earth element compositions of biogenic (conodont) apatite. *J. Sedim. Res.* (submitted).

Broecker W. S. (1982) Ocean chemistry during glacial time. *Geochim. Cosmochim. Acta* **4**, 1689–1705.

Broecker W. S. and Peng T. H. (1982) *Tracers in the Sea*. Eldigio Press, Palisades, New York, 690pp.

Brumsack H. J. (1989) Geochemistry of recent TOC-rich sediment from the Gulf of California and the Black Sea. *Geol. Rundsch.* **78**, 851–882.

Brumsack H. J. and Thurow J. (1986) The geochemical facies of black shales from the Cenomanian Turonian boundary event (CTBE). In *Biogeochemistry of Black shales—Case Studies* (eds. E. T. Degens, P. A. Meyers, and S. C. Brassell). Mitteilungen aus dem Geologisch-Palaeontologischen Institut der Universitaet Hamburg, Vol. 60, pp. 247–265.

Burdett J. W., Arthur M. A., and Richardson M. (1989) A Neogene seawater sulfate isotope age curve from calcareous pelagic microfossils. *Earth Planet. Sci. Lett.* **94**, 189–198.

Burns C. E. and Bralower T. J. (1998) Upper Cretaceous nannofossil assemblages across the Western Interior seaway: implications for the origin of lithologic cycles in the Greenhorn and Niobrara Formations. In *Stratigraphy and Paleoenvironments of the Cretaceous Western Interior Seaway*. Concepts in Sedimentology and Paleontology no. 6 (eds. M. A. Arthur and W. E. Dean). Society Economic Paleontologists Minerologists, Tulsa, pp. 35–58.

Byers C. W. (1977) Biofacies patterns in euxinic basins: a general model. In *Deep-water Carbonate Environments*. Special Publication, no. 25 (eds. H. E. Cook and P. Enos). Society Economic Paleontologists Minerologists, Tulsa, pp. 5–17.

Calvert S. E. and Karlin R. E. (1991) Relationships between sulphur, organic carbon, and iron in the modern sediments of the Black Sea. *Geochim. Cosmochim. Acta* **55**, 2483–2490.

Calvert S. E. and Karlin R. E. (1998) Organic carbon accumulation in the Holocene sapropel of the Black Sea. *Geology* **26**, 107–110.

Calvert S. E. and Pedersen T. F. (1993) Geochemistry of recent oxic and anoxic marine sediments: implications for the geological record. *Mar. Geol.* **113**, 67–88.

Calvert S. E. and Pedersen T. F. (1996) Sedimentary geochemistry of manganese: implications for the environment of formation of manganiferous black shales. *Econ. Geol.* **91**, 36–47.

Calvert S. E., Karlin R. E., Toolin L. J., Donahue D. J., Southon J. R., and Vogel J. S. (1991) Low organic carbon accumulation rates in Black Sea sediments. *Nature* **350**, 692–695.

Calvert S. E., Thode H. G., Yeung D., and Karlin R. E. (1996) A stable isotope study of pyrite formation in the late Pleistocene and Holocene sediments of the Black Sea. *Geochim. Cosmochim. Acta* **60**, 1261–1270.

Canfield D. E. (1989) Sulfate reduction and oxic respiration in marine sediments: implications for organic carbon preservation in euxinic sediments. *Deep-Sea Res.* **36**, 121–138.

Canfield D. E. (1994) Factors influencing organic carbon preservation in marine sediments. *Chem. Geol.* **114**, 315–329.

Canfield D. E. (1998) A new model for Proterozoic ocean chemistry. *Nature* **396**, 450–453.

Canfield D. E. (2001) Isotope fractionation by natural populations of sulfate-reducing bacteria. *Geochim. Cosmochim. Acta* **65**, 1117–1124.

Canfield D. E. and Raiswell R. (1991) Pyrite formation and fossil preservation. In *Taphonomy: Releasing the Data Locked in the Fossil Record*. Topics in Geobiology 9 (eds. P. A. Allison and D. E. G. Briggs). Plenum, New York, pp. 411–453.

Canfield D. E. and Teske A. (1996) Late Proterozoic rise in atmospheric oxygen concentration inferred from phylogenetic and sulphur-isotope studies. *Nature* **382**, 127–132.

Canfield D. E. and Thamdrup B. (1994) The production of ^{34}S-depleted sulfide during bacterial disproportionation of elemental sulfur. *Science* **266**, 1973–1975.

Canfield D. E., Raiswell R., and Bottrell S. H. (1992) The reactivity of sedimentary iron minerals toward sulfide. *Am. J. Sci.* **292**, 659–683.

Canfield D. E., Lyons T. W., and Raiswell R. (1996) A model for iron deposition to euxinic Black Sea sediments. *Am. J. Sci.* **296**, 818–834.

Canfield D. E., Habicht K. S., and Thamdrup B. (2000) The Archean sulfur cycle and the early history of atmospheric oxygen. *Science* **288**, 658–661.

Chambers L. A., Trudinger P. A., Smith J. W., and Burns M. S. (1975) Fractionation of sulfur isotopes by continuous cultures of *Desulfovibrio desulfuricans*. *Can. J. Microbiol.* **21**, 1602–1607.

Chester M. A., Guymer I., and Freestone R. (1997) Managing water quality in the tidal Ouse (UK), problems associated with suspended sediment oxygen demand. In *Environmental and Coastal Hydraulics: Protecting the Aquatic Habitat*. Proceedings of Congress of International Association for Hydraulic Research 27, Theme B (eds. F. M. Holly, Jr., A. Alsaffar, S. S. Wang, and T. Carstens). International Association for Hydraulic Research Congress pp. 665–670.

Chester R. (1990) *Marine Geochemistry*. Harper Collins, NY, 690pp.

Chester R. (2000) *Marine Geochemistry* 2nd edn. Blackwell, Oxford, UK, 506pp.

Chester R. and Murphy K. J. T. (1990) Metals in the marine atmosphere. In *Heavy Metals in the Marine Environment* (eds. R. Furnace and P. Rainbow). CRC Press, Boca Raton, pp. 27–49.

Claypool G. E., Holser W. T., Kaplan I. R., Sakai H., and Zak I. (1980) The age curves of sulfur and oxygen isotopes in marine sulfate and their mutual interpretations. *Chem. Geol.* **28**, 199–260.

Coveney R. M., Jr, Watney W. L., and Maples C. G. (1991) Contrasting depositional models for Pennsylvanian black shale discerned from molybdenum abundances. *Geology* **19**, 147–150.

Creaney S. and Passey Q. R. (1993) Recurring patterns of total organic carbon and source rock quality within a sequence stratigraphic framework. *Am. Assoc. Petrol. Geol. Bull.* **77**, 386–401.

Crusius J. and Anderson R. F. (1992) Inconsistencies in accumulation rates of Black Sea sediments inferred from records of laminae and ^{210}Pb. *Paleoceanography* **7**, 215–227.

Crusius J., Calvert S., Pedersen T., and Sage D. (1996) Rhenium and molybdenum enrichments in sediments as indicators of oxic, suboxic and sulfidic conditions of deposition. *Earth Planet. Sci. Lett.* **145**, 65–78.

Cullers R. L. (1994a) The controls on the major and trace element variation of shales, siltstones, and sandstones of Pennsylvanian-Permian age from uplifted continental blocks in Colorado to platform sediment in Kansas, USA. *Geochim. Cosmochim. Acta* **58**, 4955–4972.

Cullers R. L. (1994b) The chemical signature of source rocks in size fractions of Holocene stream sediment derived from metamorphic rocks in the Wet Mountains region, Colorado, USA. *Chem. Geol.* **113**, 327–343.

Davis C., Pratt L. M., Sliter W. V., Mompart L., and Murat B. (1999) Factors influencing organic carbon and trace metal accumulation in the Upper Cretaceous La Luna Formation of the western Maracaibo Basin, Venezuela. In *Evolution of the Cretaceous Ocean-Climate System*. Geological Society of America Special Paper 332 (eds. E. Barrera and C. Johnson). Geological Society of America, Boulder, pp. 203–230.

de Baar H. J. W., German C. R., Elderfield H., and van Gaans P. (1988) Rare earth element distributions in anoxic waters of the Cariaco Trench. *Geochim. Cosmochim. Acta* **52**, 1203–1219.

de Leeuw J. W., Frewin N. L., van Bergen P. F., Sinninghe Damsté J. S., and Collinson M. E. (1995) Organic carbon as a palaeoenvironmental indicator in the marine realm. In *Marine Paleoenironmental Analysis from Fossils*. Geological Society of London Special Publication 83 (eds. D. W. J. Bosence and P. A. Allison). Geological Society of London, London, pp. 43–72.

DeMenocal P. B. (1995) Plio-Pleistocene African climate. *Science* **270**, 53–59.

Dean W. E. and Arthur M. A. (1998) Geochemical expression of cyclicity in Cretaceous pelagic limestone sequences: Niobrara Formation, Western Interior Seaway. In *Stratigraphy and Paleoenvironments of the Cretaceous Western Interior Seaway USA*. Concepts in Sedimentology and Paleontology, no 6 (eds. W. E. Dean and M. A. Arthur). Society Economic Paleontologists Minerologists, Tulsa, pp. 227–255.

Dean W. E., Piper D. Z., and Peterson L. C. (1999) Molybdenum accumulation in Cariaco basin sediment over the past 24 k.y: a record of water-column anoxia and climate. *Geology* **27**, 507–510.

Degens E. T. (1965) *Geochemistry of Sediments: A Brief Survey*. Prentice Hall, Englewood Cliffs, NJ, 342pp.

Degens E. T. and Ittekkot V. (1987) The carbon cycle: tracking the path of organic particles from sea to sediment. In *Marine Petroleum Source Rocks*. Geological Society of London Special Publication 26 (eds. J. Brooks and A. J. Fleet). Geological Society of London, London, pp. 121–135.

Demaison G. I. and Moore G. T. (1980) Anoxic marine environments and oil source bed genesis. *Am. Assoc. Petrol. Geol. Bull.* **64**, 1179–1209.

DePaolo D. J. (1986) Detailed record of the Neogene Sr isotopic evolution of seawater from DSDP Site 590B. *Geology* **14**, 103–106.

Detmers J., Brüchert V., Habicht K. S., and Kuever J. (2001) Diversity of sulfur isotope fractionations by sulfate-reducing prokaryotes. *Appl. Environ. Microbiol.* **67**, 888–894.

Dickens G. R. (1999) Carbon cycle: the blast in the past. *Nature* **401**, 752–755.

Dickens G. R., O'Neil J. R., Rea D. K., and Owen R. M. (1995) Dissociation of oceanic methane hydrate as a cause of the carbon-isotope excursion at the end of the Paleocene. *Paleoceanography* **10**, 965–971.

Durand B. (1980) Sedimentary organic matter and kerogen: definition and quantitative importance of kerogen. In *Kerogen: Insoluble Organic Matter from Sedimentary Rocks* (ed. B. Durand). Technip, Paris, pp. 13–34.

Dyman T. S., Cobban W. A., Fox J. E., Hammond R. H., Nichols D. J., Perry W. J., Jr., Porter K. W., Rice D. D., Setterholm D. R., Shurr G. W., Tysdal R. G., Haley J. C., and Campen E. B. (1994) Cretaceous rocks from southwestern Montana to southwestern Minnesota, Northern Rocky Mountains and Great Plains region. In *Perspectives on the Eastern Margin of the Cretaceous Western Interior Basin*. Geological Society of America Special Paper 287 (eds. G. W. Shurr, G. A. Ludvigson, and R. H. Hammond). Geological Society of America, Boulder, pp. 5–26.

Dymond J., Suess E., and Lyle M. (1992) Barium in deep-sea sediment: a geochemical proxy for paleoproductivity. *Paleoceanography* **7**, 163–181.

Dymond J., Collier R., McManus J., Honjo S., and Manganini S. (1997) Can the aluminum and titanium contents of ocean sediments be used to determine the paleoproductivity of the oceans? *Paleoceanography* **12**, 586–593.

Eicher D. L. and Diner R. (1989) Origin of the Cretaceous Bridge Creek cycles in the western interior, United States. *Palaeogeogr. Palaeoclimatol. Palaeoecol.* **74**, 127–146.

Elder W. P. (1985) Biotic patterns across the Cenomanian-Turonian extinction boundary near Pueblo, Colorado. In *Fine-Grained Deposits and Biofacies of the Cretaceous Western Interior Seaway: Evidence of Cyclic Sedimentary Processes* (eds. L. M. Pratt, E. G. Kauffman, and F. B. Zelt). Society Economic Paleontologists and Minerologists, Tulsa, pp. 157–169.

Elder W. P. (1989) Molluscan extinction patterns across the Cenomanian-Turonian stage boundary in the Western Interior of the United States. *Paleobiology* **15**, 299–320.

Elder W. P. (1991) Molluscan paleoecology and sedimentation patterns of the Cenomanian-Turonian extinction interval in the southern Colorado Plateau region. In *Stratigraphy, Depositional Environments, and Sedimentary Tectonics of the Western Margin, Cretaceous Western Interior Seaway*. Geological Society of America Special Paper 260 (eds. J. D. Nations and J. G. Eaton). Geological Society of America, Boulder, pp. 113–137.

Elder W. P., Gustason E. R., and Sageman B. B. (1994) Basinwide correlation of parasequences in the Greenhorn Cyclothem, Western Interior US. *Geol. Soc. Am. Bull.* **106**, 892–902.

Ellwood B. B., Crick R. E., El Hassani A., Benoist S. L., and Young R. H. (2000) Magnetosusceptibility event and cyclostratigraphy method applied to marine rocks: detrital input versus carbonate productivity. *Geology* **28**, 1135–1138.

Emerson S. (1985) Organic carbon preservation in marine sediments. In *The Carbon Cycle and Atmospheric CO_2: Natural Variations Archean to Present*. American Geophysical Union, Geophysical Monograph 32 (eds. E. T. Sunquist and W. S. Broecker). American Geophysical Union, Washington, DC, pp. 78–87.

Emerson S. and Hedges J. I. (1988) Processes controlling the organic carbon content of open ocean sediments. *Paleoceanograpy* **53**, 1233–1240.

Emerson S. R. and Huested S. S. (1991) Ocean anoxia and the concentrations of molybdenum and vanadium in seawater. *Mar. Chem.* **34**, 177–196.

Engel M. H. and Macko S. (1993) *Organic Geochemistry: Principles and Applications*. Plenum, New York, 861pp.

Ettensohn F. R. (1985a) The Catskill Delta complex and the Acadian Orogeny: a model. In *The Catskill Delta*. Geological Society of America Special Paper 201 (eds. D. L. Woodrow and W. D. Sevon). Geological Society of America, Boulder, pp. 39–49.

Ettensohn F. R. (1985b) Controls on development of Catskill Delta complex basin-facies. In *The Catskill Delta*. Geological Society of America Special Paper 201 (eds. D. L. Woodrow and W. D. Sevon). Geological Society of America, Boulder, pp. 65–77.

Ettensohn F. R., Miller M. L., Dillman S. B., Elam T. D., Geller K. L., Swager D. R., Markowitz G., Woock R. D., and Barron L. S. (1988) Characterization and implications of the Devonian-Mississippian black-shale sequence, eastern and central Kentucky, USA: Pycnoclines, transgression, regression and tectonism. In *Devonian of the World: Proceedings of the Second International Symposium on the Devonian System, Volume II, Sedimentation*. (eds. N. J. McMillan, A. F. Embry, and D. J. Glass) Canadian Society of Petroleum Geologists, Memoir 14, pp. 323–345.

Faill R. T. (1985) The Acadian Orogeny and the Catskill Delta. In *The Catskill Delta*. Geological Society of America Special Paper 201 (eds. D. L. Woodrow and W. D. Sevon). Geological Society of America, Boulder, pp. 15–37.

Falkowski P. G. (1997) Evolution of the nitrogen cycle and its influence on the biological sequestration of CO_2 in the ocean. *Nature* **387**, 272–275.

Farquhar J., Bao H. M., and Thiemens M. (2000) Atmospheric influence of Earth's earliest sulfur cycle. *Science* **289**, 756–758.

Filippelli G. (1997) Controls on the phosphorus concentration and accumulation in marine sediments. *Mar. Geol.* **139**, 231–240.

Filippelli G. M., Sierro F. J., Flores J. A., Vazquez A., Utrilla R., Perez-Folgado M., and Latimer J. C. (2003) A sediment-nutrient-oxygen feedback responsible for productivity variations in Late Miocene sapropel sequences of the western Mediterranean. *Palaeogeogr. Palaeoclimatol. Palaeoecol.* **190**, 335–348.

Fisher I. J. St. and Hudson J. D. (1987) Pyrite formation in Jurassic shales of contrasting biofacies. In *Marine Petroleum Source Rocks*. Geological Society of London Special Publications 26 (eds. J. Brooks and A. J. Fleet). Geological Society of London, London, pp. 69–78.

Francois R. (1988) A study on the regulation of the concentrations of some trace metals (Rb, Sr, Zn, Pb, Cu, V, Cr, Ni, Mn and Mo) in Saanich Inlet sediments, British Columbia, Canada. *Mar. Geol.* **83**, 285–308.

Francois R., Honjo S., Manganini S. J., and Ravizza G. E. (1995) Biogenic barium fluxes to the deep sea: implications for paleoproductivity reconstruction. *Global Biogeochem. Cycles* **9**, 289–303.

Froelich P. N., Klinkhammer G. P., Bender M. L., Luedtke N., Heath G. R., Cullen D., Dauphin P., Hammond D., Hartman B., and Maynard V. (1979) Early oxidation of organic matter in pelagic sediments of the eastern equatorial Atlantic: suboxic diagenesis. *Geochim. Cosmochim. Acta* **43**, 1075–1090.

Froelich P. N., Blanc V., Mortlock R. A., Chilrud S. N., Dunstan W., Udomkit A., and Peng T. H. (1992) River fluxes of dissolved silica to the ocean were higher during glacials: Ge/Si in diatoms rivers, and oceans. *Paleoceanography* **7**, 739–767.

Froelich P. N., Kvenvolden K. A., and Torres M. (1993) Evidence for gas hydrate in the accretionary prism near the Chile triple junction: ODP Leg 141. *Eos. Trans., AGU* **74**, 369.

Fry B., Jannasch H. W., Molyneaux S. J., Wirsen C. O., Muramoto J. A., and King S. (1991) Stable isotope studies of the carbon, nitrogen and sulfur cycles in the Black Sea and the Cariaco Trench. *Deep-Sea Res.* **38**, S1003–S1019.

Garrels R. M. and Lerman A. (1981) Phanerozoic cycles of sedimentary carbon and sulfur. *Proc. Natl. Acad. Sci. USA* **78**, 4652–4656.

Garrels R. M. and Mackenzie F. T. (1971) *Evolution of Sedimentary Rocks*. W.W. Norton and Company, New York, 397pp.

Gartner S. (1977) Nannofossils and biostratigraphy: an overview. *Earth Sci. Rev.* **13**, 227–250.

Gautier D. L. (1986) Cretaceous shales from the Western Interior of North America: sulfur/carbon ratios and sulfur-isotope composition. *Geology* **14**, 225–228.

Gautier D. L. (1987) Isotopic composition of pyrite: relationship to organic matter type and iron availability in some North American Cretaceous shales. *Chem. Geol.: Isotope Geosci. Sect.* **65**, 293–303.

Gellatly A. M. and Lyons T. W. Trace sulfate in Mesoproterozoic carbonates: implications for seawater sulfate and oxygen availability. *Geochim. Cosmochim. Acta*. (submitted).

German C. R. and Elderfield H. (1989s) Rare earth elements in Saanich Inlet, British Columbia, a seasonally anoxic basin. *Geochim. Cosmochim. Acta* **53**, 2561–2571.

German C. R. and Elderfield H. (1990) Application of the Ce anomaly as a paleoredox indicator: the ground rules. *Paleoceanography* **5**, 823–833.

German C. R., Holliday B. P., and Elderfield H. (1991) Redox cycling of rare earth elements in the suboxic zone of the Black Sea. *Geochim. Cosmochim. Acta* **55**, 3553–3558.

Goldhaber M. B. and Kaplan I. R. (1974) The sulfur cycle. In *The Sea, vol. 5* (ed. E. D. Goldberg). Wiley, New York, pp. 569–655.

Goldhaber M. B. and Kaplan I. R. (1980) Mechanisms of sulfur incorporation and isotope fractionation during early diagenesis in sediments of the Gulf of California. *Mar. Chem.* **9**, 95–143.

Gorsline D. S. (1984) A review of fine-grained sediment origins, characteristics, transport and deposition. In *Fine-grained Sediments: Deep-water Processes and Facies*, Geological Society of America Special Publication 5 (eds. D. A. V. Stow and D. J. W. Piper). Geological Society of America, Boulder, pp. 17–34.

Gromet L. P., Dymek R. F., Haskin L. A., and Kortev R. L. (1984) The "North American shale composite" its compilation, major and trace element characteristics. *Geochim. Cosmochim. Acta* **48**, 2469–2482.

Habicht K. S. and Canfield D. E. (1997) Sulfur isotope fractionation during bacterial sulfate reduction in organic-rich sediments. *Geochim. Cosmochim. Acta* **6**, 5351–5361.

Habicht K. S. and Canfield D. E. (2001) Isotope fractionation by sulfate-reducing natural populations and the isotopic composition of sulfide in marine sediments. *Geology* **29**, 555–558.

Habicht K. S., Gade M., Thamdrup B., Berg P., and Canfield D. E. (2002) Calibration of sulfate levels in the Archean Ocean. *Science* **298**, 2372–2374.

Harris S. E., Mix A. C., and King T. (1997) Biogenic and terrigenous sedimentation at Ceara Rise, western tropical Atlantic, supports Pliocene-Pleistocene deep-water linkage between hemispheres. *Proc. ODP Sci. Results* **154**, 331–345.

Hartnett H. E., Keil R. G., Hedges J. I., and Devol A. H. (1998) Influence of oxygen exposure time on organic carbon preservation in continental margin sediments. *Nature* **391**, 572–574.

Hassold N., Rea D. K., and Meyers P. A. (2003) Grain size evidence for variations in delivery of terrigenous sediments to a Middle Pleistocene interrupted sapropel from ODP Site 969, Mediterranean Ridge. *Palaeogeogr. Palaeoclimatol. Palaeoecol.* **190**, 211–219.

Haug G. H., Gunther D., Peterson L. C., Sigman D. M., Hughen K. A., and Aeschlimann B. (2003) Climate and the collapse of Maya Civilization. *Science* **299**, 1731–1735.

Hayes J. M. (1993) Factors controlling ^{13}C contents of sedimentary organic compounds: principles and evidence. *Mar. Geol.* **113**, 111–125.

Hayes J. M., Popp B. N., Takigiku R., and Johnson M. W. (1989) An isotopic study of biogeochemical relationships between carbonates and organic carbon in the Greenhorn Formation. *Geochim. Cosmochim. Acta* **53**, 2961–2972.

Hayes J. M., Freeman K. H., Popp B. N., and Hoham C. H. (1990) Compound-specific isotopic analyses: a novel tool for reconstruction of ancient biogeochemical processes. *Org. Geochem.* **16**, 1115–1128.

Hayes J. M., Strauss H., and Kaufman A. J. (1999) The abundance of ^{13}C in marine organic matter and isotopic fractionation in the global biogeochemical cycle of carbon during the past 800 Ma. *Chem. Geol.* **161**, 103–125.

Haynes J. R. (1981) *Foraminifera*. Wiley, New York, 433pp.

Hedges J. I. and Keil R. G. (1995) Sedimentary organic matter preservation: an assessment and speculative synthesis. *Mar. Chem.* **49**, 81–115.

Helz G. R., Miller C. V., Charnock J. M., Mosselmans J. F. W., Pattrick R. A. D., Garner C. D., and Vaughan D. J. (1996) Mechanism of molybdenum removal from the sea and its concentration in black shales: EXAFS evidence. *Geochim. Cosmochim. Acta* **60**, 3631–3642.

Henrichs S. M. and Reeburgh W. S. (1987) Anaerobic mineralization of marine sediment organic matter: rates and the role of anaerobic processes in the oceanic carbon economy. *Geomicrobiol. J.* **5**, 191–237.

Herbert T. D. and Fischer A. G. (1986) Milankovitch climatic origin of mid-Cretaceous black shale rhythms in central Italy. *Nature* **321**, 739–743.

Herbert T. D., Stallard R. F., and Fischer A. G. (1986) Anoxic events, productivity rhythms, and the orbital signature in a mid-Cretaceous deep-sea sequence from central Italy. *Paleoceanography* **1**, 495–506.

Hem J. D. (1981) Rates of manganese oxidation in aqueous systems. *Geochim. Cosmochim. Acta* **45**, 1369–1374.

Hofmann P., Wagner T., and Beckmann B. (2003) Millennial- to centennial-scale record of African climate variability and organic carbon accumulation in the Coniacian-Santonian eastern tropical Atlantic (Ocean Drilling Program Site 959, off Ivory Coast and Ghana). *Geology* **31**, 135–138.

Holland H. D. (2002) Volcanic gases, black smokers, and the Great Oxidation Event. *Geochim. Cosmochim. Acta* **66**, 3811–3826.

Holmes M. E., Schneider R. R., Mueller P. J., Segl M., and Wefer G. (1997) Reconstruction of past nutrient utilization in the eastern Angola Basin based on sedimentary $^{15}N/^{14}N$ ratios. *Paleoceanography* **12**, 604–614.

House M. R. and Kirchgasser W. T. (1993) Devonian goniatite biostratigraphy and timing of facies movements in the Frasnian of eastern North America. In *High Resolution Stratigraphy*. Geological Society of London Special Publications 70 (eds. E. A. Hailwood and R. B. Kidd). Geological Society of London, London, pp. 267–292.

Huang Y., Freeman K. H., Wilkin R. T., Arthur M. A., and Jones A. D. (2000) Black Sea chemocline oscillations during the Holocene: molecular and isotopic studies of marginal sediments. *Org. Geochem.* **31**, 1525–1531.

Hughen K., Overpeck J. T., Peterson L. C., and Trumbore S. (1996) Rapid climate changes in the tropical Atlantic region during the last deglaciation. *Nature* **380**, 51–54.

Hughen K. A., Overpeck J. T., Lehman S. J., Kashgarian M., Southon J., Peterson L. C., Alley R., and Sigman D. M. (1998) Deglacial changes in ocean circulation from an extended radiocarbon calibration. *Nature* **391**, 65–68.

Hurtgen M. T., Lyons T. W., Ingall E. D., and Cruse A. M. (1999) Anomalous enrichments of iron monosulfide in euxinic marine sediments and the role of H_2S in iron sulfide transformations: examples from Effingham Inlet, Orca Basin, and the Black Sea. *Am. J. Sci.* **299**, 556–588.

Hurtgen M. T., Arthur M. A., Suits N. S., and Kaufman A. J. (2002) The sulfur isotopic composition of Neoproterozoic seawater sulfate: implications for a snowball Earth? *Earth Planet. Sci. Lett.* **203**(1), 413–429.

Ingall E. D. and Jahnke R. (1997) Influence of water-column anoxia on the elemental fractionation of carbon and phosphorus during sediment diagenesis. *Mar. Geol.* **139**, 219–229.

Ingall E. D., Bustin R. M., and Van Cappellen P. (1993) Influence of water column anoxia on the burial and preservation of carbon and phosphorus in marine shales. *Geochim. Cosmochim. Acta* **57**, 303–316.

Isozaki Y. (1997) Permo-Triassic boundary superanoxia and stratified superocean: records from lost deep sea. *Science* **276**, 235–238.

Ittekkot V. (1988) Global trends in the nature of organic matter in river suspensions. *Nature* **332**, 436–438.

Jacobs L. and Emerson S. (1982) Trace metal solubility in an anoxic Fjord. *Earth Planet. Sci. Lett.* **60**, 237–252.

Jacobs L., Emerson S., and Skei J. (1985) Partitioning and transport of metals across the O_2/H_2S interface in a permanently anoxic basin: Framvaren Fjord, Norway. *Geochim. Cosmochim. Acta* **49**, 1433–1444.

Jacobs L., Emerson S., and Huested S. S. (1987) Trace metal geochemistry in the Cariaco Trench. *Deep-Sea Res.* **34**, 965–981.

Joachimski M. M. and Buggisch W. (1993) Anoxic events in the late Frasnian: causes of the Frasnian-Famennian faunal crisis? *Geology* **21**, 675–678.

Johnson-Ibach L. E. (1982) Relationship between sedimentation rate and total organic carbon content in ancient marine sediments. *Am. Assoc. Petrol. Geol. Bull.* **66**, 170–188.

Johnson J. G. and Sandberg C. A. (1988) Devonian eustatic events in the Western United States and their biostratigraphic responses. In *Devonian of the World: Proceedings of the Second International Symposium on the Devonian System, Volume III, Paleontology, Paleoecology and Biostratigraphy* (eds. N. J. McMillan, A. F. Embry, and D. J. Glass) Canadian Society of Petroleum Geologists, Memoir **14**, pp. 171–178.

Johnson J. G., Klapper G., and Sandberg C. A. (1985) Devonian eustatic fluctuations in Euramerica. *Geol. Soc. Am. Bull.* **96**, 567–587.

Jones B. and Manning D. A. C. (1994) Comparison of geochemical indices used for the interpretations of paleoredox conditions in ancient mudstones. *Chem. Geol.* **111**, 111–129.

Jørgensen B. B. (1979) A theoretical model of the stable sulfur isotope distribution in marine sediments. *Geochim. Cosmochim. Acta* **43**, 363–374.

Kah L. C., Lyons T. W., and Frank T. D. Mesoproterozoic marine sulfate: Evidence for a changing biosphere. *Science* (submitted).

Kakegawa T., Kawai H., and Ohmoto H. (1998x) Origins of pyrites in the approximately 2.5 Ga Mt. McRae Shale, the Hamersley District, Western Australia. *Geochim. Cosmochim. Acta* **62**, 3205–3220.

Kaplan I. R. and Rittenberg S. C. (1964) Microbial fractionation of sulphur isotopes. *J. Gen. Microbiol.* **34**, 195–212.

Kauffman E. G. (1984) Paleobiogeography and evolutionary response dynamic in the Cretaceous Western Interior Seaway of North America. In *Jurassic-Cretaceous Biochronology and Paleogeography of North America*. Geological Association of Canada Special Paper 27 (ed. G. E. G. Westermann). Geological Association of Canada, St. John's, pp. 273–306.

Kauffman E. G. and Caldwell W. G. E. (1993) The Western Interior Basin in space and time. In *Evolution of the Western Interior Basin*. Geological Association Canada Special Paper 39 (eds. W. E. Caldwell and E. G. Kauffman). Geological Association of Canada, St. John's, pp. 1–30.

Keil R. G. and Hedges J. I. (1993) Sorption of organic matter to mineral surfaces and the preservation of organic matter in coastal marine sediments. *Chem. Geol.* **107**, 385–388.

Kennedy M. J., Pevear D. R., and Hill R. J. (2002) Mineral surface control of organic carbon in black shale. *Science* **295**, 657–660.

Koopmans M. P., Koester J., van Kaam-Peters H. M. E., Kenig F., Schouten S., Hartgers W. A., de Leeuw J. W., and Sinninghe Damste J. S. (1996) Diagenetic and catagenetic products of isorenieratene; molecular indicators for photic zone anoxia. *Geochim. Cosmochim. Acta* **60**, 4467–4496.

Kuhnt W., Herbin J. P., Thurow J., and Wiedmann J. (1990) Distribution of Cenomanian-Turonian organic facies in the western Mediterranean and along the adjacent Atlantic margin. *Am. Assoc. Petrol. Geol. Stud. Geol.* **30**, 133–160.

Kump L. R. and Arthur M. A. (1999) Interpreting carbon-isotope excursions: carbonates and organic matter. *Chem. Geol.* **161**, 181–198.

Kump L. R., Arthur M. A., Patzkowsky M. E., Gibbs M. T., Pinkus D. S., and Sheehan P. M. (1999) A weathering hypothesis for glaciation at high atmospheric pCO_2 during the Late Ordovician. *Palaeogeogr. Palaeoclimatol. Palaeoecol.* **152**, 173–187.

Kump L. R., Brantley S. L., and Arthur M. A. (2000) Chemical weathering, atmospheric CO_2, and climate. *Ann. Rev. Earth Planet. Sci.* **28**, 611–667.

Kuypers M. M. M. (2001) Mechanisms and biogeochemical implications of the mid-Cretaceous global organic carbon burial events. PhD Dissertation, Universitiet Utrecht, 135pp (unpublished).

Kuypers M. M. M., Blokker P., Erbacher J., Kinkel H., Pancost R. D., Schouten S., and Sinninghe Damsté J. S. (2001) Massive expansion of marine Archea during a Mid-Cretaceous oceanic anoxic event. *Science* **293**, 92–94.

Lantzy R. J. and Mackenzie F. T. (1979) Atmospheric trace metals: global cycles and assessment of man's impact. *Geochim. Cosmochim. Acta* **43**, 511–526.

Laurin J. and Sageman B. (2001) Tectono-sedimentary evolution of the western margin of the Colorado Plateau during the latest Cenomanian and early Turonian. In *The Geologic Transition: High Plateaus to Great Basin*. Utah Geological Association Publication 30 (eds. M. C. Erskin, J. E. Faulds, J. M. Bartley, and P. D. Rowley). Utah Geological Association, Salt Lake City, pp. 57–74.

Leckie R. M., Schmidt M. G., Finkelstein D., and Yuretich R. (1991) Paleoceanographic and paleoclimatic interpretations of the Mancos Shale (Upper Cretaceous), Black Mesa Basin, Arizona. In *Stratigraphy, Depositional Environments, and Sedimentary Tectonics of the Western Margin, Cretaceous*

Western Interior Seaway. Geological Society of America Special Paper 260 (eds. J. D. Nations and J. G. Eaton). Geological Society of America, Boulder, pp. 139–152.

Leckie R. M., Yuretich R. F., West O., Finkelstein D. B., and Schmidt M. G. (1998) Paleoceanography of the southwestern Western Interior sea during the time of Cenomanian-Turonian boundary (Late Cretaceous). In *Stratigraphy and Paleoenvironments of the Cretaceous Western Interior Seaway USA*. Concepts in Sedimentology and Paleontology, no. 6 (eds. W. E. Dean and M. A. Arthur). Society Economic Paleontologists Minerologists, Tulsa, pp. 101–126.

Leithold E. L. (1994) Stratigraphical architecture at the muddy margin of the Cretaceous Western Interior Seaway, southern Utah. *Sedimentology* **41**, 521–542.

Leithold E. L. and Blair N. E. (2001) Watershed control on the carbon loading of marine sedimentary particles. *Geochim. Cosmochim. Acta* **65**, 2231–2240.

Leventhal J. S. (1995) Carbon-sulfur plots to show diagenetic and epigenetic sulfidation in sediments. *Geochim. Cosmochim. Acta* **59**, 1207–1211.

Lewan M. D. (1984) Factors controlling the proportionality of vanadium to nickel in crude oils. *Geochim. Cosmochim. Acta* **48**, 2231–2238.

Lewan M. D. (1986) Stable carbon isotopes of amorphous kerogens from Phanerozoic sedimentary rocks. *Geochim. Cosmochim. Acta* **50**, 1977–1987.

Lewan M. D. and Maynard J. B. (1982) Factors controlling enrichment of vanadium and nickel in the bitumen of organic sedimentary rocks. *Geochim. Cosmochim. Acta* **46**, 2547–2560.

Lewis B. L. and Landing W. M. (1991) The biogeochemistry of manganese and iron in the Black Sea. *Deep-Sea Res.* **38**, S773–S803.

Lewis B. L. and Landing W. M. (1992) The investigation of dissolved and suspended-particulate trace metal fractionation in the Black Sea. *Mar. Chem.* **40**, 105–141.

Libes S. M. (1992) *An Introduction to Marine Biogeochemistry*. Wiley, New York, 734pp.

Loutit T. S., Hardenbol J., Vail P. R., and Baum G. R. (1988) Condensed sections: the key to age determination and correlation of continental margin sequences. In *Sea-Level Changes: An Integrated Approach*. Special Publication, no. 42 (eds. C. K. Wilgus, B. S. Hastings, C. A. Ross, H. Posamentier, J. Van Wagoner, and C. G. St. C Kendall). Society Economic Paleontologists Minerologists, Tulsa, pp. 183–213.

Luepke J. J. and Lyons T. W. (2001) Pre-Rodinian (Mesoproterozoic) supercontinental rifting along the western margin of Laurentia: geochemical evidence from the Belt-Purcell Supergroup. *Precamb. Res.* **111**, 79–90.

Lyons T. W. (1991) Upper Holocene sediments of the Black Sea: summary of leg 4 box cores (1988 Black Sea Oceanographic Expedition). In *Black Sea Oceanography*. NATO ASI Series (eds. E. Izdar and J. W. Murray). Kluwer, pp. 401–441.

Lyons T. W. (1997) Sulfur isotopic trends and pathways of iron sulfide formation in upper Holocene sediments of the anoxic Black Sea. *Geochim. Cosmochim. Acta* **61**, 3367–3382.

Lyons T. W. and Berner R. A. (1992) Carbon-sulfur-iron systematics of the uppermost deep-water sediments of the Black Sea. *Chem. Geol.* **99**, 1–27.

Lyons T. W., Berner R. A., and Anderson R. F. (1993) Evidence for large pre-industrial perturbations of the Black Sea chemocline. *Nature* **365**, 538–540.

Lyons T. W., Luepke J. J., Schreiber M. E., and Zieg G. A. (2000) Sulfur geochemical constraints on Mesoproterozoic restricted marine deposition: Lower Belt Supergroup, Northwestern United States. *Geochim. Cosmochim. Acta* **64**, 427–437.

Lyons T. W., Werne J. P., Hollander D. J., and Murray R. W. (2003) Contrasting sulfur geochemistry and Fe/Al and Mo/Al ratios across the last oxic-to-anoxic transition in the Cariaco Basin, Venezuela. *Chem. Geol.* **195**, 131–157.

Lyons T. W., Kah L. C., and Gellatly A. M. The Precambrian sulfur isotope record of evolving atmospheric oxygen. In *Tempos and Events in Precambrian Time* (eds. Eriksson *et al.*). Developments in Precambrian Geology Series, Elsevier (in press).

Martin J. M. and Maybeck M. (1979) Elemental mass-balance of material carried by major world rivers. *Mar. Chem.* **7**, 173–206.

Martin J. M. and Whitfield M. (1983) The significance of the river input of chemical elements to the ocean, NATO Conference Series, IV. *Mar. Sci.* **9**, 265–296.

Mayer L. M. (1994) Relationships between mineral surfaces and organic carbon concentrations in soils and sediments. *Chem. Geol.* **114**, 347–363.

Maynard J. B. (1981) Carbon isotopes as indicators of dispersal patterns in Devonian-Mississippian shales of the Appalachian Basin. *Geology* **9**, 262–265.

McDaniel D. K., Hemming S. R., McLennan S. M., and Hanson G. N. (1994) Resetting of neodymium isotopes and redistribution of REEs during sedimentary processes; the early Proterozoic Chelmsford Formation, Sudbury Basin, Ontario, Canada. *Geochim. Cosmochim. Acta* **58**, 931–941.

McGhee G. R., Jr (1982) The Frasnian-Famennian extinction event: a preliminary analysis of Appalachian marine ecosystems. In *Geological Implications of Impacts of Large Asteroids and Comets on the Earth*. Geological Society of America Special Paper (eds. L. T. Silver and P. H. Schultz). Geological Society of America, Boulder, 190, pp. 491–500.

McLennan S. M., Hemming S., McDaniel D. K., and Hanson G. N. (1993) Geochemical approaches to sedimentation, provenance, and tectonics. In *Processes Controlling the Composition of Clastic Sediments*. Geological Society of America Special Paper (eds. M. J. Johnsson and A. Basu). Geological Society of America, Boulder, 284, pp. 21–40.

McManus J., Berelson W. M., Klinkhammer G. P., Johnson K. S., Coale K. H., Anderson R. F., Kumar N., Burdige D. J., Hammond D. E., Brumsack H. J., McCorkle D. C., and Rusdi A. (1998) Geochemistry of barium in marine sediments: implications for its use as a paleoproxy. *Geochim. Cosmochim. Acta* **62**, 3453–3473.

McManus J., Berelson W. M., Hammond D. E., and Klinkhammer G. P. (1999) Barium cycling in the North Pacific: implications for the utility of Ba as a paleoproductivity and paleoalkalinity proxy. *Paleoceanography* **14**, 53–61.

Meyers P. A. (1994) Preservation of elemental and isotopic source identification of sedimentary organic matter. *Chem. Geol.* **114**, 289–302.

Meyers S., Sageman B., and Hinnov L. (2001) Integrated quantitative stratigraphy of Cenomanian-Turonian Bridge Creek Limestone Member using Evolutive Harmonic Analysis and stratigraphic modeling. *J. Sedim. Res.* **71**, 628–644.

Meyers S., Sageman B., and Lyons T. The role of sulfate reduction in organic matter degradation and molybdenum accumulation: theoretical framework and application to a Cretaceous organic matter burial event, Cenomanian-Turonian OAE II. *Paleoceanography* (in review).

Middelburg J. J., Calvert S. E., and Karlin R. (1991x) Organic-rich transitional facies in silled basins: response to sea-level change. *Geology* **19**, 679–682.

Milliman J. D. and Syvitski J. P. M. (1992) Geomorphic/tectonic control of sediment discharge to the ocean: the importance of small mountainous rivers. *J. Geology* **100**, 524–544.

Morford J. L. and Emerson S. (1999) The geochemistry of redox sensitive trace metals in sediments. *Geochim. Cosmochim. Acta* **63**, 1735–1750.

Moore W. S. and O'Neill D. J. (1991) Radionuclide distributions in recent Black Sea sediments. In *Black Sea*

Oceanography. NATO ASI Series (eds. E. Izdar and J. W. Murray). Kluwer, pp. 343–359.

Morse J. W. and Berner R. A. (1995) What determines sedimentary C/S ratios? *Geochim. Cosmochim. Acta* **59**, 1073–1077.

Mueller P. J. and Suess E. (1979) Productivity, sedimentation rate and sedimentary organic matter in the oceans: I. *Organic carbon preservation. Deep-Sea Res.* **26**, 1347–1362.

Murray R. W. and Leinen M. (1996) Scavenged excess aluminum and its relationship to bulk titanium in biogenic sediment from the central Equatorial Pacific Ocean. *Geochim. Cosmochim. Acta* **60**, 3869–3878.

Murray R. W., Leinen M., and Isern A. R. (1993) Biogenic flux of Al to sediment in the central Equatorial Pacific Ocean: evidence for increased productivity during glacial periods. *Paleoceanography* **8**, 651–670.

Murphy A. E., Sageman B. B., Hollander D. J., Lyons T. W., and Brett C. E. (2000a) Black shale deposition in the Devonian Appalachian Basin: siliciclastic starvation, episodic water-column mixing, and efficient recycling of biolimiting nutrients. *Paleoceanography* **15**, 280–291.

Murphy A. E., Sageman B. B., and Hollander D. J. (2000b) Organic carbon burial and faunal dynamics in the Appalachian basin during the Devonian (Givetian-Famennian) greenhouse: an integrated paleoecological/biogeochemical approach. In *Warm Climates in Earth History* (eds. B. Huber, K. MacLeod, and S. Wing). Cambridge University Press, Cambridge, pp. 351–385.

Murphy A. E., Sageman B. B., and Hollander D. J. (2000c) Eutrophication by decoupling of the marine biogeochemical cycles of C, N, and P: a mechanism for the Late Devonian mass extinction. *Geology* **28**, 427–430.

Obradovich J. (1993) A cretaceous time scale. In *Evolution of the Western Interior Basin*. Geological Association Canada Special Paper 39 (eds. W. G. E. Caldwell and E. G. Kauffman). Geological Association Canada, St. John's, pp. 379–396.

Oschmann W. (1991) Anaerobic-poikiloaerobic-aerobic: a new facies zonation for modern and ancient neritic redox facies. In *Cyclic and Event Stratification* (eds. G. Einsele and A. Seilacher). Springer, Berlin, pp. 565–571.

Pancost R. D., Freeman K. H., Patzkowsky M. E., Wavrek D. A., and Collister J. W. (1998) Molecular indicators of redox and marine photoautotroph composition in the late Middle Ordovician of Iowa USA. *Org. Geochem.* **29**, 1649–1662.

Pancost R. D., Baas M., van Geel B., and Sinninghe Damsté J. S. (2002) Biomarkers as proxies for plant inputs to peats: an example from a sub-boreal ombrotrophic bog. *Org. Geochem.* **33**, 675–690.

Park J. and Herbert T. (1987) Hunting for paleoclimatic periodicities in a geologic time series with an uncertain time scale. *J. Geophys. Res.* **92**, 14027–14040.

Pasley M. A., Gregory W. A., and Hart G. F. (1991) Organic matter variations in transgressive and regressive shales. *Org. Geochem.* **17**, 483–509.

Paytan A., Kastner M., and Chavez F. (1996) Glacial to interglacial fluctuations in productivity in the equatorial Pacific as indicated by marine barite. *Science* **274**, 1355–1357.

Pedersen T. F. and Calvert S. E. (1990) Anoxia vs. productivity: What controls the formation of organic-carbon-rich sediments and sedimentary rocks? *Am. Assoc. Petrol. Geol. Bull.* **74**, 454–466.

Pedersen T. F., Nielsen B., and Pickering M. (1991) Timing of late Quaternary productivity pulses in the Panama Basin and implications for atmospheric CO_2. *Paleoceanography* **6**, 657–677.

Perlmutter M. A., Radovich B. J., Matthews M. D., and Kendall C. G. St. c (1998) The impact of high-frequency sedimentation cycles on stratigraphic interpretation. In *Sequence Stratigraphy—Concepts and Application*. Special Publication 8 (eds. F. M. Gradstein, K. O. Sandvik, and N. J. Milton). Norwegian Petroleum Society, Oslo, pp. 141–170.

Peterson L. C., Overpeck J. T., Kipp N. G., and Imbrie J. (1991) A high-resolution late Quaternary upwelling record from the anoxic Cariaco Basin, Venezuela. *Paleoceanography* **6**, 99–119.

Petsch S. T., Berner R. A., and Eglinton T. I. (2000) A field study of the chemical weathering of ancient sedimentary organic matter. *Org. Geochem.* **31**, 475–487.

Petsch S. T., Edwards K. J. and Eglington, T. (2002) Interactions of chemical, biological and physical processes during weathering of black shales. In *Proceedings, 6th International Symposium on the Geochemistry of the Earth's Surface (GES-6)*: Honolulu, 47–52.

Pettijohn F. J. (1949) *Sedimentary rocks*. Harpers Inc., New York, 526pp.

Piper D. Z. (1994) Seawater as the source of minor elements in black shales, phosphorites and other sedimentary rocks. *Chem. Geol.* **114**, 95–114.

Piper D. Z. and Isaacs C. M. (1995) Minor elements in Quaternary sediment from the Sea of Japan: a record of surface-water productivity and intermediate-water redox conditions. *Geol. Soc. Am. Bull.* **107**, 54–67.

Pollastro R. M. (1980) Mineralogy and diagenesis of gas-bearing reservoirs in Niobrara Chalk. *US Geol. Surv. Prof. Pap.* **P1175**, 36–37.

Popp B. N., Laws E. A., Bidigare R. R., Dore J. E., Hanson K. L., and Wakeham S. G. (1998) Effect of phytoplankton cell geometry on carbon isotopic fractionation. *Geochim. Cosmochim. Acta* **62**, 69–77.

Potter P. E., Maynard B. J., and Pryor W. A. (1980) *Sedimentology of Shale: Study Guide and Reference Source*. Springer, New York, 270pp.

Pratt L. M. (1984) Influence of paleoenvironmental factors on preservation of organic matter in the Middle Cretaceous Green Formation, Pueblo, CO. *Am. Assoc. Petrol. Geol. Bull.* **68**, 1146–1159.

Pratt L. M. (1985) Isotopic studies of organic matter and carbonate in rocks of the Greenhorn Marine Cycle. In *Fine-Grained Deposits and Biofacies of the Cretaceous Western Interior Seaway: Evidence of Cyclic Sedimentary Processes*. (eds. L. M. Pratt, E. G. Kauffman, and F. B. Zelt). Society Economic Paleontologists and Minerologists, Tulsa, pp. 38–48.

Pratt L., Arthur M., Dean W., and Scholle P. (1993) Paleoceanographic cycles and events during the late Cretaceous in the Western Interior Seaway of North America. In *The Evolution of the Western Interior Basin*. Geological Association of Canada Special Paper 39 (eds. W. G. E. Caldwell and E. G. Kauffman) Geological Association of Canada, St. John's, pp. 333–353.

Pye K. and Krinsley D. H. (1986) Diagenetic carbonate and evaporite minerals in Rotliegend aeolian sandstones of the southern North Sea: their nature and relationship to secondary porosity. *Clay Min.* **21**, 443–457.

Raiswell R. and Berner R. A. (1985) Pyrite formation in euxinic and semi-euxinic sediments. *Am. J. Sci.* **285**, 710–724.

Raiswell R. and Berner R. A. (1986) Pyrite and organic matter in Phanerozoic normal marine shales. *Geochim. Cosmochim. Acta* **50**, 1967–1976.

Raiswell R. A. and Canfield D. E. (1996) Rates of reaction between silicate iron and dissolved sulfide in Peru margin sediments. *Geochim. Cosmochim. Acta* **60**, 2777–2787.

Raiswell R. and Canfield D. E. (1998) Sources of iron for pyrite formation in marine sediments. *Am. J. Sci.* **298**, 219–245.

Raiswell R., Buckley F., Berner R. A., and Anderson T. F. (1988) Degree of pyritization of iron as a paleoenvironmental indicator of bottom-water oxidation. *J. Sedim. Petrol.* **58**, 812–819.

Raiswell R., Bottrell S. H., Al-Biatty H. J., and Tan M. M. D. (1993) The influence of bottom water oxygenation and reactive iron content on sulfur incorporation into bitumens from Jurassic marine shales. *Am. J. Sci.* **293**, 569–596.

Raiswell R., Newton R., and Wignall P. B. (2001) An indicator of water-column anoxia: resolution of biofacies variations in the Kimmeridge Clay (Upper Jurassic UK). *J. Sedim. Res.* **71**, 286–294.

Ravizza G. (1993) Variations of the $^{187}Os/^{186}Os$ ratio of seawater over the past 28 million years as inferred from metalliferous carbonates. *Earth Planet. Sci. Lett.* **118**, 335–348.

Rea D. R. and Hovan S. A. (1995) Grain size distribution and depositional processes of the mineral component of abyssal sediments: lessons from the North Pacific. *Paleoceanography* **10**, 251–258.

Redfield A. C., Ketchum B. H., and Richards F. A. (1963) The influence of organisms on the composition or seawater. In *The Sea, vol. 2* (ed. M. N. Hill). Wiley, New York, pp. 26–77.

Repeta D. J. (1993) A high resolution historical record of Holocene anoxygenic primary production in the Black Sea. *Geochim. Cosmochim. Acta* **57**, 4337–4342.

Rhoads D. C. and Morse J. W. (1971) Evolutionary and ecologic significance of oxygen-deficient marine basins. *Lethaia* **4**, 413–428.

Rickard, L.V. (1975) *Correlation of the Devonian Rocks in New York State*. New York Museum and Science Service, Map and Chart Series, No. 24.

Roen J. B. (1984) Geologic framework and hydrocarbon evaluation of Devonian and Mississippian black shales in the Appalachian Basin. *AAPG Eastern Section meeting, Am. Assoc. Petrol. Geol. Bull.* **68**, 1927.

Roen J. B. and Kepferle R. C. (1993) Petroleum geology of the Devonian and Mississippian black shale of eastern North America. *US Geol. Surv. Bull.* **B1909**, A1–A8.

Rohl H. J., Schmid-Rohl A., Oschmann W., Frimmel A., and Schwark L. (2001) The Posidonia Shale (Lower Toarcian) of SW-Germany: an oxygen-depleted ecosystem controlled by sea level and palaeoclimate. *Palaeogeogr. Palaeoclimatol. Palaeoecol.* **165**, 27–52.

Saelen G., Tyson R. V., Talbot M. R., and Telnaes N. (1998) Evidence of recycling of isotopically light $CO_{2(aq)}$ in stratified black shale basins: contrasts between the Whitby Mudstone and Kimmeridge Clay formations, United Kingdom. *Geology* **26**, 747–750.

Sageman B. B. (1985) High-resolution stratigraphy and paleobiology of the Hartland Shale Member: analysis of an oxygen-deficient epicontinental sea. In *Fine-Grained Deposits and Biofacies of the Cretaceous Western Interior Seaway: Evidence of Cyclic Sedimentary Processes.* (eds. L. M. Pratt, E. G. Kauffman, and F. B. Zelt). Society Economic Paleontologists and Minerologists, Tulsa, pp. 110–121.

Sageman B. B. (1989) The benthic boundary biofacies model: Hartland Shale Member, Greenhorn Formation (Cenomanian), Western Interior, North America. *Palaeogeogr. Palaeoclimatol. Palaeoecol.* **74**, 87–110.

Sageman B. B. (1996) Lowstand tempestites: depositional model for cretaceous skeletal limestones, Western Interior US. *Geology* **24**, 888–892.

Sageman B. B. and Bina C. (1997) Diversity and species abundance patterns in Late Cenomanian black shale biofacies: western interior US. *Palaios* **12**, 449–466.

Sageman B. B. and Johnson C. C. (1985) Stratigraphy and paleobiology of the Lincoln Limestone Member, Greenhorn Limestone, Rock Canyon Anticline, Colorado. In *Fine-Grained Deposits and Biofacies of the Cretaceous Western Interior Seaway: Evidence of Cyclic Sedimentary Processes.* (eds. L. M. Pratt, E. G. Kauffman, and F. B. Zelt). Society Economic Paleontologists and Minerologists, Tulsa, pp. 100–109.

Sageman B. B., Rich J., Arthur M. A., Birchfield G. E., and Dean W. E. (1997) Evidence for Milankovitch periodicities in Cenomanian-Turonian lithologic and geochemical cycles, Western Interior USA. *J. Sedim. Res.* **67**, 286–302.

Sageman B. B., Murphy A. E., Werne J. P., Ver Straeten C. A., Hollander D. J., and Lyons T. W. (2003) A tale of shales: the relative roles of production, decomposition, and dilution in the accumulation of organic-rich strata, Middle-Upper Devonian, Appalachian basin. *Chem. Geol.* **195**, 229–273.

Sageman B., Rich J., Savrda C. E., Bralower T., Arthur M. A., and Dean W. E. (1998) Multiple Milankovitch cycles in the Bridge Creek Limestone (Cenomanian-Turonian), Western Interior basin. In *Stratigraphy and Paleoenvironments of the Cretaceous Western Interior Seaway USA*. Concepts in Sedimentology and Paleontology no. 6 (eds. W. E. Dean and M. A. Arthur). Society Economic Paleontologists Minerologists, Tulsa, pp. 153–171.

Sancetta C., Lyle M., Heusser L., Zahn R., and Bradbury J. P. (1992) Late-glacial to Holocene changes in winds, upwelling, and seasonal production of the Northern California current system. *Quat. Res.* **38**, 359–370.

Savrda C. E. (1998) Ichnology of the Bridge Creek Limestone: evidence for chemical and spatioal variations in paleo-oxygenation in the Western Interior Seaway. In *Stratigraphy and Paleoenvironments of the Cretaceous Western Interior Seaway USA*. Concepts in Sedimentology and Paleontology, no. 6 (eds. W. E. Dean and M. A. Arthur). Society Economic Paleontologists Minerologists, Tulsa, pp. 127–136.

Schenau S. J. and de Lange G. J. (2001) Phosphorous regeneration vs. burial in sediments of the Arabian Sea. *Mar. Chem.* **75**, 201–217.

Schieber J. (1996) Early diagenetic silica deposition in algal cysts and spores: a source of sand in black shales? *J. Sedim. Res.* **66**, 175–183.

Schieber J., Krinsley D., and Riciputi L. (2000) Diagenetic origin of quartz silt in mudstones and implications for silica cycling. *Nature* **406**, 981–985.

Schieber J., Zimmerle W., and Sethi P. S. (1998) *Shales and Mudstones.* vol. 1, E. Schweizerbart'sche Verlagsbuchhandlung, Stuttgart, 384pp.

Schieber J., Zimmerle W., and Sethi P. S. (1998) *Shales and Mudstones.* vol. 2, E. Schweizerbart'sche Verlagsbuchhandlung, Stuttgart, 296pp.

Schlanger S. O. and Jenkyns H. C. (1976b) Cretaceous Oceanic Anoxic Events: causes and consequences. *Geologie En Mijnbouw* **55**, 179–184.

Schlanger S. O., Arthur M. A., Jenkyns H. C., and Scholle P. A. (1987) The Cenomanian-Turonian oceanic anoxic event: I. Stratigraphy and distribution of organic carbon-rich beds and the marine $d^{13}C$ excursion. In *Marine Petroleum Source Rocks*. Geological Society of London Special Publication 26 (eds. J. Brooks and A. J. Fleet). Geological Society of London, London, pp. 371–399.

Schlesinger W. H. (1991) *Biogeochemistry: An Analysis of Global Change*. Academic Press, San Diego, 443pp.

Scholle P. and Arthur M. A. (1980) Carbon isotope fluctuations in Cretaceous pelagic limestones: potential stratigraphic and petroleum exploration tool. *Am. Assoc. Petrol. Geol. Bull.* **64**, 67–87.

Seilacher A. (1982) Posidonia Shales (Toarcian S. Germany)—Stagnant Basin Model Revalidated. In *Proceedings of the First International Meeting on "Paleontology, Essential of Historical Geology"* (ed. E. M. Gallitelli). Venice, Italy, pp. 25–55.

Shen Y. N., Canfield D. E., and Knoll A. H. (2002) Middle Proterozoic ocean chemistry: evidence from the McArthur Basin, northern Australia. *Am. J. Sci.* **302**, 81–109.

Silliman J. E., Meyers P. A., and Bourbonniere R. A. (1996) Record of post-glacial organic matter delivery and burial in sediments of Lake Ontario. *Org. Geochem.* **24**, 463–472.

Simons D.-J. H. and Kenig F. (2001) Molecular fossil constraints on water column structure of the Cenomanian-Turonian Western Interior Seaway, USA. *Palaeogeogr. Palaeoclimatol. Palaeoecol.* **169**, 129–152.

Sinninghe Damsté J. S. and Koester J. (1998) A euxinic southern North Atlantic Ocean during the Cenomanian Turonian oceanic anoxic event. *Earth Planet. Sci. Lett.* **165**, 173.

Sinninghe Damsté J. S., Wakeham S. G., Kohnen M. E. L., Hayes J. M., and de Leeuw J. W. (1993) A 6000-year sedimentary molecular record of chemocline excursions in the Black Sea. *Nature* **827**, 829.

Slomp C. P., Thomson J., and de Lange G. J. (2002) Enhanced regeneration of phosphorous during formation of the most recent eastern Mediterranean sapropel (S1). *Geochim. Cosmochim. Acta* **66**, 1171–1184.

Sloss L. L. (1962) Stratigraphic models in exploration. *J. Sedim. Petrol.* **32**, 415–462.

Spencer D. W. and Brewer P. G. (1971) Vertical advection diffusion and redox potentials as controls on the distribution of manganese and other trace metals dissolved in waters of the Black Sea. *J. Geophys. Res.* **76**, 5877–5892.

Stein R. (1986) Organic carbon and sedimentation rate, further evidence for anoxic deep-water conditions in the Cenomanian/Turonian Atlantic Ocean. *Mar. Geol.* **72**, 199–209.

Stevenson F. J. and Cheng C.-N. (1972) Organic geochemistry of the Argentine Basin sediments: carbon-nitrogen relationships and Quarternary correlations. *Geochim. Cosmochim. Acta* **36**, 653–671.

Strauss H. (2002) The isotopic composition of Precambrian sulphide—Seawater chemistry and biological evolution. In *Precambrian Sedimentary Environments: A Modern Approach to Ancient Depositional Systems*. International Association of Sedimentologists Special Publication no. 33 (eds. W. Altermann and P. L. Corocoran). Blackwell, Oxford, pp. 67–105.

Stumm W. and Morgan J. J. (1996) *Aquatic Chemistry*. Wiley, New York, 1022pp.

Suess E. (1980) Particulate organic carbon flux in the oceans–surface productivity and oxygen utilization. *Nature* **288**, 260–263.

Sweeney R. E. and Kaplan I. R. (1980) Stable isotope composition of dissolved sulfate and hydrogen sulfide in the Black Sea. *Mar. Chem.* **9**, 145–152.

Swift D. J. P. and Thorne J. A. (1991) Sedimentation on continental margins: I. A general model for shelf sedimentation. *International Association of Sedimentologists, (Special Publication)* **14**, 3–31.

Taylor S. R. and McLennan S. M. (1985) *The Continental Crust: Its Composition and Evolution*. Blackwell, Malden, Massachusetts, 312pp.

Thomson J., Mercone D., de Lange G. J., and van Santvoort P. J. M. (1999) Review of recent advances in the interpretation of eastern Mediterranean sapropel S1 from geochemical evidence. *Mar. Geol.* **153**, 77–89.

Thurow J., Moullade M., Brumsack H. J., Masure E., Taugourdeau-Lantz J., and Dunham K. W. (1988) The Cenomanian/Turonian boundary event (CTBE) at Hole 641A, ODP Leg 103 (compared with the CTBE interval at Site 398). *Proc. ODP Sci. Results* **103**, 587–634.

Tissot B. P. and Welte D. H. (1984) *Petroleum Formation and Occurrence*. Springer, Berlin, 699pp.

Toth D. J. and Lerman A. (1977) Organic matter reactivity and sedimentation rates in the ocean. *Am. J. Sci.* **277**, 465–485.

Trask P. D. (ed.) (1955) *Recent Marine Sediments*. Society of Economic Paleontologists and Mineralogists, Tulsa, OK, 736pp.

Tromp T. K., Van Cappellen P., and Key R. M. (1995) A global model for the early diagenesis of organic carbon and organic phosphorus in marine sediments. *Geochim. Cosmochim. Acta* **59**, 1259–1284.

Truyols-Massoni M., Montesinos R., Garcia-Alcalde J. L., and Leyva F. (1990) The Kacak-Otomari Event and its characterization in the Palentine Domain (Cantabrian Zone, NW Spain). In *Extinction Events in Earth History: Proceedings of Project 216, Global Biological Events in Earth History, 3rd International Conference on Global Bio-events*. Lecture Notes in Earth Sciences (eds. E. G. Kauffman and O. H. Walliser) 30, pp. 133–144.

Tucker R. D., Bradley D. C., Ver Straeten C. A., Harris A. G., Ebert J. R., and McCutcheon S. R. (1998) New U–Pb zircon ages and the duration and division of Devonian time. *Earth Planet. Sci. Lett.* **158**, 175–186.

Turekian K. K. and Wedepohl K. H. (1961) Distribution of the elements in some major units of the Earth's crust. *Geol. Soc. Am. Bull.* **72**, 175–192.

Twenhofel W. H. (1926) *Treatise on sedimentation*. Williams and Wilkins, Baltimore, 661pp.

Twenhofel W. H. (1950) *Principles of sedimentation* 2nd edn. McGraw Hill, New York, 673pp.

Tyrrell T. (1999) The relative influences of nitrogen and phosphorus on oceanic primary production. *Nature* **400**, 525–531.

Tyson R. V. (1995) *Sedimentary Organic Matter: Organic Facies and Palynofacies*. Chapman and Hall, London, 615pp.

Tyson R. V. and Pearson T. H. (1991) Modern and ancient continental shelf anoxia: an overview. In *Modern and Ancient Continental Shelf Anoxia*. Geological Society of London, Special Publication, no. 58 (eds. R. V. Tyson and T. H. Pearson). Geological Society of London, London, pp. 1–24.

Valley J. and Cole D. R. (2001) Stable Isotope Geochemistry, Rev. Mineral. Geochem. Mineralogical Society of America, Washington, DC, vol. 43, 662p.

Van Cappellen P. and Ingall E. D. (1994) Benthic phosphorous regeneration, net primary production, and ocean anoxia: a model of the coupled biogeochemical cycles of carbon and phosphorous. *Paleoceanography* **9**, 667–692.

Van Cappellen P., Viollier E., Roychoudhury A., Clark L., Ingall E., Lowe K., and DiChristina T. (1998) Biogeochemical cycles of manganese and iron at the oxic-anoxic transition of a stratified marine basin (Orca Basin, Gulf of Mexico). *Environ. Sci. Technol.* **32**, 2931–2939.

Van der Weijden C. H. (2002) Pitfalls of normalization of marine geochemical data using a common divisor. *Mar. Geol.* **184**, 167–187.

Van Mooy B. A. S., Keil R. G., and Devol A. H. (2002) Impact of suboxia on sinking particulate organic carbon; enhanced carbon flux and preferential degradation of amino acids via denitrification. *Geochim. Cosmochim. Acta* **66**, 457–465.

Van Santvoort P. J. M., de Lange G. J., Thomson J., Cussen H., Wilson T. R. S., Krom M. D., and Ströhle K. (1996) Active post-depositional oxidation of the most recent sapropel (S1) in sediments of the eastern Mediterranean Sea. *Geochim. Cosmochim. Acta* **60**, 4007–4024.

Ver Straeten C. A. and Brett C. E. (1995) Lower and Middle Devonian foreland basin fill in the Catskill Front: stratigraphic synthesis, sequence stratigraphy, and the Acadian Orogeny. In *67th Annual Meeting Field Trip Guidebook* (eds. J. I. Garver and J. A. Smith). New York State Geological Association, Albany, pp. 313–356.

Vincent E. and Berger W. H. (1985) Carbon dioxide and polar cooling in the Miocene: the monterey hypothesis. In *The Carbon Cycle and Atmospheric CO_2: Natural Variations Archean to Present*. American Geophysical Union, Geophysical Monograph, no. 32 (eds. E. T. Sunquist and W. S. Broecker). American Geophysical Union, Washington, DC, pp. 455–468.

Watkins D. K. (1989) Nanoplankton productivity fluctuations and rhythmically bedded pelagic carbonates of the Greenhorn Limestone (Upper Cretaceous). *Palaeogeogr. Palaeoclimatol. Palaeoecol.* **74**, 75–86.

Weisel C. P., Duce R. A., Fasching J. L., and Heaton R. W. (1984) Estimates of the transport of trace metals from the ocean to the atmosphere. *J. Geophys. Res. D Atmos.* **89**, 11607–11618.

Weldeab S., Emeis K., Hemleben C., and Siebel W. (2002) Provenance of lithogenic surface sediments and pathways of riverine suspended matter in the eastern Mediterranean Sea:

evidence from $^{143}Nd/^{144}Nd$ and $^{87}Sr/^{86}Sr$ ratios. *Chem. Geol.* **186**, 139–149.

Werne J. P., Hollander D. J., Lyons T. W., and Peterson L. C. (2000) Climate-induced variations in the productivity and planktonic ecosystem structure from the Younger Dryas to Holocene in the Cariaco Basin, Venezuela. *Paleoceanography* **15**, 19–29.

Werne J. P., Sageman B. B., Lyons T., and Hollander D. J. (2002) An integrated assessment of a "type euxinic" deposit: evidence for multiple controls on black shale deposition in the Middle Devonian Oatka Creek Formation. *Am. J. Sci.* **302**, 110–143.

Werne J. P., Lyons T. W., Hollander D. J., Formolo M. J., Formolo M. J., and Sinninghe Damsté J. S. (2003) Reduced sulfur in euxinic sediments of the Cariaco Basin: sulfur isotope constraints on organic sulfur formation. *Chem. Geol.* **195**, 159–179.

Wignall P. B. (1994) *Black Shales*. Claredon, Oxford, 127pp.

Wignall P. B. and Myers K. J. (1988) Interpreting benthic oxygen levels in mudrocks: a new approach. *Geology* **16**, 452–455.

Wignall P. B. and Newton R. (1998) Pyrite framboid diameter as a measure of oxygen deficiency in ancient mudrocks. *Am. J. Sci.* **298**, 537–552.

Wijsman J. W. M., Middelburg J. J., and Heip C. H. R. (2001) Reactive iron in Black Sea sediments: implications for iron cycling. *Mar. Geol.* **172**, 167–180.

Wilde P., Quinby-Hunt M. S., and Erdtmann B.-D. (1996) The whole-rock cerium anomaly: a potential indicator of eustatic sea-level changes in shales of the anoxic facies. *Sedim.Geol.* **101**, 43–53.

Wilkin R. T. and Arthur M. A. (2001) Variations in pyrite texture, sulfur isotope composition, and iron systematics in the Black Sea: evidence for late Pleistocene to Holocene excursions of the O_2–H_2S redox transition. *Geochim. Cosmochim. Acta* **65**, 1399–1416.

Wilkin R. T., Barnes H. L., and Brantley S. L. (1996) The size distribution of framboidal pyrite in modern sediments: an indicator of redox conditions. *Geochim. Cosmochim. Acta* **60**, 3897–3912.

Wilkin R. T., Arthur M. A., and Dean W. E. (1997) History of water-column anoxia in the Black Sea indicated by pyrite framboid size distributions. *Earth Planet. Sci. Lett.* **148**, 517–525.

Witzke B. J. and Heckel P. H. (1988) Paleoclimatic indicators and inferred Devonian paleolatitudes of Euramerica. In *Devonian of the World: Proceedings of the Second International Symposium on the Devonian System, Volume II, Sedimentation*, Canadian Society of Petroleum Geologists, Memoir 14 (eds. N. J. McMillan, A. F. Embry, and D. J. Glass), Canadian Society of Petroleum Geologists, Calgary, pp. 49–63.

Woodrow D. L. (1985) Paleogeography, paleoclimate, and sedimentary processes of the Late Devonian Catskill Delta. In *The Catskill Delta*. Geological Society of America Special Paper 201 (eds. D. L. Woodrow and W. D. Sevon). Geological Society of America, Boulder, pp. 51–63.

Wortmann U. G., Hesse R., and Zacher W. (1999) Major-element analysis of cyclic black shales: paleoceanographic implications for the early cretaceous deep western Tethys. *Paleoceanography* **114**, 525–541.

Wortmann U. G., Bernasconi S. M., and Boettcher M. E. (2001) Hypersulfidic deep biosphere indicates extreme sulfur isotope fractionation during single-step microbial sulfate reduction. *Geology* **29**, 647–650.

Wright J., Schrader H., and Holser W. T. (1987) Paleoredox variations in ancient oceans recorded by rare earth elements in fossil apatite. *Geochim. Cosmochim. Acta* **51**, 631–644.

Yarincik K. M., Murray R. W., Lyons T. W., Peterso L. C., and Haug G. H. (2000a) Oxygenation history of bottom waters in the Cariaco Basin, Venezuela, over the past 578,000 years: results from redox-sensitive metals (Mo, V, Mn, and Fe). *Paleoceanography* **15**, 593–604.

Yarincik K. M., Murray R. W., and Peterson L. C. (2000b) Climatically sensitive eolian and hemipelagic deposition in the Cariaco Basin, Venezuela, over the past 578,000 years: results from Al/Ti and K/Al. *Paleoceanography* **15**, 210–228.

Yochelson E. L. and Lindemann R. H. (1986) Considerations on the systematic placement of the Styliolines (*incertae sedis*: Devonian). In *Problematic Fossil Taxa*. Oxford Monographs in Geology and Geophysics, no. 5 (eds. Hoffman and Nitecki). Oxford University Press, Oxford, pp. 45–58.

Young G. M. and Nesbitt H. W. (1998) Processes controlling the distribution of Ti and Al in weathering profiles, siliciclastic sediments. *J. Sedim. Res.* **68**, 448–455.

Zaback D. A., Pratt L. M., and Hayes J. M. (1993) Transport and reduction of sulfate and immobilization of sulfide in marine black shales. *Geology* **21**, 141–144.

Zheng Y., Anderson R. F., van Geen A., and Kuwabara J. S. (2000) Authigenic molybdenum formation in marine sediments: a link to pore water sulfide in the Santa Barbara Basin. *Geochim. Cosmochim. Acta* **64**, 4165–4178.

7.07
Late Diagenesis and Mass Transfer in Sandstone–Shale Sequences

K. L. Milliken
The University of Texas at Austin, TX, USA

7.07.1 INTRODUCTION

Between Ca 50 °C and 300 °C, sandstones and mudrocks ("shales") undergo massive chemical and textural reorganization. In this temperature interval detrital grains, and the rock textures defined by grains, are lost by reactions with pore fluids. Chemical and physical processes in late diagenesis transform siliciclastic sediments into rocks. Predictive models of porosity evolution with depth depend upon an understanding of these processes. Because the magnitude of the mineralogical changes in late diagenesis is large, these changes also have important implications for understanding rates and mechanisms of element cycling through the crust.

Controversy regarding the scale of the elemental mobility that accompanies the mineralogical and textural reorganization has been a defining theme of research in late diagenesis. Conundrums arising from apparent conflicts between petrographic and petrophysical constraints on elemental mobility are well known to students of clastic diagenesis. Interestingly, similar paradoxes have long vexed students of low-grade metamorphism (e.g., Ague, 1991; Rumble, 1994). A related issue in late diagenesis concerns apparent subsurface weathering. Weathering during erosion and transport at the surface fails to remove high-temperature phases from sediments completely, and these detrital components arrive in the realm of late diagenesis with considerable reactive potential. However, after reaching a temperature of 200 °C, these metastable compounds have largely been lost by reaction with pore fluids. Of course, volumetrically significant weathering processes require acid. However, the source(s) of this acid remains disputed. In the context of identifying volumetrically significant sources of acid, other questions arise regarding the extent to which precipitation reactions in late diagenesis should be construed as acid-releasing reverse-weathering reactions.

Historically, late diagenesis of siliciclastic rocks was viewed as physical and isochemical, involving only compaction and dewatering of sedimentary materials. Detrital phases that survived weathering were seen as essentially inert to subsequent reaction during burial and prior to the onset of metamorphism. Metamorphism itself was viewed as isochemical and accomplished principally through solid-state reactions (see a brief summary in Ague, 1991). Notable exceptions to these views constitute the foundations of existing theory regarding the nature of late diagenesis. "Intrastratal solution" of chemically unstable detrital minerals (e.g., Pettijohn, 1941), apparent potassium metasomatism of shales (e.g., Weaver and Beck, 1971), and massive mineralogical changes during progressive burial (e.g., Coombs *et al.*, 1959) are observations that established the directions of modern research in late diagenesis. Advances in petrographic imaging techniques (e.g., backscattered electron- and cathodoluminescence-imaging) and integration of petrographic observations with both bulk and spatially resolved chemical analyses have greatly accelerated the evolution of concepts about late diagenesis and early metamorphism.

Today, basin-scale mass transfer of some materials (e.g., helium, water, and petroleum) is unquestioned (e.g., Hunt, 1996). Other materials (e.g., titanium and the REEs) are sufficiently mobile to appear within authigenic precipitates, but are likely to be "immobile" on the scale of a hand specimen. Mobilities of the major elements that make up sandstones and shales (silicon, aluminum, calcium, sodium, potassium) remain controversial. Conflicting notions about processes in rock suites across the wide range of burial conditions and alteration show that fundamental questions remain unanswered about the nature of the volumetrically significant processes within a major segment of the rock cycle. It is very likely that something is wrong, or at least inadequate, with the present concepts and/or data pertaining to the evolution of permeability, transport mechanisms, and timing of reactions during late diagenesis. This chapter summarizes the existing framework of observations relating to small- and large-scale elemental mobility in the late diagenesis of siliciclastic rocks.

7.07.2 THE REALM OF "LATE DIAGENESIS"

"Diagenesis" refers to the physical and chemical processes that affect sedimentary materials after deposition and before metamorphism and between deposition and weathering. The effects of diagenetic processes on rock properties such as porosity and the degree of lithification are progressive. It is therefore difficult to draw absolute boundaries between diagenesis and the adjacent segments of the rock cycle. Divisions within diagenesis are even less readily drawn. Diagenesis includes many chemical and physical processes that are also active during deposition, weathering, and metamorphism. Crystalline rocks are also affected by subsolidus aqueous reactions that resemble diagenesis (e.g., Sprunt and Nur, 1979; Clauer *et al.*, 1989; Ramseyer *et al.*, 1992; Kominou and Yardley, 1997).

"Late diagenesis" of siliciclastic materials here refers to processes that postdate the initial stages of consolidation through compactional grain rearrangement and early cementation that occurs in fluids that have a clear affiliation with the depositional environment. It is essentially synonymous with "mesodiagenesis" (Morad *et al.*, 2000). Late diagenesis takes place in the subsurface under conditions of elevated temperature (~50–300 °C). In general, late diagenesis occurs at depths greater than 1.5 km. As a result of fluid–rock interactions, pore fluids during late diagenesis differ in important ways from fluids associated with depositional environments (e.g., Land and Macpherson, 1992b; Hanor, 1994).

This chapter is concerned especially with chemical processes that create pervasive modifications of volumetrically significant rock components and impart chemical and textural modifications that can be detected in the bulk rock. Processes of similar type and magnitude may occur at lower temperatures and at shallower depths in zones having either enhanced fluid flow (e.g., in fault zones) or high geothermal gradients (e.g., adjacent to intrusions). Chemical and physical modifications during late diagenesis are not as severe as those during metamorphism. Processes encompassed by this review take place in rocks that have retained their textures as deposits of particulate debris. "Grain," in the context of sandstones and shales, has a specific genetic connotation that is distinct from "crystal."

Much of our knowledge about late diagenesis has come from the study of samples obtained by drilling for oil and gas. Cenozoic–Mesozoic sedimentary accumulations 10–12 km thick occur in many basins, but wells that penetrate below 6 km are rare. As a result, the full transition into greenschist metamorphism has never been studied in the context of simple burial under conditions of "normal" geothermal gradients (in the range of 20–30 °C km^{-1}). Knowledge about late diagenesis in the realm of 200–300 °C comes almost entirely from tectonically uplifted and deformed regions or from settings of anomalously high geothermal gradients. This leaves us with great uncertainty regarding the relative importance of burial effects (primarily thermal and compactional) and conditions imposed during dynamic tectonism and uplift on a variety of rock modifications.

7.07.3 ELEMENTAL MOBILITY AT THE GRAIN SCALE

7.07.3.1 Physical Processes

7.07.3.1.1 Compaction

The intergranular volume (IGV) is a key parameter in compaction studies of sand to sandstone. IGV is the space between the framework grains, more or less synonymous with the term "pre-cement porosity." At deposition, IGV constitutes 45% of the total volume of matrix-free sands (e.g., Pryor, 1973; Atkins and McBride, 1992), essentially equivalent to the primary porosity. In the average sandstone, compaction appears to have accomplished an overall greater loss of primary porosity than cementation (Lundegard, 1992). In the absence of cementation, compactional processes alone can reduce IGV to 0% (Pittman and Larese, 1991). IGV decline with burial provides a vital context for constraining material transfer during late diagenesis, because it places upper limits on the space available for fluid flow and cementation.

The decline of IGV during burial is a strong function of initial composition. Quartz-rich materials compact at a slower rate than more ductile lithic-rich materials (Pittman and Larese, 1991). Brittle crushing of quartz and feldspar is increasingly recognized as an important compaction mechanism (e.g., Milliken, 1994b; Dickinson and Milliken, 1995b; Chuhan *et al.*, 2000b; Milliken and Laubach, 2000; Makowitz and Milliken, 2001). Compaction by simple grain rearrangement can reduce IGV to ~26% (Graton and Fraser, 1935), but greater reductions are generally believed to require additional mechanisms of compaction such as ductile grain deformation, extreme brittle deformation (as in quartz-rich fault gouges), or pressure solution (e.g., Lundegard, 1992).

Pressure solution has long been recognized as a cause of IGV decline in sandstones (e.g., Heald, 1955; Thompson, 1959), but the specific mechanisms by which this highly localized dissolution of quartz is accomplished are still much debated. Most theories of pressure solution envision pressure at grain contacts as the principal driving force for dissolving quartz (e.g., de Boer, 1977; Rutter, 1983; Tada and Siever, 1989), but there are many exceptions and variations of this view (e.g., Bjørkum, 1996; Milliken and Laubach, 2000). Whatever the actual mechanism of quartz mobilization, it seems clear that the stress field ultimately controls the morphology and alignment of pressure solution features (sutured grain boundaries and stylolites) which are arranged essentially perpendicular to the maximum stress.

The concept of IGV in muddy sediments is less well established because it is difficult to petrographically discriminate between grains, cements, and porosity. As a result, compaction within mudrocks is described primarily on the basis of porosity decline curves (Giles *et al.*, 1998). The porosity of muddy sediments is very high at deposition (near 80%); it declines exponentially with depth (summarized in Giles *et al.*, 1998) in response to temperature, burial rate, and the degree of overpressuring (Bjørlykke, 1999).

7.07.3.1.2 Fracturing

Much tectonic deformation, both brittle and ductile, takes place within the temperature realm of late diagenesis. Hence deformation, apart from burial compaction, is an integral part of diagenesis. The assemblages of authigenic phases that appear as cements and grain replacements within sandstones also fill fractures, many of them microfractures (e.g., Milliken, 1994b; Eichhubl and Boles, 2000; Laubach and Milliken, 1996a,b; Laubach, 1997; Marrett and Laubach, 1997; Marrett *et al.*, 1997). Observations and measurements of actual mineral-filled fractures in shales (as opposed to slates) (e.g., Beach, 1977; Jowette, 1987; Al-Aasm *et al.*, 1993; Parnell *et al.*, 2000; Cosgrove, 2001) are surprisingly rare given the great importance assigned to shale fracturing in the context of overpressuring and the modeling of fluid flow in the deep subsurface (e.g., Nunn, 1996; Wang and Xie, 1998; Nunn and Meulbroek, 2002). Bitumen-filled microfractures in shales revealed by UV fluorescence microscopy provide tantalizing indications that additional work on the petrographic characterization of shale fractures would be fruitful (e.g., summarized in Talukdar *et al.*, 1987; Littke *et al.*, 1988; Hunt, 1996).

7.07.3.2 Chemical Processes in Late Diagenesis

Despite the elevated temperature during late diagenesis (pressure is not considered to be a critical control), chemical reactions are mechanistically similar to those occurring at the Earth's surface (cf. Berner, 1981). First, all chemical processes during late diagenesis involve interactions between solid phases and pore fluids; second, kinetic controls on reactions tend to override thermodynamic controls; and third, chemical equilibrium in the bulk material is not achieved. The mineral assemblages and rock textures that exist in sandstones and shales during late diagenesis must be interpreted in the context of these controls and observations.

Solid-state diffusion (volume diffusion) in minerals is too slow to be effective in the temperature range of late diagenesis and even throughout the early stages of metamorphism (e.g., Freer, 1981; Fortier and Giletti, 1989; Yund *et al.*, 1989; Farver and Yund, 1996). All reactions in diagenesis can be classed as either dissolution or precipitation reactions involving transfer of material into or out of aqueous solution. Since all reactions in diagenesis are fluid mediated, elemental transfer can occur in late diagenesis because diagenetic fluids are mobile (e.g., Bethke and Marshak, 1990; Deming, 1994; Ferry, 1994; Parnell, 1998).

Chemical processes in diagenesis are slow, because reactions at mineral surfaces are rate limiting ("surface-reaction controlled"); these reactions tend to be strongly temperature dependent (Berner, 1981). Microscale equilibrium between an authigenic phase and a pore fluid may occur on a transient, local basis in diagenesis, but will not tend to occur simultaneously in the bulk rock. Throughout the conditions of late diagenesis, a diversity of phases coexist which are not all in equilibrium either with pore fluids or with one another. Geothermometers based on ratios of dissolved cations are not generally applicable in the temperature range of late diagenesis (Land and Macpherson, 1992a). A lack of pervasive equilibrium during late diagenesis is not surprising since nonequilibrium persists widely throughout the early stages of metamorphism (e.g., Ferry, 2000; Carlson, 2002).

7.07.3.2.1 Dissolution

The slowness of surface-reaction-controlled rate dissolution results in metastability—the tendency of minerals to persist for geologically long periods under conditions greatly different from those under which they are thermodynamically stable. This is illustrated dramatically where detrital minerals formed at high temperatures by crystallization from melts persist even for billions of years, at low temperatures in contact with aqueous solutions with which they are not in equilibrium (Figure 1). The reactivity of metastable minerals varies tremendously with their structural defects and trace element content. Hence, grains of detrital minerals of the same bulk composition display large variations in the timing and the conditions under which they react (Milliken *et al.*, 1989; Milliken, 1992) (Figure 2). Authigenic minerals in sediment rocks can also display prominent metastability. They may form during one stage of diagenesis and then persist through a long history of subsequent burial and fluid flow without dissolving or recrystallizing.

Even detrital or authigenic minerals in bulk equilibrium under some specific diagenetic condition (e.g., detrital albite in contact with albite-saturated water) may still be at disequilibrium with respect to their trace element and isotopic composition, as well as their structural state. Such disequilibrium favors dissolution, even if fluids are near equilibrium with respect to the bulk mineral.

An important advance in the understanding of dissolution was made possible in the 1970s by the application of scanning electron microscopic (SEM) techniques to the dissolution of detrital grains during weathering (Berner and Holdren, 1977, 1979; Berner and Schott, 1982). Dissolution is not uniformly distributed across mineral surfaces. Rather, it is highly localized at crystal defects ("surface-reaction controlled"). Observations of dissolution textures in the subsurface confirm that surface-controlled dissolution is the primary dissolution mechanism throughout diagenesis (Figure 3).

A concurrent advance in understanding subsurface dissolution was made by using colored

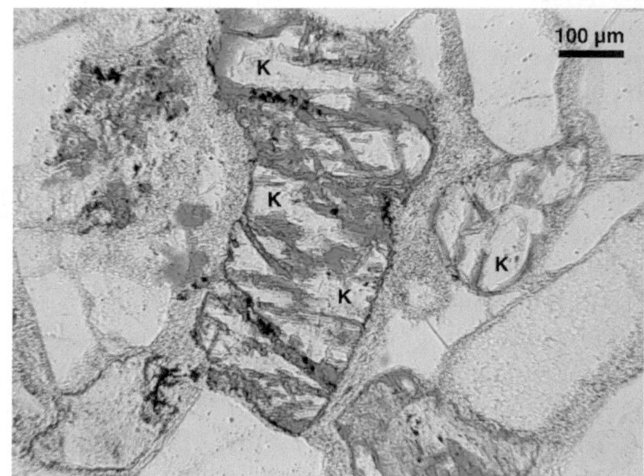

Figure 1 Detrital orthoclase grains (K) in subarkosic sandstone, from Moodies Group, Barberton Mountains, South Africa. Apart from dissolution that may relate to modern outcrop weathering (blue), these feldspars have survived since the Archean (at least 3.2 Ga).

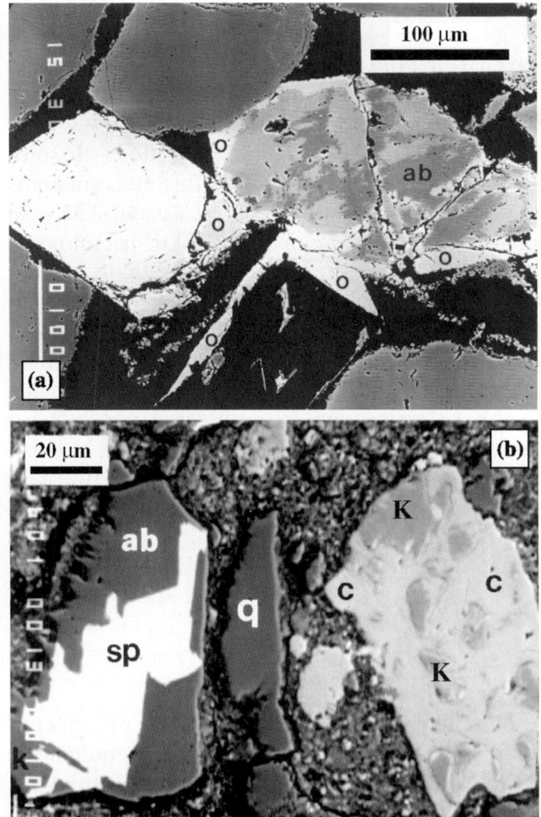

Figure 2 Heterogeneity of grain alteration reflects the highly variable nature of metastability in detrital grain populations. Backscattered electron images. (a) Overgrowths (o) of adularia rim the three central grains, signifying that all were once K-feldspar. The grain lower center is completely dissolved. Grain at middle right is Na-rich alkali feldspar, partially replaced by albite (ab) along cleavages and possibly fractures. The third grain, virtually unaltered, has a K-rich composition and contains primary exsolution lamellae of albite. K for the adularia (which predates the grain dissolution) was imported into this field of view whereas materials mobilized by the grain dissolution are missing. Frio Sandstone, Oligocene, South Texas. (b) The grain on the left has a small remaining patch of unreplaced K-feldspar (K) near the lower-left edge of the photomicrograph. The rest of this grain is replaced by albite (ab) and authigenic sphene (titanite, sp). The center grain is quartz (q). The grain on the right is also a K-feldspar (K) that is partially replaced by calcite (c). Sphene and calcite replace around 1% and 4%, respectively, of the feldspars in Frio Formation mudrocks in South Texas. Frio Formation, Oligocene, South Texas ((a) is reproduced by permission of SEPM (Society for Sedimentary Geology) from *J. Sedim. Petrol.*, **1989**, *59*, 743).

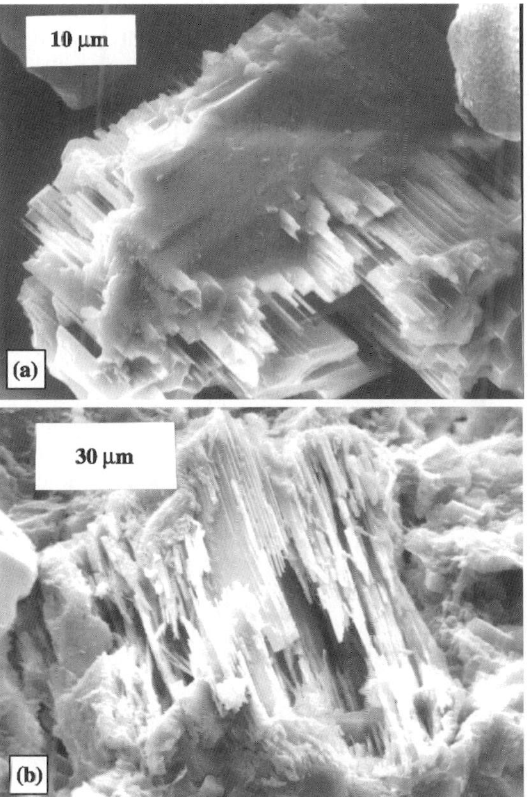

Figure 3 Dissolution is a surface-reaction controlled process throughout late diagenesis: (a) dissolution of an epidote grain from the subsurface Plio-Pleistocene sequence offshore Louisiana; SEM image and (b) Ca-plagioclase, Frio Formation, Oligocene, South Texas, SEM image ((a) is reproduced by permission of SEPM (Society for Sedimentary Geology) from *J. Sedim. Petrol.*, **1988**, *58*, 999).

impregnation media in thin section preparation. With this technique it became possible to discern whether a pore space in sandstone was natural or whether it was created during thin section preparation. Widespread use of dyed impregnation media led to a revolution in thinking about the nature of porosity in siliciclastic materials. It was quickly recognized that *many* sandstones contain secondary pores produced by the dissolution of unstable detrital components, and, in some cases, by the dissolution of authigenic minerals (Lindquist, 1977; Stanton, 1977; Schmidt and McDonald, 1979a,b; Bloch, 1994).

7.07.3.2.2 Cementation

Precipitation reactions tend to be sluggish if nucleation is very slow. At the scale of individual grains, cements are not distributed uniformly throughout the available pore spaces, but tend to be spatially localized. Cements are localized on pre-existing surfaces, probably because surface area and surface composition affect nucleation rates. Incipient crystals nucleated on a surface have a smaller total surface area than free crystals of the same volume in the pore fluid, leading to a free energy difference that favors nucleation on a substrate (heterogeneous nucleation, e.g., Kirkpatrick, 1981). In the absence of a suitable nucleation substrate, authigenic minerals may fail to precipitate even from fluids that are highly supersaturated with respect to these minerals.

Mineral precipitation within primary intergranular pore spaces leads to cementation. The existence of cements demands elemental mobility at least on the scale of a thin section, because cements reduce porosity that was present at the time of deposition. During early diagenesis of unconsolidated sediments, displacive cementation may lead to cement volumes in excess of 45%. In late diagenesis the upper limit of cement volume is constrained by the IGV at the time of cement emplacement.

Some cements are not localized and nucleation occurs uniformly on all available surfaces. Calcite cement in sandstones typically has such a distribution. Other minerals—notably quartz, albite, adularia, and ankerite—tend to nucleate almost exclusively on the surfaces of specific preexisting crystals (Figure 4). Even individual grains of the appropriate composition may not be suitable substrates if their surface is coated, e.g., by clays or oxides, early microcrystalline cements, or bitumen (e.g., Pittman and Lumsden, 1968; Heald and Larese, 1974; Boles and Hickey, 1996). Clean, "fresh" mineral surfaces created by fracturing may localize cement in preference to adjacent "normal" grain surfaces (Reed and Laubach, 1996; Chuhan *et al.*, 2000b; Milliken and Laubach, 2000). Cement distribution is therefore controlled not only by the availability and concentration of the components needed for precipitation, but also by the distribution of suitable substrates. Some authigenic minerals, kaolinite for example, do not form as overgrowths. Yet they display highly patchy distributions that suggest some as yet unidentified localizing factor in their nucleation.

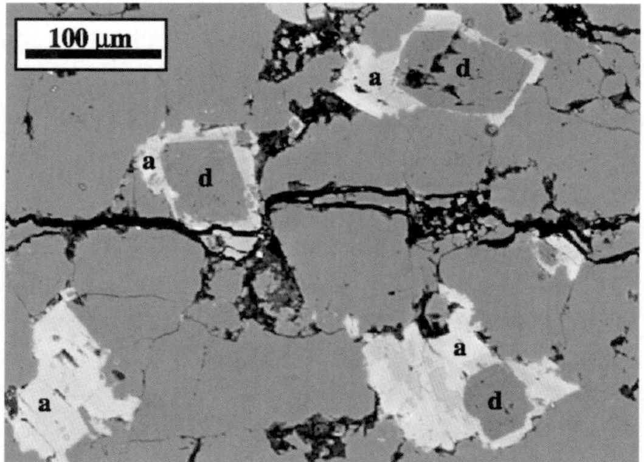

Figure 4 Ankerite (a) localized as overgrowths on grains of detrital dolomite (d). Similar high degrees of substrate control on nucleation are observed for several of the common authigenic phases in late diagenesis. Weber Formation, Pennsylvanian, western Colorado. Backscattered electron image.

Certain spatial distributions of cements are generally *not observed* (Milliken, 1995). Halo or coronal textures are common in metamorphic rocks and are generally interpreted to signify incomplete equilibrium between solid phases that participated in linked chemical reactions (dissolutions/precipitations) within diffusional gradients (e.g., Spry, 1969; Ferry, 2000; Carlson, 2002). In sandstones and shales such spatial distributions of authigenic minerals are seldom seen (Stoessell, 1987). The substrate controls described above may mask distributions related to diffusional gradients, or, advection may overwhelm diffusional controls on element distribution at the single grain scale. Coronal distributions in authigenic phases have been reported mainly in minerals that contain highly immobile elements such as titanium (Morad and Aldahan, 1982; Morad, 1988), in low-permeability rocks exposed to relatively high temperatures (Sutton and Land, 1996).

7.07.3.2.3 Replacement

Dissolution involves the transfer of material from the rock into pore fluids. Cementation involves the opposite. Replacement involves transfer in both directions. During replacement authigenic phases are precipitated within spaces formerly occupied by detrital grains. It is a special case of cementation within secondary pores. "Recrystallization" is replacement in which the gross mineralogy (but not necessarily the trace element and isotopic composition) of the original grain and its replacement are the same. Almost all cement minerals also occur as grain replacements (Milliken, 1989). This observation indicates that primary and secondary pores are essentially equivalent in terms of their propensity to be infiltrated by mineral-bearing fluids.

On a large scale, replacements may appear pseudomorphic after original detrital grains on a small scale, replacements are very imperfect. Typically they consist of microporous aggregates that only partially mimic the original crystal (Figure 5; Milliken, 1989). Many cements and replacements have a close spatial association, because they are both localized on the same substrate. For example, an ankerite crystal that replaces a detrital feldspar will often also fill an adjacent primary pore space.

Localization of replacement minerals on *dissolving* grains is demonstrated by the distribution of replacement minerals within partially replaced grains. In such cases, the distribution of the replacement phase tends to be patchy, following the same pattern of defect-related distribution that is seen in secondary pores (Milliken, 1989). Replacement only occurs at the locus of dissolution (Figure 6). Coronal or halo textures are not

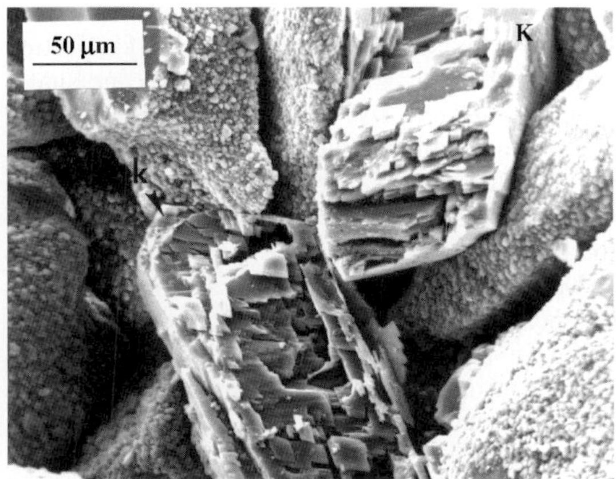

Figure 5 Highly microporous aggregate of authigenic albite replacing a K-feldspar (evidenced by the adularia overgrowth, K). Frio Formation, Oligocene, South Texas. SEM image (source Milliken, 1989) (reproduced by permission of SEPM (Society for Sedimentary Geology) from *J. Sedim. Petrol.*, **1989**, *59*, 364).

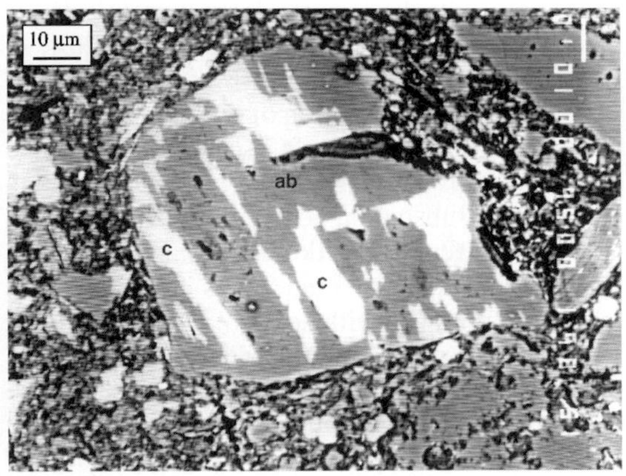

Figure 6 Patchy calcite (c) replacing detrital feldspar (now albite, ab) in shale. Similar microscale distributions of replacement phases are observed widely in sandstones and shales. Frio Formation, Oligocene, South Texas. Backscattered electron image.

generally observed in replacements during late diagenesis.

Localization of replacement at sites of dissolution implies that there is no *necessary* temporal association between the dissolution step and the subsequent precipitation. Precipitation and dissolution may be essentially simultaneous. Alternatively, a secondary pore *may* persist for geologically long periods before being filled by a secondary phase. In either case, the secondary phase is localized by precipitation on the surface of the secondary pore. In the first case, however, the dissolution step may have a mechanistic role in controlling the precipitation step, for example, by providing some portion of the elements that are then precipitated; in the second case there is no such necessary connection.

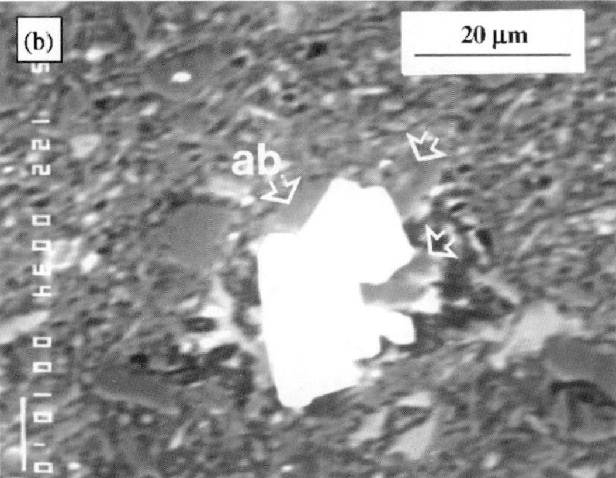

Figure 7 Replacement driven by "force-of-crystallization" is characterized by authigenic phases that develop euhedral faces that are not plausibly construed as crystal growth within pore spaces: (a) sphalerite replaces albitized detrital feldspar and adjacent portions of clay-rich matrix, Frio Formation, Oligocene, South Texas and (b) siderite crystal (s) "attacks" a detrital K-feldspar (K) in sandstone, Breathitt Formation, Pennsylvanian, eastern Kentucky.

An intriguing aspect of replacement is introduced by the "force of crystallization" (Weyl, 1959; Carmichael, 1987; Maliva and Siever, 1988a,b; Dewers and Ortoleva, 1990; Ferry, 2000). Where this force is important, the replacement phase acts as more than a passive filling localized on surfaces within secondary pores. A compelling body of petrographic and chemical evidence suggests that in certain cases, especially in late diagenesis at somewhat elevated temperatures, precipitating authigenic phases actually induce the dissolution of an adjacent pre-existing phase, presumably by enhancing the solubility of the dissolving phase by pressure exerted across a water film at the crystal boundary. The replacement phase in such a case of "aggressive" replacement typically has euhedral crystal faces adjacent to the dissolved crystal and may replace multiple adjacent phases (Figure 7). Many authigenic phases do not find ready substrates on which to nucleate; it is likely that a high degree of supersaturation with respect to the replacement phase must be achieved before this form of replacement occurs.

7.07.4 VOLUMETRICALLY SIGNIFICANT PROCESSES OF LATE DIAGENESIS

7.07.4.1 Detrital Feldspar Dissolution and Replacement

Detrital feldspars in both sandstones and shales are subject to dissolution and replacement during late diagenesis. As previously discussed, replacement processes during diagenesis commence with a dissolution step. In that sense, *all* detrital feldspar loss is due to dissolution. Only a portion of the total feldspar loss in sandstones and shales is compensated by replacement. Feldspathic components within lithic clasts are equally subject to dissolution. In lithic-rich sandstones this constitutes a volumetrically significant portion of the total feldspar reaction.

Detrital feldspars in modern sediments have a compositional range that reflects their composition in the source rocks (Trevena and Nash, 1981). During diagenesis, sodium-rich K-feldspars and more calcium-rich detrital plagioclase grains are more subject to dissolution (and replacement) than potassium-rich K-feldspar or sodium-rich plagioclases (Maynard, 1984; Milliken, 1988, 1992; Milliken *et al.*, 1989).

Small secondary pores are frequent in feldspars of modern sediments (Passaretti and Eslinger, 1987; Brantley *et al.*, 1999). Additional dissolution of detrital feldspars proceeds through the earliest stages of diagenesis (Milliken, 1988). Sandstones flushed with large quantities of meteoric water are especially prone to loss of feldspars through dissolution (Mathisen, 1984).

Petrographic observation of dissolution features in detrital grains (mostly feldspars, but also lithic clasts and heavy minerals) reveals that secondary porosity is a substantial portion of total porosity in many rocks across a wide range of burial depths and tectonic settings, and tends to increase in relative importance with depth (e.g., Loucks *et al.*, 1984). Several lines of evidence suggest that secondary pores, like primary pores, are subject to compactional collapse and filling with authigenic phases (Figure 8(a)). The most direct evidence for this is that the absolute amount of secondary

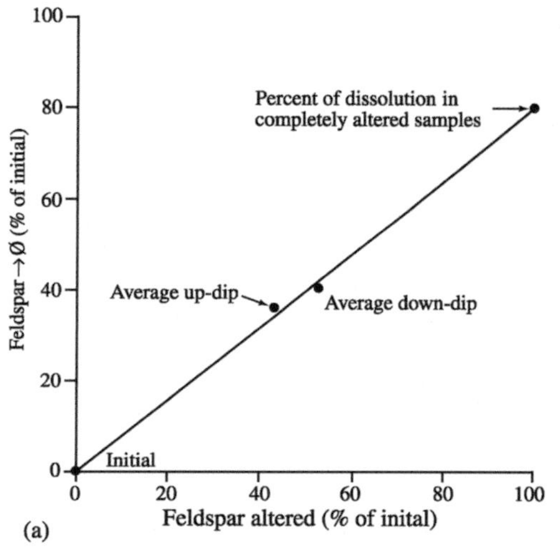

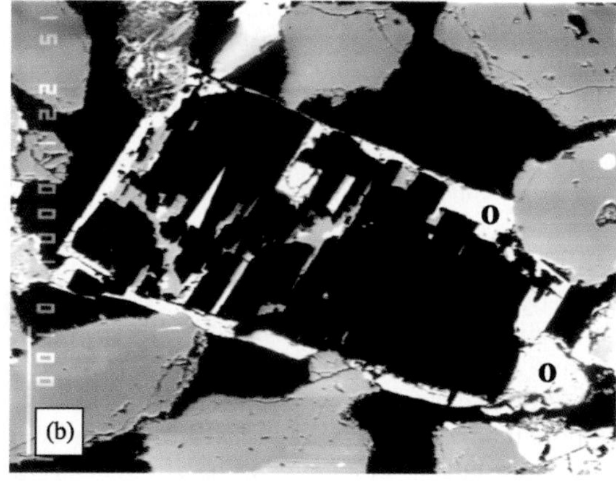

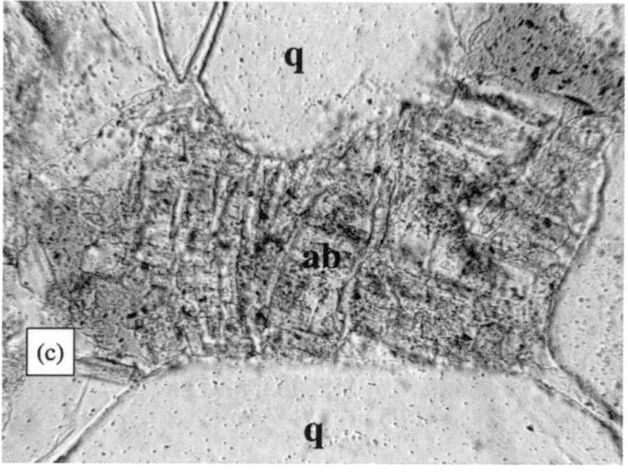

Figure 8 Alteration of detrital feldspars is dominated by dissolution. (a) Total alteration (dissolution + replacement) of the feldspar assemblage versus dissolution for partially altered Frio Formation sandstones in two contrasting structural situations (up-dip = 15 samples; down-dip = 17 samples) in south Texas. By extrapolation, completely albitized feldspar assemblages would represent, on average, dissolution of ~80% of the initial grain population, an amount of dissolution that is not reflected in the observed amounts of secondary porosity. (b) Former locus of a feldspar grain with adularia overgrowth (o). Near-total dissolution has occurred without local precipitation of Al- or Si-bearing replacement phases. Many such pores apparently collapse with further burial. Frio Formation, Oligocene, South Texas. Backscattered electron image. (c) Partially dissolved detrital feldspar (now albite) showing compactional collapse between adjacent quartz grains (q). Frio Formation, Oligocene, Brazoria County, Texas. Plane polarized light ((a) is reproduced by permission of SEPM (Society for Sedimentary Geology) from *J. Sedim. Petrol.*, **1989**, *59*, 754).

porosity declines with depth, similar to primary porosity. For example, in the data set of the Loucks *et al.* (1984) study secondary porosity declined from an average of 10.1% in sandstones between 80 °C and 120 °C to ~5.4% in sandstones above ~160 °C. Cryptic feldspar loss compromises provenance information and complicates efforts to make mass balance calculations (Milliken, 1988; Harris, 1990; McLaughlin *et al.*, 1994). Extreme loss of detrital feldspars leads to the formation of "diagenetic quartzarenites" (McBride, 1987; Harris, 1990; Glasmann, 1992; De Ros, 1998). Many petrographic assessments of secondary porosity conclude that the mobilized aluminum is not compensated by local precipitation of clays and therefore is transported from the sandstone (e.g., the above-cited studies and Siebert *et al.*, 1984; Wilkinson and Haszeldine, 1996), although a few studies indicate that aluminum is conserved within the volume of thin sections (e.g., Giles and De Boer, 1990; Hayes and Boles, 1992).

Feldspar dissolution can be quantified independently of petrographic observations by comparing compositional variations in a population of altered grains to the initial compositional variation of that population. Such an approach requires first that a reasonable approximation of the initial mineral assemblage can be identified, and second that the altered samples are not completely altered. As alteration approaches completion (complete dissolution or replacement), the range of possible paths to the near-end member composition cannot be usefully constrained.

This approach was applied by Milliken *et al.* (1989) to estimate the relative importance of dissolution and replacement during the alteration of feldspar grains within a small region of volcanically derived Oligocene Frio sandstones in south Texas. The average composition of calcium-plagioclase and K-feldspar grain populations, together with ratios of different grain types (e.g., plagioclase/total feldspar, albite/total plagioclase), were compared in an estimated initial, "least-altered" population and in partially altered samples of that same population. The number of grains of pure albite (i.e., grains replaced by authigenic albite) was compared with the amount of detrital grain loss that was required to achieve the observed average composition and grain type ratios of the surviving detrital grain population. The amount of albite grain replacement in these sandstones is too small to account for the amount of detrital grain loss required to achieve the observed average compositions and grain ratios in partially altered samples. The trend for partially altered samples suggests that, in the purely albitic feldspar assemblages at depth, on an average 80% of the initial feldspars dissolved without subsequent precipitation of local replacement phases (Figure 8).

A portion of the detrital feldspar that dissolves in sandstones is subsequently replaced by authigenic albite (e.g., Land and Milliken, 1981; Boles, 1982). The same process affects a portion of the detrital feldspars in the silt component of shales (e.g., Milliken, 1992; J. I. Lee and Y. I. Lee, 1998; Moore, 2000). Albitization affects both detrital calcium-plagioclase (e.g., Gold, 1987; Morad *et al.*, 1990) and K-feldspar grains (Walker, 1984; Saigal *et al.*, 1988; Aagaard *et al.*, 1990). Some low-rank metamorphic rocks contain detrital feldspars of very sodic composition (Trevena and Nash, 1981) and discrimination of these detrital albites from authigenic albite is a challenge. In general, the composition of authigenic albite is close to that of the sodium end member. It is largely nonluminescence (Kastner and Siever, 1979).

Inspection of albite grains with light microscopy suggests that they are highly pseudomorphic after the original detrital grains, but at the SEM scale the replacements are highly imperfect. Typically they form as highly microporous aggregates that only partially mimic the original crystals (Figure 5; Milliken, 1989). As with other common grain replacements, albite grain replacement is accompanied by albite overgrowth cements (e.g., Boles, 1982). Albite distribution within partially replaced grains is highly patchy and is typically associated with secondary porosity (Figure 2(a)) (Boles, 1982; Walker, 1984; Gold, 1987; Saigal *et al.*, 1988; Milliken, 1989). This is consistent with a dissolution/precipitation mechanism for albitization.

Figures 9(c) and (d) provide an indication of the magnitude of feldspar dissolution and replacement reactions in sedimentary basins. Because most sandstones and shales contain several volume percent of feldspar, the volumetric significance of detrital feldspar albitization is large. In the deeper portions of sedimentary basins where this process has gone to completion, a feldspar assemblage consisting of 100% albite is the end result of complete dissolution and replacement of the detrital feldspars in both sandstones and shales. Even if the total amount of replacement is far less than the total dissolution, authigenic albite in shales and sandstones is the second most abundant authigenic phase in siliciclastic rocks, exceeded only by illite, and exceeding the volume of all the common cements. Only in very quartz-rich sands does the amount of authigenic quartz (and perhaps carbonate) exceed the volume of authigenic albite in late diagenesis.

Sodium import does not necessarily accompany albitization. In south Texas sandstones across the zone of albitization show no change in whole-rock sodium content (Milliken *et al.*, 1994), consistent

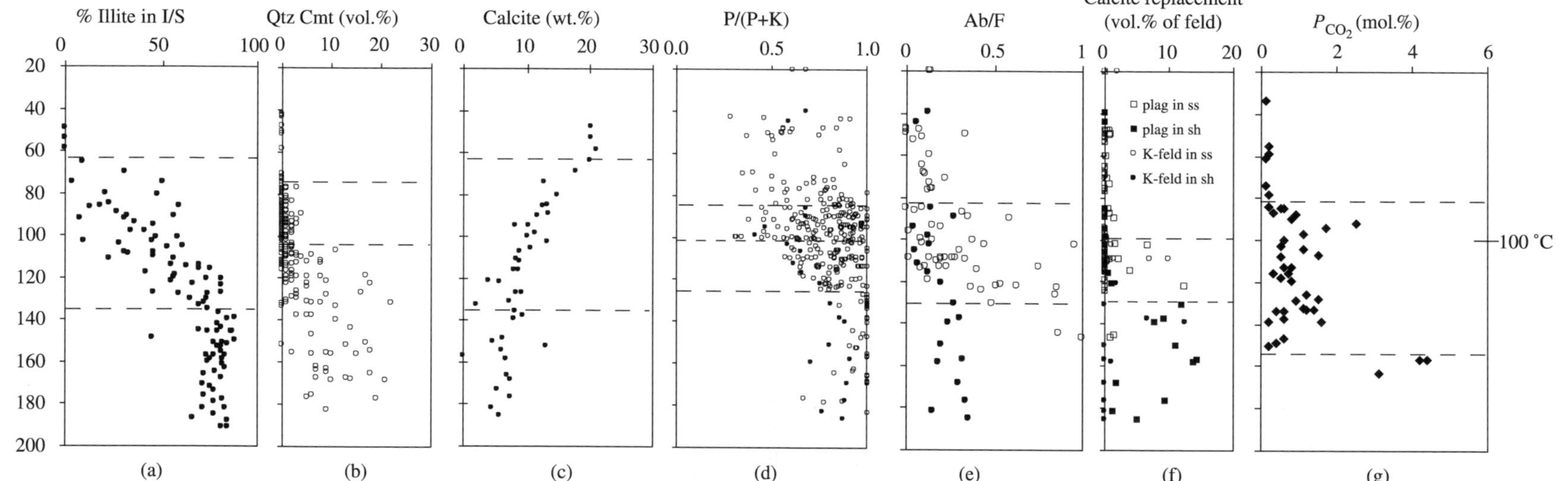

Figure 9 Temperature trends for volumetrically significant silicate and carbonate mineral reactions in late diagenesis, illustrated with data from the Oligocene Frio Formation, South Texas. Open symbols represent data for sandstones, closed symbols represent data for shales, expecting for (g) which represents the percent of CO_2 in gas. Similar trends are observed in basins worldwide. The Frio is chosen for illustrative purpose because of the large depth range of samples available, the simple burial and thermal history, the high degree of overpressuring (minimizing the complications of meteoric incursion), and the large amount of data available. Sources: Lynch (1996), (a); unpublished data from study of Loucks *et al.* (1984), (b) and (d); Milliken and Land (1993) and Land *et al.* (1997), (c); feldspar studies of Milliken *et al.* (1989), and Milliken (1992), (d)–(f); Lundegard (1985), (g). Temperatues for (a) and (g) are taken from the cited sources. Other temperatures are calculated on the basis of the regional 30 °C km^{-1} geothermal gradient. Dashed lines indicate interpreted boundaries between significant reaction zones. Correlations and miscorrelations between plots (a)–(g) suggest that relationships between volumetrically significant reactions in late diagenesis are highly complex and strongly influenced by reaction kinetics.

with the observation that alteration of the feldspar assemblage is not a constant-volume process. Constant-volume albitization of the sodium-rich K-feldspars and the calcium-rich plagioclases would require an approximate doubling of the sodium content.

A diverse assemblage of authigenic minerals in addition to albite replaces detrital feldspars in both sandstones and shales. In the Gulf Coast calcite is volumetrically second in importance to albite, constituting ~1–2% of former detrital feldspar volumes in Oligocene sandstones and ~4% in shales. Whereas calcite cements in sandstones show a poorly defined decline with depth, authigenic calcite in feldspars is highly correlated with the overall alteration of the feldspar assemblage (Figure 9(f)).

Other common, though volumetrically minor, feldspar-replacing minerals include titanite, anatase, sphalerite, barite, ankerite, siderite, and fluorite. With the exception of replacement driven by force of crystallization, feldspar replacements have intracrystalline distributions that are strongly localized at sites of surface-controlled dissolution. Interestingly, replacement of detrital feldspars by authigenic clays is rarely observed in late diagenesis.

In a general sense, dissolution of feldspars can be viewed as a kind of weathering and can be described by hydrolysis reactions. The exact formulation of these reactions is far from certain, however. In the simplest construction, dissolution of K-feldspar and calcium plagioclase consumes H^+, the plagioclase consuming slightly more:

$$(Na_{0.7}Ca_{0.3})Al_{1.3}Si_{2.7}O_8 + 5.2H^+ + 2.8H_2O = 0.7Na^+ + 0.3Ca^{2+} + 1.3Al^{3+} + 2.7H_4SiO_4 \quad (1)$$

$$(K_{0.5}Na_{0.5})AlSi_3O_8 + 4H^+ + 4H_2O = 0.5K^+ + 0.5Na^+ + Al^{3+} + 3H_4SiO_4 \quad (2)$$

Unfortunately, the speciation of aluminum in subsurface pore fluids is poorly known. Formulations of feldspar dissolution that utilize other aluminum-bearing ions require less H^+, e.g., for a K-feldspar:

$$KAlSi_3O_8 + 8H_2O = K^+ + Al(OH)_4^- + 3H_4SiO_4 \quad (3)$$

Two lines of evidence suggest that reactions (1) and (2) may represent a more accurate description of feldspar dissolution than reaction (3). First, there is the small-scale evidence relating to feldspar replacement by calcite. The most plausible link between feldspar dissolution and calcite precipitation (Figures 2(b) and 6) is pH-buffering ((4) and (5)):

Dissolving phase:

$$KAlSi_3O_8 + 4H^+ + 4H_2O = K^+ + Al^{3+} + 3H_4SiO_4 \quad (4)$$

Replacement phase:

$$Ca^{2+} + HCO_3^- = CaCO_3 + H^+ \quad (5)$$

Thus, dissolution of 0.25 mol of K-feldspar buffers sufficient acid to allow (potentially) precipitation of 1 mol of calcite. Buffering by silicates has been postulated as the primary control on subsurface pH (Hutcheon and Abercrombie, 1989, 1990; Smith and Ehrenberg, 1989; Hutcheon *et al.*, 1993), consistent with the observation that dissolution affecting feldspars typically leaves adjacent calcite cements and skeletal debris unaffected (Siebert *et al.*, 1984).

A further indication of the role of silicate buffering in late diagenesis is the depth (temperature)-related rise in subsurface P_{CO_2} (Franks and Forester, 1984; Lundegard and Land, 1989; Smith and Ehrenberg, 1989) (Figure 9(g)). The $\delta^{13}C$ of CO_2 increases with depth (Figure 10), suggesting that the mechanisms responsible for the CO_2 rise are complex, involving reactions that generate CO_2 from multiple sources in deep basins. At lower temperatures, $\delta^{13}C$ values are light, suggesting a prominent contribution from organic alkalinity (Equation (6)) (Lundegard and Land, 1989):

$$Ca^{2+} + 2HCO_3^- = CaCO_3 + H_2O + CO_2 \quad (6)$$

In the Gulf of Mexico, the increase in P_{CO_2} occurs to a greater degree (up to 16 mol.%) and at shallower depths in Eocene units (Wilcox formation) that contain smaller quantities of less-reactive detrital feldspars (of plutonic as opposed to volcanic provenance) (Lundegard and Land, 1986). Apparently, as the detrital feldspar supply is exhausted, pH buffering shifts to the carbonate assemblage. The P_{CO_2} then rises

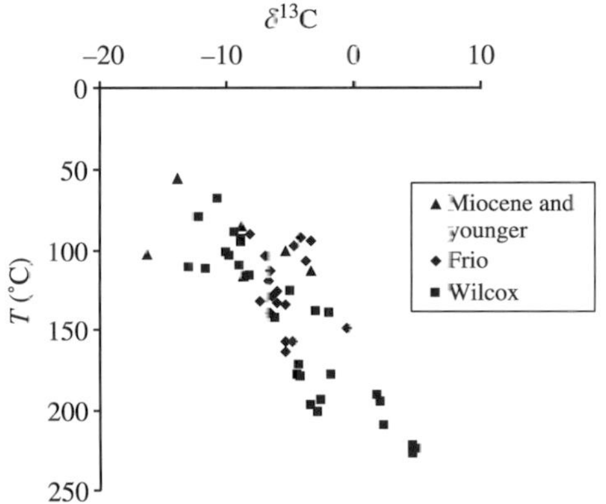

Figure 10 $\delta^{13}C$ of CO_2 in Cenozoic units of the Gulf of Mexico Basin (replotted from data of Lundegard, 1985).

dramatically as a result of carbonate dissolution (Equation (7)) (Smith and Ehrenberg, 1989), releasing carbon with the characteristic heavier isotopic signature of marine skeletal debris into the CO_2 reservoir (Figure 10):

$$CaCO_3 + 2H^+ = Ca^{2+} + H_2O + CO_2 \quad (7)$$

The temperature-related increase in the $\delta^{13}C$ of CO_2 is a trend noted in basinal gases worldwide (James, 1990; Hunt, 1996). It may reflect the evolving balance between different pathways of CO_2 generation (organic versus inorganic) as late diagenesis progresses.

7.07.4.2 Dissolution of Detrital Quartz (Pressure Solution)

Most subsurface fluids are supersaturated with respect to quartz (Figure 11). Thus, secondary porosity formed within detrital quartz in the subsurface is not observed. This is not surprising given the kinetic and nucleation limitations on quartz precipitation (Oelkers *et al.*, 1996; Walderhaug, 1996) and the fact that the silica geothermometer does not work well in sedimentary basins (Land and Macpherson, 1992a).

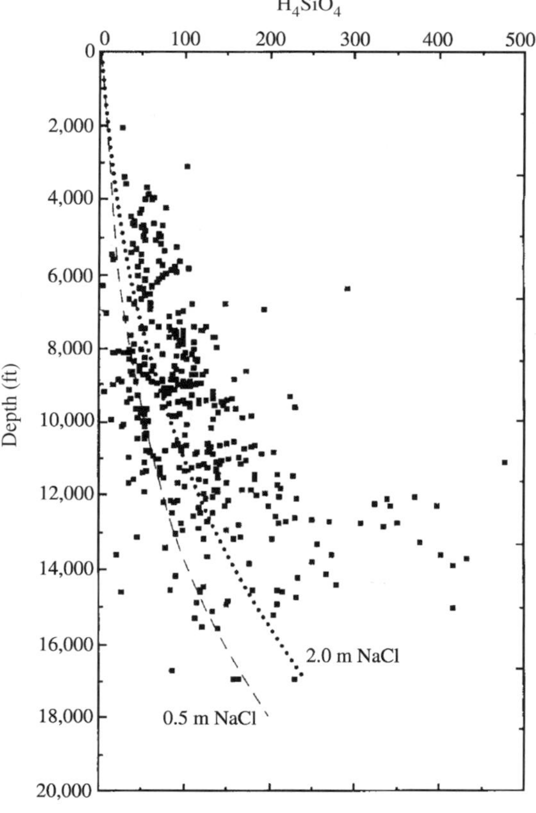

Figure 11 Silica concentration in pore fluids versus depth for Cenozoic units of the Gulf of Mexico Basin. Most fluids are oversaturated with respect to quartz (1 ft = 0.3048 m) (reproduced by permission of SEPM (Society from Sedimentary Geology) from *Basin-wide Diagenetic Patterns*, **1997**, p. 36).

Cathodoluminescence evidence supports the notion that detrital quartz (outside of pressure solution) is essentially inert to volumetrically significant reaction with pore fluids even into the early stages of metamorphism. Detrital quartz grains display heterogeneity in luminescence intensity, color, and texture, reflecting differences in trace element content and crystal defect characteristics inherited from source rocks (Zinkernagel, 1978; Ramseyer *et al.*, 1988; Houseknecht, 1991; Owen, 1991; Milliken, 1994a; Seyedolali *et al.*, 1997; Boggs *et al.*, 2002). Stable isotopic heterogeneity within quartz grain assemblages also documents the detrital diversity of quartz grain assemblages (Blatt, 1987; Hervig *et al.*, 1995; Williams *et al.*, 1997). Chemical heterogeneity of detrital quartz grains, and also differences in cathodoluminescence between grains and cements, are ultimately homogenized during metamorphism (Sprunt *et al.*, 1978) as quartz undergoes massive recrystallization, the last of the volumetrically significant detrital phases to do so. Recrystallization of the detrital quartz imparts a major textural reorganization as textures related to grains, cements, and their intervening pore spaces are erased (Holness and Watt, 2001). This textural transition is a key step in the progression from "diagenesis" to "metamorphism," but the exact conditions of temperature and fluid pressure under which the detrital quartz fraction ultimately re-equilibrates with pore fluids are poorly known.

Pressure solution at grain contacts and along stylolites is the only documented process whereby silica is mobilized from detrital quartz in late diagenesis. It has been long recognized that pressure solution of quartz is enhanced at grain surfaces in contact with potassium-rich phyllosilicates (Heald, 1955). A mechanism for mobilizing quartz by chemical means rather than directly by pressure has been proposed as the primary cause for the dissolution of quartz along pressure solution seams (Bjørkum, 1996; Renard *et al.*, 1997). Many apparently sutured boundaries between quartz grains are observed on close inspection by cathodoluminescence imaging to consist of highly comminuted detrital quartz particles that have been cemented by authigenic quartz (Milliken, 1994b, 2001; Dickinson and Milliken, 1995a). This indicates that grain loss is far smaller than estimates based on light microscopy. A possible role for brittle processes in pressure solution is an intriguing possibility that remains to be investigated. Such a role for highly comminuted particles is supported by theoretical work that suggests a role for small-scale surface roughness and fluid channelization in silica mobilization (Wood, 1981; Wiltschko and Sutton, 1982; Gratz, 1991; den Brok, 1998).

It is clear that the relative importance of pressure solution as a mechanism of IGV decline increases markedly with depth. It also varies in importance among units of different bulk composition. For example, pressure solution appears to be widespread in sandstones of plutonic derivation (Thomas *et al.*, 1993; Oelkers *et al.*, 1996; Spötl *et al.*, 2000), but not in the largely volcanogenic Cenozoic sandstones of the Gulf of Mexico basin (Land *et al.*, 1987; Land and Milliken, 2000). Controls that appear to favor the development of pressure solution include abundant potassium-rich micaceous debris, a history of meteoric water incursion, and elevated temperature (>100 °C?).

If surface interactions with potassium-rich phyllosilicates are important in silica mobilization, it would not be surprising to find that this process is active in shales. This possibility has received little attention since it was suggested by Füchtbauer (1967). Evidence that silt particles in shale undergo changes in aspect ratio during progressive burial has been reported (Evans, 1990). If phyllosilicate-induced pressure solution affects quartz in shale, it is of great importance for the overall silica budget in diagenesis, because most detrital quartz resides in shales (Blatt and Schultz, 1976; Potter *et al.*, 1980).

The role of pressure solution as a cause of volume loss in late diagenetic shales deserves closer examination in view of the huge volume loss from pelitic rocks (up to 50%) that is postulated at slightly higher temperature in connection with slaty cleavage development and the early stages of metamorphism (e.g., Wright and Platt, 1982; Ague, 1991).

7.07.4.3 Dissolution of Detrital Carbonate

Little has been reported regarding the petrographic characterization of carbonate in late diagenetic shales. Most carbonate (~99%?) in Cenozoic shales of the Gulf of Mexico basin is detrital and marine allochemical in origin (Milliken and Land, 1982, 1993). Carbonate decline is a commonly observed feature of progressive burial (increasing temperature) and early metamorphism in shales (Hower *et al.*, 1976; Boles, 1978; Freed, 1981, 1982; Calvert and Klimentidis, 1986; Jennings and Thompson, 1986; Lundegard and Land, 1986; Wintch and Kvale, 1991, 1994; Milliken and Land, 1993; Land *et al.*, 1997; Land and Milliken, 2000) (Figure 9(c)). The dominant mechanism of carbonate loss in Gulf Coast shales appears to be pressure solution (Milliken and Land, 1993). In the Frio Formation of the Texas Gulf Coast, carbonate loss proceeds mostly within the zone of smectite illitization (Figure 9(a)) and across the zone of feldspar reaction (Figures 9(d)–(f)).

Calcite decline in shale sequences of the Gulf of Mexico during late diagenesis varies between geographic regions (Land and Milliken, 2000), but on the whole this reaction is important (~10 vol.%). There is a substantial conundrum regarding the fluid volumes responsible for transporting this material and the source(s) of the acid responsible for carbonate mobilization (assuming this trend is not in part depositional). If trends observed in Gulf Coast shales are typical for basins worldwide, the depth-related decline in shale carbonate is the most significant mass transfer process in late diagenesis.

Loss of detrital and marine skeletal carbonate by pressure solution is also observed in sandstones (e.g., Cavazza and Gandolfi, 1992; Hesse and Abid, 1998; Milliken *et al.*, 1998). Convincing evidence of authigenic carbonate removal (deeply etched cleavages, detached remnants of once-continuous poikilotopic crystals) is found in deep, hot basinal settings. In these the supply of reactive feldspars has been exhausted. This allows carbonates to react with acids generated by other processes (e.g., McDowell and Paces, 1985; Taylor and Land, 1997). The inverse correlation between the dissolved bicarbonate and the calcium content of Gulf Coast basinal brines argues that dissolution of carbonate is probably due to reaction with HCl, similar to reaction (7) rather than by reaction with CO_2 (Land and Macpherson, 1992b).

7.07.4.4 Illitization

The precipitation of authigenic illite is recognized in basins worldwide. It is the single volumetrically most significant reaction in the late diagenesis of siliciclastic materials (Lynch, 1997) (Figure 9(a)). In sandstones, illite forms largely as cement in primary pores (Macchi, 1987; Ehrenberg and Nadeau, 1989). In shales, it forms primarily as replacements of the detrital mineral smectite (e.g., Hower *et al.*, 1976; Pearson and Small, 1988; Furlan *et al.*, 1996; Lynch, 1997). Precipitation of illite is strongly temperature dependent (e.g., Eslinger and Glasmann, 1993; Huang *et al.*, 1993), and thus, increases with depth in relatively young basins (Hower *et al.*, 1976; Pearson and Small, 1988; Furlan *et al.*, 1996).

Smectite forms during weathering of silicates in a wide range of climates. It is the volumetrically most significant detrital clay mineral in Cenozoic basins worldwide (Potter *et al.*, 1980). On an average, shales contain between 30% and 50%

nonclay minerals (e.g., Blatt and Schultz, 1976; Potter *et al.*, 1980; Blatt and Totten, 1981), mostly as silt-size particles. Among the clay minerals, smectite (including its mixed-layer variations) is by far the most abundant. It averages 40–50% in shallow Gulf Coast shales (Lynch, 1997). On a volumetric basis, smectite is the most abundant detrital sedimentary mineral. Replacement of the weathering product smectite by the diagenetic mineral illite during late burial diagenesis is therefore a reaction of great interest for understanding the mass balance of weathering and reverse weathering in the crust.

It is generally agreed on the basis of TEM evidence (e.g., Rask *et al.*, 1997) and on the basis of chemical substitution in the several structural sites of illite (e.g., Lanson and Champion, 1991; Awwiller, 1993; Lynch *et al.*, 1997) that the illitization of smectite, like other replacement reactions in late diagenesis, proceeds through a dissolution/precipitation mechanism. Similar to other replacement minerals, illite also occurs as cements in sandstones as well as in shales, where it forms overgrowths on detrital illite particles and discrete crystals (cements) (e.g., Lanson and Champion, 1991; Rask *et al.*, 1997).

Numerous reactions describing the illitization of smectite within shales have been proposed. However, few attempts have been made to balance such reactions using accurately determined compositions of naturally occurring smectite and illite or to treat these reactions in a manner consistent with TEM evidence for a dissolution/precipitation reaction mechanism. Compared to smectite, illite is enriched in aluminum and potassium and deficient in silicon. (Ammonium derived from concurrent organic maturation also substitutes into illite, e.g., Lindgreen, 1994.) Thus, precipitation of illite requires quantities of potassium and aluminum in excess of those supplied by smectite, and dissolution of smectite supplies dissolved silica in excess of that needed to precipitate illite. (Hower *et al.*, 1976; Pearson and Small, 1988; Awwiller, 1993; Lynch, 1997; Lynch *et al.*, 1997). Hower *et al.* (1976) described the general reaction, (8), as

$$\text{smectite} + Al^{3+} + K^{+} = \text{illite} + Si^{4+} \quad (8)$$

acknowledging that the reaction also involves loss of iron and magnesium. Such a model for the replacement of smectite by illite requires a supply of aluminum and potassium from other mineral dissolution reactions. Dissolution of detrital illite and K-feldspar are the most commonly suspected sources (e.g., Hower *et al.*, 1976; Chuhan *et al.*, 2000a). A similar reaction, (9), proposed by Lynch *et al.* (1997) is expanded to take into account a greater spectrum of the observed depth-related mineralogical changes observed in Oligocene Gulf Coast shales between 7,000 and 15,000 ft (1 ft = 0.3048 m)

$$\begin{aligned} &10.93\,\text{illite/smectite (20\% illite)} \\ &\quad + 0.91\,\text{discrete illite} + 2.75\,\text{kaolinite} \\ &\quad + 0.86\,\text{potassium feldspar} + 1.46\,\text{plagioclase} \\ &\quad \text{feldspar} + 2.11Al_2O_3 + 2.66K_2O \\ &\qquad = 12.59\,\text{I/S(85\% illite)} + 0.27\,\text{chlorite} \\ &\qquad\quad + 3.16\,\text{quartz} + 1.98\,\text{albite} + 4.7SiO_2 \end{aligned} \quad (9)$$

Alternative constructions of this reaction are conservative with respect to aluminum (Boles and Franks, 1979b; Land *et al.*, 1987). Such reactions require less input of potassium from minerals other than smectite, but result in a diminution of the total amount of clay by ~25% and the generation of excess silicon, iron, and magnesium in amounts (near 15% of the total shale volume) that are not readily sequestered in known authigenic phases in either shales or sandstones (Awwiller, 1993).

Illite precipitation can also be construed as an acid-generating reverse-weathering reaction by invoking the observed reduction of iron in illite and precipitation of additional chlorite (e.g., Land *et al.*, 1987; Lynch *et al.*, 1997). Illite replacement of kaolinite is widely observed in sandstones (e.g., Kisch, 1983; Ehrenberg and Nadeau, 1989; Chuhan *et al.*, 2000a), and also requires potassium derived from a dissolution reaction and can also be written as reverse weathering reactions (Awwiller, 1993):

$$\begin{aligned} &0.48K^{+} + 1.08Al_2Si_2O_5(OH)_4 + 1.29SiO_2 \\ &\quad + 0.11X_2O + 0.1Fe_2O_3 + 0.19MgO \\ &\qquad = K_{0.48}X^{+}_{1.22}Mg_{0.19}Fe_{0.20}Al_{2.16}Si_{3.45}O_{10}(OH)_2 \\ &\qquad\quad + 1.16H_2O + 0.48H^{+} \end{aligned}$$

7.07.4.5 Quartz Cement

A dispute over silica mobility in quartz cementation was the opening volley of what has become a wide-ranging discussion of elemental mobility in late diagenesis. Silica concentrations in subsurface fluids are typically higher than saturation (Land, 1997), but still low enough so that large quantities of fluid are apparently required to account for quartz cementation. Land and Dutton (1978) showed that a minimum of 10^5 pore volumes of fluid were required to transport the amount of silica required to provide the quartz cement in a Pennsylvanian sandstone in north Texas. This is more fluid than could be accounted for by simple burial compaction in the absence of incursions of meteoric water (Land and Dutton, 1978; Bjørlykke, 1979; Boles and Franks, 1979a; Land and Dutton, 1979). The volume of quartz cement in a wide variety of sandstones poses similar problems

(Blatt, 1979; McBride, 1989), as do quartz veins in low-grade metamorphic pelites (Yardley, 1986).

Numerous studies have shown that quartz cementation in basins worldwide is prominently temperature correlated (e.g., Land and Fisher, 1987; Land *et al.*, 1987; Bjørlykke and Egeberg, 1993; Walderhaug, 1994) (Figure 9(b)), suggesting that precipitation is the rate-limiting step in the formation of quartz cement (Oelkers *et al.*, 1996; Walderhaug, 1996; Lander and Walderhaug, 1999). Sandstones that have been heated above a threshold temperature of ~70 °C have a high probability of becoming quartz cemented. The ultimate degree of quartz cementation is highly predictable with a classic Arrhenius equation, based empirically on the observed correlation of quartz cementation with thermal history and corrected for the amount of quartz surface that is available as a precipitation substrate (Lander and Walderhaug, 1999; Walderhaug *et al.*, 2000; Bloch *et al.*, 2002). Thus, the overall amount of quartz cement in sandstones should correlate in a general way with the content of detrital quartz. As previously mentioned quartz nucleation is strongly localized at the grain scale and several types of surface coatings inhibit the nucleation of quartz on detrital grain surfaces (Pittman and Lumsden, 1968; Heald and Larese, 1974; Aase *et al.*, 1996; Boles and Hickey, 1996). This adds considerable complexity to the prediction of porosity in the subsurface and requires substantial empirical adjustments in kinetic models of quartz cementation (Oelkers *et al.*, 2000).

A simple kinetic rate formulation for quartz cementation predicts that finer-grained detrital materials will become cemented at a faster rate than coarser materials because the surface area available on finer-grained particles is larger (e.g., Walderhaug, 1996). However, most large data sets for quartz cement volumes do not indicate a strong correlation between the amount of quartz cement and grain size (e.g., Dutton and Diggs, 1990, 1992) and many data sets indicate a tendency toward greater quartz cementation in coarser sandstones (McBride, 1989).

Application of the Lander–Walderhaug model tends to overestimate the amount of quartz cement in fine-grained sandstones, and underestimates it in coarse-grained sandstones (Walderhaug, 2000; Milliken, 2001). An intriguing observation that may explain this phenomenon is the tendency for larger quartz grains to have larger overgrowths than smaller quartz grains in the same sample (Makowitz and Sibley, 2001). Recent experiments by Larese and colleagues show that incipient quartz nuclei initiate as complex polyhedral aggregates that grow more rapidly than crystals dominated by a single set of faces. As crystal growth progresses, smaller nuclei (initiated on smaller substrates) advance to simpler crystal face sets more quickly and consequently reach smaller final sizes (Rob Lander, personal communication, 2002). Thus, controls on nucleation and crystal growth govern the spatial distribution and also the local rate of quartz cementation.

That nucleation and growth rate are the limiting steps in quartz cementation has no particular implications with respect to the ultimate source of the silica or the mechanism of transport. Potential sources of silica for quartz cementation are numerous (McBride, 1989) and include all documented silicate dissolution reactions in sandstones and shales.

With the execution of the early diagenetic silcretes, the depth distribution of quartz cement in young basins precludes extremely early quartz cementation during burial (Bjørlykke and Egeberg, 1993). Yet petrographic evidence regarding the timing of quartz cementation relative to other reactions within sandstones is somewhat hindered by the strong localization of quartz cement on detrital quartz surfaces. Quartz cement does not occur uniformly throughout sandstones. Physical evidence for the timing of its formation relative to other diagenetic processes may therefore be difficult to locate. For example, quartz cement and albite grain replacements are so strongly localized on mutually exclusive substrates that determination of their relative age is difficult.

Most quartz cementation appears to predate most feldspar dissolution (Land *et al.*, 1987). This observation removes local detrital feldspar dissolution as a likely source of silica for quartz cementation in many sandstones. This conclusion is somewhat enigmatic, because temperature trends for quartz cementation (Figure 9(b)) and detrital feldspar loss (Figures 9(c) and (d)) indicate that the zone of most intense quartz emplacement is deeper than that of most intense feldspar reaction. Perhaps this is yet another indication of the degree of secondary porosity collapse that accompanies burial. The surviving secondary pores represent only the latest (post-quartz cement) stages of the total feldspar loss, formed essentially at the same depth at which they are observed. Quartz precipitation within grain dissolution pores does occur, but rarely, and mostly in cases of very high temperature diagenesis (>150 °C?) (e.g., Morad and Aldahan, 1987; Dutton and Diggs, 1990). Trace aluminum concentrations in authigenic quartz show that mobilization of an aluminum-bearing phase does accompany quartz cementation (Kraishan *et al.*, 2000). Smectite dissolution is one possibility (Abercrombie *et al.*, 1994) that has been widely advocated as a source for quartz cement for over 40 years (e.g., Towe, 1962; Lynch *et al.*, 1997). Interestingly, the zones of major reaction progress in illitization (Figure 9(a)) and silicification (Figure 9(b)) correspond only partially. Pressure solution of detrital quartz in sandstone has widely been called upon as a local source for silica for quartz cementation. However, stylolitization typically cuts across zones within quartz cement, suggesting that these two processes are separated in time (Land, 1997). Thus, chemical and petrographic evidence largely support the notion that silica for quartz cementation is imported from sources external to sandstones (Land *et al.*, 1987, 1997; Land, 1997).

Enhanced quartz cementation within highly permeable sandstone units has been cited as evidence for the importance of fluid flow in quartz cementation (Haszeldine *et al.*, 1984, 1992; McBride, 1989; Lynch, 1996). The correlation of quartz cement with other late diagenetic phases that appear to require advective transport supports this idea (Burley *et al.*, 1989). However, the success of kinetic growth rate models for predicting quartz cement suggests that fluid flow may be, at most, a second-order control on the occurrence of quartz cement, because fluid flow controls would impose large spatial variations on the occurrence of quartz cement that are incompatible with the predictability of quartz cement content on the basis of thermal history (Bjørlykke and Egeberg, 1993). The strong control of growth rate combined with difficulty of isolating the variables affecting quartz cement distribution such as grain size (= nucleation substrate size) and grain coatings (including detrital matrix) make the evaluation of secondary external controls on quartz cementation (e.g., local permeability, proximity to shale contacts) extremely challenging.

A growing body of evidence suggests that precipitation of quartz in sandstones induces lithification that leads to brittle behavior at scales larger than grains (Laubach and Milliken, 1996a). Many sandstones appear to have a deformational history in which the progressive opening of transgranular fractures partially overlaps the timing of quartz cementation. This leads to the development of dramatic crack-seal textures within fracture-filling quartz (Figure 12) (Milliken and Laubach, 2000). Quartz cement has also been observed intergrown with bitumen in fractures in shales (Parnell *et al.*, 1996).

The difficulties imposed by clay coatings on quartz nucleation probably explain why there are few reports of quartz cementation in shales. The oxygen in quartz silt is isotopically heavy ($\delta^{18}O = 19‰$) compared to sand-size quartz ($\delta^{18}O = 14‰$) (Blatt, 1987) suggesting that a substantial component of relatively low-temperature quartz resides in shales. The application of cathodoluminescence microscopy to shales has been quite limited. In Frio Formation shales of south Texas no convincing quartz cement is observed across a range of depths in which quartz cementation becomes important in associated

Figure 12 Quartz precipitation that overlaps the timing of fracture opening results in dramatic crack-seal fabrics. Here, a pale blue-luminescing quartz grain has been repeatedly fractured to yield "rafted" grain slivers encased in blue- and red-luminescing authigenic quartz. Quartz precipitation ultimately did not keep pace with fracture opening as considerable fracture porosity survives (green). Cozzette Sandstone, Iles Formation (upper Cretaceous), Eastern Piceance Basin, Colorado. Photomicrograph by Rob Reed.

sandstones (Milliken, 1994a). Quartz silt particles in shales from the Eocene Wilcox formation do show evidence of tiny quartz overgrowths (Land and Milliken, 2000) (Figure 13). Interestingly, the content of smectite is higher in the Frio, and so the occurrence of quartz overgrowth in shales does not appear to reflect the local availability of silica from illitization reactions. Devonian black shales in the eastern US apparently contain large quantities of authigenic quartz formed at very low temperatures ($\delta^{18}O = 28‰$) (Schieber *et al.*, 2000). This material is probably related to the early diagenetic alteration of opaline skeletal debris. Nonetheless, it is clear that the cathodoluminescence imaging has great potential for documenting the origins of quartz in shale.

7.07.4.6 Authigenic Calcite

In contrast to diagenetic processes involving silicates, calcite cement in sandstones shows neither a strong temperature dependence (e.g., Milliken *et al.*, 1981; Land, 1984; Land and Fisher, 1987; Land *et al.*, 1987; Lynch and Land, 1996; Morad, 1998) nor strong spatial localization at the thin section scale (Milliken, 2001). Petrographic evidence indicates that calcite in sandstones precipitates over the entire diagenetic spectrum from near-seafloor conditions to deep burial (Morad, 1998). Some calcites date early and preserve IGVs near their depositional values; calcite also postdates quartz cement and other late diagenetic phases such as ankerite and is emplaced in

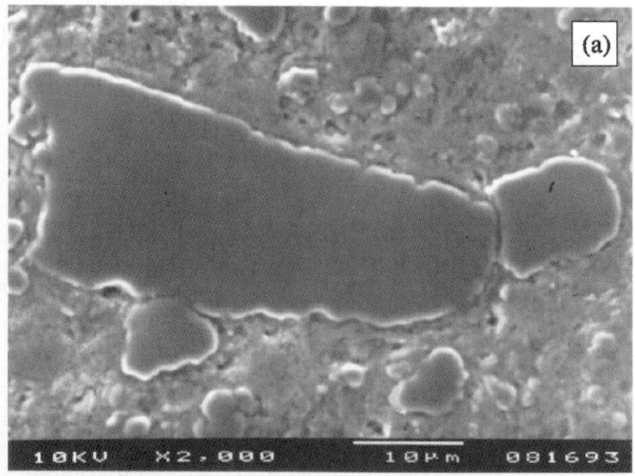

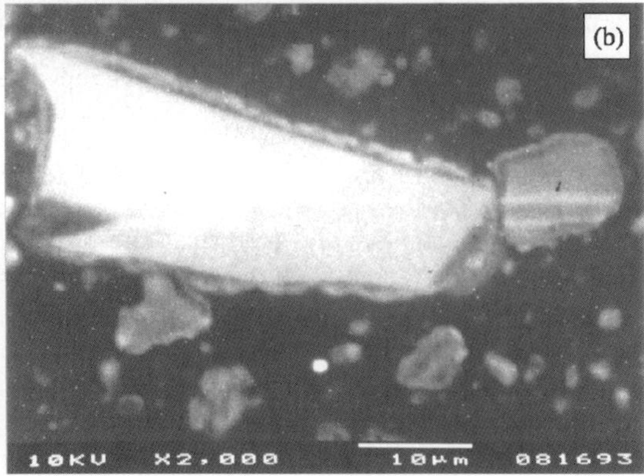

Figure 13 Quartz on silt grains in Wilcox Formation shale, Eocene, Texas Gulf Coast: (a) secondary electron image and (b) panchromatic cathodoluminescence image.

sandstones that have undergone significant compaction (IGVs in the ranges of 20% or less).

The heterogeneity of calcite cement distribution at scales larger than a thin section is pronounced. Both early (e.g., Irwin, 1980; Scotchman, 1991; Bjørkum *et al.*, 1992) and late (e.g., McBride *et al.*, 1995; Girard, 1998) calcite cements tend to form as concretions or isolated cemented horizons. At the field scale, or on a formation-wide basis, calcite cement typically accounts for only a few percent of the total sandstone volume (1–3%); locally, cemented zones may contain 20–30% calcite. The cause of intense local calcite emplacement (zones are typically on the order of a few cm to 1 m thick) is uncertain. Isolated shell lags have been proposed as a localizing influence on carbonate precipitation (Bjørkum and Walderhaug, 1990), but there is no evidence to suggest that calcite requires special nucleation substrates as is typical of authigenic silicates. Concretions are observed in units containing no skeletal debris (Girard, 1998), including nonmarine sandstones (e.g., Beckner and Mozley, 1998; Milliken, 1998).

Chemical evidence supports the petrographic observation that calcite in sandstones precipitates over a wide range of conditions (Morad, 1998). Stable isotopic analysis ($\delta^{18}O$, $\delta^{13}C$, $^{87}Sr/^{86}Sr$) of authigenic calcite in sandstone typically reveals large compositional heterogeneities between samples at any given depth (Milliken *et al.*, 1981; Lynch and Land, 1996; Macaulay *et al.*, 1998). Partial replacement (recrystallization) of early formed calcite creates daunting (for the use of bulk methods of analysis) trace element and isotopic heterogeneities at the scale of a few tens of micrometers, but does not homogenize calcite chemistry during progressive burial (Lynch and Land, 1996). Thus, calcites in deeply buried sandstones appear to record a complex history of multiple precipitation events.

The amount of calcite cement in sandstones varies widely, as mentioned, but formations as a whole tend to contain 1–5% calcite. Some formations contain sufficient detrital carbonate to supply the formation of calcite cement (Milliken *et al.*, 1998). Some formations, however, appear to require an import of calcium (and CO_2) from external sources. The values of $\delta^{18}O$, $\delta^{13}C$, and $^{87}Sr/^{86}Sr$ have widely been used to infer the possible sources of calcite cement. A common finding is that marine skeletal debris supplies much of the carbon and, therefore, presumably, the calcium. In late diagenesis, the $\delta^{13}C$ values for calcite suggest that organic carbon generally supplies less than 50% of the total carbon in calcite, and that this amount declines with depth (Milliken *et al.*, 1981; Land, 1984; Lynch and Land, 1996) in a pattern reminiscent of the heavier carbon isotopic values observed for CO_2 (described in the feldspar section) and for dissolved bicarbonate (Land and Macpherson, 1992b). Strontium isotope data for late diagenetic calcite show the competing influences of marine skeletal debris and feldspar reactions on the evolution of strontium in pore fluids. Earlier calcites tend to reflect marine strontium signatures, whereas later calcites contain radiogenic strontium released by dissolution of old feldspars (Land and Fisher, 1987; Mozley and Hoernle, 1990; Girard, 1998). In a few instances, feldspar reactions involving large quantities of young volcanogenic debris produce the opposite effect, imparting very low strontium isotope values to carbonate cement (Mack, 1990; Boles, 1998).

The trace element content of late diagenetic calcite contrasts markedly with that of calcite formed early in diagenesis. Often these are relatively depleted in iron and magnesium and enriched in manganese (Lynch and Land, 1996; Milliken, 1998; Milliken *et al.*, 1998). It is common for iron and magnesium to covary in late diagenetic calcite. The reason for this is unclear.

Several well-known mechanisms account for precipitation of authigenic calcite in early diagenesis (Morad, 1998), but the sources of calcium in authigenic calcite in late diagenesis are less certain. It is difficult to account for the amount of calcite cement observed in sandstones from internal sources of calcium and CO_2 alone (Land and Macpherson, 1992b). Possible internal sources include calcium plagioclase, early formed calcite cements, marine skeletal debris, and carbonate rock fragments. Similar potential sources exist in shales. Detrital carbonate particles are, of course, attractive as a source, because they dissolve in significant amounts (as described in the previous section) and because calcite cements have isotopic compositions (δ^{13}C and ^{87}Sr/^{86}Sr) that are consistent with such a source. Calcium plagioclase in Gulf Coast sandstones contains insufficient calcium to serve as a significant source of calcite cement (Land and Macpherson, 1992b). These sandstones contain unusually calcic plagioclase. It is therefore unlikely that calcium plagioclase is a significant source for calcium in calcite cement in general (Morad, 1998).

Two difficulties beset the notion of a mudrock source for calcite cement in sandstones. First, the lack of a strong depth trend for calcite cement in sandstones is difficult to reconcile with the prominent depth-related trend for carbonate loss in mudrocks. Second, there is a tremendous volume imbalance between the large amounts of $CaCO_3$ that appear to be missing from shales compared to the observed amount of calcite cement in sandstone (Land and Macpherson, 1992b). Unfortunately, quantitative data for depth trends of calcite in associated shales and sandstones outside the Gulf of Mexico are few.

The patterns of calcite grain replacements in sandstones and mudrocks contrast with those observed for cements in sandstones. As described in connection with feldspar dissolution, authigenic calcite is strongly localized as feldspar replacements and increases prominently below 3 km (Milliken, 1992; Milliken and Land, 1993) (Figure 9(f)). Unfortunately, the scarcity of similar petrographic studies in mudrocks leaves great uncertainty regarding the generality of these observations. Authigenic calcite in the form of cements is notably absent in shales (O'Brien and Slatt, 1990; Milliken and Land, 1993) with the exception of very early carbonate concretions (Irwin, 1980; Gluyas, 1984; Raiswell and Fisher, 2000; Hudson *et al.*, 2001).

7.07.4.7 Kaolinite and Chlorite

Kaolinite and chlorite occur both in sandstones and shales. It is clear that in sandstones these minerals are primarily authigenic. They form largely as cements, but also as grain replacements. While these minerals may constitute a few percent locally, they usually amount to less than a percent on a formation-wide basis (~0.86% in the Frio Formation), though notable exceptions exist (e.g., Trevena and Clark, 1986).

Prominent kaolinization in sandstone is mostly a feature of early diagenesis related to meteoric influx, e.g., at unconformities (e.g., Bjørlykke and Aagaard, 1992; Khanna *et al.*, 1997). In many cases, they are related to acidic fluids generated in coals (e.g., Huggett, 1984; Cookenboo and Bustin, 1999). Organic complexation has been implicated as an important mechanism in the mobilization of aluminum for the precipitation of kaolinite in limestones (Maliva *et al.*, 1999), and the same may hold true for sandstones (e.g., Platt, 1993). Kaolinite typically displays a highly patchy distribution at the thin section scale. Its distribution does not necessarily coincide with that of leached feldspars (Milliken, 2002). As previously mentioned, illite may form as a replacement of kaolinite in late diagenesis.

Chlorite in sandstones forms as both early (~20 °C) and late cements and grain replacements (Grigsby, 2001). As described in Section 7.07.4.5, early chlorite may exert a profound effect on porosity in late diagenesis as a result of its inhibitory effect on quartz cementation.

In shales, these minerals are volumetrically significant; the relative proportion of detrital and authigenic grains is very uncertain. It seems likely that both detrital and authigenic varieties are usually present. Isolated, anomalously large pores provide an interesting window into the behavior of these minerals in shales (Primmer and Shaw, 1987). Both phases precipitate as cements (Figure 14). Lynch (1997) reports an average content of ~17% kaolinite and 2% chlorite in

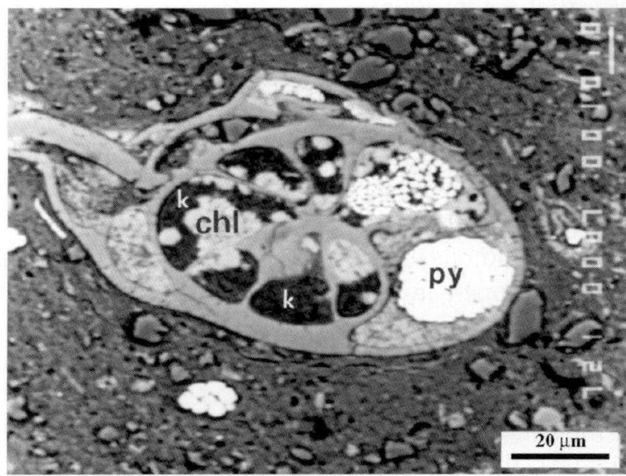

Figure 14 Partially crushed foraminifer with chambers filled by chlorite cement (chl), kaolinite (k), and pyrite (py) cements. Reactions evident in these anomalously large pores are indicative of reactions that may also proceed in smaller pores throughout the shale. Frio Formation, Oligocene, South Texas. Backscattered electron image.

Oligocene Gulf Coast shales at 2 km depth; at 4.5 km depth the kaolinite in these rocks is reduced to ~10% whereas the chlorite has increased to 4%. This suggests that the same illitization reaction that affects kaolinite in sandstones may be active in shales as well. Alternatively, the kaolinite may be converted into chlorite.

Dunoyer de Segonzac (1970, p. 282) wrote: "Chlorites are the least well-known clay minerals in diagenesis." The persistence of this status is unfortunate, because chlorite precipitation is especially important to understanding the overall acid balance in late diagenesis. The chlorite structure includes a brucite layer with a high concentration of hydroxyl, and thus precipitation of chlorite, is potentially a potent reverse-weathering reaction as proposed by numerous workers (Hower *et al.*, 1976; Boles and Franks, 1979b; Land *et al.*, 1987; Lundegard and Land, 1989):

$$\begin{aligned}&2.88Al^{3+} + 3.27Fe^{2+} + 0.96Mg^{2+}\\&\quad + 2.89H_4SiO_4 + 6.72H_2O\\&\qquad = (Al_{1.77}Fe_{3.27}Mg_{0.96})Si_{2.89}Al_{1.11})O_{10}(OH)_8\\&\qquad\quad + 17.1H^+ \qquad (10)\end{aligned}$$

(chlorite formula of Genuise, 1991).

Chloritization is particularly important in shales containing magnesium-rich smectites. It proceeds there progressively with increasing depth in a manner highly analogous to illitization (Dunoyer de Segonzac, 1970; Chang *et al.*, 1986).

7.07.4.8 Minor Cement and Replacement Phases

The foregoing sections have considered the volumetrically significant reactions in siliciclastic diagenesis with regard to reaction mechanisms and the apparent constraints on the mobility of elements required to accomplish these reactions. Other widely distributed late diagenetic authigenic minerals are ferroan dolomite ("ankerite") (e.g., Boles, 1978; Spötl and Pitman, 1998) and the zeolite laumontite (Helmold and van de Kamp, 1984). These phases may be quite abundant locally (up to 20%), but they typically constitute less than 1 vol.% on a formation-wide basis. Both are aggressive replacement phases (reverse weathering reactions?) that are prone to attack the detrital feldspars.

Especially relevant to an understanding of elemental mobility in late diagenesis are reactions that are volumetrically minor on a formation scale (though, locally these may be major) that involve elements whose minerals have extremely low solubilities. All of these phases are known as both cements and grain replacements and are observed as a ubiquitous part of the diagenetic component in both sandstones and shales, entirely apart from ore deposits. The appearance of such elements in authigenic phases demonstrates that pore fluids have a capacity for element transport that clashes somewhat with our current understanding of element complexing and fluid flow.

Examples of these authigenic phases include sphalerite (Woronick and Land, 1985; Løndøy *et al.*, 1986; Taylor and Land, 1997), barite (Breit *et al.*, 1990; Ramm and Ryseth, 1996; Fishman, 1997; Gluyas, 1997; Krzysztof *et al.*, 1997; Taylor and Land, 1997; Fisher *et al.*, 2000), anatase (Morad and Aldahan, 1982; Yau *et al.*, 1987; Milliken and Mack, 1990; Ramm and Ryseth, 1996), titanite (Merino, 1975; Richmann *et al.*, 1980; Morad and Aldahan, 1982; Helmold and van de Kamp, 1984; Aldahan and Morad, 1986; Yau *et al.*, 1987; Van Panhuys and Trewin, 1990; Milliken, 1992; Ramm and Ryseth, 1996), and a variety of phases that contain thorium, zirconium and the rare earth elements (Mellon, 1964; Cathelineau, 1987; Rubin *et al.*, 1989, 1993; Kisch, 1991; Milodowski and Zalasiewicz, 1991; Sutton and Maynard, 1993; Awwiller, 1994; Bouch *et al.*, 1995; Rasmussen, 1996; Lev *et al.*, 1998, 1999; Cuney and Mathieu, 2000; Mathieu *et al.*, 2001). In general, sandstones that contain such phases are also characterized by abundant quartz cement, highly altered feldspar assemblages, and associated fluids high in CO_2.

7.07.4.9 Summary: Massive Reorganization by Fluid-mediated Reactions

Table 1 summarizes the major dissolution and precipitation reactions of late diagenesis in sandstones and shales in nonvolcanogenic basins. In volcanic basins, zeolites play a more prominent role. All of the reactions listed in the table proceed over a wide range of temperature in late diagenesis and are, in large part, synchronous.

In contrast to diagenesis in carbonates rocks, the volumetrically significant stabilization reactions in sandstones and shales take place at elevated temperature. Beginning at a temperature ~80 °C, surface-reaction-controlled dissolution, cementation, and replacement processes induce a major shift towards a mineral assemblage that, by 200 °C, approaches (but does not reach) equilibrium amongst mineral phases and pore fluids. The largest step toward the quartz–albite–illite–chlorite–(±calcite or ankerite) assemblage that is characteristic of greenschist metamorphism takes place during late diagenesis. Of the components present at the time of deposition, only detrital quartz has a high probability of surviving

Table 1 Petrographically documented reactions of volumetric significance in late diagenesis of sandstones and shales. As written, all reactions only proceed to the right in the nonequilibrium realm of late diagenesis. Compositions are from Lynch (1996) for clay minerals and from Milliken *et al.* (1989) for feldspars.

	(°C)	
Dissolution reactions: sources of dissolved components		
In shales:		
Smectite	60–140	$K_{0.12}Na_{0.25}(Al_{1.41}Fe_{0.22}Mg_{0.41})(Si_{3.88}Al_{0.12})O_{10}(OH)_2 + 9.96H_2O = 0.12K^+ + 0.25Na^+ + 0.22Fe^{3+} + 0.41Mg^{2+} + 1.53Al^{3+} + 3.88H_4SiO_4 + 6.44OH$
Organic matter	40–100	$CH_2O + H_2O = CO_2 + 4H^+$
In sandstones and shales:		
Kaolinite	75–150	$Al_2Si_2O_5(OH)_4 + 5H_2O = 2Al^{3+} + 2H_4SiO_4 + 6OH^-$
Ca-plagioclase[c]	70–130	$(Na_{0.7}Ca_{0.3})Al_{1.3}Si_{2.7}O_8 + 5.2H^+ + 2.8H_2O = 0.7Na^+0.3Ca^{2+} + 1.3Al^{3+} + 2.7H_4SiO_4$
K–Na alkali feldspars[c]	70–120	$(K_{0.5}Na_{0.5})AlSi_3O_8 + 4H^+ + 4H_2O = 0.5K^+0.5Na + Al^{3+} + 3H_4SiO_4$
K-feldspar[c]	80–160	$KAlSi_3O_8 + 4H^+ + 4H_2O = K^+ + Al^{3+} + 3H_4SiO_4$
Calcite[a] (weathering + thermal decomposition?)	>60?	$CaCO_3 + 2H^+ = Ca^{2+} + CO_2 + H_2O$
?quartz (by pressure solution)	>110?	$SiO_2 + H_2O = H_4SiO_4$
Precipitation reactions: sinks for dissolved components		
In shales:		
Illite[a,d]	60–200+	$65K^+ + 0.08Na^+ + 0.14Fe^{2+} + 0.2Mg^{2+} + 3.4H_4SiO_4 + 2.27Al^{3+} = K_{0.65}Na_{0.08}(Al_{1.68}Fe_{0.14}Mg_{0.2})(Si_{3.41}Al_{1.59})O_{10}(OH)_2 + 8.22H^+ + 2.69H_2O$
Chlorite[d]	60–200+	$2.88Al^{3+} + 3.27Fe^{2+} + 0.96Mg^{2+} + 2.89H_4SiO_4 + 6.72H_2O = (Al_{1.77}Fe_{3.27}Mg_{0.96})Si_{2.89}Al_{1.11})O_{10}(OH)_8 + 17.1H^+$
In sandstones and shales:		
Kaolinite/dickite[e]	80–150	$2Al^{3+} + 2H_4SiO_4 + 6OH^- = Al_2Si_2O_5(OH)_4 + 5H_2O$
Albite[d]	75–200+	$Na^+ + 3H_4SiO_4 + Al^{3+} = NaAlSi_3O_8 + 4H^+ + 4H_2O$
In sandstones:		
Quartz	>80	$H_4SiO_4 = SiO_2 + 2H_2O$
Calcite[b]	100–160?	$Ca^{2+} + CO_2 + H_2O = CaCO_3 + 2H^+$ (in the presence of active feldspar buffers)
	40–100?	$Ca^{2+} + 2HCO_3^- = CaCO_3 + CO_2 + H_2O$ (in the presence of organic alkalinity?)

[a] Locally in sandstones. [b] Locally in shales as feldspar replacement. [c] Weathering reactions. [d] Reverse weathering reactions. [e] In organic-rich materials.

into the early stages of metamorphism with its primary chemical makeup intact. This means that, with the exception of detrital quartz zircon, and monazite every cation–oxygen bond existing in the crystals of the primary detrital assemblage has been broken by the end of late diagenesis (Land, 1997). Including the cements, a portion of the quartz and all the other minerals in the ultimate assemblage are precipitated from pore fluids (authigenic). The average sandstone contains 65% detrital quartz and the average shale contains 30% (Blatt *et al.*, 1980). The authigenic volumes in these rocks at the end of late diagenesis are approximately 35% and 70%, respectively (ignoring for the moment the possible volume losses involved). Given the magnitude of these dissolution and precipitation reactions, it is not surprising that fluids in the deep portions of sedimentary basins have compositions that are strongly rock buffered (Land and Macpherson, 1992b; Hanor, 1994).

7.07.5 WHOLE-ROCK ELEMENTAL DATA AND LARGER-SCALE ELEMENTAL MOBILITY

At the scale of thin sections, dissolution, cementation, and replacement are all *prima facie* evidence of elemental mobility. It is clear that late diagenesis involves complex history of material transfer at the scale of a few millimeters because detrital materials immediately adjacent to cements and grain replacements may show no evidence of dissolution, or contain no elements in common with the authigenic minerals. For minor elements that are concentrated in authigenic phases in amounts far in excess of their concentration in the bulk rock (e.g., barite cement), extraction from some relatively larger rock volume is necessary.

The scale of material transport at distances larger than a thin section is difficult to discern from petrographic evidence alone. The principal reason for this is that the largest volume of material in sedimentary basins resides in shales that are not amenable to the same level of petrographic scrutiny as sandstones. If a database for sandstone is large and representative, it may be possible to determine that an element has, on average, been exported at the hand specimen scale; it is not possible to determine petrographically if that same element has been sequestered locally in shales or lost from the basin entirely. This is because the origin of a substantial portion of the authigenic component (cements and grain replacements) in shales is very likely to be cryptic. For example, it is difficult to decide whether shale kaolinite belongs to the dissolution or to the precipitation side of the overall diagenetic "reaction" if we cannot ascertain which portion of the kaolinite is authigenic and which detrital. Petrographic data to assess quantitatively the products of many diagenetic reactions in shale are lacking.

Bulk compositional data also have certain disadvantages and limitations for interpreting mass transfer (summarized by Milliken *et al.*, 1994). First, there is the formidable challenge of discriminating primary variations related to sorting and provenance from diagenetically induced variations (de Caritat *et al.*, 1997; Bloch *et al.*, 1998; Hutcheon *et al.*, 2000). In sandstones, even subtle variations in grain size lead to compositional variations in the primary grain assemblage that impact diagenetic heterogeneity (Milliken, 2001). A far greater textural and compositional range is encompassed by "shale." Thus, it is extremely challenging to identify "primary" or "least-altered" compositions against which elemental gains and losses can be assessed across a range of thermal conditions.

One strategy for constraining initial variation is to sample a single stratigraphic layer at a range of tectonic burial depths (Cavazza and Gandolfi, 1992), at progressive distances from an intrusion (Summer and Ayalon, 1995), or within and without early concretions (Knoke, 1966; Evans, 1989; Gluyas and Coleman, 1992; Potdevin and Hassouta, 1997). The more common strategy in bulk elemental analysis, and the only one for which a substantial body of data exists, is to sample across progressive burial in relatively young basins, an approach that is necessarily fraught with the complication of depositional heterogeneity due to spatial and temporal variations in environment and provenance.

In the late diagenesis of sandstones and shales, volumetrically significant bulk compositional transfers have been proposed for only a few materials. Not surprisingly, these elements are major components of volumetrically significant diagenetic reactants. Water and organic matter (petroleum) are generally agreed to be mobile at the basinal scale because many petroleum reservoirs are far removed from their source rocks (Hunt, 1996) and because fluids vented on the seafloor show evidence of long-distance transport (e.g., papers in Parnell, 1998). Major transfers of CO_2, Ca, Si, and K have also been proposed, but they remain controversial, largely because the apparent permeability of thick shale-dominated sequences appears very inadequate to permit a large amount of advective transport (Bjørlykke, 1989, 1993; Bjørlykke and Lynch, 1997).

Bulk elemental trends for CO_2, Ca, Si, and K are of particular interest because for each of these there is:

(i) petrographic evidence for reaction processes consistent with the observed mobility trends (previously discussed);

(ii) recurrence of the same trends in many basins, in different formations, and across different depositional environments;

(iii) a matching temporal trend in average rock composition;

(iv) persistence of the trend into low-grade metamorphism; and

(v) balancing trends across different rock types (sandstones and shales) or across different crustal levels.

7.07.5.1 SiO_2

The notion that silica is transferred from shales to sandstones during late diagenesis (Towe, 1962) provides a mechanism to link the observed depth trends of at least three (possibly four) of the major diagenetic reactions listed in Table 1 (smectite dissolution, illite precipitation, quartz precipitation in sandstones, and quartz pressure solution in shales). Large-scale silica transfer is to be expected as fluids flow from regions of higher to lower temperature (Ferry and Dipple, 1991).

Land *et al.* (1997) normalized SiO_2 in shale samples against six "immobile" elements (Al_2O_3, TiO_2, HREE, Zr, Hf, and Th) and found that it declines 6 g SiO_2 per 100 g of shale between approximately 2 km and 4.5 km, whereas the ratios of the various reference elements remain invariant. Land and Milliken (2000) note similar SiO_2 declines in 9 of 10 additional Gulf Cost wells, including the wells reported by Awwiller (1993) and Hower *et al.* (1976). These trends for SiO_2 are observed for shales in the Eocene Wilcox Formation and the Oligocene Frio Formation from the central and southern portions of the Texas Gulf Coast.

Similar silica losses from shales (Knoke, 1966; Evans, 1989, 1990) and gains by sandstones (Gluyas and Coleman, 1992; Gluyas *et al.*, 2000) are documented in concretion studies. Silica gained by Gulf Coast sandstones does not approach the amount that is apparently lost from shales, however, as sandstones are nearly balanced for silicon on a whole-rock basis (though not petrographically) (Milliken *et al.*, 1994), an observation also made for North Sea sandstones (Giles and De Boer, 1990).

Silica loss of considerable magnitude is also reported during early metamorphism (e.g., Wright and Platt, 1982). Ague (1991) reports that SiO_2 content declines from an average of 60.3% in shales and slates to ~56% in amphibolites.

7.07.5.2 CaO (and CO_2)

Whole-rock elemental data for calcium in Gulf Coast shales (Hower *et al.*, 1976; Land *et al.*, 1997; Land and Milliken, 2000) support the calcite decline that is widely observed by XRD (Freed, 1982; Milliken, 1981 and Land, 1982) and by chemical analysis (Figure 9(c)). Ten of the 11 wells studied by Land and Milliken (2000) show a depth-related decline in CaO relative to TiO_2, though only 7 of 11 do so relative to Al_2O_3. The carbonate content of shales, in general, decreases with age (Garrels and MacKenzie, 1971, 1974; van Moort, 1974) and degree of metamorphism. The loss of CaO from Gulf Coast shales is not at all balanced by an increase of CaO in sandstones (Land and Macpherson, 1992b; Milliken *et al.*, 1994).

7.07.5.3 K_2O

Progressive increase in K_2O is a third commonly observed trend in the bulk elemental composition of shales, as noted in the Gulf Coast (Hower *et al.*, 1976; Calvert and Klimentidis, 1986; Awwiller, 1993; Wintsch and Kvale, 1994; Land *et al.*, 1997; Berger *et al.*, 1999; Land and Milliken, 2000) and elsewhere (van Moort, 1972; Pearson and Small, 1988; Furlan *et al.*, 1996). Corresponding, though not balanced, losses of K_2O are observed in Gulf Coast sandstones (Milliken *et al.*, 1994). Potassium metasomatism is widely observed in shales and slates in a variety of settings (Hardy, 1989; Sutton and Maynard, 1990, 1993; Ague, 1991, 1997; Fedo *et al.*, 1995, 1997; Beratan, 1999; Ennis *et al.*, 2000) and is linked to secular trends in shale composition (Garrels and MacKenzie, 1971, 1974; van Moort, 1972). K_2O variability in shales is greater than trends in calcium and silicon, as loss of K_2O is observed in some Gulf Coast shale sequences (Land and Milliken, 2000), as well as in some cases of low-grade metamorphism (Sutton and Land, 1996).

7.07.6 FLUID FLOW

Transport during late diagenetic reactions is not rate limiting (surface-reaction controls are rate limiting). The spatial distribution of mineral dissolution and precipitation therefore yields little information about mechanisms of transport. The absence of halo textures reflects this fact on a small scale. At larger scales, element transfer between sandstones and shales that is diffusion controlled would be expected to impart spatial distributions on authigenic features with respect to lithologic boundaries. However, few studies have documented such patterns (Fothergill, 1955; Moncure *et al.*, 1984; Sullivan and McBride, 1991), and those that have been reported have a heterogeneity that is generally within expected variations due to grain size and sorting (Milliken, 2001). The general absence of spatial patterns with respect to lithologic boundaries probably reflects the overriding influence of surface reaction and nucleation substrate controls on the spatial distribution of diagenetic features. From a purely petrographic standpoint it is clear that import and export of material is required to accomplish the observed shift towards a quartz–albite–illite–chlorite assemblage. Petrography does not answer the question whether diffusion or advection controls element transport. Unlike chemical reactions in late diagenesis, which are controlled by the slowest of the reactions involved, element transport is limited by the fastest process, and this may differ from element to element.

Diffusional transfers of potassium and silicon between sandstones and shales may be sufficient to accomplish feldspar dissolution, illitization, and quartz cementation (Thyne, 2001; Thyne *et al.*, 2001). Losses of the magnitude observed for detrital carbonates in shales exceed the capacity of diffusion-mediated transfer. Large-scale advection seems required, although our understanding of shale permeabilities seems to preclude this (Bjørlykke, 1989, 1993; and Lynch, 1997). The possibility of convection driven by salinity heterogeneity within thick shale sequences has been demonstrated by Sharp *et al.* (2001), who note that more information for rock properties and fluid compositions within deep basinal shales is needed before the generality of their results can be assessed.

7.07.7 REVERSE WEATHERING AND CONCLUDING COMMENTS

A transition between the completion of weathering and the advent of reverse weathering *must* occur in the crust in order to maintain the acid balance within the rock cycle (e.g., Urey, 1956; Krauskopf, 1979) and to maintain levels of atmospheric and oceanic CO_2 within observed secular ranges (e.g., Berner *et al.*, 1983; Berner, 1991). Early work on reverse weathering investigated the degree to which early marine diagenesis of fluvial detritus might lead to the release of CO_2 (e.g., Mackenzie and Garrels, 1966a,b; Mackenzie *et al.*, 1981). More recently processes releasing CO_2 have been advocated in late diagenetic settings of very high geothermal gradient, evolving amounts of CO_2 that equal or exceed those released from volcanic provinces (Kerrick *et al.*, 1995). The volumetric potential of feldspar diagenesis to influence the crustal CO_2 cycle was noted (Kastner, 1974), even prior to the recognition of the importance of albitization in sedimentary basins. In continental margin basins such as the Gulf of Mexico and rift basins such as the North Sea, the magnitude of the mineralogical shift toward equilibrium that is seen in late diagenesis, together with the elemental transfers that are suggested (though not entirely proven), implies that such typical basins of normal geothermal gradient also host reactions that may enter significantly into the global CO_2 cycle.

Weathering and reverse weathering provide a useful conceptual framework for organizing the many synchronous dissolution, cementation, and replacement reactions in the late diagenesis of sandstones and shales. In Table 1 the dissolution reactions that affect weathering products such as smectite and early diagenetic kaolinite are not construed as acid-consuming reactions, because an acid imbalance of immense (and implausible) proportions would be created (Land *et al.*, 1987). Dissolution reactions that affect metastable igneous detritus (feldspars), however, are plausibly construed as weathering hydrolysis. Most precipitation reactions can be construed as acid-releasing reverse-weathering reactions.

Interestingly, quartz dissolution and precipitation reactions do not enter directly into the mass balance of reverse weathering, because protons are neither consumed nor released. Nonetheless, silica activity may play an active role in driving reactions toward reverse weathering (Abercrombie *et al.*, 1994). In a sense, quartz "competes" with metal-bearing silicates in the acid balance of late diagenesis. The balance of dissolved silicon that goes into quartz precipitation (acid-neutral) versus into precipitation of other metal-bearing silicates (acid-releasing) is potentially a major control on the acid balance. A grossly simplified reaction for organic matter is also included in Table 1. Although this reaction is locally important (in some form) as evidenced by the mix of organic and inorganic carbon in calcites, dissolved C, and CO_2, it is not volumetrically sufficient to control the acid balance of late diagenesis (Lundegard and Land, 1986, 1989).

Reactions involving carbonates present a confusing situation. Precipitation and dissolution reactions involving carbonates can both be written to evolve CO_2. The retrograde solubility of calcite and the role of P_{CO_2} in controlling carbonate solubility are further complications. Across the depth range of the observed P_{CO_2} trend, CO_2 released from both carbonate precipitation and dissolution reactions may contribute in proportions that vary with temperature.

The balance among the reactions in Table 1 evolves during late diagenesis as clay mineral reactions proceed, feldspars are consumed, and cements are precipitated. If local dissolution and precipitation reactions are out of balance at any point in this evolution, then transport of acid (and possibly other components) at scales larger than thin sections will be required. However, quantifying the acid balance among all the reactions in Table 1 requires information on shale composition and petrography that is not currently available.

Localized action of aggressive metal-bearing brines faults (e.g., Taylor and Soule, 1993) and around salt domes (Kyle and Saunders, 1997) attests to the likely presence of fluids with the imprint of acid-generating mineralogical reactions (Hemley and Jones, 1964) in deep sedimentary basins at, and perhaps in large measure, just beyond the range of our deepest sampling.

Certainly, CO_2 evolved during late diagenesis must ultimately return to the atmosphere/ocean. It also seems clear that transport of major components such as silicon and potassium between sandstones and shales at a scale of a few meters is required and can perhaps be accomplished by diffusion (Thyne *et al.*, 2001). New data, especially for shales, must be obtained before simultaneous quantitative balances can be proposed for the reactions in Table 1. The speciation of aluminum in pore fluids, the initial and final quantities of the reactants and products in both sandstones and shales, and the precise volumes of sandstones and shales in the sequences in question are key data needed to ascertain the scale of mobility for the major elements in late diagenesis. Our ability to answer basic questions about the rock cycle falls short, in large part, for lack of information about the major mineral components of shale, the most common type of sedimentary rock.

ACKNOWLEDGMENTS

My pursuit of diagenetic studies has greatly benefited from the remarkable good fortune of having an extraordinary trio of experts in this subject just down the hall. Robert Folk taught me petrography; Lynton Land advocated that microscopical observations are more powerful with something numerical attached (and vice versa!), and Earle McBride, enthusiastic for petrography and chemistry alike, has endlessly been creative in finding and sharing opportunities to apply these approaches. Of course, all errors and any misconceptions that turn up in this paper are my own fault. Sally Sutton is acknowledged for braving tick fever to collect the sample shown in Figure 1. Rob Reed kindly provided the image that appears in Figure 12. Jim Boles, Knut Bjørlykke, Stuart Haszeldine, Ian Hutcheon, and Sadoon Morad all pitched in to help me cut a navigable path through the jungle of diagenesis literature. Fred Mackenzie directed me towards some of the early literature on reverse weathering and also made numerous editorial suggestions. I thank Rob Lander, Leo Lynch, and Earle McBride for comments and discussions during the construction of this paper.

REFERENCES

Aagaard P., Egeberg P. K., Saigal G. C., Morad S., and Bjørlykke K. (1990) Diagenetic albitization of detrital K-feldspars in Jurassic, Lower Cretaceous, and Tertiary clastic reservoir rocks from offshore Norway: II. Formation water chemistry and kinetic considerations. *J. Sedim. Petrol.* **60**, 575–581.

Aase N. E., Bjørkum P. A., and Nadeau P. H. (1996) The effect of grain-coating microquartz on preservation of reservoir quality. *Am. Assoc. Petrol. Geologists* **80**, 1654–1673.

Abercrombie H. J., Hutcheon I. E., Bloch J. D., and De Caritat P. (1994) Silica activity and the smectite–illite reaction. *Geology* **22**, 539–542.

Ague J. J. (1991) Evidence for major mass transfer and volume strain during regional metamorphism of pelites. *Geology* **19**, 855–858.

Ague J. J. (1997) Crustal mass transfer and index mineral growth in Barrow's garnet zone, northeast Scotland. *Geology* **25**, 73–76.

Al-Aasm I. S., Muir I., and Morad S. (1993) Diagenetic conditions of fibrous calcite vein formation in black shales: petrographic, chemical and isotopic evidence. *Bull. Can. Petrol. Geol.* **41**, 46–57.

Aldahan A. A. and Morad S. (1986) Authigenic sphene in sandstones from the Brottum Formation (Norway) and the Dala Sandstone (Sweden). *Neues Jahrbuch fuer Mineralogie. Monatschefte* **1986**(3), 135–144.

Atkins J. E. and McBride E. F. (1992) Porosity and packing of Holocene river, dune, and beach sands. *Am. Assoc. Petrol. Geologists Bull.* **76**, 339–355.

Awwiller D. N. (1993) Illite/smectite formation and potassium transfer during burial diagenesis of mudrocks: a study from the Texas Gulf Coast Paleocene-Eocene. *J. Sedim. Petrol.* **63**, 501–512.

Awwiller D. N. (1994) Geochronology and mass transfer in Gulf Coast mudrocks (south-central Texas, USA): Rb–Sr, Sm–Nd, and REE systematics. *Chem. Geol.* **116**, 61–84.

Beach A. (1977) Vein arrays, hydraulic fractures, and pressure solution structures in a deformed flysch sequence, southwest England. *Tectonophysics* **40**, 201–225.

Beckner J. R. and Mozley P. S. (1998) Origin and spatial distribution of early vadose and phreatic calcite cements in the Zia Formation, Albuquerque Basin, New Mexico, USA. In *Carbonate Cementation in Sandstones. Distribution Patterns and Geochemical Evolution* (ed. S. Morad). International Association of Sedimentologists, Oxford, vol. 26, pp. 27–52.

Beratan K. K. (1999) Miocene potassium metasomatism, Whipple Mountains, southeastern California: a datable tracer of extension-related fluid transport. *Geology* **27**, 259–262.

Berger G., Velde B., and Aigouy T. (1999) Potassium sources and illitization in Texas Gulf Coast shale diagenesis. *J. Sedim. Res.* **69**, 151–157.

Berner R. A. (1981) Kinetics of weathering and diagenesis. In *Kinetics of Geochemical Processes* (eds. A. C. Lasaga and R. J. Kirkpatrick). Mineralogical Society of America, Princeton, NJ, vol. 8, pp. 111–134.

Berner R. A. (1991) A model for atmospheric CO_2 over Phanerozoic time. *Am. J. Sci.* **291**, 339–376.

Berner R. A. and Holdren G. R., Jr. (1977) Mechanism of feldspar weathering: some observational evidence. *Geology* **5**, 369–372.

Berner R. A. and Holdren G. R., Jr. (1979) Mechanism of feldspar weathering: II. Observations of feldspars from soils. *Geochim. Cosmochim. Acta* **43**, 1173–1186.

Berner R. A. and Schott O. (1982) Mechanism of pyroxene and amphibole weathering: II. Observations of soil grains. *Am. J. Sci.* **282**, 1214–1231.

Berner R. A., Lasaga A. C., and Garrels R. M. (1983) The carbonate-silicate geochemical cycle and its effect on atmospheric carbon dioxide over the past 100 million years. *Am. J. Sci.* **283**, 641–683.

Bethke C. M. and Marshak S. (1990) Brine migrations across North America—the plate tectonics of groundwater. *Ann. Rev. Earth Planet. Sci.* **18**, 287–315.

Bjørkum P. A. (1996) How important is pressure in causing dissolution of quartz in sandstones? *J. Sedim. Res.* **66**, 147–154.

Bjørkum P. A. and Walderhaug O. (1990) Geometrical arrangement of calcite cementation within shallow marine sandstones. *Earth Sci. Rev.* **29**, 145–161.

Bjørkum P. A., Walderhaug O., and Aase N. E. (1992) The concretions of the Berreraig sandstone formation: geometry and geochemistry. *Sedimentology* **39**, 517–520.

Bjørlykke K. (1979) Cementation of sandstones: discussion. *J. Sedim. Petrol.* **49**, 1358–1359.

Bjørlykke K. (1989) Sandstone diagenesis and porosity modification during basin evolution. *Geol. Rundsch.* **78**, 243–268.

Bjørlykke K. (1993) Fluid flow in sedimentary basins. *Sedim. Geol.* **86**, 137–158.

Bjørlykke K. (1999) Principal aspects of compaction and fluid flow in mudstones. In *Muds and Mudstone: Physical and Fluid-flow Properties* (eds. A. C. Aplin, A. J. Fleet, and J. S. Macquaker). Geological Society of London, vol. 158, pp. 73–78.

Bjørlykke K. and Aagaard P. (1992) Clay minerals in North Sea Sandstones. In *Origin, Diagenesis, and Petrophysics of Clay Minerals in Sandstones* (eds. D. W. Houseknecht and E. D. Pittman). Society for Sedimentary Geology (SEPM), Tulsa, OK, vol. 47, pp. 65–80.

Bjørlykke K. and Egeberg P. K. (1993) Quartz cementation in sedimentary basins. *Am. Assoc. Petrol. Geologists* **77**, 1538–1548.

Bjørlykke K. and Lynch F. L. (1997) Mineral/water interaction, fluid flow, and Frio sandstone diagenesis: evidence from the rocks—Discussion and Reply. *Am. Assoc. Petrol. Geologists Bull.* **81**, 1534–1537.

Blatt H. (1979) Diagenetic processes in sandstones. In *Aspects of Diagenesis* (eds. P. Scholle and P. A. Schluger). Society of Economic Paleontologists and Mineralogists, Tulsa, OK, vol. 26, pp. 141–157.

Blatt H. (1987) Oxygen isotopes and the origin of quartz. *J. Sedim. Petrol.* **57**, 373–377.

Blatt H. and Schultz D. J. (1976) Size distribution of quartz in mudrocks. *Sedimentology* **23**, 857–866.

Blatt H. and Totten M. W. (1981) Detrital quartz as an indicator of distance from shore in marine mudrocks. *J. Sedim. Petrol.* **51**, 1259–1266.

Blatt H., Middleton G., and Murray R. (1980) *Origin of Sedimentary Rocks*. Prentice-Hall, Englewood Cliffs, NJ.

Bloch J., Hutcheon I. E., and de Caritat P. (1998) Tertiary volcanic rocks and the potassium content of Gulf Coast shales—the smoking gun. *Geology* **26**, 527–530.

Bloch S. (1994) Secondary porosity in sandstones: significance, origin, relationship to subaerial unconformities, and effect on predrill reservoir quality prediction. In *Reservoir Quality Assessment and Prediction in Clastic Rocks* (ed. M. D. Wilson). Society for Sedimentary Geology (SEPM), vol. 30, pp. 137–159.

Bloch S., Lander R. H., and Bonnell L. (2002) Anomalously high porosity and permeability in deeply buried sandstone reservoirs: origin and predictability. *Am. Assoc. Petrol. Geologists Bull.* **86**, 301–328.

Boggs S., Jr., Kwon Y.-I., Goles G. G., Rusk B. G., Krinsley D., and Seyedolali A. (2002) Is quartz cathodoluminescence color a reliable provenance tool? A quantitative examination. *J. Sedim. Res.* **72**, 408–415.

Boles J. R. (1978) Active ankerite cementation in the subsurface Eocene of southwest Texas. *Contrib. Mineral. Petrol.* **68**, 13–32.

Boles J. R. (1982) Active albitization of plagioclase, Gulf Coast Tertiary. *Am. J. Sci.* **282**, 165–180.

Boles J. R. (1998) Carbonate cementation in Tertiary sandstones, San Joaquin Basin, California. In *Carbonate Cementation in Sandstones, Distribution Patterns and Geochemical Evolution* (ed. S. Morad). International Association of Sedimentologists, Oxford, vol. 26, pp. 261–284.

Boles J. R. and Franks S. G. (1979a) Cementation of sandstones: reply. *J. Sedim. Petrol.* **49**, 1362.

Boles J. R. and Franks S. G. (1979b) Clay diagenesis in the Wilcox sandstones of southwest Texas: implications of smectite diagenesis on sandstone cementation. *J. Sedim. Petrol.* **49**, 55–70.

Boles J. R. and Hickey J. J. (1996) Inhibition of quartz cementation by the presence of hydrocarbons, Point McIntyre Field, North Slope, Alaska. In *Geofluids II '97. Contributions to the second international conference on fluid evolution, migration and interaction in sedimentary basins and orogenic belts* (eds. J. P. Hendry, P. F. Carey, J. Parnell, A. H. Ruffell and R. H. Worden). The Queen's University, Belfast, pp. 272–274.

Bouch J. E., Hole M. J., Trewin N. H., and Morton A. C. (1995) Low-temperature aqueous mobility of the REE during sandstone diagenesis. *J. Geol. Soc. London* **152**, 895–898.

Brantley S. L., White A. F., and Hodson M. E. (1999) Surface area of primary silicate minerals. In *Growth, Dissolution, and Pattern Formation in Geosystems* (eds. B. Jamtveit and P. Meakin). Kluwer Academic, Dordrecht, pp. 291–326.

Breit G. N., Goldhaber M. B., Shawe D. R., and Simmons E. C. (1990) Authigenic barite as an indicator of fluid movement through sandstones within the Colorado Plateau. *J. Sedim. Petrol.* **60**, 884–896.

Burley S. D., Mullis J., and Matter A. (1989) Timing diagenesis in the Tartan reservoir (UK North Sea): constraints from combined cathodoluminescence and fluid inclusion studies. *Mar. Petrol. Geol.* **6**, 98–120.

Calvert C. S. and Klimentidis R. E. (1986) The generation of silicate cements during burial diagenesis of shales. *Clay Minerals Society 23rd Annual Meeting*, 27.

Carlson W. D. (2002) Scales of equilibrium and rates of equilibrium during metamorphism. *Am. Mineral.* **87**, 185–204.

Carmichael D. M. (1987) Induced stress and secondary mass transfer: thermodynamic basis for the tendency toward constant-volume constraint in diffusion metasomatism. In *Chemical Transport in Metasomatic Processes* (ed. H. C. Helgeson). Reidel, Dordrecht, pp. 239–264.

Cathelineau M. (1987) U–Th–REE mobility during albitization and quartz dissolution in granitoids: evidence from south-east French Massif Central. *Bull. Mineral.* **110**, 249–259.

Cavazza W. and Gandolfi G. (1992) Diagenetic processes along a basin-wide marker bed as a function of burial depth. *J. Sedim. Petrol.* **62**, 261–272.

Chang H. K., Mackenzie F. T. and Schoonmaker J. (1986) Comparisons between the diagenesis of dioctahedral and trioctahedral smectite, Brazilian offshore basins. *Clays Clay Min.* **34**, 407–423.

Chuhan F. A., Bjørlykke K., and Lowrey C. (2000a) The role of provenance in illitization of deeply buried reservoir sandstones form Haltenbanken and north Viking Graben, offshore Norway. *Mar. Petrol. Geol.* **17**, 673–689.

Chuhan F. A., Kjeldstad A., Bjørlykke K., and Høeg K. (2000b) Porosity loss in sand by grain crushing—experimental evidence and relevance to reservoir quality. *Mar. Petrol. Geol.* **19**, 39–53.

Clauer N., Fraper S. K., and Fritz B. (1989) Calcite veins of the Stripa Granite (Sweden) as records of the origin of the groundwaters and their interactions with the granite body. *Geochim. Cosmochim. Acta* **53**, 1777–1781.

Cookenboo H. O. and Bustin R. M. (1999) Pore water evolution in sandstones of the Groundhog Coalfield, northern Bowser Basin, British Columbia. *Sedim. Geol.* **123**, 129–146.

Coombs D. S., Ellis A. J., Fyfe W. S., and Taylor A. M. (1959) The zeolite facies, with comments on the interpretation of hydrothermal synthesis. *Geochim. Cosmochim. Acta* **17**, 53–107.

Cosgrove J. W. (2001) Hydraulic fracturing during the formation and deformation of a basin: a factor in the dewatering of low-permeability sediments. *Am. Assoc. Petrol. Geologists Bull.* **85**, 737–748.

Cuney M. and Mathieu R. (2000) Extreme light rare earth element mobilization by diagenetic fluids in the geological environment of the Oklo natural reactor zones, Franceville Basin, Gabon. *Geology* **28**, 743–746.

de Boer R. B. (1977) On the thermodynamics of pressure solution—interaction between chemical and mechanical forces. *Geochim. Cosmochim. Acta* **41**, 249–256.

de Caritat P., Bloch J., Hutcheon I., Longstaffe F. J., and Abercrombie H. (1997) Comparison of the mineralogical and chemical composition of 2 shales from the Western Canada sedimentary basin and the United States Gulf Coast. *Clays Clay Min.* **45**, 327–332.

Deming D. (1994) Fluid flow and heat transport in the upper continental crust. In *Geofluids: Origin, Migration, and Evolution of Fluids in Sedimentary Basins* (ed. J. Parnell). Geological Society of London, Special Publication, London, vol. 78, pp. 27–42.

den Brok S. W. J. (1998) Effect of microcracking on pressure-solution strain rate: the Gratz grain-boundary model. *Geology* **26**, 915–918.

De Ros L. F. (1998) Heterogeneous generation and evolution of diagenetic quartz arenites in the Silurian–Devonian Furnas Formation of the Paraná Basin, southern Brazil. *Sediment. Geol.* **116**, 99–128.

Dewers T. and Ortoleva P. (1990) Force of crystallization during the growth of siliceous concretions. *Geology* **18**, 204–207.

Dickinson W. W. and Milliken K. L. (1995a) Diagenetic role of brittle deformation in compaction and pressure solution, Etjo Sandstone, Namibia. *J. Geol.* **103**, 339–347.

Dickinson W. W. and Milliken K. L. (1995b) The diagenetic role of brittle deformation in compaction and pressure solution, Etjo Sandstone, Namibia. *J. Geol.* **103**, 339–347.

Dunoyer de Segonzac G. (1970) The transformation of clay minerals during diagenesis and low-grade metamorphism: a review. *Sedimentology* **15**, 281–346.

Dutton S. P. and Diggs T. N. (1990) History of quartz cementation in the Lower Cretaceous Travis Peak Formation, East Texas. *J. Sedim. Petrol.* **60**, 191–202.

Dutton S. P. and Diggs T. N. (1992) Evolution of porosity and permeability in the Lower Cretaceous Travis Peak Formation, East Texas Basin. *Am. Assoc. Petrol. Geologists Bull.* **76**, 252–269.

Ehrenberg S. N. and Nadeau P. H. (1989) Formation of diagenetic illite in sandstones of the Garn Formation, Haltenbanken area, mid-Norweigien Continental Self. *Clay Min.* **24**, 233–253.

Eichhubl P. and Boles J. R. (2000) Focused fluid flow along faults in the Monterey Formation, coastal California. *Geol. Soc. Am. Bull.* **112**, 1667–1679.

Ennis D. J., Dunbar N. W., Campbell A. R., and Chapin C. E. (2000) The effects of K-metasomatism on the mineralogy and geochemistry of silicic ignimbrites near Socorro, New Mexico. *Chem. Geol.* **167**, 285–312.

Eslinger E. and Glasmann J. R. (1993) Geothermometry and geochronology using clay minerals—an introduction. *Clays Clay Min.* **41**, 117–118.

Evans I. J. (1989) Geochemical fluxes during shale diagenesis, an example from the Ordovician of Morocco. *Water–Rock Interact.* **WRI-6**, 219–222.

Evans J. (1990) Quartz dissolution during shale diagenesis: implications for quartz cementation in sandstones. *Chem. Geol.* **84**, 239–240.

Farver J. R. and Yund R. A. (1996) Volume and grain boundary diffusion of calcium in natural and hot-pressed calcite aggregates. *Contribut. Mineral. Petrol.* **123**, 77–91.

Fedo C. M., Nesbitt H. W., and Young G. M. (1995) Unraveling the effects of potassium metasomatism in sedimentary rocks and paleosols, with implications for paleoweathering conditions and provenance. *Geology* **23**, 921–924.

Fedo C. M., Young G. M., Nesbitt H. W., and Hanchar J. M. (1997) Potassic and sodic metasomatism in the Southern Province of the Canadian Shield: evidence from the Paleoproterozoic Serpent Formation. *Precamb. Res.* **84**, 17–36.

Ferry J. M. (1994) A historical review of metamorphic fluid flow. *J. Geophys. Res.* **99**(B8), 15487–15498.

Ferry J. M. (2000) Patterns of mineral occurrence in metamorphic rocks. *Am. Mineral.* **85**, 1573–1588.

Ferry J. M. and Dipple G. M. (1991) Fluid flow, mineral reactions, and metasomatism. *Geology* **19**, 211–214.

Fisher Q. J., Knipe R. J., and Worden R. H. (2000) Microstructures of deformed and non-deformed sandstones from the North Sea: implications for the origins of quartz cement in sandstones. In *Quartz Cementation in Sandstones* (eds. R. H. Worden and S. Morad). International Association of Sedimentologists, Oxford, vol. 29, pp. 129–146.

Fishman N. S. (1997) Basin-wide fluid movement in a Cambrian paleoaquifer: evidence from the Mt. Simon sandstone, Illinois and Indiana. In *Basin-wide Diagenetic Patterns: Integrated Petrologic, Geochemical, and Hydrologic Considerations* (eds. I. Montañez, J. M. Gregg, and K. L. Shelton). Society for Sedimentary Geology (SEPM), Tulsa, OK, vol. 57, pp. 221–234.

Fortier S. M. and Giletti B. J. (1989) An empirical model for predicting diffusion coefficients in silicate minerals. *Science* **245**, 1481–1484.

Fothergill C. A. (1955) The cementation of oil reservoir sands and its origin. *Proc. World Petrol. Cong.* **59**, 301–314.

Franks S. G. and Forester R. W. (1984) Relationships among secondary porosity, pore-fluid chemistry, and carbon dioxide, Texas Gulf Coast. In *Clastic Diagenesis* (eds. D. A. McDonald and R. Surdam). American Association of Petroleum Geologists, vol. 37, pp. 63–80.

Freed R. L. (1981) Shale mineralogy and burial diagenesis of Frio and Vickburg formations in two geopressured wells, McAllen Ranch area, Hidalgo County, Texas. *Gulf Coast Assoc. Geol. Soc. Transact.* **31**, 289–293.

Freed R. L. (1982) Clay mineralogy and depositional history of the Frio Formation in two geopressured wells, Brazoria County, Texas. *Gulf Coast Assoc. Geol. Soc. Transac.* **32**, 459–463.

Freer R. (1981) Diffusion in silicate minerals and glasses: a data digest and guide to the literature. *Contrib. Mineral. Petrol.* **76**, 440–454.

Füchtbauer H. (1967) Influence of different types of diagenesis on sandstone porosity. *7th World Petrol. Cong.*, 353–369.

Furlan S., Clauer N., Chauduri S., and Sommer F. (1996) K transfer during burial diagenesis in the Mahakam Delta basin (Kalimantan, Indonesia). *Clays Clay Min.* **44**, 157–199.

Garrels R. M. and MacKenzie F. T. (1971) *The Evolution of Sedimentary Rocks*. W. W. Norton, New York.

Garrels R. M. and MacKenzie F. T. (1974) Chemical history of the oceans deduced from post-depositional changes in sedimentary rocks. In *Studies in Paleo-Oceanography,* SEPM Special Publication 20 (ed. W. W. Hey). Society of Economic Paleontologists and Mineralogists, Tulsa, OK, vol. 20, pp. 193–204.

Genuise J. J. (1991) Petrography and geochemistry of authigenic chlorite from Cretaceous and Oligocene sandstones of the Texas/Louisiana Gulf Coast. University of Texas at Austin (unpublished).

Giles M. R. and De Boer R. B. (1990) Origin and significance of redistributional secondary porosity. *Mar. Petrol. Geol.* **7**, 378–397.

Giles M. R., Indrelid S. L., and James D. M. D. (1998) Compaction, the great unknown in basin modelling. In *Basin Modelling: Practice and Progress* (eds. S. J. Duppenbecker and J. E. Iliffe). Geological Society of London, London, vol. 14, pp. 15–43.

Girard J.-P. (1998) Carbonate cementation in the Middle Jurassic Oseberg reservoir sandstone, Oseberg field, Norway: a case of deep burial-high temperature poikilotopic calcite. In *Carbonate Cementation in Sandstones. Distribution Patterns and Geochemical Evolution* (ed. S. Morad). International Association of Sedimentologists, Oxford, vol. 26, pp. 285–308.

Glasmann J. R. (1992) The fate of feldspar in Brent Group reservoir, North Sea: a regional synthesis of diagenesis in shallow, intermediate, and deep burial environments. In *Geology of the Brent Group* (eds. A. C. Morton, R. S. Haszeldine, M. R. Giles, and S. Brown). Geological Society of London, London, vol. 61, pp. 329–350.

Gluyas J. (1997) Element mobility during diagenesis: sulphate cementation of Rotliegend sandstones, Southern North Sea. *Mar. Petrol. Geol.* **14**, 1001.

Gluyas J., Garland C., Oxtoby N. H., and Hogg A. J. C. (2000) Quartz cement: the Miller's tale. In *Quartz Cementation in Sandstones* (eds. R. H. Worden and S. Morad). International Association of Sedimentologists, Oxford, vol. 29, pp. 119–218.

Gluyas J. G. (1984) Early carbonate diagenesis within Phanerozoic shales and sandstones of the NW European shelf. *Clay Min.* **19**, 309–321.

Gluyas J. G. and Coleman M. L. (1992) Material flux during sediment diagenesis: a contradiction to fluid flow prediction. *Nature* **356**, 22–23.

Gold P. B. (1987) Textures and geochemistry of authigenic albite from Miocene sediments, Louisiana Gulf Coast. *J. Sedim. Petrol.* **57**, 353–362.

Graton L. C. and Fraser H. J. (1935) Systematic packing of spheres—with particular relation to porosity and permeability. *J. Geol.* **43**, 709–785.

Gratz A. J. (1991) Solution-transfer compaction in quartzites: progress toward a rate law. *Geology* **19**, 901–904.

Grigsby J. D. (2001) Origin and growth mechanism of authigenic chlorite in sandstones of the lower Vicksburg Formation, South Texas. *J. Sedim. Res.* **71**, 27–36.

Hanor J. S. (1994) Origin of saline fluids in sedimentary basins. In *Geofluids: Origin, Migration, and Evolution of Fluids in Sedimentary Basins* (ed. J. Parnell). Geological Society of London, London, vol. 78, pp. 151–174.

Hardy L. S. (1989) Hydrothermal potassium feldspar at the Haile gold mine, South Carolina. *Econ. Geol.* **84**, 2307–2310.

Harris N. B. (1990) Diagenetic quartz arenite and destruction of secondary porosity: an example from the Middle Jurassic

Brent Sandstone of Northwest Europe. *Geology* **17**, 361–364.

Haszeldine R. S., Samson I. M., and Cornford C. (1984) Quartz diagenesis and convective fluid movement: Beatrice oilfield, UK North Sea. *Clay Min.* **19**, 391–402.

Haszeldine R. S., Brint J. F., Fallick A. E., Hamilton P. J., and Brown S. (1992) Open and restricted hydrologies in Brent Group diagenesis: North Sea. In *Geology of the Brent Group* (eds. A. C. Morton, R. S. Haszeldine, M. R. Giles, and S. Brown). Geological Society of London, London, vol. 61, pp. 401–419.

Hayes M. J. and Boles J. R. (1992) Volumetric relations between dissolved plagioclase and kaolinite in sandstones: implications for aluminum mass transfer in the San Joaquin Basin, California. In *Origin, Diagenesis, and Petrophysics of Clay Minerals in Sandstones* (eds. D. W. Houseknecht and E. D. Pittman). Society for Sedimentary Geology (SEPM), Tulsa, OK, vol. 47, pp. 111–124.

Heald M. T. (1955) Stylolites in sandstones. *J. Geol.* **63**, 101–114.

Heald M. T. and Larese R. E. (1974) Influence of coatings on quartz cementation. *J. Sedim. Petrol.* **44**, 1269–1274.

Helmold K. P. and van de Kamp P. C. (1984) Diagenetic mineralogy and controls on albitization and laumontite formation in Paleogene arkoses, Santa Ynez Mountains, California. In *Clastic Diagenesis* (eds. D. A. McDonald and R. C. Surdam). American Association of Petroleum Geologists, Tulsa, OK, vol. 37, pp. 239–276.

Hemley J. J. and Jones W. R. (1964) Chemical aspects of hydrothermal alteration with emphasis on hydrogen metasomatism. *Econ. Geol.* **59**, 538–569.

Hervig R. L., Williams L. B., Kirkland I. K., and Longstaffe F. J. (1995) Oxygen isotope microanalysis of diagenetic quartz: possible low temperature occlusion of pores. *Geochim. Cosmochim. Acta* **59**, 2537–2543.

Hesse R. and Abid I. A. (1998) Carbonate cementation—the key to reservoir properties of four sandstone levels (Cretaceous) in the Hibernia oil field, Jeanne d'Arc Basin, Newfoundland, Canada. In *Carbonate Cementation in Sandstones: Distribution Patterns and Geochemical Evolution* (ed. S. Morad). International Association of Sedimentologists, Oxford, vol. 26, pp. 363–394.

Holness M. B. and Watt G. R. (2001) Quartz recrystallization and fluid flow during contact metamorphism: a cathodoluminescence study. *Geofluids* **1**, 215–228.

Houseknecht D. W. (1991) Use of cathodoluminescence petrography for understanding compaction, quartz cementation, and porosity in sandstones. In *Luminescence Microscopy and Spectroscopy: Quantitative and Qualitative Applications* (eds. C. E. Barker and O. C. Kopp). Society for Sedimentary Geology (SEPM), Tulsa, OK, Short Course vol. 25, pp. 59–75.

Hower J., Eslinger E. V., Hower M. E., and Perry E. A. (1976) Mechanism of burial metamorphism of argillaceous sediment: 1. Mineralogical and chemical evidence. *Geolog. Soc. Am. Bull.* **87**, 725–737.

Huang W.-L., Longo T. M., and Pevear D. (1993) An experimentally derived kinetic model for smectite to illite conversion and its use as a geothermeter. *Clays Clay Min.* **41**, 162–177.

Hudson J. D., Coleman M. L., Barreiro B. A., and Hollingworth N. T. J. (2001) Septarian concretions from the Oxford Clay (Jurassic, England, UK): involvement of original marine and multiple external pore fluids. *Sedimentology* **48**, 507–531.

Huggett J. M. (1984) Controls on mineral authigenesis in coal measures sandstones of the East Midlands, UK. *Clay Min.* **19**, 343–357.

Hunt J. M. (1996) *Petroleum Geochemistry and Geology*. Freeman, San Francisco.

Hutcheon I. and Abercrombie H. (1990) Carbon dioxide in clastic rocks and silicate hydrolysis. *Geology* **18**, 541–544.

Hutcheon I. and Abercrombie H J. (1989) The role of silicate hydrolysis in the origin of CO_2 in sedimentary basins. *Water-Rock Interact.* **WRI-6**, 321–324.

Hutcheon I., Bloch J., and Modus S. (2000) Potassium enrichment in shale: fluid transport or provenance? *J. Geochem. Explor.* **69–70**, 17–22.

Hutcheon I., Shevalier M., and Abercrombie H. J. (1993) pH buffering by metastable mineral-fluid equilibria and the evolution of carbon dioxide fugacity during burial diagenesis. *Geochim. Cosmochim. Acta* **57**, 1017–1027.

Irwin H. (1980) Early diagenetic carbonate precipitation and pore fluid migration in the Kimmeridge Clay of Dorset, England. *Sedimentology* **27**, 577–591.

James A. T. (1990) Correlation of reservoired gases using the carbon isotopic compositions of wet gas components. *Am. Assoc. Petrol. Geologists Bull.* **74**, 1441–1458.

Jennings S. and Thompson G. R. (1986) Diagenesis of Plio-Pleistocene sediments of the Colorado River Delta, southern California. *J. Sedim. Petrol.* **56**, 89–98.

Jowette E. C. (1987) Formation of sulfide-calcite veinlets in the Kuperschiefer Cu–Ag deposits in Poland by natural hydrofracturing during basin subsidence. *J. Geol.* **95**, 513–526.

Kastner M. (1974) The contribution of authigenic feldspars to the geochemical balance of alkali metals. *Geochim. Cosmochim. Acta* **38**, 650–653.

Kastner M. and Siever R. (1979) Low temperature feldspars in sedimentary rocks. *Am. J. Sci.* **279**, 435–479.

Kerrick D. M., McKibben M. A., Seward T. M., and Caldeira K. (1995) Convective hydrothermal CO_2 emission from high heat flow regions. *Chem. Geol.* **121**, 17–27.

Khanna M., Saigal G. C., and Bjørlykke K. (1997) Kaolinization of Upper Triassic–Lower Jurassic sandstones of the Tampen Spur Area, North Sea: implications for early diagenesis and fluid flow. In *Basin-wide Diagenetic Patterns: Integrated Petrologic, Geochemical, and Hydrologic Considerations* (eds. I. Montañez, J. M. Gregg, and K. L. Shelton). Society for Sedimentary Geology (SEPM), Tulsa, OK, vol. 57, pp. 253–268.

Kirkpatrick R. J. (1981) Kinetics of crystallization in igneous rocks. In *Kinetics of Geochemical Processes* (eds. A. C. Lasaga and R. J. Kirkpatrick). Mineralogical Society of America, Chelsea, MI, vol. 8, pp. 321–398.

Kisch H. J. (1983) Mineralogy and petrology of burial diagenesis (burial metamorphism) and incipient metamorphism in clastic rocks. In *Diagenesis in Sediments and Sedimentary Rocks* (eds. G. A. C. Larsen and G. V. Chilinger). Elsevier, Amsterdam, vol. 2, pp. 289–493.

Kisch H. J. (1991) Development of slaty cleavage and degree of very low-grade metamorphism. *J. Metamorph. Geol.* **6**, 735–750.

Knoke R. (1966) Untersuchen zur diagenese an Kalkkonkretionen und umgebenden Tonschiefern. *Beitrage zur Mineralogie und Petrographie* **12**(2), 139–167.

Kominou A. and Yardley B. W. D. (1997) Fluid–rock interaction in the Rhine Graben: a thermodynamic model of the hydrothermal alteration observed in deep drilling. *Geochim. Cosmochim. Acta* **61**, 515–531.

Kraishan G. M., Rezaee M. R., and Worden R. H. (2000) Significance of trace element composition of quartz cement as a key to reveal the origin of silica in sandstones: an example from the Cretaceous of the Barrow Sub-basin, Western Australia. In *Quartz Cementation in Sandstones* (eds. R. H. Worden and S. Morad). International Association of Sedimentologists, Oxford, vol. 29, pp. 317–331.

Krauskopf K. B. (1979) *Introduction to Geochemistry*. McGraw-Hill, New York.

Krzysztof M. W., Goldstein R. H., and Walton A. W. (1997) Regional and local controls on diagenesis driven by basin-wide flow system: Pennsylvanian sandstones and limestones, Cherokee Basin, southeastern Kansas. In *Basin-wide Diagenetic Patterns: Integrated Petrologic, Geochemical, and Hydrologic Considerations* (eds. I. Montañez, J. M. Gregg, and K. L. Shelton). Society for Sedimentary Geology (SEPM), Tulsa, OK, vol. 57, pp. 235–252.

Kyle J. R. and Saunders J. A. (1997) Metallic deposits of the Gulf Coast Basin: diverse mineralization styles in a young sedimentary basin. In *Carbonate-hosted Lead–Zinc Deposits* (ed. D. F. Sangster). Society of Economic Geologists, Littleton, CO, vol. 4, pp. 218–229.

Land L. S. (1984) Frio sandstone diagenesis, Texas Gulf coast: a regional isotopic study. In *Clastic Diagenesis* (eds. D. A. MacDonald and R. C. Surdam). American Association of Petroleum Geologists, Tulsa, OK, vol. 37, pp. 47–62.

Land L. S. (1997) Mass-transfer during burial diagenesis in the Gulf of Mexico sedimentary basin. In *Basin-wide Diagenetic Patterns,* SEPM Special Publication 57 (eds. I. Montañez, J. M. Gregg, and K. L. Shelton). Society for Sedimentary Geology (SEPM), Tulsa, OK, pp. 29–39.

Land L. S. and Dutton S. P. (1978) Cementation of a Pennsylvanian deltaic sandstone: isotopic data. *J. Sedim. Petrol.* **48**, 1167–1176.

Land L. S. and Dutton S. P. (1979) Cementation of sandstones: Reply. *J. Sedim. Petrol.* **49**, 1359–1361.

Land L. S. and Fisher R. S. (1987) Wilcox sandstone diagenesis, Texas Gulf Coast: a regional isotopic comparison with the Frio Formation. In *Diagenesis of Sedimentary Sequences* (ed. J. D. Marshal). Geological Society of London, London, vol. 36, pp. 219–235.

Land L. S. and Macpherson G. L. (1992a) Geothermometry from brine analyses: lessons from the Gulf Coast, USA. *Appl. Geochem.* **7**, 333–340.

Land L. S. and Macpherson G. L. (1992b) Origin of saline formation waters, Cenozoic section, Gulf of Mexico sedimentary basin. *Am. Assoc. Petrol. Geologists Bull.* **76**, 1344–1362.

Land L. S. and Milliken K. L. (1981) Feldspar diagenesis in the Frio Formation, Brazoria County, Texas Gulf Coast. *Geology* **9**, 314–318.

Land L. S. and Milliken K. L. (2000) Regional loss of SiO_2 and $CaCO_3$ and gain of K_2O during burial diagenesis of Gulf Coast mudrocks, USA. In *Quartz Cementation in Sandstones* (eds. R. H. Worden and S. Morad). International Association of Sedimentologists, Oxford, vol. 29, pp. 183–198.

Land L. S., Milliken K. L., and McBride E. F. (1987) Diagenetic evolution of Cenozoic sandstones, Gulf of Mexico sedimentary basin. *Sedim. Geol.* **50**, 195–225.

Land L. S., Mack L. E., Milliken K. L., and Lynch F. L. (1997) Burial diagenesis of argillaceous sediment, south Texas Gulf of Mexico sedimentary basin: a reexamination. *Geol. Soc. Am. Bull.* **109**, 2–15.

Lander R. H. and Walderhaug O. (1999) Predicting porosity through simulating sandstone compaction and quartz cementation. *Am. Assoc. Petrol. Geologists Bull.* **83**, 433–449.

Lanson B. and Champion D. (1991) The I/S to illite reaction in the late stage diagenesis. *Am. J. Sci.* **41**, 280–287.

Laubach S. E. (1997) A method to detect natural fracture strike in sandstone. *Am. Assoc. Petrol. Geologists Bull.* **81**, 604–623.

Laubach S. E. and Milliken K. L. (1996a) New fracture characterization methods for siliciclastic rocks. In *Rock Mechanics Tools and Techniques, Proceedings of the 2nd North American Rock Mechanics Symposium* (eds. M. Aubertin and others). Balkema, Rotterdam, 2050pp.

Laubach S. E. and Milliken K. L. (1996b) Using CL observations to evaluate fractures in siliciclastic petroleum reservoirs. *International Conference on Cathodoluminescence and Related Techniques in Geosciences and Geomaterials*, Nancy, France, pp. 83–84.

Lee J. I. and Lee Y. I. (1998) Feldspar albitization in Cretaceous non-marine mudrocks, Gyeongsang Basin, Korea. *Sedimentology* **45**, 745–754.

Lev S. M., McLennan S. M., Meyers W. J., and Hanson G. N. (1998) A petrographic approach for evaluating trace-element mobility in a black shale. *J. Sedim. Res.* **68**, 970–980.

Lev S. M., McLennan S. M., and Hanson G. N. (1999) Mineralogic controls on REE mobility during black-shale diagenesis. *J. Sedim. Res.* **69**, 1071–1082.

Lindgreen H. (1994) Ammonium fixation during illite–smectite diagenesis in upper Jurassic shale, North Sea. *Clay Min.* **29**, 527–537.

Lindquist S. J. (1977) Secondary porosity development and subsequent reduction, overpressured Frio Formation sandstone (Oligocene), South Texas. *Gulf Coast Assoc. Geol. Soc. Transact.* **27**, 99–107.

Littke R., Baker D. R., and Leythaeuser D. (1988) Microscopic and sedimentologic evidence for the generation and migration of hydrocarbons in Toarcian sources rocks of different maturities. *Org. Geochem.* **13**, 549–559.

Lønøy A., Akelson J., and Rønning K. (1986) Diagenesis of a deeply buried sandstone reservoir, Hild Field, northern North Sea. *Clay Min.* **21**, 497–511.

Loucks R. G., Dodge M. M., and Galloway W. E. (1984) Regional controls on diagenesis and reservoir quality in Lower Tertiary sandstones. In *Clastic Diagenesis* (eds. D. A. McDonald and R. C. Surdam). American Association of Petroleum Geologists, Tulsa, OK, vol. 37, pp. 15–45.

Lundegard P. D. (1985) Carbon dioxide and organic acids: origins and role in burial diagenesis (Texas Gulf Coast Tertiary). Ph.D. thesis, University of Texas at Austin, Austin, Texas, 145pp.

Lundegard P. D. (1992) Sandstone porosity loss—a "big picture" view of the importance of compaction. *J. Sedim. Petrol.* **62**, 250–260.

Lundegard P. D. and Land L. S. (1986) Carbon dioxide and organic acids: their role in porosity enhancement and cementation, Paleogene of the Texas Gulf Coast. In *Roles of Organic Matter in Sediment Diagenesis* (ed. D. L. Gautier). Society of Economic Paleontologists and Mineralogists, Tulsa, OK, vol. 38, pp. 129–146.

Lundegard P. D. and Land L. S. (1989) Carbonate equilibria and pH buffering by organic acids—response to changes in P_{CO_2}. *Chem. Geol.* **74**, 277–287.

Lynch F. L. (1996) Mineral/water interaction, fluid flow, and Frio sandstone diagenesis: evidence from the rocks. *Am. Assoc. Petrol. Geologists Bull.* **80**, 486–504.

Lynch F. L. (1997) Frio shale mineralogy and the stoichiometry of the smectite-to-illite reaction—the most important reaction in clastic sedimentary diagenesis. *Clays Clay Min.* **45**, 618–631.

Lynch F. L. and Land L. S. (1996) Diagenesis of calcite cement in Frio Formation sandstones and its relationship to formation water chemistry. *J. Sedim. Res.* **A66**, 439–446.

Lynch F. L., Mack L. E., and Land L. S. (1997) Burial diagenesis of illite/smectite in shales and the origins of authigenic quartz and secondary porosity in sandstones. *Geochim. Cosmochim. Acta* **66**, 439–446.

Macaulay C. I., Fallick A. E., McLaughlin O. M., Haszeldine R. S., and Person M. J. (1998) The significance of d13C of carbonate cements in reservoir sandstones: a regional perspective from the Jurassic of the northern North Sea. In *Carbonate Cementation in Sandstones* (ed. S. Morad). International Association of Sedimentologists, Oxford, vol. 26, pp. 395–408.

Macchi L. (1987) A review of sandstone illite cements and aspects of their significance to hydrocarbon exploration and development. *Geol. J.* **22**, 333–345.

Mack L. E. (1990) Sr as a tracer of diagenesis in Cenozoic sediments of the northern Gulf of Mexico sedimentary basin. Dissertation, University of Texas at Austin.

Mackenzie F. T. and Garrels R. M. (1966a) Chemical mass balance between rivers and oceans. *Am. J. Sci.* **264**, 507–525.

Mackenzie F. T. and Garrels R. M. (1966b) Silica-bicarbonate balance in the ocean and early diagenesis. *J. Sedim. Petrol.* **36**, 1075–1084.

Mackenzie F. T., Ristvet B. L., Thorstenson D. C., Lerman A., and Leeper R. H. (1981) Reverse weathering and chemical mass balance in a coastal environment. In *River Inputs to Ocean Systems* (eds. J. M. Marten, J. D. Burton, and D. Eisma). UNEP and UNESCO, pp. 152–187.

Makowitz A. and Milliken K. L. (2001) New data on the role of mechanical burial compaction in diagenesis: Fro Formation, Gulf of Mexico Basin. Am. Assoc. Petrol. Geologists Ann. Meet, pp. 123–124.

Makowitz A. and Sibley D. F. (2001) Crystal growth mechanisms of quartz overgrowths in a Cambrian quartz arenite. *J. Sedim. Res.* **71**, 809–816.

Maliva R. G. and Siever R. (1988a) Diagenetic replacement controlled by force of crystallization. *Geology* **16**, 688–691.

Maliva R. G. and Siever R. (1988b) Mechanisms and controls of silicification of fossils in limestones. *J. Geol.* **96**, 387–398.

Maliva R. G., Dickson J. A. D., and Fallick A. E. (1999) Kaolin cements in limestones: potential indicators of organic-rich pore waters during diagenesis. *J. Sedim. Res.* **69**, 158–163.

Marrett R. and Laubach S. E. (1997) Diagenetic controls on fracture permeability and sealing. *Int. J. Rock Mech. Mineral Sci.* **34**(on CD) (3–4) Paper 204.

Marrett R., Ortega O., Reed R., and Laubach S. (1997) Predicting macrofracture permeability from microfractures. *AAPG Annual Meeting*, A76.

Mathieu R., Zetterstrom L., Cuney M., Gauthier-Lafaye F., and Hidaka H. (2001) Alteration of monazite and zircon and lead migration as geochemical tracers of fluid paleocirculations around the Oklo-Okelobondo and Bangombe natural nuclear reaction zones (Francville Basin, Gabon). *Chem. Geol.* **171**, 147–171.

Mathisen M. E. (1984) Diagenesis of Plio-Pleistocene nonmarine sandstones, Cagayan Basin, Phillipines: early development of secondary porosity in volcanic sandstones. In *Clastic Diagenesis* (eds. D. A. McDonald and R. C. Surdam). American Association of Petroleum Geologists, Tulsa, OK, vol. 37, pp. 177–193.

Maynard J. B. (1984) Composition of plagioclase feldspar in modern deep-sea sands: relationship to tectonic setting. *Sedimentology* **31**, 493–501.

McBride E. F. (1987) Diagenesis of the Maxon Sandstone (Early Cretaceous), Marathon region, Texas: a diagenetic quartzarenite. *J. Sedim. Petrol.* **37**, 98–107.

McBride E. F. (1989) Quartz cementation in sandstones: a review. *Earth Sci. Rev.* **26**, 69–112.

McBride E. F., Milliken K. L., Cavazza W., Cibin U., Fontana D., Picard M. D., and Zuffa G. G. (1995) Inhomogenous distribution of calcite cement at the outcrop scale in Tertiary sandstones, northern Apennines, Italy. *Am. Assoc. Petrol. Geologists Bull.* **79**, 1044–1063.

McDowell S. D. and Paces J. B. (1985) Carbonate alteration minerals in the Salton Sea geothermal system, California, USA. *Min. Mag.* **49**(3), 469–479.

McLaughlin O. M., Haszeldine R. S., Fallick A. E., and Rodgers G. (1994) The case of the missing clay, aluminium loss and secondary porosity, South Brae oilfield, North Sea. *Clay Min.* **29**, 651–663.

Mellon G. B. (1964) Discriminatory analysis of calcite- and smectite-cemented phases of the Mountain Park Sandstone. *J. Geol.* **72**, 786–809.

Merino E. (1975) Diagenesis in Tertiary sandstones from Kettleman North Dome, California: I. Diagenetic mineralogy. *J. Sedim. Petrol.* **45**, 320–336.

Milliken K. L. (1988) Loss of provenance information through subsurface diagenesis in Plio-Pleistocene sandstones, Gulf of Mexico. *J. Sedim. Petrol.* **58**, 992–1002.

Milliken K. L. (1989) Petrography and composition of authigenic feldspars, Oligocene Frio Formation, South Texas. *J. Sedim. Petrol.* **59**, 361–374.

Milliken K. L. (1992) Chemical behavior of detrital feldspars in mudrocks versus sandstones, Frio Formation (Oligocene), South Texas. *J. Sedim. Petrol.* **62**, 790–801.

Milliken K. L. (1994a) Cathodoluminescent textures and the origin of quartz silt in Oligocene mudrocks, South Texas. *J. Sedim. Res.* **64A**, 567–571.

Milliken K. L. (1994b) The widespread occurrence of healed microfractures in siliciclastic rocks: evidence from scanned cathodoluminescence imaging. In *Rock Mechanics, Models and Measurements, Challenges from Industry, Proceedings of the 1st North American Rock Mechanics Symposium* (eds. P. P. Nelson and S. E. Laubach). Balkema, Rotterdam, 1155pp.

Milliken K. L. (1995) Petrographic evidence of open-system chemical reactions in mudrocks in the Gulf of Mexico sedimentary basin. *Geological Society of American Annual Meeting* **27**(6), p. 461.

Milliken K. L. (1998) Carbonate diagenesis in non-marine foreland sandstones at the western edge of the Alleghanian overthrust belt, southern Appalachians. In *Carbonate Cementation in Sandstones: Distribution Patterns and Geochemical Evolution* (ed. S. Morad). International Association of Sedimentologists, Oxford, vol. 26, pp. 87–105.

Milliken K. L. (2001) Diagenetic heterogeneity in sandstone at the outcrop scale, Breathitt Formation (Pennsylvanian), eastern Kentucky. *Am. Assoc. Petrol. Geologists Bull.* **85**, 795–815.

Milliken K. L. (2002) Microscale distribution of kaolinite in Breathitt Formation sandstones (middle Pennsylvanian): implications for mass balance. In *Clay Cements in Sandstones* (eds. R. Worden and S. Morad). International Association of Sedimentalogists, Blackwell, Oxford, vol. 34 pp. 343–360.

Milliken K. L. and Land L. S. (1982) Fluid dynamics for cap-rock formation in Gulf Coast: Discussion. *Am. Assoc. Petrol. Geologists Bull.* **66**, 2585–2687.

Milliken K. L. and Land L. S. (1993) The origin and fate of silt-sized carbonate in subsurface Miocene–Oligocene mudrocks, South Texas Gulf Coast. *Sedimentology* **40**, 107–124.

Milliken K. L. and Laubach S. E. (2000) Brittle deformation in sandstone diagenesis as revealed by scanned cathodoluminescence imaging with application to characterization of fractured reservoirs. In *Cathodoluminescence in Geosciences* (eds. M. Pagel, V. Barbin, P. Blanc, and D. Ohnenstetter). Springer, Berlin, pp. 225–243.

Milliken K. L. and Mack L. E. (1990) Subsurface dissolution of heavy minerals, Frio Formation sandstones of the ancestral Rio Grande province, South Texas. *Sedim. Geol.* **68**, 187–199.

Milliken K. L., Land L. S., and Loucks R. G. (1981) History of burial diagenesis determined from isotopic geochemistry, Frio Formation, Brazoria County, Texas. *Am. Assoc. Petrol. Geologists Bull.* **65**, 1397–1413.

Milliken K. L., McBride E. F., and Land L. S. (1989) Numerical assessment of dissolution versus replacement in the subsurface destruction of detrital feldspars, Oligocene Frio Formation, South Texas. *J. Sedim. Petrol.* **59**, 740–757.

Milliken K. L., Mack L. E., and Land L. S. (1994) Elemental mobility in sandstones during burial: whole-rock chemical and isotopic data, Frio Formation, South Texas. *J. Sedim. Res.* **A64**, 788–796.

Milliken K. L., McBride E. F., Cavazza W., Cibin U., Fontana D., Picard M. D., and Zuffa G. (1998) Geochemical history of calcite precipitation in Tertiary sandstones, Northern Apennines, Italy. In *Carbonate Cementation in Sandstones. Distribution Patterns and Geochemical Evolution* (ed. S. Morad). International Association of Sedimentologists, Oxford, vol. 26, pp. 213–240.

Milodowski A. E. and Zalasiewicz J. A. (1991) Redistribution of REE during diagenesis of turbidite/hemipelagite mudrock sequences of Llandovery age from central Wales. *Geol. Soc. London Spec. Publ.* **57**, 101–124.

Moncure G. K., Lahan R. W., and Siebert R. M. (1984) Origin of secondary porosity and cement distribution in a sandstone/shale sequence from the Frio Formation (Oligocene). In *Clastic Diagenesis* (eds. D. A. MacDonald and R. C. Surdam). American Association of Petroleum Geologists, Tulsa, OK, vol. 37, pp. 151–161.

Moore D. M. (2000) Diagenesis of the Purington Shale in the Illinois Basin and implications for the diagenetic state of sedimentary rocks of shallow Paleozoic basins. *J. Geol.* **108**, 553–567.

Morad S. (1988) Diagenesis of titaniferous minerals in Jurassic sandstones from the Norwegian Sea. *Sedim. Geol.* **57**, 17–40.

Morad S. (1998) Carbonate cementation in sandstones: distribution patterns and geochemical evolution. In *Carbonate Cementation in Sandstones. Distribution Patterns and Geochemical Evolution* (ed. S. Morad). International Association of Sedimentologists, Oxford, vol. 26, pp. 1–26.

Morad S. and Aldahan A. A. (1982) Authigenesis of titanium minerals in two Proterozoic sedimentary rocks from southern and central Sweden. *J. Sedim. Petrol.* **52**, 1295–1305.

Morad S. and Aldahan A. A. (1987) Diagenetic replacement of feldspars by quartz in sandstones. *J. Sedim. Petrol.* **57**, 488–493.

Morad S., Bergan M., Knarud R., and Nystuen J. P. (1990) Albitization of detrital plagioclase in Triassic reservoir sandstones from the Snorre Field, Norwegian North Sea. *J. Sedim. Petrol.* **60**, 411–425.

Morad S., Ketzer J. M., and De Ros L. F. (2000) Spatial and temporal distribution of diagenetic alterations in siliciclastic rocks: implications for mass transfer in sedimentary basins. *Sedimentology* **47**(Suppl. 1), 95–120.

Mozley P. S. and Hoernle K. (1990) Geochemistry of carbonate cements in the Sag River and Shublik Formations (Triassic/Jurassic), North Slope, Alaska: implications for the geochemical evolution of formation waters. *Sedimentology* **37**, 817–836.

Nunn J. A. (1996) Buoyancy-driven propagation of isolated fluid-filled fractures: implications for fluid transport in Gulf of Mexico geopressured sediments. *J. Geophys. Res.* **101**, 2963–2970.

Nunn J. A. and Meulbroek P. (2002) Kilometer-scale upward migration of hydrocarbons in geopressured sediments by buoyancy-driven propagation of methane-filled fractures. *Am. Assoc. Petrol. Geologists Bull.* **86**, 907–918.

O'Brien N. R. and Slatt R. M. (1990) *Argillaceous Rock Atlas*. Springer, New York.

Oelkers E. H., Bjørkum P. A., and Murphy W. M. (1996) A petrographic and computational investigation of quartz cementation and porosity reduction in North Sea sandstones. *Am. J. Sci.* **296**, 420–452.

Oelkers E. H., Bjørkum P. A., Walderhaug O., Nadeau P. H., and Murphy W. M. (2000) Making diagenesis obey thermodynamics and kinetics: the case of quartz cementation in sandstones from offshore mid-Norway. *Appl. Geochem.* **15**, 295–309.

Owen M. R. (1991) Application of cathodoluminescence to sandstone provenance. In *Luminescence Microscopy and Spectroscopy: Quantitative and Qualitative Applications* (eds. C. E. Barker and O. C. Kopp). SEPM, Tulsa, OK, Short Course vol. 25, pp. 67–75.

Parnell J. (1998) Dating and duration of fluid flow and fluid–rock interaction. In *Special Publication*. Geological Society of London, London, vol. 144, 284pp.

Parnell J., Carey P. F., and Monson B. (1996) Fluid inclusion constraints on temperatures of petroleum migration from authigenic quartz in bitumen veins. *Chem. Geol.* **129**, 217–226.

Parnell J., Honghan C., Middleton D., Haggan T., and Carey P. (2000) Significance of fibrous mineral veins in hydrocarbon migration: fluid inclusion studies. *J. Chem. Explor.* **69–70**, 623–627.

Passaretti M. L. and Eslinger E. V. (1987) Dissolution and relic textures in framework grains of Holocene sediments from the Brazos River and Gulf Coast of Texas. *J. Sedim. Petrol.* **57**, 94–97.

Pearson M. J. and Small J. S. (1988) Illite–smectite diagenesis and paleotemperature in northern North Sea Quaternary to Mesozoic shale sequence. *Clay Min.* **23**, 109–132.

Pettijohn F. J. (1941) Persistence of heavy minerals with geologic age. *J. Geol.* **49**, 612–625.

Pittman E. D. and Larese R. E. (1991) Compaction of lithic sands: experimental results and applications. *Am. Assoc. Petrol. Geologists Bull.* **75**, 1279–1299.

Pittman E. D. and Lumsden D. N. (1968) Relationship between chlorite coatings on quartz grains and porosity, Spiro Sand, Oklahoma. *J. Sedim. Petrol.* **38**, 668–670.

Platt J. D. (1993) Controls on clay mineral distribution and chemistry in the Early Permian Rotliegend of Germany. *Clay Min.* **28**, 393–416.

Potdevin J.-L. and Hassouta L. (1997) Bian de matière des processus d'illitization et de surcroissance de quartz dans un réservoir pétrolier du champ d'Ellon (zone Alwyn, mer du Nord). *Bull. Sociétie Géologie Français* **168**, 219–229.

Potter P. E., Maynard J. B., and Pryor W. A. (1980) *Sedimentology of Shale*. Springer, New York.

Primmer T. J. and Shaw H. F. (1987) Diagenesis in shales: evidence from back-scattered electron microscopy and electron microprobe analyses. *Proc. Int. Clay Conf.*, Denver, CO, vol. 8, pp. 135–143.

Pryor W. A. (1973) Permeability–porosity patterns and variations in some Holocene sand bodies. *Am. Assoc. Petrol. Geologists Bull.* **57**, 162–189.

Raiswell R. and Fisher Q. J. (2000) Mudrock-hosted carbonate concretions: a review of growth mechanisms and their influence on chemical and isotopic composition. *J. Geol. Soc. London* **157**, 239–251.

Ramm M. and Ryseth A. E. (1996) Reservoir quality and burial diagenesis in the Statfjord Formation, North Sea. *Petrol. Geosci.* **2**, 313–324.

Ramseyer K., Baumann J., Matter A., and Mullis J. (1988) Cathodoluminescence colours of α-quartz. *Mineral. Mag.* **52**, 669–677.

Ramseyer K., AlDahan A. A., Collini B., and Landström O. (1992) Petrological modifications in granitic rocks from the Siljan impact structure: evidence from cathodoluminescence. *Tectonophysics* **216**, 195–204.

Rask J. H., Bryndzia L. T., Braunsdorf N. R., and Murray T. E. (1997) Smectite illitization in Pliocene-age Gulf of Mexico mudrocks. *Clays Clay Min.* **45**, 99–109.

Rasmussen B. (1996) Early diagenetic REE-phosphate minerals (florencite, gorceixite, crandallite, and xenotime) in marine sandstones: a major sink for oceanic phosphorus. *Am. J. Sci.* **296**, 601–632.

Reed R. M. and Laubach S. E. (1996) The role of microfractures in the development of quartz overgrowth cements in sandstones: new evidence from cathodoluminescence studies. *Geolog. Soc. Am. Annual Meeting*, **28**(7), 280p.

Renard F., Ortoleva P., and Gratier J. P. (1997) Pressure solution in sandstones: influence of clays and dependence on temperature and stress. *Tectonophysics* **280**, 257–266.

Richmann D. L., Milliken K. L., Loucks R. G., and Dodge M. M. (1980) Mineralogy, diagenesis, and porosity in Vicksburg sandstones, McAllen Ranch Field, Hidalgo County, Texas. *Gulf Coast Assoc. Geol. Soc. Transact.* **30**, 473–481.

Rubin J. N., Henry C. D., and Price J. G. (1989) Hydrothermal zircons and zircon overgrowths, Sierra Blanca Peaks, Texas. *Am. Mineral.* **74**, 865–869.

Rubin J. N., Henry C. D., and Price J. G. (1993) The mobility of Zr and other "immobile" elements during hydrothermal alteration. *Chem. Geol.* **110**, 29–47.

Rumble D., III (1994) Water circulation in metamorphism. *J. Geophys. Res.* **99**(B8), 15499–15502.

Rutter E. H. (1983) Pressure solution in nature, theory, and experiment. *J. Geol. Soc. London* **140**, 725–740.

Saigal G. C., Morad S., Bjørlykke K., Egeberg P. K., and Aagaard P. (1988) Diagenetic albitization of detrital K-feldspars in Jurassic, Lower Cretaceous, and Tertiary clastic reservoir rocks from offshore Norway: 1. Textures and origin. *J. Sedim. Petrol.* **58**, 1003–1013.

Schieber J., Krinsley D., and Riciputi L. (2000) Diagenetic origin of quartz silt in mudstones and implications for silica cycling. *Nature* **406**, 981–985.

Schmidt V. and McDonald D. A. (1979a) The role of secondary porosity in the course of sandstone diagenesis. In *Aspects of Diagenesis* (eds. P. A. Scholle and P. R. Schluger). Society of Economic Paleontologists and Mineralogists, Tulsa, OK, vol. 26, pp. 175–207.

Schmidt V. and McDonald D. A. (1979b) Texture and recognition of secondary porosity in sandstones. In *Aspects of Diagenesis* (eds. P. A. Scholle and P. R. Schluger). Society of Economic Paleontologists and Mineralogists, Tulsa, OK, vol. 26, pp. 209–225.

Scotchman I. C. (1991) The geochemistry of concretions from the Kimmeridge Clay Formation of southern and eastern England. *Sedimentology* **38**, 79–106.

Seyedolali A., Krinsley D. H., Boggs S., Jr., O'Hara P. F., Dypvik H., and Goles G. G. (1997) Provenance interpretation of quartz by scanning electron microscope—cathodoluminescence fabric analysis. *Geology* **25**, 787–790.

Sharp J. M., Fenstemaker T. R., Simmons C. T., McKenna T. E., and Dickinson J. K. (2001) Potential of salinity-driven free convection in a shale-rich sedimentary basin: example from the Gulf of Mexico basin in South Texas. *Am. Assoc. Petrol. Geologists Bull.* **85**, 2089–2110.

Siebert R. M., Moncure G. K., and Lahann R. W. (1984) A theory of framework grain dissolution in sandstones. In *Clastic Diagenesis* (eds. D. A. MacDonald and R. C. Surdam). American Association of Petroleum Geologists, Tulsa, OK, vol. 37, pp. 163–175.

Smith J. T. and Ehrenberg S. N. (1989) Correlation of carbon dioxide abundance with temperature in clastic hydrocarbon reservoirs: relationship to inorganic chemical equilibrium. *Mar. Petrol. Geol.* **6**, 129–135.

Spötl C. and Pitman J. K. (1998) Saddle (baroque) dolomite in carbonates and sandstones: a reappraisal of a burial-diagenetic concept. In *Carbonate Cementation in Sandstones. Distribution Patterns and Geochemical Evolution* (ed. S. Morad). International Association of Sedimentologists, Oxford, vol. 26, pp. 437–460.

Spötl C., Houseknecht D. W., and Riciputi L. R. (2000) High-temperature quartz cement and the role of stylolites in a deep gas reservoir, Spiro Sandstone, Arkoma Basin, USA. In *Quartz Cementation in Sandstones* (eds. R. H. Worden and S. Morad). International Association of Sedimentologists, Oxford, vol. 29, pp. 281–297.

Sprunt E., Dengler L. A., and Sloan D. (1978) Effect of metamorphism on quartz cathodoluminescence. *Geology* **6**, 305–308.

Sprunt E. S. and Nur A. (1979) Microcracking and healing in granites: new evidence of cathodoluminescence. *Science* **205**, 405–497.

Spry A. (1969) *Metamorphic Textures*. Pergamon, Oxford.

Stanton G. D. (1977) Secondary porosity in sandstones of the lower Wilcox (Eocene), Karnes County. *Gulf Coast Assoc. Geol. Soc. Trans.* **27**, 147–207.

Stoessell R. K. (1987) Mass transport in sandstones around dissolving plagioclase grains. *Geology* **15**, 295–298.

Sullivan K. B. and McBride E. F. (1991) Diagenesis of sandstones at shale contacts and diagenetic heterogeneity, Frio Formation, Texas. *Am. Assoc. Petrol. Geologists Bull.* **75**, 121–138.

Summer N. S. and Ayalon A. (1995) Dike intrusion into unconsolidated sandstone and the development of quartzite contact zones. *J. Struct. Geol.* **17**, 997–1010.

Sutton S. J. and Land L. S. (1996) Postdepositional chemical alteration of Ouachita shales. *Geol. Soc. Am. Bull.* **108**, 978–991.

Sutton S. J. and Maynard J. B. (1990) Multiple alteration events in the history of a sub-Huronian regolith at Lauzon Bay, Ontario. *Can. J. Earth Sci.* **29**, 432–445.

Sutton S. J. and Maynard J. B. (1993) Petrology, mineralogy, and geochemistry of sandstones of the lower Huronian Matinenda Formation, resemblance to underlying basement rocks. *Can. J. Earth Sci.* **30**, 1209–1223.

Tada R. and Siever R. (1989) Pressure solution during diagenesis. *Ann. Rev. Earth Planet. Sci.* **17**, 89–118.

Talukdar S., Gallango O., Vallejos C., and Ruggiero A. (1987) Observations on the primary migration of oil in the La Luna source rocks of the Maracaibo Basin, Venezuela. In *Migration of Hydrocarbons in Sedimentary Basins* (ed. B. Doligez). Editions Technip, Paris, France, pp. 59–78.

Taylor T. R. and Land L. S. (1997) Association of allochthonous waters and reservoir enhancement in deeply buried Miocene sandstones: Picaroon Field, Corsair Trend, Offshore Texas. In *Siliciclastic Diagenesis and Fluid Flow: Concepts and Applications* (eds. L. J. Crossey, R. G. Loucks, and M. W. Totten). Society for Sedimentary Geology, Tulsa, OK, vol. 55, pp. 37–48.

Taylor T. R. and Soule C. H. (1993) Reservoir characterization and diagenesis of the Oligocene 64-zone sandstone, North Belridge Field, Kern County, California. *Am. Assoc. Petrol. Geologists Bull.* **77**, 1549–1566.

Thomas A. R., Dahl W. M., Hall C. M., and York D. (1993) $^{40}Ar/^{39}Ar$ analyses of authigenic muscovite, timing of stylolitization, and implications for pressure solution mechanisms: Jurassic Norphlet Formation, offshore Alabama. *Clays Clay Min.* **41**, 269–279.

Thompson A. (1959) Pressure solution and porosity. In *Silica in Sediments* (ed. A. E Ireland). vol. 7, pp. 92–100.

Thyne G. (2001) A model for diagenetic mass transfer between adjacent sandstone and shale. *Mar Petrol. Geol.* **18**, 743–755.

Thyne G., Boudreau B. P., Ramm M., and Midtbø R. E. (2001) Simulation of potassium feldspar dissolution and illitization in the Statfjord Formation, North Sea. *Am. Assoc. Petrol. Geologists Bull.* **85**, 621–635.

Towe K. M. (1962) Clay mineral diagenesis as a possible source of silica cement in sedimentary rocks. *J. Sedim. Petrol.* **32**, 26–28.

Trevena A. S. and Clark R. A. (1986) Diagenesis of sandstone reservoirs of Pattani Basin, Gulf of Thailand. *Am. Assoc. Petrol. Geologists Bull.* **70**, 299–308.

Trevena A. S. and Nash W. P. (1981) An electron microprobe study of detrital feldspar. *J. Sedim. Petrol.* **51**, 137–150.

Urey H. C. (1956) Regarding the early history of the Earth's atmosphere. *Geol. Soc. Am. Bull.* **67**, 1125–1128.

van Moort J. C. (1972) The K_2O, CaO, MgO and CO_2 contents of shales and related rocks and their importance for sedimentary evolution since the Proterozoic. *24th Int. Geolog. Cong.*, Cong. **24**(10), 427–439.

Van Panhuys M. and Trewin N. H. (1990) Authigenic sphene cement in Permian sandstones from Arran. *Scott. J. Geol.* **26**, 139–144.

Walderhaug O. (1994) Temperatures of quartz cementation in Jurassic sandstones from the Norwegian continental shelf—evidence from fluid inclusions. *J. Sedim. Petrol.* **64**, 311–323.

Walderhaug O. (1996) Kinetic modeling of quartz cementation and porosity loss in deeply buried sandstone reservoirs. *Am. Assoc. Petrol. Geologists Bull.* **80**, 731–745.

Walderhaug O. (2000) Modeling quartz cementation and porosity in Middle Jurassic Brent Group sandstones of the Kvitebjorn Field, northern North Sea. *Am. Assoc. Petrol. Geologists Bull.* **84**, 1325–1339.

Walderhaug O., Lander R. H., Bjørkum P. A., Oelkers E. H., Bjørlykke K., and Nadeau P. (2000) Modelling quartz cementation and porosity in reservoir sandstones: examples from the Norwegian continental shelf. In *Quartz Cementation in Sandstones* (eds. R. H. Worden and S. Morad). International Association of Sedimentologists, Oxford, vol. 29, pp. 39–49.

Walker T. R. (1984) Diagenetic albitization of potassium feldspar in arkosic sandstones. *J. Sediment. Petrol.* **54**, 3–16.

Wang C.-Y. and Xie X. (1998) Hydrofracturing and episodic fluid flow in shale-rich basins: a numerical study. *Am. Assoc. Petrol. Geologists Bull.* **82**, 1857–1869.

Weaver C. E. and Beck K. C. (1971) Clay water diagenesis during burial: how mud becomes gneiss. *Geolog. Soc. Am. Spec. Pap.* **134**, 1–78.

Weyl P. K. (1959) Pressure solution and force of crystallization—a phenomenological theory. *J. Geophys. Res.* **64**, 2001–2025.

Wilkinson M. and Haszeldine R. S. (1996) Aluminum loss during sandstone diagenesis. *J. Geol. Soc. London* **153**, 657–660.

Williams L. B., Hervig R. L., and Dutton S. P. (1997) Constraints on paleofluid compositions in the Travis Peak Formation, East Texas: evidence from microanalysis of oxygen isotopes in diagenetic quartz. In *Basin-wide Diagenetic Patterns: Integrated Petrologic, Geochemical, and Hydrologic Considerations* (eds. I. Montañez, J. M. Gregg, and K. L. Shelton). Society for Sedimentary Geology (SEPM). Tulsa, OK, vol. 57, pp. 269–280.

Wiltschko D. V. and Sutton S. J. (1982) Deformation by overburden of a coarse quartzite conglomerate. *J. Geol.* **90**, 725–733.

Wintch R. P. and Kvale C. M. (1991) Open-system, constant-volume, development of slaty cleavage, and strain-induced replacement reactions in the Martinsburg Formation, Lehigh Gap, Pennsylvania. *Geol. Soc. Am. Bull.* **103**, 916–927.

Wintsch R. P. and Kvale C. M. (1994) Differential mobility of elements in burial diagenesis of siliciclastic rocks. *J. Sedim. Res.* **64**, 349–361.

Wood J. R. (1981) Fracture pressure solution: experiment and theory. *Geological Association of Canada, Mineralogical Association of Canada Meeting,* Abstracts of papers, A63.

Woronick R. W. and Land L. S. (1985) Late burial diagenesis, Lower Cretaceous Pearsall and Lower Glen Rose formations, South Texas. In *Carbonate Cements* (eds. N. Schneidermann and P. M. Harris). Society of Economic Paleontologists and Mineralogists, Tulsa, OK, vol. 36, pp. 265–275.

Wright T. O. and Platt L. B. (1982) Pressure dissolution and cleavage in the Martinsburg Shale. *Am. J. Sci.* **282**, 122–135.

Yardley B. W. D. (1986) Fluid migration and veining in the Connemara schists, Ireland. In *Fluid-rock Interactions during Metamorphism* (eds. J. V. Walther and B. J. Wood). Springer, New York, pp. 109–131.

Yau Y.-C., Peacor D. R., and Essene E. J. (1987) Authigenic anatase and titanite in shales from the Salton Sea geothermal field, California. *Neues Jahrbuch fuer Mineralogie. Monatshefte* **1987**(10), 441–452.

Yund R. A., Quigley J., and Tullis J. (1989) The effect of dislocations on bulk diffusion in feldspars during metamorphism. *J. Metamorph. Geol.* **7**, 337–341.

Zinkernagel U. (1978) In Cathodoluminescence of Quartz and its Application to Sandstone Petrology, E. Schweizerbart'sche Verlagsbuchandlung, Stuttgart, Contributions to Sedimentology 8.

7.08
Coal Formation and Geochemistry

W. H. Orem and R. B. Finkelman
US Geological Survey, Reston, VA, USA

7.08.1 INTRODUCTION

Coal is one of the most complex and challenging natural materials to analyze and to understand. Unlike most rocks, which consist predominantly of crystalline mineral grains, coal is largely an assemblage of amorphous, degraded plant remains metamorphosed to various degrees and intermixed with a generous sprinkling of minute syngenetic, diagenetic, epigenetic, and detrital mineral grains, and containing within its structure various amounts of water, oils, and gases. Each coal is unique, having been derived from different plant sources over geologic time, having experienty -45
ced different thermal histories, and having been exposed to varying geologic processes. This diversity presents a challenge to constructing a coherent picture of coal geochemistry and the processes that influence the chemical composition of coal.

Despite the challenge coal presents to geochemists, a thorough understanding of the chemistry and geology of this complex natural substance is essential because of its importance to our society. Coal is, and will remain for sometime, a crucial source of energy for the US and for many other countries (Figure 1). In the USA, more than half of the electricity is generated by coal-fired power plants, and almost 90% of the coal mined in the USA is sold for electricity generation (Pierce *et al.*, 1996). It is also an important source of coke for steel production, chemicals, pharmaceuticals, and even perfumes (Schobert, 1987). It may also, in some cases, be an economic source of various mineral commodities. The utilization of coal through mining, transport, storage, combustion, and the disposal of the combustion by-products, also presents a challenge to geochemists because of the wide range of environmental and human health problems arising from these activities. The sound and effective use of coal as a natural resource requires a better understanding of the geochemistry of coal, i.e., the chemical and mineralogical characteristics of the coal that control its technological behavior, by-product characteristics, and environmental and human health impacts. In this chapter, we will try to make geochemical sense of this wonderfully complex and important resource.

7.08.2 COAL FORMATION

We will begin this discussion by examining the processes leading to the formation of coal deposits. Coal is an organic "rock" derived from chemical and physical transformations of plant biopolymers due to: (i) microbially mediated enzymatic processes (biodegradation) occurring during early diagenesis and peat formation and (ii) the effects of pressure and temperature acting over long period of time following burial of the peat (coalification); (Stach *et al.*, 1982). Thus, coal is a product of both biological and geologic processes acting on plant remains over time. Biopolymers in vascular plants are the most important starting materials in the formation of coal (Figure 2). Especially important is the lignocellulose biopolymer, a structural component of vascular plants, which is thought to provide the basic aromatic framework for coal (Hatcher, 1990). The cuticle of leaves is also known to contain resistant biopolymers (cutan, suberan) that form part of the aliphatic structure of coal (Nip *et al.*, 1986; Tegelaar *et al.*, 1989, 1995). Resinites, the fossil remains of higher plant resins, also contribute to the aliphatic structure of coal (Langenheim, 1969, 1995; Anderson and Winans, 1991; Anderson and Crelling, 1995). Algal and microbial biopolymers were historically considered to constitute a minor portion of the material forming coal deposits (Stach *et al.*, 1982). Recent work, however, suggests that resistant biopolymers in algae (algaenan), as well as bacterial biopolymers may also contribute to the organic components in coal and kerogen (de Leeuw and Largeau, 1993; Hatcher and Clifford, 1997; Gelin *et al.*, 1999; Allard *et al.*, 2002).

Present-day coals today generally began their formation tens of thousands to hundreds of millions of years ago in ancient swamps. The oldest known coals are anthracites from Michigan, USA, dated to be ~2.3 Gyr old (Taylor *et al.*, 1998). These oldest known coals are probably of algal and fungal origin, and are extremely rare. The extensive coal deposits found around the world are principally the product of peats formed in swamps dominated by vascular plants (e.g., land plants). Vascular plants first evolved in the Middle Ordovician (~475 Myr ago), but extensive

Figure 1 Photograph of a low rank coal bed (lignite of Pliocene age) from southwestern Romania.

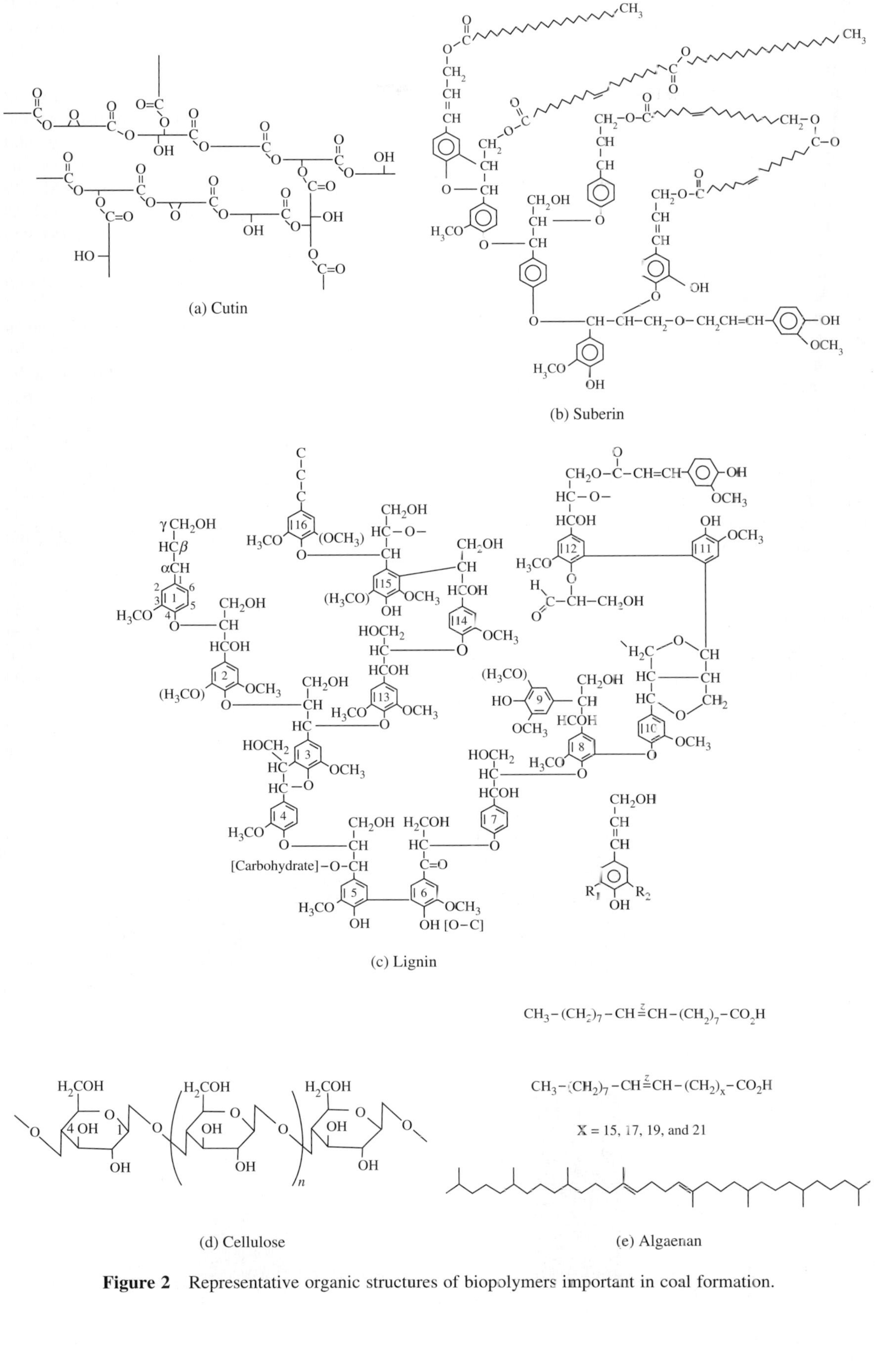

(a) Cutin

(b) Suberin

(c) Lignin

$CH_3-(CH_2)_7-CH\overset{z}{=}CH-(CH_2)_7-CO_2H$

$CH_3-(CH_2)_7-CH\overset{z}{=}CH-(CH_2)_x-CO_2H$

X = 15, 17, 19, and 21

(d) Cellulose

(e) Algaenan

Figure 2 Representative organic structures of biopolymers important in coal formation.

vascular plant-dominated peat deposits did not occur until the Carboniferous (354–298 Myr ago), when vascular plant evolution and favorable environmental conditions (climatic and tectonic) permitted the development of thick peats. The age distribution of coals is uneven. Some periods of earth history were more favorable than others for the development of thick peats that later produced coal deposits. Periods favorable for extensive peat formation leading to coal deposits occurred during the Late Carboniferous, the Late Permian, the Jurassic to Early Cretaceous, and the Late Cretaceous to Early Tertiary (Stach *et al.*, 1982; Figure 3). Similarly, the worldwide geographic distribution of coals is uneven (Landis and Weaver, 1993; Figure 3). Relatively more coal deposits occur in Asia, North America, and Europe; relatively fewer occur in Africa and South America. This geographic distribution of coal reflects the movement of continents on the earth's surface. Continents with extensive coal deposits occupied geographic positions favorable for peat formation during periods when environmental conditions favored extensive peat accumulation and preservation.

7.08.2.1 Peat Formation and Early Diagenesis

The first stage in the conversion of plant biopolymers into coal involves early diagenesis and the formation of peat (peatification) (Bouska, 1981). Peat is an organic sediment derived from the fungal and bacterial degradation of plant biopolymers (Hatcher and Spiker, 1988). Peat accretes in places where the rate at which organic matter accumulates from senescent biomass (primarily from vascular plants) exceeds the rate of microbial degradation of this organic matter. These conditions are generally found in shallow, aquatic environments, such as swamps, fens, and marshes, where waterlogged conditions limit the availability of oxygen for biodegradation, and where anaerobic microbial processes dominate.

Both climate and geology may be involved in establishing the conditions favorable for the formation of peat (McCabe, 1991). For example, geologic features that restrict the flow of water can lead to broad, shallow, flooded areas, promoting the accumulation of peat. Peat deposits formed in this setting are often referred to as topogeneous peats (Figure 4). Examples of topogeneous peats include the Okefenokee Swamp and the Everglades in the southeastern USA, where a ridge and relatively flat terrain allows the ponding of water, and where water flow is broad, shallow, and slow moving (Spackman *et al.*, 1974; Cohen, 1984; Gleason and Stone, 1994). Many extensive peat deposits are formed at high and a low latitudes. For example, moderate rainfall and a low evaporation

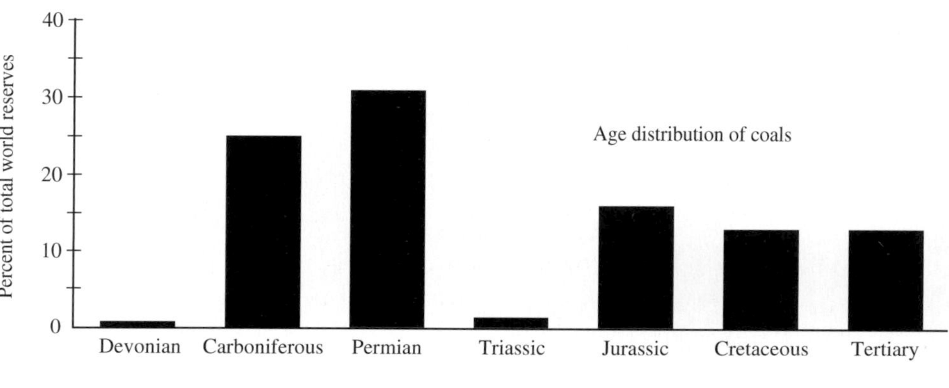

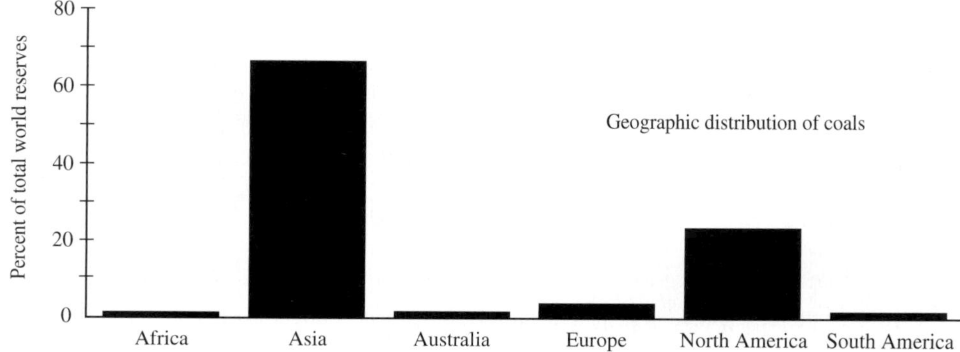

Figure 3 Distributions of world coal by age (upper plot) and geographic location (lower plot).

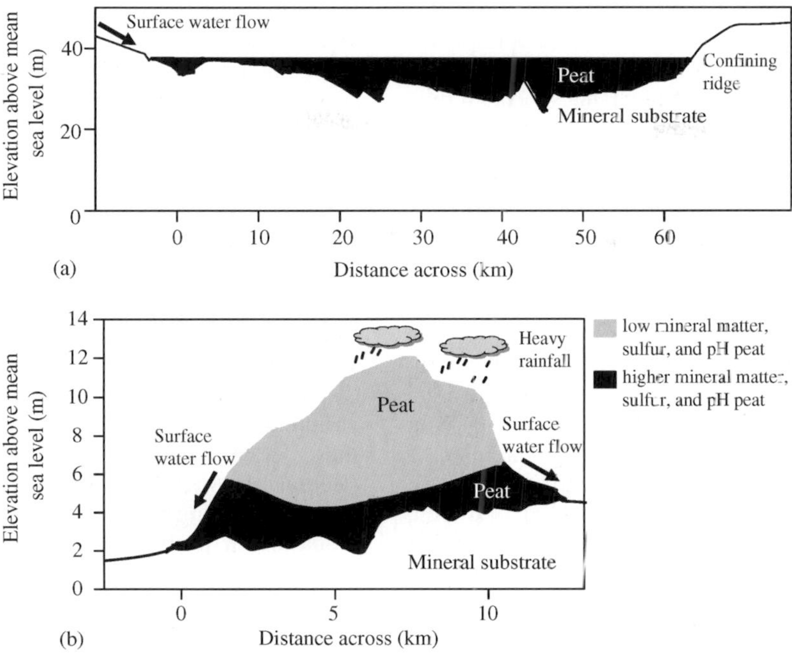

Figure 4 Schematic diagrams of different types of peat-forming environments: (a) a planar or topogenous peat, such as the Okefenokee Swamp, Georgia, USA and (b) a domed or ombrogenous peat deposit, such as in Kalimantan, Indonesia. Note the much greater peat depth in the domed peat compared to the topogenous peat.

rate of water at high latitudes can give rise to waterlogged conditions favorable for peat development (Aaby and Tauber, 1975; Ovenden, 1990; Barber, 1993; Halsey *et al.*, 1998). In tropical regions, high rainfall and warm year-round temperatures result in the accumulation of water and high rates of primary production, creating an abundant source of plant material and leading to the formation of extensive peat deposits (Anderson, 1964, 1983; Rieley *et al.*, 1997). Examples of tropical peats include the extensive "domed" peat deposits of the Indo-Malay region in southeast Asia (Morley, 1981; Dehmer, 1993; Staub and Esterle, 1994; Rieley *et al.*, 1997; Figure 4). These peats are termed ombrogenous, in reference to the excessive rainfall that allows the peats to develop (Anderson, 1983; Bragg, 1997). Ombrogenous peats, whose water supplies are generally rainfall-derived, have lower sulfur and mineral matter contents than topogenous peats, which obtain water from surface runoff (Neuzil *et al.*, 1993). The Carboniferous coals of the Appalachian Basin in the eastern USA are thought to have formed under environmental conditions similar to those of the present-day ombrogenous peat deposits of the Indo-Malay region (Cecil *et al.*, 1993; Grady *et al.*, 1993).

From a biogeochemical standpoint, the principal factors in the accumulation of peat deposits are the nature of the plants growing in the peatland and the absence of oxygen (Tissot and Welte, 1984). The nature of the plants is important, as different plant types biodegrade at different rates (Hatcher and Spiker, 1988; Cowie and Hedges, 1992). For example, wood is more resistant to biodegradation than most other plant tissues (Hatcher and Breger, 1981; Spiker and Hatcher, 1987; Stout *et al.*, 1988), and woody plants are generally better preserved in peats than nonwoody plants (Benner *et al.*, 1985). The waterlogged conditions present in peat-forming environments limit diffusion of oxygen from the atmosphere, resulting in anoxic or low-oxygen conditions in the underlying sediments (Wieder, 1985; Shotyk, 1988). Thus, fungal biodegradation is generally limited to the near-surface zone in peats where oxygen is available (Herlihy, 1973; Benner *et al.*, 1984, 1986), but bacterial biodegradation within the peat is dominated by anaerobic metabolic processes (Risatti, 1978; Given *et al.*, 1983; Yavitt and Lang, 1990). Anaerobic microbial metabolic processes (fermentation, methanogenesis, sulfate, nitrate, iron, and manganese reduction) are fundamentally less efficient in degrading organic substances than aerobic processes, allowing a greater abundance of accumulating organic matter to survive destruction (Berner, 1971; Tissot and Welte, 1984). In places where oxygenated conditions prevail, soils develop instead of peats. Compared to peats, soils contain relatively little organic matter, with only the most refractory organic compounds (resistant to attack by aerobic microbes) remaining (Flaig, 1966; Hatcher *et al.*, 1981, 1983a).

Changes in the chemical structural composition of peats occur over time as microorganisms (continually) degrade plant biopolymers. A useful way of examining chemical changes through time is the chemical analysis of peat cores; progressively deeper sections of these cores represent older and more biodegraded plant material. Changes in the elemental composition of peat with depth (i.e., over time) typically involve loss of oxygen and nitrogen, and an increase in organic carbon content (Orem and Hatcher, 1987). The loss of oxygen suggests that oxygen-rich organic compounds such as cellulose are selectively degraded; leaving behind an organic residue (peat) that is relatively enriched in organic carbon. Compounds in peats containing nitrogen also are preferentially degraded during early diagenesis, resulting in the (important) release of nitrogen. The release of nitrogen is important for sustaining the plants living in the wetland peat environment (Kalbitz and Geyer, 2002).

The complexity of the macromolecules in peat (and coal) has challenged organic geochemists attempting to decipher the mechanism of peat formation. Early studies of the transformation of plant biopolymers into humic substances and other geopolymers in peat were limited by the analytical tools available at the time (Waksman and Stevens, 1928; Given, 1972; Spackman *et al.*, 1974; Schnitzer and Levesque, 1979). Recent progress in understanding the transformation of plant biopolymers into geopolymers is largely due to advances in analytical capabilities. Instrumentation such as solid state ^{13}C nuclear magnetic resonance (NMR) spectroscopy, pyrolysis gas chromatography/mass spectrometry (PY-GC/MS), electrospray ionization mass spectrometry (ESI/MS), and other techniques have enabled researchers to better probe the complex, macromolecular structures present in peat and coal (Schenck, 1986; Wilson, 1987; Ikeda *et al.*, 1999; van Grass *et al.*, 1979; Schulten and Gleixner, 1999).

A number of studies have used solid-state ^{13}C NMR to examine changes in the chemical structure of peat with increasing depth (time) in peat profiles (Hammond *et al.*, 1985; Hatcher *et al.*, 1985, 1986; Orem and Hatcher, 1987). These changes result primarily from microbial activity, including fungal and bacterial degradation under oxic conditions at the peat surface, and bacterial processes under primarily anoxic conditions within the peat. A major change observed among the ^{13}C NMR spectra of peats from many environments is the systematic loss of oxygen-bonded aliphatic carbons (representing cellulose and other complex carbohydrates) with increasing depth (Wilson, 1987); (Figure 5). As the carbohydrates are biodegraded, other organic moieties in the peat are concentrated. Thus, ^{13}C NMR spectra often show increasing intensities due to aromatic carbons (from lignin) and aliphatic carbons (contributions from vascular plant waxes and resistant algal and microbial aliphatic substances) with increasing depth (Figure 5). Other techniques, such as PY-GC/MS have also shown that peat decomposition results in the preferential loss of carbohydrates, and the selective preservation of resistant aromatic and aliphatic organic molecules (Saiz-Jimenez *et al.*, 1987; Stout *et al.*, 1988). The preferential loss of carbohydrates and the selective preservation of lignin, resistant waxes and, algaenans is consistent with the loss of oxygen from peats revealed by elemental analysis, and results from the limited ability of anaerobic microorganisms to degrade these more resistant substances enzymatically (Benner *et al.*, 1984, 1985, 1986).

An interesting aspect of the process of biodegradation in peats is that although most of the cellulose in the lignocellulose biopolymer of woody tissue is lost, the wood retains most of its original morphology (Hatcher and Clifford, 1997). The retention of morphology in wood during peatification appears to be due to the selectivity of the exo-enzymes used by the anaerobic bacteria. These degrade the cellulosic portion of the wood but leave the lignin portion relatively intact. Nonwoody tissues (mostly leaves or grass) apparently are more susceptible to morphologic change as cellulose is degraded; this nonwoody material becomes part of the matrix material in the peat (Hatcher and Spiker, 1988).

In addition to lignin, certain components of the leaf cuticles of vascular plants are also selectively preserved. Cutin, a polyester-like biopolymer that is a major component of modern cuticles, is apparently biodegraded during peatification (Nip *et al.*, 1986; Tegelaar, 1990), but cutan, a minor component of the leaf cuticle, is more readily preserved in peats (Tegelaar *et al.*, 1989). The chemical structure of cutan is still subject to debate, but may consist of polymethylenic structures ester-bonded to cellulose or to an aromatic core (Tegelaar *et al.*, 1989; Nip *et al.*, 1989; McKinney *et al.*, 1996). Whatever its chemical structure, cutan is sufficiently resistant to survive peatification.

The largest pools of degraded organic matter in peat, soil, and sediments are humic substances (humic acid, fulvic acid, and humin). Humic substances constitute a category of organic matter that is operationally defined by its mode of isolation (Schnitzer and Khan, 1972). Humic substances are the degraded organic matter in sediments. One theory for the formation of humic substances, often referred to as the consideration or melanoidin hypothesis, proposes that biopolymers are first degraded to low-molecular-weight substances such as amino acids and sugars, and that these substances subsequently condense to

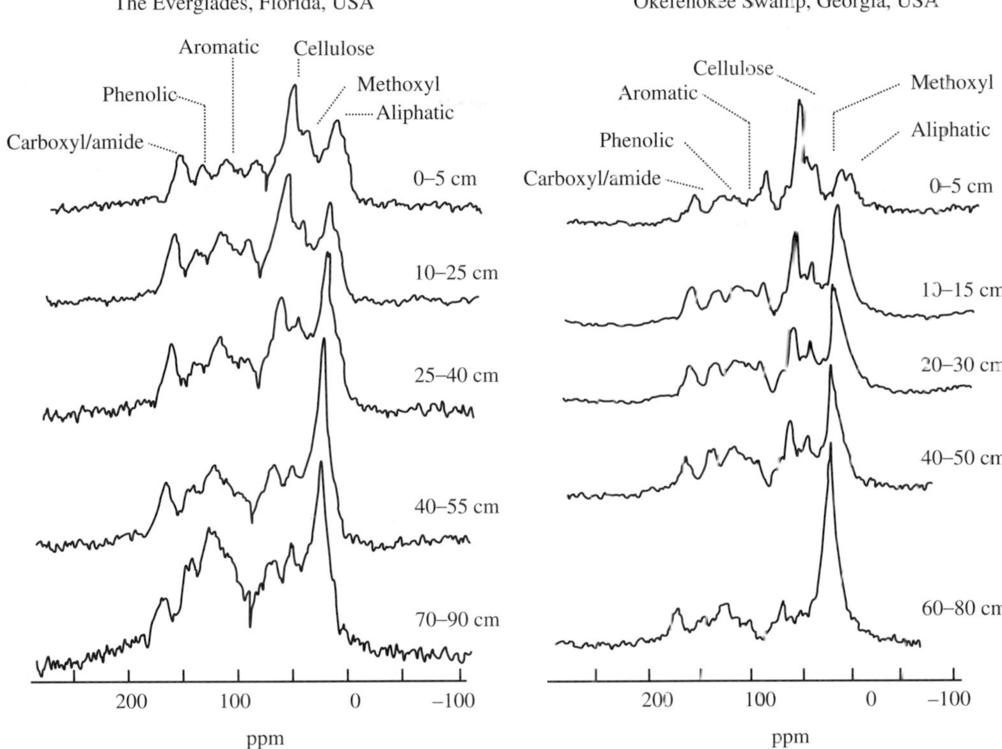

Figure 5 Solid state ^{13}C NMR spectra of peat from the Everglades, and Okefenokee Swamp. Major organic structural components originating from plant biopolymers are identified in the spectra. Note the preferential loss of cellulose with depth (time) in the peats, and the selective preservation of resistant biomolecules, such as lignin (aromatic) and vascular plant waxes (aliphatic).

form humic substances (Manskaya and Drozdova, 1968; Young *et al.*, 1977; Ertel and Hedges, 1980; Yamamoto and Ishiwatari, 1989, 1992; Poirier *et al.*, 2000). Others suggest that the more labile fraction of the plant-derived organic matter is preferentially degraded by microorganisms, leaving the more refractory components selectively preserved in the peat as humic substances (Fischer and Schrader, 1921; Hatcher *et al.*, 1983a,b; Hatcher and Spiker, 1988). Which of these competing hypotheses best explains the formation of humic substances is still subject to debate; however, the preponderance of evidence appears to favor the second hypothesis (often called the selective preservation hypothesis) (Hatcher and Clifford, 1997; Garcette-Lepecq *et al.*, 2000); (see Figure 6).

The principal degradation mechanism of organic matter in the formation of peats (humic substances included) and leading to the formation of coal, therefore, appears to involve the preferential breakdown of labile biomolecules (carbohydrates, etc.), and the selective preservation of inert and relatively inert (i.e., somewhat degraded) biomolecules (Hatcher and Clifford, 1997). Melanoidin formation may also play some role in the production of humic substances in peats, soils, and sediments, but appears to be a secondary process (Zhang *et al.*, 2001).

7.08.2.2 Coalification, Early Stage (Lignite, Brown Coal)

The transformation of peat to coal (coalification) generally occurs in subsiding basins under a range of conditions: thermal gradients that depend on local heat flow, lithostatic pressures that are proportional to the thickness of the sediment overburden, and fluid pressures that depend on the distribution of the hydraulic framework and permeability (Tissot and Welte, 1984). In an open system where fluids can move freely, as in a peat deposit buried under a porous overburden, the fluid pressure is proportional to the weight of the water column in the overburden. Temperature and time are the most important factors controlling the transformation of organic matter in peat to coal. Pressure resulting from the sediment overburden is often considered important only as a means of controlling sediment temperature (Karwell, 1955; Stanov, 1972; Peters *et al.*, 1981; Rohrback *et al.*, 1984; Venkatesan *et al.*, 1993).

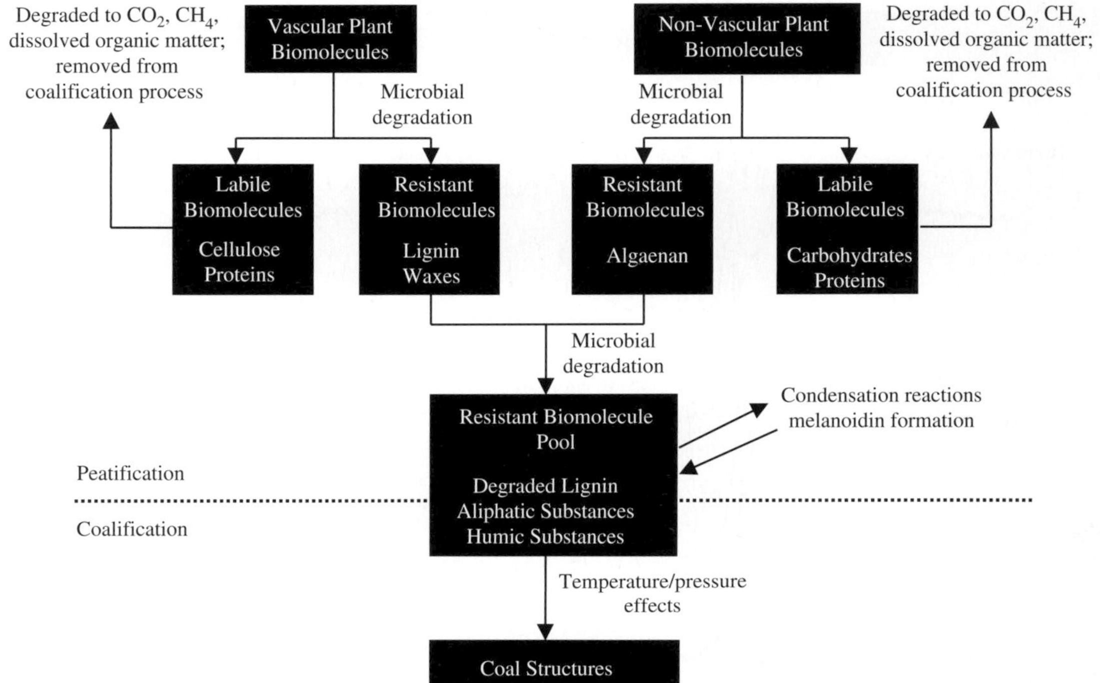

Figure 6 Schematic outline showing the major changes occurring in organic matter during peatification and coalification.

The magnitude of hydrostatic pressure relative to lithostatic pressure, however, may influence some chemical reactions during coalification and coal maturation by maintaining contact between reaction products and the starting material (Monthioux, 1988), and/or by allowing reaction products to escape.

The major changes in the elemental composition of coal occurring during early-stage coalification involve continued loss of oxygen, and to a lesser extent the loss of hydrogen accompanying the condensation of aromatic ring structures derived from lignin. Elemental H/C versus O/C plots (van Krevelen diagrams) of peats and coals of different degrees of alteration illustrate the loss of oxygen and hydrogen from organic matter normalized to carbon as coalification proceeds (Figure 7), and are a useful means of examining the maturation of organic matter in coals and in source rocks for petroleum hydrocarbons (Tissot and Welte, 1984). Nitrogen-containing compounds (especially amides) are also lost during early stage coalification, continuing the loss of nitrogen that began during the peatification stage. Pyrrolic nitrogen-containing compounds are apparently selectively preserved in coals while amides are lost (Knicker *et al.*, 1996).

Solid-state ^{13}C NMR has proved to be an extremely useful tool for examining the major changes in organic matter occurring during the coalification process (Hatcher *et al.*, 1982, 1989a, b; Botto, 1987; Wilson, 1987). Studies of coals of different degrees of alteration or rank (see Section 7.08.3 for a discussion of the concept of rank) using ^{13}C NMR have shown that coalification to the lignite or brown coal stage is accompanied by: (i) loss of residual cellulose, (ii) loss of methoxyl functional groups on lignin, and (iii) the formation of catechol-like structures from lignin (Hatcher and Clifford, 1997) (Figure 4). Lignites vary in the amount of oxygen functionality associated with the lignin molecule. Some lignites show pronounced peaks for methoxyl and phenolic carbons in their ^{13}C NMR spectra, while other lignites contain virtually no methoxyl and reduced phenolic functionality compared to peats (Figure 8). This variability in the oxygen functionality of lignites probably reflects differences in their thermal histories.

The major changes in lignin during the early stages of coalification include: (i) cleavage of β-O-4 aryl ethers, (ii) demethylation of methoxyl functional groups, and (iii) dehydroxylation reactions (Hatcher and Clifford, 1997). The β-O-4 cleavage reaction produces catechol-like structures in low-rank coals, and a carbocation on the three-carbon side chain of lignin, which can alkylate the aromatic ring on adjacent catechol structures (Botto, 1987; Hatcher *et al.*, 1989a,b; Buchanan *et al.*, 1997). This reaction accounts for many of the observed chemical changes in lignin during the early stage of coalification, and explains how the structural integrity of the macromolecule is maintained. Cleavage of aryl-O bonds through

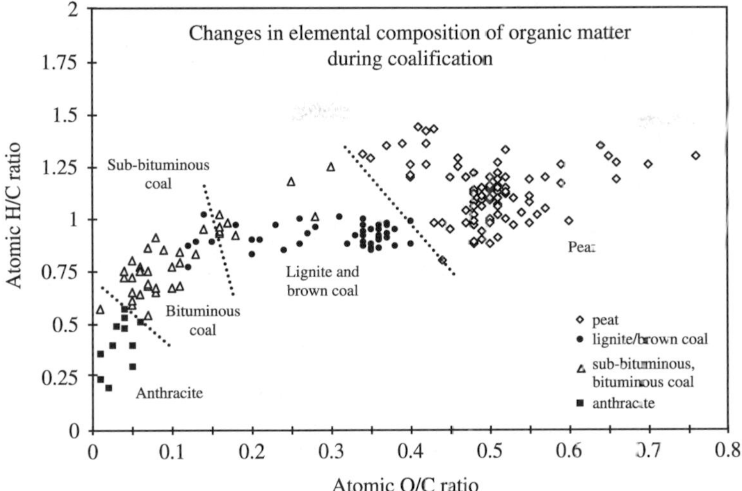

Figure 7 An H/C versus O/C (van Krevelen plot) plot of peat and coals of different rank. The plot illustrates that loss of oxygen (change in O/C) is most important during peatification and early stage coalification, with H loss (change in H/C) becoming more important during coalification to higher rank.

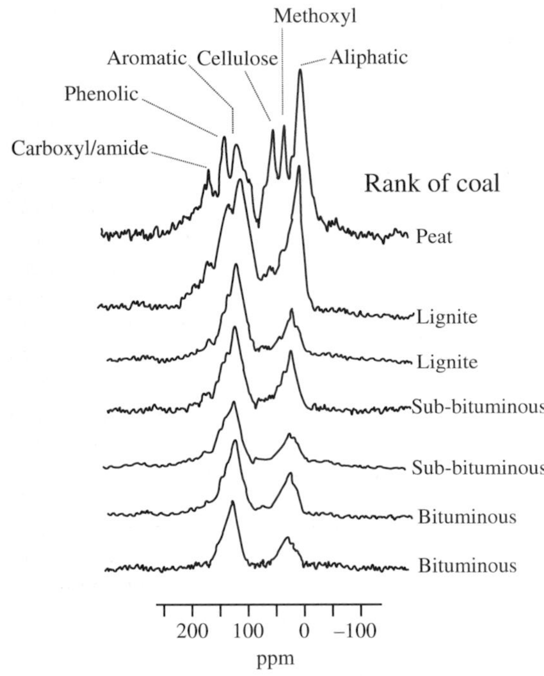

Figure 8 Solid state ^{13}C NMR spectra of a peat and a series of ranked coals (lignite to bituminous), showing the changes in organic structure occurring during coalification. Note the increasing simplicity of the spectra as rank increases, reflecting the loss of oxygen-containing organic matter such as cellulose, phenols, and methoxyls. At the highest rank, only condensed aromatic structures derived from lignin, and resistant aliphatic structures remain.

the demethylation of methoxyl groups is also an important change occurring in lignin during early stage coalification. This reaction results in the loss of methoxyl functionality and an increase in phenolic content in the form of catechol-like structures. Dehydroxylation reactions in the alkyl sidechains of the lignin during the early stage of coalification are also suggested by ^{13}C NMR data (Hatcher, 1988; van der Heijden and Boon, 1994). This process probably involves simple reduction of the hydroxyl groups on the sidechain in order to maintain the physical integrity of the lignin structure in low-rank coals (Hatcher and Clifford, 1997).

Experimental studies have also examined changes in the chemistry of organic matter during coalification (experimental coalification). These studies normally involve the heating of peat, low-rank coal, or plant material in a pressure vessel under conditions simulating natural coalification. However, the detailed conditions of the experiments vary greatly (Huck and Patteisky, 1964; Rohrback *et al.*, 1984; Monthioux *et al.*, 1985, 1986; Landais *et al.*, 1989, 1990; Lo, 1991; Rollins *et al.*, 1991; Shearer *et al.*, 1996; Orem *et al.*, 1996). Results from experimental coalification studies have generally supported the model of early stage coalification developed from organic geochemical studies of peat and coal deposits (Lo, 1991). For example, experimental coalification typically results in loss of oxygen and hydrogen, and an increase in organic carbon content (Venkatesan *et al.*, 1993; Orem *et al.*, 1996). Loss of oxygen-rich cellulose is also supported by the results of experimental coalification (Rollins *et al.*, 1991; Orem *et al.*, 1996). Changes in lignin observed during experimental coalification include: the loss of methoxyl functionality by a demethylation process, β-O-4 aryl ether cleavage, and the formation of catechol-like structures (Ohta and Venkatesan, 1992; Venkatesan *et al.*, 1993; Orem *et al.*, 1996).

This is consistent with the changes in lignin observed in studies of peats and low-rank coals (Hatcher and Clifford, 1997).

7.08.2.3 Coalification, Late Stage (Subbituminous, Bituminous, Anthracite)

Later stage coalification (formation of subbituminous coal, bituminous coal, and anthracite) results from deeper burial and exposure of organic matter to more extreme temperatures and pressures compared to those experienced by brown coals and lignite. The more intense heating and pressure results in further chemical changes in the organic molecules, as well as (physical) changes in the appearance of the coal. In bituminous coals, specific plant components in the coal (macerals) become more homogenous, and the cell structures of woody material within the coal (still visible in lignite and subbituminous coals using scanning electron microscopy) become significantly deformed (Hatcher *et al.*, 1994; Hatcher and Clifford, 1997). As a result of these physical changes during late-stage coalification, the appearance of coal changes from a brown, dull-colored substance that resembles mixtures of wood and plant remains (brown coal and lignite), to a lustrous, shiny, black organic "rock" (bituminous coals and anthracite). Bituminous coals and anthracites are often referred to as hard coal because of their physical structure, while lignite and brown coal are commonly referred to as soft coal.

The elemental composition of organic matter in coal continues to change during late-stage coalification (Figure 7). In H/C versus O/C van Krevelen plots, bituminous coals and anthracites show a somewhat smaller drop in O/C and a sharper drop in H/C values compared to brown coals and lignites (Tissot and Welte, 1984). In bituminous coals and anthracites, much of the oxygen present in the original plant organic matter has been lost. Virtually all of the cellulose is gone, and much of the oxygen functionality on lignin (especially methoxyl groups and hydroxyl groups on the alkyl sidechains) is also lost. Thus, the change in O/C ratio is slower in bituminous coals and anthracites. The H/C ratio shows a sharp decline in bituminous coals and anthracites. This change in H/C ratio reflects increased aromatization of the lignin structure, as heating and pressure effects decrease the number of functional groups on lignin and increase the number of polynuclear aromatic structures. In addition, loss of low-molecular-weight hydrocarbons from the cracking of resistant cuticular biopolymers (cutan), and algal biopolymers (algaenan) also reduces the H/C ratio.

The nitrogen content also continues to decrease during late stage coalification. Nitrogen is formed from the thermal decomposition of nitrogen-containing organic compounds in the coal at temperatures above 600 °C (Boudou and Espitalie, 1995).

As described earlier, chemical changes in the lignin molecule that occur during early diagenesis and coalification to the lignite and brown coal stage include: the loss of cellulose from the lignocellulose biopolymer, the loss of methoxyl functionality on the aromatic ring, β-O-4 ether cleavage and alkylation of the aromatic ring, the formation of catechol-like structures, and loss of hydroxyl functionality on alkyl sidechains (Hatcher and Clifford, 1997). Despite these extensive chemical changes, the physical structure of lignin is maintained through early-stage coalification, and intact cell structure of woody tissue is clearly visible in scanning electron micrographs. Chemical changes occurring during coalification to the bituminous coal stage, however, result in physical changes in the structure of lignin, including loss of cellular morphology. The loss of hydroxyl functionality on the alkyl sidechains of lignin that begins during early-stage coalification continues during later stage coalification. ^{13}C NMR spectra of bituminous coals show no indication of any alkyl hydroxyl or alkyl ether groups on the lignin sidechains, although the alkyl sidechains themselves are apparently retained at higher rank (Hatcher and Clifford, 1997). Catechol-like structures are dominant components of the lignin in low-rank coals; however, phenols dominate at higher rank (Hatcher *et al.*, 1989a; van Bergen *et al.*, 1994). This change in structural components reduces the number of oxygen-bonded carbons in the aromatic ring structure of lignin from two in lignites to one in subbituminous and bituminous coal. The mechanism by which catechol-like structures are converted to phenols is not confirmed, but may involve loss of water from catechol-like structures to form hydroxylated diaryl ethers as intermediates (Behar *et al.*, 1995; Behar and Hatcher, 1995). If correct, this model is a self-perpetuating mechanism whereby subsequent heating and cleavage of the diaryl ether bonds produces a phenol and a catechol, with the catechol available for further reaction to form additional hydroxylated diaryl ether intermediates and phenols (Hatcher and Clifford, 1997).

Another characteristic change in the geochemistry of lignin during late-stage coalification is aromatization (Straka *et al.*, 1999). The aromaticity of coal and coalified wood is greater in bituminous coals than in subbituminous coals (Hatcher, 1988; Cody, 1992). Analysis of bituminous coals using PY-GC/MS also typically shows the presence of polynuclear aromatic ring systems that are absent in lignites (Hatcher and Clifford, 1997). The mechanism of aromatization is not well established, but may involve condensation and pyrolysis of the phenol structures produced during coalification to the subbituminous coal stage, to form benzene-like structures through aryl ether intermediates (Hatcher and Clifford, 1997). These benzene-like structures with their associated alkyl sidechains may then condense to form polynuclear aromatic structures characteristic of bituminous coals and anthracites.

7.08.3 COAL RANK

Coal is typically classified by "rank" a term referring to the degree of alteration of the coal from the initial organic material (plant biopolymers). The higher the rank of the coal, the greater the degree of alteration. The major categories of coal rank are presented in Table 1, along with a description of the chemical and physical changes occurring at each stage of rank. Rank is an important parameter because of its relation to key industrial characteristics of coal. For example, the energy produced per unit weight of coal (calorific value in kcal lb^{-1} or kJ kg^{-1}) is dependent on rank. Higher-rank coals typically have higher calorific values, reflecting loss of moisture and volatiles and the progressive increase in aromatization of the coal structure. Rank also plays a role as a predictor of oil and natural gas generation from coal beds, as discussed later (Section 7.12.5). Rank is also a useful means of categorizing coals, which can vary greatly in their physical and chemical characteristics.

Although microorganisms play a role in the alteration of plant biopolymers to form peat and coal, temperature and (to a lesser degree) pressure are the principal factors controlling the degree of alteration (rank) attained by a particular coal. The rank of coal is not consistently correlated with its age, and there are examples of Carboniferous coals of lower rank than Tertiary coals. The absence of correlation between coal age and rank reflects differences in the geologic history of different coals, such as depth of burial, temperature exposure, and duration of exposure. Depth of burial is frequently correlated with rank, reflecting the effects of the local geothermal gradient (temperature generally increasing with depth). Geothermal gradients vary regionally due to differences in heat flow and thermal conductivity of rocks. Heat flow in continents typically averages about 50–60 mW m^{-2} (Chapman, 1985), but can be much higher (hundreds of mW m^{-2}) in areas of hydrothermal activity (Haenel and Staroste, 1988) and magmatic intrusion (Barker, 1979; Taylor *et al.*, 1998; Stadler and Teichmuller, 1971). The duration of heating can also play a role in determining the rank of a coal, but is less important than the maximum temperature experienced during heating. Duration of heating seems to play little role in controlling

Table 1 Classifications of coal according to rank, and chemical and physical parameters characteristic of rank.

Rank	*Sub category*	*Refl. Rm_{oil}*	*Org. C (%) Dry-ash free*	*Atomic H/C*	*Atomic O/C*	*Characteristics*
Peat		0.2	50–60	0.80–1.40	0.4–0.7	Identifiable plant fragments; abundant cellulose
Lignite		0.3	60–70	0.80–1.20	0.2–0.5	Some identifiable plant fragments; only a residue of cellulose present; dehydration
Sub-bituminous	C	0.4				
	B	0.4–0.5	70–80	0.75–1.00	0.15	Gelification; no cellulose present; loss of methoxyl functionality on lignin
	A	0.5				
High volatile	C	0.6				
bituminous	B	0.7	80–90	0.60–0.75	0.10	Condensation of aromatic structures and loss of H and O; peak of oil, window from coal
	A	0.8–1.1				
Medium volatile bituminous		1.1–1.5	80–90	0.50–0.75	0.05–0.10	
Low volatile bituminous		1.5–1.9	85–90	0.50–0.75	0.05–0.10	
Semi-anthracite		2.0	90	0.50	0.05	
Anthracite		3.0	>90	0.25–0.50	<0.05	Condensation of aromatic structures and loss of H and O nearly complete; graphitization
Meta-anthracite		>4.0	>90	0.25	<0.05	

rank at temperatures below 50 °C (Lopatin, 1971), or at temperatures above 130 °C (Gretener and Curtis, 1982), but at temperatures between 50 °C and 130 °C the length of heating may influence coal rank (Gretener and Curtis, 1982).

A number of different parameters have been used to define coal rank, including: moisture and volatile matter content, reflectance, calorific value, and organic carbon, hydrogen, and oxygen contents. Unfortunately, no single rank parameter is useful over the entire coalification range. Moisture content, for example, changes rapidly during the earliest stages of coalification, and is often used to differentiate peat, lignite, and subbituminous coal. At higher rank, however, changes in moisture content are much smaller, and this parameter becomes less useful in classifying coal by type. Vitrinite reflectance is the most commonly used parameter to classify coals over the main range of hard coals (bituminous coals). This technique measures the proportion of light reflected from specific components of coalified organic matter (the so-called vitrinite macerals) in the polished surface of a coal, compared to that of a standard material of known reflectance. Vitrinite reflectance depends on both the refraction and absorption of light in coal. Refraction and absorption are physical parameters that change based on the electron density in the coal, which varies depending on the content of condensed aromatic structures in the coal. Vitrinite reflectance values (R_0 values, see Table 1) increase with coal rank due to the increase in condensed aromatic structures in coal at higher rank, and the resulting delocalization of electrons (McCartney and Teichmuller, 1972).

7.08.4 STRUCTURE OF COAL

Coal is a heterogeneous organic “rock.” The heterogeneity arises from the fact that coal is composed of components derived from a number of different vascular plant types and organs, as well as organic material derived from algae, fungi, and bacteria. In addition, heterogeneity arises from differences in the degree and manner in which these components are modified prior to burial, and during diagenesis and coalification. The heterogeneous nature of coal is readily apparent when viewing a coal bed. It becomes even more apparent when coal is viewed under a light microscope or when using electron microscopy. Under the microscope, individual constituents or fragments of organic matter are recognizable. These fragments are referred to as macerals (Stopes, 1935).

Three primary maceral groups are recognized: vitrinite, liptinite (also referred to as exenite in the older literature), and inertinite. Each maceral group contains a number of different maceral types. These are grouped together on the basis color, shape, reflectivity, and structure (Taylor *et al.*, 1998). The three primary maceral groups are thought to be derived from different sources. Vitrinite is thought to represent microbially and thermally altered remnants of lignocellulose biopolymers in wood. Liptinite macerals in coal are thought to originate from the lipid-rich fractions of vascular plants: surface waxes (cutan, suberan), resins, sporopollenin, and fats. Inertinite macerals originate from the same plant source materials as vitrinite and liptinite, but the source materials were altered during an early stage (e.g., prior to the onset of coalification) by oxidation processes such as fire. The different maceral groups have distinct chemical compositions, reflecting differences in their origin (Taylor *et al.*, 1998). Liptinite has a higher hydrogen content than other macerals, reflecting the high aliphatic organic matter content of the waxes and other lipid-rich biopolymers that are concentrated in this maceral group. Vitrinite has a high aromatic carbon and oxygen content due to its origin in the lignocellulose of woody tissue in plants. Inertinite has a higher carbon content than the other maceral groups due to carbonization by combustion processes. A number of macerals are created during coalification, and these are termed secondary macerals to distinguish them from the primary macerals that derive from the various plant structural components or early oxidation processes. Macerals may contain inorganic substances bound to or otherwise imbedded in the organic structure at a molecular level. These inorganic substances differ from distinct mineral phases that constitute the bulk of the inorganic portion of coal (see Section 7.12.6).

The nonmineral (i.e., organic) structure of coal is thought to consist of two basic parts: (i) A three-dimensional, cross-linked, macromolecular matrix composed of altered biopolymers such as lignin, and (ii) lower-molecular-weight, more mobile organic substances that are either entrapped within or weakly bonded to the macromolecular matrix. The lower-molecular-weight substances within the coal matrix may include degradation products of lignocellulose, waxes, and other biopolymers, as well as more mobile liquid hydrocarbons and gases (methane, CO_2, N_2). A conceptual model for coal proposes that physically associated interactions of moderate molecular weight polar molecules with the macromolecular three-dimensional matrix are important structural components (Kuangzong *et al.*, 1998). In this model, the organic structure of coal consists of three parts: (i) a rigid three-dimensional structure of macromolecules (altered biopolymers) chemically bonded together, (ii) polar molecules of high and moderate molecular weights (e.g., asphaltenes and resins), and (iii) hydrocarbons of lower molecular weights. The polar molecules of moderate molecular weight are bound to each other and to the macromolecular three-dimensional structure by physical intermolecular forces (e.g., hydrogen bonding, charge-transfer interactions, etc.) to form a fine network structure within which lower molecular weight, nonpolar molecules (primarily hydrocarbons) are sequestered.

7.08.5 HYDROCARBONS FROM COAL

7.08.5.1 Liquid Hydrocarbons (Oil)

Coal can act as a source rock for the generation of liquid hydrocarbons. The production of liquid hydrocarbons from coal is similar to that of type III kerogen, reflecting the origin of the organic matter in coal and type III kerogen, primarily from hydrogen-poor (lipid-poor) and oxygen-rich vascular plants (Tissot and Welte, 1984). Type III kerogens plot at the high end of the O/C ratio in van Krevelen plots. Worldwide, most petroleum source rocks and petroleum deposits are derived from type I and II kerogens. The organic matter in these rocks is derived primarily from more lipid-rich algae (Tissot and Welte, 1984). In some cases, however, coal beds may produce commercial petroleum fields, and a number of oil accumulations worldwide are believed to be derived primarily from coal (Clayton, 1993). One example of coal-derived petroleum commonly cited is the Kutai Basin (Mahakam delta), Indonesia (Durand and Oudin, 1979); other examples have been reported from Australia, China, USA, Canada, and New Zealand (Clayton, 1993). A number of basins in the Far East that contain economic petroleum and gas accumulations have been correlated with Tertiary source rocks containing coal and carbon-rich shale (Lin *et al.*, 1997).

Most coal-derived liquid hydrocarbon deposits are of Late Cretaceous or Tertiary age. Only four known deposits are of Paleozoic age (Clayton, 1993). The age distribution of coal-derived petroleum does not correspond to the worldwide age distribution of coal deposits (Figure 3), as ~60% of the world's coals are Permian or older (Bouska, 1981). Evolutionary changes in the biopolymers characteristic of vascular plants may have contributed to the greater abundance of coal-derived petroleum from Late Cretaceous and younger coals.

The lipid-rich portions of coal are generally regarded as the source of the liquid hydrocarbons derived from coal (Snowdon, 1991). This lipid-rich fraction includes higher plant waxes and resins, as well as algal lipids admixed with the vascular plant materials. The element composition of coal is a key factor in determining its oil-generating potential. Generally, coals are considered oil-prone if their atomic H/C value is 0.8 or higher, corresponding to a hydrogen index (HI) value of >200 in Rock-Eval analysis (Powell, 1988; Hunt, 1991). Lipid-rich components of coal are usually higher in liptinite macerals, but there is no definitive correlation between the maceral composition of coal and its oil potential, because the composition of liptinite macerals in coals is rather variable (Boreham and Powell, 1993). Still, coals with abundant liptinite (15–25% relative abundance) and liptinite bands expel liquid hydrocarbons most efficiently (Pearson and Moore, 2000). Some vitrinite macerals have also been observed to produce oil (Khorasani, 1987; Boreham and Powell, 1991; Powell *et al.*, 1991), but their chemistry is still unclear (Boreham and Powell, 1993). Algal and bacterial biopolymers in inertinites have also been shown to produce small quantities of oil (Taylor *et al.*, 1988). In general, coals containing a significant proportion of microbial remains (the so-called perhydrous coals) also generate more liquid hydrocarbons than average coals.

The initiation of the petroleum-forming potential of coals begins during peatification, when extractable hydrocarbons increase due to the microbial degradation and the transformation of biopolymers in peat. This process continues during the early stages of coalification. The primary liquid hydrocarbon-generating phase for coal (the petroleum/oil window) typically occurs between the subbituminous and the low-volatile bituminous coal rank (vitrinite reflectance R_0 range of 0.5–1.6%), (Figure 9). Rock-Eval pyrolysis, which examines the volatiles emitted from organic matter in order to estimate the hydrocarbon-generating potential of coal, sediments, or source rocks (see Katz, 1983; Hartman-Stroup, 1987; Espitalie and Bordenave, 1993), suggests that coals do not generate significant amounts of hydrocarbons before they reach the high volatile A bituminous rank, and that a significant generative potential remains even in higher-rank, medium-volatile bituminous coals (Newman *et al.*, 1994).

Despite the well-known association and trapping of liquid hydrocarbons within the coal structure, liquid hydrocarbons apparently are continuously generated and expelled from coal. Nevertheless, residual quantities of liquid hydrocarbons are normally trapped within the coal macromolecular matrix. These trapped liquid hydrocarbons are probably converted to gas during continued coalification.

Liquid hydrocarbons derived from coal have a composition that is somewhat distinctive, and that distinguishes them from oils derived from algal-dominated type I and II source rocks. For example, petroleum derived from coal tends to have high pristane/phytane ratios; a ratio >4 is

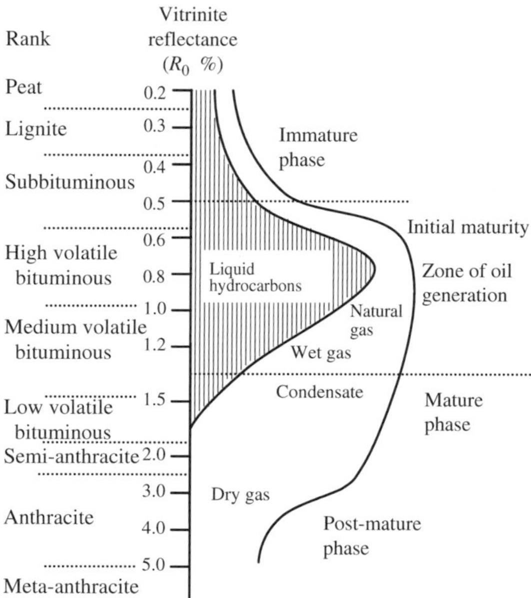

Figure 9 Zones of hydrocarbon generation from coals (oil and natural gas), in relation to rank and other coal properties.

considered diagnostic of a coal source (Clayton, 1993). The high pristane/phytane ratio in coal-derived petroleum may reflect the alteration of the phytol sidechain of chlorophyll during peat formation, producing more of the oxidized isoprenoid product (pristane), than oils from algal-derived source rocks. Petroleum derived from coal also tends to have a dominance of odd-carbon-number *n*-alkanes (i.e., a CPI or carbon preference index >1), reflecting the well-documented dominance of odd-carbon-number molecules in waxes from higher plants (Clayton, 1993). Polycyclic aromatic structures, especially substituted naphthalenes and phenanthrenes, are also more abundant in coal-derived petroleum compared to algal-derived petroleum (Strachan *et al.*, 1988; Alexander *et al.*, 1988). Several other compositional characteristics of petroleum-derived coals distinguish them from those derived from algal-dominated sources (Clayton, 1993).

7.08.5.2 Natural Gas

Coals can produce large reservoirs of natural gas with commercial potential. In the USA alone, coal-bed gas resources are estimated to be 400 trillion cubic feet (Tcf) (1 cf = 0.0283 m^3); 90 Tcf are considered to be recoverable (ICF Resources, 1990). 399 Tcf of natural gas from other conventional sources are considered recoverable (Rice, 1993). Most coals of sufficient rank have generated large volumes of gas that may be trapped by adsorption. Gas production from high-rank coal beds in the Appalachian Basin dates back to the early 1920s (Kuuskraa *et al.*, 1997). Recently, extensive coal bed methane (CBM) resources have been shown to occur in coals of subbituminous rank in the Powder River Basin, WY, USA (Ayers, 2000).

Natural gas is generated in coal beds both by biogenic and thermogenic processes. Gas derived from biogenic processes consists mostly of methane and CO_2, and is produced as a by-product during the anaerobic microbial decomposition of organic matter (i.e., methanogenesis). The various species of methanogenic bacteria (Kingdom Archea) responsible for methane production are the terminal organisms in a consortium of bacteria that degrade biopolymers and geopolymers, first to low-molecular-weight organic compounds via fermentation, and finally to methane and CO_2. Bacteria produce methane via two types of metabolic pathways: CO_2 reduction and fermentation (demethylation of low-molecular-weight organic acids to produce CO_2 and methane). Both CO_2 reduction and fermentation produce natural gas in recent peat deposits, but CO_2 reduction is normally the dominant metabolic pathway for bacterial methanogenesis in older (more decomposed) organic deposits. Large amounts of biogenic natural gas may be generated where conditions are favorable for microbes to flourish and methane to accumulate. The major requirements for biogenic methane production and accumulation include: water, anoxic conditions, a source of metabolizable organic matter, relatively low temperature, a low sulfate concentration, and adequate pore space for natural gas retention.

Large amounts of methane are produced in wetlands during peatification, but most of this methane is probably lost to the atmosphere prior to entrapment during burial. After peat deposits are buried, and during the early stages of coalification (to subbituminous rank), significant amounts of methane can be produced. At low temperatures, rapid burial, and anoxic conditions, economic amounts of biogenic methane may be produced from peat and coal deposits for tens of thousands of years (Claypool and Kaplan, 1974). These early-stage processes probably account for much of the biogenic methane in coal beds (Rice and Claypool, 1981; Rice, 1992).

Biogenic methane may also be produced from coal beds long after burial and coalification. Late-stage biogenic methane production occurs during a relatively short period (tens of thousands to a few million years) following the intrusion of groundwater into coal beds. Intrusion of groundwater into coals is a common occurrence, and coal beds act as regional aquifers in some areas. Oxic groundwater may enter the coal bed, sustaining aerobic metabolism (bacteria and fungi) of the geopolymers in the coal and generating a reservoir of lower-molecular-weight organic material suitable for supporting anaerobic metabolism. As the oxygen in the groundwater is consumed, anoxic conditions develop, and methanogenesis can begin. Late-stage biogenic methane production can occur in coal of any rank where anoxic conditions, a suitable reservoir of metabolizable organic substrate, and relatively low temperature conditions exist. Late-stage biogenic methane production is thought to occur primarily by bacterial CO_2 reduction.

Thermogenic production of methane from coal occurs during coalification as increasing temperature and pressure cause devolatilization of methane, CO_2, water, and other gases from the coal. The generation of methane and other hydrocarbon gases from coal occurs at the high volatile bituminous rank and higher (Figure 9). Gas production increases rapidly with increasing rank. The amount of natural gas produced by thermogenic processes depends on the composition of the coal, and on regional variations in coal rank (Levine, 1987). Other factors influencing natural gas production from coal include the growth of structural features, differences in regional temperature gradients, and basin hydrology (Carroll and Pashin, 1997). The components of coal thought to produce natural gas are liptinite macerals and lipid-rich vitrinite macerals, based on the known ability of these macerals to produce oil from coal. Natural gas can either be produced directly from these macerals in the coal, or by cracking the heavy hydrocarbons present within the coal matrix. Natural gas yields from coal are typically in the range of 100–300 $cm^3\ g^{-1}$ of coal (Rice, 1993; Figure 10).

Retention of natural gas in coal beds takes place via adsorption by the coal surface, absorption within the molecular structure of the coal, and storage within pore spaces or cleats in the coal. The large surface area of the three-dimensional macromolecular structure of coal promotes molecular attraction of natural gas to the coal surface. The presence of liquid hydrocarbons in coal usually reduces its natural gas sorption capacity and natural gas production in coal beds. Liquid hydrocarbons plug the micropore space within coal and reduce the space available for natural gas storage (Levine, 1991). At higher temperatures, however, liquid hydrocarbons in coal may be cracked to produce natural gas. This process opens up micropore space for gas storage (Thomas and Damberger, 1976). The capacity of coal to store methane decreases with increasing rank, largely due to a decrease in pore space with increasing rank (Meissner, 1984; Figure 11). As mentioned earlier, methane production from coal increases with rank (Figure 9), and in higher

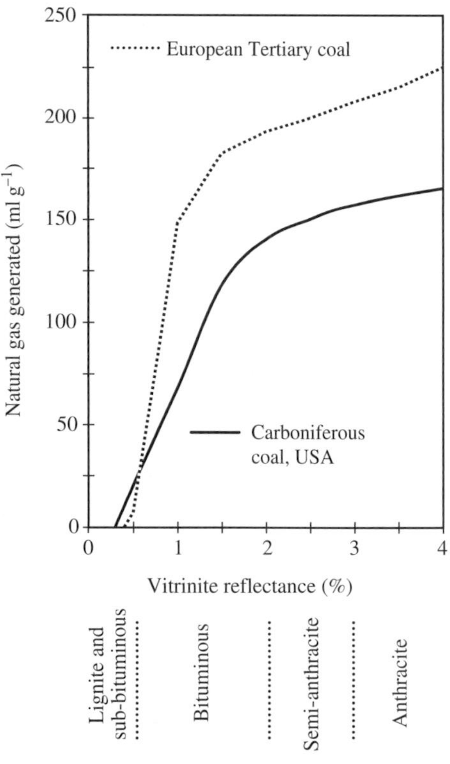

Figure 10 Plot of natural gas yields as a function of rank from a European Tertiary coal and a Carboniferous coal from the eastern USA.

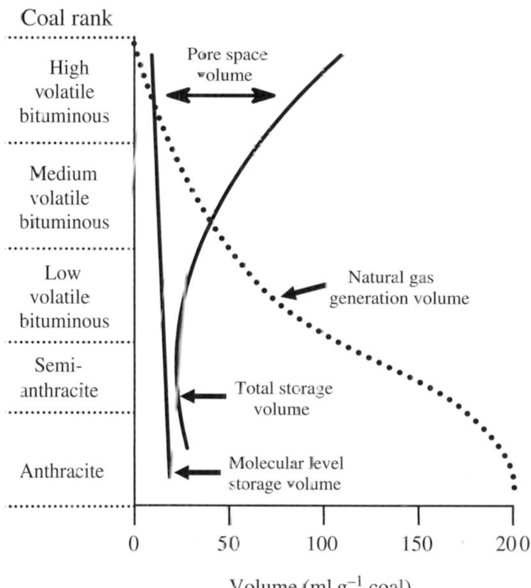

Figure 11 Schematic plot showing natural gas yield and changes in storage volume (molecular storage and pore space storage) as a function of coal rank.

rank coals methane production may exceed the storage capacity of the coal (Rice, 1993). However, the intense heat from igneous intrusions may increase the gas reservoir potential of coal by creating microscopic pores that enhance gas adsorption capacity, gas desorption, and permeability (Gurba and Weber, 2001).

The pressure within coal beds (i.e., the reservoir pressure) helps to retain natural gas within the microstructure of the coal (Rice, 1993). Gas is released from the coal as pressure is reduced, either by natural uplift and erosion of the coal bed or by coal mining or drilling wells into the coal. Coal-gas production can take place only when the reservoir pressure is reduced sufficiently to allow the gas to desorb. Gas flow to wells drilled into the coal takes place through natural fractures, not through the relatively impermeable coal matrix. Economic production of natural gas from coal beds depends on a number of factors, including: coal structure, gas content of the coal, stratigraphy, the distribution of fractures, hydrologic conditions, reservoir pressure, and the presence or absence of "available" gas. The high water content of coals also means that after the wells are drilled, water must be pumped to depressurize the reservoir. The produced water often contains high concentrations of salts and other organic and inorganic substances solubilized from the coal bed. The disposal of these waters can present environmental problems.

The molecular and isotopic composition of natural gas from coal beds is quite varied. Coal beds often contain a mixture of gases, but methane is usually the dominant component (Rice, 1993). Carbon dioxide is the major nonhydrocarbon gas found in coal beds, and can range from 0% to 99% of the total gas in coal beds (Rice, 1993). Most of this CO_2 is formed recently (late-stage) from microbial activity, thermal degradation of carbonates, or from underlying magma chambers. Most of the CO_2 derived from devolatilization during coalification is lost due to the high solubility of CO_2 in formation waters. Some coal beds contain significant amounts (20% or more) of ethane (C_2) and higher hydrocarbons (C_{2+}) (Rice, 1993). The source of the gas (biogenic or thermogenic) plays a key role in determining its composition. Biogenic gases and thermogenic gases produced from higher rank coals consist primarily of methane, while thermogenic gases at lower rank can include variable amounts of C_{2+} hydrocarbons. Most coal beds contain a mixture of biogenic and thermogenic gases. For example, gases from bituminous Pennsylvanian coals of the Appalachian Plateau appear to be a mixture of thermogenic and biogenic gas (Laughrey and Baldassare, 1998). The isotopic signatures (δ^{13}C, δD) of coal-bed gases have been used to examine their origin, but this analytical approach can be complicated by factors such as mixed sources and the method of gas collection (Rice, 1993). In general, biogenic sources produce isotopically lighter methane (δ^{13}C values of -55 to -90 per mil, and δD values of -150 to -400 per mil), due to fractionation of the lighter isotopes during microbial methanogenesis (Rice and Claypool, 1981; Rice, 1992). Thermogenic methane is generally isotopically heavier (more positive δ^{13}C and δD values). The isotopic values generally increase (i.e., become heavier) with increasing rank.

7.08.6 INORGANIC GEOCHEMISTRY OF COAL

7.08.6.1 Abundance of Elements

Although coal consists largely of organic matter, the inorganic constituents in coal commonly attract a disproportionate amount of attention and may ultimately determine how a coal will be used. Except for a few extremely rare elements (actinium, astatine, francium, polonium, protactinium), every element has been found in coal. In ash (the inorganic residue from the complete incineration of coal), elemental concentrations range from the parts per trillion level to more than 50 wt%. Table 2 contains estimates of the arithmetic and geometric means for 79 elements reported in US coal. Although few, if any, coal samples have compositions similar to that of "average" US coal, these values are useful in establishing norms for comparisons. In most coal samples silicon, aluminum, sulfur, iron, and calcium are present at levels of several percent by weight. All other elements (excluding carbon, oxygen, hydrogen, and nitrogen), are usually present at concentrations below 1 wt.% (Bragg *et al.*, 1998).

The abundance of the inorganic constituents in coal varies at every level—between coal basins, between coal beds within a basin, and within coal beds—over distances from micrometers to kilometers. These variations are due to differences in the geologic and geochemical processes acting on the peat and the coal over geologic time (Swaine, 1990; Bouska, 1981). Variations in the plant communities, source material, detrital influx, diagenetic processes, and epigenesis all influence the type and abundance of the inorganic chemical constituents of coal.

The plants that were the precursors of the coal required a range of elements as nutrients or for structural support. Elements essential to plant metabolism include phosphorus, potassium, sulfur, calcium, and magnesium. Some plants also require boron, chlorine, copper, iron, manganese, molybdenum, and zinc (Severson and Shacklette, 1988). Plants may contribute inorganic constituents

Table 2 Arithmetic and geometric means for chemical elements in US coal.

Component	*Arithmetic*		*Geometric*		*Max.*	*Num.*
	Mean	*SD*	*Mean*	*SD*		
Ash (%)	13.1	8.3	10.9	1.9	50.0	7,976
Aluminum (Al) (%)	1.5	1.1	1.1	2.1	10.6	7,882
Antimony (Sb) (ppm)	1.2	1.6	0.61	3.6	35	7,473
Arsenic (As) (ppm)	24	60	6.5	5.5	2,200	7,676
Barium (Ba) (ppm)	170	350	93	3.0	22,000	7,836
Beryllium (Be) (ppm)	2.2	4.1	1.3	3.5	330	7,484
Bismuth (Bi) (ppm)	(<1.0)	ND	ND	ND	14	920
Boron (B) (ppm)	49	54	30	3.1	1,700	7,874
Bromine (Br) (ppm)	17	19	9.1	4.1	160	4,999
Cadmium (Cd) (ppm)	0.47	4.6	0.02	18	170	6,150
Calcium (Ca) (%)	0.46	1.0	0.23	3.3	72	7,887
Carbon (C) (%)	63	15	62	1.3	90	7,154
Cerium (Ce) (ppm)	21	28	5.1	7.1	700	5,525
Cesium (Cs) (ppm)	1.1	1.1	0.70	3.2	15	4,972
Chlorine (Cl) (ppm)	614	670	79	41	8,800	4,171
Chromium (Cr) (ppm)	15	15	10	2.7	250	7,847
Cobalt (Co) (ppm)	6.1	10	3.7	2.9	500	7,800
Copper (Cu) (ppm)	16	15	12	2.1	280	7,911
Dysprosium (Dy) (ppm)	1.9	2.7	0.008	35	28	1,510
Erbium (Er) (ppm)	1.0	1.1	0.002	73	11	1,792
Europium (Eu) (ppm)	0.40	0.33	0.12	5.8	4.8	5,268
Fluorine (F) (ppm)	98	160	35	15	4,000	7,376
Gadolinium (Gd) (ppm)	(1.8)	ND	ND	ND	39	2,376
Gallium (Ga) (ppm)	5.7	4.2	4.5	2.1	45	7,565
Germanium (Ge) (ppm)	5.7	14	0.59	16	780	5,689
Gold (Au) (ppm)	(<0.05)	ND	ND	ND	ND	ND
Hafnium (Hf) (ppm)	0.73	0.68	0.04	33	18	5,120
Holmium (Ho) (ppm)	(0.35)	ND	ND	ND	4.5	1,130
Hydrogen (H) (%)	5.2	0.09	5.2	1.2	9.5	7,155
Indium (In) (ppm)	(<0.3)	ND	ND	ND	ND	ND
Iodine (I) (ppm)	(<1.0)	ND	ND	ND	ND	ND
Iridium (Ir) (ppm)	(<0.001)	ND	ND	ND	ND	ND
Iron (Fe) (ppm)	1.3	1.5	0.75	2.9	24	7,882
Lanthanum (La) (ppm)	12	16	3.9	6.0	300	6,235
Lead (Pb) (ppm)	11	37	5.0	3.7	1,900	7,469
Lithium (Li) (ppm)	16	20	9.2	3.3	370	7,848
Lutetium (Lu) (ppm)	0.14	0.10	0.06	4.7	1.8	5,008
Magnesium (Mg) (%)	0.11	0.12	0.07	2.7	1.5	7,887
Manganese (Mn) (ppm)	43	84	19	3.9	2,500	7,796
Mercury (Hg) (ppm)	0.17	0.24	0.10	3.1	10	7,649
Molybdenum (Mo) (ppm)	3.3	5.6	1.2	6.5	280	7,107
Neodymium (Nd) (ppm)	(9.5)	ND	ND	ND	230	4,749
Nickel (Ni) (ppm)	14	15	9.0	2.8	340	7,900
Niobium (Nb) (ppm)	2.9	3.1	1.0	7.7	70	6,843
Nitrogen (N) (%)	1.3	0.4	1.3	1.4	13	7,153
Osmium (Os) (ppm)	(<0.001)	ND	ND	ND	ND	ND
Oxygen (O) (%)	16	12	12	2.0	60	7,151
Palladium (Pd) (ppm)	(<0.001)	ND	ND	ND	ND	ND
Phosphorus (P) (ppm)	430	1,500	20	20	58,000	5,079
Platinum (Pt) (ppm)	(<0.001)	ND	ND	ND	ND	ND
Potassium (K) (%)	0.18	0.21	0.10	3.5	2.0	7,830
Praseodymium (Pr) (ppm)	(2.4)	ND	ND	ND	65	1,533
Rhenium (Re) (ppm)	(<0.001)	ND	ND	ND	ND	ND
Rhodium (Rh) (ppm)	(<0.001)	ND	ND	ND	ND	ND
Rubidium (Rb) (ppm)	21	20	0.62	41	140	2,648
Ruthenium (Ru) (ppm)	(<0.001)	ND	ND	ND	ND	ND
Samarium (Sm) (ppm)	1.7	1.4	0.35	13	18	5,151
Scandium (Sc) (ppm)	4.2	4.4	3.0	2.3	100	7,803
Selenium (Se) (ppm)	2.8	3.0	1.8	3.1	150	7,563

(continued)

Table 2 (continued).

Component	Arithmetic		Geometric		Max.	Num.
	Mean	SD	Mean	SD		
Silicon (Si) (%)	2.7	2.4	1.9	2.4	20	7,846
Silver (Ag) (ppm)	(<0.1)	0.35	0.01	9.1	19	5,038
Sodium (Na) (%)	0.08	0.12	0.04	3.5	1.4	7,784
Strontium (Sr) (ppm)	130	150	90	2.5	2,800	7,842
Sulfur (S) (%)	1.8	1.8	1.3	2.4	25	7,214
Tantalum (Ta) (ppm)	0.22	0.19	0.02	13	1.7	4,622
Tellurium (Te) (ppm)	(<0.1)	ND	ND	ND	ND	ND
Terbium (Tb) (ppm)	0.30	0.23	0.09	7.7	3.9	5,024
Thallium (Tl) (ppm)	1.2	3.4	0.00004	205	52	1,149
Thorium (Th) (ppm)	3.2	3.0	1.7	5.0	79	6,866
Thulium (Tm) (ppm)	[0.15]	ND	ND	ND	1.9	365
Tin (Sn) (ppm)	1.3	4.3	0.001	54	140	3,004
Titanium (Ti) (%)	0.08	0.07	0.06	2.2	0.74	7,653
Tungsten (W) (ppm)	1.0	7.6	0.10	14	400	4,714
Uranium (U) (ppm)	2.1	16	1.1	3.5	1,300	6,923
Vanadium (V) (ppm)	22	20	17	2.2	370	7,924
Ytterbium (Yb) (ppm)	[0.95]	ND	ND	ND	20	7,522
Yttrium (Y) (ppm)	8.5	6.7	6.6	2.2	170	7,897
Zinc (Zn) (ppm)	53	440	13	3.4	19,000	7,908
Zirconium (Zr) (ppm)	27	32	19	2.4	700	7,913

All values are on a coal basis. Data are exclusively from the US Geological Survey (USGS) except for estimated values in parenthesis which are based on USGS and literature data. Values in brackets are calculated from cerium and lanthanum data and assuming a chondrite normalized rare-earth-element distribution pattern. (ND = no data; SD = standard deviation; Max. = maximum; Num. = number of samples).

to coal in other ways. Several authors have noted the presence of crystalline and noncrystalline biological material, such as phytoliths and sponge spicules in peats (Andrejko and Cohen, 1984; Raymond *et al.*, 1990). Some of this biogenically derived material may be preserved in coal as discrete particles.

Detrital input (air- and water-borne particulates and dissolved species) is one of the more important sources of the inorganic constituents in coal. Many of the minerals in the moderate- to high-ash coals (5 wt.% to more than 20 wt.%) occur intermixed with fragments of organic matter in distinct bands. Although many of the chemical elements may have been originally associated with the plants or with detrital particles, some of these elements are remobilized during the coalification process. Chemical and mineralogical evidence (Finkelman, 1981b) indicates that some of the remobilized elements are precipitated in the coal as authigenic minerals, whereas other elements, such as calcium, sodium, and magnesium can be lost during the coalification process.

The largest and most obvious minerals in coal are the epigenetic minerals that have precipitated in the cleat and fractures subsequent to coal formation. These minerals include sulfides, carbonates, and kaolinite. They are common but volumetrically minor constituents. Nevertheless, the epigenetic sulfide minerals can have a profound impact on the utility of the coal, because they contain potentially hazardous trace elements.

Table 3 Major minerals in world coals.

Major species

- Quartz (SiO_2)
- Clays
  - Illite (K, Al, Silicate)
  - Kaolinite ((Al_2 Si_2)5 $(OH)_4$)
  - Mixed-layer clays (K, Al, Mg, Fe silicates)
- Carbonates
  - Calcite ($CaCO_3$)
  - Siderite ((Fe, Mn)CO_3)
- Pyrite (FeS_2)

Accessories

- Galena (PbS)
- Sphalerite ((Zn, Cd)S)
- Clausthalite (PbSe)
- Chalcopyrite ($CuFeS_2$)
- Crandallite Group (Ca, Ba, Sr)Al_3 $(PO_4)_2$ $(OH)_5H_2O$
- Monazite (REE, Th) PO_4
- Apatite (Ca_5 $(PO_4)_5$ (F, OH))
- Barite ($BaSO_4$)
- Rutile (TiO_2)
- Zircon ($ZrSiO_4$)
- Feldspars (Ca, Na, K, Al silicates)
- Zeolites (Ca, Na, K, Al silicates)
- Ankerite (Ca (Fe, Mg, Mn)$(CO_3)_2$)
- Micas (K, Fe, Mg, Ti, Al silicates)

7.08.6.2 Mineralogy

More than 100 minerals have been identified in coal. However, only a few are common in most coal samples. Table 3 lists these together with some of the more important minor and trace mineral constituents in coal.

The minerals in a coal clearly reflect its geochemical environment of deposition and diagenesis rather than the composition of the source rocks. Ferromagnesian minerals such as pyroxenes, amphiboles, and olivines are very rare in coal; feldspars are also uncommon. These minerals are unstable in low-pH peat–swamp environments and are apparently destroyed in the early stages of coalification. In contrast, more resistant accessory minerals such as rutile, zircon, and rare-earth phosphates are relatively common in coal.

When ferromagnesian minerals are dissolved in acidic swamp waters, they release metals such as nickel, copper, cobalt, chromium, and lead. The anoxic conditions in the swamp and the abundant sulfur liberated by the putrefaction of the decaying plant matter are ideal for the precipitation of sulfide minerals. Generally, the most abundant of these is pyrite. Syngenetic pyrite, commonly in the form of framboids (Figure 10), typically contains small amounts of chalcophile elements such as arsenic, mercury, lead, selenium, cadmium, and thallium. Later generations of pyrite typically contain higher concentrations of these minor elements. This is especially true for late stage, epigenetic pyrite (Figure 12). Other sulfide minerals that are common diagenetic products in coal include chalcopyrite, sphalerite, galena, and the lead selenide chlausthalite (Figure 13). The accessory sulfide minerals may be inconspicuous, but they are not insignificant. These minerals generally contain the bulk of the environmentally deleterious elements such as arsenic, mercury, lead, selenium, cadmium, and thallium. Oxidation, leaching, or heating releases these elements into the environment, sometimes with devastating consequences (see Section 7.12.10.4).

Clays are volumetrically the most abundant mineral group in coal. They can be authigenic or detrital in origin. Kaolinite is the most common clay and the most common authigenic mineral in coals. The silicon and aluminum in kaolinite are, perhaps, residual from the dissolution of ferromagnesian minerals and feldspars. Illite and mixed layer clays in coal are almost exclusively detrital in origin. Chlorites, smectites, and other clay minerals may be abundant locally.

Quartz, one of the most common minerals in most coals, can have a detrital, authigenic, or epigenetic origin. Carbonate minerals, generally, calcite, siderite, and ankerite occur most commonly as syngenetic nodules or as epigenetic cleat and fracture fillings. The carbonate minerals in coal can have a beneficial effect by neutralizing sulfuric acid produced during the oxidation of the sulfide minerals.

Information on coal mineralogy has been used primarily to assess the technological performance

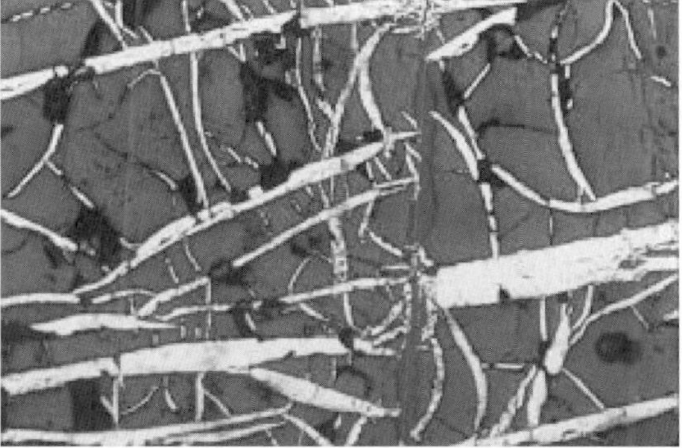

Figure 12 Late stage (epigenetic) pyrite fracture fillings in a bituminous coal. Reflected light photomicrograph.

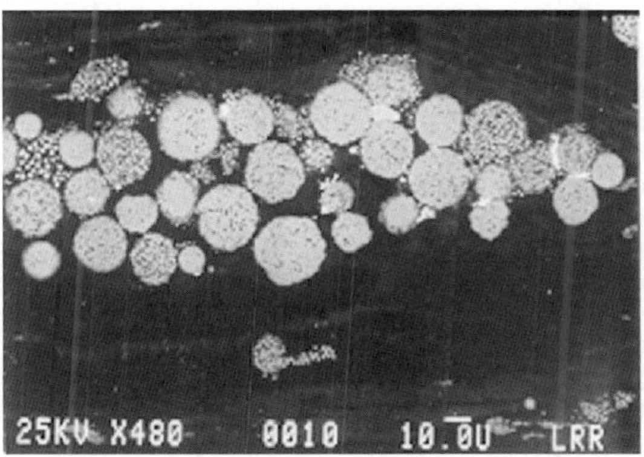

Figure 13 A cluster of syngenetic pyrite framboids in a bituminous coal. Micrometer-sized bright grains dispersed in the cluster are crystals of clausthalite (PbSe). Scanning Electron photomicrograph, backscattered electron image (scale bar = 10 μm).

of coal in slagging, fouling, corrosion, erosion, and abrasion (Gupta *et al.*, 1999), and the potential environmental impact due to acid generation caused by pyrite oxidation (Kolker *et al.*, 2001). Ward *et al.* (2001) have used the mineral abundances in Australian coals as an aid to seam correlation.

7.08.6.3 Modes of Occurrence

Information on the element and mineral content of coals provides only a partial picture of the potential behavior of the trace elements in coal. The picture can be brought into sharper focus if we know an element's mode of occurrence (chemical form). The mode of occurrence of an element in coal determines its technological behavior, environmental impact, and economic potential, and provides insight into the geochemical processes that occurred during coalification.

There has been considerable work undertaken to determine the modes of occurrence of elements in coal. Table 4 contains a list of the more likely modes of occurrence for selected elements in coal. It is apparent from the variety of mineral species and organic associations that determining the modes of occurrence can be a challenging task. The low concentration levels of some elements, the intimate intermixtures of organic and inorganic materials, and the very fine grain sizes of the minerals (most are less than 2 μm in diameter) further complicate the task. For these reasons,

Table 4 Modes of occurrence of selected elements in coal in decreasing order of probability.

Element	*Modes of occurrence*
Aluminium	Clays, feldspars, organic association, Al-oxides
Antimony	Accesory sulfide, organic association
Arsenic	Solid solution in pyrite, rare organic association
Barium	Barite, crandallite, organic association
Beryllium	Clay, organic association
Bismuth	Accessory sulfide
Boron	Organic association, illite
Bromine	Organic association
Cadmium	Sphalerite, clay, pyrite
Calcium	Calcite, sulphates, organic association, silicates, phosphates
Cesium	Clays, feldspar, mica
Chlorine	Organic association
Chromium	Clays (especially illite), organic association, spinels
Cobalt	Accessory sulfides, pyrite, clays, organic association (?)
Copper	Chalcopyrite, organic association, clays
Fluorine	Perhaps apatite, clays, mica, amphiboles (rare)
Gallium	Clays, organics, sulfides
Germanium	Organic association
Gold	Native gold, organic association
Hafnium	Zircon, clays
Indium	Probably sulfides
Iodine	Probably organic association
Iron	Pyrite, siderite, sulfates, oxides, organic association
Lead	Galena, PbSe, clays, organic association (?)
Lithium	Clays
Magnesium	Clays, organic association
Manganese	Siderite, calcite, pyrite, clays, organic association
Mercury	Solid solution with pyrite, rare organic association
Molybdenum	Clays, sulfides, organic association
Nickel	Mono-sulfides, pyrite, organic association, and clay
Niobium	Oxides
Phosphorus	Phosphates
Potassium	Clays, feldspars
Platinum	Native alloys, perhaps some organic association
Rare earths	Phosphates, some organic association
Rubidium	Probably illite
Scandium	Clays, phosphates, organic association
Selenium	Organic association, pyrite, PbSe
Silicon	Quartz, clays, silicates
Silver	Perhaps silver sulfides
Sodium	Organic association, clays, zeolites, silicates
Strontium	Carbonates, phosphates, organic association
Tantalum	Oxides, silicates
Tellurium	Unclear
Thorium	Rare-earth phosphates, clays
Tin	Tin oxides, sulfides
Titanium	Clays, oxides, organic association
Tungsten	Oxides, clays, organic association
Uranium	Organic association, zircon, phosphates
Vanadium	Clays, some organic association
Yttrium	Rare-earth phosphates
Zinc	Sphalerite
Zirconium	Zircon

Note: Elements in **bold type** have more significant organic association in low-rank coal.

most efforts to determine the modes of occurrence of trace elements in coal have been qualitative.

Modes of occurrence of the elements in coal can be determined using a variety of procedures. Perhaps the most effective method is the use of scanning electron microscopy-energy dispersive X-ray analysis (SEM-EDX). This method can detect and analyze minerals as small as 1 μm in diameter (Figure 14). The SEM-EDX also provides useful information on the textural relationships of the minerals. Other microbeam techniques, such as the electron microprobe analyzer, ion microprobe, laser mass analyzer, and transmission electron microscopy, have also been used to determine modes of occurrence of elements in coal.

Mineralogical techniques such as X-ray diffraction, differential thermal analysis, Fourier transform infrared analysis, and Mössbauer spectroscopy may be useful for determining modes of occurrence of major elements in coal, but the ability of these techniques to determine the modes of occurrence of the minor- and trace-elements is quite limited. X-ray absorption–fine structure (XAFS) spectroscopy has been used to determine the modes of occurrence of several important minor and trace elements (e.g., arsenic and chromium) in coal (Huggins *et al.*, 2000; Kolker *et al.*, 2000).

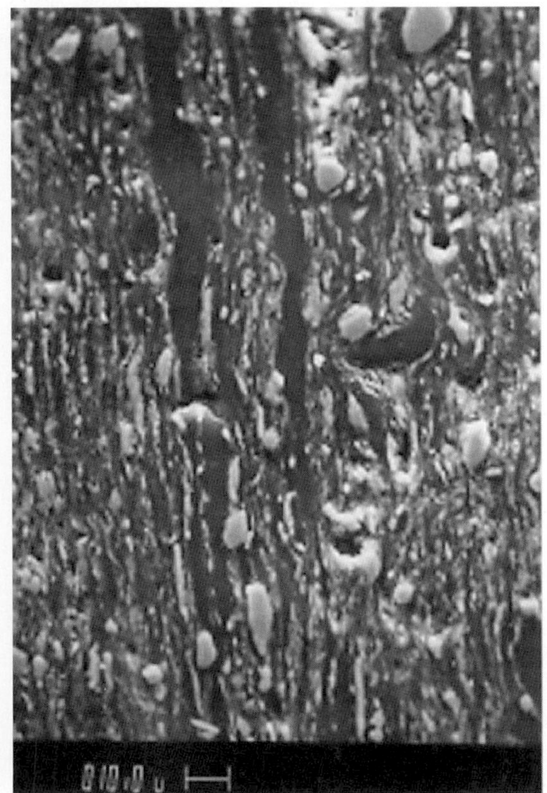

Figure 14 Backscattered SEM image depicting detrital clay and quartz grains in a bituminous coal. Most minerals in coal are only (1–2 μm) in diameter (scale bar = 10 μm).

The modes of occurrence of an element can be inferred from indirect evidence such as the chemical composition of density separates, from statistical correlation with other elements, from other coal characteristics such as correlation with ash yield, from an element's geochemical characteristics, or from its behavior during heating or leaching of the coal. These indirect methods are generally relatively rapid, but they are prone to error (Finkelman, 1981a), and they are best used in conjunction with one of the direct methods. Palmer *et al.* (1998) have attempted to quantify the modes of occurrence by combining results from sequential leaching, scanning electron microscopy, electron microprobe analysis, and X-ray diffraction analysis. From the sequential leaching results, the modes of occurrence of ~40 elements can be inferred from their response to different solvents (Palmer *et al.*, 1998). Microprobe analyses, generally sensitive to about 100 ppm, provide information regarding the concentration of chromium (in illite), arsenic, selenium, cobalt, nickel, lead (in pyrite), and manganese and strontium (in carbonates). In the leaching procedure, generally more than 50% of the arsenic and mercury are leached by HNO_3, indicating an association of these elements and pyrite. With few exceptions, more than 50% of the Be and 40% of the chromium are leached by HF, indicating an association with aluminosilicates. More than 35% of the selenium is associated with disulfides in most coals, but as much as 95% of the selenium was not leached in several low-rank western US coal samples, indicating an organic association. Cobalt, nickel, antimony, thorium, and uranium have multiple modes of occurrence. Commonly, more than 55% of the cadmium and up to 75% of the lead is associated with mono- and disulfides.

7.08.7 GEOCHEMISTRY OF COAL UTILIZATION

One of the costliest problems of coal utilization is the buildup of sintered ash deposits on the heat-exchange surfaces of coal-fired boilers (Reid, 1981). These deposits not only reduce the efficiency of the boiler drastically, but also promote corrosion and erosion (Honea *et al.*, 1982). The size and strength of these deposits depends on the configuration of the boiler, the operating conditions, and the inorganic composition of the coal being combusted. The primary compositional factors that influence the size and strength of these deposits are the ash yield and the concentration of sodium, calcium, and magnesium in the coal (Raask, 1985). A high silicon:aluminum ratio, a function of coal mineralogy, is also important. Slagging, the buildup of deposits on the boiler walls is particularly sensitive to the concentration and mineralogy of iron and sulfur.

Abrasion of the mining and grinding equipment can also be a costly problem. Excess abrasion is caused by high concentrations or large particle size of hard minerals such as quartz. During combustion, chlorine in the coal may form hydrochloric acid and contribute to corrosion of metal surfaces in coal-fired boilers.

Numerous algorithms incorporating element concentrations have been developed to predict the propensity of a coal to foul and slag, but none have won general acceptance (Vaninetti and Busch, 1982). Among the reasons for the difficulty of developing accurate predictive models are the variations in element modes of occurrence, differences in boiler configuration, variable composition of the feed coal, and the complexity of the nonequilibrium chemical reactions that occur during combustion.

7.08.8 ECONOMIC POTENTIAL OF METALS FROM COAL

Some inorganic constituents in coal have an important economic potential as by-products. Extraordinarily high concentrations of some trace elements in coal have been reported, some many times higher than the maximum values in Table 2. For example, Jenny (1903) described several pre-1900 mining operations in which minerals were extracted from coal and coal-bearing rocks, including vanadium from lignite in Argentina, antimonial silver from coal in Peru, and sphalerite (ZnS) and galena (PbS) from Missouri coal. In the 1960s there was limited commercial production of uranium from North and South Dakota lignites (Noble, 1973; Schnable, 1975). In one part of the Dakotas, molybdenum was enriched to such a degree that it was recovered together with the uranium. However, burning of the molybdenum-rich lignite allegedly caused molybdenosis in cattle (E. A. Noble, personal communication, 1990). In the Russian Far East, coals have been found that contain potentially economic concentrations of germanium, platinum group elements, rare earth elements, and other elements (Seredin, 1995).

The association of minerals and coal can be used in mineral exploration. Although the concentration of elements in coal may not be economic, it may indicate the presense of nearby mineralization. Coal may thus be used as a geochemical prospecting tool to locate economic mineral deposits. Finkelman and Brown (1991) noted the remarkable high silver content (500 ppm in the ash) in a central Texas bituminous coal and suggested that this anomalously high silver content might indicate the presence of nearby economic deposits (Figure 15). Similarly, Ruppert *et al.* (1996) used data for the concentration and distribution of nickel in the Kosovo Lignite Basin in the former Yugoslavia to identify a nearby potential economic nickel deposit.

7.08.9 INORGANICS IN COAL AS INDICATORS OF DEPOSITIONAL ENVIRONMENTS

There have been numerous attempts to use trace-element concentrations in coal as indicators of depositional environments. Most commonly, these studies have sought evidence of a marine influence on the coal (Goodarzi, 1987, 1988; Swaine, 1983; Chou, 1984; Hart and Leahy, 1983). The elements cited as indicators of marine influence include molybdenum, magnesium, boron, chlorine, bromine, sodium, yttrium, and uranium. However, problems such as mixing in brackish environments, the reworking of sediments, and postdepositional enrichment or leaching, make the data equivocal, and there is no consensus regarding reliable indicators.

Several authors have used trace-element concentrations to correlate coal beds (Butler, 1953; Alpern and Morel, 1968; Nichols and D'Auria, 1981; Swaine, 1983; O'Connor, 1988). Significant vertical and lateral variations in the concentration of trace elements within coal beds has limited the success of these attempts. The proper selection of elements for this purpose is critical. Elements that have a tendency to form complexes with organic matter (e.g., boron, germanium, beryllium) or with sulfides (zinc, copper, lead) or that are soluble in water or dilute acids (sodium, magnesium, potassium, manganese, calcium) are poor candidates for correlating coal beds. Elements that are constituents of relatively inert minerals (zirconium, hafnium, niobium, tantalum, scandium) should be more useful (Finkelman, 1981a). The concentrations or ratios of these elements are not significantly affected by changes in Eh and pH, rate of detrital influx, changes in plant or bacterial communities, or availability of sulfide ions. Palmer *et al.* (2002) used the chemistry of coal particles recovered from the site of the Titanic wreck to indicate the source of the coal used by the ill-fated ocean liner. The Cr/Ni ratio of most of the 20 coal particles analyzed were more consistent with coal from Great Britain than with coal from the United States.

7.08.10 ENVIRONMENTAL IMPACTS

7.08.10.1 Coal Combustion

About 90% of the coal mined in the US is burned, principally to generate electricity. The coal

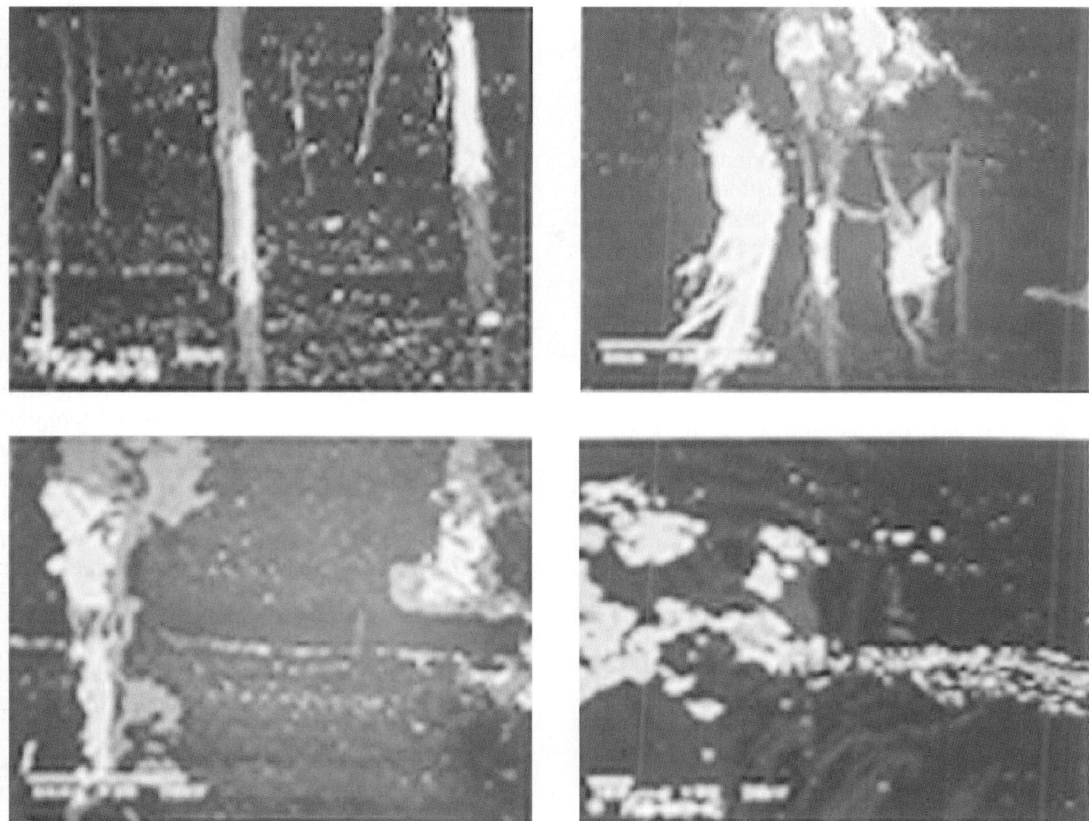

Figure 15 Composite of backscattered SEM images depicting large fracture filling silver-bearing sphalerite crystals in a bituminous coal from Texas.

combustion process produces large quantities of waste products that may be released into the environment. Most of what is seen emitted from the smoke stacks of utility boilers is harmless water vapor (steam). However, the emissions do contain small but significant amount of pollutants. Among these pollutants are sulfur and nitrogen oxides, carbon dioxide, mineral and coal particulates, trace elements, and trace amounts of organic compounds. Each of these pollutants has different effects on human health and the environment.

The environmental impact of coal combustion has been of concern for nearly one thousand years. In the thirteenth century, complaints were raised in London about the deterioration of air quality from the extensive burning of coal (Schobert, 1987). Since then, from time to time, coal combustion has been restricted or banned in London, England, because of the environmental and health problems caused primarily by the release of sulfur gases.

7.08.10.1.1 Sulfur emissions

Sulfur in coal occurs principally as pyritic sulfur and organically bound sulfur. Other sulfur forms (sulfate and elemental) are common trace constituents of coal. Regardless of the form, sulfur is oxidized during coal combustion to form various gaseous sulfur oxide compounds.

These sulfur oxides have a broader environmental impact than any other combustion product of coal. Sulfur's undesirable effects have been known for a long time. In 1880, the death of approximately 1,000 people in London, England were attributed to sulfurous gases produced by coal combustion (Schobert, 1987). One of the more obvious, but less significant, impact is the obnoxious odor of gaseous sulfur compounds. Sulfur dioxide can be absorbed in the lining of nasal passages where it is converted to sulfuric acid which causes a burning sensation. Sulfuric acid can also coat respirable dust particles that are taken into the lungs, causing severe breathing problems.

In the atmosphere, sulfur oxides can combine with water and oxygen to form sulfurous and sulfuric acids. The deposition of these acids causes corrosion or decomposition of materials such as limestone, marble, iron, and steel. The deterioration of building facades and monuments is one result of this worldwide problem. Flushing of the sulfur oxides from the air by precipitation (acid rain) can lead to acidification of lakes and soils, weakening or killing plants and animals.

The US has been very successful in reducing sulfur emissions from coal combustion. There are several ways in which sulfur emissions can be reduced. These include cleaning coals by physically removing the pyritic sulfur prior to combustion; combustion gas scrubbing to remove gaseous sulfur compounds after combustion but prior to release of the gases to the atmosphere; and switching from high-sulfur to low-sulfur coals or other fuels.

7.08.10.1.2 Nitrogen emissions

Virtually all of the nitrogen in coal is organically bound. During coal combustion, nitrogen is oxidized to form several gaseous oxides. However, a large part of the nitrogen oxides produced and emitted during coal combustion comes from nitrogen in the air reacting with oxygen at high temperatures. These products contribute to the formation of smog and react with oxygen in the presence of light to produce a variety of materials that cause eye and respiratory irritation. Like sulfur oxides, nitrogen oxides combine with oxygen and air moisture to form acids. Nitric acid that is flushed from the air by precipitation contributes to the acidification of surface water and soil. Technology is available to reduce the amount of nitrogen oxides produced during coal combustion and to enhance their decomposition prior to release.

7.08.10.1.3 Carbon dioxide emissions

The combustion of fossil fuel produces substantial amount of carbon dioxide. The accumulation of carbon dioxide and other "greenhouse gases" in the atmosphere affects the global climate (McCabe *et al.*, 1993). Heat radiated from the Earth is emitted in the infrared wavelength region. Carbon dioxide and water vapor in the atmosphere absorb infrared energy. The energy is emitted uniformly in all directions. The downward directed IR must make at least one more pass before it escapes from the atmosphere. This process warms the atmosphere and may have profound effects on precipitation, crop yields, and sea level. However, there is still no concensus regarding the quantitative aspects of these consequences.

As there is no politically and economically attractive alternative to energy generation by fossil fuels during the next several decades, research is being conducted to reduce the environmental impact of fossil fuel burning by reducing or sequestering the CO_2 produced by fossil fuel combustion.

7.08.10.1.4 Emission of particulates

Coal generally contains a mineral fraction that amounts to 5–20 wt.%. During combustion, most of the minerals are transformed into dust-sized, glassy particles and, along with some unaltered mineral grains and unburned carbon, are emitted from the smoke stacks. These particles contribute to the smog problem, are eye- and respiratory irritants, and act as substrates for the deposition of sulfuric and nitric acid.

Modern pollution control technology can remove ~99% of the particulates in the combustion gases. Improved coal cleaning procedures and pollution control technology promise to be even more efficient in reducing the amount of particulates produced by coal combustion.

7.08.10.1.5 Emission of trace elements

Most trace elements in coal are associated with the mineral fraction although some are organically bound (Finkelman, 1981a). At sufficiently high exposures, some of these elements (e.g., antimony, arsenic, beryllium, cadmium, chromium, mercury, nickel, lead, selenium, and uranium) can be harmful to human health and to the environment. Although these elements are present in small concentrations in coal (generally a few ppm), the vast amount of coal that is burned annually mobilizes tons of these pollutants. Deposition of these pollutants downwind from power plants can lead to high trace-element concentrations in soil and uptake by plants. The pollutants can retard plant growth or enter the food chain, causing adverse health effects in animals and humans (Adriano, 1986).

Existing pollution control devices can remove as much as 99% of the trace elements (except mercury and selenium) from the combustion gases. Substantial amounts of mercury and, to a lesser extent, selenium are emitted with the combustion gases. There are a number of ways to reduce trace-element emissions due to coal combustion. These pollution control options include switching to coals with a lower trace-element content, selectively mining those parts of coal beds with lower trace-element contents, cleaning the mined coal, and using pollution control devices such as fabric filters, electrostatic precipitators, and combustion gas scrubbers.

7.08.10.1.6 Emission of organic compounds

The combustion of coal is not 100% efficient. Several percent of the carbon are not burned and this residue if commonly is carried along with the combustion gases. This carbon can react to form small amounts of polycyclic aromatic hydrocarbons (PAH) and other organic compounds. Certain PAH compounds are known to be carcinogens. The temperatures achieved by coal-burning utility and industrial boilers are sufficient to destroy most of these organic compounds. They do, however, persist in the less efficient residential coal-combustion systems.

7.08.10.2 Environmental Impacts of Mining and Disposal of Coal Wastes

Coal mining exposes sulfides (primarily pyrite) in coal and associated rocks to oxygen and moisture. These oxidize the sulfides and form sulfuric acid. The resulting acidic waters (referred to as acid mine drainage (AMD)) adversely impact the biota in watersheds downstream from active and abandoned mines. Oxidation of the sulfides also releases chalcophyllic trace elements into the water. Many of these elements precipitate in oxygenated surface waters and are concentrated in stream sediments (Goldhaber *et al.*, 2001).

Many state regulatory agencies recognize the potentially harmful effects of trace elements released during coal mining and utilization. Boon *et al.* (1987) noted that most western US coal-mining states require baseline information for several trace elements, especially selenium, boron, and molybdenum. The release of selenium from coal combustion wastes has caused significant environmental damage (extensive fish kills) at two sites in North Carolina and one in Texas (Shepard, 1987).

7.08.10.3 Disposal of Solid Wastes from Coal Utilization

Coal mining, coal cleaning, and coal combustion generate large quantities of solid by-products. The ash generated from coal combustion alone produces about 100 Mt of solid residue. About 30% of this material is used, primarily in construction materials; but the remaining 70% is stored in impoundments and landfills. Concerns about the leaching of toxic elements from coal ash has led to federal and state guidelines for disposal practices to limit the potential of leachates to contaminate surface and ground waters.

7.08.10.4 Human Health

Health problems caused by coal combustion are due to either the use of poor quality coal (high ash, high sulfur, or high content of toxic trace elements) or to the improper use of coal. When poor quality coal is used improperly, health problems due to coal burning can be widespread and severe. In millions of houses in many developing countries coal is burned in unvented stoves, causing severe indoor air pollution. In Guizhou Province, Southwest China, the situation is exacerbated by the use of coal in which toxic elements such as arsenic are concentrated to an extraordinary degree (Finkelman *et al.*, 1999). Thousands of people in this region are suffering from severe arsenic poisoning. The primary source of the arsenic appears to be the consumption of chili peppers dried over fires fueled with high-arsenic coal. Coal samples in the region were found to contain up to 35,000 ppm arsenic compared to a US average of 24 ppm (Table 2). Chili peppers dried over high-arsenic coal fires adsorbed 500 ppm arsenic on average. In addition, more than 10 million people in Guizhou Province and surrounding areas suffer from dental and skeletal fluorosis. The excess fluorine intake occurs during the ingestion of corn that has been dried over burning briquettes made from high-fluorine coals and high-fluorine clay binders.

Data from Bencko *et al.* (1977) indicate that arsenic released from coal-burning power plants in Czechoslovakia may have affected the hearing of children in the nearby communities. The coal cleaning and effluent treatment practices in the US and in most developed countries greatly reduce the level of toxic emissions from coal-burning power plants.

An unusual situation exists in the Balkans, where health problems may stem from undisturbed coal deposits. Well waters containing toxic organic compounds (aromatic amines, heterocyclic compounds) leached from low-rank coals (Pliocene lignites) may be the cause of (or a contributing factor to) the disease Balkan Endemic Nephropathy (BEN) (Orem *et al.*, 1999, 2002; Tatu *et al.*, 1998, 2000a,b). BEN is a kidney disease (interstitial nephropathy) with a very high co-incidence of renal/pelvic cancer (uroepithelial carcinomas). It is believed to have killed more than 100,000 people in Yugoslavia alone. Adverse health effects associated with toxic organic compounds derived from coal may be very widespread, but few studies have addressed this issue (Orem, 2001).

Perhaps the most widely known health impact of coal is "black lung disease" also known as coal workers' pneumoconiosis. This was a widespread respiratory problem of coal miners caused by inhalation of dust particles during mining. The mineralogical content of the coal may play a role in the incidence and severity of the disease (Finkelman *et al.*, 2002).

Not all of the allegations of health problems caused by coal are legitimate. Concerns expressed about exposure to radioactivity from coal and coal combustion products are misplaced. The products of commercial coal combustion (fly ash, bottom ash) do contain uranium and thorium in concentrations ~5–10 times higher than that of the coal. But the uranium and the thorium in the coal by-products are not a cause for concern, because they are mostly in insoluble forms at concentration levels similar to those in most soils.

7.08.11 CONCLUSIONS

Coal is arguably the most important energy resource worldwide. It produces electricity, heat for warmth and cooking in third world countries, and petroleum and natural gas (the latter in great abundance). Understanding the complex geochemistry of coal is important so that we can make efficient and intelligent use of this economically critical (but often unappreciated) natural resource.

The organic geochemistry of coal is complex. Myriad sources of organic matter contribute to its formation, and a large variety of biological and physical processes affect changes in the biomolecules that are the starting material for coal. Early work on the organic geochemistry of coal was hampered by the absence of analytical tools for probing the complex organic structures of this material. Since the mid-1970s, however, significant advances in our understanding of coal formation, and of the organic geochemistry of coal have been made, largely due to the application of new analytical techniques.

The basic framework of coal formation begins during peatification. Microorganisms preferentially degrade certain plant biomolecules (e.g., carbohydrates), leaving other biomolecules selectively preserved (e.g., lignin and waxes). This observation has replaced earlier ideas regarding peatification. Studies of early-stage coalification have shown that numerous chemical changes in biomolecules occur due to increasing temperature (e.g., loss of oxygen functionality and changes in lignin), but that the overall structural characteristics of low-rank coals remain intact. Later-stage coalification (subbituminous rank and higher) involves more dramatic changes to the structure of coal, including the transformation of coal from a soft, brown substance to a hard, black, lustrous, organic rock. Both oxygen and hydrogen are lost during late stage coalification, and aromatic structures become more condensed.

Despite these advances, our knowledge of the details of coal formation remains quite sketchy. For example, we do not yet have a full knowledge of the biopolymers from plants that contribute to coal formation. Furthermore, while the selective preservation hypothesis provides the basis outlines of coal formation and organic geochemistry, many details of the processes determine the formation of coal are relatively poorly understood.

Coal is an important source of oil and natural gas (especially natural gas), but our understanding of the process of hydrocarbon production from coal remains quite primitive. Natural gas is produced from coal by both thermal and microbial processes, but the specific organic moieties in coal that are responsible for oil and natural gas production are poorly understood. A better understanding of natural gas production by microbial activity (methanogenesis of coal) may allow a better prediction of the coal beds likely to produce quantities of natural gas. It would be useful to produce a modified methanogenic bacterium capable of degrading organic substances in coal to produce natural gas.

Virtually all chemical elements can be found in coal. They occur in a wide variety of forms, most as constituents of minerals. The concentration of the elements and their modes of occurrence are influenced by various factors, including the amount and source of detrital input, organic complexing, climatic changes, water chemistry, diagenetic processes, and epigenetic mineralization. This diversity of influences makes the study of coal geochemistry challenging and rewarding.

Trace and minor elements and the minerals with which they are associated play crucial roles in the utilization of coal. They are the cause of technological problems, such as abrasion of mining equipment and the corrosion, fouling, and slagging of utility boilers. They also contribute to environmental pollution and health problems. Not all of their impact is detrimental; inorganic constituents offer the potential of by-product recovery from coal and their use in geochemical exploration. Moreover, we may learn how to use their distribution as an aid in correlating coal beds and deciphering depositional environments and diagenetic histories.

If we are to mitigate the problems caused by the inorganic constituents in coal, or realize their potential, we must conduct more detailed research. We must better understand the ways in which the inorganic components are incorporated in coal and what factors dictate their modes of occurrence. We must quantify the modes of occurrence and determine how these various forms behave during the processes (e.g., cleaning, combustion, gasification, liquefaction, oxidation, leaching) to which the coal will be subjected. We can then construct models that will lead us to the appropriate coal for each particular process, as well as to models that will tell us the most appropriate use for the readily available coal resources. When we accomplish these objectives, coal will be a more efficient, cost-effective, and environmentally compatible energy source.

REFERENCES

Aaby B. and Tauber H. (1975) Rates of peat formation in relation to degree of humification and local environment, as shown by studies of a raised bog in Denmark. *Boreas* **4**, 1–17.

Adriano D. C. (1986) *Trace Elements in the Terrestrial Environment*. Springer, Berlin.

Alexander R., Larcher A. V., Kagi R. I., and Price P. L. (1988) The use of plant-dervied biomarkers for correlation of oils with source rocks in the Cooper/Eromanga Basin system, Australia. *APEA J.* **28**, 310–324.

Allard B., Rager M.-N., and Templier J. (2002) Occurrence of high molecular weight lipids (C^{80+}) in the trilaminar outer cell walls of some freshwater microalgae. A reappraisal of algaenan structure. *Org. Geochem.* **33**, 789–801.

Alpern B. and Morel P. (1968) Examen. dans Ie cadre du basin houiller lorrain, des possibilitiés stratigraphiques de la géochimie. *Ann. Soc. Geol. Nord.* **88**(4), 185–202.

Anderson J. A. R. (1964) The structure and development of the peat swamps of Sarawak and Brunei. *J. Tropical Geogr.* **18**, 7–16.

Anderson J. A. R. (1983) The tropical peat swamps of western Malesia. In *Mires: Swamps, Bog, Fen, and Moor (Ecosystems of the World 4A)* (ed. A. J. P. Gore). Elsevier, pp. 181–199.

Anderson K. B. and Crelling J. (1995) *Amber Resinite, and Fossil Resins*, American Chemical Society Symposium Series 617. American Chemical Society, Washington, DC.

Anderson K. B. and Winans R. E. (1991) The nature and fate of natural resins in the geosphere: II. Identification, classification, and nomenclature of resinites. *Org. Geochem.* **18**, 829–841.

Andrejko M. J. and Cohen A. D. (1984) Scanning electron microscopy of silicophytoliths from the Okefenokee swamp-marsh complex. In *The Okefenokee Swamp* (eds. A. D. Cohen, D. J. Casagrande, M. J. Anderjko, and G. R. Best). Wetland Surveys, pp. 468–491.

Ayers W. B. (2000) Methane production from thermally immature coal, Fort Union Formation, Powder River Basin. *Am. Assoc. Petrol. Geol. Ann. Meet. Exp. Abstr.* **2000**, 8.

Barber K. E. (1993) Peatlands as scientific archives of past biodiversity. *Biodiver. Conser.* **2**, 474–489.

Barker C. E. (1979) Vitrinite reflectance geothermometry in the Cerro Prieto geothermal system, Baja California, Mexico. PhD Dissertation, University of California, 127pp.

Behar F. and Hatcher P. G. (1995) Artificial coalification of a fossil wood from brown coal by confined system pyrolysis. *Energy Fuels* **9**, 984–994.

Behar F., Vandenbroucke M., Teermann S. C., Hatcher P. G., Leblond C., and Lerat O. (1995) Experimental simulation of gas generation from coals and a marine kerogen. *Chem. Geol.* **126**, 247–260.

Bencko V., Symon K., Chladek V., and Pihrt (1977) Health aspects of burning coal with a high arsenic content: II. Hearing changes in exposed children. *Environ. Res.* **13**, 386–395.

Benner R., Newell S. Y., Maccubbin A. E., and Hodson R. E. (1984) Relative contributions of bacteria and fungi to rates of degradation of lignocellulosic detritus in salt-marsh sediments. *Appl. Environ. Microbiol.* **48**, 36–40.

Benner R., Morgan M. A., and Hodson R. E. (1985) Effects of pH and plant source on lignocellulose biodegradation rates in two wetland ecosystems, the Okefenokee Swamp and a Georgia salt marsh. *Limnol. Oceanogr.* **30**, 489–499.

Benner R., Morgan M. A., and Hodson R. E. (1986) Biogeochemical cycling of lignocellulosic carbon in marine and freshwater ecosystems: relative contributions of procaryotes and eucaryotes. *Limnol. Oceanogr.* **31**, 89–100.

Berner R. A. (1971) *Principle of Chemical Sedimentology*. McGraw Hill, New York.

Boon D. Y., Munshower F. F., and Fisher S. E. (1987) Overburden chemistry: a review and update. In *Billings Symposium on surface Mining and Reclamation in the Great Plains*, American Society for Surface Mining and Reclamation, Reclamation Research Unit Report No. 8704, pp. A1.1–A1.18.

Boreham C. J. and Powell T. G. (1991) Variation in pyrolysate composition of sediments from the Jurassic Walloon coal measures, eastern Australia as a function of thermal maturation. *Org. Geochem.* **17**, 723–733.

Boreham C. J. and Powell T. G. (1993) Petroleum source rock potential of coal and associated sediments: qualitative aspects. In *Hydrocarbons from Coal* (eds. B. E. Law and D. D. Rice). American Association of Petroleum Geologists, AAPG Studies in Geology #38, American Association of Petroleum Geologists, Tulsa, OK, pp. 133–157.

Botto R. E. (1987) Solid ^{13}C NMR tracer studies to probe coalification. *Energy Fuels* **1**, 228–230.

Boudou J. P. and Espitalie J. (1995) Molecular nitrogen from coal pyrolysis: kinetic modeling. *Chem. Geol.* **126**, 319–333.

Bouska V. (1981) *Geochemistry of Coal, Coal Science and Technology 1*. Elsevier, Amsterdam.

Bragg L. J., Oman J. K., Tewalt S. J., Oman C. L., Rega N. H., Washington P. M., and Finkelmann R. B. (1998) *US Geological Survey Coal Quality (COALQUAL) Database: Version 2.0*. US Geological Survey Open-File Report 97-134.

Bragg O. M. (1997) Understanding ombrogenous mires of the temperate zone and the tropics: an ecohydrologists viewpoint. In *Biodiversity and Sustainability of Tropical Peatlands* (eds. J. O. Rieley and S. E. Page). Samara Publishing, Cardigan, UK, pp. 135–146.

Buchanan A. C., III, Britt P. F., and Struss J. A. (1997) Investigation of reaction pathways involved in lignin maturation. *Energy Fuels* **11**, 247–248.

Butler J. R. (1953) Geochemical affinities of some coals from Svalbard. *Nor. Polarinst. Skr.* 96.

Carroll R. E. and Pashin J. C. (1997) Tectonic and hydrologic influences on coalification and hydrocarbon generation in the Warrior and Cahaba coal fields, Alabama. *AAPG Eastern Section and the Society for Organic Petrology Joint Meeting, Abstracts, AAPG Bulletin* **81**, 1546.

Cecil C. B., Dulong F. T., Cobb J. C., and Supardi (1993) Allogenic and autogenic controls on sedimentation in the central Sumatra Basin as an analogue for Pennsylvanian coal-bearing strata in the Appalachian Basin. In *Modern and Ancient Coal forming Environments* (eds. J. C. Cobb and C. B. Cecil). Geological Society of America Special Publication. Washington, DC, vol. 286, pp. 3–22.

Chapman D. S. (1985) Continental heat flow data. In *Geophysics of the Solid Earth, the Moon, and the Planets*. (eds. K. Fuchs and H. Stoffel). Springer, Landolt-Bornstein, N. Ser., Group V, 2b, pp. 1–19.

Chou L. (1984) Relationship between geochemistry of coal and the nature of strata overlying the Herrin coal in the Illinois Basin, USA. *Memoir Geol. Soc. China* **6**, 269–280.

Claypool G. E. and Kaplan I. R. (1974) The origin and distribution of methane in marine sediments. In *Natural Gases in Marine Sediments*. Plenum, New York, pp. 99–139.

Clayton J. L. (1993) Composition of crude oils generated from coals and coaly organic matter in shales. In *Hydrocarbons from Coal*, AAPG Studies in Geology #38 (eds. B. E. Law and D. D. Rice), American Association of Petroleum Geologists, Tulsa, OK, pp. 185–201.

Cody G. D. (1992) The macromolecular geochemistry of bituminous vitrain. Doctoral Dissertation, Pennsylvania State University, University Microfilms.

Cohen A. D. (1984) The Okeefenokee Swamp: a low-sulfur end-member of a shoreline-related depositional model for coastal plain coals. In *The Okeefenokee Swamp: Its Natural History, Geology, and Geochemistry* (eds. A. D. Cohen, D. J. Casagrande, M. J. Andrejko, and G. R. Best). Wetland Surveys, Las Alamos, NM, pp. 668–680.

Cowie G. L. and Hedges J. I. (1992) The role of anoxia in organic matter preservation in coastal sediments: relative stabilities of the major biochemicals under oxic and anoxic depositional conditions. *Org. Geochem.* **19**, 229–234.

Dehmer J. (1993) Petrology and organic geochemistry of peat samples from a raised bog in Kalimantan (Borneo). *Org. Geochem.* **20**, 340–362.

de Leeuw J. W. and Largeau C. (1993) A review of macromolecular organic compounds that comprise living organisms and their role in kerogen, coal, and petroleum formation. In *Organic Geochemistry Principles and Applications* (eds. M. H. Engel and S. A. Macko). Plenum, New York, pp. 23–72.

Durand B. and Oudin J. L. (1979) Exemple de migration des hydrocarbures dans une serie deltaique: le delta de la Mahakam, Indonesie. *Proc. 10th World Petrol. Congr. Bucharest* **2**, 3–11.

Ertel J. R. and Hedges J. I. (1980) A test of the melanoidin hypothesis. *AAPG Bull.* **64**, 704.

Espitalie J. and Bordenave M. L. (1993) Screening techniques for source rock evaluation: tools for source rock routine analysis: Rock-Eval pyrolysis. In *Applied Petroleum Geochemistry* (ed. M. L. Bordenave). Editions Technip, Paris, France, pp. 237–261.

Finkelman R. B. (1981a) *Modes of Occurrence of Trace Elements in Coal.* US Geological Survey Open-File Report 81-99.

Finkelman R. B. (1981b) Recognition of authigenic and detrital minerals in coal. *Geol. Soc. Am. Abstr. Progr.* **13**(7), 451.

Finkelman R. B. and Brown R. D., Jr. (1991) Coal as a host and as an indicator of mineral resources. In *Geology in Coal Resource Utilization* (ed. D. C. Peters). TechBooks, pp. 471–481.

Finkelman R. B., Belkin H. E., and Zheng B. (1999) Health impacts of domestic coal use in China. *Proc. Nat. Acad. Sci. USA* **96**, 3427–3431.

Finkelman R. B., Orem W., Castranova V., Tatu C. A., Belkin H. E., Zheng B., Lerch H. E., Maharaj S. V., and Bates A. L. (2002) Health impacts of coal and coal use: possible solutions. *Int. J. Coal Geol.* **50**, 425–443.

Fischer F. and Schrader H. (1921) The origin and chemical structure of coal. *Brennstoff Chem.* **2**, 37–45.

Flaig, W. (1966) The chemistry of humic substances. In *The Use of Isotopes in Soil Organic Matter Studies.* Pergamon, London, pp. 103–127.

Garcette-Lepecq A., Derenne S., Largeau C., Bouloubassi I., and Saliot A. (2000) Origin and formation pathways of kerogen-like organic matter in Recent sediments off the Danube Delta (northwestern Black Sea). *Org. Geochem.* **31**, 1663–1683.

Gelin F., Volkman J. K., Largeau C., Derenne S., Sinninghe Damste J. S., and de Leeuw J. W. (1999) Distribution of aliphatic nonhydrolysable biopolymers in marine microalgae. *Org. Geochem.* **30**, 147–159.

Given P. H. (1972) Biological aspects of the geochemistry of coal. In *Advances in Organic Geochemistry 1971* (eds. H. R. von Gaertner and H. Wehner). Pergamon, London, pp. 69–92.

Given P. H., Spackman W., Imbalzano J. R., Casagrande D. J., Lucas A. J., Cooper W., and Exarchos C. (1983) Physicochemical characteristics and levels of microbial activity in some Florida peat swamps. *Int. J. Coal Geol.* **3**, 77–99.

Gleason P. J. and Stone P. (1994) Age, origin, and landscape evolution of the Everglades peatland. In *Everglades, the Ecosystem and its Restoration* (eds. S. M. Davis and J. C. Ogden). St. Lucie Press, Delray Beach, FL, pp. 149–197.

Goldhaber M. B., Irwin E., Atkins B., Lee L., Black D. D., Zappia H., Hatch J., Pashin J., Barwick L. H., Cartwright W. E., Sanzolone R., Ruppert L., Kolker A., and Finkelman R. (2001) *Arsenic in Stream Sediments of Northern Alabama.* US Geological Survey Miscellaneous Field Studies Map MF-2357.

Goodarzi F. (1987) Concentration of elements in lacustrine coals from zone A Hat Creek deposit No. 1, British Columbia, Canada. *Int. J. Coal Geol.* **8**, 247–268.

Goodarzi F. (1988) Elemental distribution in coal seams at the Fording coal mine, British Columbia, Canada. *Chem. Geol.* **68**, 129–154.

Grady W. C., Eble C. F., and Neuzil S. G. (1993) Brown coal maceral distributions in a modern domed tropical Indonesian peat and a comparison with maceral distributions in middle Pennsylvanian-age Appalachian bituminous coal beds. In *Modern and Ancient Coal-forming Environments*, Geological Society of America Special Paper 286, (eds. J. C. Cobb and C. B. Cecil). Geological society of America, Tulsa, OK, pp. 63–82.

Gretener P. E. and Curtis C. D. (1982) Role of temperature and time on organic metamorphism. *AAPG Bull.* **66**, 1124–1149.

Gupta R. P., Wall T. F., and Baxter L. A. (eds.) (1999) *The Impact of Mineral Impurities in Solid Fuel Combustion.* Plenun, New York, 768pp.

Gurba L. W. and Weber C. R. (2001) Effects of igneous intrusions on coalbed methane potential, Gunnedah Basin, Australia. *Int. J. Coal Geol.* **46**, 113–131.

Haenel R. and Staroste E. (1988) *Atlas of Geothermal Resources in the European Community, Austria, and Switzerland.* Publication EUR 11026 of the Commission of the European Communities, ESC, EEC, EAEC.

Halsey L. A., Vitt D. H., and Bauer I. E. (1998) Peatland initiation during the Holocene in continental western Canada. *Climat. Change* **40**, 315–342.

Hammond T. E., Cory D. G., Ritchey W. M., and Morita H. (1985) High resolution solid state ^{13}C n.m.r. of Canadian peats. *Fuel* **64**, 1687–1695.

Hart R. J. and Leahy R. M. (1983) The geochemical characterization of coal seams from the Witbank Basin. *Spec. Publ. Geol Soc. S. Afr.* **7**, 169–174.

Hartman-Stroup C. (1987) The effect of organic matter type and organic content on Rock-Eval hydrogen index in oil shales and source rocks. *Org. Geochem.* **11**, 351–369.

Hatcher P. G. (1988) Dipolar dephasing ^{13}C NMR studies of decomposed wood and coalified xylem tissue: evidence for chemical structural changes associated with defunctionalization of lignin structural units during coalification. *Energy Fuels* **2**, 48–58.

Hatcher P. G. (1990) Chemical structural models for coalified wood (vitrinite) in low rank coal. *Org. Geochem.* **16**, 959–968.

Hatcher P. G. and Breger I. A. (1981) Nuclear magnetic resonance studies of ancient buried wood: I. Observations on the origins of coal to the brown coal stage. *Org. Geochem.* **3**, 49–55.

Hatcher P. G. and Clifford D. J. (1997) The organic geochemistry of coal: from plant materials to coal. *Org. Geochem.* **27**, 251–274.

Hatcher P. G. and Spiker E. C. (1988) Selective degradation of plant biomolecules. In *Humic Substances and their Role in the Environment* (eds. F. M. Frimmel and R. F. Christman). Wiley, New York, pp. 59–74.

Hatcher P. G., Schnitzer M., Dennis L. W., and Maciel G. E. (1981) Aromaticity of humic substances in soils. *Soil Sci. Soc. Am. J.* **45**, 1089–1094.

Hatcher P. C., Breger I. A., Szeverenyi N., and Maciel G. E. (1982) Nuclear magnetic resonance studies of ancient buried wood: II. Observations on the origin of coal from lignite to bituminous coal. *Org. Geochem.* **4**, 9–18.

Hatcher P. G., Breger I. A., Dennis L. W., and Maciel G. E. (1983a) Solid-state ^{13}C NMR of sedimentary humic substances: new revelations on their chemical composition. In *Aquatic and Terrestrial Humic Substances* (eds. R. F. Christman and E. T. Gjessing). Ann Arbor Science Publishers, Ann Arbor, MI, pp. 37–82.

Hatcher P. G., Spiker E. C., Szeverenyi N. M., and Maciel G. E. (1983b) Selective preservation and the origin of petroleum-forming aquatic kerogen. *Nature* **305**, 498–501.

Hatcher P. G., Breger I. A., Maciel G. E., and Szeverenyi N. M. (1985) Geochemistry of humin. In *Humic Substances in Soil, Sediment, and Water: Geochemistry, Isolation, and Characterization* (eds. G. R. Aiken, D. M. McKnight, R. L. Wershaw, and P. MacCarthy). Wiley, New York, pp. 275–302.

Hatcher P. G., Spiker E. C., and Orem W. H. (1986) Organic geochemical studies of the humification process in low-moor peat. In *Peat and Water, Aspects of Water Retention and Dewatering in Peat* (ed. C. H. Fuchsman). Elsevier, London, pp. 195–213.

Hatcher P. G., Lerch H. E., III, and Verheyen T. V. (1989a) Organic geochemical studies of the transformation of gymnosperm xylem during peatification and coalification to subbituminous coal. *Int. J. Coal Geol.* **13**, 65–97.

Hatcher P. G., Wilson M. A., Vassallo A. M., and Lerch H. E. (1989b) Studies of angiospermous wood in Australian brown coal by nuclear magnetic resonance and analytical pyrolysis: new insights into the early coalification process. *Int. J. Coal Geol.* **13**, 99–126.

Hatcher P. G., Wenzel K. A., and Cody G. D. (1994) The coalification reactions of vitrinite derived from coalified wood: transformations to the rank of bituminous coal. In *Vitrinite Reflections as a Maturity Parameter* (eds. P. K. Mukhopadhyay and W. G. Dow). American Chemical Society, Washington, DC, pp. 112–135.

Herlihy M. (1973) Distribution of nitrifying and heterotrophic microorganisms in cutover peats. *Soil Biol. Biochem.* **5**, 621–628.

Honea F. I., Montgomery G. G., and Jones M. L. (1982) Recent research on ash fouling in combustion of low rank coals. In *Technology and Use of Lignites. Volume 1: Proceedings, 11th Biennial Lignite Symposium*, Grand Forks Energy Technology Center IC-82/1, (eds. W. R. Kube, E. A. Sondreal, and D. M. White), pp. 504–545.

Huck G. and Patteisky K. (1964) Coalification reactions under pressure. *Fortschritte in der Geologie von Rheinland und Westfalen* **12**, 551–558.

Huggins F. E., Shah N., Huffman G. P., Kolker A., Crowley S. S., Palmer C. A., and Finkelman R. B. (2000) Mode of occurrence of chromium in four US coals. *Fuel Process. Technol.* **63**, 79–92.

Hunt J. M. (1991) Generation of gas and oil from coal and other terrestrial organic matter. *Org. Geochem.* **17**, 673–680.

ICF Resources (1990) The United States coalbed methane resource. *Quart. Rev. Methane Coal Seams Technol.* **7**, 10–28.

Ikeda K., Arimura R., Echigo S., Shimizu Y., Minear R. A., and Matsui S. (1999) The fractionation/concentration of aquatic humic substances by the sequential membrane system and their characterization with mass spectrometry. *Water Sci. Technol.* **42**, 383–390.

Jenny W. P. (1903) The chemistry of ore-deposition. *Am. Inst. Min. Eng. Trans.* **33**, 445–498.

Kalbitz K. and Geyer S. (2002) Different effects of peat degradation on dissolved organic carbon and nitrogen. *Org. Geochem.* **33**, 319–326.

Karwell J. (1955) The metamorphism of mineral coal in terms of physical chemistry. *Zeitschrift* **107**, 132–139.

Katz B. J. (1983) Limitations of "Rock-Eval" pyrolysis for typing organic matter. *Org. Geochem.* **4**, 195–199.

Khorasani G. K. (1987) Oil-prone coals of the Walloon coal measures, Surat basin, Australia. In *Coal and Coal-bearing Strata: Recent Advances*, Society Special Publication No. 32 (ed. A. C. Scott). Blackwell, London, pp. 303–310.

Knicker H., Hatcher P. G., and Scaroni A. W. (1996) A solid-state ^{15}N NMR spectroscopic investigation of the origin of nitrogen structures in coal. *Int. J. Coal Geol.* **32**, 255–278.

Kolker A., Palmer C. A., Crowley S. S., Finkelman R. B., Huggins F. E., Sha N., and Huffman G. P. (2000) Mode of occurrence of arsenic in four US coals. *Fuel Process. Technol.* **63**, 167–178.

Kolker A., Cecil B. C., Dulong F. T., and Fedarko N. (2001) Effect of pyrite composition texture, and form on acid mine drainage potential in coal-bearing strata of the central Appalachian Basin. *Geol. Soc. Am. Ann. Meet. Abstr. Progr. Pap. 172-0.*

Kuangzong Q., Shaohui G., and Shuyuan L. (1998) New concept on coal structure and new consideration for the generation mechanism of oil from coal. *Chinese Sci. Bull.* **43**, 2025–2034.

Kuuskraa V. A., Kelafant J., and Kuuskraa J. A. (1997) A critical look at the geologic and reservoir controls on producing Appalachian Basin coalbed methane. *AAPG Eastern Section and the Society for Organic Petrology Joint Meeting*. Abstracts, *AAPG Bulletin.* **81**, 1556.

Landis E. R. and Weaver J. N. (1993) Global coal occurrence. In *Hydrocarbons from Coal*, AAPG Studies in Geology #38 (eds. B. E. Law and D. D. Rice). American Association of Petroleum Geologists, Tulsa, OK, pp. 1–12.

Landais P., Muller J. F., Michels R., Oudin J. L., and Zaugg P. (1989) Comparative behaviour of coal and maceral concentrates during artificial coalification. *Fuel* **68**, 1616–1619.

Landais P., Monthioux M., and Poty B. (1990) Simulations of natural coalification by high-pressure pyrolysis. *Int. J. Coal Geol.* **16**, 230–234.

Langenheim J. H. (1969) Amber: a botanical enquiry. *Science* **163**, 1157–1169.

Langenheim J. H. (1995) Biology of amber-producing trees: focus on case studies of Hymenaea and Agathis. In *Amber, Resinite, and Fossil Resins* (eds. K. B. Anderson and J. C. Crelling). American Chemical Society, Washington, DC, pp. 1–31.

Laughrey C. D. and Baldassare F. J. (1998) Geochemistry and origin of some natural gases in the Plateau Province, central Appalachian Basin, Pennsylvania and Ohio. *AAPG Bull.* **82**, 317–335.

Levine J. R. (1987) Influence of coal composition on the generation and retention of coalbed natural gas. *Proceed. 1987 Coalbed Methane Symp.* 15–18.

Levine J. R. (1991) The impact of oil formed during coalification on generation and storage of natural gas in coalbed reservoir systems. *Proc. 1991 Coalbed Methane Symp.* 307–315.

Lin R., Dzou L., Doyle K., and Jardine E. (1997) Far-east Tertiary coal-related petroleum systems. *AAPG Bull.* **81**, 1557.

Lo H. B. (1991) How well can artificial coalification simulate the natural maturation trend? *Org. Geochem.* **17**, 415–420.

Lopatin N. V. (1971) Temperature and geologic time as factors in coalification. *Izvestiya Akademii Nauk SSSR. Seriya Geologicheskaya* **3**, 95–106.

Manskaya S. M. and Drozdova T. V. (1968) *Geochemistry of Organic Substances*. Pergamon, London.

McCabe P. J. (1991) Geology of coal: environments of deposition. In *Economic Geology, US, The Geology of North America* (eds. H. J. Gluskoter, D. D. Rice, and R. B. Taylor). The Geological Society of America, Tulsa, OK, pp. 469–482.

McCabe P. J., Gautier D. L., Lewan M. D., Turner C., and others (1993) *The Future of Energy Gases*. US Geological Survey Circular, 1115.

McCartney J. T. and Teichmuller M. (1972) Classification of coals according to degree of coalification by reflectance of the vitrinite component. *Fuel* **51**, 64–68.

McKinney D. E., Bortiatynski J. M., Carson D. M., Clifford D. J., de Leeuw J. W., and Hatcher P. G. (1996) Tetramethylammonium hydroxide (TMAH) thermochemolysis of the aliphatic biopolymercutan: insights into its chemical structure. *Org. Geochem.* **24**, 641–650.

Meissner F. F. (1984) Cretaceous and lower Tertiary coals as sources for gas accumulations in the Rocky Mountain area. In *Hydrocarbon Source Rocks of the Greater Rocky Mountain Region* (eds. J. Woodward, F. F. Meissner, and

J. L. Clayton). Rocky Mountain Association of Geologists, pp. 401–431.

Monthioux M. (1988) Expected mechanisms in nature and in confined-system pyrolysis. *Fuel* **67**, 843–847.

Monthioux M., Landais P., and Monin J.-C. (1985) Comparison between natural and artificial maturation series of humic coals from the Mahakam delta, Indonesia. *Org. Geochem.* **8**, 275–292.

Monthioux M., Landais P., and Durand B. (1986) Comparison between extracts from natural and artificial series of Mahakam delta coals. *Org. Geochem.* **10**, 299–311.

Morley R. J. (1981) Development and vegetation dynamics of a lowland ombrogenous peat swampin Kalimantan Tengah, Indonesia. *J. Biogeogr.* **8**, 383–404.

Neuzil S. G., Supardi, Cecil C. B., Kane J. S., and Soedjono K. (1993) Inorganic geochemistry of domed peat in Indonesia and its implication for the origin of mineral matter in coal. In *Modern and Ancient Coal-forming Environments*, Geological Society of America Special Paper 286 (eds. J. C. Cobb and C. B. Cecil), pp. 23–44.

Newman J., Price L., and Johnston J. H. (1994) Source potential of New Zealand coals, based on relationships between conventional coal chemistry, Rock-Eval pyrolysis, and GCMS biomarkers. *1994 New Zealand Petroleum Conference Proceedings: The Post Maui Challenge: Investment and Development Opportunities*, p. 47 (abstract).

Nichols C. L. and D'Auria J. M. (1981) Seam and location differentiation of coal specimens using trace element concentrations. *Analyst* **106**, 874–882.

Nip M., Tegelaar E. W., de Leeuw J. W., Schenck P. A., and Holloway P. J. (1986) A new non-saponifiable highly aliphatic, resistant biopolymer in plant cuticles: evidence from pyrolysis, ^{13}C NMR analysis of present day, fossil plants. *Naurwissenschaften* **73**, 579–585.

Nip M., de Leeuw J. W., Schenck P. A., Windig W., Meuselaar H. L. C., and Crelling J. C. (1989) A flash pyrolysis and petrographic study of cutinite from the Indiana paper coal. *Geochim. Cosmochim. Acta* **53**, 671–683.

Noble E. A. (1973) Uranium in coal. In *Mineral and Water Resources of North Dakota*, North Dakota Geological Survey Bulletin 63, pp. 80–83.

O'Connor J. T. (1988) The Campbell Creek/No. 2 *Gas/ Peerless* Powellton coal bed correlation from the middle of the Kanawha Formation of the central Appalachian Basin. In *USGS Research on Energy Resources-1988 Program and Abstracts*. US Geological Survey Circular 1025 (ed. L. M. H. Carter), pp. 39.

Ohta K. and Venkatesan M. I. (1992) Pyrolysis of wood specimens with and without minerals: implications for lignin diagenesis. *Energy Fuels* **6**, 271–277.

Orem W. H. and Hatcher P. G. (1987) Early diagenesis of organic matter in a sawgrass peat from the Everglades, Florida. *Int. J. Coal Geol.* **8**, 33–54.

Orem W. H., Neuzil S. G., Lerch H. E., and Cecil C. B. (1996) Experimental early-stage coalification of a peat sample and a peatified wood sample from Indonesia. *Org. Geochem.* **24**, 111–125.

Orem W. H., Feder G. L., and Finkelman R. B. (1999) A possible link between Balkan Endemic Nephropathy and the leaching of toxic organic compounds from Pliocene lignites by groundwater. *Int. J. Coal Geol.* **40**, 237–252.

Orem W. H., Tatu C. A., Feder G. L., Finkelman R. B., Lerch H. E., Maharaj S., Szilagyi D., Dumitrascu V., Paunescu V., and Margineanu F. (2002) Environment, geochemistry and the eitiology of Balkan endemic nephropathy: lessons from Romania. *Facta Universitatis, Med. Biol.* **9**(1), 39–48.

Ovenden L. (1990) Peat accumulation in northern wetlands. *Quat. Res.* **33**, 377–386.

Palmer C. A., Mroczkowski S. J., Finkelman R. B., Crowley S., and Bullock J. (1998) The use of sequential laboratory leaching to quantify the modes of occurrence of trace elements in coal. In *Pittsburgh Coal Conference, Pittsburgh, PA, 1998*, CD-ROM, pp. 1–18.

Palmer C. A., Finkelman R. B., and Luttrell G. H. (2002) Coal from a mid-Atlantic Ocean shipwreck: the source of the coal in the Titanic and effects of exposure to seawater. *Nineteenth Annual International Pittsburgh Coal Conference*, CD-ROM.

Pearson D. L. and Moore T. R. (2000) Coal as a thermal insulator and oil source rock. *Am. Assoc. Petrol. Geol., Annu. Meet. Exp. Abstr.* 113 pp.

Peters K. E., Rohrback B. G., and Kaplan I. R. (1981) Geochemistry of artificially heated humic and sapropelic sediments: I. Protokerogen. *AAPG Bull.* **65**, 688–705.

Pierce B., Ruppert L. F., Hatch J., Warwick P. D., Flores R. M., and Kirschbaum M. A. (1996) *Assessing the Coal Resources of the United States*. US Geological Survey Fact Sheet FS-157-96.

Poirier N., Derenne S., Rouzaud J.-N., Largeau C., Mariotti A., Balesdent J., and Marquet J. (2000) Chemical structure and sources of the macromolecular, resistant, organic fraction isolated from a forest soil (Lacadee, South-West France). *Org. Geochem.* **31**, 813–827.

Powell T. G. (1988) Developments in concepts of hydrocarbon generation from terrestrial organic matter. In *Petroleum Resources of China and Related Subjects*, Circum-Pacific Council for Energy and Mineral Resources, Earth Science Series (eds. H. C. Wagner, L. C. Wagner, F. F. H. Wang, and F. L. Wong), vol. 10, pp. 807–824.

Powell T. G., Boreham C. J., Smyth M., Russel N., and Cook A. C. (1991) Petroleum source rock assessment in non-marine sequences: pyrolysis and petrographic analysis of Australian coals and carbonaceous shales. *Org. Geochem.* **17**, 375–394.

Raask E. (1985) *Mineral Impurities in Coal Combustion*. Hemisphere Publ. Corp.

Raymond R., Jr., Bish D. L., and Cohen A. D. (1990) Inorganic contents of peats. In *Mineral Matter and Ash Deposition from Coal*, Engineering Foundation Conference (eds. R. W. Breyers and K. S. Vorres), pp. 23–37.

Reid W. T. (1981) Coal ash-its effects on combustion systems. In *Chemistry of Coal Utilization, 2nd Supplementary Volume* (ed. M. A. Elliot). Wiley, pp. 1389–1445.

Rice D. D. (1992) Controls, habitat, and resource potential of ancient bacterial gas. In *Bacterial Gas* (ed. R. Vialy), Editions Technip, pp. 91–118.

Rice D. D. (1993) Composition and origins of coalbed gas. In *Hydrocarbons from Coal*, AAPG Studies in Geology #38 (eds. B. E. Law and D. D. Rice). American Association of Petroleum Geologists, Tulsa, OK, pp. 159–184.

Rice D. D. and Claypool G. E. (1981) Generation, accumulation, and resource potential of biogenic gas. *Am. Assoc. Petrol. Geol. Bull.* **65**, 5–25.

Rieley J. O., Page S. E., Limin S. H., and Winarti S. (1997) The peatland resource of Indonesia and the Kalimantan peat swamp forest research project. In *Biodiversity and Sustainability of Tropical Peatlands* (eds. J. O. Rieley and S. E. Page). Samara Publishing, Cardigan, UK, pp. 37–44.

Risatti J. B., Jr. (1978) Geochemical and microbial aspects of Volo Bog, Lake County, Illinois. PhD Dissertation, Illinois University at Urbana-Champaign.

Rohrback B. G., Peters K. E., and Kaplan I. R. (1984) Geochemistry of artificially heated humic and sapropelic sediments: II. Oil and gas generation. *AAPG Bull.* **68**, 961–970.

Rollins M. S., Cohen A. D., Bailey A. M., and Durig J. R. (1991) Organic chemical and petrographic changes induced by early-stage artificial coalification of peats. *Org. Geochem.* **17**, 451–465.

Ruppert L., Finkelman R., Boti E., Milosavljevic M., Tewalt S., and Dulong F. (1996) Origin and significance of high nickel and chromium in the Pliocene Kosovo Basin lignite, Central Yugoslavia. *Int. J. Coal Geol.* **29**, 235–258.

Saiz-Jimenez C., Boon J. J., Hedges J. I., Hessels J. K. C., and de Leeuw J. W. (1987) Chemical characterization of recent and buried woods by analytical pyrolysis: comparison of pyrolysis data with ^{13}C NMR and wet chemical data. *J. Analyt. Appl. Pyrol.* **11**, 437–450.

Schenck P. A. (1986) Curie point pyrolysis; mass spectrometry and Curie point pyrolysis; gas chromatography; mass spectrometry in organic geochemistry. In *Workshop on Advances in Biomarkers and Kerogens*. Academia Sinica, Inst. Geochem., Guiyang, China, pp. 83–87.

Schnable R. W. (1975) Uranium. In *Mineral and Water Resources of South Dakota*, Report prepared by the US Geological Survey for the Committee on Interior and Insular Affairs, United States Senate, pp. 172–176.

Schnitzer M. and Khan S. U. (1972) *Humic Substances in the Environment*. Dekker, New York.

Schnitzer M. and Levesque M. (1979) Electron spin resonance as a guide to the degree of humification of peats. *Soil Sci.* **127**, 140–145.

Schobert H. H. (1987) *Coal—The Energy Source of the Past and Future*. American Chemical Society, Washington, DC.

Schulten H.-R. and Gleixner G. (1999) Analytical pyrolysis of humic substances and dissolved organic matter in aquatic systems: structure and origin. *Water Res.* **33**, 2489–2498.

Seredin V. (1995) New types of REE and Au-RGE mineralization of Russian coal-bearing depressions. *Mineral Deposits, Rotterdam, Balkema*, 799–801.

Severson R. C. and Shacklette H. T. (1988) *Essential Elements and Soil Amendments for Plants: Sources and Use for Agriculture. US Geol. Surv. Circ.* 1017.

Shearer J. C., Moore T. A., and Cecil C. B. (1996) Experimental coalification of Indonesian peat: petrography. *Org. Geochem.* **24**, 127–140.

Shepard M. (1987) Toxic resources and the real world. *Elect. Power Res. Inst. J.* **September 1987**, 17–21.

Shotyk W. (1988) Review of the inorganic geochemistry of peats and peatland waters. *Earth Sci. Rev.* **25**, 95–176.

Snowdon L. R. (1991) Oil from Type III organic matter: resinite revisited. *Org. Geochem.* **17**, 743–747.

Spackman W., Cohen A. D., Given P. H., and Casagrande D. J. (1974) *The Comparative Study of the Okefenokee Swamp and the Everglades-Mangrove Swamp-Marsh Complex of Southern Florida*, Field Guide Book, Geological Society of America.

Spiker E. C. and Hatcher P. G. (1987) The effects of early diagenesis on the chemical and stable carbon isotopic composition of wood. *Geochim. Cosmochim. Acta* **51**, 1385–1391.

Stach E., Mackowsky M.-Th., Teichmuller M., Taylor G. H., Chandra D., and Teichmuller R. (1982) *Stach's Textbook of Coal Petrology* 3rd edn. Gebruder Borntraeger, Berlin.

Staub J. R. and Esterle J. S. (1994) Peat-accumulating depositional systems of Sarawak. *East Malaysia. Sedim. Geol.* **89**, 91–106.

Stadler G. and Teichmuller M. (1971) The change of coal and the diagenesis of the clay- and sandstones in Subsurface Borehole 150 of the anthracite mines at Ibbenbueren. *Fortschritte in der Geologie von Rheinland und Westfalen* **18**, 125–146.

Stanov V. V. (1972) Equation for metamorphism of coal. *Doklady Akademii Nauk SSSR* **207**, 1200–1203.

Stopes M. C. (1935) On the petrology of banded bituminous coal. *Fuel* **14**, 4–13.

Stout S. A., Boon J. J., and Spackman W. (1988) Molecular aspects of the peatification and early coalification of angiosperm and gymnosperm woods. *Geochim. Cosmochim. Acta* **52**, 405–414.

Strachan M. G., Alexander R., and Kagi R. I. (1988) Trimethylnaphthalenes in crude oils and sediments: effects of source on maturity. *Geochim. Cosmochim. Acta* **52**, 1255–1264.

Straka P., Buchtele J., and Nahunkova J. (1999) Chemical structure of maceral fractions of coal. *Acta Montana. Serie B: Fuel, Carbon, Mineral Processing* **9**(112), 47–53.

Swaine D. J. (1983) Geological aspects of trace elements in coal. In *The Significance of Trace Elements in Solving Petrogenetic Problems and Controversies* (ed. S. S. Augustithis). Theophrastus Publications, pp. 521–532.

Swaine D. J. (1990) *Trace Elements in Coal*. Butterworths.

Tatu C. A., Orem W. H., Finkelman R. B., and Feder G. L. (1998) The etiology of Balkan Endemic Nephropathy: still more questions than answers. *Environ. Health Perspect.* **106**, 689–700.

Tatu C. A., Orem W. H., Feder G. L., Finkelman R. B., Szilagyi D. N., Dumitrascu V., Margineanu F., and Paunescu V. (2000a) Additional support for the role of the Pliocene lignite derived organic compounds in the etiology of Balkan endemic nephropathy. *J. Med. Biochem.* **4**, 95–101.

Tatu C. A., Orem W. H., Feder G. L., Paunescu V., Dumitrascu V., Szilagyi D. N., Finkelman R. B., Margineanu F., and Schneider F. (2000b) Balkan endemic nephropathy etiology: a link to the geological environment. *Cent. Euro. J. Occupat. Environ. Med.* **6**(2–3), 138–150.

Taylor G. H., Liu S. Y., and Smyth M. (1988) New light on the origin of Cooper basin oil. *APEA J.* **28**, 303–309.

Taylor G. H., Teichmuller M., Davis A., Diessel C. F. K., Littke R., and Robert P. (1998) *Organic Petrology*. Gebruder Borntraeger, Berlin.

Tegelaar E. W. (1990) Resistant biomacromolecules in morphologically characterized constituents of kerogen: a key to the relationship between biomass and fossil fuels. PhD Dissertation, Delft University of Technology.

Tegelaar E. W., de Leeuw J. W., Derenne S., and Largeau C. (1989) A reappraisal of kerogen formation. *Geochim. Cosmochim. Acta* **53**, 3103–3106.

Tegelaar E. W., Hollman G., van der Vegt P., de Leeuw J. W., and Holloway P. J. (1995) Chemical characterization of the periderm tissue of some angiosperm species: recognition of an insoluble, non-hydrolysable, aliphatic biomacromolecule (suberan). *Org. Geochem.* **23**, 239–251.

Thomas J., Jr. and Damberger H. H. (1976) *Internal Surface Area, Moisture Content, and Porosity in Illinois Coals: Variations with Coal Rank*. Illinois Geological Survey Circular 493.

Tissot B. P. and Welte D. H. (1984) *Petroleum Formation and Occurrence*. Springer, Berlin.

Van Bergen P. F., Goni M., Collinson M. E., Barrie P. J., Sinninghe Damste J. S., and de Leeuw J. W. (1994) Chemical and microscopic characterization of outer seed coats of fossil and extant water plants. *Geochim. Cosmochim. Acta* **58**, 3823–3844.

van der Heijden E. and Boon J. J. (1994) A combined pyrolysis mass spectrometric and light microscopic study of peatified Calluna wood isolated from raised bog peat deposits. *Org. Geochem.* **22**, 903–919.

van Grass G., de Leeuw J., and Schenck P. A. (1979) Analysis of coals of different rank by Curie-point pyrolysis–mass spectrometry and Curie-point pyrolysis–gas chromatography—mass spectrometry. *Phys. Chem. Earth* **12**, 485–494.

Vaninetti G. E. and Busch C. F. (1982) Mineral analysis of ash data—a utility perspective. *J. Coal Qual.* **Spring 1982**, 22–31.

Venkatesan M. I., Ohta K., Stout A., Steinberg S., and Oudin J. L. (1993) Diagenetic trends of lignin phenols in Mahakam Delta coals: correlation between laboratory thermal maturation and natural samples. *Org. Geochem.* **20**, 463–473.

Waksman S. A. and Stevens K. R. (1928) Contribution to the chemical composition of peat: I. chemical nature of organic complexes in peat and methods of analysis. *Soil Sci.* **26**, 113–137.

Ward C. R., Bocking M., and Ruan C.-D. (2001) Mineralogical analysis of coals as an aid to seam correlation in the Gloucester Basin, New South Wales, Australia. *Int. J. Coal Geol.* **47**, 31–49.

Wieder R. K. (1985) Peat and water chemistry at Big Run Bog, a peatland in the Appalachian mountains of West Virginia, USA. *Biogeochemistry* **1**, 277–302.

Wilson M. A. (1987) *NMR Techniques and Applications in Geochemistry and Soil Chemistry*. Pergamon.

Yamamoto S. and Ishiwatari R. (1989) A study of the formation mechanism of sedimentary humic substances: II. Protein-based melanoidin model. *Org. Geochem.* **14**, 479–489.

Yamamoto S. and Ishiwatari R. (1992) A study of the formation mechanism of sedimentary humic substances: III. Evidence for the protein-based melanoidin model. *Sci. Total Environ.* **117–118**, 279–291.

Yavitt J. B. and Lang G. E. (1990) Methane production in contrasting wetland sites: response to organic-chemical components of peat and to sulfate reduction. *Geomicrobiol. J. GEJODG* **8**, 27–46.

Young D. K., Sprang S. R., and Yen T. F. (1977) Preliminary investigation on the precursors of the organic components in sediments-melanoidin formations. In *Chemistry of Marine Sediments* (ed. T. F. Yen). Ann Arbor Science, pp. 101–110.

Zhang X., Nguyen R. T., Harvey H. R., Knicker H., and Hatcher P. G. (2001) Preservation of proteinaceous material during the degradation of the green alga Botryococcus braunii: a solid-state 2D ^{15}N ^{13}C NMR spectroscopy study. *Geochim. Cosmochim. Acta* **65**, 3299–3305.

7.09
Formation and Geochemistry of Oil and Gas

R. P. Philp
University of Oklahoma, Norman, OK, USA

7.09.1 INTRODUCTION

It is commonly agreed that oil and gas are derived from the debris of living organic matter which has been deposited in some type of aquatic environment, and subjected to diagenesis, burial, and the effects of increasing temperature and pressure. From time to time, the concept of an abiogenic origin for oil and gas rises to the surface and is vigorously debated, but supporters of this concept are certainly in the minority at this time. It is not intended to cover the abiogenic origin of petroleum in this chapter except to say it was originally conceived by Russian geochemists in the early part of the twentieth century and was resurrected to some degree by the work of Thomas Gold and his ideas of a deep hot origin for hydrocarbons (Gold, 1999). Recent studies have suggested that abiogenic formation of alkanes in the Earth's crust would represent only a minor contribution to the global hydrocarbon reservoirs (Sherwood-Lollar *et al*., 2002). A reader interested in obtaining additional information on the topic of the abiogenic origin of oil and gas is referred to books by Hunt (1979) and Gold (1999). The remainder of this chapter will focus on the organic origin of oil and gas and review the major developments and advances that have occurred in this area during the past several decades.

7.09.2 THE EARLY STEPS IN OIL AND GAS FORMATION: WHERE DOES IT ALL BEGIN?

Crude oils are complex mixtures of organic compounds covering a wide range of polarity, molecular weights, size, shape, solubility, and elemental composition (Hunt, 1979, 1996; Tissot and Welte, 1978, 1984). The organic compounds in crude oils are produced through one or two major pathways from organic compounds present in living systems, such as higher plants, algae, and bacteria. One pathway permits a small percentage of the compounds in crude oils to be derived directly from compounds synthesized by living organisms and readily converted to hydrocarbons. The predominant pathway for the formation of the bulk of the compounds present in crude oils is via the ubiquitous kerogen intermediate (Figure 1; McIver, 1967; Dow, 1977; Durand, 1980). Prior to discussing these pathways it is necessary to go back a few steps in the process to discuss the origin of the organic matter.

The process responsible for the origin of crude oil can be thought of as starting with photosynthesis, which indirectly plays a very significant role in determining the geochemical characteristics of oils and, to some extent, natural gas. Photosynthesis starts with the assimilation of CO_2 from the atmosphere by primary photosynthesizers, such as higher plants on land and phytoplankton in aquatic environments. From a very simplistic point of view, we can think of photosynthesis as a transfer of hydrogen from water to carbon dioxide to produce organic matter and oxygen. The simplest organic molecule formed in this reaction is glucose, from which autotrophic organisms can synthesize polysaccharides, such as cellulose and starch, and other necessary constituents. Collectively these compounds can be thought of as the basic building blocks of life; they include carbohydrates, proteins, lipids, and lignin, which are present in varying proportions in different types of living organic matter. They all play a specific role in the lifecycle of the particular plant or organism with which they are associated. For example, higher plants contain much higher proportions of lignin than grasses, since lignin gives trees the rigidity necessary for them to remain upright. Lipid contents in marine organisms are relatively high, since they provide buoyancy and a source of energy for the organisms.

Plants on land and phytoplankton in the oceans are primary photosynthesizers and provide a source of food and energy for secondary organisms to utilize for the process of

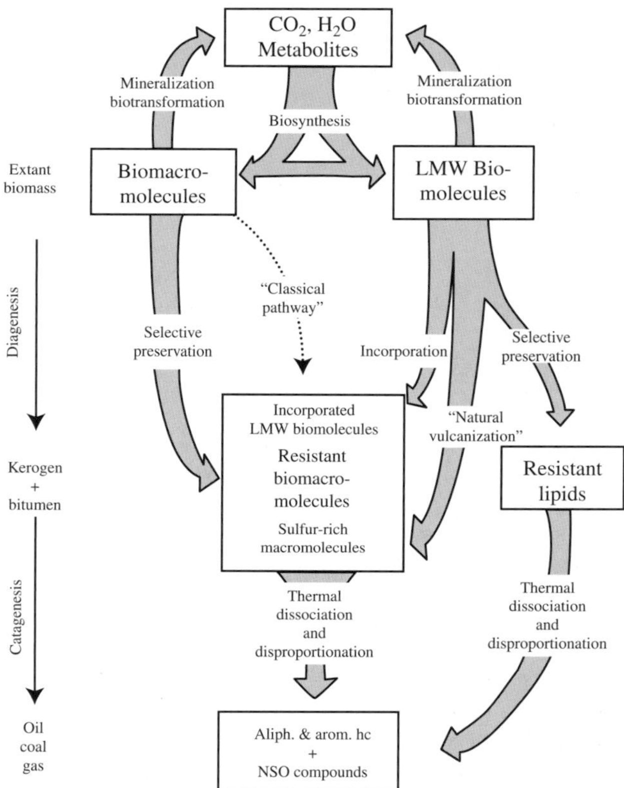

Figure 1 Kerogen is a key intermediate in the formation of oil and gas. Numerous schemes have been proposed both for the formation and structure of kerogen. This diagram shows one of the schemes which favors the "selective preservation" mechanism for kerogen formation.

biosynthesizing compounds needed for their own growth, survival, and lifecycle. Ultimately, all living creatures, both primary and secondary photosynthesizers, die and many are deposited in some type of aquatic environment. The nature of the depositional environment will clearly play an important role in determining the amount of organic matter that will ultimately be preserved. For example, organic matter deposited in a very oxic, high-energy environment will be rapidly oxidized and converted to CO_2 and water, with little opportunity for preservation of significant amounts of organic material. Hence a large amount of carbon in this material is recycled back into the atmosphere, where it is again available for photosynthesis. Organic matter deposited in a highly anoxic environment, and perhaps accompanied by rapid sedimentation or burial, will be preserved in far greater concentrations than in the oxic situation. Tissot and Welte (1978) estimated that $\sim 0.1\%$ of the organic carbon from the photosynthetic cycle escaped and was fixed in the geological record, where an even smaller percentage would ultimately be converted into oil and gas.

Differences in preservation relate to the efficiency of microbial reworking of organic matter. In brief, aerobic degradation of organic matter will proceed at a more rapid pace than anaerobic degradation. This very general interpretation of the situation for the formation of organic-rich source rocks was fairly widely accepted following the classic paper of Demaison and Moore (1980). The process described in their paper is fairly simplistic and in reality the situation is far more complex microbially than represented by these two extremes of oxicity. The complexity of the problem has been further enhanced to some extent, by the work of Pedersen and Calvert (1990), who proposed that the driving force in the formation of organic-rich source rocks was not so much the oxicity of the depositional environment but rather the level of productivity. In other words, an environment with a very high rate of productivity would be more likely to produce organic-rich sediments than one where the rate of productivity was low. In reality it would appear both concepts, productivity and preservation, are important in the formation of organic-rich rocks. The microbial degradation of the organic matter is an on-going process from the moment the organism dies and continues as the organic matter settles through the water column and is deposited into the sediments. It may experience episodes of both oxic and anoxic degradation during deposition. Microbial alteration will be the most important factor involved in the reworking of the organic matter at the relatively low temperatures of deposition.

An interesting proposal to differentiate the importance of productivity versus preservation was published by Hollander *et al.* (1993), who used the extent of carbon isotope fractionation ($\delta^{13}C$) between CO_2 from carbonate, and primary organic matter in ancient organic rich sediments to assess productivity. During periods of high productivity, the difference between the $\delta^{13}C$ values for the carbonates and organic matter increases as a result of enhanced CO_2 demand. In times of low productivity with lower CO_2 demand, the difference decreases. Hydrogen-rich sediments deposited during periods of high productivity show an inverse correlation between the $\delta^{13}C$ values of the carbonates and organic matter and hydrogen index (HI) values, while hydrogen-rich sediments deposited under anoxic and stagnant conditions are characterized by a positive correlation between these two parameters. In a more recent application, Gonclaves (2002) plotted HI versus $\delta^{13}C$ to illustrate the importance of increasing productivity in the formation of the Rio de Contas Formation of the Camamu Rift sequence on the eastern Brazilian continental margin, and increasing stagnation in the Morro do Barro Formation.

Productivity and preservation are two of the main driving forces behind the formation of organic-rich rocks that may ultimately become source rocks. The nature and composition of the organic material are other important factors that contribute to the extent of preservation of the organic matter. Carbohydrates, for example, are highly susceptible to biodegradation due to their high oxygen content. In addition, the initial degradation products are relatively water soluble, thus facilitating their rate of degradation. Proteins are susceptible to degradation to peptides, amino acids, which are very water soluble, and ultimately carbon dioxide, water, and nitrogen. Other components of living systems are somewhat more recalcitrant and may be selectively preserved to form the basic building blocks of kerogen. The idea of selective preservation was originally proposed by Philp and Calvin (1976) but refined by various groups in the 1980s and has now become an accepted part of the mechanism of kerogen formation. Insoluble and bacterially resistant highly aliphatic biopolymers have been found in plant cuticles and referred to as cutans and suberans (Nip *et al.*, 1986; Tegelaar *et al.*, 1995) and algaenans in the cells walls of several algae (Gelin *et al.*, 1999). In the 1970s the focus of attention was the molecular structures of kerogens; more recently, attention has been directed towards determining the structures of algaenans. de Leeuw and Largeau (1993) proposed a three-dimensional network composed of ether-linked long alkyl chains but a recent paper by Allard *et al.* (2002) has proposed that at least part of the algaenan structure is comprised of long-chain (di)carboxylic acids up to C_{120}. Once again, from the viewpoint of petroleum exploration, this may be looked at as a somewhat academic exercise. However, as discussed below, high-molecular-weight hydrocarbons up to C_{120} have been found in oils from a wide variety of source materials and, in all probability, a certain proportion of these hydrocarbons originated from the algaenans or structurally similar materials (Hsieh and Philp, 2001).

7.09.3 INSOLUBLE ORGANIC MATERIAL—KEROGEN

From an oil and gas perspective, the precise molecular structures of kerogens are not so important, but general structural characteristics are important since they determine the nature of the product of a particular kerogen generates. At one time it was thought that kerogens had a homogeneous structure and showed little structural variation from basin to basin. However, it is now clear that kerogen is a very heterogeneous and complex agglomerate of macerals such that one can anticipate variations in the kerogen structures over relatively small distances within a source facies. (Macerals are discrete particles of insoluble organic material which can be identified under the microscope and represent residual detritus from various sources of organic material.) Once the organic origin for oil and gas and the importance of kerogen as an intermediate became accepted (McIver, 1967), it was clear that a routine method was necessary to characterize and classify source rocks and their associated kerogens.

In the 1970s kerogens were often thought of as polymeric structures with repeating subunits. Significant time and effort were invested in unraveling the structure of kerogen through stepwise degradation studies (e.g., Burlingame and Simoneit, 1969; Vitorovic *et al.*, 1974). However, the degradation products obtained in this manner, typically carboxylic acids, were rarely successfully reconstructed into viable kerogen structures. Around the same time, another important paper was published by Tissot *et al.* (1974) extrapolating a concept long applied by the coal geochemists to the characterization of coals and petroleum source rocks. Coal chemists typically characterized or classified coals in terms of their maceral content and elemental composition. Macerals represent various types of organic debris in the coals that can be identified microscopically and related to definitive sources. Classification of coals in this manner is a relatively complex topic but, for example, alginate and exinite are macerals commonly associated with an algal input; vitrinite with a higher plant input; sporonite with spores. Detailed descriptions of these classifications can be found in the classic coal petrology text by Stach *et al.* (1982). Tissot and others took this idea and realized that kerogen, like coal, was a very heterogeneous material comprised of macerals in the same manner as coal and not simply a homogeneous polymeric-type material as suggested earlier. Kerogens could be characterized petrographically along with their elemental compositions in a manner very similar to the classification of coals. This led to the concept of classifying kerogens as types I–IV. In the original papers, different kerogen types were constrained by well-defined boundaries on the H/C versus O/C plot as shown in Figure 2(a). The publication of an important paper by Jones (1987) made it clear that such boundaries did not in reality exist and it was more meaningful to think of kerogens as a continuum going from the type I to type IV with various degrees of mixing between the extremes. This permitted a better understanding of why a kerogen classified as a type III kerogen was still able to produce significant amounts of oil. It also explained variations in the products generated from the different kerogen types. The Tissot–Welte diagram, derived from the coal petrographers' van Krevelen diagram, is still used to assign kerogen types, provide an indication of maturity levels, and nature of products that a particular kerogen may be expected to generate at appropriate levels of maturity.

A change to the original diagram resulted from the fact that, for several years, kerogen types have been determined not by elemental analyses but by use of the Rock Eval pyrolysis system (Espitalie *et al.*, 1977; Peters, 1986). Determination of the elemental composition of kerogens is a relatively time-consuming process. Development of the Rock Eval technique for characterization of the source rocks provided a relatively rapid alternative method for determination of two indices that could be used to replace the H/C and O/C parameters. This technique is a pyrolysis method whereby a sample is exposed to a temperature-programmed pyrolysis from ambient to 600 °C and the pyrolysis products detected immediately without any chromatographic separation (Espitalie *et al.*, 1977). The result is basically three peaks: S_1, S_2, and S_3 (modern versions of the Rock Eval technique produce some additional peaks but, for the purpose of this discussion, the presence of the three aforementioned peaks is sufficient). S_1 corresponds to the material which is normally solvent extracted from a source rock; S_2 corresponds to the products formed from the thermal breakdown of the kerogen; and S_3 is derived from oxygen-containing moieties within the kerogen. From these three parameters, plus the total organic carbon content of the sample, two important parameters are developed, namely, the so-called HI, which is the S_2 peak normalized to

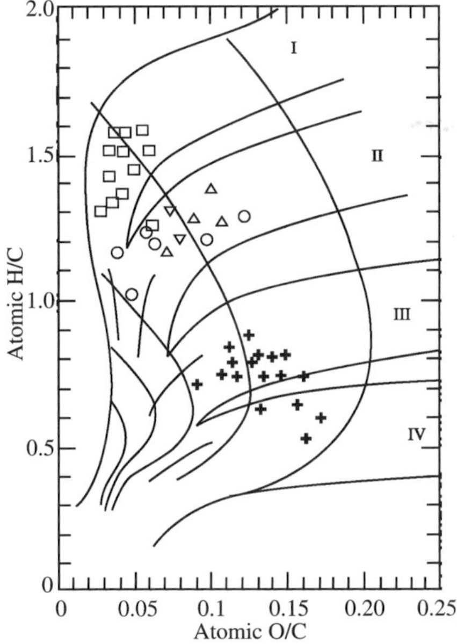

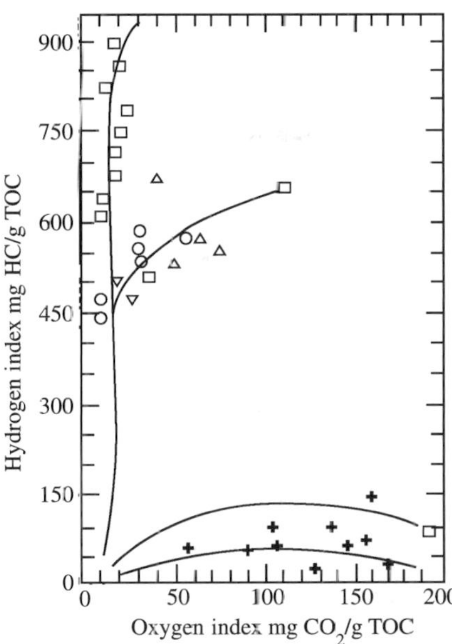

Figure 2 (a) An early method used to characterize kerogens was based on the elemental composition of the kerogen. Initially kerogens were divided into four major types as illustrated on this diagram with well-defined boundaries. (b) With the development of the Rock Eval pyrolysis system it was found that the HI and OI indicies were directly proportional to the H/C and O/C ratios and therefore a plot of HI to OI could be used to replace the H/C and O/C values on the Tissot–Welte diagram (Hunt, 1996) (reproduced by permission of Freeman from *Petroleum Geochemistry and Geology*, **1996**).

the TOC, and the oxygen index (OI), which is the S_3 peak normalized to the TOC. It has been shown that the HI and OI are directly proportional to the H/C and O/C ratios and, therefore, a plot of HI to OI can be used to replace the H/C and O/C values on the Tissot–Welte diagram (Figure 2(b)).

Another extremely important feature related to the generation of oil or gas is the maturity level of the source rock. Organic matter has to reach a certain level of maturity before it starts to thermally degrade and is converted into liquid or gaseous hydrocarbons. The threshold level for oil generation varies depending upon kerogen type. For example, a type IIS kerogen enriched in sulfur generates oil at lower temperatures than a type II kerogen that is not enriched in sulfur. Information such as this is critical in any exploration study and for modeling basins (Orr, 1986).

Determination of maturity levels is critical to the success of any exploration program. Recovery of immature, but organic-rich, source rocks would indicate good source potential for such rocks if buried more deeply in other parts of the basin. At the other extreme, an overmature source rock would indicate a mature part of the basin not capable of generating additional liquid hydrocarbons, but possibly gas. There are several indicators available that can be used to estimate the relative maturity of a source rock. The traditional method is measuring the maturity of vitrinite. The chemical composition of the maceral vitrinite, derived from higher plant debris, changes as the level of maturity increases. With increasing maturity, the ability of vitrinite to reflect light increases and hence a vitrinite reflectance scale has been developed which correlates the degree of reflectance with maturity. Maturity changes of vitrinite have been studied by coal chemists for a long period of time (Teichmuller, 1958). A similar approach was adopted by the petroleum geochemists (Dow, 1977). The vitrinite reflectance scale is summarized in Figure 3, which indicates the important values of this scale representing the onset of oil generation, peak oil generation, onset of gas generation, and, for very high levels of maturity, the onset of graphitization. There is no complete agreement within the exploration community as to the values associated with these different processes, although those cited here provide useful general guidelines. In addition, certain factors, such as a high sulfur content, lower the levels of reflectance at which oil generation occurs.

The second important point is that not all samples contain vitrinite. Pre-Devonian samples do not contain vitrinite since higher plants did not evolve until around this time and vitrinite is derived from higher plant remains. However, there are a number of alternative maturity scales that can be used. For example, the thermal alteration index (Staplin, 1969) and spore coloration index (Burgess, 1974) have been used

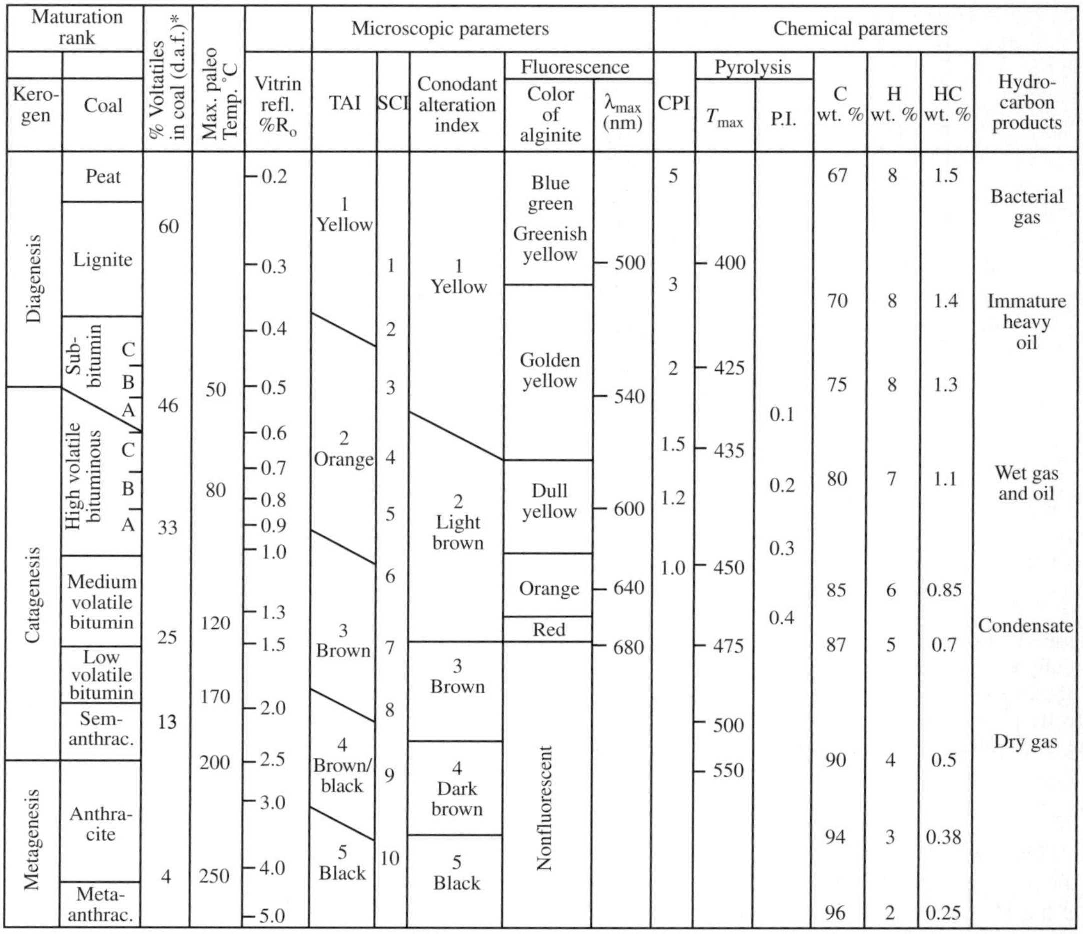

Maturation rank: Kerogen	Maturation rank: Coal	% Volatiles in coal (d.a.f.)*	Max. paleo Temp. °C	Vitrin refl. $\%R_o$	Microscopic parameters: TAI	SCI	Conodant alteration index	Fluorescence: Color of alginite	Fluorescence: λ_{max} (nm)	Chemical parameters: CPI	Pyrolysis: T_{max}	Pyrolysis: P.I.	C wt. %	H wt. %	HC wt. %	Hydrocarbon products
Diagenesis	Peat			0.2				Blue green		5			67	8	1.5	Bacterial gas
		60			1 Yellow			Greenish yellow								
	Lignite			0.3		1	1 Yellow		500	3	400					
													70	8	1.4	Immature heavy oil
	Sub-bitumin C			0.4		2										
	Sub-bitumin B							Golden yellow		2	425					
	Sub-bitumin A	46	50	0.5		3			540				75	8	1.3	
												0.1				
Catagenesis	High volatile bituminous C			0.6	2 Orange					1.5	435					
				0.7		4										
	High volatile bituminous B		80	0.8			2 Light brown	Dull yellow	600	1.2		0.2	80	7	1.1	Wet gas and oil
	High volatile bituminous A	33		0.9		5										
				1.0								0.3				
	Medium volatile bitumin					6				1.0	450					
								Orange	640				85	6	0.85	
				1.3								0.4				
		25	120					Red								Condensate
				1.5	3 Brown	7			680		475		87	5	0.7	
	Low volatile bitumin		170				3 Brown									
				2.0		8										
	Sem-anthrac.	13									500					Dry gas
Metagenesis			200	2.5	4 Brown/black	9	4 Dark brown	Nonfluorescent			550		90	4	0.5	
	Anthra-cite			3.0												
													94	3	0.38	
		4	250	4.0	5 Black	10	5 Black									
	Meta-anthrac.			5.0									96	2	0.25	

*Dry ash free

Figure 3 Vitrinite reflectance is used to provide an indication of the maturity of a source rock or potential source rock. The approximate values for the onset of oil generation, peak oil generation, and onset of gas generation are also shown in this diagram. It should be noted that there is some disagreement as to the precise values for these different processes (Hunt, 1996) (reproduced by permission of Freeman from *Petroleum Geochemistry and Geology*, **1996**).

widely to provide an alternative indication of maturity, as shown in Figure 3. With the development of the Rock Eval system, one of the most widely measured maturation parameters by the oils companies today is T_{max}.

The initial result from Rock Eval pyrolysis is a chromatogram with the two major peaks described above, S_1 and S_2 (Figure 4), along with the S_3 peak. As the maturity of a sample increases, the temperature at which the S_1 peak appears remains relatively constant; however, the temperature at which S_2 maximizes increases. S_3 is not used directly for maturity determinations. The increase in the temperature at which S_2 maximizes results from the fact that what is being measured here is the temperature at which the residual material in the rock breaks down. As the maturity level of the rock increases, the temperature required to degrade the residual material also increases. It is important to note that a direct correlation between T_{max} and vitrinite reflectance is not necessary. These parameters result from measuring two different processes, both of which increase with increasing maturity, but not necessarily at the same rate. Care also needs to be taken when comparing maturity parameters for samples thought to be at similar levels of maturity but from different basins. Burial and heating rates will differ and therefore relative changes in maturity parameters will also differ.

In summary, kerogen is a key intermediate in the formation of oil and gas. Kerogen types and maturity levels play an important role in determining the characteristics of the products that will be generated from a specific kerogen (Tissot *et al.*, 1972, 1974). In other words, will a source rock produce oil or gas; light oil or heavy oil; dry gas or wet gas? It is also important to remember that the bulk of organic carbon in the sedimentary record is in the form of kerogen. Kerogen is made up of varying proportions of different types of macerals bound together in a variety of ways. It is not a

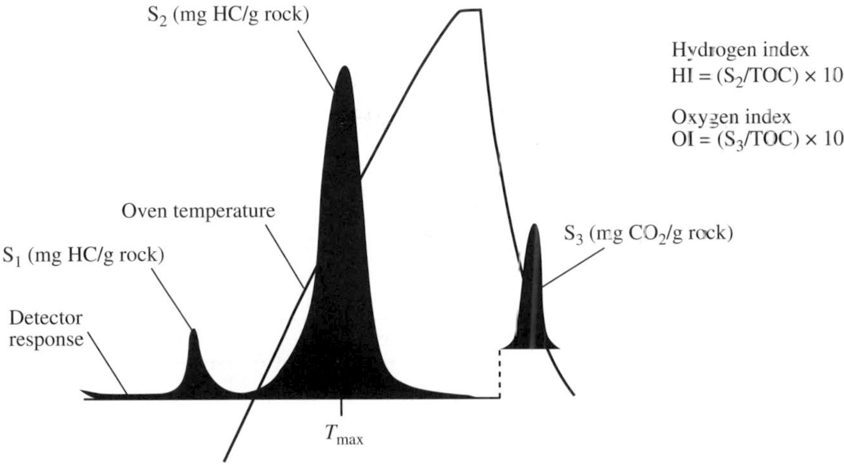

Figure 4 Rock Eval pyrolysis typically produces three major peaks that are used extensively in geochemical characterization of source rocks. The most recent version of the system produces additional data but for most routine analyses the peaks of interest are the S_1, S_2, and S_3 peaks as shown here. Various production parameters can be derived from these peaks plus values for HI and OI indices.

polymer, but is derived from a variety of "building blocks" preserved from insoluble material present in living systems. The role of coal as a potential source rock has not been discussed in detail in this chapter, but it is accurate to say that our understanding of the relative importance of coals versus shales as sources for oil and gas is certainly incomplete and requires a great deal of additional work (Scott and Fleet, 1994). The expulsion efficiency of coaly rocks is also difficult to assess due to their tendency to adsorb hydrocarbons (Saxby and Shibaoka, 1986; Powell and Boreham, 1991). These are questions that will continue to be addressed in the future as exploration continues in areas such as the Mahakam Delta, and other areas of South East Asia, where coals or dispersed coal-like materials clearly play an important role as a source for much of the oil produced in that area.

7.09.4 SOLUBLE ORGANIC MATERIAL

As mentioned in the preceding section, most crude oil and gas is derived from organic matter with kerogen as the major intermediate. In most source rocks, kerogen comprises the bulk of the organic carbon (Figure 5). A cursory examination of the petroleum geochemistry literature quickly reveals that the bulk of published papers have been disproportionately concerned with a minor fraction of the source rock, namely, the extractable organic material. The most significant part of geochemistry has been based on improving our understanding of the origin and fate of the extractable organic matter present in the source rock. Geochemistry made some dramatic advances during 1960–1980 in parallel with developments in analytical technology particularly gas chromatography–mass spectrometry (GCMS; Eglinton *et al.*, 1964; Henderson *et al.*, 1969; Eglinton and Calvin, 1967; Eglinton and Murphy, 1969; Balogh *et al.*, 1973). It was observed, basically for the first time, that organic-rich rocks, such as the Green River and Messel shales, contained mixtures of relatively small amounts of complex molecules such as steranes and terpanes (Henderson *et al.*, 1969). In order to identify those compounds, it was necessary to go through many steps to isolate and purify individual compounds. The combination of the GC and MS plus the development of automated data systems greatly improved our capability to characterize trace amounts of components in very complex mixtures. The ancillary technique of multiple ion detection (MID) also became a key part of this characterization process (Hites and Bieman, 1970). This technique makes use of the fact that when a compound is ionized in a mass spectrometer, it produces certain ions characteristic of its structure. By monitoring the characteristic ion, rather than the complete spectrum, it is possible to greatly improve the sensitivity and specificity of the system. There are tables of characteristic ions (used for monitoring various classes of biomarkers) in the books by Philp (1985) and Peters and Moldowan (1992). To illustrate the approach, Figure 6 illustrates m/z 191 and 217 chromatograms for a typical crude oil. In the conventional gas chromatogram the presence of steranes and terpanes would not be evident due to their relatively low concentrations. The MID chromatograms clearly show these distributions which are widely used in correlation studies and for purposes of source and maturity related determination as described below. GCMS is probably the

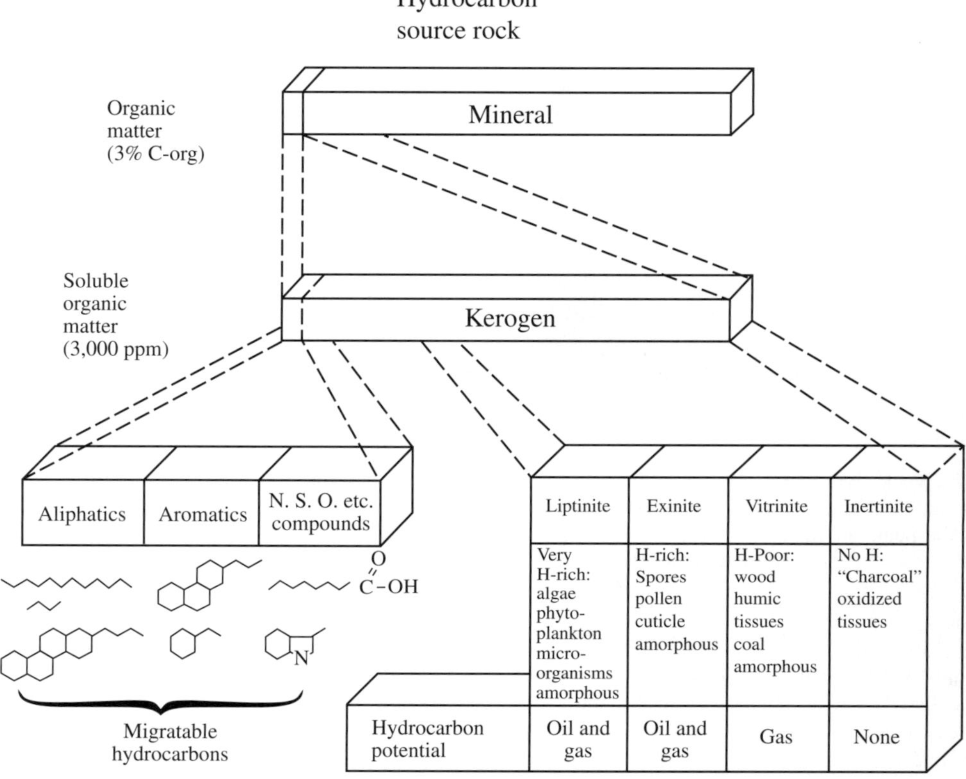

Figure 5 The majority of the organic carbon in a source rock is in the form of kerogen. This schematic diagram provides an indication of the relative amounts and form of organic matter in a typical source rock (Brooks *et al.*, 1987) (reproduced by permission of Blackwell from *Marine Petroleum Source Rocks*, **1987**).

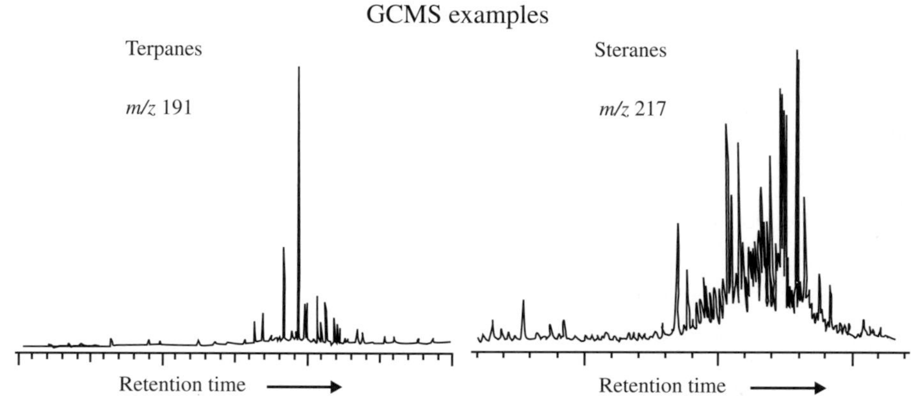

Figure 6 GCMS is a widely used technique in petroleum exploration and production studies. The use of multiple ion detection to obtain the fingerprints for the various classes of biomarkers is an extremely important ancillary GCMS technique. In this diagram the *m/z* 191 and 217 chromatograms show the distributions of the terpanes and steranes, respectively.

most widely used analytical technique in petroleum geochemistry today, and additional applications of this technique will be shown below (Philp, 1985; Peters and Moldowan, 1992).

Compounds such as steranes and terpanes in crude oils are referred to as biomarkers. The biomarker concept was developed by Eglinton and Calvin (1967), who recognized the presence of compounds in ancient sediments with carbon skeletons that could be related to precursor molecules present in organisms and plants living at the time of deposition. For example, sterols are very abundant in many living organisms and plants and, upon burial, diagenesis, and thermal maturation, are ultimately converted into steranes. The only significant change that occurs is loss of the hydroxyl group and the double bond(s) present in the sterols. Therefore, a precursor/product relationship can be established between the sterol and the sterane. Similar precursor–product relationships can be established for a large number of other compounds or compound classes. Specific precursors may not always be known for a particular hydrocarbon in a crude oil, but it may still be possible to associate a particular compound with a specific input or type of organic matter. When discussing biomarkers, it is important to recall the fact that several decades earlier Alfred Treibs had recognized structural similarities between chlorophyll in plants and porphyrins in crude oils (Treibs, 1934, 1936). As a result, he was able to infer an organic origin for crude oils. What is remarkable about his discovery is that it was made long before the availability of GC, GCMS, and other sophisticated analytical tools available today. He proposed a pathway for the conversion of chlorophyll to porphyrins based on compounds identified using classical organic techniques. This pathway was verified years later by the work of Maxwell *et al.* (1981) and others using more sophisticated techniques.

The development and commercial availability of GCMS systems accelerated the search for biomarkers in samples ranging in age from Precambrian to very recent sediments. The work of Kimble (1972) was a major contribution to the understanding of stereochemical changes in many of these biomarkers with increasing maturity. Kimble *et al.* (1974) observed that the Green River and Messel shale contained similar distributions of terpenoids and steranes differing in relative proportions of stereoisomers as a result of their different thermal histories. Wolfgang Seifert, a petroleum geochemist with Chevron, was one of the first to recognize the potential of these stereochemical variations and applied the biomarker concept to petroleum exploration problems. Along with his co-workers, he published many landmark papers in the 1970s and the early 1980s (Seifert, 1977, 1978; Seifert and Moldowan, 1978, 1979, 1981).

The concentration of individual biomarkers in a crude oil are relatively low (Peters and Moldowan, 1992; Figure 7), but their information content is significant and related to source, depositional environments, maturity, age, migration, and extent of biodegradation. The complete biomarker fingerprints obtained by MID are used extensively for the purpose of undertaking oil/oil or oil/source rock correlation studies. There are numerous papers and books which have documented the utilization of biomarkers in petroleum exploration; only a few examples of the major applications will be

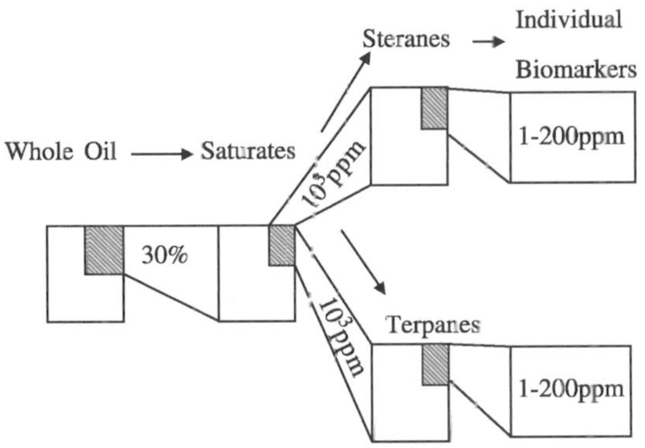

Figure 7 The concentrations of individual biomarkers in a crude oil are relatively low but the complex nature of their structures means their information content is high. This diagram provides an indication of the approximate concentrations of individual steranes and terpanes in a typical crude oil.

given here; additional references are provided for further reading.

7.09.4.1 Source

Biomarkers evolved from our ability to associate the origin of certain hydrocarbons with specific functionalized molecules that in turn could be associated with certain types of organic matter. Many of these precursor–product relationships evolved from studies of recent sediments, where the inputs of organic matter could be easily documented (Maxwell *et al.*, 1971; Eglinton *et al.*, 1974). For example, it was recognized many years ago that plant waxes were dominated by *n*-alkanes with a marked odd/even predominance. If a recent sediment was examined and observed to have a marked odd/even distribution in the C_{25}–C_{35} region, it could be associated with a higher plant input to the sediments. With increasing maturity, the odd/even predominance is reduced as a result of an increase in even-numbered and decrease in odd-numbered *n*-alkanes. However, the carbon number range can still be used to differentiate between a higher plant versus marine input to the sediments. The presence of individual compounds can be used in the same way. 18α(H)-Oleanane is a classic example of a compound associated with a higher plant input, and more specifically an angiosperm source (Whitehead, 1971, 1974; Grantham *et al.*, 1983; ten Haven and Rullkotter, 1988; Figure 8). Mistakes were often made in these assignments. For example, it was widely reported prior to the mid-1980s that the presence of C_{29} steranes in a crude oil represented contributions from higher plant material in the oil. It is true that there are abundant C_{29} sterols in higher plant materials but, unfortunately, petroleum geochemists were perhaps not aware

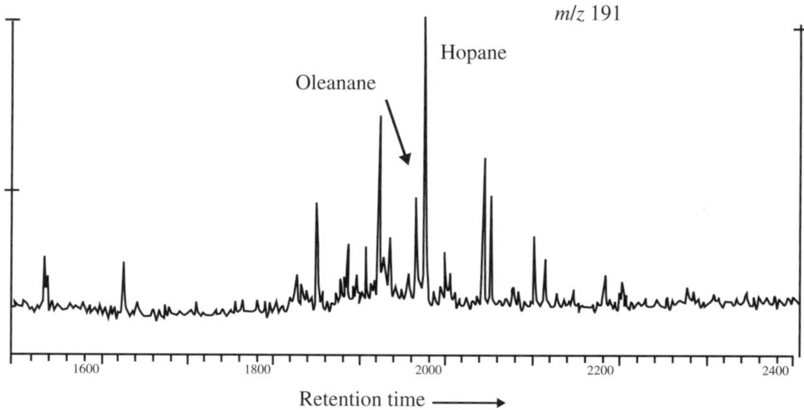

Figure 8 Various individual biomarkers have been used as age-dating compounds. A classic example of this is 18α(H)-Oleanane, a compound associated with a higher plant input, more specifically an angiosperm source. Angiosperms generally are believed to have evolved at the end of the Cretaceous/Early Tertiary.

of the fact that C_{29} sterols had also been reported to occur in various algae (Volkman, 1986). The specificity associated with the presence of C_{29} steranes was therefore lost. The overall sterane fingerprint could still be used for correlation purposes and specific steranes, such as the dinosteranes, were still valuable as specific source indicators. There are many more source indicators than the few described here as an illustration. An excellent source of additional information on this topic and many of the topics described below is the "Biomarker Guide" by Peters and Moldowan (1992). Although the book was published over 10 years ago, there has not been a significant number of novel biomarkers discovered since that time. This is somewhat in contrast to the exponential rise in the number of biomarkers discovered in the 1970s–1980s during a period of rapid expansion in geochemistry (Mackenzie *et al.*, 1982).

7.09.4.2 Depositional Environments

The utility of biomarkers as indicators of depositional environments arises from the fact that certain types of compounds are associated with organisms, or plants that grow in specific types of depositional environments. It has been possible, in some situations, to characterize depositional environments through a composite of biomarker parameters, e.g., the case of offshore Brazil (Mello *et al.*, 1988). However, the majority of work in this area has utilized single compounds. For example, gammacerane has long been used as an indicator of hypersalinity (de Leeuw and Sinninghe Damste, 1990), although, as noted in Peters and Moldowan (1992), gammacerane is present in relatively low concentrations in virtually all oils from marine depositional environments. However, there is a general consensus that with increasing salinity the ratio of the concentration of gammacerane to that of hopane increases quite significantly. Oils or extracts from evaporitic environments, particularly lacustrine environments, typically have terpane chromatograms where gammacerane is the most abundant component in the chromatogram (Figure 9). It is also noteworthy that for many years the exact nature of the precursor of gammacerane was unclear until the work of Venkatesen (1989) established that gammacerane was derived from tetrahymanol precursor. It is of interest to note that while gammacerane is an indicator of salinity, its precursor, tetrahymanol, is widespread in freshwater and marine ciliates. Sinninghe Damste *et al.* (1995) proposed that in certain Miocene and Upper Jurassic formations, gammacerane is derived from bacterivorous ciliates that fed on green sulfur bacteria. If their diet is deprived of sterols, these anaerobic ciliates synthesize tetrahymanol, and hence gammacerane can be thought of as an indicator of water column stratification and not simply hypersalinity, as suggested by Schoell *et al.* (1994).

Botryococcane, an irregular C_{34} isoprenoid, is another good example of a compound associated with a specific alga, *Botryococcous braunii*, a green alga which only grows in fresh/brackish water environments (Moldowan and Seifert, 1980). This compound is unique, in that it is both a source indicator and an indicator of depositional environments. Hence, it is extremely useful for paleoreconstructing depositional environments, and the fact it is not ubiquitous makes it extremely useful for correlation purposes. Seifert and Moldowan (1981) reported relatively high concentrations of

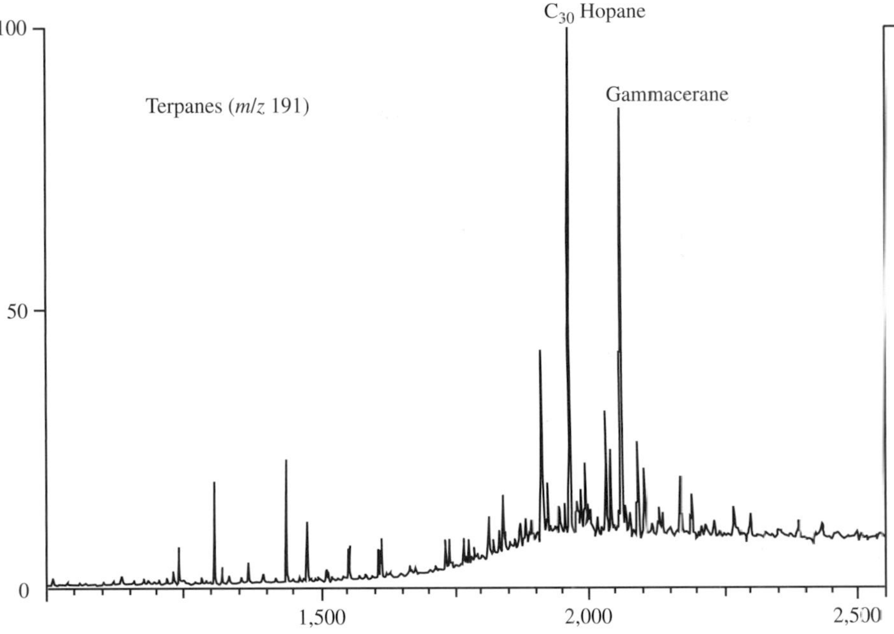

Figure 9 Depositional environments can be characterized by individual biomarkers or groups of biomarkers. The classic example is gammacerane, whose relative concentration has been shown to vary with salinity in many well-defined cases.

this compound in certain oils from Indonesia, restricting the possible sources for these oils and indicating that the source rocks for these particular oils would be expected to have concentrations of botryococcane.

Other compounds that were earlier proposed to be associated with particular types of environments have been re-evaluated. For example, Brooks *et al.* (1969) noted the presence of the regular isoprenoids pristane (Pr) and phytane (Ph) in crude oils and coal extracts. This led Powell and McKirdy (1973) to propose a mechanism for the production of relatively high concentrations of pristane in oxic type environments and high concentrations of phytane in reducing type environments. Thus, the Pr/Ph ratio evolved as an indicator of the oxicity of the depositional environment. It was used in this way for many years, but as more and more samples were investigated and reaction mechanisms were more clearly understood, it became apparent that this ratio was not universally applicable. Reasons include the fact that any depositional environment is not totally homogeneous in terms of oxicity (ten Haven *et al.*, 1987). An environment thought to be aerobic has microenvironments which are anaerobic: so, what does the Pr/Ph ratio mean in that situation? Further, a number of alternative sources of pristane and phytane have also been discovered in addition to chlorophyll (Goosens *et al.*, 1984; Rowland, 1990). Hence, an organism containing an alternative source for phytane may thrive in an oxic environment, producing Pr/Ph ratio that is very low for an oxic environment. In brief, a great deal of caution needs to be exercised when using the Pr/Ph ratio as an indicator of depositional environment. In the past few years with the development of combined gas chromatography isotope ratio mass spectrometry (GCIRMS), it is now possible to measure the stable isotope composition of the pristane and phytane (Freeman *et al.*, 1990). If they have similar isotopic compositions, they are probably derived from the same source; the Pr/Ph ratio can then be used as an indicator of the depositional environment. In spite of these uncertainties, the Pr/Ph ratio is still a useful correlation parameter, in conjunction with many of the other parameters used for correlating oils to source rocks and oils with oils.

In a slightly different approach, and one which is part of reservoir geochemistry, Dahl *et al.* (1994) and McCaffery *et al.* (1994a) proposed that vital source rock properties can be predicted from biomarker parameters measured in reservoired oils, and that prediction of source properties in vertically drained basins can prove extremely useful in oil-rich regions such as the US Gulf Coast, the North Sea, the Beaufort Sea, and the Mahakam and Niger Deltas. If oils can be used to predict source rock properties, this will permit the explorers to know where to look for the source rocks and to predict migration pathways. Basically, it was observed that many of the characteristic biomarker parameters show a good correlation with bulk source rock properties such as kerogen H/C ratios, HI, and maceral compositions. The homohopane ratio was not quite as useful as some of the others, since it is maturity dependent and is less useful for determining the source rock qualities of expelled oils than the sterane parameters. The Ni/(Ni + V) ratio, shown by Moldowan *et al.* (1985) to be a good indicator of paleoenvironment, also trended in the expected direction. Steranes showed a good correlation for the percentage of C_{30} steranes (24-n-propyl) to the 4-desmethyl steranes in the extracts to the percentage of marine organic matter in the rock. Methyl sterane contents correlated strongly with kerogen HI and H/C ratios. 3β-Methyl-24-ethylcholestanes dominated in kerogens with low H/C ratios, whereas 4α,23,24-methyl cholestanes (dinosteranes) dominated in rocks with high H/C ratios. One might expect the highest dinosterane concentrations in areas where algal blooms occurred, which in turn could have yielded rocks with higher H/C and HI values. Relative resistance of C_{29} sterane precursors from terrestrial versus marine sources may also reflect why these sterane precursors correlate with the percentage of terrestrial organic matter. The marine C_{29} steranes may be more prone to oxidation and thus the terrestrial contribution may be bound in plant membranes and more resistant to degradation.

The data described above can be used to predict the location of better source rocks in vertically drained basins, especially in deltaic-type environments with relatively young source rocks. With long-distance vertical migration, some of the biomarker parameters may become skewed. A number of factors must first be considered before applying this approach: first, some of the parameters vary with maturity; second, C_{30} steranes are not present in lacustrine samples and so the approach will not work in that situation; and finally, it will not work where the oils were deposited prior to land plant evolution, since no vitrinite was present at that time. Oils from mixed source rocks also complicate the issue. The ability to predict source rock properties on the basis of biomarker distributions in crude oils is a very interesting concept, since most exploration efforts try to discover oil and not source rocks.

7.09.4.3 Maturity

In the same way that certain biomarkers have been used to characterize source materials and depositional environments, selected biomarkers have been used to evaluate the relative maturity of suspected source rocks and the oils they may have generated. As more samples have been examined and our understanding of changes resulting from differences in maturity has improved, many of the maturity ratios and their associated interpretations have changed, although the end result is often the same. Again, it should be noted that the use of biomarkers as maturity indicators really started with the work of Kimble (1972) and Kimble *et al.* (1974), who noticed that the distribution of individual stereoisomers differed between shales of differing maturity levels. This observation was noted by Seifert at Chevron and was applied to a far wider range of samples. It then became apparent that steranes and hopanes had a number of stereoisomers and optical isomers whose configurations changed with increasing maturity (Seifert *et al.*, 1978; Seifert and Moldowan, 1980; Mackenzie *et al.*, 1982).

The most commonly used parameters which have survived scrutiny over an extended period of time are summarized in Table 1. The maturity parameters can be divided into a number of groups. The first group includes those based on changes in distributions of stereoisomers or optical isomers; the second includes those resulting from thermal degradation of certain molecules in a predictable manner, e.g., the cleavage of side chains from the triaromatic steroid hydrocarbon; and the third group contains various porphyrin ratios that depend on ring opening of the porphyrin which occurs with increasing maturity. The most recent set of biomarker maturity parameters are those based on changes in the relative distribution of various methylated adamantane structures (Chen *et al.*, 1996). These parameters are still being investigated and there is discussion over their maturity range; nevertheless, the ubiquitous nature and stability of these compounds make them ideal maturity parameters.

While not all of the maturity parameters will be discussed in detail, it is useful to make a few comments concerning those that are most commonly used. Ratios based on changes in sterane

Table 1 Common biomarker maturity parameters.

1. %Tricyclic terpanes/(Tricyclics + 17α-Hopanes)
2. %22S/(22S + 22R)-Hopanes(C_{32})
3. %$\beta\alpha$/($\alpha\beta$ + $\beta\alpha$)-Hopanes(C_{30})
4. %Ts/(Ts + Tm)
5. %18β/(α + β)-Oleananes
6. %20S/(20S + 20R)-Steranes (C_{29})
7. %$\beta\beta$/($\beta\beta$ + $\alpha\alpha$)-Steranes (C_{29})
8. %Diasteranes/(Diasteranes + Regular Steranes)
9. %TA-C_{28}/(TA-C_{28} + MA-C_{29})
10. %MA(I)/MA(I + II)
11. %TA(I)/TA(I + II)
12. %ETIO/(ETIO + DPEP) Nickel
13. %ETIO/(ETIO + DPEP) Vanadyl
14. %C_{28}/(C_{28}E + C_{32D}) (PMP)
15. Ratio of 4,6-DMDBT/1,4-DMDBT, 2,4-DMDBT/1,4-DMDBT, and 4-MDBT/1-MDBT

This table only provides a list of these maturity parameters. The majority of these parameters are discussed in detail in the Peters and Moldowan (1992) monograph and do not need to be explained in more detail in this chapter. The dibenzothiophene (DBT) parameters are described in Chakhmakhchev *et al.* (1997).

stereochemistry depend on the fact that in the initial conversion process of functionalized sterols to steranes, the so-called kinetic isomer is the first to be formed, but it is not necessarily the most stable isomer. This kinetic isomer is slowly transformed into the more thermodynamically stable isomer. As a result, ratios can be developed which monitor changes with increasing maturity.

The two most commonly used sterane maturity parameters are the C_{29} 14α(H),17α(H)(20S/(20S + 20R)) ratio and the C_{29} 14α(H),17α(H) (20S + 20R)/14α(H),17α(H)(20S + 20R) + 14β(H)17β (H)(20S + 20R) (Seifert and Moldowan, 1986). There have been numerous changes in the interpretation of the significance of these ratios. First it should be emphasized that, whilst these various ratios are extremely useful, most of them are basin specific or even formation specific. In other words, simply because oils from two different basins have a ratio of, say, 0.4 for the 20S/(20S + 20R) ratio, it does not mean they are at the same maturity level. These ratios can be influenced by factors such as the heating rate, the presence of clay minerals, and for oils, biodegradation and the nature of the source material and the depositional environment. There have also been a number of attempts to interpret the mechanisms responsible for these changes. For example, does the ratio 20S/(20S + 20R) increase as a result of the epimerization reaction between these two epimers, or does it change because one epimer is being degraded, or aromatized, more rapidly than the other? Requejo (1992) carried out very elegant quantification studies and provided detailed evidence to show that both epimers decreased in concentration with increasing maturity but with the 20R decreasing faster than the 20S. Regardless of whether it is preferential destruction or conversion, the end result is the same, namely, a ratio that increases with maturity and can be used to compare relative maturity for source rocks or oils from the same basin. Another process that may affect this particular ratio is biodegradation. In extremely heavily biodegraded oils, isomers with the "naturally" occurring stereochemistry, the 20R configuration, are removed preferentially over the 20S configuration, leading to an erroneously high value for the 20S/(20S + 20R) ratio.

The second sterane ratio, also used extensively for many years, is the 14α(H),17α(H)/(14α(H), 17α(H) + 14β(H),17β(H)), or $\beta\beta/(\alpha\alpha + \beta\beta)$ ratio. Initially, it was proposed that this ratio changed as a result of the $\alpha\alpha$-isomers being converted to the $\beta\beta$-isomers with increasing maturity (Mackenzie *et al.*, 1982; Seifert and Moldowan, 1986). It was also observed in certain samples, particularly those from hypersaline evaporitic environments, that very immature samples contained relatively high concentrations of the $\beta\beta$-isomers as a result of a specific source input. Hence, samples from hypersaline environments may have a 20S/(20S + 20R) ratio, suggesting that the sample is immature. However, these same samples could have a high $\beta\beta/(\alpha\alpha + \beta\beta)$ ratio, suggesting high levels of maturity. The latter results from a specific source input associated with that particular environment (ten Haven *et al.*, 1986, 1988). In view of these anomalies, a certain amount of care needs to be exercised in the interpretation of maturity ratios, and factors such as source and depositional environments also need to be taken into consideration when interpreting the results.

Another commonly cited maturity parameter is the ratio of the 22S/(22S + 22R) diastereomers of the extended hopanes (Ensminger *et al.*, 1977). As the maturity level increases, the concentration of the 22S epimer increases relative to the 22R. It was initially proposed to be an epimerization reaction but, as the reaction was investigated in more detail, it appeared that the change in the ratio probably reflected differences in the rate of degradation for the two isomers. This ratio is only useful in the early stages of oil generation until it reaches its maximum value of ~0.64 and does not increase beyond that with increasing maturity. Lewan (1985) has noted that these commonly used maturity parameters often undergo reversal and start to decrease at higher levels of maturity.

Biomarker maturity measurements are very useful for comparing relative maturities in different regions of a basin to determine whether or not, all other things being equal, a particular formation has entered the oil or gas generation window. Biomarker ratios can be particularly important in samples that may not contain vitrinite, preventing the determination of a vitrinite reflectance value. In addition, these parameters permit measurements of relative maturity for oils as well as rocks, something not possible with vitrinite, TAI, or SCI. Whilst the majority of the commonly used parameters have been in use for many years, it should be noted that several new maturity parameters have recently been introduced based on the adamantane structures. As noted above, there is still some debate concerning the range over which these parameters are valid, but their ubiquity and stability have the potential to make them extremely useful as maturity indicators. One of the earliest papers published on the use of these compounds as maturity parameters was that of Chen *et al.* (1996), who discussed the use of two parameters based on these compounds for use as high-maturity indices in a number of Chinese basins. One of the main attractions of these parameters as proposed by Chen *et al.* (1996) was the claim that they operated over a wide range of maturities from immature ($<$0.6% R_o) to over mature ranges (~2.0% R_o). A more recent paper by Li *et al.* (2000) suggested that there was no linear correlation between these adamantane maturity parameters with maturity at the higher levels of maturity and, therefore, they may have a limited range just like many of the biomarker parameters. As with any new parameter, clearly additional work needs to be done to clarify this issue.

7.09.4.4 Biodegradation

Biodegraded tar sand accumulations in North America and Canada account for a large percentage of the world's oil inventory. A general and empirical understanding of the mechanisms for the formation of these tar sands and other heavy oil accumulations has been available for many decades. More recently, several novel hypotheses concerning mechanisms of crude oil biodegradation have been introduced. These include: (i) the concept of paleopasteurization of reservoirs; (ii) anaerobic, rather than exclusively aerobic degradation of crude oils in reservoirs; and (iii) different mechanisms for degradation of surface seeps versus deep reservoirs. Organic geochemistry continues to play a major role in furthering our understanding of the mechanisms of biodegradation and whether crude oils represent mixtures of degraded and nondegraded crude oils. Determination of the presence of degraded and nondegraded oils in a reservoir can provide valuable information on whether a reservoir represents a single filling episode or a filling episode followed by degradation and influx of fresh oil.

The most significant changes which occur in crude oil upon biodegradation were comprehensively reviewed by Connan (1984). Much of the discussion in this chapter remains unchanged and observations have consistently supported the information presented in that review. Additional observations have improved upon our understanding concerning the relative rate of removal of various individual and classes of biomarkers, which can be particularly important in judging the relative extent of degradation of a crude oil accumulation (Alexander *et al.*, 1983; Philp and Lewis, 1987; Volkman *et al.*, 1983, 1984). The basic premise is that linear alkanes are removed most readily followed by compounds of increasing structural complexity such as isoprenoids, monocyclic, tricyclic, tetracyclic, and pentacyclic terpanes. Within each group of cyclic compounds, the naturally occurring stereoisomers are removed more rapidly than the thermally stable isomers, e.g., with steranes 20R epimers are removed more rapidly than the 20S epimers. The most recent summary of changes resulting from biodegradation was published by Peters and Fowler (2002).

An extensive review of previously published information will not be given in this chapter, but it is extremely important to briefly review some of the more recent developments related to biodegradation. For many years, a great deal of the information in the geochemical literature has focused on the notion of biodegradation occurring only under aerobic conditions. Few papers were published supporting the concept of anaerobic biodegradation in reservoirs (Wilkes *et al.*, 1995) and only in the late 1990s did the anaerobic mechanism gain the same level of support as aerobic degradation (Zengler *et al.*, 1999; Wilkes *et al.*, 2001). One of the reasons was the somewhat circumstantial and tenuous nature of the evidence presented in support of such a mechanism, typically culturing anaerobic bacteria from oil field waters. A number of papers have appeared recently that describe laboratory experiments, performed under very carefully controlled conditions, where anaerobic degradation of typical petroleum hydrocarbons was observed. A number of groundwater studies have clearly demonstrated anaerobic degradation of aromatic compounds (Elshahed *et al.*, 2001) through the presence of various intermediates, such as succinyl derivatives. These derivatives are formed through the addition of the fumaryl entity to the methyl group being removed or the end of the hydrocarbon chain undergoing degradation. Succinyl derivatives have been found in groundwater samples, but to date the corresponding compounds have not been found in oilfield waters, only in the water associated with the laboratory studies. However, their absence in the oilfield waters is not necessarily a major problem, since these compounds are intermediates and could undergo further degradation. Conclusive evidence, in the form of intermediates, exists that demonstrates the anaerobic degradation of toluene and short straight chain *n*-alkanes in groundwater samples. A more extensive search is, however, necessary to determine whether or not these intermediates are also present in crude oils and associated waters and can be used to indicate anaerobic degradation of the crude oils.

The second major development has been the evolution of the paleopasteurization concept (Wilhems *et al.*, 2001). Paleopasteurization explains why it is possible to observe reservoirs of nondegraded crude oils at relatively shallow depths in the subsurface where one might expect the oils to be degraded. The explanation makes use of the fact that, in general, biodegradation of crude oils has not been reported in reservoirs that have experienced temperatures above 80 °C. For paleopasteurization to occur, a reservoir will have to have been exposed to temperatures above 80 °C at some time during its history. Above that temperature, the reservoir is effectively sterilized and subsequent uplift, or overburden erosion, leading to the temperature falling below 80 °C, will mean that the oil which enters the reservoirs at lower temperatures will not experience any additional biodegradation. In other words, the sterilized sediments are not recolonized by hydrocarbon degrading bacteria. Paleopasteurization can be used to explain the presence of nondegraded oils at shallow depths and also the presence of both regular and 25-norhopanes (thought to be derived by biodegradation of the regular hopanes) in oils which appear to be nondegraded. Co-occurrence of these compounds has long been attributed to mixing of degraded and nondegraded oils (Alexander *et al.*, 1983; Philp, 1983) but paleopasteurization provides a plausible explanation for the timing of mixing event(s).

These relatively new developments represent significant advances in our understanding of the mechanisms involved in biodegradation. Study in these areas should continue to improve our ability to predict the occurrence and extent of biodegraded oils plus an improved picture on the filling histories of certain reservoirs.

7.09.4.5 Age Dating

One problem of major interest to explorationists is: when was an oil produced or what is the geologic age of an oil? In most exploration efforts, oil companies drill for oil, not source rocks, and in some basins the source rocks may never be penetrated, or even discovered. If the age of the potential source rocks can be determined directly from the oil, it greatly assists in the elimination of certain formations as the source of an oil, and it will permit potential migration pathways to be mapped more accurately, filling histories to be determined, and additional reserves to be possibly discovered. Several attempts have been made to age-date oils through the use of geochemical parameters. The main idea behind these attempts has been to find a compound, or class of compounds, that can be associated with a particular plant or organism, based on the fossil record. The earliest example based on this approach is probably the work of Grantham and Wakefield (1988), who measured the relative proportions of steranes in crude oils derived from marine source rocks with known ages. A calibration curve was constructed based on the ratio of the C_{27}/C_{28} steranes versus geologic time. An exponential increase was noted in this ratio as the samples became progressively younger, as a result of the increase in the concentration of the C_{28} steranes. Comparison with information on the increase in diversity of phytoplankton with time showed a good correlation with the sterane ratio and provided an explanation for this increase. This age-dating approach based on these sterane ratios has not received a great deal of attention in the recent literature, probably because the nature of the calibration curve was such that over certain time ranges significant errors could be introduced in the age determined.

More recently, the work of Moldowan *et al.* (1996) in particular has led to a number of specific compounds being used as age-related biomarkers, and a brief overview of the more important parameters will be given below to provide an insight into these developments. It has been well documented that the biomarker 18α(H)-Oleanane is associated with the flowering plants or angiosperms, which did not evolve until the end of the Cretaceous and into the Early Tertiary. Hence, the presence of 18α(H)-Oleanane is generally used as an indication that the oil was sourced from a rock of Late Cretaceous age or younger (ten Haven and Rullkotter, 1988; Moldowan *et al.*, 1994).

A second class of compounds used for age-dating purposes are the dinosteranes, which are derived from dinosterols (Moldowan *et al.*, 1996). These compounds were initially associated with the dinoflagellates that evolved in the Early Triassic. Examination of the source rock record clearly showed an increase in the concentration of various dinosterane derivatives at that time. Through the use of various ratios, for the most part based on aromatized dinosteranes, it was possible to associate certain dinosterane ratios with oils younger in age than the Triassic. This approach worked reasonably well until a number of oils older than the Triassic were also found to contain relatively high concentrations of these dinosteranes. Ultimately, it was concluded that the reason for this was that Acritarchs, distant relatives of the dinoflagellates, also contained high concentrations of the dinosteranes. Hence, whilst the dinosteranes may not be quite as successful as age-dating compounds as originally anticipated, there are regions in the timescale where the very low values for this ratio make it such that it can still be used to age-date oils in that region.

Holba *et al.* (1998) have developed another age-dating parameter based on variations in the distributions of certain regular and rearranged C_{26} steranes. Through the examination and characterization of a large number of oils and rock samples, a dramatic increase in the nordiacholestane and norcholestane ratios was observed as the oils became geologically younger in age. A plot of this ratio against geologic age has led to the development of a calibration chart that has been applied to age-date oils on a worldwide basis.

McCaffery *et al.* (1994b) discovered the presence of the novel C_{30} 24-isopropylcholestanes in oils and bitumens from Early Proterozoic (~1,800 Ma) to Miocene (~15 Ma). It was observed that the abundance of these compounds relative to the 24-n-propylcholestanes varied with source rock age. Late Proterozic (Vendian) and Early Cambrian oils and/or bitumens from Siberia, the Urals, Oman, Australia, and India had a high ratio of 24-isopropylcholestanes to 24-n-propylcholestanes (>1), while younger and older samples have a lower ratio (<0.4), permitting this ratio to be used as an age-dating parameter. Temporal changes in this parameter may reflect the corresponding relative abundance of certain Porifera (sponges) and certain marine algae, which are believed to contain the precursors of these steranes.

In summary, the potential for the use of age-related biomarkers is tremendous. For precise age-dating, however, it is necessary to establish the relationship between certain compounds and the evolution of certain plants or organisms. This is clearly a very time consuming process and requires a great deal of work and correlation between biomarkers and the fossil record. Once the ratios are established and successfully calibrated for the oils, they can be potentially very powerful. As mentioned above, it is generally oils that are recovered from drilling operations and not source rocks, so the more information that can be recovered from oils the more useful the results.

7.09.4.6 Migration

The fact that crude oil compositions will change with migration distance was proposed many years ago and has been documented in many publications since that time. Changes occur as a result of geochromatography, or basically the fact that smaller molecules move faster and further than the large molecules. All other things being equal, oils that have migrated the greatest distance will be enriched in the lighter components and numerous attempts have been made to quantify such changes. Early work by Seifert and Moldowan (1978) suggested that since the 20S/(20S + 20R) and the $\alpha\alpha/(\alpha\alpha + \beta\beta)$ sterane parameters were both maturity-related parameters, the extent of any deviance between these parameters could be related to a migration effect for oils derived from similar source materials. This idea was published back in the 1970s and represented the first attempt to use molecular ratios for measuring either relative or absolute migration distances. Prior to that, Silverman (1965) had noticed changes in bulk isotopic changes for crude oils with increasing migration distances. These changes were not directly due to isotopic fractionation but resulted from the enrichment of the lighter components with increasing migration distance, which was accompanied by a change in the isotopic composition. The lighter components of a crude oil are typically isotopically heavier and, therefore, if the molecular distribution of a crude oils changes one can expect isotopic changes to occur. Few developments were made following this observation until 1996 when Larter *et al.* (1996) suggested that variations in the ratio of benzo(a)-carbazole/(benzo(a)carbazole+benzo(c)carbazole) could be calibrated to determine absolute migration distances. The carbazole isomers have slightly different shapes, but the same structure, and are equally affected by maturation; therefore, any changes in this ratio can be attributed to migration effects. The benzo(a)carbazole, being more linear, would migrate faster than the benzo(c)carbazole due to less interaction with the surrounding mineral phases. Initially, this application was met with great enthusiasm. However, as with any new parameter, it was also subject to great scrutiny and, with time, it was realized that there were a number of factors, such as source input, thermal maturity, or depositional environments (Li *et al.*, 1997; Clegg *et al.*, 1998), which could influence this parameter. Changes in lithology and geometry of migration conduits along the migration pathway can also affect the calculated migration distances for oils of similar maturity and source (Li *et al.*, 1998). Research in this area is continuing and several additional molecular parameters are currently being investigated as potential indicators of migration distances. For example, Taylor *et al.* (1997) noted a decrease in phenol concentrations with increasing migration distance for some North Sea oils, and Galimberti *et al.* (2000) developed a molecular migration index based on *o*-cresol/phenol to infer a migration trend for some North Sea oils.

7.09.5 GEOCHEMISTRY AND SEQUENCE STRATIGRAPHY

To improve their degree of exploration success in recent years, petroleum companies have integrated sequence stratigraphy with geochemistry to define the prospective areas for petroleum source rocks and reservoirs. The connection between organic-rich rocks and petroleum generation was recognized many years ago and more recently it was recognized that particular facies produced organic-rich rocks. Organic facies can be defined as "mappable subdivisions of a designated stratigraphic unit, distinguished from adjacent subdivisions on the basis of the character of organic constituents, without regard to the inorganic aspects of the sediment" (Mann and Stein, 1997). Organic facies have been differentiated on the basis of decreasing oxygen content of the water bottom, which controls the quantitative and qualitative preservation of the organic matter and, therefore, the petroleum source rock potential. The facies deposited in the most anoxic environment will most likely contain the richest amount of organics. Chemical parameters associated with the different organic facies have been described in detail by Jones (1987). The recognition of the organic facies integrates well with sequence stratigraphy, which is basically the "study of rock relationships within a chronostratigraphic framework of repetitive, genetically related strata bounded by surfaces of erosion or nondeposition, or their correlative conformities" (Curiale *et al.*, 1992). In the petroleum industry, sequence stratigraphy uses large-scale geology based on the resolution of seismic exploration. In sequence stratigraphy, the largest unit is a sequence that is divided into system tracts, which are a linkage of contemporaneous depositional systems. This means that, in the same system tract, a sandstone would be included with a shale if they were deposited at the same time but they might be located in different areas. The sandstone would be closer and the shale would be farther away from the paleoshore. There are three system tracts: lowstand system tract (LST), transgressive system tract, and highstand system tract. Each system tract has unique characteristics which can be summarized in the following manner.

7.09.5.1 Lowstand System Tract

When the relative sea level falls, causing intense erosion and infill of the basin by sediment bypass, the area experiences a (LST), which rests on a subaerially produced unconformity (Curiale *et al.*, 1992). During an LST, organic material is degraded in the oxic water column and sediments are reworked, allowing biodegradation of organics because the oxic conditions let organisms graze at depth (Steffen and Gorin, 1993).

7.09.5.2 Transgressive System Tract

A transgressive system tract (TST) occurs during a relative sea level rise and is characterized by an upward decrease in abundance, size, and angularity of terrigenous sediments, and by the upward increase in organic material (Steffen and Gorin, 1993). The top of a TST is bounded by a maximum flooding surface that marks the change from the TST's retrogradational deposition to the HST's aggradational deposition (Curiale *et al.*, 1992).

7.09.5.3 Condensed Section

Condensed sections occur within a TST and the lowermost section of a HST if the sediment supply is low (Curiale *et al.*, 1992; Wignall and Maynard, 1993). The deposits of a condensed section are generally fine-grained rocks that usually have a high concentration of organic matter and are formed in sediment starved, deep-water conditions (Wignall and Maynard, 1993).

7.09.5.4 Highstand System Tract

During the latter part of relative sea level rise and part of the following relative sea level fall, the HST occurs and is characterized by an upward increase in abundance, size, and angularity of terrigenous sediments, and by an upward decrease in organic material. The sediments are more bioturbated than TST sediments (Robison *et al.*, 1996).

Several attempts have been made to integrate geochemical parameters into this framework of system tracts to assist in differentiating one facies from another. Potential source rocks can be found in all three system tracts, with the richest source rocks generally found in the condensed section within a TST. The best petroleum reservoirs are found in LST. With the use of geochemical data, the various system tracts in a sequence can be identified and therefore used to distinguish the location of both potential petroleum source rocks and reservoirs.

LSTs have several unique geochemical characteristics, including their total organic carbon content (Bohacs, 1993) and hydrogen indices typically lower than those expected for a TST (Creasey and Passey, 1993). The kerogens associated with a LST are generally type III or type IV (Lambert, 1993). Organic-sulfur content in the kerogen can be higher in LSTs than HSTs (Bohacs, 1993). In general, source rocks within an LST will usually be gas-prone if present (Robinson *et al.*, 1996) and the best reservoir rocks will be found in an LST due to the porous and permeable terrigenous deposits that dominate LSTs. A TST has more unique geochemical parameters than any other system tract, with high TOCs in the range 2–10% and high hydrogen indices ranging from 300 to 600 (Robison *et al.*, 1996) and typically oil-prone type II kerogens. The rise of relative sea level that causes a TST is recorded in the geochemical data by an increase in the sulfur content, decrease in the relative concentration of 24-ethylcholestane, an increase in the ratio of desmethyl C_{30}/C_{29} 24-n-alkylcholestanes, decrease in relative diasterane concentration for each carbon number, and an increase in the ratio of desmethylated to methylated C_{30}.

HSTs have a variety of geochemical parameters with TOC values (0–4%) decreasing due to increasing sediment dilution and the pyrolyzable hydrocarbons lower in a HST than those in a TST. The average value for the HI in an HST 0–150 and the kerogens typically contain type III to type IV kerogen with low organic-sulfur content. In general, HST source rocks are oil-rich at the beginning of a highstand and become more gas-rich as the highstand continues.

With this type of information the geochemical characteristics of an organic facies can be interpreted basically in terms of HST versus LST, etc. In certain cases, individual biomarkers can be used to aid in interpreting the sequence stratigraphy of a particular section. Several papers have appeared in the literature where geochemistry has been integrated with sequence stratigraphy in order to interpret the history of a particular basin (Horsfield *et al.*, 1994). The most recent comprehensive example on the integration of geochemistry and sequence stratigraphy is that reported by Peters *et al.* (2000), who revised early models of the Mahakam–Makassar region in Indonesia and upgraded the potential of the outer shelf area. Applying their new model and the associated interpretations led to several new discoveries in the area. Type III organic matter was recognized as the dominant source of organic matter in the area and four petroleum systems could be recognized in the area: an HST, two LSTs, and one TST. The highstand oils were waxy oils derived from middle-upper Miocene coal and shale source rocks deposited in a coastal plain highstand kitchen; one group of lowstand oils were less waxy and derived from middle-upper Miocene coaly source rocks deposited in deep-water lowstand kitchens; the other lowstand oils occur onshore and are generated from coaly source rocks; finally, the nonwaxy transgressive oils occur mainly onshore in middle to upper Miocene reservoirs. However, by incorporating geochemical and sequence stratigraphic data in this model, previous models which downgraded source potential in deep-water offshore areas were revised.

In brief, the integration of geochemistry in conjunction with sequence stratigraphy is continuing to develop. There are a limited number of geochemical parameters that can be used to identify the different system tracts at this time. However, care must be used when employing the parameters because of the lack of extensive testing. With more research and testing, more reliable geochemical parameters will be available to identify system tracts that are so easily seen in seismic.

7.09.6 FLUID INCLUSIONS

Another relatively recent development has been the study of fluid inclusions in reservoir rocks and along migration fairways. The information gained from these studies can be utilized to evaluate the evolution and migration of petroleum in sedimentary basins, as well as information on the filling history of the reservoir (Roedder, 1984; Karlsen *et al.*, 1993). Oil in fluid inclusions retains the composition of fluids at the time of trapping and is not subject to the same effects of weathering as the free oil. Light-hydrocarbon information present in fluid inclusions is often lost in the free crude oils. It should be noted that certain methods used to evaluate fluid inclusions that involve crushing lead to loss of the light components. Closed-system decrepitation combined with GC and/or GCMS prevent loss of these lighter components (Jones and Macleod, 2000; Volk *et al.*, 2000). Initially, one of the major concerns with fluid inclusion studies was contamination of the fluid with the free oil. However, methods developed by Jones and Macleod (2000) have shown that this contamination can basically be eliminated by various cleanup procedures. Sample selection is also of critical importance in ensuring that the inclusions are of a single period of geological time. In certain circumstances, the data generated can be combined with microthermometry and paleo-*PVT* data in order to give a detailed picture of the source, maturity, and physical properties of paleopetroleum.

The basic premise behind these studies is the fact that inclusions are constantly being formed as oil migrates along the migration fairway into the reservoir. Once the inclusions are formed, a fingerprint is trapped in the inclusions and represents whatever was migrating at that time. If another pulse of oil comes along prior to sample collection, then the rock sample will basically contain two fingerprints—one which represents the original pulse trapped in the inclusion and the other being the free product. Techniques are available that make it possible to remove the free product and decrepitate the fluid inclusions such that the product in the inclusions can be analyzed in the same manner as the free product. The initial trapped pulse is actually unaltered phase since, once it is trapped in the inclusion, it cannot be altered by bacterial action, or water washing. The second, extractable free phase represents a second pulse of oil currently migrating into the reservoir. The techniques have evolved to the level that it is possible to analyze the product in a small number of inclusions such that the biomarker parameters normally determined from the crude oil itself can be determined providing information available from the inclusion on the various topics described above.

A recent study by Volk *et al.* (2000) in the Prague Basin characterized fluid inclusions, and was able to show that the dominant processes during petroleum migration in this basin were gas migration, gas flushing, migration in gaseous solution, and petroleum fractionation. George *et al.* (1997) published several papers concerned with the application of fluid inclusion technology to constraining migration and filling histories of various fields in Australia and surrounding areas.

7.09.7 RESERVOIR GEOCHEMISTRY

The major part of the information provided above has been concerned with the characteristics of the oil and source rock extracts. Another very important part of geochemistry is concerned with the fate of the oil after it leaves the source rock (primary migration) and moves along the migration fairway (secondary migration) and is emplaced in the reservoir itself. Many years ago, Hunt (1961) reported on the compositional differences between source rock extracts and related oils in the reservoir rocks. These differences were a result of primary and secondary migration processes. Most of the changes occurring along the migration fairway can be related to migration distances, as described in the previous section. What happens once an oil has migrated and started to fill the reservoir? It is of interest to note how the answers to this, and other reservoir-related questions, have evolved over the past decade. A search of the literature will show that, prior to 1990, there was very little, if any, mention in the literature of reservoir geochemistry. Most geochemists in the oil business and in academia were then devoting their attention to the application of geochemistry to exploration problems. The early 1990s saw yet another downturn in the oil business and the geochemists involved in the exploration business needed to diversify. One way of doing this was to apply geochemical concepts to reservoir and production problems. Landmark papers published by England (1990) and England and Mackenzie (1989a,b) provided the necessary spark to get geochemists involved in this area of reservoir geochemistry to understand filling mechanisms to determine the presence of barriers preventing communication between blocks and to address various problems associated with production that had not been clearly understood in the past. Additional reports by the Chevron group provided the impetus for the widespread development of geochemical applications to continuity studies (Baskin and Jones, 1993; Baskin *et al.*, 1995).

Studies of reservoir and production problems were once the unique domain of the engineers. However, there were, and still are, certain problems that cannot be completely answered by engineering concepts alone. Petroleum geochemistry offers a low-cost means of approaching issues that ultimately can stimulate production from reservoirs thought to have produced all of their recoverable oil. Significant areas where geochemistry might play a role include continuity studies (Ross and Ames, 1988; Hwang *et al.*, 1994; Halpern, 1995), proportions of commingled production from multiple pay zones (Kaufman *et al.*, 1990), oil quality (Karlsen and Larter, 1990), and gas/oil and gas/water contacts (Baskin *et al.*, 1995).

From all the above topics, the most widely used geochemical approach is the determination of continuity. For example, two fields may be connected hydrostatically but the oil column might not be continuous between the two fields. Engineering data alone cannot confirm this, but geochemical data can. Is there communication between fault blocks or do faults act as seals? Are there barriers within a field that prevent communications, thus affecting the location of production wells? The geochemical approach to answering these questions is relatively simple and is based on high-resolution gas chromatography. In the chromatogram of a crude oil, there are a large number of small peaks between the major n-alkane components in a crude oil (Figure 10). Pairs of peaks are selected and ratios of these peaks calculated and plotted on a polar or star diagram. For a family of oils derived from a common source, these ratios will have very similar values if the oils are in communication with each other; if they are not, then there will be significant differences in the values for the ratios. The most successful result is one where differences are observed between these ratios, since such a result permits one to assume, with a fairly high degree of certainty, that the oils are not in communication with each other. When the oils have similar values for these ratios, this may simply be a coincidence, having little to do with whether or not the samples are in communication with each other. The success of the method depends having upon a highly reproducible GC technique, since the variations in the ratios are often quite small and one does not want the standard deviation from one analysis to another to be greater than differences from sample to sample.

In the 1990s the validity of this approach was demonstrated with numerous examples published in the literature and presented at international conferences. It now forms a routine part of any field development operation. A number of modifications have been introduced, such as those described by Halpern (1995). In the typical fingerprinting approach described above, the peaks are selected pretty much at random since their identity is not crucial as long as we know the same peaks are being utilized for each analyses. In the Halpern approach, peaks of known identity were selected and two sets of parameters developed. The first set consists of correlation

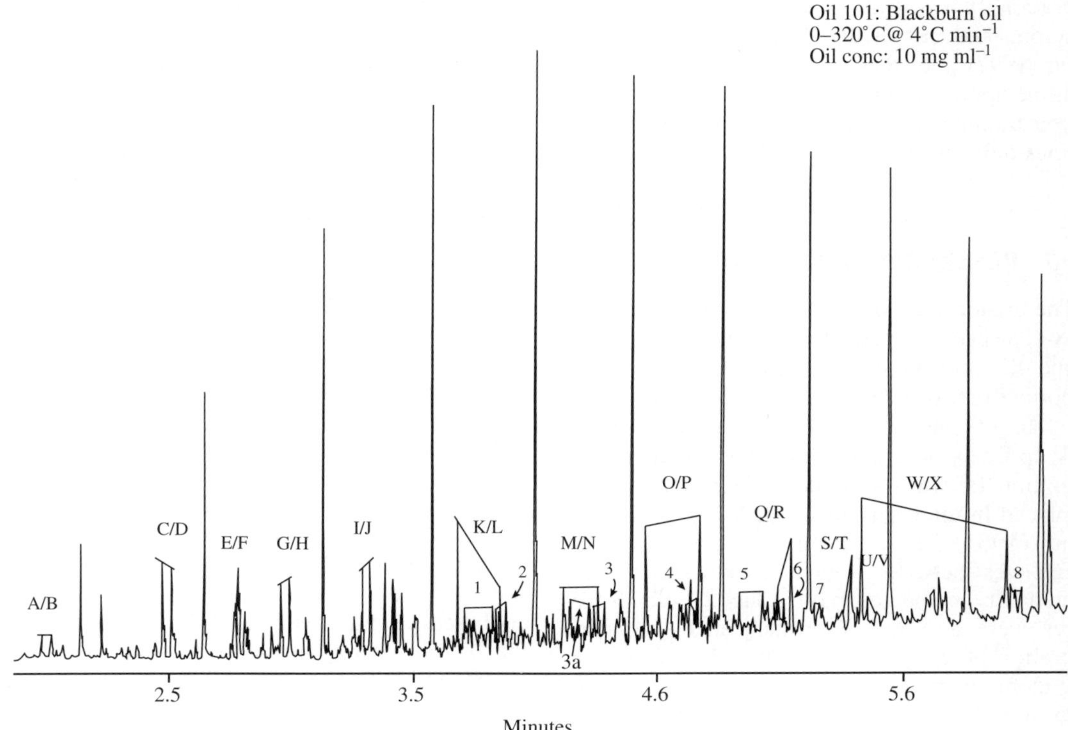

Figure 10 A routine tool for the determination of continuity between fault blocks makes use of high-resolution GC traces. Pairs of peaks between the *n*-alkanes are selected and peak ratios measured. These ratios are plotted on a star diagram and if two samples from the same source are not in communication with each other it can be expected that their plots on the star diagram will not be coincident. The situation is a little less clear when two plots are coincident since this might mean the samples are in communication or it might simply be a coincidence.

parameters, which demonstrate whether the samples actually correlate with each other. The second set consists of transformation ratios, which provide one with the ability to predict why two samples may not be correlated with each other if they were thought to be related. For example, if two samples differed on the basis of a transformation ratio containing toluene, this would probably be due to the effects of water washing. If the transformation ratio containing the *n*-alkane was significantly different then one could say this resulted from biodegradation. By using the transformation ratios in this manner, not only can one say whether two samples were in communication with each other, but if there were differences between them, whether the differences might be related to a weathering effect.

Other reservoir and production areas where geochemistry has played an important role include topics related to the study of tar mats and paraffin accumulations. Both of these are problems which can have a significant impact on production, but have not received a great deal of attention. However, with advances in analytical technology and a better understanding of processes occurring in the reservoir, some progress has been made. It should be emphasized that the geochemists are not necessarily going to make these problems go away, but at least it should be possible to predict when these problems may be anticipated, and more importantly understand whether they are being caused by production practices. For example, tar mats are known to be enriched in asphaltenes and other high-molecular-weight components. They are also known to form barriers in reservoirs that can impede production and drainage of a field, but how do they form? Numerous ideas have been advanced including: biodegradation, deasphaltening as a result of gas production or leakage, water washing, and a number of other processes. One question which still seems difficult to answer is: do the tar mats form as a result of production or were they already in place prior to production? In reality, it is probably a combination of both effects. However, another important aspect of this work is trying to determine or predict where one might expect to find tar mats. Do they always occur at the oil/water contact, and, if so why are the occurrences not more uniform throughout the field? These are certainly questions which deserve more attention in the near future.

Paraffins are another major problem in oil and gas production. As a result of the classic

Hedberg (1968) paper, it was believed for many years that paraffins were only associated with oils derived from source rocks containing an abundance of higher plant material. This is clearly not the case anymore; oils from most source rock of all ages contain higher-molecular-weight (HMW) hydrocarbons ($>C_{40}$), which can be deposited as oils migrate through a temperature gradient to the reservoir (Hsieh and Philp, 2001). The amounts precipitated depend on the nature of the original oil plus the temperature conditions or gradient between the reservoir and ultimately the storage facility. Geochemistry cannot prevent paraffin deposition but can more accurately characterize the HMW composition of the oils and predict when these problems may occur. In addition, it can provide valuable information on the concentrations of these HMW compounds, which can be used in various models designed to simulate paraffin deposition.

Hedberg (1968) proposed that waxy crude oils were associated with terrigenous source materials and generally originated from stratigraphic sequences of nonmarine origin, shale-sandstone lithology, water salinities less than that of marine waters, and ages ranging from the Devonian to Pliocene. Kinghorn (1983) also suggested that petroleum waxes were restricted to terrigenous sources. However, since those studies were made, evidence has appeared which suggests that waxy crude oils may also be derived from marine or lacustrine source materials (Moldowan *et al.*, 1985; Tegelaar *et al.*, 1989; Carlson *et al.*, 1993; Heath *et al.*, 1995; Philp *et al.*, 1995; Hsieh, 1999; Hsieh and Philip, 2001).

High-temperature gas chromatography (HTGC) has led to the determination of hydrocarbons extending as high as C_{120} in crude oils (del Rio and Philp, 1992; del Rio *et al.*, 1992; Carlson *et al.*, 1993; de Aquino Neto *et al.*, 1994; Heath *et al.*, 1995; Philp *et al.*, 1995). HMW hydrocarbons present in crude oils include long-chain aliphatic and alkyl-aromatic hydrocarbons, and represent a significant fraction of the whole oil. Burger *et al.* (1981) measured an oil sample composed of 14% wax, of which 58% included hydrocarbons in the C_{46+} range. Barker (1995) determined that microcrystalline waxes (i.e., alkanes with melting points >60 °C) in a refined petroleum sample were composed of 20–40% *n*-alkanes, 15–40% isoalkanes, and ~35% cycloalkanes. Simulated distillation (SIMDIST) of crude oils has typically indicated the presence of HMW hydrocarbons in crude oils, but the method lacks the chromatographic resolution obtained by HTGC necessary to demonstrate the complexity of these HMW hydrocarbon fractions (Trestianu *et al.*, 1985; Durand *et al.*, 1998). Recent modifications of the extraction techniques have demonstrated that the HMW hydrocarbons are also present in source rock extracts and extend into the C_{60+} region of the chromatograms (Mueller and Philp, 1998).

The high-molecular-weight hydrocarbons in crude oils are comprised of several homologous series of compounds. One series, identified as alkylcyclopentanes, extends into the C_{60} region of crude oils, and the distribution of these compounds can be correlated with the depositional environment (Carlson *et al.*, 1993; Wavrek and Dahdah, 1995). Carlson *et al.* (1993) observed that in marine oils this series of compounds had a distinct odd/even predominance pattern (Figure 11(a)), whereas oils from saline lacustrine environments demonstrate a high even/odd predominance pattern (Figure 11(b)), and freshwater lacustrine oils are characterized by a low even/odd to no clear predominance pattern (Figure 11(c)). HMW hydrocarbons in the C_{40+} range have been observed to be quite stable towards biodegradation (Heath *et al.*, 1997).

The qualitative and quantitative composition of waxes is highly variable in oils. A comprehensive knowledge of this variability is extremely beneficial for improving methods used for the remediation of wax precipitation during oil production. Variations in the carbon-number distribution of wax components have a significant effect on the solubility of the wax, but the variability is not necessarily restricted to differences in geographical location or source material. Chromatograms of wax fractions from two oils collected from the same formation and equivalent depths of production, but different well locations, are shown in Figures 12(a) and (b). The shallower sample shown in Figure 12(a) has a bimodal distribution of macrocrystalline (C_{20}–C_{35}) and microcrystalline (C_{35}–C_{65}) waxes. The deeper sample contains only the macrocrystalline wax and is probably located further from the source than the shallower sample. It is suggested that as the oil moves further from the source material, most of the microcrystalline waxes precipitate along the migration pathway and the residual oil (Figure 12(b)) is depleted in HMW hydrocarbons.

7.09.8 BASIN MODELING

Basin modeling, in one form or another, has been an integral part of petroleum exploration studies since the 1970s. Geochemical parameters have played a role in the development of these models as they have become more sophisticated. The major role of basin models is to reconstruct the history of sedimentary basins in an effort to predict how the processes of generation, expulsion, migration, trapping, and preservation control the volumetrics, quality, and distribution of oil and gas in a basin. There are two aspects to basin modeling: thermal and fluid flow modeling. As

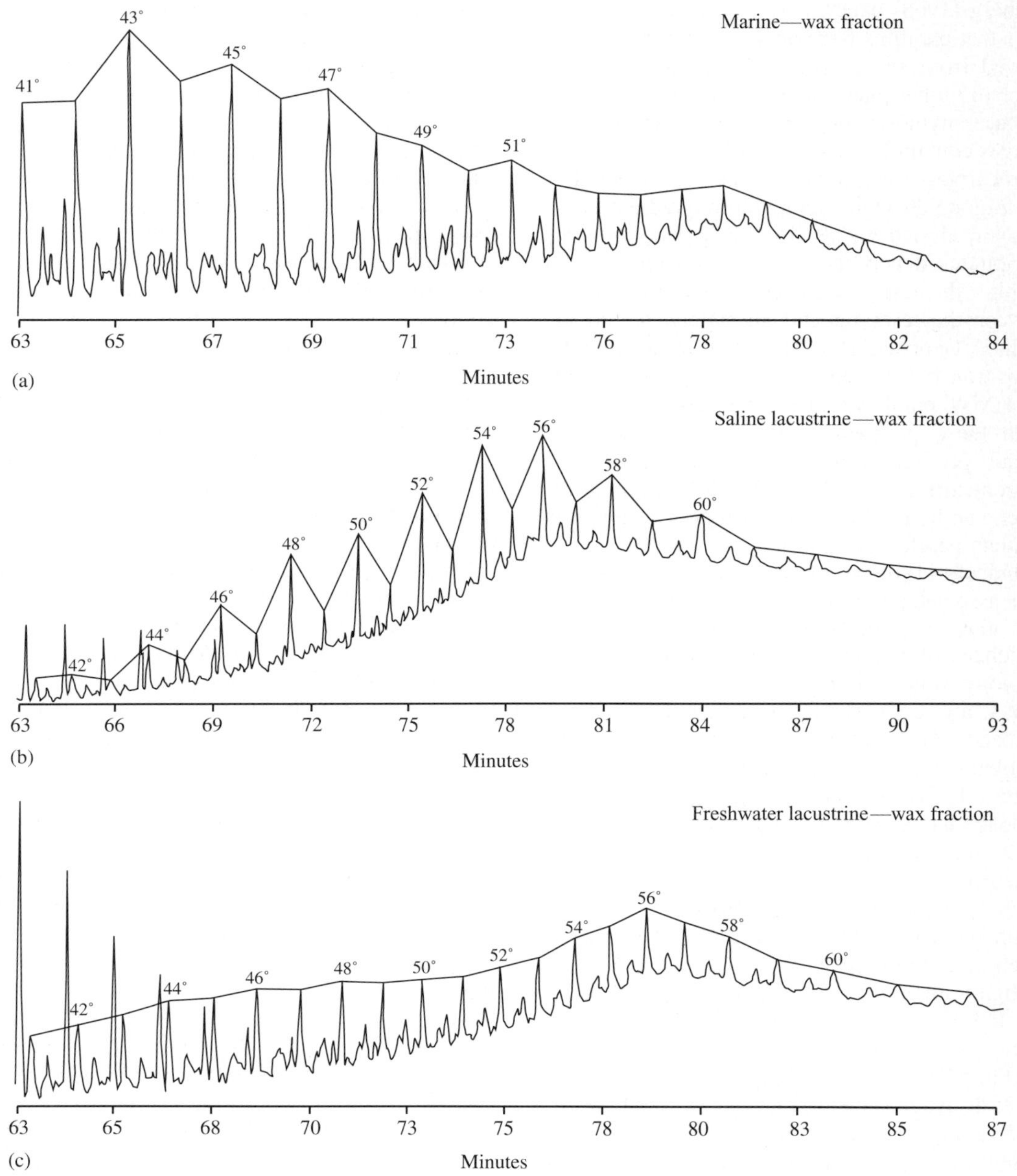

Figure 11 Long-chain alkylcyclopentanes are relatively abundant in the C_{40+} region of the chromatograms of most oils. Oils from marine (a), saline lacustrine (b), and freshwater lacustrine (c) environments can be differentiated on the basis of these alkylcyclopentane distributions. Marine oils have a distinct odd/even predominance pattern; oils from saline lacustrine environments demonstrate a high even/odd predominance pattern, and freshwater lacustrine oils are characterized by a low even/odd to no clear predominance pattern.

one might anticipate, thermal modeling deals with maturation, generation, and cracking. Many parameters are required for the three-dimensional models in use today. But, from a geochemical point of view, probably the most important factors are related to generation kinetics and type of organic material in the source rock. Numerous geological parameters are also required, such as timing of geological events with respect to the source, carrier, reservoir, and overburden rock. This also includes information on the extent of deposition, nondeposition, uplift, erosion, and subsidence.

It has been recognized for some time that as organic matter is buried and subject to progressively higher temperatures, thermal degradation of the kerogen occurs to yield petroleum range hydrocarbons under reducing conditions. Among the earliest experiments demonstrating these effects were those of Engler (1913), who heated

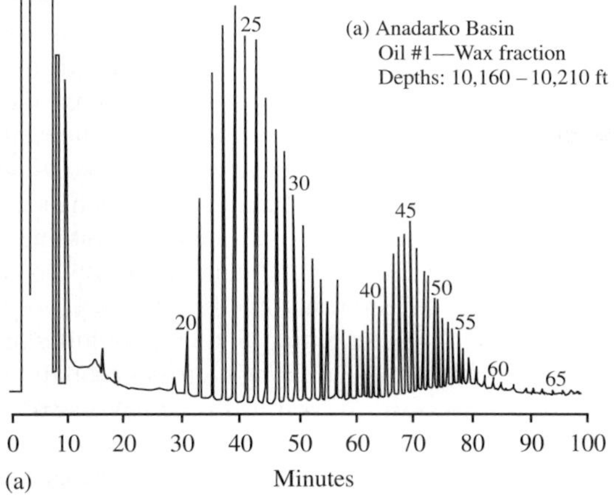

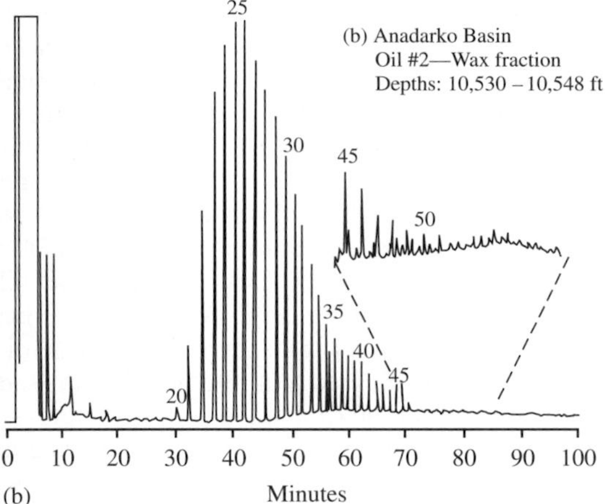

Figure 12 Chromatograms of wax fractions from two oils collected from the same formation and equivalent depths of production, but different well locations. The shallower sample shown (a) has a bimodal distribution of macrocrystalline (C_{20}–C_{35}) and microcrystalline (C_{35}–C_{65}) waxes. The deeper sample (b) contains only the macrocrystalline wax and is probably located further from the source than the shallower sample. As the oil moves further from the source material, much of the microcrystalline waxes precipitate along the migration pathway and the residual oil becomes depleted in HMW hydrocarbons.

oleic acid and other organic matter to produce an entire range of petroleum hydrocarbons. Hunt (1979) described a number of additional experiments that involved heating various types of organic matter to produce petroleum hydrocarbons. It is now generally accepted that the major mechanism for the formation of most, but not all, oil is the following reaction derived directly from these early heating experiments:

$$\text{kerogen} \rightarrow \text{bitumen} \rightarrow \text{oil} + \text{gas} + \text{residue}$$

The timing of generation is important, since oil accumulation does not occur if reservoirs or traps are not available when the source rock has started to generate oil. Timing of petroleum generation is also important in relation to the formation of faults, which act as migration pathways. This information can be obtained by modeling the time–temperature history of the source rock. The timing of petroleum generation in relation to the presence of appropriate traps and migration pathways led to the development of the so-called petroleum system concept originally described in papers by Magoon and Dow (1994). A petroleum system basically includes all the geological elements and processes that are essential for an oil and gas deposit to occur in nature, including source rock, reservoir rock, seals, migration pathways, and the geological processes that created them.

It was not until 1964 that Habicht made the first attempt at modeling petroleum generation. A burial history curve was constructed for a Jurassic source rock, and Arrhenius equation kinetics were used to determine time and depth of oil generation in the Gifhorn trough of northwestern Germany. Philippi (1965) documented the increase in the yield of hydrocarbons from source rocks in the Los Angeles and Ventura basins of California with increased time and temperature. A few years later, Poulet and Roucache (1969) used a series of source rock burial history curves to describe the origin of hydrocarbons in the Northern Sahara Hassi Massoud area. The main finding of these early studies was that it was not temperature alone that was necessary for the generation of the hydrocarbons but that it was also necessary to take time into account. In other words, rapid burial with high geothermal gradients does not result in the same level of maturation as slow burial with low geothermal gradients.

In his early studies, Habicht (1964) used Arrhenius kinetics to determine time and depth of oil generation, and the first mathematical model using Arrhenius kinetic theory along a source rock burial curve was published by Tissot (1969). The model did not find widespread use since it was quite rigorous. At the time when these studies were taking place, Teichmuller (1958) demonstrated the relationship between vitrinite reflectance and the occurrence of oil. The reflectance of vitrinite increases exponentially with a linear increase in temperature and generally plots as a straight line on a semilog plot (Dow, 1977).

Hood at Shell Oil developed a simplified method of predicting oil generation of a source rock from its maximum temperature and effective heating time. The latter was defined as the time during which a specific rock was within 15 °C of its maximum temperature where most of the generation occurs (Hood *et al.*, 1975). He later correlated vitrinite reflectance with the level of organic maturation (LOM). Larskaya and Zhabrev (1964) were the first geochemists to demonstrate that the generation of oil from the kerogen of shales increases exponentially with temperature. A time–temperature Arrhenius plot was constructed to define the threshold of intense generation for several sedimentary basins, which could be used to evaluate drilling prospects (Connan, 1974).

In 1971 Lopatin developed a model based on the empirical relationship between vitrinite reflectance and petroleum formation. This model was simpler than that proposed by Tissot and used both time and temperature to calculate the thermal maturity of organic matter in sediments. Lopatin (1971) (see also Waples, 1985) developed the time–temperature index (TTI) which was a measure of the total thermal exposure for a source rock since deposition, based on the old chemical rule that reaction rates double for each 10 °C rise in temperature (Bergius, 1913). As it was applied to more problems, it was found that the Lopatin method tended to underestimate maturity at higher heating rates. This led to the development of models that combined the Lopatin approach with Arrhenius kinetics. These pioneering methods ultimately gave rise to very sophisticated models which could be combined with data on evolution of migration pathways, traps, and structures to better evaluate the risk in prospecting for economic hydrocarbon accumulations.

One of the crucial aspects of the models that have been developed is the determination of the kinetic parameters to be used in the Arrhenius equation, namely, the activation energy E and the pre-exponential or frequency factor A. These parameters can be determined experimentally by heating the source rock in the laboratory at various temperatures and measuring the yields of hydrocarbons. The most commonly used techniques are an open nonisothermal dry programmed temperature pyrolysis system such as Rock Eval, and a closed isothermal wet system such as hydrous pyrolysis. There are many documented advantages and disadvantages with all the methods and each has its supporters and detractors. Open-system pyrolysis provides a range or distribution of activation energies which is more realistic than a single value, since kerogen breakdown represents a whole series of reactions and not a single reaction. Detractors argue that it does not monitor the rate-controlling step of bitumen to oil but rather all reactions involved in the formation of oil and gas. Dry open pyrolysis also shows a decrease in yield with decrease in heating temperature, and Lewan (1985) showed that extrapolation of these yields will reach zero before reaching geological heating rates. The products that are generated do not closely resemble crude oils, since they contain higher amounts of polar compounds and unsaturate compared to a regular crude oil, possibly due to the absence of water. Hydrous pyrolysis performed isothermally in a closed system produces a liquid which more closely resembles a crude oil than produced by the open system. It has been proposed that "the closed system and the presence of water" more closely resembles the natural burial process by developing internal pressures inside the reaction vessel and dissolving water in the bitumen, which in turn enhances hydrogen exchange between water and bitumen phases. A significant disadvantage of hydrous pyrolysis is the time involved. It is not very practical for characterization of a large number of samples in this way, although cross-correlation with open pyrolysis is possible through the use of a standard set of hydrous pyrolysis analyses.

There have been a number of studies comparing open and closed, hydrous and anhydrous, pyrolysis experiments, worth mentioning among them are by Qin *et al.* (1994) and Burnham *et al.* (1987). Qin *et al.* (1994) compared thermal maturation results from a brown coal from the Lower Tertiary of East China and showed that, whilst the hydrous pyrolysis results closely followed the natural pathway of coal, the anhydrous pathway deviated significantly from it. Burnham *et al.* (1987) showed that using Rock Eval kinetics, the oil window covers a wider temperature range than that obtained from hydrous pyrolysis, where oil generation is considered, but bitumen plus oil and gas generation are not included.

The method used to determine the kinetic parameters is important but one also has to consider the kerogen type and burial history and geothermal gradient. In view of the constraints associated with the kinetic parameters, it is not possible to assume a single set of kinetic parameters that can describe all petroleum formation, as was often done in the past. Hunt (1996, see Table 6-1, p. 147) has summarized the variability in kinetic parameters for the different kerogen types, initially described by Tissot and Espitalie (1975) and later modified by Lewan (1985) and others. The impact the sulfur content has on the kinetic parameters for the type IIS kerogens means that these kerogens will generate oil at much lower levels of thermal maturity than kerogens with lower sulfur contents. The importance of activation energies for each kerogen type is further exemplified by Hunt (1996; Figure 13), who showed oil generation curves for different kerogen types in a pull apart basin and a cratonic basin with geothermal gradients of 45 °C km^{-1} and 25 °C km^{-1}, respectively. The differences in burial depths for complete oil generation are substantial, e.g., kerogen type IID in the cratonic basin requires 2,800 m more of burial than type IIA for oil generation. This clearly emphasizes the importance of using different kinetic parameters for different kerogen types in defining the petroleum generation intervals in sedimentary basins.

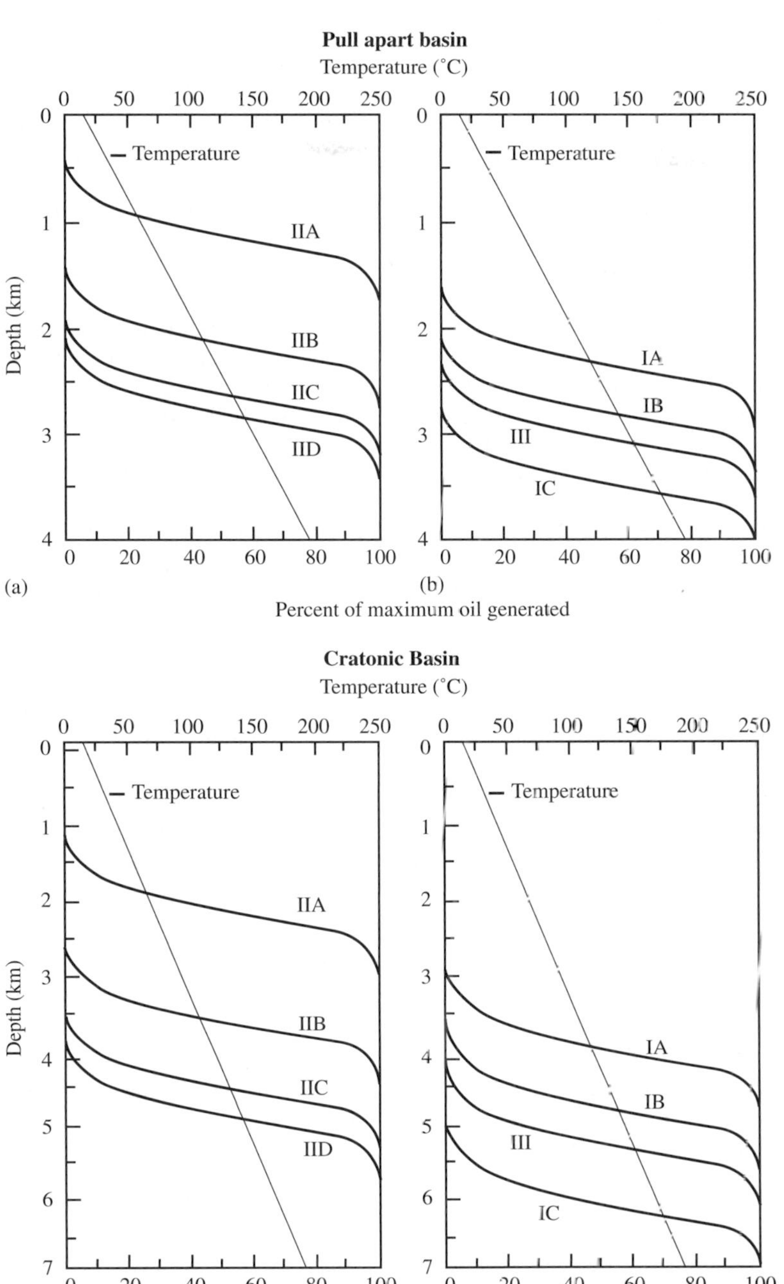

Figure 13 The importance of activation energies for each kerogen type is further exemplified by Hunt (1996), who showed oil generation curves for different kerogen types in a pull apart basin and a cratonic basin with geothermal gradients of 45 °C km^{-1} and 25 °C km^{-1}, respectively (Hunt, 1996) (reproduced by permission of Freeman from *Petroleum Geochemistry and Geology*, **1979**).

From a geochemical point of view, one of the most important contributions to basin modeling would be the determination of kinetic parameters, followed by characterization of organic matter types. More comprehensive reviews can be found on these topics in articles by Waples (1984, 1994), Hunt (1996), and Welte *et al.* (1997).

7.09.9 NATURAL GAS

The major part of this chapter has been directed towards the formation and accumulation of crude oil; however, geochemistry plays an equally important role in understanding the formation of natural gas (Schoell *et al.*, 1993). Unlike crude oils or source rock extracts, where we have very complex mixtures containing a large variety of compounds and compound classes, natural gases may contain methane, or methane plus a few other components, in much lower abundance, in the C_2–C_5 range. There are a number of sources for hydrocarbons and nonhydrocarbons in natural gas including: (i) methanogenic bacteria, (ii) all types of kerogen, (iii) coal, and (iv) oil in source and reservoir rocks. Generation of methane from the thermal cracking of coal or kerogen is known as primary cracking, whereas generation from oil is known as secondary cracking. Other components of natural gas such as CO_2, H_2S, and N_2 can be formed by both organic and inorganic processes. In general, all gases are originally of biogenic origin in the sense that they are derived from organic matter deposited in various sedimentary environments. The quantities of natural gas derived from microbial reworking of organic material are relatively small, and, as noted by Schoell and Clayton (2002), are not of commercial significance at this time. Accumulation of abiogenic hydrocarbon gases are also not significant from a commercial standpoint (Sherwood-Lollar *et al.*, 2002). The yields and relative proportions of various components in a natural gas accumulation vary, depending on source materials. Gases derived from humic type source material tend to produce larger quantities of CO_2 and N_2 than those derived from sapropelic sources, whereas the latter generate higher quantities of C_{2+} components.

Microbial methane generation tends to cease as bacterial populations start to decrease at temperatures exceeding 70 °C. As the burial temperature starts to increase above 50 °C, thermally formed methane plus higher hydrocarbons and CO_2 start to increase. CO_2 production from inorganic sources continues to much higher temperatures. The major source of H_2S associated with natural gas reservoirs is thermochemical sulfate reduction, not diagenesis. Very little of the H_2S formed at low temperatures survives long-term diagenesis and maturation. Peak generation of H_2S is probably at depths beyond 5 km.

With natural gas being a mixture of such structurally simple compounds, options available for correlation purposes are limited. Gas chromatography provides an indication of the distribution of the components present in the natural gas sample from which the ratio of methane/C_{2+} fraction can be determined. This is measure of the wetness of the gas and in certain cases can provide a measure of the maturity of the gas. The ratio needs to be combined with carbon and isotope ratio data to obtain the most useful information. A gas can be dominated by methane and, from the chromatogram alone, it is impossible to determine whether this is biogenic gas or a high maturity gas. As a result of thousands of analyses over time, it can be predicted that dry bacterial gas will have carbon isotope values in the range −110‰ to −60‰; wet thermogenic gas from −60‰ to −30‰; and dry thermogenic gas from −40‰ to −15‰. The hydrogen values show a similar trend. It should be noted that these values should serve only as guidelines, since there are going to be anomalies. For example, bacterial conversion of CO_2 to methane causes a fractionation of ~−70‰. Hence if the CO_2 is −15‰, the resulting methane will be −85‰, which is in the bacterial range. However, if the CO_2 is +30‰, it will form methane of −40‰, which is not in the microbial range even though the gas is microbial.

Combination of the isotopic values with the methane/C_{2+} ratio provides a very useful tool to differentiate gases of biogenic versus thermal origin, as recognized many years ago (Bernard *et al.*, 1976). More recent studies have also seen much greater use of the hydrogen and deuterium isotopes for natural gas components combined with the carbon values to differentiate gases derived at different levels of thermal maturities, as documented in many of the papers by Schoell *et al.* (1993).

A second geochemical application in natural gas studies is the use of carbon and hydrogen isotopes in reservoir geochemistry. As with crude oils, it is very important to be able to determine the extent of continuity between fault blocks, to determine whether there are barriers present within a reservoir and whether the gases are derived from the same source. The simplicity of the natural gas mixture means that the most successful method for gas continuity studies involves the use of carbon and hydrogen isotopes. A few examples of this approach have been published. An excellent example is given by Schoell *et al.* (1993), for a Gulf Coast gas field. A number of samples were taken from different formations with the idea of determining whether or not the samples were in communication with each other. On the basis of the carbon and hydrogen isotope data, it was possible to determine which two samples were in contact with each other and which ones were totally isolated from each other.

Another disadvantage of working with gases rather than crude oils and source rocks is that it is difficult to say very much directly concerning the maturity of the source rocks from which the gas was generated. In the case of source rocks, we have the vitrinite reflectance methods, and with the oils we can get maturity information from the biomarker ratios. In the case of gases, again we do not have that luxury. However, a number of attempts have been made, based on carbon isotope data, to determine these values. Although this is not discussed in detail in this chapter, there are several papers where this information is readily available (Berner and Faber, 1988, 1996; Clayton, 1991).

7.09.10 SURFACE PROSPECTING

The majority of early oil discoveries up until the 1950s resulted from drilling wells on or near active seeps or other surface features thought to be related to underlying hydrocarbon accumulations (DeGolyer, 1940). This led to the development of techniques referred to as surface prospecting tools. Oil seeps have always been relatively easy to find, and with the several satellite imagery techniques available today, they are far more easily detected. Surface prospecting has had varying degrees of success over the years, and has been received with varying levels of skepticism. Many of the early publications only reported successes, gave no indication of the rates of success, and certainly no indication of the spectacular failures. Others implied that a successful discovery of a surface anomaly would indicate a drilling target at the center of the anomaly, completely ignoring the fact that volatile components do not necessarily migrate vertically but instead take pathways of least resistance to the surface via faults and fissures (summarized in Philp and Crisp, 1982). Used as a regional tool in conjunction with other data, such as geophysics, regional geology and stratigraphy, surface prospecting can provide an extremely powerful indicator for the presence of hydrocarbons, particularly in frontier basins, but not the exact location of the trap.

A petroleum seep is defined as visible evidence at the Earth's surface of the present or past leakage of oil, gas, or bitumen from the subsurface. Seeps have been utilized at least since ancient tribes of the Near East recovered blocks of asphalt from the Dead Sea. Many studies have demonstrated the worldwide correlation between seeps and earthquakes, with most visible seeps being near past or present areas of tectonic activity. The presence of seeps in a basin can considerably reduce the exploration risk because these seeps indicate that petroleum-forming processes have been active in the subsurface. In a situation where the major pathway for seeps is along a fault, the intensity and shape of the seep are significantly different from those situations where vertical leakage has occurred, and this difference permits their distinction.

Early investigators assumed that hydrocarbons would migrate vertically, and any resulting surface anomaly would be directly over the location of the trap. This concept has changed considerably over the years, and it is now known that a number of factors may affect the location of any surface anomaly relative to the underlying hydrocarbon accumulation. Overlying geology, groundwater, depth of accumulation, degree of fracturing, and many other factors will ultimately determine the relative position of any surface feature or anomaly. Weathering effects such as evaporation of volatiles, leaching of water soluble constituents, microbial degradation, polymerization, and auto-oxidation may also affect the appearance of the seep and ultimately lead to the oil being converted to solid bitumen.

A systematic development of geochemical methods for oil and gas prospecting was first initiated by Laubmeyer (1933) in Germany, Horvitz (1939, 1972) and Rosaire (1940) in the United States. Reviews of the early Russian work can be found in Kartsev *et al.* (1954). Rosaire (1940) discussed early prospecting techniques in the United States, particularly those applied to gases in soils and groundwaters. Both DeGolyer (1940) and Link (1952) emphasized the importance of the relationship between oil and gas seeps and the underlying accumulations of hydrocarbon reserves. Reviews of surface prospecting by Davis (1967) included discussions of microbial prospecting methods, infrared analyses of soil extracts (Bray, 1956a), and the utilization of carbon isotopes as a means of characterizing soil lipids (Bray, 1956b).

A major problem with surface prospecting in the past has been the interpretation of the results, in particular to determine if the gases are biogenic (microbial and near-surface origin) or thermogenic. A partial solution to this problem made use of the concentration of methane relative to the other gases. Biogenic gases typically contain mostly methane, likely the most abundant and widespread of all hydrocarbon gases held within the sedimentary rocks of the crust (Hedberg, 1980). Another approach is to use the isotopic composition of the methane, along with the composition of the light hydrocarbon gases, ethane and propane, to distinguish microbial from thermal gas (Bernard *et al.*, 1976).

Initial attempts at surface prospecting were confined to onshore studies, but in the late 1960s and 1970s similar applications offshore, including the development of various sniffing devices, were developed. Sniffers used various means to strip the gases, typically those in the C_1–C_4 range, from water samples collected close to the sediment–water interface prior to their analyses by GC. One of the advantages of this type of offshore sniffing, compared to onshore studies, was that background interference from gases coming from plant debris and roots was nonexistent; however, as with onshore studies, interference from microbial methane was a potential problem. Offshore sediment samples were also being collected over grid patterns, and gases were isolated and analyzed using many of the same techniques described for onshore sediments. The advantage of this type of sampling was again the lack of background interference compared to that in onshore sampling. In selected cases, gas bubbles from seeps could be observed on the seafloor and collected directly into sample bottles and returned to the laboratory for analyses (Brooks *et al.*, 1973; Bernard *et al.*, 1976).

In many situations, the use of methods to detect indirect surface expression or alteration of surface features resulting from the seep was a more viable option. Such indirect approaches ranged from the extremes of changes in the mineralogy of the overlying soil to changes in the vegetation type as a result of a gas seep. Near-surface carbonate cements can be formed from carbon dioxide produced by bacterial degradation of hydrocarbons seeping to the surface from deep petroleum deposits. Horovitz (1981) noted that ferrous carbonate anomalies, which could be distinguished on the basis of their unusual thermal decomposition pathways, showed high concentrations above the edges of hydrocarbon accumulations. Their origin was verified, in some cases, through determination of their carbon isotope composition (Behrens, 1988). Donovan and Dalziel (1977) measured the isotopic composition of carbonate cements in outcrops and detected anomalous isotopic compositions indicative of hydrocarbon seepage. Donovan (1981) reported anomalous isotope compositions for the carbonates and also noted the associated occurrences of pyrite. Other indirect methods of detecting surface anomalies included iron and manganese leaching from surface rocks (Kartsev *et al.*, 1954; Donovan *et al.*, 1975), ions associated with petroleum (Kartsev *et al.*, 1954; Hitchon, 1974; Davis, 1967), helium associated with gas seeps (Dyck, 1976; Roberts *et al.*, 1976), radon 222 (Gates and McEldowney, 1977), and dissolved nitrogen in groundwaters (Zorkin *et al.*, 1976). LANDSAT imagery techniques (Marrs and Kimansky, 1977) became available in the late 1970s. Various microbiological methods have also been reported and involve detection of enhanced levels of methane- and ethane-consuming bacteria in the sediments. Some of the more notable early reviews of this method include Davis (1967), Brisbane and Ladd (1968), Sealey (1974a, b), and Miller (1976).

7.09.11 SUMMARY

Organic geochemistry has played a pivotal role in the continued development of oil and gas exploration and production. It has consistently made useful contributions, from the earliest days when the presence of blocks of asphalt floating in the Dead Sea was an indication of oil and gas in the area to the latest four-dimensional basin models for reservoir production modeling. Recognition of the organic origin for oil and gas followed by the development of the biomarker concept provided a quantum leap forward in terms of exploration. More recently, similar and related concepts have been applied to furthering our understanding of the filling histories of reservoirs and the continuity between fault blocks within a reservoir.

Organic geochemistry applied to oil and gas related problems is a far more mature science than it was in the 1980s. It is an integral part of any exploration package and the intensity of new discoveries in the field is not as great as it was even in the 1990s. What can we expect to see from geochemistry in this area in the next few years? Clearly, the routine searching for novel biomarkers is almost a thing of the past. However, as more and more samples are examined, then recognition of specific compounds that can be related to a specific source or depositional environment or oil of a certain age will continue to be of significance. Integration of geochemical parameters with sequence stratigraphic models will continue to assume greater importance. I feel we have only scratched the surface of his combination, and such integration as demonstrated by the recently published study in the Mahakam Delta indicates the success of this process. With greater emphasis on production today than in the past, it is clear that geochemistry will continue to play a significant role in this area as well, not only for continuity studies but for providing a better understanding of the mechanisms responsible for problems in reservoirs such as formations of tar mat barriers. The origin of high-molecular-weight hydrocarbons is another area that has recently started to be investigated in significant detail but one which will be very important for exploration purposes. What is the structure of these various series of compounds and where do they come from? As of early 2000s, the utilization of stable carbon and hydrogen isotopes in both oil and gas exploration and reservoirs studies has certainly not been fully exploited. I think we will see many more applications of this concept to exploration and production problems.

In brief, despite various changes over the past few years, geochemistry will continue to have an important role in oil and gas production and exploration. It may not be at the same level of visibility as it had previously, but it will be integrated with many established engineering concepts and provide valuable information for use in many areas related to exploration and production.

REFERENCES

Alexander R., Kagi R., Woodhouse G. W., and Volkman J. K. (1983) The geochemistry of some biodegraded Australian crude oils. *Austral. Petrol. Explor. Assoc. J.* **23**, 53–63.

Allard B., Rager M.-N., and Templier J. (2002) Occurrence of high molecular weight lipids (C_{80+}) in the trilaminar outer cell walls of some freshwater microalgae. A reappraisal of the algaenan structure. *Org. Geochem*, **33**, 789–802.

de Aquino Neto F. R., Cardoso J. N., dos Santos Pereira A., Zupo Fernandez M. C., Caetano C. A., and de Castro Machado A. L. (1994) Application of high temperature high resolution gas chromatography to paraffinic deposits in petroleum production pipelines. *J. H. Res. Chrom.* **17**, 259–263.

Balogh B., Wilson D. M., Christiansen P., and Burlingame A. L. (1973) 17α(H)-Hopane identified in oil shale of the Green River Formation (Eocene) by carbon 13 NMR. *Nature* **242**, 603–605.

Barker A. (1995) The chromatographic analysis of refined and synthetic waxes. In *Chromatography in the Petroleum Industry J. Chrom. Lib. Ser.* (ed. E. R. Adlard). Elsevier, Amsterdam, **56**, 55–93.

Baskin D. K. and Jones R. W. (1993) Prediction of oil gravity prior to drill stem testing in Monterey Formation reservoirs, Offshore California. *Am. Assoc. Petrol. Geol. Bull.* **77**, 1479–1487.

Baskin D. K., Hwang R. J., and Purdy R. K. (1995) Prediction of gas, oil, and water intervals in Niger Delta reservoirs using gas chromatography. *Am. Assoc. Petrol. Geol. Bull.* **79**, 337–350.

Behrens E. W. (1998) Geology of a continental slope oil seep, northern Gulf of Mexico. *Am. Assoc. Petrol. Geol. Bull.* **72**, 105–114.

Bergius F. (1913) Production of hydrogen from water and coal from cellulose at high temperatures and pressures. *J. Soc. Chem. Ind.* **32**, 462–467.

Bernard B. B., Brooks J. M., and Sackett W. M. (1976) Natural gas seepage in the Gulf of Mexico. *Earth Planet. Sci. Lett.* **31**, 48–54.

Berner U. and Faber E. (1988) Maturity related mixing model for methane, ethane, and propane based on carbon isotopes. *Org. Geochem.* **13**(1), 67–72.

Berner U. and Faber E. (1996) Empirical carbon isotope/ maturity relationships for gases from algal kerogens and terrigenous organic matter, based on dry, open system pyrolysis. *Org. Geochem.* **24**(10/11), 947–995.

Bohacs K. M. (1993) Source quality variations tied to sequence development in the monterey and associated formations, southwestern California. In *Source Rocks in a Sequence Stratigraphic Framework*, AAPG Studies in Geology #37. American Association of Petroleum Geologists, Tulsa, OK.

Bray E. E. (1956a) Geochemical exploration methods. *US Patent 2,742,575.*

Bray E. E. (1956b) Method of geochemical prospecting. *US Patent 2,773,991.*

Brisbane P. G. and Ladd J. N. (1968) The utilization of methane, ethane, and propane by soil micro-organisms. *J. Gen. Appl. Microbiol.* **19**, 351–364.

Brooks J. D., Gould K., and Smith J. W. (1969) Isoprenoid hydrocarbons in coal and petroleum. *Nature* **222**, 257–259.

Brooks J. M., Frederick A. D., Sackett W. M., and Swinnerton zJ. W. (1973) Baseline concentrations of light hydrocarbons in Gulf of Mexico. *Environ. Sci. Technol.* **7**, 639–642.

Brooks J., Cornford C., and Archer R. (1987) The role of hydrocarbon source rocks in petroleum exploration. In *Marine Petroleum Source Rocks* (eds. J. Brooks and A. J. Fleet). Blackwell, Oxford, pp. 17–46.

Burger E. D., Perkins T. K., and Striegler J. H. (1981) Studies of wax deposition in Trans Alaska Pipeline. *J. Petrol. Technol.* **33**, 1076–1086.

Burgess J. (1974) Microscopic examination of kerogen (dispersed organic matter) in petroleum exploration. Washington, DC. *Geol. Soc. Am. Spec. Publ.* **153**, 19–30.

Burlingame A. L. and Simoneit B. R. (1969) High resolution mass spectrometry of Green River kerogen oxidations. *Nature* **222**, 741–747.

Burnham A. K., Braun R. L., Gregg H. R., and Samoun A. M. (1987) Comparison of methods for measuring kerogen pyrolysis rates and fitting kinetic parameters. *Energy Fuels* **1**, 452–458.

Carlson R. M. K., Teerman S. C., Moldowan J. M., Jacobson S. R., Chan E. I., Dorrough K. S., Seetoo W. C., and Mertani B. (1993) High temperature gas chromatography of high-wax oils. In *Indonesian Petroleum Association, 22nd Annual Convention Proceedings, Jakarta*, Indonesian Petroleum Association, Indonesia, pp. 483–507.

Chakhmakhchev A., Suzuki M., and Takayama K. (1997) Distribution of akylated dibenzothiophenes in petroleum as a tool for maturity assessments. *Org. Geochem.* **26**, 483–490.

Chen J., Fu J., Sheng G., Liu D., and Zhang J. (1996) Diamondoid hydrocarbon ratios: novel maturity indices for highly mature crude oils. *Org. Geochem.* **25**(3/4), 179–190.

Clayton C. (1991) Carbon isotope fractionation during natural gas generation from kerogen. *Mar. Petrol. Geol.* **8**, 232–240.

Clegg H., Horsfield B., Wilkes H., Sinninghe Damste J., and Koopmans M. P. (1998) Effect of artificial maturation on carbazole distribution as revealed by the hydrous pyrolysis of an organic sulfur rich source rock (Ghareb Formation-Jordan). *Org. Geochem.* **29**, 1953–1960.

Connan J. (1974) Time-temperature relation in oil genesis. *Am. Assoc. Petrol. Geol. Bull.* **58**, 2516–2521.

Connan J. (1984) Biodegradation of crude oils in reservoirs. In *Advances in Petroleum Geochemistry* (eds. J. Brooks and D. H. Welte). Academic Press, London, vol. 1, pp. 299–335.

Creasey S. and Passey Q. R. (1993) Recurring patterns of total organic carbon and source rock quality within a sequence stratigraphy framework. *Am. Assoc. Petrol. Geol. Bull.* **77**(3), 386–401.

Curiale J. A., Cole R. D., and Witmer R. J. (1992) Application of organic geochemistry to sequence stratigraphy analysis: Four Corners Platform area, New Mexico, USA. *Org. Geochem.* **19**(1–3), 53–75.

Dahl J. E. P., Moldowan J. M., Teerman S. C., McCaffrey M. A., and Sundararaman P. (1994) Source rock quality determination from oil biomarkers: I. A new geochemical technique. *Am. Assoc. Petrol. Geol. Bull.* **78**(10), 1507–1526.

Davis J. B. (1967) *Petroleum Microbiology*. Elsevier, Amsterdam, 604pp.

DeGolyer E. (1940) Future position of petroleum geology in the oil industry. *Am. Assoc. Petrol. Geol. Bull.* **24**, 1389–1399.

Demaison G. J. and Moore G. T. (1980) Anoxic environments and oil source bed genesis. *Am. Assoc. Petrol. Geol. Bull.* **64**, 1179–1209.

Donovan T. J. (1981) Geochemical prospecting for oil and gas from orbital and suborbital altitudes. In *Unconventional Methods in Exploration for Petroleum and Natural gas II* (ed. B. M. Gottlieb). Southern Methodist University Press, Dallas, pp. 96–115.

Donovan T. J. and Dalziel M. C. (1977) Late diagenetic indicators of buried oil and gas. *US Geol. Surv. Open File Rep.* 77-817, 38pp.

Donovan T. J., Friedman I., and Gleason J. D. (1975) Recognition of petroleum-bearing traps by unusual isotopic compositions of carbonate-cemented surface rocks. *Geology* **2**, 351–354.

Dow W. G. (1997) Kerogen studies and geological interpretations. *J. Geochem. Explor.* **7**(2), 77–79.

Durand B. (ed.) (1980) *Kerogen-Insoluble Organic Matter from Sedimentary Rocks.* Editions Technip, Paris, 520pp.

Durand J. P., Bré A., Béboulène J. J., Ducrozet A., and Carbonneaux S. (1998) Simulated distillation methods for petroleum fractions with minimal residue in the boiling range of 35–700 °C. *J. Chromatogr. Sci.* **36**, 431–434.

Dyck W. (1976) The use of helium in mineral exploration. *J. Geochem. Explor.* **5**, 3–20.

Eglinton G. and Calvin M. (1967) Chemical fossils. *Sci. Am.* **216**, 32–43.

Eglinton G. and Murphy M. T. J. (eds.) (1969) *Organic Geochemistry: Methods and Results.* Springer, New York.

Eglinton G., Scott P. M., Belsky T., Burlingame A. L., and Calvin M. (1964) Hydrocarbons of biological origin from a one billion year old sediment. *Science* **145**, 263–264.

Eglinton G., Maxwell J. R., and Philp R. P. (1974) Organic geochemistry of sediments from contemporary aquatic environments. In *Advances in Organic Geochemistry 1973* (eds. B. Tissot and F. Biener). Editions Technip, Paris, pp. 942–961.

Elshahed M. S., Gieg L. M., McInerney M. J., and Suflita J. M. (2001) Signature metabolites attesting to the *in-situ* attenuation of alkylbenzenes in anaerobic environments. *Environ. Sci. Technol.* **35**, 682–689.

England W. A. (1990) The organic geochemistry of petroleum reservoirs. *Org. Geochem.* **16**, 419–426.

England W. A. and Mackenzie A. S. (1989a) Some aspects of the organic geochemistry of petroleum fluids. *Geol. Rundsch.* **78**(1), 291–303.

England W. A. and Mackenzie A. S. (1989b) Geochemistry of petroleum reservoirs. *Geol. Rundsch.* **78**, 214–237.

Engler K. O. V. (1931) *Die Chemie und Physik des Erdols*, S. Hirzel, Leipzig, vol.1.

Ensminger A., Albrecht P., Ourisson G., and Tissot B. (1977) Evolution of polycyclic alkanes under the effect of burial (Early Toarcian shales, Paris Basin). In *Advances in Organic Geochemistry 1975* (eds. R. Campos and J. Goni). ENADIMSA, Madrid, pp. 45–52.

Espitalie J., La Porte J. L., Madec M., Marquis F., Le Plat P., Paulet J., and Boutefeu A. (1977) Methode rapide de caracterisation des roches meres de leur potential petrolier et de leur degre d'evolution. *Rev.l'Inst. Fran.du Petr.* **32**(1), 23–42.

Freeman K. H., Hayes J. M., Trendel J.-M., and Albrecht P. (1990) Evidence from carbon isotope measurements for diverse origins of sedimentary hydrocarbons. *Nature* **343**, 254–256.

Galimberti R., Ghiselli C., and Chiaramonte M. A. (2000) Acidic polar compounds in petroleum: a new analytical methodology and applications as molecular migration indices. *Org. Geochem.* **31**, 1375–1386.

Gates T. M. and McEldowney R. C. (1977) Uranium exploration method may help find gas and oil. *World Oil* **184**, 55–57.

Gelin F., Volkman J. K., Largeau C., Derenne S., Sinninghe Damsté J. S., and de Leeuw J. W. (1999) Distribution of aliphatic, nonhydrolyzable biopolymers in marine microalgae. *Org. Geochem.* **30**, 147–159.

George S. C., Krieger F. W., Eadington P. J., Quezada R. A., Greenwood P. F., Eisenberg L. I., Hamilton P. J., and Wilson M. A. (1997) Geochemical comparison of oil-bearing fluid inclusions and produced oil from the Toro sandstone, Papua New Guinea. *Org. Geochem.* **26**, 155–173.

Gold T. (1999) *The Deep Hot Biosphere*. Springer, New York, 235pp.

Gonclaves F. T. T. (2002) Organic and isotope geochemistry of the Early Cretaceous rift sequence in the Camamu Basin., Brazil: paleolimnolgical inferences and source rock models. *Org. Geochem.* **33**(1), 67–81.

Goosens H., de Leeuw J. W., Schenck P. A., and Brassell S. C. (1984) Tocopherols as likely precursors of pristane in ancient sediments and crude oils. *Nature* **312**, 440–442.

Grantham P. J. and Wakefield L. L. (1988) Variations in the sterane carbon number distributions of marine source rock derived crude oils through geological time. *Org. Geochem.* **12**, 61–73.

Grantham P. J., Pesthwma J. and Baak A. (1983) Triterpanes in a number of Far-Eastern crude oils. In *Advances in Organic Geochemistry 1981* (eds. M. Bjoroy *et al.*). Wiley, New York, pp. 675–683.

Habicht J. K. A. (1964) Comment on the history of migration in the Gifhorn Trough. *Proceedings of the 6th World Petroleum Congress*, Paper 19-PD2, sec. 1, 480.

Halpern H. I. (1995) Development and applications of light-hydrocarbon based star diagrams. *Am. Assoc. Petrol. Geol. Bull.* **79**, 801–815.

ten Haven H. L. and Rullkotter J. (1988) The diagenetic fate of taraxer-14-ene and oleanane isomers. *Geochem. Cosmochim. Acta* **52**, 2543–2548.

ten Haven H. L., de Leeuw J. W., Peakman T. M., and Maxwell J. R. (1986) Anomalies in steroid and hopanoid maturity indices. *Geochim. Cosmochim. Acta* **50**, 853–855.

ten Haven H. L., de Leeuw J. W., Rullkötter J., and Sinninghe Damsté J. S. (1987) Restricted utility of the pristane/phytane ratio as a palaeoenvironmental indicator. *Nature* **330**, 641–643.

ten Haven H. L., de Leeuw J. W., Sinninghe Damste J. S., Schenck P. A., Palmer S. E., and Zumberge J. E. (1988) Application of biological markers in the recognition of paleohypersaline environments. In *Lacustrine Petroleum Source Rocks*, Geological Society Special Publication No. 40 (eds. A. J. Fleet, K. Kelts, and M. R. Talbot), Blackwell, Oxford, pp. 123–130.

Heath D., Moffatt R. L., and Rowland S. (1995) Quantification of the C_{30+} fraction of North Sea gas condensates by high temperature capillary gas chromatography. *Anal. Proc.* **32**, 485–487.

Heath D. J., Lewis C. A., and Rowland S. J. (1997) The use of high temperature gas chromatography to study the biodegradation of high molecular weight hydrocarbons. *Org. Geochem.* **26**, 769–785.

Hedberg H. D. (1968) Significance of high-wax oils with respect to genesis of petroleum. *Am. Assoc. Petrol. Geol. Bull.* **52**, 736–750.

Hedberg H. D. (1980) Methane generation and petroleum migration. In *Problems of Petroleum Migration*, AAPG Studies in Geology 10 (eds. W. H. Roberts, III and R. J. Cordell). American Association of Petroleum Geologists, Tulsa, pp. 79–206.

Henderson W., Wollrab V., and Eglinton G. (1969) Identification of steranes and triterpanes from a geological source by capillary gas liquid chromatography and mass spectrometry. In *Advances in Organic Geochemistry 1968* (eds. P. A. Schenck and I. Havenaar). Pergamon, London, pp. 181–207.

Hitchon B. (1974) Application of geochemistry to the search for crude oil and natural gas. In *Introduction to Exploration Geochemistry* (ed. A. A. Levinson). Applied Publishing, Calgary, pp. 509–545.

Hites R. A. and Bieman K. (1970) Computer evaluation of continuously scanned mass spectra of gas chromatographic effluents. *Anal. Chem.* **42**, 855–860.

Holba A. G., Tegelaar E. W., Huizinga B. J., Moldowan J. M., Singletary M. S., McCaffrey M. A., and Dzou L. I. P. (1998) 24-Norcholestanes as age sensitive molecular fossils. *Geology* **26**(9), 783–786.

Hollander D. J., McKenzie J. A., Hsu K. J., and Huc A. Y. (1993) Application of an eutrophic lake model to the origin of ancient organic carbon-rich sediments. *Global Biogeochem. Cycles* **7**, 157–179.

Hood A., Gutjahr C. C. M., and Heacock R. L. (1975) Organic metamorphism and the generation of petroleum. *Am. Assoc. Petrol. Geol. Bull.* **59**, 986–996.

Horsfield B., Curry D., Bohacs J. K., Littke R., Rullkotter J., Schenck H. J., Radke M., Schaefer R. G., Carroll A. R., Isaksen G., and Witte E. G. (1994) Organic geochemistry of freshwater and alkaline lacustrine sediments in the Green River Formation of the Washakie Basin, Wyoming, USA. *Org. Geochem.* **22**(3–5), 415–440.

Horvitz L. (1939) On geochemical prospecting. *Geophysics* **4**, 210–225.

Horvitz L. (1972) Vegetation and geochemical prospecting for petroleum. *Am. Assoc. Petrol. Geol. Bull.* **56**, 925–940.

Horvitz L. (1981) Hydrocarbon geochemical prospecting after forty years. In *Unconventional Methods in Exploration for Petroleum and Natural Gas II* (ed. B. M. Gottlieb). Southern Methodist University Press, Dallas, pp. 83–95.

Hsieh M. (1999) Characterization of waxes in high pour-point crude oils. MS Thesis, University of Oklahoma, 113p.

Hsieh, and Philp R. P. (2001) Ubiquitous occurrence of high molecular weight hydrocarbons in crude. *Org. Geochem.* **32**, 955–966.

Hunt J. M. (1961) Distribution of hydrocarbons in sedimentary rocks. *Geochim. Cosmochim. Acta* **22**, 37–49.

Hunt J. M. (1979) *Petroleum Geochemistry and Geology*. Freeman, San Francisco.

Hunt J. M. (1996) *Petroleum Geochemistry and Geology,* 2nd edn. Freeman, New York.

Hwang R. J., Ahmed A. S., and Moldowan J. M. (1994) Oil composition variation and reservoir continuity: Unity Field, Sudan. *Org. Geochem.* **21**(2), 171–188.

Jones R. W. (1987) Organic facies. In *Advance in Petroleum Geochemistry* (eds. J. Brooks and D. Welte). Academic Press, New York, pp. 1–90.

Jones D. and Macleod G. (2000) Molecular analysis of petroleum in fluid inclusions: a practical methodology. *Org. Geochem.* **31**, 1163–1173.

Karlsen D. A. and Larter S. (1990) A rapid correlation method for petroleum population mapping within individual petroleum reservoirs: applications to petroleum reservoir description. In *Correlation in Hydrocarbon Exploration* (ed. J. D. Collinson). Norwegian Petroleum Society, Graham and Trotman, pp. 75–85.

Karlsen D. A., Nedkvitne T., Larter S. R., and Bjorlykke K. (1993) Hydrocarbon composition of authegenic inclusions: application to elucidation of petroleum filling reservoir history. *Geochim. Cosmochim. Acta* **57**, 3641–3659.

Kartsev A. A., Tabasaranskii Z. A., Subbota M. I., and Mogilevskii G. A. (1954) *Geochemical Methods for Prospecting and Exploration for Petroleum and Natural Gas.* State Scientific and Technical Publishing House of Petroleum and Mineral Fuel Literature, Moscow (English translation by P. A. Witherspoon and W. D. Romey (1959), University of California Press, Berkeley).

Kaufman R. L., Ahmed A. S., and Elsinger R. J. (1990) Gas chromatography as a development and production tool for fingerprinting oils from individual reservoirs: applications in the Gulf of Mexico. *GCSSEPM Found. 9th Ann. Res. Conf. Proc.* 263–282.

Kimble B. (1972) The geochemistry of triterpenoid hydrocarbons. PhD Thesis, University of Bristol, 302pp.

Kimble B. J., Maxwell J. R., Philp R. P., Eglinton G., Albrecht P., Ensminger A., Arpino P., and Ourisson G. (1974) Tri- and tetraterpenoids in Messel Oil Shale. *Geochim. Cosmochim. Acta* **38**, 1165–1181.

Kinghorn R. R. F. (1983) *An Introduction to the Physics and Chemistry of Petroleum.* Wiley, Chichester, 420p.

Lambert M. W. (1993) Internal stratigraphy and organic facies of the Devonian–Mississippian Chattanooga (Woodford) Shale in Oklahoma and Kansas. In *Source Rocks in a Sequence Stratigraphic Framework*, AAPG Studies in Geology #37. American Association of Petroleum Geologists, Tulsa, OK. pp. 163–176.

Larskaya Ye. S. and Zhabrev D. V. (1964) Effects of stratal temperatures and pressures on the composition of dispersed organic matter. *Dokl. Akad. Nauk SSSR* **157**(4), 135–139.

Larter S. R., Bowler B. F. J., Li M., Chen M., Brincat D., Bennett B., Noke K., Donohoe P., Simmons D., Kohnen M., Allan J., Telnaes N., and Horstad I. (1996) Molecular indicators of secondary oil migration distances. *Nature* **383**, 593–597.

Laubmeyer G. (1933) A new geophysical prospecting method, especially for deposits of hydrocarbons. *Petroleum* **29**, 1–4.

de Leeuw J. W. and Largeau C. (1993) A review of macromolecular organic compounds that comprise living organisms and their role in kerogen, coal, and petroleum formation. In *Organic Geochemistry Principles and Applications* (eds. M. H. Engel and S. A. Macko). Plenum, New York, pp. 23–72.

de Leeuw J. W. and Sinninghe Damste J. S. (1990) Organic sulfur compounds and other biomarkers as indicators of paleosalinity. In *Geochemistry of Sulfur in Fossil Fuels*, American Chemical Society Symposium Series 429 (eds. W. L. Orr and C. M. White), American Chemical Society, Washington, DC, pp. 417–443.

de Leeuw J. W., van Bergen P. F., van Aarssen B. G. K., Gateller J.-P. L. A., Sinninghe Damste J. S., and Collinson M. E. (1991) Resistant biomarkers as major contributors to krogen. In *Molecules through Time-Fossil Molecules and Biochemical Systematics* (eds. G. Eglinton and G. B. Curry). The Royal Society, London, pp. 329–337.

Lewan M. D. (1985) Evaluation of petroleum generation by hydrous pyrolysis experimentation. *Phil. Trans. Roy. Soc. London A* **315**, 123–134.

Li M., Yao H., Stasiuk L. D., Fowler M. G., and Larter S. R. (1997) Effect of maturity and petroleum expulsion on pyrrolic nitrogen compound yields and distributions in the Duvernay Formation petroleum source rocks in central Alberta, Canada. *Org. Geochem.* **26**, 731–744.

Li J., Philp R. P., and Cui M. (2000) Methyl diamantane index (MDI) as a maturity parameter for Lower Paleozoic carbonate rocks at high maturity and over maturity. *Org. Geochem.* **31**, 267–272.

Link W. K. (1952) Significance of oil and gas seeps in world oil exploration. *Am. Assoc. Petrol. Geol. Bull.* **36**, 1505–1540.

Lopatin N. V. (1971) Temperature and geological time as factors of carbonifaction. *Akad. Nauk SSSR. Izv. Ser. Geol.* **3**, 95–106.

Mackenzie A. S., Brassell S. C., Eglinton G., and Maxwell J. R. (1982) Chemical fossils: the geological fate of steroids. *Science* **217**, 491–504.

Magoon L. B. and Dow W. G. (eds.) (1994) The Petroleum System. The Petroleum System–From Source to Trap. In *AAPG Memoir 60.* American Association of Petroleum Geologists, Tulsa.

Mann U. and Stein R. (1997) Organic facies variations, source rock potential, and sea level changes in Cretaceous black shales of the Quebrada Ocal, upper Magdalena Valley, Columbia. *Am. Assoc. Petrol. Geol. Bull.* **81**(4), 556–576.

Marrs R. W. and Kaminsky B. (1977) Detection of Petroleum-related soil anomalies from LANDSAT. *Am. Assoc. Petrol. Geol. Bull.* **61**, 1560–1561.

Maxwell J. R., Pillinger C. T., and Eglinton G. (1971) Organic geochemistry. *Quart. Rev. Chem. Soc.* **25**, 628.

Maxwell J. R., Quirke J. M. E., and Eglinton G. (1981) Aspects of modern porphyrin geochemistry and the Treibs hypothesis. In *The Impact of the Treibs Porphyrin Concept on the Modern Organic Geochemistry* (ed. A. A. Prashowsky). Halbigdruck, Wurburg, pp. 37–56.

McCaffery M. A., Dahl J. E., Sundararaman P., Moldowan J. M., and Schoell M. (1994a) Source rock quality determination from oil biomarkers: II. A case study using Tertiary-reservoired Beaufort Sea oils. *Am. Assoc. Petrol. Geol. Bull.* **78**(10), 1527–1540.

McCaffery M. A. Moldowan J. M., Lipton P. A., Summons R. E., Peters K. E., Jeganathan A., and Watt D. S. (1994b) Paleoenvironmental implications of novel C_{30} steranes in Precambrian to Cenozoic age petroleum and bitumen. *Geochim. Cosmochim. Acta* **58**, 529–532.

McIver R. D. (1967) Composition of kerogen-clue to its role in the origin of petroleum. In *Proceedings of the 7th World Petroleum Congress in Mexico City*. Elsevier, London, vol. 2, pp. 26–36.

Mello M. R., Gaglianone P. C., Brassell S. C., and Maxwell J. R. (1988) Geochemical and biological marker assessment of depositional environments using Brazilian offshore oils. *Mar. Petrol. Geol.* **5**, 205–223.

Miller G. H. (1976) Microbial surveys help evaluate geological, geophysical prospects. *Oil Gas J.* **74**, 192–202.

Moldowan J. M. and Seifert W. K. (1980) First discovery of botryococcane in petroleum. *J. Chem. Soc. Chem. Comm.* 912–914.

Moldowan J. M., Seifert W. K., and Gallegos E. J. (1985a) Relationship between petroleum composition and depositional environment of petroleum source rocks. *Am. Assoc. Petrol. Geol. Bull.* **69**, 1255–1268.

Moldowan J. M., Sundararaman P., and Schoell M. (1985b) Sensitivity of biomarker properties to depositional environment and/or source input in the Lower Toarcian of SW-Germany. *Adv. Org. Geochem.* **10**, 915–926.

Moldowan J. M., Dahl J., Huizinga B. J., Fago F. J., Hickey L. J., Peakman T. M., and Taylor D. W. (1994) The molecular fossil record of oleanane and its relation to angiosperms. *Science* **265**, 768–771.

Moldowan J. M., Dahl J., Jacobson S. R., Huizinga B. J., Fago F. J., Shetty R., Watt D. S., and Peters K. E. (1996) Chemostratigraphic reconstruction of biofacies: molecular evidence linking cyst-forming dinoflagellates with per-Triassic ancestors. *Geology* **24**, 159–162.

Mueller E. and Philp R. P. (1998) Extraction of high molecular weight hydrocarbons from source rocks: an example from the Green River Formation, Uinta Basin, Utah. *Org. Geochem.* **28**, 625–631.

Nip M., Tegelaar E. W., de Leeuw J. W., Schenck P. A., and Holloway P. J. (1986) A new non-saponifiable highly aliphatic, resistant biopolymer in plant cuticles: evidence from pyrolysis, ^{13}C NMR analysis of present day, fossil plants. *Naturwissenschaften* **73**, 579–585.

Orr W. (1986) Kerogen/asphaltene/sulfur relationships in sulfur rich Monterey oils. *Org. Geochem.* **10**, 499–516.

Pedersen T. F. and Calvert S. E. (1990) Anoxia vs. productivity: what controls the formation of organic-carbon-rich sediments and sedimentary rocks? *Am. Assoc. Petrol. Geol. Bull.* **74**, 454–466.

Peters K. E. (1986) Guidelines for evaluating petroleum source rocks using programmed pyrolysis. *Am. Assoc. Petrol. Geol. Bull.* **70**, 318–329.

Peters K. E. and Fowler M. (2002) Applications of petroleum geochemistry to exploration and reservoir management. *Org. Geochem.* **33**(1), 5–37.

Peters K. E. and Moldowan J. M. (1992) *The Biomarker Guide: Interpreting Molecular Fossils in Petroleum and Ancient Sediments*. Prentice Hall, Englewood Cliffs, NJ.

Peters K. E., Snedden J. W., Sulemaeman A., Sarg J. F., and Enrico R. J. (2000) A New Geochemical–sequence stratigraphic model for the Mahakam Delta and Makassar Slope, Kalimantan, Indonesia. *Am. Assoc. Petrol. Geol. Bull.* **84**(1), 12–44.

Philippi G. T. (1965) On the depth, time and mechanism of petroleum generation. *Geochim. Cosmochim. Acta* **29**, 1021–1049.

Philp R. P. (1983) Correlation of crude oils from the San Jorges Basin, Argentina. *Geochim. Cosmochim. Acta* **47**(2), 267–277.

Philp R. P. (1985) *Fossil Fuel Biomarkers. Applications and Spectra*. Elsevier, Amsterdam, 294p.

Philp R. P. and Calvin M. (1976) Possible origin for insoluble organic (kerogen) debris in sediments from insoluble cell-wall materials of aglae and bacteria. *Nature* **262**, 134–136.

Philp R. P. and Crisp P. T. (1982) Surface geochemical methods used for oil and gas prospecting: a review. *J. Geochem. Expl.* **17**, 1–34.

Philp R. P. and Lewis C. A. (1987) Organic geochemistry of biomarkers. *Ann. Rev. Earth Planet. Sci.* **15**, 363–395.

Philp R. P., Bishop A. N., del Rio J.-C., and Allen J. (1995) Characterization of high molecular weight hydrocarbons ($>C40$) in oils and reservoir rocks. In *The Geochemistry of Reservoirs*. Geol. Soc. Spec. Publ. No. 86 (eds. J. M. Cubitt and W. A. England). The Geological Society, London, pp. 71–85.

Poulet M. and Roucache J. (1969) Etude geochemique des gisements du Nord-Sahara (Algerie). *Rev. l'Inst Fran. du petrol.* **24**, 615–644.

Powell T. G. and Boreham C. J. (1991) Petroleum generation and source rock assessment in terrigenous sequences: an update. *APEA J.* **31**, 297–311.

Powell T. G. and McKirdy D. M. (1973) Relationship between ratio of pristane to phytane, crude oil composition and geological environment in Australia. *Nature* **243**, 37–39.

Qin K., Yang Q., Guo S., and Shu W. (1994) Chemical structure and hydrocarbon formation of the Huanxian brown coal. *Org. Geochem.* **21**(3–4), 333–341.

Requejo A. G. (1992) Quantitative analysis of triterpane and sterane biomarkers: methodology and applications in molecular maturity studies. In *Biological Markers in Sediments and Petroleum* (eds. J. M. Moldowan, P. Albrecht, and R. P. Philp). Prentice Hall, Englewood Cliffs, NJ, pp. 223–240.

del Rio J. C. and Philp R. P. (1992) High molecular weight hydrocarbons: a new frontier in organic geochemistry. *Trends Anal. Chem.* **11**, 187–193.

del Rio J. C., Philp R. P., and Allen J. (1992) Nature and geochemistry of high molecular weight hydrocarbons (above C_{40}) in oils and solid bitumens. *Org. Geochem.* **18**, 541–553.

Roberts A. A., Dalziel M., Pogorski L. A., and Quirt S. G. (1976) A possible helium anomaly in the soil gas, Boulder and Weld Counties, Colorado. *US Geol. Surv. Open-File Report*, 76–544.

Robison V. D., Liro L. M., Robison C. R., Dawson W. C., and Russo J. W. (1996) Integrated geochemistry, organic petrology, and sequence stratigraphy of the Triassic Shublik Formation, Tenneco Phoenix well, North Slope, Alaska, USA. *Org. Geochem.* **24**(2), 257–272.

Roedder E. (1984) Fluid inclusions. *Rev. Min.* 12.

Rosaire E. E. (1940) Symposium on geochemical exploration. Geochemical prospecting for petroleum. *Am. Assoc. Petrol. Geol. Bull.* **24**, 1400–1433.

Ross L. M. and Ames R. L. (1988) Stratification of oils in Colombus Basin off Trinidad. *Oil Gas J.* **26**, 72–76.

Rowland S. J. (1990) Production of acyclic isoprenoid hydrocarbons by laboratory maturation of methanogenic bacteria. *Org. Geochem.* **15**, 9–16.

Saxby J. D. and Shibaoka M. (1986) Coal and coaly macerals as source rocks for oil and gas. *Appl. Geochem.* **1**, 25–36.

Schoell M. and Clayton C. (2002) Formation and occurrence of bacterial gas. *American Association of Petroleum Geologists Annual Meeting*, Houston, March 2002, Abstract A157.

Schoell M., Jenden P. D., Beeunas M. A., and Coleman D. D. (1993) Isotope analyses of gases in gas field and gas storage operations. Richardson, TX: Society of Petroleum Engineers, *SPE 26171*, pp. 334–337.

Schoell M., Hwang R. J., Carlson R. M. K., and Welton J. E. (1994) Carbon isotopic composition of individual biomarkers in gilsonites (Utah). In *Compound-specific Analysis in Biogeochemistry and Petroleum Research* (eds. M.Schoell and J. M. Hayes). *Org. Geochem.* **21**, 673–683.

Scott A. C. and Fleet A. J. (eds.) (1994) In *Coal and Coal-bearing Strata as Oil-prone Source Rocks?* Geological Society Special Publication No. 77. The Geological Society, London, 213pp.

Sealey J. Q. (1974a) A geomicrobiological method of prospecting for petroleum: Part I. *Oil & Gas J.* **72**(14), 142–146.

Sealey J. Q. (1974b) A geomicrobiological method of prospecting for petroleum, Part II. *Oil & Gas J.* **72**(15), 98–102.

Seifert W. K. (1977) Source rock/oil correlations by C_{27}–C_{30} biological marker hydrocarbons. In *Advances in Organic Geochemistry 1975* (eds. R. Campos and J. Goni). ENADIMSA, Madrid, pp. 21–44.

Seifert W. K. (1978) Steranes and terpanes in kerogen pyrolysis for correlation of oils and source rocks. *Geochim. Cosmochim. Acta* **42**, 473–484.

Seifert W. K. and Moldowan J. M. (1978) Applications of steranes, terpanes and monoaromatics to the maturation, migration and source of crude oils. *Geochim. Cosmochim. Acta* **42**, 77–95.

Seifert W. K. and Moldowan J. M. (1974) The effect of biodegradation on steranes and terpanes in crude oils. *Geochim. Cosmochim. Acta* **43**, 111–126.

Seifert W. K. and Moldowan J. M. (1980) The effect of thermal stress on source rock quality as measured by hopane stereochemistry. *Phy. Chem. Earth* **12**, 229–237.

Seifert W. K. and Moldowan J. M. (1981) Paleoreconstruction by biological markers. *Geochim. Cosmochim. Acta* **45**, 783–794.

Seifert W. K. and Moldowan J. M. (1986) Use of biological markers in petroleum exploration. In *Methods in Geochemistry and Geophysics* (ed. R. B.Johns), Elsevier, Amsterdam, **24**, 261–290.

Seifert W. K., Moldowan J. M., Smith G. W., and Whitehead E. V. (1978) First proof of a C_{28} pentacyclic triterpane in petroleum. *Nature* **271**, 436–437.

Sherwood-Lollar B., Westgate T. D., Ward J. A., Slater G. F., and Lacrampe-Couloume G. (2002) Abiogenic formation of alkanes in the Earth's crust as a minor source for global hydrocarbon reservoirs. *Nature* **416**, 522–524.

Silverman S. R. (1965) Migration and segregation of oil and gas. In *AAPG Memoir 4* (eds. A. Young and G. E. Galley). American Association of Petroleum Geologists, Tulsa, pp. 54–65.

Sinninghe Damste J. S., Kenig F., Koopmans M. P., Koster J., Schouten S., Hayes J. M., and de Leeuw J. (1995) Evidence for gammacerane as an indicator of water column stratification. *Geochim. Cosmochim. Acta* **59**(9), 1895–1900.

Stach E., Mackowsky M.-Th., Teichmeuller M., Taylor G. H., Chandra D., and Teichmeuller R. (1982) *Textbook of Coal Petrology,* 3rd edn. Bebruder Borntraeger, Berlin.

Staplin F. L. (1969) Sedimentary organic matter, organic metamorphism, and oil and gas occurrence. *Can. Petrol. Geol. Bull.* **17**(1), 47–66.

Steffen D. and Gorin G. E. (1993) Sedimentology of organic matter in upper Tithonian–Berriasian Deep–Sea Carbonates of Southeast France: evidence of eustatic control. In *Source Rocks in a Sequence Stratigraphic Framework*, AAPG Studies in Geology #37. America Association of Petroleum Geologists, Tulsa, OK, pp. 49–66.

Taylor P., Larter S. R., Jones D. M., Dale J., and Horstad I. (1997) The effect of oil-water partioning of alkylphenols in petroleum systems. *Geochim. Cosmochim. Acta* **61**, 1899–1910.

Tegelaar E. W., Matthezing R. M., Jansen J. B. H., Horsfield B., and de Leeuw J. W. (1989) Possible origin of n-alkanes in high-wax crude oils. *Nature* **342**, 529–531.

Tegelaar E. W., Hollman G., van der Vegt P., de Leeuw J. W., and Holloway P. J. (1995) Chemical characterization of the periderm tissue of some angiosperm species: recognition of an insoluble, non-hydrolysable aliphatic biomacromolecule (suberan). *Org. Geochem.* **23**, 239–251.

Teichmuller M. (1958) Metamorphism du carbon et propection du petrole. *Rev. Indus. Min.* (Special Issue) 1–15.

Tissot B. (1969) Primieres donnees sur le mecanismes et la cinetique de la formation du petrole dans les sediments: simulation d'un schema reactionnel sur ordinateur. *Revue l'Inst. Francais du Petrole* **24**(4), 470–501.

Tissot B. and Espitalie J. (1975) L'evolution thermique de la matiere organiques des sediments: application d'une simulation mathematique. *Rev. l'Inst. Francais Petrole* **30**, 743–777.

Tissot B. and Welte D. H. (1978) *Petroleum Formation and Occurrence*. Springer, Heidlberg.

Tissot B. and Welte D. H. (1984) *Petroleum Formation and Occurrence,* 2nd edn. Springer, Berlin.

Tissot B., Oudin J. L., and Pelet R. (1972) Criteria of the origin and evolution of petroleum application to the geochemical study of sedimentary basins. In *Advances in Geochemistry 1971* (eds. H. v. Gaertner and H. Wehner). Pergamon, New York, pp. 113–134.

Tissot B., Durand B., Espitalie J., and Combaz A. (1974) Influence of nature and diagenesis of organic matter in formation of petroleum. *Am. Assoc. Petrol. Geol. Bull.* **58**, 499–506.

Treibs A. (1934) The occurrence of chlorophyll derivatives in an oil shale of the upper Triassic. *Annalen* **517**, 103–114.

Treibs A. (1936) Chlorophyll and hemin derivatives in organic materials. *Angew Chem.* **49**, 682–686.

Trestianu S., Zilioli G., Sironi A., Saravalle C., and Munari F. (1985) Automatic simulated distillation of heavy petroleum fractions up to 800 °C TBP by capillary gas chromatography. Part I: Possibilities and limits of the method. *J. High Res. Chrom. Comm.* **8**, 771–781.

Venkatesan M. I. (1989) Tetrahymanol: its widespread occurrence and geochemical significance. *Geochim. Cosmochim. Acta* **53**(11), 3095–3101.

Vitorovic D., Djuricic M. V., and Ilic B. (1974) New structural information obtained by stepwise oxidation of kerogen from the Aleksinac (Yugoslavia) shale. In *Advances in Organic Geochemistry 1973*, Editions Technip, Paris, pp. 180–189.

Volk H., Mann U., Burde O., Horsfield B., and Suchy V. (2000) Petroleum inclusions and residual oils: constraints for deciphering petroleum migration. *J. Geochem. Explor.* **71**, 307–311.

Volkman J. K. (1986) A review of sterol markers for marine and terrigenous organic matter. *Org. Geochem.* **9**, 84–99.

Volkman J. K., Alexander R., Kagi R. I., and Woodhouse G. W. (1983) Demethylated hopanes in crude oils and their applications in petroleum geochemistry. *Geochim. Cosmochim. Acta* **47**, 1033–1040.

Volkman J. K., Alexander R., Kagi R. I., Rowland S. J., and Sheppard P. N. (1984) Biodegradation of aromatic hydrocarbons in crude oils from the Barrow SubBasin of western Australia. *Org. Geochem.* **6**, 619–632.

Waples D. W. (1984) Thermal models for oil generation. In *Advances in Petroleum Geochemistry* (eds. J. Brooks and D. H. Welte). Academic Press, London, vol. 1, pp. 7–67.

Waples D. W. (1985) *Geochemistry in Petroleum Exploration.* International Human Resources Development Corporation, Boston.

Waples D. W. (1994) Maturity modelling: thermal indicators, hydrocarbon generation and oil cracking. In *The Petroleum System-from Source to Trap*, AAPG Memoir 60 (eds. L. Magoon and W. G. Dow). American Association of Petroleum Geologists, Tulsa, pp. 285–306.

Wavrek D. A., and Dahdah N. F. (1995) Characterization of high molecular weight compounds: implications for advanced-recovery technologies. *SPE 28965*, pp. 207–210.

Welte D. H., Horsfield B., and Baker D. R. (1997) *Petroleum and Basin Evolution*. Springer, New York.

Whitehead E. V. (1971) Chemical clues to petroleum origin. *Chem. Indus.* No. 27. pp. 1116–1118.

Whitehead E. V. (1974) The structure of petroleum pentacyclanes. In *Advances in Organic Geochemistry 1973* (eds. B. Tissot and F. Bienner). Editions Technip, Paris, pp. 225–243.

Wignall P. B. and Maynard J. R. (1993) The sequence stratigraphy of transgressive black shales. In *Source Rocks in a Sequence Stratigraphic Framework*, AAPG Studies in Geology #37. American Association of petroleum Geologists, Tulsa, OK, p.35.

Wilhelms A., Larter S., Head I., Farrimond P., di-Primo R., and Zwach C. (2001) Biodegradation of oil in uplifted basins prevented by deep burial sterilization. *Nature* **411**, 1034–1037.

Wilkes H., Willsch H., Rabus R., Aeckersberg F., Rueter P., and Widdel F. (1995) Compositional changes of crude oils upon anaerobic degradation by sulphate reducing bacteria.

In *Organic Geochemistry, Developments and Applications to Energy, Climate, Environment and Human History* (eds. J. O. Grimalt and C. Dorronsoro). 17th International Meeting on Organic Geochemistry, San Sebastian, Spain, pp. 321–324.

Wilkes H., Rabus R. Fischer Th., Armstroff A., Behrends A., Pierik A. J., and Widdel F. (2001). The anaerobic degradation of n-alkanes as pure substrates and in crude oil: first direct insights into the molecular mechanism of the initial C–H-activation and the subsequent degradation pathway. Paper present at the 20th International Organic Geochemistry Meeting, Nancy France, September, 2001. Abstract p. 218.

Zengler K., Richnow H. H., Rosselo-Moura R., Michaelis W., and Widdel F. (1999) Methane formation from long chain alkanes by anaerobic microorganisms. *Nature* **401**, 266–269.

Zorkin L. M., Stadnik Ye. V., and Yurin G. A. (1976) Geochemistry of nitrogen in ground water of oil and gas bearing basins. *Int. Geol. Rev.* **19**, 1404–1410.

7.10
Sulfur-rich Sediments

M. B. Goldhaber
US Geological Survey, Denver, CO, USA

7.10.1. INTRODUCTION

7.10.1.1 Overview

Marine sediments with more than a few tenths of a percent of organic carbon, as well as organic-matter-bearing, nonmarine sediments with significant concentrations of sulfate in the depositional waters contain the mineral pyrite (FeS_2). Pyrite, along with sulfur-bearing organic compounds, form indirectly through the metabolic activities of sulfate-reducing microorganisms. The geochemical transformations of sulfur in sediments leading to these products significantly impact the pathway of early sedimentary diagenesis, conditions for the localization of mineral deposits (Ohmoto and Goldhaber, 1997), the global cycling of sulfur and carbon, the abundance of oxygen in the Earth's atmosphere, and perhaps even the emergence of life on Earth (e.g., Russell and Hall, 1997). This chapter provides an overview of sedimentary-sulfur geochemistry from its microbial and abiologic pathways to the global consequences of these processes.

The geochemistry of sulfur is complicated by its wide range of oxidation states (Table 1). Under oxidizing conditions (e.g., in the presence of atmospheric oxygen) sulfate, with sulfur in the +6 valence state, is the stable form of sulfur. Under reducing conditions (e.g., in the presence of H_2), sulfide ($S = -2$ valent) is the stable oxidation state. However, a range of additional aqueous and solid-phase sulfur species exist with valences between these two end-members. What makes the study of sulfur geochemistry so exciting and challenging is that many of these intermediate-valent forms play key roles in sedimentary-sulfur transformations. Furthermore, many of these reactions are microbially mediated. As detailed below, these complex biogeochemical pathways are now yielding to research whose scope ranges from molecular to global level.

7.10.1.2 History of the Study of Sedimentary-sulfur Geochemistry

The study of sulfur diagenesis has a long history. Scientists have long accepted that microorganisms play a major role in geochemical sulfur transformations (Baas Becking, 1925). They also recognized at an earlier time that removal of sulfate occurs in the pore waters of marine mud (Murry, 1895). Subsequent work established that depletion of sulfate from marine pore water is a microbial process that results in formation of sedimentary pyrite (Berner, 1964a; Emery and Rittenberg, 1952; Hartmann and Nielsen, 1969; Kaplan *et al.*, 1963; ZoBell and Rittenberg, 1950). The abundance of this pyrite, together with its morphological and isotopic characteristics, provides clues to the details of these transformations.

There are a number of useful summary articles on sulfur geochemistry (e.g., Belyayev *et al.*, 1981; Berner, 1973; Ohmoto *et al.*, 1990, #1031; Bottrell and Raiswell, 2000; Canfield and Raiswell, 1991; Chambers and Trudinger, 1979; Goldhaber and Kaplan, 1974; Grinenko and Ivanov, 1983; Ivanov, 1981; Krouse and McCready, 1979a,b; Migdisov *et al.*, 1983; Morse *et al.*, 1987; Ohmoto and Goldhaber, 1997; Skyring, 1987; Strauss, 1997). In addition to this important body of work, there have been a number of recent advances in the study of sedimentary-sulfur diagenesis that make a new review timely. These advances are linked to a much deeper understanding of the complex microbial processes that dominate sulfur transformations in sediments, coupled with advances in analytical technology, enabling processes to be studied with increasing specificity and sophistication.

7.10.2 BACTERIAL SULFATE REDUCTION

7.10.2.1 Biochemistry of Bacterial Sulfate Reduction

The key reaction in the global sulfur cycle is the reduction of sulfate (SO_4^{2-}) to hydrogen sulfide (H_2S). In this chapter H_2S, unless otherwise specified, is taken to indicate the sum of the reduced sulfur species including H_2S_{aq} and HS^-_{aq}.

Table 1 Forms of sulfur in marine sediments and their oxidation states.

Aqueous species or mineral	*Formula*	*Oxidation state(s) of sulfur*
Sulfide	$H_2S(aq)$, $HS^-(aq)$	−2
Iron sulfide[a]	$FeS(s)$	−2
Greigite	$Fe_3S_4(s)$	−2, 0
Pyrite	$FeS_2(s)$	−2, 0
Polysulfide	$S_{X^{2-}}(aq)$	−2, 0
Sulfur	$S_8(s)$	0
Hyposulfite	$S_2O_4^{2-}(aq)$	+3
Sulfite	$SO_3^{2-}(aq)^-$	+4
Thiosulfate	$S_2O_3^{2-}(aq)$	−1, +5
Dithionate	$S_2O_6^{2-}(aq)$	+5
Trithionate	$S_3O_6^{2-}(aq)$	−2, +6
Tetrathionate	$S_4O_6^{2-}(aq)$	−2, +6
Pentathionate	$S_5O_6^{2-}(aq)$	−2, +6
Sulfate	$SO_4^{2-}(aq)$	+6

After Kasten and Jorgensen (2000) (reproduced by permission of Springer-Verlag from *Marine Geochemistry*, **2000**, 263–281).
[a] Includes troilite, mackinawite, and pyrrhotite.

The stability of SO_4^{2-} towards naturally occurring reducing agents is so great that either elevated temperatures or a microbially (enzymatically) catalyzed process is required for reduction to occur. Thermodynamic calculations indicate that reduction of SO_4^{2-} to H_2S should occur at Earth-surface temperatures by a wide range of organic compounds. Yet below temperatures ~110–150 °C this reaction is strongly kinetically inhibited, and may not be observable even on geologic timescales (Goldhaber and Orr, 1994; Ohmoto and Goldhaber, 1997). This kinetic inhibition of sulfate reduction is effectively overcome in nature by two types of microbially catalyzed processes: assimilatory and dissimilatory sulfate reduction (Madigan *et al.*, 2000).

Many organisms, including higher plants, fungi, algae, and most prokaryotes, reduce sulfate and incorporate it as a sulfur source for biosynthesis of proteins. This reaction, termed assimilative sulfate reduction, although widely distributed in nature, is neither a major component of the global sulfur cycle, nor is it a major mechanism of sulfur transformation in sediments. In contrast, dissimilatory reduction, in which sulfate is an electron acceptor for energy generating bacterial processes, is of global importance (Equation (1)):

$$SO_4^{2-} + 2CH_2O \rightarrow H_2S + 2HCO_3^- \quad (1)$$

In this equation, CH_2O represents a generic form of organic matter with the oxidation state of carbohydrate. In effect, sulfate-reducing prokaryotes (including both bacteria and archea) respire (breathe in) sulfate (as we do oxygen), and breathe out H_2S and carbon dioxide. As a result, they release large amounts of H_2S into their environment. Dissimilatory reduction is restricted to the sulfate-reducing bacteria and archea.

The biochemical pathway of both assimilatory and dissimilatory sulfate reduction is illustrated in Figure 1. The details of the dissimilatory reduction pathway are useful for understanding the origin of bacterial stable isotopic fractionations. The overall pathways require the transfer of eight electrons, and proceed through a number of intermediate steps. The reduction of sulfate requires activation by ATP (adenosine triphosphate) to form adenosine phosphosulfate (APS). The enzyme ATP sulfurylase catalyzes this reaction. In dissimilatory reduction, the sulfate moiety of APS is reduced to sulfite (SO_3^{2-}) by the enzyme APS reductase, whereas in assimilatory reduction APS is further phosphorylated to phospho-adenosine phosphosulfate (PAPS) before reduction to the oxidation state of sulfite and sulfide. Although the reduction reactions occur in the cell's cytoplasm (i.e., the sulfate enters the cell), the electron transport chain for dissimilatory sulfate reduction occurs in proteins that are periplasmic (within the bacterial cell wall). The enzyme hydrogenase requires molecular hydrogen supplied either from the external environment or by the oxidation of organic compounds such as lactate. Hydrogenase supplies eight electrons to cytochrome c_3 (cytochromes are proteins with iron-containing porphyrin rings), which in turn transfers electrons to a second cytochrome complex (Hmc). The electrons from this periplasmic electron transport system are transported across the cytoplasmic membrane to an iron–sulfur protein in the cytoplasm that, in turn, supplies electrons for the reduction reactions shown in Figure 1.

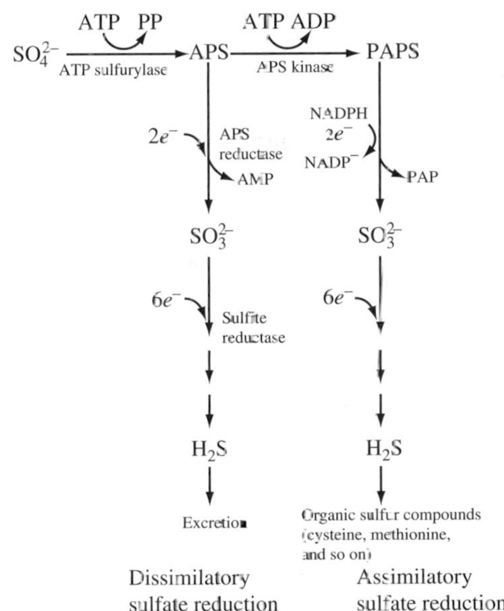

Figure 1 Biochemical pathway of dissimilatory and assimilatory SO_4^{2-} reduction (after Madigan *et al.*, 2000).

7.10.2.2 Ecology of Sulfate-reducing Bacteria

The impact of sulfate-reducing bacteria (SRB) on marine sediments can be extensive. In organic-rich sediments, large quantities of reactant organic matter are consumed and dissolved products produced (Equation (1)). The impact of SRB metabolic activities can modify the overall geochemistry of the sedimentary package. For example, in many environments, more than 50% of the total carbon mineralization (oxidation) is due to SRB (Canfield and Des Marais, 1993; Jorgensen, 1982). The ecology of these organisms determines to what extent they modify the sediments. The most important ecological requirement is that sulfate-reducing microorganisms are strict anaerobes. In the presence of organic matter and absence of oxygen, SRB can grow in a wide range of environments spanning the spectrum of pressure, temperature, salinity, and pH values found in the Earth's upper crust.

SRB must compete, however, for the available food supply in sediments. In addition to sulfate reducers, other microorganisms gain energy by anaerobic respiration using electron acceptors other than sulfate. In marine sediments, a well-defined succession of microbial ecosystems closely follows the energy yield available from oxidation of organic matter with the available electron acceptors. The order is aerobic respiration followed by respiration using nitrate, manganese oxide, iron oxide, sulfate, and carbonate (Figure 2). The presence of significant amounts of one of these electron acceptors, more favored than sulfate, suppresses the activities of SRB. For example, addition of Fe(III) oxyhydroxides to sediments in which sulfate reduction is active will suppress SRB nearly completely (Lovely and Phillips, 1987).

Oxygen respiration is by far the most efficient respiratory process, but because oxygen is present in relatively low concentration in seawater, and because rates of organic oxidation with oxygen are rapid, aerobic respiration is self-limiting. The penetration distance of oxygen by diffusion may only be millimeters in sediments with abundant organic matter. However, oxygen is also physically mixed (advected) or pumped into sediments by the life activities of macroorganisms, a process termed bioturbation (Aller, 1980a, 1988; R. C. Aller and J. Y. Aller, 1998; Goldhaber *et al.*, 1977). Bioturbation may increase the thickness of the oxygenated zone, or may create a series of heterogeneous oxic and anoxic volumes within the upper sediment layers. Following oxygen removal,

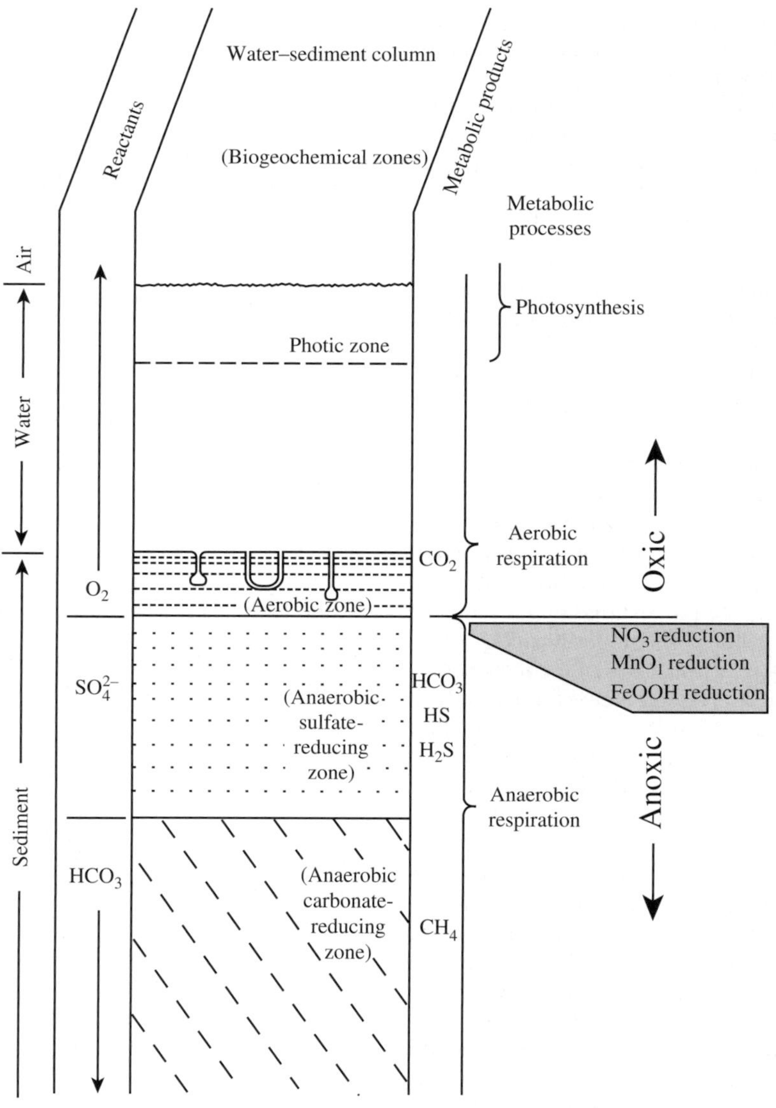

Figure 2 Schematic depth zonation of biogeochemical processes in marine sediments.

a series of microorganisms utilize terminal electron acceptors in the sequence shown in Figure 2. However, with the exception of sulfate, the other electron acceptors commonly do not occur in high concentrations in the marine environment (although exceptions exist). By one estimate, the concentration of sulfate is more than 50 times higher than that of the other more energetically favored terminal electron acceptors (Froelich *et al.*, 1979). Sulfate is the second most abundant anion in seawater, so seawater overlying marine sediments represents an enormous reservoir of this anion. Therefore, anaerobic respiration using sulfate tends to dominate anaerobic respiratory processes in organic-matter-rich marine sediments.

A critical control on sulfate reduction in sediments is the amount and type of food available for the bacteria. Sulfate reducers can only metabolize a relatively limited range of organic compounds. Although the recognized range of genera/species of sulfate reducers and the compounds they can metabolize has increased dramatically in recent years (Madigan *et al.*, 2000, table 13.21), the typical substrates consist of small organic monocarboxylic and dicarboxylic acids, alcohols, and amino acids. However, the organic matter that persists in sediments through the preceding zones of aerobic and anaerobic metabolism is present as complex polymerized biological residues. For this reason, sulfate reducers rely on a precursor group of organisms to break down organic matter to simpler compounds by fermentation. The "quality" (susceptibility to microbial attack) of organic matter in sediments may have a dramatic effect on overall sulfate reduction rates in sediments. These rates can vary over many orders of magnitude (see Section 7.10.5.2).

A number of other ecological factors beyond abundance and type of food may influence the activities of SRB. These factors include temperature, salinity, sulfate concentration, and pH. Specific species of SRB are adapted to specific temperature ranges. These ranges can be divided into psychrophilic (thermal optimum near 4 °C), mesophilic (optimum near 39 °C), thermophilic (optimum near 60 °C), and hyperthermophilic (optimum near 88 °C) (Madigan *et al.*, 2000). It is evident that sulfate-reducing microorganisms are viable over a temperature range spanning at least 0–100 °C. The effect of temperature on a specific sedimentary setting is nonlinear and cannot be simplified to a single temperature factor. This complexity may be related to the highly variable nature of organic matter (Westrich and Berner, 1988). In summary, temperature may influence the species present in a sedimentary environment, and along with food source may influence rates of reduction of sulfate.

Neither variation in salinity nor sulfate concentration seems to have a major impact on the viability of SRB, although very high salinity may impact the sulfate reduction rate. SRB are able to metabolize sulfate down to very low concentrations. Half saturation constants (k_m values) of 70 μM and 200 μM for marine strains have been reported (Ingvorsen and Jorgensen, 1984). SRB have been recognized in the entire spectrum of water types from dilute freshwater up to saturation with halite. For example, one study of a series of salt-evaporation ponds in San Francisco Bay (Klug *et al.*, 1985) reported sulfate-reduction rates from sediments with pore-water salinities ranging from 33‰ up to 300‰. Maximum sulfate reduction rates, which occurred in the upper 1 cm of sediment, tended to decrease with increasing salinity from values as high as 16 μmol sulfate reduced per gram wet sediment per day, to values of 0.04 μmol reduced per cm or less in the sediment with salinity of 150‰ and 300‰.

Below the top 1–2 cm, however, reduction rates were similar at all salinities. The Dead Sea with a total salt content of ~310 g L^{-1} is a second example of a highly saline environment where microbial sulfate reduction is clearly occurring, and, in fact, dominating the geochemistry of sulfur (Nissenbaum and Kaplan, 1976). Very high rates of sulfate reduction have been recorded in microbial mats forming in high salinity environments (Habicht and Canfield, 1997). Like temperature, salinity may influence the species of SRB present in the sediment (Pfennig *et al.*, 1979).

The influence of pH on SRB has not been systematically studied. Most Earth-surface environments tend to fall in a rather narrow pH range of 5–8.5, and SRB are not markedly impacted by pH changes within this range (Fauque, 1995). There is research indicating that at least some species of SRB are active at the extremes of pH. One study looked in detail at sulfate reduction in an acid (pH 3) strip mine lake in Indiana (Konopka *et al.*, 1985). They found that sulfate reduction was occurring microbiologically at rates comparable to those in pH neutral settings. At the other extreme, H_2S occurs at concentrations of 20 mM in the bottom waters of Soap Lake in the state of Washington. This lake is saline and has a pH of 9.8. Sulfur isotopic data indicate that the sulfide formed microbiologically (Tuttle *et al.*, 1990).

7.10.2.3 Sulfur Isotopic Fractionation during Bacterial Reduction

7.10.2.3.1 Laboratory studies

One of the characteristics of sulfate-reducing microorganisms that has proved most useful for documenting their activity in the natural environment is their ability to fractionate stable sulfur isotopes during the reduction process. The aqueous sulfide produced is enriched in the light stable isotope of sulfur (^{32}S) compared to the heavy isotope (^{34}S) during reduction. In this chapter, the term "isotopically light" indicates relative enrichment in this light isotope. This fractionation was first demonstrated by Thode *et al.* (1951), who recognized a 10‰ enrichment in ^{32}S relative to the starting sulfate. A number of subsequent laboratory research efforts demonstrated variable isotope fractionation effects associated with dissimilatory sulfate reduction ranging from +3‰ to −46‰ (see an earlier review by Chambers and Trudinger, 1979). These studies reported that isotope enrichment is inversely proportional to sulfate reduction rate. This inverse relation to rate was addressed conceptually by Rees (1973), who pointed out that in order for fractionation to occur, there must be a back reaction involving release of isotopically heavy intermediates in the reduction pathway. Otherwise, if all sulfate that enters the cell undergoes reduction, mass balance requires there can be no net isotopic enrichment of ^{32}S in the product sulfide. Rapid rates of reduction imply rapid throughput of intermediates along the reduction pathway with little opportunity for buildup and back reaction of these intermediates. Conversely, slow rates of reduction allow isotopically heavy intermediates in the pathway to accumulate and back-react so that they may be ultimately released.

Older experimental work utilized a limited number of species of microorganisms and electron donors. A more recent survey (Detmers *et al.*, 2001) evaluated isotopic fractionation by 32 strains of sulfate-reducing microorganisms spanning a wide range of temperatures and electron donors (Table 2). The overall fractionations observed ranged from −3‰ to −42‰. Organisms that oxidize the organic substrate completely to bicarbonate typically produced larger fractionations (between −15‰ and −42‰, average 25‰) compared to organisms that incompletely oxidize the substrate, and excrete acetate (−2‰ to −18.7‰, average −9.5‰). There was no trend in isotope fractionation with optimum growth temperature of the organism.

Table 2 Cell-specific fractionation factors of sulfate-reducing prokaryotes.

Microorganism	*Isolated from*	*Electron donor* (mM)	*Fraction factor* (‰)
Complete oxidizing			
Desulfonema magnum	Marine mud	Benzoate (3)	42.0
Desulfobacula phenolica	Marine mud	Benzoate (3)	36.7
Desulfobacterium autotrophicum	Marine mud	Butyrate (20)	32.7
Desulfobacula toluolica	Marine mud	Benzoate (3)	28.5
Desulfotomaculum gibsoniae	Freshwater mud	Butyrate (20)	27.8
Desulfospira joergensenii	Marine mud	Pyruvate (20)	25.7
Desulfotignum balticum	Marine mud	Butyrate (20)	23.1
Desulfofrigus oceanense	Arctic sediment	Acetate (20)	22.0
Desulfobacter sp. ASv20	Arctic sediment	Acetate (20)	18.8
Desulfobacca acetoxidans	Anaerobic sludge	Acetate (20)	18.0
Desulfococcus sp.	Marine mud	Pyruvate (20)	16.1
Desulfosarcina variabilis	Marine mud	Benzoate (3)	15.0
Incomplete oxidizing			
Desulfonatronum lacustre	Alkaline lake mud	Ethanol (20)	18.7
Archaeoglobus fulgidus strain Z	Submarine hot spring	Lactate (20)	17.0
Thermodesulfovibrio yellowstonii	Thermal vent water	Lactate (20)	17.0
Desulfotomaculum thermocistemum	Oil reservoir	Lactate (20)	15.0
Desulfohalobium redbaense	Saline sediment	Lactate (20)	10.6
Desulfocella halophila	Great salt lake	Pyruvate (20)	8.1
Desulfobulbus "marinus"	Marine mud	Propionate (20)	6.8
Desulfotalea arctica	Arctic sediment	Lactate (20)	6.1
Desulfovibrio sp. strain X	Hydrothermal vent	Lactate (20)	5.4
Themtodesulfobacterium commune	Thermal spring	Lactate (20)	5.0
"*Desulfovibrio oxyclinae*"	Hypersaline mat	Lactate (20)	4.5
Desulfotalea psychrophila	Arctic sediment	Lactate (20)	4.3
Desulfovibrio profundus	Deep sea sediment	Lactate (20)	4.1
Desulfovibrio halophilus	Hypersaline microbial mat	Lactate (20)	2.0
Hydrogen and formate			
Desulfobactetium autotrophicum	Marine mud	H_2	14.0

After Detmers *et al.* (2001).

Recent work has also addressed the impact of elevated temperature and decreased SO_4^{2-} concentration on sulfur isotopic fractionation factors during sulfate reduction. Canfield *et al.* (2000) measured fractionation factors in marine sediment whose *in situ* temperature was up to 90 °C. Sediments maintained at temperatures up to 85 °C were found to reduce SO_4^{2-} with fractionation factors of 13–28‰. Habicht *et al.* (2002) studied isotopic fractionation factors as a function of SO_4^{2-} concentration. They found high fractionations of up to 32‰ at SO_4^{2-} concentrations of 200 μM or greater, whereas fractionation dropped dramatically below this concentration. The low SO_4^{2-} limit for large fractionation is less than 1% of the modern seawater concentration of 28 mM.

7.10.2.3.2 Field studies

The laboratory studies cited above have been compared to measurements made in natural populations of SRB (Canfield, 2001; Habicht and Canfield, 1996, 1997). The two studies by Habicht and Canfield (1996) determined isotope fractionation in natural populations of SRB for microbial mats. These mats are quite organic rich and rates of sulfate reduction are very high compared to most other natural settings. Nonetheless, the measured isotope fractionations were similar to those measured in the laboratory. A subsequent paper by the same authors (Habicht and Canfield, 1997) looked at isotopic fractionation in marine mud. Figure 3 is a comparison of the field and laboratory determined fractionations plotted as a function of rate of reduction from that paper. The two sets of data are similar except that the fractionations from natural populations have a minimum near 20‰, whereas the lab cultures show occasional smaller values (2–20‰), particularly when grown on H_2. Habicht and Canfield (1997) took this difference to indicate that H_2 is not a major substrate for SRB.

Canfield (2001) undertook a comprehensive study of isotope fractionation by natural populations of SRB in sediment from a small marine lagoon located off the Danish coast. He used a flowthrough reactor and controlled the sulfate concentration, temperature, and ultimately the nature of the organic substrate. High fractionations of 30‰ and 40‰ were found when the SRB metabolized with natural organic substrate at environmental temperatures of 25 °C. The experiment continued until depletion of the natural substrate occurred, and then acetate, ethanol, and lactate were added from an external source. The isotopic fractionations observed ranged from 7‰ to 40‰, and increased fractionations correlated with decreased rates of sulfate reduction as noted in the laboratory studies. An exception occurred at low temperatures, where decreased reduction rates

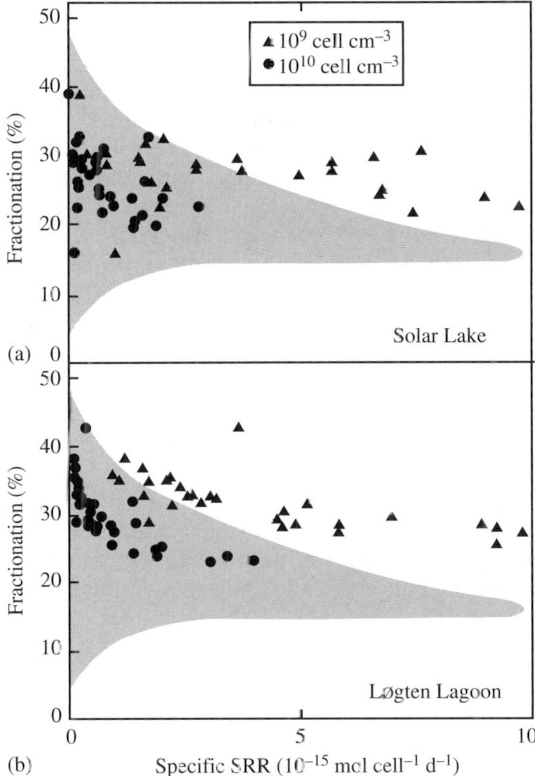

Figure 3 Fractionation of SO_4^{2-} by SRB versus specific sulfate reduction rate (SRR). The shaded area is the range of values observed in laboratory experiments. The points are from field data assuming two different densities of cells for comparison with lab data (Habicht and Canfield, 1997) (reproduced by permission of Elsevier from *Geochim. Cosmochim. Acta*, **1997**, *61*, 5351–5361).

correlated with decreased fractionations. Canfield (2001) attributed that low-temperature behavior to increased viscosity of the cell membrane that inhibits transport of sulfate across it.

7.10.3 FORMS OF SULFUR IN MARINE SEDIMENTS

7.10.3.1 Dissolved Species

The dominant dissolved form of sulfur in seawater is sulfate (including the free form, SO_4^{2-}, plus ion pairs with the major cations of seawater). In pore water of near-surface marine sediments, sulfate also dominates, but it is progressively removed with increasing depth by bacterial reduction and supplanted by dissolved H_2S. Pore-water SO_4^{2-} and H_2S profiles are discussed more fully below.

Other dissolved forms of sulfur in sediments are, by comparison, present at much lower concentrations, but may nonetheless play a significant role in diagenetic processes. These include many of the species summarized in

Table 1, but specifically thiosulfate ($S_2O_3^{2-}$), sulfite (SO_3^{2-}), polysulfides (S_X^{2-}), and elemental sulfur (S^0_{aq}, which also occurs in the solid phase). The presence of this spectrum of sulfur species and their cyclic interconversion in a redox gradient (such as is present in organic-matter-bearing marine sediments with oxygenated depositional waters) were recognized long ago by Bass Becking (1925) (see also Truper, 1982).

As will be discussed in a subsequent section, thiosulfate, $S_2O_3^{2-}$, likely plays a significant role as a reactive intermediate in sulfur diagenesis. It has a wide variety of sources and sinks (Fossing and Jorgensen, 1990; Habicht *et al.*, 1998; Jorgensen and Bak, 1991). Equivalent reactions can be written for sulfite and polysulfides. Thiosulfate forms by oxidation of both iron sulfide minerals and aqueous sulfide (Equations (2)–(6)). The oxidants are dissolved oxygen or iron/manganese-oxide minerals. These reactions occur both abiologically and microbiologically:

$$FeS_2 + 1.5O_2(aq) \rightarrow S_2O_3^{2-} + Fe^{2+} \quad (2)$$

$$FeS + O_2(aq) + H^+ \rightarrow 5S_2O_3^{2-} + Fe^{2+} + 0.5H_2O \quad (3)$$

$$H_2S(aq) + O_2(aq) \rightarrow 0.5S_2O_3^{2-} + H^+ + 0.5H_2O \quad (4)$$

$$H_2S(aq) + 7H^+ + 4FeOOH(\text{goethite}) \rightarrow 0.5S_2O_3^{2-} + 6.5H_2O + 4\,Fe^{+2} \quad (5)$$

$$2H_2S(aq) + 3H_2O \rightarrow S_2O_3^{2-} + 5H_2(aq)$$
(photosynthetic oxidation by *Calothrix*. sp. × *and Oscillatoria* sp.) (6)

Thiosulfate may be removed by reoxidation to sulfate in the uppermost (oxygenated) portions of sediments (Equation (7)):

$$S_2O_3^{2-} + 2O_2(aq) + H_2O \rightarrow 2H^+ + 2SO_4^{2-} \quad (7)$$

It may also be reduced microbiologically to H_2S, in which it serves as an electron acceptor analogous to sulfate (e.g., Equation (1)):

$$2\,\text{lactate} + S_2O_3^{2-} + 2H^+ \rightarrow 2H_2S + 2\,\text{acetate} + 2CO_2 + H_2O \quad (8)$$

Finally, it may undergo disproportionation, a microbial reaction (Canfield and Thamdrup, 1994) that is the inorganic equivalent of an organic fermentation reaction:

$$S_2O_3^{2-} + H_2O \rightarrow H_2S + SO_4^{2-} \quad (9)$$

Thamdrup *et al.* (1994a) made detailed measurements of thiosulfate and sulfite concentrations in a Danish salt marsh and subtidal marine sediments. Thiosulfate occurred at levels from less than 0.05 μM to 0.45 μM and sulfite at 0.1–0.8 μM. They reported that concentrations of both species increased with depth; lowest values were in the oxic and suboxic zones and highest were in the zone of sulfate reduction. They also determined rapid turnover times for thiosulfate of 4 h in the upper 1 cm of sediment, and attributed this rapid turnover to the competing production and consumption processes summarized above.

Luther *et al.* (1985) determined thiosulfate concentrations of 400–1,300 μM and sulfite contents of 6.8–7.9 μM from subtidal sediments off the coast of Massachusetts, USA. Later work (Luther *et al.*, 2001) also confirmed the presence of polysulfides in a microbial mat from Great Marsh, Delaware, and subtidal sediments from Rehoboth Bay, Delaware at concentrations of up to 40 μM. Elemental sulfur (S_8^0) dominated near the surface of both environments, polysulfide, was detected at intermediate depth (6.5 cm in the case of the subtidal sediments), and only H_2S was present at greater depth.

7.10.3.2 Pyrite

Pyrite is the dominant solid-phase form of sulfur in organic-matter-bearing marine sediments. Pyrite is cubic FeS_2. It can have a wide range of crystal morphologies (Gait, 1978; Love, 1967) although only a subset of these are recognized from sedimentary environments. Canfield and Raiswell (1991) review pyrite morphology in modern and ancient sediments. They recognize framboidal, aggregated, bladed, and equant morphologies. Framboids (from the French framboisraspberry) consist of densely packed generally spheroidal aggregates of discrete equigranular submicron-sized pyrite crystals (Love, 1967). The individual microcrysts are usually cubes or pyritohedra. Framboidal pyrite forms during the early stages of diagenesis, and retains a size distribution characteristic of its depositional environment during later stages of diagenesis (Wignall and Newton, 1998; Wilkin *et al.*, 1996). Along with framboids, the most abundant morphology of pyrite in sediments is small (generally <2 μm) single crystals (Love, 1967).

Aggregated pyrite consists of small pyrite crystals forming irregular rounded clots, which may partly coalesce. According to Canfield and Raiswell (1991), aggregates are less regular in size and shape than framboids. Bladed crystals elongate and tend to occur as (rare) fillings in chambered organisms. Equant pyrite consists of single crystals with well-developed crystal faces. These range in size from minute, micrometer-sized crystals up to large millimeter-sized varieties. The smaller-sized variety is by far the most common. Framboids and small euhedra are the dominant morphologies found in sediments. The coarser varieties are generally only recognized in ancient sediments, suggesting that they may arise during later diagenesis.

The mineral chemistry of pyrite has been extensively described by Vaughan and Craig (1978). Pyrite has a dimorph, Marcasite that also has the nominal formula FeS_2, but is orthorhombic rather than cubic. Marcasite is metastable with respect to pyrite, and will convert to pyrite over time (Murowchick, 1992). Marcasite has not been reported in recent marine sediments. Likewise, the mineral pyrrhotite ($Fe_{1-x}S$) is extremely uncommon in low-temperature sedimentary environments, although it is known from some ancient lake beds (Tuttle and Goldhaber, 1993), and in some oil-bearing strata (Goldhaber and Reynolds, 1991; Reynolds *et al.*, 1990b)

7.10.3.3 Metastable Iron Sulfides

In addition to pyrite, there are two additional iron-sulfide minerals commonly found in recent sediments. These phases have been termed metastable iron sulfides, because they are thermodynamically unstable with respect to transformation to pyrite, or to a mixture of pyrite plus pyrrhotite (Berner, 1967). They are also known as acid-volatile iron sulfides, because in contrast to pyrite they are soluble in nonoxidizing mineral acids such as HCl. The dominant acid-volatile sulfides found in both natural and experimental systems are mackinawite and greigite.

Mackinawite is a tetragonal, sulfur-deficient Fe(II) sulfide with the formula FeS_{1-x}. The initial precipitate formed during the rapid mixing of H_2S and Fe(II) is an amorphous iron sulfide that transforms very rapidly to a poorly crystalline form of mackinawite. The properties of mackinawite have been extensively reviewed (Morse *et al.*, 1987; Rickard *et al.*, 1994; Vaughan and Craig, 1978). The black color seen below the sediment–water interface in many marine and nonmarine sediments is probably largely due to mackinawite, although direct confirmation of the presence of this phase has proven difficult because it is extremely fine grained and oxidizes rapidly.

Greigite is the thiospinel of iron (Skinner *et al.*, 1964). It has the formula Fe_3S_4. It is the sulfur analogue of magnetite. Both natural and synthetic greigite are sooty black powders that are strongly ferromagnetic (Dekkers and Schoonen, 1996). Its properties have been reviewed by Morse *et al.* (1987). Magnetotactic bacteria may form greigite (Konhauser, 1998; Neal *et al.*, 2001; Posfai *et al.*, 1998, 2001). Greigite dominates the magnetic properties of some modern and even some ancient sedimentary rocks (Krs *et al.*, 1990; Reynolds *et al.*, 1990a). Laboratory synthesis of the metastable iron sulfides is described below. Both sulfides play an important role in the pathway of pyrite formation.

7.10.3.4 Organic Sulfur

In some sedimentary environments such as sapropels, sulfur formed by the assimilatory processes described above may be significant (Canfield *et al.*, 1998). In marine plankton, the observed S/C weight ratio is ~0.02, and similar ratios may be observed in surface marine sediments (Suits and Arthur, 2000). More typically, however, the diagenetic formation of organosulfur compounds from microbially produced H_2S is the dominant pathway of organosulfur formation, and can be a significant overall sink for reduced sulfur in marine sediments. S/C ratios tend to increase with depth in the sediment column (Francois, 1987). In most environments, authigenic organosulfur compounds are second only to pyrite as products of sulfur diagenesis. In some iron-poor sediments (where pyrite formation is inherently minor), the formation of organosulfur may dominate (Canfield *et al.*, 1998; Ferdelman *et al.*, 1991). For example, in diatomaceous sediments of the Peru margin, S/C weight ratios at depth in the sediment column fall in the range 0.06–0.33 (Mossmann *et al.*, 1991; Suits and Arthur, 2000). The detailed pathways of formation of specific organosulfur compounds or compound classes are complex and are outside the scope of this review, and the reader is referred to published literature (e.g., Adam *et al.*, 2000; Eglinton *et al.*, 1994; Gelin *et al.*, 1998; Kohnen *et al.*, 1991, 1989, 1990; Nissenbaum and Kaplan, 1972; Sinninghe Damste *et al.*, 1998; Vairavamurthy and Anonymous, 1993; Vairavamurthy *et al.*, 1997; van Kaam-Peters *et al.*, 1998; Wakeham, 1995; Werne *et al.*, 2000).

7.10.3.5 Elemental Sulfur

Elemental sulfur is an oxidation product of dissolved sulfide. It may form inorganically, but in near-surface marine sediments where dissolved sulfide exists close to the sediment–water interface, it commonly forms because of the metabolic activities of sulfide oxidizing bacteria. Bacterial mats of filamentous sulfur producing *Beggiatoa* spp. have been observed in the surface of nearshore marine sediments (Schimmelmann and Kastner, 1993; Troelsen and Jorgensen, 1982). Elemental sulfur can be stored within invaginations of the bacterial cell wall in the form of quasi-liquid minute droplets of metastable polysulfur ($^{-}O_3S{-}S\ldots S{-}S{-}O_3^{-}$) (Steudel *et al.*, 1987). Schimmelmann and Kastner (1993) gave evidence that this polysulfur is converted to solid elemental sulfur on the timescale of a few years.

7.10.4 MECHANISM OF PYRITE FORMATION

7.10.4.1 Evidence from Experimental Studies

Experimental investigations of pyrite formation provide important constraints on sedimentary geochemistry. Beginning with the pioneering work of Allen *et al.* (1912) and later that of Berner (1962, 1964b,c), the formation of pyrite at Earth surface conditions has been the subject of a number of detailed laboratory studies. Much of the earlier literature is reviewed by Morse *et al.* (1987). These laboratory investigations have resulted in the recognition of several potential mechanisms for formation of pyrite at $T < 100\,°C$. All these mechanisms involve redox reactions because the ultimate source of sulfur in pyrite, H_2S, is more reduced than the disulfide in pyrite. These redox reactions include:

(i) direct precipitation of pyrite from homogeneous polysulfide (S_x^{2-}) solutions:

$$Fe(HS)^+(aq) + S_x^{2-}(aq) \rightarrow FeS_2 + S_{x-1}^{2-} + H^+ \quad (10)$$

(ii) progressive sulfidation and conversion of solid iron monosulfide to pyrite by an oxidized aqueous sulfur species such as polysulfide:

$$FeS(s) + S_n^{2-} \rightarrow FeS_2(s) + S_{n-1}^{2-} \quad (11)$$

(iii) direct reaction of H_2S with a precursor solid monosulfide releasing $H_2(g)$:

$$FeS(s) + 2H_2S(aq) \rightarrow FeS_2(s) + 2H_2(g) \quad (12)$$

(iv) iron loss from a precursor solid monosulfide by an oxidation mechanism:

$$2FeS(s) + \tfrac{1}{2}H_2O + \tfrac{3}{4}O_2(aq) \rightarrow FeS_2(s) + FeOOH(s) \quad (13)$$

With the exception of Equation (10), all these pathways involve a precursor iron monosulfide phase. Direct precipitation of pyrite from solution (Equation (10)) is strongly inhibited. This is due to the difficulty of direct nucleation of pyrite, leading to very large supersaturation with respect to pyrite in experimental and natural solutions (Schoonen and Barnes, 1991a). Experimental studies have thus focused on the role of one or more iron monosulfide precursors to pyrite, which have long been recognized as intermediates in sedimentary pyrite formation (see review by Morse *et al.*, 1987). Poorly crystalline mackinawite is the initial product of reaction of H_2S with aqueous or solid iron sources at Earth-surface temperature (Morse *et al.*, 1987; Rickard, 1969). Thus, the fate of the initial mackinawite precipitate is key to understanding pyrite genesis.

Rickard (1997) and Rickard and Luther (1997) studied the direct reaction of H_2S with mackinawite (Equation (12)) and presented experimental and theoretical data in support of this pathway. In contrast, a group at Pennsylvania State University (Benning *et al.*, 2000; Wilkin and Barnes, 1996) argued that in the absence of an oxidizing agent or prior exposure to air, mackinawite is stable in the presence of H_2S over a range of temperature, time, and chemical conditions, thus implying that reaction (4) is not a dominant pathway of pyrite formation. They argued that iron monosulfide will transform to pyrite in the presence of an oxidizing agent such as dissolved O_2 and/or in the presence of an aqueous or solid sulfur species more oxidized than H_2S such as polysulfide, thiosulfate, or elemental sulfur.

Reactions involving mackinawite and an oxidized sulfur species have been repeatedly shown to lead to pyrite formation (e.g., Berner, 1969; Rickard, 1969, 1975). In addition, Wilkin and Barnes (1996) and Benning *et al.* (2000) have shown that pyrite formation is exceptionally rapid when the mackinawite is "pre-oxidized" (e.g., exposed briefly to air) prior to the experiment. Based partly on X-ray photoelectron and Auger spectroscopy results of pyrrhotite oxidation (Mycroft *et al.*, 1995), Wilkin and Barnes (1996) hypothesized that this oxidative exposure initiates an iron-loss pathway similar to Equation (13). In sulfidic solutions, Fe(II) oxyhydroxides, shown as a product in this reaction, would not accumulate, but instead would undergo reductive dissolution by a reaction similar to Equation (14):

$$4FeOOH + \tfrac{1}{2}H_2S + 7H^+ \rightarrow 4Fe^{2+} + \tfrac{1}{2}SO_4^{2-} + 6H_2O \quad (14)$$

The loss of one-fourth of the iron from mackinawite with simultaneous oxidation of one-half of the initial iron leads to formation of greigite (Wilkin and Barnes, 1996; Equation (15)). Iron loss (as opposed to sulfur gain—Equation (11)) is energetically favored, because mackinawite and greigite share the same close-packed sulfur sublattice:

$$4FeS(s) + \tfrac{1}{2}O_2(aq) + 2H^+ \rightarrow Fe_3S_4 + Fe^{2+} + H_2O \quad (15)$$

Further work may be required to define the full spectrum of reactions forming pyrite in sediments. The pathways involving oxidation such as in Equation (13) seem appropriate in the upper portions of marine sediments, where most pyrite actually forms (see below) and where oxidants such as O_2 or Fe/Mn oxides are abundant. However, transformation of iron monosulfide to pyrite also occurs in deeper sediment strata in the presence of excess H_2S. At these deeper levels, access to oxidants is more problematic, and a reaction such as Equation (12) may be required.

Greigite has been repeatedly implicated as an intermediate in pyrite formation (Benning *et al.*, 2000; Schoonen and Barnes, 1991b; Sweeney and Kaplan, 1973; Wang and Morse, 1994; Wilkin and Barnes, 1996). Evidence that the greigite to pyrite transformation can be a solid-state reaction comes from observations that intermediate stages of this reaction lead to grains with greigite cores with pyrite rims (Benning *et al.*, 2000; Sweeney and Kaplan, 1973). Greigite may also be a required intermediate in the generation of specific morphological variants of pyrite (see below).

7.10.4.2 Isotope Effects during Experimental Pyrite Formation

Experimental studies have demonstrated that inorganic reactions during pyrite formation produce minimal isotopic fractionation. Precipitation of iron monosulfide by reaction of H_2S with an iron source results in fractionations generally less than 1.5‰ (Bottcher *et al.*, 1998; Price and Shieh, 1979), and subsequent transformation of monosulfide to pyrite shows similarly small fractionations (Price and Shieh, 1979).

7.10.4.3 Origin of Morphological Variations in Pyrite

As noted above, pyrite in sediments occurs in a variety of morphologies. Of particular interest, however, is the origin of framboids. This unusual morphology is of very widespread occurrence in sediments, but has been surprisingly difficult (although not impossible) to synthesize in the laboratory (Berner, 1969; Butler and Rickard, 2000; Farrand, 1970; Sweeney and Kaplan, 1973). The most convincing explanation of the origin of framboids is based on the strongly ferromagnetic properties of greigite. Framboids may arise by aggregation of uniformly sized magnetic greigite microcrystals to form greigite framboids, followed by replacement of greigite by pyrite (Wilkin and Barnes, 1997).

More generally, pyrite morphology is a function of the degree of supersaturation of the solution (Murowchick and Barnes, 1987; Wang and Morse, 1994). At very low degrees of supersaturation and mineral growth, ions are added to the growing surface in the most energetically favorable sites. For pyrite, the lowest energy morphology is the cube. Very rapid growth from highly supersaturated solutions is characterized by low selectivity for the site of addition of ions. This leads to spherulitic forms. Wang and Morse (1994) recognized the following sequence of pyrite morphologies as a function of solution supersaturation:

Cube	Combination of cube and octahedron	Octahedron	Spherulite
		Increasing pyrite supersaturation →	

7.10.5 SULFUR DIAGENETIC PROCESSES IN MARINE SEDIMENTS

7.10.5.1 Depth Distribution of Diagenetic Sulfur Products

Diagenetic processes in sediments occur over varying periods of time. For nearshore sediments, sulfur diagenetic processes may occur over months to centuries, whereas in deep-sea sediments, similar reactions may occur over times exceeding 10^6 yr. For this entire spectrum of sedimentary environments, a useful concept is steady state. This is a time-invariant distribution of diagenetic products. At steady state, pore-water and solid-phase profiles do not change as sediment accumulates. Achieving this steady-state condition requires that a large number of variables such as sedimentation rate, organic matter deposition, bottom-water conditions, and many others are held constant. The discussions that follow are predicated upon this concept except where noted.

There are three major fates for microbiologically produced aqueous sulfide. Listed in order of their relative importance, these are: loss to the overlying water by oxidation, fixation during formation of pyrite, and fixation by formation of organosulfur compounds. In addition, two minor products of sulfur diagenesis, elemental sulfur and metastable iron sulfides, tend to be transient and are most commonly recognized in the uppermost part of the sulfate reduction zone. These conclusions are based on a very large number of studies of the diagenesis of sulfur in a variety of depositional settings (Table 3). Sulfate reduction has been recognized from the most rapidly deposited nearshore environments, to portions

Table 3 Studies on sedimentary-sulfur geochemistry for selected depositional settings.

Location	*Water depth* (m)	*S parameters measured*[a]	*References*
Great Marsh (Delaware USA Salt Marsh)	0–1	Spy, Sorg, S^0, AVS	Ferdelman *et al.* (1991)
Limfjorden, Kysing Fjord, Denmark	0.5–12	SO_4, H_2S, Spy, AVS, SRR	(Detmers *et al.*, 2001; Howarth and Jorgensen, 1984; Jorgensen, 1977b)
Kysing Fjork Denmark	<1	SO_4, H_2S, S^0, Spy, AVS, $\sum S$, SRR	(Fossing and Jorgensen, 1990)
Cape Lookout Bight, North Carolina	8	SO_4, H_2S, $\sum S$, AVS, δSO_4, δSpy	(Chanton *et al.*, 1987a,b)
Long Island Sound, USA	8–34	SO_4, H_2S, Spy, AVS, SRR	(Aller, 1980a; Goldhaber *et al.*, 1977)
Gulf of Mexico outer shelf	20–112	SO_4, $\sum S$	(Filipek and Owen, 1980)
Baltic Sea	100–200	$\sum S$, Spy, Sorg, S^0, δSO_4, δSpy, SRR	(Hartmann and Nielsen, 1969; Sternbeck and Sohlenius, 1997; Yu *et al.*, 1982)
Peru Margin	252–457	SO_4, H_2S Spy AVS, Sorg, S^0	(Mossmann *et al.*, 1991; Suits and Arthur, 2000)
Southern California Borderland	100–1,500	SO_4, H_2S, Spy, S^0 ,Sorg, AVS, δSpy	(Kaplan *et al.*, 1963; Schimmelmann and Kastner, 1993; Sweeney and Kaplan, 1980)
Jervis Inlet, British Columbia	650	SO_4, H_2S, S^0, Spy, Sorg, δSH_2S, δSorg	(Francois, 1987)
Gulf of California	476–3,361	SO_4, H_2S, $\sum S$, Spy, AVS, δSO_4, δSH_2S, δSpy	(Berner, 1964a; Goldhaber and Kaplan, 1980)
Arabian Sea	>500	$\sum S$, Spy, δSpy	(Lueckge *et al.*, 1999; Passier *et al.*, 1997; Schenau *et al.*, 2002)
Black Sea	100–2,000	$\sum S$, Spy, Sorg, S^0, δSO_4, SRR	Various including (Jorgensen *et al.*, 2001; Lyons, 1997; Lyons and Berner, 1992; Wilkin and Arthur, 2001) among many others
Northeast Atlantic	158–4,920	SO_4, SRR	(Battersby *et al.*, 1985)
Cariaco Basin	900–1,400	$\sum$Spy, δSpy,	(Lyons *et al.*, 2003; Werne *et al.*, 2000)
Western North Atlantic	1,291–4,595	$\sum S$, Spy, SO_4	(Cagatay *et al.*, 2001)
New Jersey Margin; DSDP Hole 603	4,796	$\sum S$, Spy, δSpy	(Bonnell and Anderson, 1987; Dean and Arthur, 1987)
Atlantic Ocean Basin	Various	$\sum S$, δSpy	(Lew, 1981)

[a] $\sum S$: content of total solid-phase sulfur. Spy: pyrite sulfur content. δSpy: isotopic composition of pyrite. Sorg: organic sulfur. S^0: elemental sulfur. SO_4: pore-water sulfate concentration. δSO_4: isotopic composition of pore-water sulfate. H_2S: pore-water sulfide concentration. δSH_2S: isotopic composition of pore-water sulfide. SRR: sulfate reduction rate.

of the central ocean basins (Canfield, 1991; D'hondt *et al.*, 2002; Jorgensen, 1982). Despite this range in environmental settings, it is possible to generalize from the large number of studies to recognize overall similarities in the diagenetic pathway. Specifically, chemical analyses of sediments and pore fluids from core samples from a variety of marine settings indicate that there is a characteristic distribution of reactions and products with depth (time). Diagenetic reactions occur in a series of depth zones. Figure 4 shows these zones schematically. The uppermost zone (I) is nearly continuously oxic. The next zone (II) is strongly influenced by the life activities of macroinfauna (bioturbation). Zone III is characterized by evolving pore-water profiles of sulfur species (decreasing SO_4^{2-} and variable H_2S). The deepest zone (IV) is characterized by the near absence of diagenetic sulfur processes. Note that this section focuses only on reactions involving sulfur geochemistry, whereas as shown in Figure 4 there are competing reactions involving other electron acceptors. In some sediments, these other acceptors, particularly the iron and manganese oxides, may dominate organic remineralization (Aller *et al.*, 1986; Canfield *et al.*, 1993), although more commonly sulfate dominates over all but aerobic processes. It should also be noted that Figure 4 shows an idealized steady-state sequence. Individual zones may be absent or modified in certain circumstances. An obvious example is that zones I and II, which are directly or indirectly related to the presence of oxygen, are absent in anoxic depositional environments, and zone II may be absent or insignificant under dysoxic (low-oxygen) depositional waters.

Zone I is characterized by the presence of oxygen in the pore waters. This aerobic zone may extend a few millimeters into nearshore organic-matter-rich sediments and considerably deeper in hemi-pelagic sediments. In most environments, this upper layer is extensively mixed by the activities of macroorganisms (bioturbated) and probably stirred by waves and currents as well. Because of this activity, pore-water SO_4^{2-} concentrations are essentially those of the overlying seawater. Iron oxides are abundant, giving these sediments a brown color. Manganese oxides may be present as well. Free H_2S is absent due to its loss by aerobic oxidation discussed above and/or buffering of H_2S to very low concentration by reaction with iron oxides (Goldhaber and Kaplan, 1974). Elemental sulfur may be present (Troelsen and Jorgensen, 1982). Within this zone (or the next), assimilatory sulfur may be lost by hydrolysis reactions. Surprisingly, pyrite may also be present in this zone, 10% or more of the amount found at depth. This pyrite may form in anaerobic "microenvironments" (Jorgensen, 1977a) within the overall oxic setting. This possibility is reinforced by measured sulfate reduction rates in zone I surface sediment, which indicate substantial activity by SRB (Canfield, 1993; Jorgensen, 1977a; Jorgensen and Bak, 1991). SRB can withstand exposure to O_2 for extended time periods and remain viable (Cypionka *et al.*, 1985). It is also likely that some portion of the shallow pyrite may have actually formed in deeper

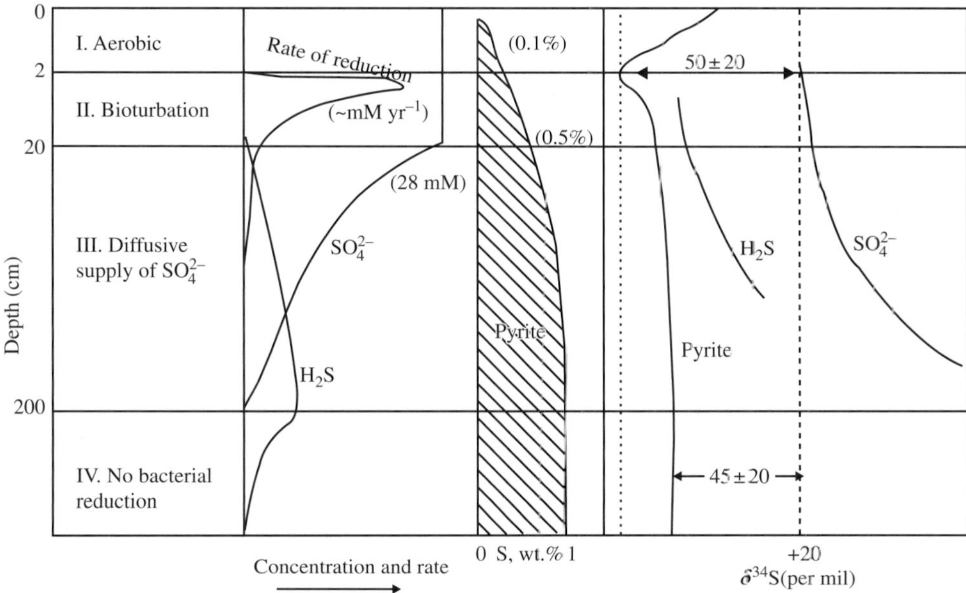

Figure 4 Idealized depth dependence of pore-water and solid-phase sulfur constituents and their sulfur isotopic values. The zones labeled I, II, III, and IV are classified according to processes of transport of sulfur and redox processes within each zone. See text for discussion.

layers and then transported to the surface through biogenic mixing processes. However, oxidation by a variety of reactants, chiefly O_2, but also NO_3^- and MnO_2, is the primary fate for reduced sulfur in this zone (Luther *et al.*, 1982; Schippers and Jorgensen, 2001, 2002; Thamdrup *et al.*, 1994b).

Zone II is the zone of bioturbation. Near the top of this interval, sulfate reduction rates are at their maximum and decrease with depth, iron monosulfide phases reach a maximum and then decline, whereas pyrite increases systematically with depth reflecting net fixation of reduced sulfur. If the relative rates of sulfate reduction are less than sulfate resupply by the activities of macrofaunal organisms, sulfate concentrations may differ only slightly from bottom-water values due to mixing and irrigation activities of macroorganisms (Aller, 1980b, 1988; R. C. Aller and J. Y. Aller, 1998; Goldhaber *et al.*, 1977; Goldhaber and Kaplan, 1980). Zone II may be nearly fully open to addition of overlying seawater sulfate, depending upon the relative rates of bioturbation and sulfate reduction. Extensive reoxidation of solid-phase pyrite and iron monosulfides also occurs within this zone (Berner and Westrich, 1985; Sommerfield *et al.*, 2001). H_2S is generally not measurable (<1 μM) but is occasionally present in very low concentrations <60 μM (Goldhaber and Kaplan, 1980). In either case, the low values reflect oxidation, buffering by reaction with sedimentary iron, and biogenic mixing with sulfide-free overlying water. The net effect of extensive reoxidation of reduced-sulfur phases is that only a fraction of the total sulfate reduced to sulfide is retained within the sediments. Nonetheless, most pyrite is typically formed in the upper two zones (Berner, 1970; Hartmann and Nielsen, 1969; Mossmann *et al.*, 1991).

The high rates of sulfate reduction in this zone reflect a number of impacts of macrofaunal activity, many of which enhance the rate and/or extent of organic decomposition. These effects are summarized in Table 4 (after R. C. Aller and J. Y. Aller, 1998). These authors give a qualitative evaluation of the impact of bioturbation on diagenetic processes involving organic matter.

Zone III is the zone of diffusive transport. Sulfate in this interval decreases with depth, as does the rate of sulfate reduction. The steepness of the SO_4^{2-} gradient reflects the balance between rate of sulfate removal by SRB and rate of sulfate addition by diffusion plus a smaller contribution from net burial of pore water. Zone III is only partially open to sulfate addition from overlying seawater. The steepness of the sulfate gradient varies dramatically as a function of depositional setting. In the most organic-matter-rich rapidly deposited sediments (e.g., near man-made sewage outfalls), SO_4^{2-} may be totally removed from pore waters within the uppermost few centimeters. Where evaporites or migrating sulfate-bearing brines occur at depth, sulfate may be supplied from below (e.g., Kastner *et al.*, 1990). More typically, in nearshore sediments with organic carbon contents in the range 1–3 wt.%, total SO_4^{2-} removal occurs between a depth of several tens of centimeters to a few meters. In hemi-pelagic sediments with a few tenths of a percent organic carbon, total sulfate removal may require hundreds of meters or not occur at all. The steepness of the SO_4^{2-} gradient is related to the overall sedimentation rate (Berner, 1978; Canfield, 1991; Goldhaber and Kaplan, 1975), because sedimentation rate, in turn, controls the metabolizability of organic matter and hence reduction rate (see below). The sulfur isotopic composition of SO_4^{2-} in this zone increases systematically with depth as the light isotope is preferentially removed to form H_2S.

Table 4 Macrofaunal effects on organic carbon decomposition/remineralization.

Macrofaunal activity	*Effect*	*Effect on rate/extent of organic matter decomposition*
Particle manipulation	Substrate exposure, surface area increase	+
Grazing	Microbe consumption, bacterial growth stimulation	+
Excretion/secretion	Mucus substrate, nutrient release, bacterial growth/stimulation	+
Construction/secretion	Synthesis of refractory or inhibitory structural products (tube linings, halophenols, body structural products)	−
Irrigation	Soluble reactants supplied, metabolite buildup lowered, increased reoxidation	+
Particle transport	Transfer between major redox zones, increased reoxidation, redox oscillation	+

Reproduced by permission of USGS from *J. Marine Res.*, **1998**, *56*, 905–936.

Dissolved sulfide in this zone builds up because of its bacterial production. Its maximum concentration is less than the total sulfate reduced because a portion of the H_2S reacts with iron and organic matter to form insoluble products. At any depth, the concentration gradient of H_2S is kinetically controlled and reflects the balance between the rate of these removal processes, the rate of gain or loss by diffusion, and the rate of its formation by reduction. One of these processes, the production of H_2S, must cease when all SO_4^{2-} has been consumed. The net result is a concentration maximum that falls in a range from 1 μM to >10 mM. The depth of maximum pore-water H_2S commonly correlates closely with the depth of total SO_4^{2-} depletion. In most environments, H_2S persists at measurable concentrations (i.e., greater than a few micromolar) in pore waters to depths of a few centimeters to several meters below the point at which SO_4^{2-} is removed. The essentially total removal of pore-water H_2S is a reflection of the availability of excess iron over sulfide sulfur in most sediments (see below). Pyrite content may increase gradually within zone III, but the rate of this increase is most rapid at the top of this zone. Frequently, increases in pyrite cannot confidently be distinguished from scatter in the data within this zone.

Zone IV lies below the depth of removal of pore-water H_2S. The absence of sulfur as sulfate or sulfide in pore waters precludes further pyrite formation, and signals the cessation of the typical early diagenetic sulfur reactions. However, during deeper burial and even after lithification, many additional processes can impact sedimentary-sulfur geochemistry. With increasing temperature, release of H_2S from organosulfur compounds takes place. As diagenesis passes into metamorphism, pyrite converts to pyrrhotite plus H_2S (Ohmoto and Goldhaber, 1997). Both these reactions provide additional sulfide for late pyrite formation. In addition, some sedimentary environments are open to migration of external fluids that may be H_2S bearing. Marine mud interbedded sandstones are candidates for addition of epigenetic fluids. In extreme cases, whole geographic regions may be impacted by migrating sulfide- (and metal)-bearing fluids (Goldhaber *et al.*, 2002, 1995). The result of these processes may range from the development of coarse overgrowths on earlier diagenetic pyrite, growth of sulfide nodules and layers, all the way to development of massive sulfide-ore bodies with tens of millions of tons of metal sulfide precipitates.

7.10.5.2 Rates of Sulfate Reduction

Sulfate reduction rates have been extensively quantified in marine sediments. There are a number of methods for measuring sulfate reduction rates in marine sediments, although addition of radioactive sulfate ($^{35}SO_4^{2-}$) and recovery of radiolabled ^{35}S reduced sulfur compounds (Jorgensen, 1978; Thode-Anderson and Jorgensen, 1989) is the most commonly utilized technique in coastal sediments. Incubation of sediments over time has also been successfully employed (Aller and Mackin, 1989). Mathematical modeling of sulfate profiles is an alternative that is widely utilized, particularly in deep-sea sediments, where sulfate reduction rates are too low for ratio-tracer studies (Canfield, 1991), and a combination of direct measurement in upper sediment layers (where rates are high) and modeling in deeper layers may be appropriate in many situations (Jorgensen *et al.*, 2001). A striking result of these rate measurements is the wide range of sulfate reduction rates that are recognized. Figure 5, from Canfield (1989b), illustrates this point. Sulfate reduction rates in this plot, integrated over the depth of the sediment column, vary by nearly seven orders of magnitude. It is important to note that these reduction rates are proportional to (scale with) sedimentation rate. Sedimentation rate is an important master variable for sulfate reduction rate. With decreasing sedimentation rate, the exposure of organic matter to aerobic oxidation during transit through the water column, as well as metabolism by macro- and microorganisms increases relative to rapidly deposited inshore sediments (Berner, 1978; Canfield, 1991; D'honct *et al.*, 2002; Goldhaber and Kaplan, 1975).

7.10.6 SULFUR ABUNDANCE IN RECENT MARINE SEDIMENTS

7.10.6.1 Controls on Sulfur Abundance

Figure 6 is a histogram of the total sulfur concentration in weight percent in a cross-section of normal marine siliciclastic sedimentary environments (i.e., exclusive of sediments deposited under anoxic conditions, and those dominated by a carbonate matrix). The data come from a range of depths in the sediment column. As noted above, the dominant form of sulfur in these sediments is pyrite, typically >80%. The total sulfur analyses in Figure 6 have a mean of ~0.6%, with very few analyses registering valuesgreater than 2%. The measured total sulfur abundances displayed in Figure 6 are greater than what would form from the simple burial of pore-water sulfate if the accumulating sediment pile were closed to additional SO_4^{2-} input. This "closed system" formation of pyrite would amount to only 0.1–0.3% by weight of pyrite sulfur (depending upon the porosity of the sediments). Addition of sulfur

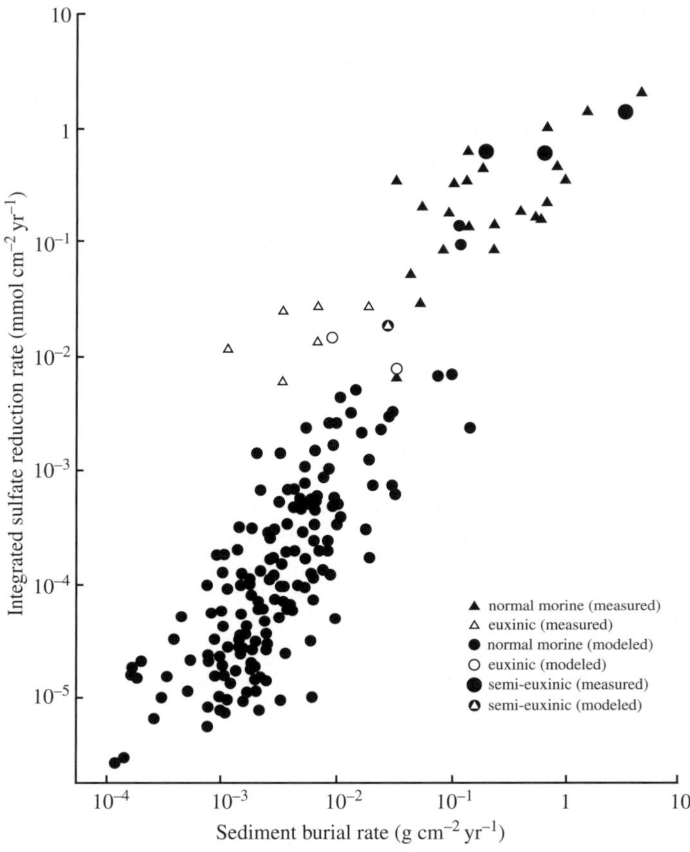

Figure 5 Rate of SO_4^{2-} reduction integrated over the entire depth of the sulfate profile plotted versus sedimentation rate (Canfield, 1989b) (reproduced by permission of Elsevier from *Deep-Sea Res. I*, **1989b**, *36*, 121–138).

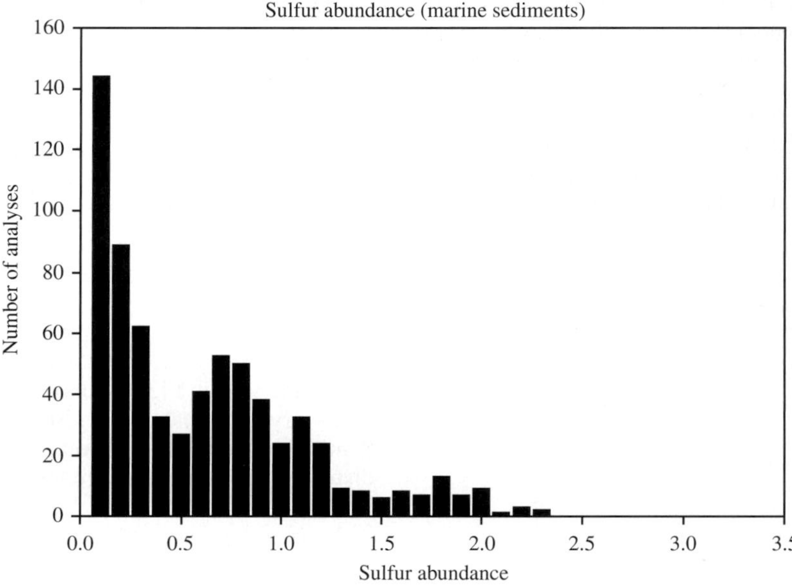

Figure 6 Histogram showing sulfur abundance in marine sediments. Data are from a variety of sources but exclude euxinic environments.

in excess of this closed system amount results from advection of sulfate during bioturbation and diffusion of sulfate from the overlying water.

Three variables are of potential importance in controlling the total amount of pyrite formed. These are the availability of: sulfate, iron, and organic matter. The effect of the first is relatively minor in marine sediments, whereas last two are more critical.

Sedimentation rate would, at first glance, seem to be an important control of total sulfur incorporation into marine sediments in view of the correlation between this parameter and sulfate reduction rate (Figure 5). This is not the case (Goldhaber and Kaplan, 1975; Volkov and Rozanov, 1983). The reason is that the total sulfate reduced (sulfide produced) is equal to the product of the sulfate reduction rate and the time over which the reduction occurs. The slower reduction rates occur for a longer time in more slowly deposited sediments, and the total sulfate reduced is not as dramatically affected. An exception is in very slowly deposited pelagic sediments, which lack adequate metabolizable organic matter to support SRB. Thus, red pelagic oozes and other open-ocean sediments do not typically contain appreciable amounts of diagenetic sulfides.

As discussed above, sulfate concentration in the range found in marine pore water has essentially no effect on sulfate reduction rates until very low sulfate concentrations are reached. It is also true that most pyrite forms before pore-water sulfate concentrations reach such low levels (Figure 4). Thus, the influence of sulfate on the abundance of sedimentary pyrite is relatively minor in marine depositional settings. Reinforcing this conclusion are studies of modern marine environments with somewhat reduced salinities such as portions of the Sea of Azov (salinity 11–15 per mil), and the Baltic Sea (salinity 10–13 per mil). Sediments from these settings have total reduced-sulfur contents ranging well over 1% by weight (compare with Figure 6 for normal marine sediments), and thus show no evidence of reduced pyrite contents (Volkov *et al.*, 1983). Of course, in freshwater settings with very low sulfate concentrations, sedimentary-sulfur accumulation is severely limited by sulfate availability (Berner and Raiswell, 1984).

Iron availability may be a limitation on pyrite formation in many marine environments. However, the situation is complex, because iron availability is a relative concept that depends on a competition between rate of H_2S addition and rate of iron mineral reaction with H_2S. Various iron oxides (e.g., ferrihydrite, lepidocrocite, goethite, hematite, and magnetite) and silicates (e.g., chlorite, biotite, amphiboles, etc.) supply iron for pyrite formation. However, the dominant sources of iron for pyrite formation are the amorphous or poorly crystalline iron oxides (Canfield *et al.*, 1992; Goldhaber and Kaplan, 1974; Morse and Wang, 1997; Pyzik and Sommer, 1981). Table 5 shows a series of rate constants and half-lives compiled by Raiswell and Canfield (1996), which illustrate the wide spectrum of reaction rates of H_2S (assumed to be at a concentration of 1 mM) with iron minerals (see also Morse and Wang, 1997). Iron oxides such as ferrihydrite, lepidocrocite, and goethite react so rapidly that excess H_2S is removed from solution to levels below 1 μM on diagenetic timescales. Dissolved sulfide does not accumulate in pore waters until these reactive oxides are removed. At the other extreme, some iron silicates react with half-lives of over 10^5 yr. As discussed above, the exposure age of marine sediments to diagenetic reactants and products including H_2S may range from tens of years to over a million years. At the short-exposure end of this spectrum, iron-bearing silicate minerals will be essentially unreactive, whereas at the long-exposure end, the iron in such silicate minerals is at least partially converted to pyrite (Raiswell and Canfield, 1996).

A useful parameter tied to the reactivity of iron is the degree of pyritization (DOP) of iron. This parameter (Equation (16)) has been widely measured in both modern and ancient sediments

Table 5 Rate constants and half-lives of sedimentary iron minerals with respect to their sulfidation.

Iron mineral	*Rate constant* (yr^{-1})	*Half-life*
Ferrihydrite	2,200	2.8 h
Lepidocrocite	>85	<3 d
Goethite	22	11.5 d
Hematite	12	31 d
Magnetite (uncoated)	6.6×10^{-3}	105 yr
"Reactive silicates"	3.0×10^{-3}	230 yr
Sheet silicates	8.2×10^{-6}	84,000 yr
Ilmenite, garnet, augite, amphibole	$\ll 8.2 \times 10^{-6}$	$\gg$ 84,000 yr

Reproduced by permission of USGS from *Am. J. Sci.*, **1992**, *292*, 659–683.

(Raiswell, 1993; Raiswell *et al*., 1988):

$$\text{DOP} = \text{pyrite Fe}/(\text{pyrite Fe} + \text{``reactive iron''}) \quad (16)$$

Berner (1970) pointed out that the reservoir of iron for pyrite formation may be smaller than the total sediment iron concentration, because some iron-bearing phases are essentially unreactive towards H_2S on the timescale of sediment diagenesis. He operationally defined "reactive" (towards H_2S) iron as that dissolved during a short exposure to hot 12N HCl (Berner, 1970). However, a further analysis of the reactive iron issue has shown that the concentrated HCl technique removes a considerable amount of iron that is unreactive to H_2S on diagenetic timescales. More gentle treatments using dithionite (Canfield, 1989a) and 1N HCl (Leventhal and Taylor, 1990) remove an iron fraction that more closely corresponds to the reactive iron oxides. The highly reactive iron in sediments is thus defined as the sum of dithionate or 1N HCl soluble iron plus that tied up in pyrite plus acid volatile sulfur (AVS). Thus defined, the fraction of highly reactive iron in a wide range of nearshore and deep-sea sediments from aerobic environments, as well as sediments accumulating in dysaerobic environments, is low (based on 219 analyses, average 25–28% of the total iron; 1σ uncertainly $\pm 10\%$).

Carbonate and opaline-silica-rich sediments are exceptions to these generalizations. They are very low in detrital iron, because calcareous and siliceous skeletal debris is much lower in iron than terrigenous material. In these dominantly biogenic sediments, iron may become limiting; the degree of pyritization is very high (>80%) and values tend to be constant. Plots of either total or reactive iron versus total sulfur generally result in a linear relationship. In these low iron sediments, pore-water H_2S builds up to very high concentrations (Thorstenson and Mackenzie, 1974). These low iron, high H_2S environments are favorable for the formation of high relative proportions of authigenic organosulfur compounds.

The primary control on the abundance of sedimentary sulfur in many normal marine siliciclastic marine sediments is related to the abundance of organic matter. This conclusion is based, in part, on a linear relationship between organic carbon and total sulfur that is commonly (although not always) observed for normal marine sediments (Figure 7). This correlation was first noted by Sweeney (1972) (see also Goldhaber and Kaplan, 1974). This important relationship has subsequently been examined by a number of researchers (Berner, 1982; Leventhal, 1983a,b, 1995; Lin and Morse, 1991; Morse and Berner, 1995; Raiswell and Berner, 1985, 1986) among others, and has been verified for Phanerozoic normal marine sediments (Raiswell and Berner, 1986). Berner (1982) determined that the slope of the C/S line for normal marine siliciclastic sediments is 2.8 ± 0.8. In effect, nearly all marine siliciclastic sediments deposited in oxic environments fall within a range of a factor of 2 in C/S ratio. An analysis of the origin of this correlation (Morse and Berner, 1995; Sweeney, 1972) shows that three factors controlling this ratio are closely coupled to each other. These are (Morse and Berner, 1995): (i) the fraction of the total organic carbon deposited that is metabolized; (ii) the fraction of metabolized organic carbon that is metabolized by SRB; and (iii) the fraction of

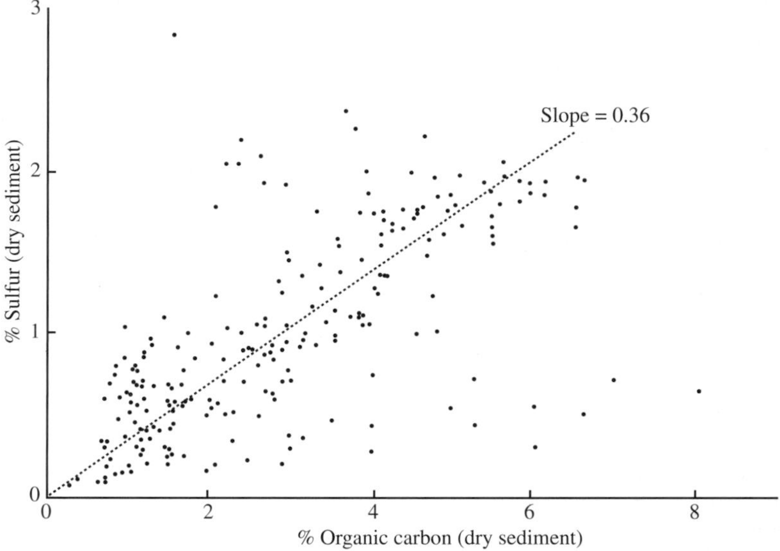

Figure 7 Plot of $\sum$S versus organic carbon (after Sweeney, 1980).

the total sulfide formed by SRB that is buried as pyrite. It is truly remarkable that these factors either remain constant or scale with each other in a way that allows the observed constancy of the C/S ratio.

Tim Lyons (written communication 2002) has suggested that these three factors do scale with each other, and has provided a rationale for understanding the relation between carbon and iron limitation. He proposes that the Raiswell and Canfield (1998) paper defines a window for reactive iron. We can expect concentrations of pyrite up to ~1–2 wt.% where there is enough highly reactive iron present such that total organic carbon may be limiting. Note that in Figure 6, total sulfur in many marine sediments fall at or below this concentration range of total sulfur. He further suggests that in oxic settings where a fixed C/S ratio is observed, the required scaling of factors controlling the constancy of the C/S ratio is due to the coupling of iron and organic-carbon delivery to sediments. In offshore environments, this coupling may occur through stimulation of surface productivity (and hence organic carbon transport to sediments) by iron. In nearshore environments, that coupling may be through direct adsorption of organic matter to iron-bearing detritus (Keil and Hedges, 1993)

7.10.6.2 Consequences of Non-steady-state Diagenesis and Impact of Euxinic Environments

There are some environments or diagenetic settings where exceptions occur to the above generalizations on abundance and depth distribution of sedimentary sulfur. Dominant among these are sites with prominent non-steady-state effects. These effects may arise from periods of nondeposition or very rapid deposition (e.g., turbidities). Combinations of these processes may lead to iron-oxide-rich layers interlaminated with organic-rich layers. The iron-oxide-rich layers will be sulfidized by H_2S diffusing from the organic-rich sediments resulting in very small C/S ratios in the iron-rich layers (Passier *et al.*, 1999). A related situation is marine sediments overlying nonmarine sediments. The nonmarine sediments would not have been subjected to sulfidization reactions because of low bottom-water sulfate concentrations, and thus would potentially contain high concentrations of reactive iron. Again, diffusion of H_2S into the underlying sediments can lead to extreme values of C/S ratio. This type of non-steady-state effect is relatively common and must be borne in mind when applying steady-state diagenetic concepts such as those pictured in Figure 4.

Another example of an atypical environment is when the overlying depositional waters contain H_2S (i.e., euxinic environments). These euxinic settings have been very extensively studied. In part, this is because such settings, though rare at present, may have been much more extensive (at times) in the past (Arthur and Sageman, 1994; Berry and Wilde, 1978; Canfield, 1998). By far the most work has been done on the Black Sea (Berner, 1974; Calvert *et al.*, 1996; Canfield *et al.*, 1996; Jorgensen *et al.*, 2001; Leventhal, 1983a; Lyons, 1997; Lyons and Berner, 1992; Lyons *et al.*, 1993; Morozov, 1995a,b; Ostroumov, 1953; Weber *et al.*, 2001; Wijsman *et al.*, 2001a,b; Wilkin and Arthur, 2001). These studies show that pyrite formation in modern Black Sea sediments is iron limited. It is also evident that most of the pyrite in the sediments forms in the water column, although sulfate reduction does continue within the sediments.

7.10.7 SULFUR ISOTOPE SYSTEMATICS OF MARINE SEDIMENTS

7.10.7.1 Overview

A characteristic feature of sedimentary sulfide in recent marine sediments is enrichment in the light isotope of sulfur. This enrichment is largely a result of the isotopic fractionation introduced during sulfate reduction by SRB. However, the isotopic composition of sedimentary sulfide is frequently found to be lighter than predicted based on the documented fractionation by SRB discussed above. Thus, additional processes beyond single step kinetic reduction of SO_4^{2-} to H_2S are required to explain the data.

The isotopic composition of sedimentary sulfide is summarized in Figure 8, which is a histogram of isotope data for solid-phase forms of sulfur from marine environments. The information is generally true for pyrite sulfur although in some cases metastable iron-sulfide minerals are included as well. The data are exclusive of anoxic depositional environments. The mean value of the isotope data is approximately −18‰ and the median is −17.3‰, but the data span a very wide range, nearly +10‰ to −60‰. The mean value of −18‰ represents enrichment in ^{32}S of ~40‰ compared with seawater SO_4^{2-}, which for the modern ocean has a $\delta^{34}S$ value of about +20‰. For comparison, the equilibrium isotopic fractionation between SO_4^{2-} and H_2S is 75‰ at 25 °C. The large average isotopic separation between sulfate and pyrite ($\Delta_{SO_4-FeS_2}$) is noteworthy. Many of the analyses represent fractionations greater than 45‰, which is the maximum observed value for SRB. Some $\Delta_{SO_4-FeS_2}$ values exceed 70‰.

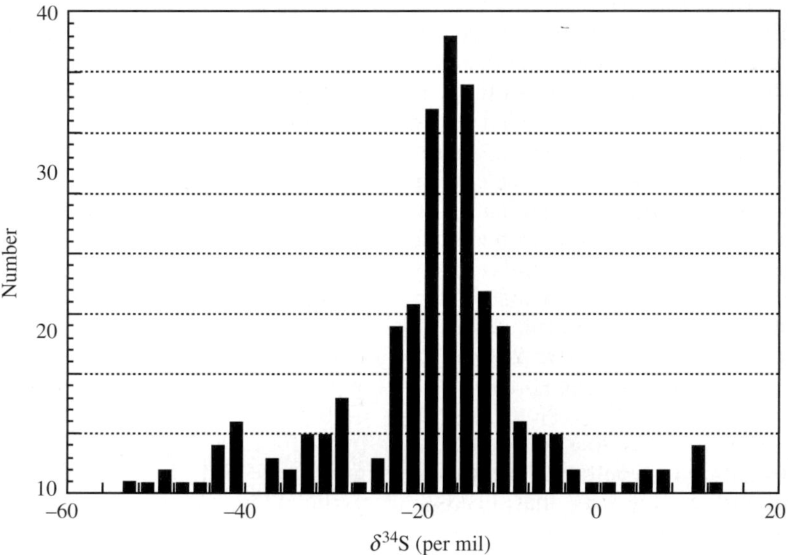

Figure 8 Histogram showing the distribution of sulfur isotopes in marine sediments (after Ohmoto and Goldhaber, 1997).

The overall widespread of $\delta^{34}S$ values shown in Figure 8, with variable but generally large negative fractionation compared to seawater sulfate, is a characteristic of sedimentary sulfur formed as a consequence of bacterial sulfate reduction. Furthermore, a wide range of isotopic values may also occur in a single depositional environment. This variability can be expressed as isotopic differences occurring between pyrite grains or size fractions that presumably formed during different stages of diagenesis (Canfield *et al.*, 1992; Goldhaber and Kaplan, 1974; Kohn *et al.*, 1998). Kohn *et al.* (1998) recognized isotopic differences of up to 35‰ between grains in the same sediment, and up to 15‰ within a single pyrite grain. Where temporal trends are observed (i.e., from core to rim in an overgrowth sequence), the most commonly observed direction is towards heavier isotopic values with increasing relative age in all these studies. This is the expected trend given the evolution of sulfur isotopes towards heavier values with depth in the sediment column (Figure 4). Isotopic variability is also commonly observed with depth in the total pyrite sulfur. The variability with depth in sediments can, in part, be due to ongoing diagenesis, but in many cases it is due to non-steady-state diagenetic effects such as depositional hiatuses, turbidite input, changes in the flux of organic carbon or iron to the sediment, etc.

For steady-state sedimentary processes, the isotopic composition of recent sedimentary sulfur and the reason for its large range of $\delta^{34}S$ values is primarily controlled by three factors: (i) the magnitude of the instantaneous bacterial isotopic fractionation factor between SO_4^{2-} and H_2S; (ii) isotope effects associated with cyclic oxidation– reduction reactions; and (iii) the mechanism of addition of sulfur to sediments. The last two of these factors, and possibly the first as well, are depth dependent. For this reason it is logical to divide this discussion by diagenetic zone.

7.10.7.2 Isotopic Processes in the Upper (Bioturbated) Sediment Regime

Bioturbation and other physical processes associated with the upper portions of marine sediments may lead to rapid exchange between pore-water and overlying depositional water. Depending on the intensity of bioturbation, sulfate in depth zones I and II and the uppermost part of zone III (Figure 4) may be effectively in contact with an infinite reservoir of seawater sulfate. When this is the case, pore-water SO_4^{2-} will have a nearly constant $\delta^{34}S$ value with depth regardless of the withdrawal of isotopically light sulfur to form H_2S. The initial isotopic composition of H_2S produced by SRB in zones I and II will be equal to the instantaneous isotopic separation between seawater sulfate and bacterial sulfide (i.e., up to about $\Delta_{SO_4-H_2S} = 45‰$). Metastable iron sulfides and pyrite formed from this H_2S will have an isotopic composition very close to this initial H_2S because of the small fractionation observed during sulfidization of iron minerals.

Characteristically, a high proportion of the aqueous and solid-phase sulfide produced in the upper sediment layer is reoxidized owing to the availability of oxidants such as oxygen as well as iron and manganese oxides and to the vigor of advective processes. The research summarized above has stressed that a portion of the reoxidation leads not only to direct loss of SO_4^{2-}, but also to cyclic oxidation–reduction reactions. It is now believed (see below) that these cyclic oxidation–reduction reactions are required to explain the large isotopic differences (i.e., $\Delta_{SO_4-FeS_2} > 45‰$) that are commonly observed in marine sediments. This cyclic redox mechanism contrasts with a previous proposal that the large values of $\Delta_{SO_4-FeS_2}$ arose because microbial metabolism in sediments occurred at slower rates than were reproduced in laboratory culture studies (Goldhaber and Kaplan, 1975). The evidence supporting the cyclic redox pathway consists of extensive studies of the isotopic fractionation exhibited during bacterial reduction (see above). Similar studies document that disproportionation of $S_2O_3^{2-}$ and other intermediate sulfide oxidation products can result in significant isotopic fractionation (Canfield and Thamdrup, 1994; Habicht *et al.*, 1998). For example, the disproportionation of $S_2O_3^{2-}$ (Equation (9)) can produce isotopic fractionations between product SO_4^{2-} and H_2S of between 3‰ and 15‰. Equivalent reactions of sulfite can produce fractionation of 28‰ (see e.g., Habicht and Canfield (2001) for a summary). Furthermore, direct measurements of isotopic fractionation processes in sediments have been conducted. These studies support the concept that oxidation of isotopically light H_2S will produce isotopically light intermediates such as $S_2O_3^{2-}$ or SO_3^{2-}. Subsequent microbial processing of these intermediates by reduction (e.g., Equation (8)) or disproportionation (e.g., Equation (9)) will lead to sulfide (and thus potentially iron–sulfide minerals) that are incrementally lighter than the original H_2S produced by SRB. Cyclic redox processes of this type have previously been proposed in the context of the formation of sedimentary ore deposits (Granger and Warren, 1969; Reynolds and Goldhaber, 1983)

Habicht and Canfield (2001) provide evidence that the pyrite forming in marine sediments is influenced by such a pathway by simultaneously determining the isotopic composition of H_2S produced by SRB and that of iron-sulfide minerals coexisting in the same sediments. Their measured bacterial fractionations could only explain 41–85% of the observed ^{34}S depletion in iron sulfides from a range of marine environments. They calculated a closed-system model showing that microbial-disproportionation fractionations involving partially oxidized sulfur species, as documented in the laboratory, could explain their data. Their calculations assumed that all H_2S initially produced by SRB was cycled through the disproportionation pathway. Based on such an assumption, very large isotopic enrichments in regenerated H_2S of 20‰ to over 60‰ were calculated. This assumption does not account for the continuous exchange of reactants and products that occurs in near-surface marine sediments, and the actual pathways are probably much more dynamic than the closed-system calculations indicate. It is also not clear why redox cycling gives rise to a frequently observed $\delta^{34}S$ "ceiling" of 60–70‰ (Lyons *et al.*, in press-a).

Isotopes of organosulfur in marine sediments are typically heavier than coexisting pyrite (Anderson and Pratt, 1994). Figure 9 from the paper by Anderson and Pratt summarizes the data. The organic sulfur is ~5–15‰ heavier than pyrite from the same samples. Anderson and Pratt suggested that the isotopically heavy component is, in part, derived from biosynthetic (assimilatory) sulfur. They also suggested that early diagenetic H_2S was not the immediate precursor. Instead, they invoke a component such as polysulfide or a sulfur oxyanion (e.g., $S_2O_3^{2-}$) as the sulfidizing agent. However, the evidence summarized above indicates that these oxidation products may be lighter than H_2S rather than being heavier. Thus, an alternative explanation for the isotopic composition of organosulfur is that H_2S is the sulfidizing agent for organic matter, and that the intermediates cycled through the

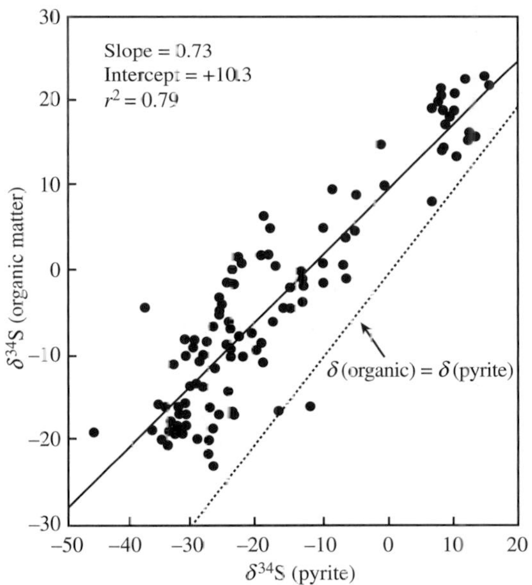

Figure 9 Plot of the sulfur isotopic composition of organic sulfur in marine sediments versus the sulfur isotopic composition of pyrite in the same sediments (reproduced by permission of American Chemical Society from *Geochemical Transformations of Sedimentary Sulfur*, **1994**, *612*, 378–396).

oxidation–reduction–disproportionation pathway go to form pyrite.

7.10.7.3 Isotopic Processes in the Deeper Diffusion-dominated Sediment Regime

Sulfur isotope systematics in the diffusion zone (zone III) reflect the condition that this interval is only partially open to SO_4^{2-} addition. This restriction on sulfate transport results in enrichment in ^{34}S of the residual sulfate. Pore-water H_2S likewise becomes heavier, like its parent SO_4^{2-} (Figure 4). The shift in H_2S and particularly in SO_4^{2-} isotope values as total sulfate depletion is approached can be dramatic—greater than 50‰ compared to that in initial seawater SO_4^{2-}. A corresponding increase in pyrite-sulfur isotope values with depth in zone III, if it occurs at all, is much smaller than that for SO_4^{2-}. In large part, this small change in pyrite $\delta^{34}S$ values with depth results from formation of most FeS_2 under more open-system conditions in the bioturbation zone where reduction rates are high and reactive iron is available. Clearly some solid-phase sulfur will form within the diffusion-dominated regime, even if the amounts are trivial compared to those formed during earlier stages of diagenesis. The existence of a maximum in pore-water H_2S profiles (e.g., Figure 4) requires that below this maximum, sulfide will diffuse deeper into the sediments and will be retained there. Examples are recognized, however, where clear shifts in $\delta^{34}S$ of pyrite towards heavier values occur with depth in the diffusion-dominated interval requiring significant formation of pyrite in zone III (Goldhaber and Kaplan, 1980). This buildup of pyrite can occur when reactive iron is available in zone III.

Calculation of the isotopic composition of sulfur formed in the diffusion zone requires recognition of some effects specific to the diffusion process. The flow of matter by diffusion, i.e., the flux, is a function of the product of the concentration gradient and diffusion coefficient according to Equation (17):

$$J(SO_4^{2-}) = -\phi D_{SO_4^{2-}} \frac{\partial[SO_4^{2-}]}{\partial x} \quad (17)$$

For sulfate diffusion, this overall flux may be separated into two distinct terms, one for ^{32}S sulfate and one for ^{34}S sulfate (Equations (18a) and (18b)):

$$J(^{32}SO_4^{2-}) = -\phi D_{SO_4^{2-}} \frac{\partial[^{32}SO_4^{2-}]}{\partial x} \quad (18a)$$

$$J(^{34}SO_4^{2-}) = -\phi D_{SO_4^{2-}} \frac{\partial[^{34}SO_4^{2-}]}{\partial x} \quad (18b)$$

Because SRB preferentially utilize $^{32}SO_4^{2-}$, the gradient in this species will be increased in a relative sense compared to $^{34}SO_4^{2-}$ and compared to a hypothetical situation in which SO_4^{2-} is reduced without isotope fractionation. Therefore, diffusion of $^{32}SO_4^{2-}$ will be more rapid than would be the case without isotope fractionation, and isotopically light SO_4^{2-} will be added by diffusion (Chanton *et al.*, 1987a,b; Goldhaber and Kaplan, 1980; Jorgensen, 1979). It can also be shown, by a similar line of reasoning, that diffusion of H_2S upwards in the direction of the sediment–water interface leads to a net transport of isotopically heavy sulfur upwards (Chanton *et al.*, 1987b; Jorgensen, 1979). The magnitude of these effects can be large. The isotopic composition of the SO_4^{2-} diffusing into the sediment has been calculated to be 24.5‰ (Goldhaber and Kaplan, 1980) and 21‰ (Chanton *et al.*, 1987b) lighter than the bottom-water value. Figure 10 shows a plot of pore-water SO_4^{2-} concentration, sulfur isotopic composition of pore-water SO_4^{2-}, and the isotopic composition of the diffusing SO_4^{2-} at the same depth. Note the striking enrichment in ^{32}S of the diffusing SO_4^{2-} compared to its value at the same depth in the pore water. The impact of this differential isotopic transport leads to burial of sulfur that is isotopically lighter than would otherwise be predicted. Both Goldhaber and Kaplan (1980) and Chanton *et al.* (1987a,b) were able to show, based on detailed mass and isotopic balances for aqueous and solid-phase sulfur in two differing depositional settings, that differential diffusion of the several distinct isotopic species had a significant effect on the final isotopic composition of the buried pyrite (Figure 10).

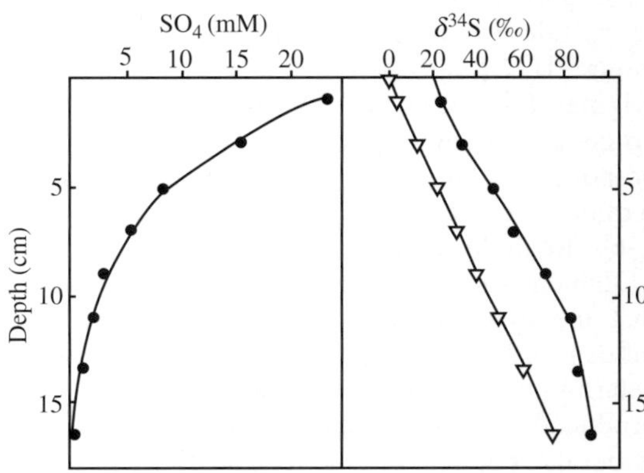

Figure 10 Plot of the concentration of SO_4^{2-} and its isotopic composition versus depth in a core from Cape Lookout Bight, NC, USA. The open triangles show the isotopic composition of the SO_4^{2-} diffusing into the sediment. Note that the diffusing sulfate is approximately 20‰ lighter than its isotopic composition at the same depth (Chanton *et al.*, 1987b) (reproduced by permission of Elsevier from *Geochim. Cosmochim. Acta*, **1987b**, *51*, 1201–1208).

Special consideration of diffusion processes is also required to calculate the fractionation factor between SO_4^{2-} and H_2S in this diffusion zone. If it is simply assumed that diffusion does not occur, closed system Rayleigh distillation equations can be used to calculate an isotopic fractionation factor for this process (Goldhaber and Kaplan, 1974; Sweeney and Kaplan, 1980). However, in a sediment column open to diffusion, differential isotopic diffusion effects such as those described above must be accounted for. This can be done using a mathematical model incorporating diffusion, burial, and chemical reaction written for the isotopically distinct species and then combined to model the isotopic depth distribution of sulfur species: either SO_4^{2-} or H_2S (Goldhaber and Kaplan, 1980; Jorgensen, 1979). It was shown by these authors that the assumption of a closed system yields a calculated instantaneous fractionation factor between SO_4^{2-} and H_2S that is less than the true value. Goldhaber and Kaplan (1980) applied such an open-system mathematical model to pore-water profiles in zone III for sediments of the Gulf of California. Closed-system calculation of the fractionation factor between SO_4^{2-} and H_2S in these sediments yielded values from 16‰ to 35‰. In contrast, the open-system calculation yielded values of ~60‰. Wortmann *et al.* (2001) applied a similar model to a deep sulfidic core from the Great Australian Bight, and calculated an open-system fractionation factor of 65‰. Claypool (2003, in press) has interpreted a similarly large fractionation factor by modeling pore-water $\delta^{34}S$ values in SO_4^{2-} profiles from extensive Ocean Drilling Program/Deep Sea Drilling Project data. Thus, the calculated open-system fractionation is larger than fractionations measured for SRB in laboratory and field studies as described above. A possible explanation for this large fractionation is that it is a net value, as the difference $\Delta_{FeS_2-H_2S}$ is a net value. Thus, it may incorporate the effects of additional redox fractionations beyond single-step reduction of SO_4^{2-} to H_2S. What is troubling about this explanation is that the additional fractionation effects involve reoxidation of H_2S and other reduced sulfur species to $S_2O_3^{2}/SO_3^{2-}$, and subsequent microbial reduction or disproportionation of these intermediate oxidation state species. The deeper, more reduced portion of the sediment column represented by zone III would seem to be a less favorable environment for oxidation reactions than the upper sediment layers where reactive Fe/Mn oxides and other oxidants such as O_2 are available (Wortmann *et al.*, 2001). The origin of the very large apparent fractionations between SO_4^{2-} and H_2S in the deeper portions of marine sediments is presently unresolved.

7.10.8 ANCIENT MARINE SEDIMENTS

Sulfur diagenesis in lithified sediments deposited under normal marine conditions would have been controlled by the same range of processes as those discussed above for their modern analogues. However, some additional factors have been recognized, which will influence interpretation of the observed sulfur abundances and isotope ratios of these ancient sediments. These factors are: (i) loss of organic matter during burial diagenesis; (ii) changes in the nature of organic matter with geologic age; (iii) changes in the sulfur isotopic composition of seawater sulfate; and (iv) later diagenetic processes such as epigenetic addition of H_2S.

Diagenetic loss of organic matter continues following the total depletion of SO_4^{2-} in lower zone III and in zone IV (Figure 4), via the microbial production of methane. Methane production continues over a considerable depth interval, up to burial temperatures in the range of 60–80 °C. Eventually, with sufficient depth and temperature increase above this range, diagenesis passes into a thermocatalytic regime in which organic maturation and attendant loss of organic matter to pore fluids occurs via thermally driven processes (catagenesis). This loss of organic matter is not associated with a corresponding mobilization of FeS_2 until much higher temperatures are attained (Ohmoto and Goldhaber, 1997). Consequently, carbon–sulfur ratios of sediments that experienced one or both of these later diagenetic regimes may have smaller organic carbon to pyrite–sulfur ratios than younger sediments (Berner and Raiswell, 1983). Because the shift is due solely to the loss of carbon, the total sulfur contents of these diagenetically altered sediments will not be different from their recent normal-marine equivalents, and will consequently be less than ~2 wt.%.

An important control on sulfur diagenesis over time is a change in the origin of organic matter incorporated into sediments. Organic matter accumulating in modern marine sediments is a mixture of materials produced on land and materials originating in the marine realm. Of the two types, the terrestrially derived organics are less readily metabolizable by the sulfate-reducing community of microorganisms (Lyons and Gaudette, 1979). Terrestrial organic materials contain a high proportion of polyaromatic derivatives of woody tissue (lignin), which are rather stable during anaerobic diagenesis. These land-derived organics also have a very long transport path and only the most refractory substances survive to be incorporated in marine sediments. These considerations, of course, hold only subsequent to the development of land plants in the Devonian. Pre-Devonian marine environments would have had a higher proportion of the total organics in metabolizable form undiluted by the more refractory land-derived organic matter (Berner and Raiswell, 1983), and a higher percentage of the deposited organic matter may have participated in sulfur diagenesis. For this reason, C/S ratios of Cambrian, Ordovician, and Silurian normal marine sediments are smaller than in younger sediments. The magnitude of this effect is larger than would be expected for the diagenetic loss of organic material due to methanogenesis and thermocatalytic reactions discussed above. Cambrian and Ordovician shale has C-organic/S-total ratios ~0.6 (Raiswell and Berner, 1986) compared to Pleistocene and Holocene values of 2.8 (see Section 6.1).

Sulfur is incorporated in sedimentary organic matter at low temperatures during early diagenesis as discussed above. The re-release of this organic sulfur as H_2S can occur. This release is tied to the overall process of thermal maturation of sedimentary kerogen (the dominant form of organic matter in sediments). Initially CO_2 is released to pore fluids followed by overlapping episodes of petroleum generation, and N_2, H_2S, and CH_4 release. The temperature of onset of this thermocatalytic sequence is ~70 °C. However, this onset temperature is dramatically affected by a variety of variables such as the nature of the organic matter and the interplay of time and temperature during burial. One important variable effecting organic maturation is the sulfur content of the kerogen. High-sulfur kerogens generate hydrocarbons at much lower thermal exposures than do normal kerogen (Orr, 1986; Tannenbaum and Aizenshtat, 1985).

7.10.9 ROLE OF SEDIMENTARY SULFUR IN THE GLOBAL SULFUR CYCLE

7.10.9.1 The Phanerozoic Time Period

The mixing time of SO_4^{2-} in the ocean is short compared to its residence time. Thus, the isotopic composition of SO_4^{2-} in seawater is constant over geologically short time periods. One pathway for removal of this seawater sulfate is by precipitation of sulfate-bearing evaporite minerals. When marine evaporite minerals such as gypsum ($CaSO_4{\cdot}2H_2O$) precipitate from seawater, the sulfur isotope fractionation between the aqueous and solid-phase sulfate is very small, about +1.7‰ (Thode and Monster, 1965). The combination of these factors implies that contemporaneous marine evaporites have the same isotopic composition worldwide, and they are nearly identical isotopically to seawater SO_4^{2-} of that age. When the sulfur isotopic composition of marine evaporite SO_4^{2-} is plotted as a function of time (Figure 11; Claypool *et al.*, 1980; Strauss, 1997), the resulting age curves are quite different for different evaporites. For example, Cambrian evaporites have $\delta^{34}S$ values of about +30‰, whereas for Permian evaporites $\delta^{34}S$ is of about +10‰. More recently, other recorders of marine SO_4^{2-} $\delta^{34}S$ in past oceans including barite ($BaSO_4$) (Paytan *et al.*, 1998) and sulfate trapped in the lattice of carbonate minerals (Lyons *et al.*, in press; Staudt and Schoonen, 1994) have begun to expand the data available from evaporites.

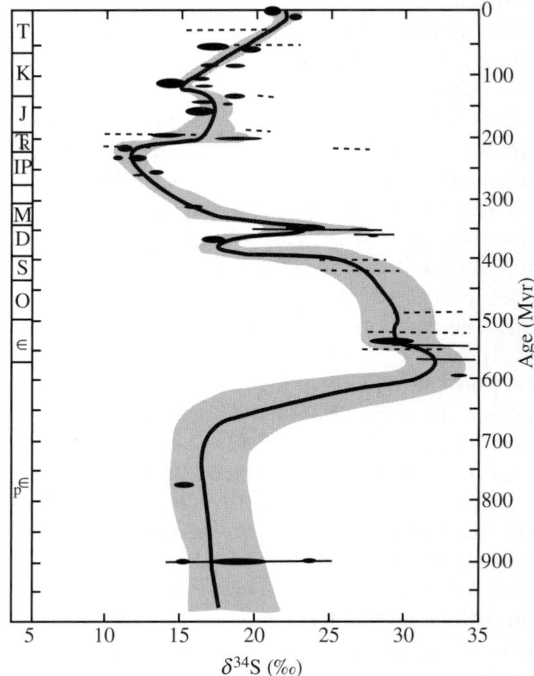

Figure 11 Sulfur isotopic composition of marine evaporite SO_4^{2-} over geologic time (Claypool *et al.*, 1980) (reproduced by permission of Elsevier from *Chem. Geol.*, **1980**, *28*, 199–260).

The age curve of Figure 11 is a critical constraint on the geochemical history of the Earth. Over geologic timescales, removal of pyrite sulfur from ocean water plus formation of sulfate-bearing evaporite minerals is required to balance the addition of sulfate in rivers and much smaller sources (and sinks) from intra-oceanic igneous processes. The river-borne sulfate is derived from terrestrial weathering of pyrite-bearing sediments and evaporites (Holser and Kaplan, 1966). Because pyrite formation involves significant isotopic fractionation and evaporite formation does not, the balance between net burial versus oxidative weathering of pyrite controls the isotopic composition of seawater SO_4^{2-}.

This global cycling of sulfur is closely linked to other geochemical cycles as well. In particular, it is tied to the abundance of atmospheric oxygen.

This linkage has been described by Berner and Petsch (1998). Weathering (oxidation) of pyrite and organic carbon are the dominant sinks for atmospheric oxygen. Unless this oxidation is balanced by a reciprocal burial of organic carbon and pyrite, there will be a net change in atmospheric oxygen. These coupled processes can be summarized in terms of two net reactions involving organic matter and pyrite (Berner, 1999):

$$CO_2 + H_2O \leftrightarrow CH_2O + O_2 \quad (19)$$

$$2Fe_2O_3 + 16HCO_3^- + 8SO_4^{2-} + 16Ca^{2+} \leftrightarrow 4FeS_2 + 15O_2 + 16CaCO_3 + 8H_2O \quad (20)$$

In this equation, CH_2O represents a generic form of organic matter. Reaction (19) going from left to right represents a net long-term photosynthesis, and from right to left represents oxidation of sedimentary organic matter. Reaction (20) going from left to right represents formation of sedimentary pyrite, and from right to left pyrite oxidation. Mass balance models have been derived, which utilize isotopic age curves such as that in Figure 11 to evaluate the role of the global sulfur and carbon cycles on the evolution of atmospheric oxygen (Petsch and Berner, 1998 and references therein).

7.10.9.2 The Precambrian

The evolution of global cycling of sulfur as exemplified by Equations (19) and (20) above is closely tied to the evolution of atmospheric oxygen. Thus, sulfur geochemistry, particularly sulfur isotope data, has proved to be an important probe into the overall evolution of the Earth atmosphere system (Canfield *et al.*, 2000; Canfield and Raiswell, 1999; Canfield and Teske, 1996; Knoll *et al.*, 1998; Lyons *et al.*, in press; Schidlowski, 1979; Schidlowski *et al.*, 1983).

The earliest isotopic record of sedimentary pyrite in the Archean lacks the large spread between coeval sulfate and sulfide sulfur that characterizes the Phanerozoic data. Sedimentary sulfides formed between 3.4 Ga and 2.7 Ga have isotopic compositions within ±5‰ of contemporaneous seawater SO_4^{2-}. This change in isotopic pattern compared to Phanerozoic data is to be expected if atmospheric O_2 was low, and consequently no widespread mechanism existed for oxidizing reduced sulfur to form SO_4^{2-} in seawater. Sulfate concentrations <200 μM are required to suppress isotope fractionation by SRB (Habicht *et al.*, 2002). Most researchers hold to a scenario that photosynthetic production of O_2 began by 2.7 Ga, but that there was a lag of several hundred million years before atmospheric O_2 began to accumulate (2.2 Ga–2.4 Ga) (although compare Ohmoto (1992) and Ohmoto *et al.* (1993). The offset between isotopic values for sedimentary sulfide and coexisting sulfate minerals increases to values indicating the impact of SRB after ~2.7 Ga, but the overall fractionation remained smaller than for Phanerozoic values. Sulfate reduction in rocks of this age may have occurred preferentially in localized SO_4^{2-}-rich environments (Canfield *et al.*, 2000). Only after ~2.3 Ga the full modern expression of large $\Delta_{SO_4-FeS_2}$ values is recognized.

The conclusion that both atmospheric O_2 and oceanic SO_4^{2-} were very low prior to ~2.4 Ga is strengthened by recently discovered mass-independent isotope fractionation recognized in sedimentary sulfate and sulfide minerals. The isotopic fractionation processes discussed previously are mass dependant; they result in predictable relations among the sulfur isotopic values for $\delta^{33}S$, $\delta^{34}S$, and $\delta^{36}S$. Mass-independent sulfur fractionation is exhibited by sedimentary sulfate and sulfide minerals prior to 2.45 Ga (Farquhar *et al.*, 2000). The most likely mechanism for producing such mass-independent fractionations involves atmospheric processes at very low values of P_{O_2}. Levels of O_2 greater than trace levels would inhibit the mass-independent fractionation processes, because it adsorbs UV radiation at wavelengths required for these reactions to occur. The mass-dependent fractionations are not recognized after ~2.45 Ga, suggesting a significant increase in Earth's O_2 content after that time (see Chapter 6.21).

The accumulation of O_2 in Earth's atmosphere and its influence on the global sulfur cycle between ~2.4 Ga and near modern levels during the Phanerozoic has been the subject of much recent research. One model (Lyons *et al.*, in press) calls for generally low oceanic SO_4^{2-} concentrations (~10% of modern value) in the Mesoproterozoic ocean (1.6–0.9 Ga). However, by ~1.8 Ga, sufficient SO_4^{2-} was delivered to the ocean to potentially result in pervasive euxinic conditions in the world ocean as proposed by Canfield (1998). Levels of atmospheric oxygen and seawater SO_4^{2-} may not have been reached until the Neoproterozoic (~900 Ma).

7.10.10 CLOSING STATEMENT

This is an exciting time for researchers studying sedimentary sulfur geochemistry. Although the broad outlines of processes leading to the incorporation of sulfur into marine sediments have been recognized for over half a century, the detailed pathways and mechanisms were not clear. It is only since the advent of methodology to investigate the component microbiological metabolic reactions and quantify low concentrations of inorganic and organic reactive intermediates in these pathways that the complexity of these processes has started to emerge. We can anticipate that further research will continue to reveal additional layers of complexity. Some particularly fruitful areas of research emerge from the study of the mechanism(s) of incorporation of sulfur into organic matter, and the detailed pathways of formation and consumption of reactive sulfur oxyanions in both near-surface and deeper layers of marine sediments. We can also anticipate that as the depth of understanding of modern processes advances, these insights will continue to be applied in understanding the geologic history of sulfur geochemistry and the key role that it has played in shaping our planet's geochemistry.

REFERENCES

Adam P., Schneckenburger P., Schaeffer P., and Albrecht P. (2000) Clues to early diagenetic sulfurization processes from mild chemical cleavage of labile sulfur-rich geomacromolecules. *Geochim. Cosmochim. Acta* **64**(20), 3485–3503.

Allen E. T., Crenshaw J. L., Johnston J., and Larsen E. S. (1912) The mineral sulphides of iron with crystallographic study. *Am. J. Sci.* **23**, 169–236.

Aller R. C. and Mackin J. E. (1989) Open-incubation, diffusion methods for measuring solute reaction rates in sediments. *J. Mar. Res.* **47**, 411–440.

Aller R. C., Mackin J. E., and Cox R. T., Jr. (1986) Diagenesis of Fe and S in Amazon inner shelf muds: apparent dominance of Fe reduction and implications for the genesis of ironstones. *Continent. Shelf Res.* **6**, 263–289.

Aller R. C. (1980a) Diagenetic processes near the sediment-water interface of Long Island Sound: I. Decomposition and nutrient element geochemistry (S, N, P). *Estuarine Physics and Chemistry; Studies in Long Island Sounds* (ed. B. Salzman). Academic Press, New York, vol. 22, pp. 237–350.

Aller R. C. (1980b) Quantifying solute distributions in the bioturbated zone of marine sediments by defining an average microenvironment. *Geochim. Cosmochim. Acta* **44**(12), 1955–1966.

Aller R. C. (1988) Benthic fauna and biogeochemical processes in marine sediments; the role of burrow structures. In *Nitrogen Cycling in Coastal Marine Environments* (eds. T. H. Blackburn and J. P. Sorensen). Wiley, pp. 301–338.

Aller R. C. and Aller J. Y. (1998) The effect of biogenic irrigation intensity and solute exchange on diagenetic reaction rates in marine sediments. *J. Mar. Res.* **56**, 905–936.

Anderson T. F. and Pratt L. M. (1994) Isotopic evidence for the origin of organic sulfur and elemental sulfur in marine sediments. In: *Geochemical Transformations of Sedimentary Sulfur*, American Chemical Society Symposium Series (eds. M. A. Vairavamurthy, M. A. A. Schoonen, T. I. Eglinton, and G. W. Luther, III, B. Manowitz). American Chemical Society, Washington, DC, vol. 612, pp. 378–396.

Arthur M. A. and Sageman B. B. (1994) Marine black shales; depositional mechanisms and environments of ancient deposits. *Ann. Rev. Earth Planet. Sci.* **22**, 499–551.

Baas Becking K. G. M. (1925) Studies on the sulfur bacteria. *Ann. Bot.* **39**, 613–650.

Battersby N. S., Malcolm S. J., Brown C. M., and Stanley S. O. (1985) Sulphate reduction in oxic and sub-oxic North–East Atlantic sediments. *FEMS Microbiol. Ecology* **31**, 225–228.

Belyayev S. S., Lein A. Y., and Ivanov M. V. (1981) Roles of methane-producing and sulfate-reducing bacteria in the destruction of organic matter. *Geochem. Int.* **18**(2), 59–67.

Benning L. G., Wilkin R. T., and Barnes H. L. (2000) In *Reaction pathways in the Fe–S system below 100 degree C.* Chemical Geology (eds. D. J. Wesolowski and T. M. Seward). *Chem. Geol.* **167**, 25–51.

Berner R. A. (1962) Experimental studies of the formation of sedimentary iron sulfides. In: *Biogeochemistry of Sulfur Isotopes: Symposium* (ed. M. LeRoy), Yale University, New Haven.

Berner R. A. (1964a) Distribution and diagenesis of sulfur in some sediments from the Gulf of California. *Mar. Geol.* **1**(2), 117–140.

Berner R. A. (1964b) Iron sulfides formed from aqueous solution at low temperatures and atmospheric pressure. *J. Geol.* **72**(3), 293–306.

Berner R. A. (1964c) Stability fields of iron minerals in anaerobic marine sediments. *J. Geol.* **72**(6), 826–834.

Berner R. A. (1967) Thermodynamic stability of sedimentary iron sulfides. *Am. J. Sci.* **265**(9), 773–785.

Berner R. A. (1969) The synthesis of framboidal pyrite. *Econ. Geol. and Bull. Soc. Econ. Geol.* **64**(4), 383–384.

Berner R. A. (1970) Sedimentary pyrite formation. *Am. J. Sci.* **268**(1), 1–23.

Berner R. A. (1973) Pyrite formation in the oceans. In: *Symposium on Hydrogeochemistry and Biogeochemistry:* vol. 1. Hydrogeochemistry, Clarke Co., Washington, DC, pp. 402–417.

Berner R. A. (1974) Iron sulfides in Pleistocene Deep Black Sea sediments and their paleo-oceanographic significance. In: *The Black Sea: Geology, Chemistry, Geochemistry, and Biology* (eds. E. Degens T and D. Ross A). American Association of Petroleum Geologists, vol. 20, pp. 524–531.

Berner R. A. (1978) Sulfate reduction and the rate of deposition of marine sediments. *Earth Planet. Sci. Lett.* **37**(3), 492–498.

Berner R. A. (1982) Burial of organic carbon and pyrite sulfur in the modern ocean; its geochemical and environmental significance. *Am. J. Sci.* **282**(4), 451–473.

Berner R. A. (1999) Atmospheric oxygen over Phanerozoic time. *Proc. Natl. Acad. Sci. USA* **96**, 10955–10957.

Berner R. A. and Petsch S. T. (1998) The sulfur cycle and atmospheric oxygen. *Science* **282**(5393), 1426–1427.

Berner R. A. and Raiswell R. (1983) Burial of organic carbon and pyrite sulfur in sediments over Phanerozoic time; a new theory. *Geochim. Cosmochim. Acta* **47**(5), 855–862.

Berner R. A. and Raiswell R. (1984) C/S method for distinguishing freshwater from marine sedimentary rocks. *Geology* **12**(6), 365–368.

Berner R. A. and Westrich J. T. (1985) Bioturbation and the early diagenesis of carbon and sulfur. *Am. J. Sci.* **285**(3), 193–206.

Berry W. B. N. and Wilde P. (1978) Progressive ventilation of the oceans; an explanation for the distribution of the lower Paleozoic black shales. *Am. J. Sci.* **278**(3), 257–275.

Bonnell L. M. and Anderson T. F. (1987) Sulfur isotopic variations in nodular and disseminated pyrite; Hole 603B. In: *Initial Reports of the Deep Sea Drilling Project* (eds. J. E. van Hinte and S. W. Wise). US Government Printing Office, Washington, DC, vol. XCII, pp. 1257–1262.

Bottcher M. E., Smock A. M., and Cypionka H. (1998) Sulfur isotope fractionation during experimental precipitation of iron(II) and manganese(II) sulfide at room temperature. *Chem. Geol.* **146**, 127–134.

Bottrell S. H. and Raiswell R. (2000) Sulphur isotopes and microbial sulphur cycling in sediments. In *Microbial Sediments* (eds. R. E. Riding and S. M. Awramik). Springer, Berlin, pp. 96–104.

Butler I. B. and Rickard D. (2000) Framboidal pyrite formation via the oxidation of iron(II) monosulfide by hydrogen sulphide. *Geochim. Cosmochim. Acta* **64**(15), 2665–2672.

Cagatay M. N., Borowski W. S., and Ternois Y. G. (2001) Factors affecting the diagenesis of Quaternary sediments at ODP Leg 172 sites in western North Atlantic; evidence from pore water and sediment geochemistry. *Chem. Geol.* **175**(3–4), 467–484.

Calvert S. E., Thode H. G., Yeung D., and Karlin R. E. (1996) A stable isotope study of pyrite formation in the late Pleistocene and Holocene sediments of the Black Sea. *Geochim. Cosmochim. Acta* **60**(7), 1261–1270.

Canfield D. E. (1989a) Reactive iron in marine sediments. *Geochim. Cosmochim. Acta* **53**(3), 619–632.

Canfield D. E. (1989b) Sulfate reduction and oxic respiration in marine sediments; implications for organic carbon preservation in euxinic environments. *Deep-Sea Res. I: Oceanogr. Res. Pap.* **36**(1A), 121–138.

Canfield D. E. (1991) Sulfate reduction in deep-sea sediments. *Am. J. Sci.* **291**(2), 177–188.

Canfield D. E. (1993) Organic matter oxidation in marine sediments. In *Interactions of C, N, P and S Biogeochemical Cycles and Global Change* (eds. R. Wollast, F. T. Mackenzie, and L. Chou). Springer, Berlin, pp. 333–363.

Canfield D. E. (1998) A new model for Proterozoic ocean chemistry. *Nature* **396**(6710), 450–453.

Canfield D. E. (2001) Isotope fractionation by natural populations of sulfate-reducing bacteria. *Geochim. Cosmochim. Acta* **65**(7), 1117–1124.

Canfield D. E. and Des Marais D. J (1993) Biogeochemical cycles of carbon, sulfur, and free oxygen in a microbial mat. *Geochim. Cosmochim. Acta* **57**(16), 3971–3984.

Canfield D. E. and Raiswell R. (1991) Pyrite formation and fossil preservation. In *Taphonomy: Releasing the Data Locked in the Fossil Record* (eds. P. A. Allison, D. E. G. Briggs, F. G. Stehli, and D. S. Jones). Plenum, New York, pp. 337–387.

Canfield D. E. and Raiswell R. (1999) The evolution of the sulfur cycle. *American Journal of Science, Special Issue: Biogeochemical Cycles and their Evolution over Geologic Time: A tribute to the Career of Robert A. Berner* **299**(7–9), 697–723.

Canfield D. E. and Teske A. (1996) Late Proterozoic rise in atmospheric oxygen concentration inferred from phylogenetic and sulphur-isotope studies. *Nature* **382**(6587), 127–132.

Canfield D. E. and Thamdrup B. (1994) The production of (super 34) S-depleted sulfide during bacterial disproportionation of elemental sulfur. *Science* **266**, 1973–1975.

Canfield D. E., Raiswell R., and Bottrell S. H. (1992) The reactivity of sedimentary iron minerals toward sulfide. *Am. J. Sci.* **292**(9), 659–683.

Canfield D. E., Jorgensen B. B., Fossing H., Glud R., Gundersen J., Ramsing N. B., Thamdrup B., Hansen J. W., and Nielsen L. P. (1993) Pathways of organic carbon oxidation in three continental margin sediments. *Marine Sediments, Burial, Pore Water Chemistry, Microbiology and Diagenesis* (eds. R. J. Parkes, P. Westbroek, and J. W. de Leeuw). Elsevier, Amsterdam, vol. 113, pp. 27–40.

Canfield D. E., Lyons T. W., and Raiswell R. (1996) A model for iron deposition to euxinic Black Sea sediments. *Am. J. Sci.* **296**(7), 818–834.

Canfield D. E., Boudreau B. P., Mucci A., and Gundersen J. K. (1998) The early diagenetic formation of organic sulfur in the sediments of Mangrove Lake, Bermuda. *Geochim. Cosmochim. Acta* **62**(5), 767–781.

Canfield D. E., Habicht K. S., and Thamdrup B. (2000) The Archean sulfur cycle and the early history of atmospheric oxygen. *Science* **288**(5466), 658–661.

Chambers L. A. and Trudinger P. A. (1979) Microbiological fractionation of stable sulfur isotopes; a review and critique. *Geomicrobiol. J.* **1**(3), 249–293.

Chanton J. P., Martens C. S., and Goldhaber M. B. (1987a) Biogeochemical cycling in an organic-rich coastal marine basin; 7, Sulfur mass balance, oxygen uptake and sulfide retention. *Geochim. Cosmochim. Acta* **51**(5), 1187–1199.

Chanton J. P., Martens C. S., and Goldhaber M. B. (1987b) Biogeochemical cycling in an organic-rich coastal marine basin; 8, A sulfur isotopic budget balanced by differential diffusion across the sediment-water interface. *Geochim. Cosmochim. Acta* **51**(5), 1201–1208.

Claypool G. E. (2003) Ventilation of marine sediments indicated by depth profiles of pore-water sulfate and $\delta^{34}S$. In *Geochemical Investigations in Earth and Space Sciences: A Tribute to Issac R. Kaplan*, Elsevier, Amsterdam.

Claypool G. E., Holser W. T., Kaplan I. R., Sakai H., and Zak I. (1980) The age curves of sulfur and oxygen isotopes in marine sulfate and their mutual interpretation. *Chem. Geol.* **28**(3–4), 199–260.

Cypionka H., Widdel F., and Pfennig N. (1985) Survival of sulfate-reducing bacteria after oxygen stress, and growth in sulfate-free oxygen-sulfide gradients. *FEMS Microbiol. Ecol.* **31**, 39–45.

Dean W. E. and Arthur M. A. (1987) Inorganic and organic geochemistry of Eocene to Cretaceous strata recovered from the lower continental rise, North American basin, site 603, deep sea drilling project leg 93. In *Initial Reports of the Deep Sea Drilling Project* (eds. J. E. van Hinte, Jr., S. W. Wise, B. N. M. Biart, J. M. Covington, D. A. Dunn, J. A. Haggerty, M. W. Johns, P. A. Meyers, M. R. Moullade, J. P. Muza, J. G. Ogg, M. Okamura, and M. Sarti, U. von Rad). Texas A & M University, Ocean Drilling Program, College Station, TX, vol. 93, pp. 1093–1137.

Dekkers M. J. and Schoonen M. A. A. (1996) Magnetic properties of hydrothermally synthesized greigite (F (sub 3) S (sub 4)): I. Rock magnetic parameters at room temperature. *Geophys. J. Int.* **126**(2), 360–368.

Detmers J., Bruchert V., Habicht K., and Kuever J. (2001) Diversity of sulfur isotope fractionations by sulfate-reducing prokaryotes. *Appl. Environ. Microbiol.* **67**(2), 888–894.

D'hondt S., Rutherford S., and Spivack A. J. (2002) Metabolic activity of subsurface life in deep-sea sediments. *Science* **295**, 2067–2070.

Eglinton T. I., Irvine J. E., Vairavamurthy A., Zhou W., and Manowitz B. (1994) Formation and diagenesis of macromolecular organic sulfur in Peru margin sediments. In: *Organic Geochemistry* (eds. N. Telnaes, G. van Graas, and K. Oygard). Pergamon, Oxford, vol. 22, pp. 781–799.

Emery K. O. and Rittenberg S. C. (1952) Early diagenesis of California Basin sediments in relation to origin of oil. *Bull. Am. Assoc. Petrol. Geol.* **36**(5), 735–806.

Farquhar J., Huiming B., and Thiemens M. H. (2000) Atmospheric influence of earth's earliest sulfur cycle. *Science* **289**, 756–758.

Farrand M. (1970) Framboidal sulphides precipitated synthetically. *Mineralium Deposita* **5**(3), 237–247.

Fauque G. D. (1995) Ecology of sulfate-reducing bacteria. In: *Sulfate Reducing Bacteria* (ed. L. L. Barton). Plenum, New York, vol. 8, pp. 217–241.

Ferdelman T. G., Church T M., and Luther G. W., III (1991) Sulfur enrichment of humic substances in a Delaware salt marsh sediment core. *Geochim. Cosmochim. Acta* **55**(4), 979–988.

Filipek L. H. and Owen R. M. (1980) Early diagenesis of organic carbon and sulfur in outer shelf sediments from the Gulf of Mexico. *Am. J. Sci.* **280**(10), 1097–1112.

Fossing H. and Jorgensen B B. (1990) Oxidation and reduction of radiolabeled inorganic sulfur compounds in an estuarine sediment, Kysing Fjord, Denmark. *Geochim. Cosmochim. Acta* **54**(10), 2731–2742.

Francois R. (1987) A study of sulphur enrichment in the humic fraction of marine sediments during early diagenesis. *Geochim. Cosmochim. Acta* **51**(1), 17–27.

Froelich P. N., Klinkhammer G. P., Bender M. L., Luedtke N., Heath G. R., Cullen D., Dauphin P., Hammond D., Hartman B., and Maynard V. (1979) Early oxidation of organic matter in pelagic sediments of the eastern equatorial Atlantic; suboxic diagenesis. *Geochim. Cosmochim. Acta* **43**(7), 1075–1090.

Gait R. I. (1978) The crystal forms of pyrite. *Mineral. Rec.* **9**(4), 219–229.

Gelin F., Kok M. D., de Leeuw J. W., and Sinninghe Damste J. S. (1998) Laboratory sulfurisation of the marine microalga Nannochloropsis salina. *Org. Geochem.* **29**(8), 1837–1848.

Goldhaber M. B. and Kaplan I. R. (1974) The sulfur cycle. In: *The Sea* (ed. E. D. Goldberg). Wiley, New York, vol. 5, pp. 569–654.

Goldhaber M. B. and Kaplan I. R. (1975) Controls and consequences of sulfate reduction rates in recent marine sediments. *Soil Sci.* **119**(1), 42–55.

Goldhaber M. B. and Kaplan I. R. (1980) Mechanisms of sulfur incorporation and isotope fractionation during early diagenesis in sediments of the Gulf of California. *Mar. Chem.* **9**(2), 95–143.

Goldhaber M. B. and Orr W. L. (1994) Kinetic controls on thermochemical sulfate reduction as a source of sedimentary H (sub 2) S. In *Geochemical Transformations of Sedimentary Sulfur* (eds. M. A. Vairavamurthy, M. A. A. Schoonen, T. I. Eglinton, and G. W. Luther, III, B. Manowitz). American Chemical Society, vol. 612, pp. 412–425.

Goldhaber M. B. and Reynolds R. L. (1991) Relations among hydrocarbon reservoirs, epigenetic sulfidization, and rock magnetization; examples from the South Texas coastal plain. *Geophysics* **56**(6), 748–757.

Goldhaber M. B., Aller R. C., Cochran J. K., Rosenfeld J. K., Martens C. S., and Berner R. A. (1977) Sulfate reduction, diffusion, and bioturbation in Long Island Sound sediments; report of the FOAM Group. *Am. J. Sci.* **277**(3), 193–237.

Goldhaber M. B., Church S. E., Doe B. R., Aleinikoff J. N., Brannon J. C., Podosek F. A., Mosier E. L., Taylor C. D., and Gent C. A. (1995) Lead and sulfur isotope investigation of Paleozoic sedimentary rocks from the southern Midcontinent of the United States: implications for paleohydrology and ore genesis of the Southeast Missouri lead belts. *Econ. Geol.* **90**(7), 1875–1910.

Goldhaber M., Lee R. C., Hatch J. C., Pashin J. C., and Treworgy J. (2002) Role of large scale fluid-flow in subsurface arsenic enrichment. In *Arsenic in Groundwater, Geochemistry and Occurrence* (ed. A. K. S. Alan Welch). Kluwer Academic, pp. 127–176.

Granger H. C. and Warren C. G. (1969) Unstable sulfur compounds and the origin of roll-type uranium deposits. *Econ. Geol. and Bull. Soc. Econ. Geol.* **64**(2), 160–171.

Grinenko V. A. and Ivanov M. V. (1983) Principal reactions of the global biogeochemical cycle of sulphur. In *The Global Biogeochemical Sulphur Cycle* (eds. M. V. Ivanov and J. R. Freney). Wiley, Chichester, pp. 1–23.

Habicht K. S. and Canfield D. E. (1996) Sulphur isotope fractionation in modern microbial mats and the evolution of the sulphur cycle. *Nature* **382**(6589), 342–343.

Habicht K. S. and Canfield D. E. (1997) Sulfur isotope fractionation during bacterial sulfate reduction in organic-rich sediments. *Geochim. Cosmochim. Acta* **61**(24), 5351–5361.

Habicht K. S. and Canfield D. E. (2001) Isotope fractionation by sulfate-reducing natural populations and the isotopic composition of sulfide in marine sediments. *Geology* **29**(6), 555–558.

Habicht K. S., Canfield D. E., and Rethmeier J. (1998) Sulfur isotope fractionation during bacterial reduction and disproportionation of thiosulfate and sulfite. *Geochim. Cosmochim. Acta* **62**(15), 2585–2595.

Habicht K. S., Gade M., Thamdrup B., Berg P., and Canfield D. (2002) Calibration of Sulfate Levels in the Archean Ocean. *Science* **298**, 2372–2374.

Hartmann M. and Nielsen H. (1969) Delta 34S-Werte in rezenten Meeressedimenten und ihre Deutung am Beispiel einiger Sedimentprofile aus der westlichen Ostsee. Delta-34S values in recent marine sediments and their significance, with examples from sediment profiles from the western Baltic sea. *Geol. Rundsch.* **58**(3), 621–655.

Holser W. T. and Kaplan I. R. (1966) Isotope geochemistry of sedimentary sulfates. *Chem. Geol.* **1**(2), 93–135.

Howarth R. W. and Jorgensen B. B. (1984) Formation of S-35-labelled elemental sulphur and pyrite in coastal marine sediments (Limfjorden and Kysing Fjord Denmark) during short-term S-35 sulphate reduction measurements. *Geochim. Cosmochim. Acta* **48**(9), 1807–1818.

Ingvorsen K. and Jorgensen B. B. (1984) Kinetics of sulfate uptake by freshwater and marine species of *Desulfovibrio*. *Arch. Microbiol.* **139**, 61–66.

Ivanov M. V. (1981) The global biogeochemical sulphur cycle. In *SCOPE 17: Some Perspectives of the Major Biogeochemical Cycles* (ed. G. E. Likens). Wiley, Chichester, pp. 61–78.

Jorgensen B. B. (1977a) Bacterial sulfate reduction within reduced microniches of oxidized marine sediments. *Mar. Biol.* **41**, 7–17.

Jorgensen B. B. (1977b) *The Sulfur Cycle of a Coastal Marine Sediment*. Limfjorden, Denmark.

Jorgensen B. B. (1978) A comparison of methods for the quantification of bacterial sulfate reduction in coastal marine sediments. *Geomicrobiol. J.* **1**(1), 11–27.

Jorgensen B. B. (1979) A theoretical model of the stable isotope distribution in marine sediments. *Geochim. Cosmochim. Acta* **43**(3), 363–374.

Jorgensen B. B. (1982) Mineralization of organic matter in the sea bed—the role of sulphate reduction. *Nature* **296**(5858), 643–645.

Jorgensen B. B. and Bak F. (1991) Pathways and microbiology of thiosulphate transformations and sulphate reduction in a marine sediment (Kattegat, Denmark). *Appl. Environ. Microbiol.* **57**(3), 847–856.

Jorgensen B. B., Weber A., and Zopfi J. (2001) Sulfate reduction and anaerobic methane oxidation in Black Sea sediments. *Deep-Sea Res. I: Oceanogr. Res. Pap.* **48**(9), 2097–2120.

Kaplan I. R., Emery K. O., and Rittenberg S. C. (1963) The distribution and isotopic abundance of sulphur in recent marine sediments off southern California. *Geochim. Cosmochim. Acta* **27**(4), 297–331.

Kasten S. and Jorgensen B. B. (2000) Sulfate reduction in marine sediments. In *Marine Geochemistry* (eds. H. D. Schulz and M. Zabel). Springer, Berlin, pp. 263–281.

Kastner M., Elderfield H., Martin J. B., Suess E., Kvenvolden K. A., and Garrison R. E. (1990) Diagenesis and interstitial-water chemistry at the Peruvian continental margin; major constituents and strontium isotopes. In *Proceedings of the Ocean Drilling Program, Peru Continental Margin; Covering Leg 112 of the Cruises of the Drilling Vessel JOIDES Resolution, Callao, Peru to Valparaiso, Chile, Sites 679-688, 20 October 1986–25 December 1986* (ed. S. Stewart). vol. 112, pp. 413–440.

Keil R. G. and Hedges J. I. (1993) Sorption of organic matter to mineral surfaces and the preservation of organic matter in coastal marine sediments. *Chem. Geol.* **107**(3–4), 385–388.

Klug M., Boston P., Francois R., Gyure R., Javor B., Tribble G., and Vairavamurthy A. (1985) Sulfur reduction in sediments of marine and evaporite environments. In *The Global Sulfur Cycle* (ed. D. Sagan). NASA Technical Memorandum 87570, NASA, Washington, DC.

Knoll A. H., Canfield D. E., Norris R. D., and Corfield R. M. (1998) Isotopic inferences on early ecosystems. In *Isotope Paleobiology and Paleoecology* (eds. W. L. Manger and L. K. Meeks). The Paleontological Society, Lawrence, KS, pp. 212–243.

Kohn M. J., Riciputi L. R., Stakes D., and Orange D. L. (1998) Sulfur isotope variability in biogenic pyrite; reflections of heterogeneous bacterial colonization? *Am. Mineral.* **83**(11–12 (Part 2)), 1454–1468.

Kohnen M. E. L., Sinninghe Damste J. S., ten Haven H. L., and de Leeuw J. W. (1989) Early incorporation of polysulphides in sedimentary organic matter. *Nature* **341**(6243), 640–641.

Kohnen M. E. L., Sinninghe-Damste J. S., Kock-Van Dalen A. C., Ten Haven H. L., Rullkoetter J., and de Leeuw J. W.

(1990) Origin and diagenetic transformations of C (sub 25) and C (sub 30) highly branched isoprenoid sulphur compounds; further evidence for the formation of organically bound sulphur during early diagenesis. *Geochim. Cosmochim. Acta* **54**(11), 3053–3063.

Kohnen M. E. L., Sinninghe Damste J. S., Baas M., Schouten S., de Leeuw J. W., and Anonymous. (1991) Sulphur quenching of functionalised lipids in marine sediments; its consequences for (palaeo)environmental reconstruction. In *Abstracts of Papers—American Chemical Society, National Meeting*. American Chemical Society, vol. 201, pp. GEOC 31.

Konhauser K. O. (1998) Diversity of bacterial iron mineralization. *Earth Sci. Rev.* **43**(3–4), 91–121.

Konopka A., Gyure R. A., Doemel W., and Brooks A. (1985) *Microbial sulfate reduction in extremely acid lakes*. Purdue University Water Resources Research Center, Technical Report 173, West Lafayette, pp. 1–50.

Krouse H. R. and McCready R. G. L. (1979a) Biogeochemical cycling of sulfur. In *Biogeochemical Cycling of Mineral-forming Elements* (eds. P. A. Trudinger and D. J. Swaine). Elsevier, Amsterdam, pp. 401–425.

Krouse H. R. and McCready R. G. L. (1979b) Reductive reactions in the sulfur cycle. In *Biogeochemical Cycling of Mineral-forming Elements* (eds. P. A. Trudinger and D. J. Swaine). Elsevier, Amsterdam, pp. 315–358.

Krs M., Krsova M., Pruner P., Zeman A., Novak F., and Jansa J. (1990) A petromagnetic study of Miocene rocks bearing micro-organic material and the magnetic mineral greigite (Sokolov and Cheb basins, Czechoslovakia). *Phys. Earth Planet. Inter.* **63**, 98–112.

Leventhal J. and Taylor C. (1990) Comparison of methods to determine degree of pyritization. *Geochim. Cosmochim. Acta* **54**, 2621–2625.

Leventhal J. S. (1983a) An interpretation of carbon and sulfur relationships in Black Sea sediments as indicators of environments of deposition. *Geochim. Cosmochim. Acta* **47**(1), 133–137.

Leventhal J. S. (1983b) Organic carbon, sulfur, and iron relationships as an aid to understanding depositional environments and syngenetic metals in recent and ancient sediments. In *US Geological Survey Circular, Report: C 0822* (eds. T. M. Cronin, W. F. Cannon, and R. Z. Poore). US Geological Survey, Reston, VA, pp. 34–36.

Leventhal J. S. (1995) Carbon-sulfur plots to show diagenetic and epigenetic sulfidation in sediments. *Geochim. Cosmochim. Acta* **59**(6), 1207–1211.

Lew M. (1981) The distribution of some major and trace elements in sediments of the Atlantic Ocean (DSDP samples). *Chem. Geol.* **33**, 205–224.

Lin S. and Morse J. W. (1991) Sulfate reduction and iron sulfide mineral formation in Gulf of Mexico anoxic sediments. *Am. J. Sci.* **291**, 55–89.

Love L. G. (1967) Early diagenetic iron sulphide in recent sediments of the Wash (England). *Sedimentology* **9**(4), 327–352.

Lovely D. R. and Phillips E. J. P. (1987) Competitive mechanisms for inhibition of sulfate reduction and methane production in the zone of ferric iron reduction in sediments. *Appl. Environ. Microbiol.* **53**(11), 2636–2641.

Lueckge A., Ercegovac M., Strauss H., and Littke R. (1999) Early diagenetic alteration of organic matter by sulfate reduction in Quaternary sediments from the northeastern Arabian Sea. *Mar. Geol.* **158**(1–4), 1–13.

Luther G. W., III, Giblin A., Howarth R. W., and Ryans R. A. (1982) Pyrite and oxidized iron mineral phases formed from pyrite oxidation in salt marsh and estuarine sediments. *Geochim. Cosmochim. Acta* **46**(12), 2667–2671.

Luther G. W., III, Giblin A. E., and Varsolona R. (1985) Polarographic analysis of sulfur species in marine porewaters. *Limnol. Oceanogr.* **30**(4), 727–736.

Luther G. W., III, Glazer B. T., Hohmann L., Popp J. I., Taillefert M., Rozan T. F., Brendel P. J., Theberge S. M., and Nuzzio D. B. (2001) Sulfur speciation monitored *in situ* with solid state gold amalgam voltammetric microelectrodes: polysulfides as a special case in sediments, microbial mats and hydrothermal vent waters. *J. Environ. Monitor.* **3**, 61–66.

Lyons T. W. (1997) Sulfur isotopic trends and pathways of iron sulfide formation in upper Holocene sediments of the anoxic Black Sea. *Geochim. Cosmochim. Acta* **61**(16), 3367–3382.

Lyons T. W. and Berner R. A. (1992) Carbon–sulfur–iron systematics of the uppermost deep-water sediments of the Black Sea. In: *Chemical Geology* (eds. P. A. Meyers, L. M. Pratt, and B. Nagy). Elsevier, vol. 99, pp. 1–27.

Lyons T. W., Kah L. C., and Gellatly A. M. The Precambrian sulfur isotope record of evolving atmospheric oxygen. In *Tempos and Events in Precambrian Time* (eds. P. G. Eriksson, W. Altermann, D. R. Nelson, W. Mueller, O. Catuneanu, and Strand). Elsevier (in press).

Lyons T. W., Raiswell R., and Anonymous. (1993) Carbon-sulfur-iron geochemistry of modern Black Sea sediments; a summary. In *Geological Society of America, 1993 Annual Meeting*. Geological Society of America (GSA), vol. 25, pp. 239.

Lyons T. W., Werne J. P., Hollander D. J., and Murry R. W. (2003) Contrasting sulfur geochemistry and Fe/Al and Mo/Al ratios across the last oxic-to-anoxic transition in the Cariaco Basin, Venezuela. *Chem. Geol.* **195**, 131–157.

Lyons W. B. and Gaudette M. E. (1979) Sulfate reduction and the nature of organic matter in estuarine sediments. *Org. Geochem.* **1**, 151–155.

Madigan M., Martinko J., and Parker J. (2000) *Biology of Microorganisms*. Prentice Hall, New Jersey.

Migdisov A. A., Ronov A. B., and Grinenko V. A. (1983) In *The sulphur cycle in the lithosphere: Part I. Reservoirs*. SCOPE (Chichester) (eds. M. V. Ivanov and J. R. Freney). Wiley, Chichester, vol. 19, pp. 25–95.

Morozov A. A. (1995a) Fe and S in the sedimentary process, oxygen-rich Black Sea Zone: Part II. Early sediment diagenesis and its role in Holocene shelf sedimentation. *Lithol. Min. Resour.* **29**(5), 437–448.

Morozov A. A. (1995b) Iron and sulfur in the sedimentary process; oxygen-enriched Black Sea zone: Part 1. Fe- and S-species in Holocene deposits of certain shelf areas. *Lithol. Min. Res.* **29**(4), 311–322.

Morse J. W. and Berner R. A. (1995) What determines sedimentary C/S ratios? *Geochim. Cosmochim. Acta* **59**(6), 1073–1077.

Morse J. W., Millero F. J., Cornwell J. C., and Rickard D. (1987) The chemistry of the hydrogen sulfide and iron sulfide systems in natural waters. *Earth Sci. Rev.* **24**(1), 1–42.

Morse J. W. and Wang Q. (1997) Pyrite formation under conditions approximating those in anoxic sediments: II. Influence of precursor iron minerals and organic matter. *Mar. Chem.* **57**(3–4), 187–193.

Mossmann J.-R., Aplin A. C., Curtis C. D., and Coleman M. L. (1991) Geochemistry of inorganic and organic sulphur in organic-rich sediments from the Peru margin. *Geochim. Cosmochim. Acta* **55**(12), 3581–3595.

Murowchick J. B. (1992) Marcasite inversion and the petrographic determination of pyrite ancestry. *Econ. Geol. and Bull. Soc. Econ. Geol.* **87**(4), 1141–1152.

Murowchick J. B. and Barnes H. L. (1987) Effects of temperature and degree of supersaturation on pyrite morphology. *Am. Mineral.* **72**(11–12), 1241–1250.

Murry J. I. R. (1895) On the chemical changes which take place in the composition of seawater associated with blue muds on the floor of the ocean. *Trans. Roy. Soc. Edinburgh* **37**, 481–508.

Mycroft J. R., Nesbitt H. W., and Pratt A. R. (1995) X-ray photoelectron and Auger electron spectroscopy of air-oxidized pyrrhotite; distribution of oxidized species with depth. *Geochim. Cosmochim. Acta* **59**(4), 721–733.

Neal A. L., Techkarnjanaruk S., Dohnalkova A., McCready D., Peyton B. M., and Geesey G. G. (2001) Iron sulfides and sulfur species produced at hematite surfaces in the presence of sulfate-reducing bacteria. *Geochim. Cosmochim. Acta* **65**(2), 223–235.

Nissenbaum A. and Kaplan I. R. (1972) Chemical and isotopic evidence for the in situ origin of marine humic substances. *Limnol. Oceanogr.* **17**(4), 570–582.

Nissenbaum A. and Kaplan I. R. (1976) Sulfur and carbon isotopic evidence for biogeochemical processes in the Dead Sea ecosystem. In: *Environmental Biogeochemistry: Carbon, Nitrogen, Phosphorus, Sulfur and Selenium Cycles* (ed. J. O. Nriagu). Ann Arbor Sci. Publ, Ann Arbor, vol. 1, pp. 309–325.

Ohmoto H. (1992) Biogeochemistry of sulfur and the mechanisms of sulfide–sulfate mineralization in Archean oceans. In *Early Organic Evolution: Implications for Mineral and Energy Resources* (eds. M. Schidlowski, S. Golubic, M. M. Kimberly, and P. A. Trudinger). Springer, Berlin.

Ohmoto H. and Goldhaber M. B. (1997) Sulfur and carbon isotopes. In *Geochemistry of Hydrothermal Ore Deposits* (ed. H. L. Barnes). Wiley, New York, pp. 517–612.

Ohmoto H., Kaiser C. J., and Geer K. A. (1990) Systematics of sulphur isotopes in recent marine sediments and ancient sediment-hosted basemetal deposits. In: *Stable Isotopes and Fluid Processes in Mineralization* (eds. H. K. Herbert and S. E. Ho). University of Western Australia, Crawley, vol. 23, pp. 70–120.

Ohmoto H., Kakegawa T., and Lowe D. R. (1993) 3.4-billion-year-old biogenic pyrites from Barberton, South Africa; sulfur isotope evidence. *Science* **262**(5133), 555–557.

Orr W. L. (1986) Kerogen/asphaltene/sulfur relationships in sulfur-rich Monterey oils. In: *Organic Geochemistry* (eds. D. Leythaeuser and J. Ruellkotter). Pergamon, Oxford, vol. 10, pp. 499–516.

Ostroumov E. A. (1953) Different forms of combined sulfur in the sediments of the Black Sea. *Trud. Inst Okeanol, Akad. Nauk. SSSR* **7**, 70–90.

Passier H. F., Luther G. W., III, and de Lange G. J. (1997) Early diagenesis and sulphur speciation in sediments of the Oman Margin, northwestern Arabian Sea. *Deep-Sea Res. II: Top. Stud. Oceanogr.* **44**(6–7), 1361–1380.

Passier H. F., Middelburg J. J., de Lange G. J., and Boettcher M. E. (1999) Modes of sapropel formation in the eastern Mediterranean; some constraints based on pyrite properties. *Mar. Geol.* **153**(1–4), 199–219.

Paytan A., Kastner M., Campbell D., and Thiemens M. H. (1998) Sulfur isotopic composition of Cenozoic seawater sulfate. *Science* **282**(5393), 1459–1462.

Petsch S. T. and Berner R. A. (1998) Coupling the geochemical cycles of C, P, Fe, and S; the effect on atmospheric O (sub 2) and the isotopic records of carbon and sulfur. *Am. J. Sci.* **298**(3), 246–262.

Pfennig N., Widdel F., and Truper H. G. (1979) The Dissimilatory Sulfate-reducing Bacteria. In: *Biogeochemical Cycling of Mineral-forming Elements* (eds. P. A. Trudinger and D. J. Swaine). Elsevier, Amsterdam, vol. 3, pp. 926–940.

Posfai M., Buseck P. R., Bazylinski D. A., and Frankel R. B. (1998) Iron sulfides from magnetotactic bacteria; structure, composition, and phase transitions. *Am. Mineral.* **83**(11–12 (Part 2)), 1469–1481.

Posfai M., Cziner K., Marton E., Marton P., Buseck P. R., Frankel R. B., and Bazylinski D. A. (2001) Crystal-size distributions and possible biogenic origin of Fe sulfides. Biogenic Iron Minerals Symposium and Workshop, Tihany, Hungary, May 20–23, 2000. *Euro. J. Mineral.* **13**, 691–703.

Price F. T. and Shieh Y. N. (1979) Fractionation of sulfur isotopes during laboratory synthesis of pyrite at low temperatures. *Chem. Geol.*, 27.

Pyzik A. J. and Sommer S. E. (1981) Sedimentary iron monosulfides; kinetics and mechanism of formation. *Geochim. Cosmochim. Acta* **45**(5), 687–698.

Raiswell R. (1993) *Iron mineralogy; influence on degree of pyritization.* 17th Meeting of the European Union of Geosciences: Abstract Supplement, Blackwell, Oxford, pp. 691–692.

Raiswell R. and Berner R. A. (1985) Pyrite formation in euxinic and semi-euxinic sediments. *Am. J. Sci.* **285**(8), 710–724.

Raiswell R. and Berner R. A. (1986) Pyrite and organic matter in Phanerozoic normal marine shales. *Geochim. Cosmochim. Acta* **50**(9), 1967–1976.

Raiswell R. and Canfield D. E. (1996) Rates of reaction between silicate iron and dissolved sulfide in Peru margin sediments. *Geochim. Cosmochim. Acta* **60**(15), 2777–2787.

Raiswell R. and Canfield D. E. (1998) Sources of iron for pyrite formation in marine sediments. *Am. J. Sci.* **298**(3), 219–245.

Raiswell R., Buckley F., Berner R. A., and Anderson T. F. (1988) Degree of pyritization of iron as a paleoenvironmental indicator of bottom-water oxygenation. *J. Sedim. Petrol.* **58**(5), 812–819.

Rees C. E. (1973) A steady-state model for sulphur isotope fractionation in bacterial reduction processes. *Geochim. Cosmochim. Acta* **37**(5), 1141–1162.

Reynolds R. L. and Goldhaber M. B. (1983) Iron disulfide minerals and the genesis of roll-type uranium deposits. *Econ. Geol. and Bull. Soc. Econ. Geol.* **78**(1), 105–120.

Reynolds R. L., Fishman N. S., Wanty R. B., and Goldhaber M. B. (1990a) Iron sulfide minerals at cement oil field, Oklahoma; implications for magnetic detection of oil fields. *Geol. Soc. Am. Bull.* **102**(3), 368–380.

Reynolds R. L., Nicholson A., Goldhaber M. B., Colman S. M., King J. W., Rice C. A., Tuttle M. L., and Sherman D. M. (1990b) Diagnosis for greigite (Fe (sub 3) S (sub 4)) in Cretaceous beds, North Slope, Alaska, and Holocene sediments, Lake Michigan. *EOS, Trans., AGU* **71**(43), 1282–1283.

Rickard D. (1997) Kinetics of pyrite formation by the H_2S oxidation of iron(II) monosulfide in aqueous solutions between 25 and 125 degrees C; the rate equation. *Geochim. Cosmochim. Acta* **61**(1), 115–134.

Rickard D. and Luther G. W., III (1997) Kinetics of pyrite formation by the H_2S oxidation of iron(II) monosulfide in aqueous solutions between 25 and 125 degrees C; the mechanism. *Geochim. Cosmochim. Acta* **61**(1), 135–147.

Rickard D., Schoonen M. A. A., and Luther G. W., III (1994) Chemistry of iron sulfides in sedimentary environments. In *Geochemical Transformations of Sedimentary Sulfur* (eds. M. A. Vairavamurthy, M. A. A. Schoonen, T. I. Eglinton, and G. W. Luther, III, B. Manowitz). American Chemical Society, Washington, DC, pp. 168–193.

Rickard D. T. (1969) The chemistry of iron sulphide formation at low temperatures. *Stockholm Contributions in Geology* **20**, 67–95.

Rickard D. T. (1975) Kinetics and mechanism of pyrite formation at low temperatures. *Am. J. Sci.* **275**(6), 636–652.

Russell M. J. and Hall A. J. (1997) The emergence of life from iron monosulphide bubbles at a submarine hydrothermal redox and pH front. *J. Geol. Soc. London* **154**(3), 377–402.

Schenau S. J., Passier H. F., Reichart G. J., and de Lange G. J. (2002) Sedimentary pyrite formation in the Arabian Sea. *Mar. Geol.* **185**(3–4), 393–402.

Schidlowski M. (1979) Antiquity and evolutionary status of bacterial sulfate reduction; sulfur isotope evidence. In: *Origins of Life*. D. Reidel, Dordrecht, vol. 9, pp. 299–310.

Schidlowski M., Hayes J. M., and Kaplan I. R. (1983) Isotopic inferences of ancient biochemistries; carbon, sulfur, hydrogen, and nitrogen. In *Earth's Earliest Biosphere; its Origin and Evolution* (ed. J. W. Schopf). Princeton University Press, Princeton, pp. 149–186.

Schimmelmann A. and Kastner M. (1993) Evolutionary changes over the last 1000 years of reduced sulfur phases and organic carbon in varved sediments of the Santa

Barbara Basin, California. *Geochim. Cosmochim. Acta* **57**(1), 67–78.

Schippers A. and Jorgensen B. B. (2001) Oxidation of pyrite and iron sulfide by manganese dioxide in marine sediments. *Geochim. Cosmochim. Acta* **65**(6), 915–922.

Schippers A. and Jorgensen B. B. (2002) Biogeochemistry of pyrite and iron sulfide oxidation in marine sediments. *Geochim. Cosmochim. Acta* **66**(1), 85–92.

Schoonen M. A. A. and Barnes H. L. (1991a) Reactions forming pyrite and marcasite from solution: I. Nucleation of FeS (sub 2) below 100 degrees C. *Geochim. Cosmochim. Acta* **55**(6), 1495–1504.

Schoonen M. A. A. and Barnes H. L. (1991b) Reactions forming pyrite and marcasite from solution: II. Via FeS precursors below 100 degrees C. *Geochim. Cosmochim. Acta* **55**(6), 1505–1514.

Sinninghe Damste J. S., Kok M. D., Koester J., and Schouten S. (1998) Sulfurized carbohydrates; an important sedimentary sink for organic carbon? *Earth Planet. Sci. Lett.* **164**(1–2), 7–13.

Skinner B. J., Erd R. C., and Grimaldi F. S. (1964) Greigite, the thio-spinel of iron; a new mineral. *Am. Mineral.* **49**, 543–555.

Skyring G. W. (1987) Sulfate reduction in coastal ecosystems. *Geomicrobiol. J.* **5**(3–4), 295–374.

Sommerfield C. K., Aller R. C., and Nittrouer C. A. (2001) Sedimentary carbon, sulfur, and iron relationships in modern and ancient diagenetic environments of the Eel River basin (USA.). *J. Sedimen. Res.* **71**(3), 335–345.

Staudt W. J. and Schoonen M. A. A. (1994) Sulfate incorporation into sedimentary carbonates. In: *Geochemical Transformations of Sedimentary Sulfur* (eds. M. A. Vairavamurthy, M. A. A. Schoonen, T. I. Eglinton, and G. W. Luther, III, B. Manowitz). American Chemical Society, Washington, DC, vol. 612, pp. 332–345.

Sternbeck J. and Sohlenius G. (1997) Authigenic sulfide and carbonate mineral formation in Holocene sediments of the Baltic Sea. *Chem. Geol.* **135**(1–2), 55–73.

Steudel R., Holdt G., Gobel T., and Hazeu W. (1987) Chromatographic separation of higher polythionates (Sn)621 (n = 3…22) and their detection in cultures of Thiobacilus ferrooxidans: molecular composition of bacterial sulfur secretions. *Angew. Chemie. Int. Ed. Engl.* **26**, 151–153.

Strauss H. (1997) The isotopic composition of sedimentary sulfur through time. *Paleogeogr. Paleoclimatol. Paleoecol.* **132**(1–4), 97–118.

Suits N. S. and Arthur M. A. (2000) Sulfur diagenesis and partitioning in Holocene Peru shelf and upper slope sediments. *Chem. Geol.* **163**(1–4), 219–234.

Sweeney R. E. (1972) Pyritization during diagenesis of marine sediments. PhD, University of California, LA.

Sweeney R. E. and Kaplan I. R. (1973) Pyrite framboid formation; laboratory synthesis and marine sediments. *Econ. Geol. and Bull. Soc. Econ. Geol.* **68**(5), 618–634.

Sweeney R. E. and Kaplan I. R. (1980) Diagenetic sulfate reduction in marine sediments. *Mar. Chem.* **9**(3), 165–174.

Tannenbaum E. and Aizenshtat Z. (1985) Formation of immature asphalt from organic-rich carbonate rocks: I. Geochemical correlation. *Org. Geochem.* **8**(2), 181–192.

Thamdrup B., Finster K., Fossing H., Hansen J. W., and Jorgensen B. B. (1994a) Thiosulfate and sulfite distributions in porewater of marine sediments related to manganese, iron, and sulfur geochemistry. *Geochim. Cosmochim. Acta* **58**(1), 67–73.

Thamdrup B., Fossing H., and Jorgensen B. B. (1994b) Manganese, iron, and sulfur cycling in a coastal marine sediment, Aarhus Bay, Denmark. *Geochim. Cosmochim. Acta* **58**(23), 5115–5129.

Thode H. C., Kleerekoper H., and McElcheran D. (1951) Isotopic fractionation in the bacterial reduction of sulphate. *Research (London)* **4**, 581–582.

Thode H. G. and Monster J. (1965) Sulfur-isotope geochemistry of petroleum, evaporites, and ancient seas. In *Fluids in Sub Surface Environments—A Symposium of American Association of Petroleum Geologists*, Tulsa, OK, Memoir, pp. 367–377.

Thode-Anderson S. and Jorgensen B. B. (1989) Sulphate reduction and the formation of sulphur-35 labelled FeS, FeS_2 and elemental Sulphur in coastal marine sediments. *Limnol. Oceanogr.* **34**(5), 793–806.

Thorstenson D. C. and Mackenzie F. T. (1974) Time variability of pore water chemistry in recent carbonate sediments, Devil's Hole, Harrington Sound, Bermuda. *Geochim. Cosmochim. Acta* **38**(1), 1–19.

Troelsen H. and Jorgensen B. B. (1982) Seasonal dynamics of elemental sulfur in two coastal sediments. *Estuar. Coast. Shelf Sci.* **15**(3), 255–266.

Truper H. G. (1982) Microbial process in the sulfur cycle through time. In *Mineral Deposits and the Evolution of the Biosphere* (eds. S. H. Holland and M. Schidlowski). Springer, Berlin, pp. 5–30.

Tuttle M. L. and Goldhaber M. B. (1993) Sedimentary sulfur geochemistry of the Paleogene Green River Formation, Western USA; implications for interpreting depositional and diagenetic processes in saline alkaline lakes. *Geochim. Cosmochim. Acta* **57**(13), 3023–3039.

Tuttle M. L., Rice C. A., and Goldhaber M. B. (1990) Geochemistry of organic and inorganic sulfur in ancient and modern lacustrine environments; case studies of freshwater and saline lakes. In *Geochemistry of Sulfur in Fossil Fuels* (eds. W. L. Orr and C. M. White). American Chemical Society, Washington, DC, pp. 114–148.

Vairavamurthy A. and Anonymous (1993) *Geochemical incorporation of sulfur into organic matter; importance of hydrogen sulfide oxidation product.* Abstracts with Programs—Geological Society of America, Geological Society of America (GSA), vol. 25, pp. 19.

Vairavamurthy M. A., Maletic D., Wang S., Manowitz B., Eglinton T. I., Lyons T., and Anonymous (1997) Characterization of sulfur-containing functional groups in sedimentary humic substances by X-ray absorption near-edge structure spectroscopy. *Energy and Fuels* **11**, 546–553.

van Kaam-Peters H. M. E., Schouten S., Koester J., and Sinninghe Damste J. S. (1998) Controls on the molecular and carbon isotopic composition of organic matter deposited in a Kimmeridgian euxinic shelf sea; evidence for preservation of carbohydrates through sulfurisation. *Geochim. Cosmochim. Acta* **62**(19–20), 3259–3283.

Vaughan D. J. and Craig J. R. (1978) *Mineral Chemistry of Metal Sulfides*. Cambridge University Press, Cambridge.

Volkov I. I. and Rozanov A. G. (1983) The sulphur cycle in oceans: Part I. Reservoirs and fluxes. SCOPE (Chichester) (eds. M. V. Ivanov and J. R. Freney). Wiley, Chichester, vol.19, pp. 357–423.

Volkov I. I., Rozanov A. G., and Zhabina N. N. (1983) Sulfur compounds in sediments of the Gotland Basin (Baltic Sea). *Lithology and Mineral Resour.* **18**(6), 584–598.

Wakeham S. G., Sinninghe Damste J. S., Kohnen M. E. L., and de Leeuw J. W. (1995) Organic sulfur compounds formed during early diagenesis in Black Sea sediments. *Geochim. Cosmochim. Acta* **59**(3), 521–533.

Wang Q. and Morse J. W. (1994) Laboratory simulation of pyrite formation in anoxic sediments. In *Geochemical Transformations of Sedimentary Sulfur* (eds. M. A. Vairavamurthy, M. A. A. Schoonen, T. I. Eglinton, and G. W. Luther, III, B. Manowitz). American Chemical Society, Washington, DC, pp. 206–223.

Weber A., Riess W., Wenzhoefer F., and Jorgensen B. B. (2001) Sulfate reduction in Black Sea sediments; *in situ* and laboratory radiotracer measurements from the shelf to 2000m depth. *Deep-Sea Res.: Part I. Oceanogr. Res. Pap.* **48**(9), 2073–2096.

Werne J. P., Hollander D. J., Behrens A., Schaeffer P., Albrecht P., and Sinninghe Damste J. S. (2000) Timing of early diagenetic sulfurization of organic matter; a precursor-product relationship in Holocene sediments of the anoxic Cariaco Basin, Venezuela. *Geochim. Cosmochim. Acta* **64**(10), 1741–1751.

Westrich J. T. and Berner R. A. (1988) The effect of temperature on rates of sulfate reduction in marine sediments. *Geomicrobiol. J.* **6**(2), 99–117.

Wignall P. B. and Newton R. (1998) Pyrite framboid diameter as a measure of oxygen deficiency in ancient mudrocks. *Am. J. Sci.* **298**(7), 537–552.

Wijsman J. W. M., Middelburg J. J., and Heip C. H. R. (2001a) Reactive iron in Black Sea sediments; implications for iron cycling. *Mar. Geol.* **172**(3/4), 167–180.

Wijsman J. W. M., Middelburg J. J., Herman P. M. J., Boettcher M. E., and Heip C. H. R. (2001b) Sulfur and iron speciation in surface sediments along the northwestern margin of the Black Sea. *Mar. Chem.* **74**(4), 261–278.

Wilkin R. T. and Arthur M. A. (2001) Variations in pyrite texture, sulfur isotope composition, and iron systematics in the Black Sea; evidence for Late Pleistocene to Holocene excursions of the O (sub 2)–H (sub 2) s redox transition. *Geochim. Cosmochim. Acta* **65**(9), 1399–1416.

Wilkin R. T. and Barnes H. L. (1996) Pyrite formation by reactions of iron monosulfides with dissolved inorganic and organic sulfur species. *Geochim. Cosmochim. Acta* **60**(21), 4167–4179.

Wilkin R. T. and Barnes H. L. (1997) Formation processes of framboidal pyrite. *Geochim. Cosmochim. Acta* **61**(2), 323–339.

Wilkin R. T., Barnes H. L., and Brantley S. L. (1996) The size distribution of framboidal pyrite in modern sediments; an indicator of redox conditions. *Geochim. Cosmochim. Acta* **60**(20), 3897–3912.

Wortmann U. G., Bernasconi S. M., and Boettcher M. E. (2001) Hypersulfidic deep biosphere indicates extreme sulfur isotope fractionation during single-step microbial sulfate reduction. *Geology* **29**(7), 647–650.

Yu A., Lein M. B., Varyshteyn B. B., Namsarayev Y. V., Kashparova A. G., VBondar V. A., and Ivanov M. V. (1982) Biogeochemistry of aneroobic diagenesis of recent Baltic sediments. *Geochem. Int.* **19**(2), 90–103.

ZoBell C. E. and Rittenberg S. C. (1950) Sulfate-reducing bacteria in marine sediments. *J. Mar. Res.* **7**(3), 602–617.

7.11
Manganiferous Sediments, Rocks, and Ores

J. B. Maynard
University of Cincinnati, OH, USA

Manganese is the ninth or tenth most abundant element in the Earth's crust (depending on the model used for crustal composition). Most of its industrial use is in steel making with a much lesser amount going into the production of batteries. It is very similar to iron in its chemical properties. Both are commonly found with +2 and +3 valences with high spin states for the 3*d* electrons and with similar ionic radii. Mn and Fe +2 ions have radii 0.83 Å and 0.78 Å, while the +3 ions have 0.70 Å and 0.65 Å, respectively (Li, 2000, tables 1–4). Accordingly, manganese is commonly found substituted in small amounts in iron minerals. Manganese, however, also has access to a higher valence state, +4, which gives rise to a plethora of complex manganese oxide minerals. Both elements are mined from supergene-enriched sedimentary deposits of a great variety of ages. Iron, however, is dominantly produced from Archean to very early Paleoproterozoic deposits, whereas manganese ores come from post-Archean rocks.

7.11.1 CHEMICAL FUNDAMENTALS

As with iron, manganese geochemistry in sedimentary environments is governed by oxidation and reduction. Eh–pH relations show a relatively large field of stability for dissolved Mn^{2+} compared to the solid oxides (Figure 1). At the pH value of 8 for seawater, or of 5–7 for fresh surface waters, manganese should be soluble except under strongly oxidizing conditions. The addition of carbonate species to the system creates a large region in which solid manganese, in the form of rhodochrosite, is stable under reducing conditions (Figure 2). Unlike FeS or FeS_2, the sulfides of manganese (alabandite and hauerite) are very soluble under reducing conditions, so manganese behavior at low Eh is controlled by carbonate minerals, in contrast to iron, which is controlled more by the sulfides. In freshwater, the pH is normally too low for rhodochrosite precipitation, but for seawater, a slight increase in the amount of CO_3^{2-} should lead to rhodochrosite

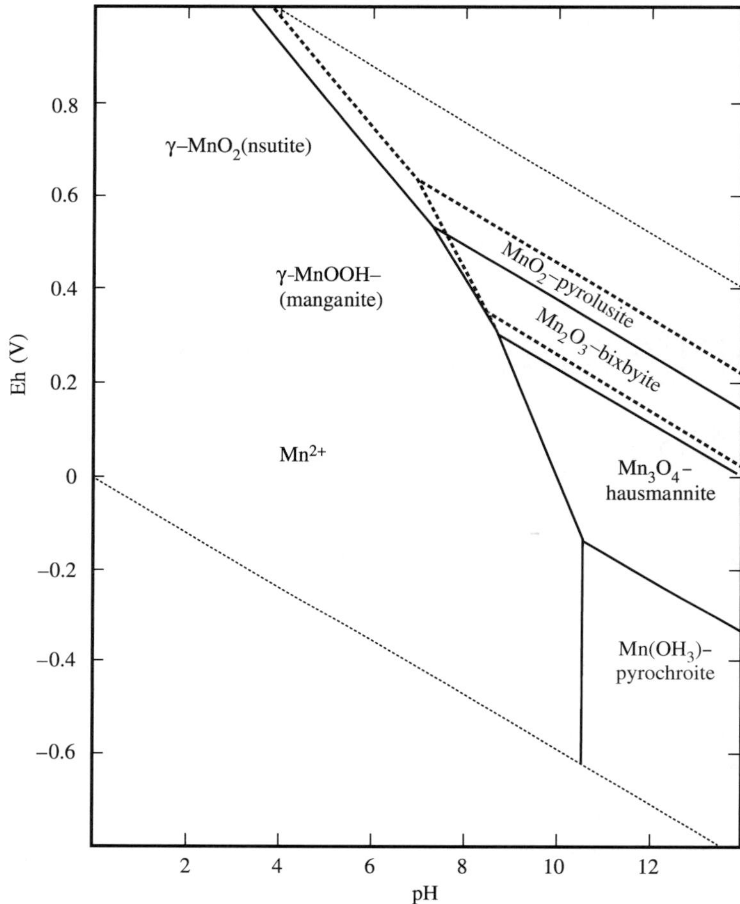

Figure 1 Eh–pH diagram showing the relationships among Mn oxides. Note the large field of stability for dissolved Mn^{2+} and low Eh and pH.

formation if there is a supply of Mn. $MnCO_3$ minerals have, in fact, been reported from a variety of low Eh marine environments.

As first pointed out by Krauskopf (1957), Eh–pH diagrams suggest a mechanism for separating manganese from iron. Figure 3 shows that soluble manganese has a considerably larger stability field than soluble iron under moderately reducing conditions. Because almost all sediments become at least slightly reducing a few centimeters below the sediment–water interface, Mn^{2+} is commonly mobilized into the pore water, while iron remains fixed as an oxide or hydroxide. Under conditions of low Eh and high sulfur content, such as in reducing marine sediments, the iron is fixed as a sulfide, but manganese is still mobile. Dissolved Mn^{2+} diffuses upward, and is either precipitated at the sediment–water interface by the oxygen in the bottom water, or dispersed into the overlying water and carried to other parts of the basin, depending on the oxygen content of the water.

In soils, the greater mobility of manganese will lead to a greater tendency to move downwards in profiles, leaving behind an iron-rich crust. If the manganese is not lost from the soil altogether, its downward movement can produce significant enrichments (e.g., Sivaprakash, 1980; Roy, 1981, pp. 124–132). The great majority of mineable manganese ores have undergone supergene enrichment (Varentsov, 1996).

The variety of manganese phases found in sediments and the common presence of metastable phases suggests that kinetic factors are important in manganese geochemistry. Unlike most other metals, considerable work has been done on the geochemical kinetics of manganese (Hem, 1972, 1981; Glasby, 1974; Stumm and Morgan, 1996). For most manganese oxides, precipitation involves oxidation of the Mn^{2+} in solution to Mn^{4+}. Although thermodynamically favored, this transition is strongly inhibited kinetically. Oxidation of Fe^{2+} can be modeled by the expression

$$-\mathrm{d}[\mathrm{Fe}^{2+}]/\mathrm{d}t = k[\mathrm{Fe}^{2+}][\mathrm{O}_2][\mathrm{OH}^-]^2 \qquad (1)$$

which involves only solution species. Mn^{2+} oxidation is a heterogeneous reaction that follows

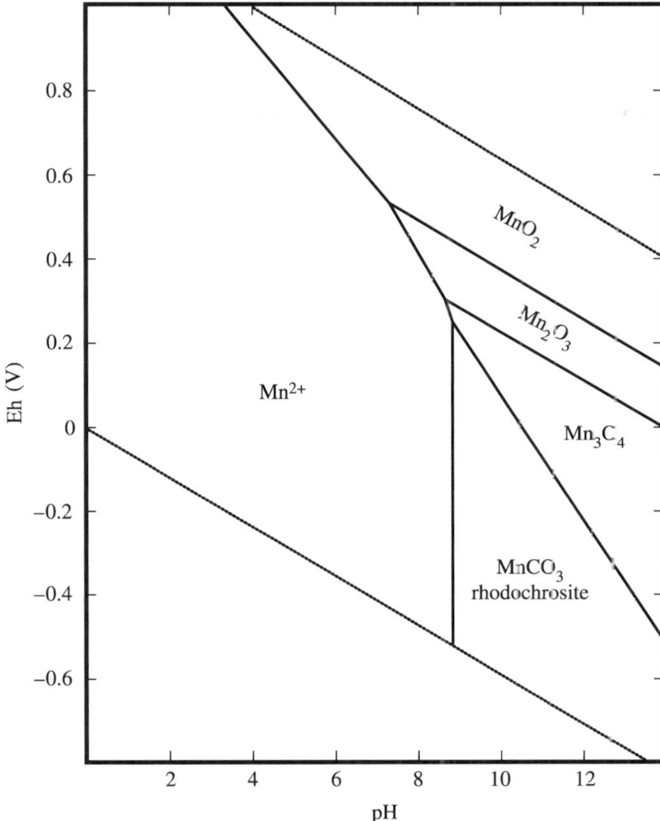

Figure 2 Eh–pH diagram for Mn carbonate and oxides. Rhodochrosite should be the dominant mineral at high pH under low Eh conditions.

an autocatalytic relationship

$$-\mathrm{d}[\mathrm{Mn}^{2+}]/\mathrm{d}t = k_0[\mathrm{Mn}^{2+}] + k_1[\mathrm{Mn}^{2+}] \times [\mathrm{MnO_2}][\mathrm{O_2}][\mathrm{OH}^-]^2 \qquad (2)$$

This expression tells us that the precipitation of manganese oxides is favored by the presence of a pre-existing surface of manganese oxide or, as it turns out, of iron oxide. These oxide surfaces have the ability to sorb appreciable quantities of ions from solution, particularly favoring the cations of the transition metals. Oxidation of Mn^{2+} involves, first, a sorption of the ion onto the oxide surface, followed by the oxidation step, hence the importance of a pre-existing surface.

Manganese and iron oxidation also differ in their pH dependence: iron oxide forms at an appreciable rate at pH values above 6, whereas the equivalent rate for manganese oxide is not reached until pH = 8.5 (Stumm and Morgan, 1996, figure 11.6). Thus, in seawater (pH = 8), there will be a tendency for iron to precipitate, but for manganese to remain in solution, even when thermodynamic considerations suggest that both should precipitate. The catalytic effect of oxide surfaces is therefore especially important in seawater. In fact, most manganese nodules appear to have nucleated around a grain such as a shark's tooth, and these nuclei commonly have a rim of iron oxide or hydroxide that precedes the deposition of the manganese oxide layers (Burns and Brown, 1972). The constants in Equations (1) and (2) were determined in batch experiments and are difficult to apply to natural systems with flowing water, but Hem (1981) has calculated that, at a pH of 8.5, it would take nearly a million years to form an oxide layer 0.1 mm thick. Note that these kinetic constraints do not apply to manganese carbonates, which form directly from Mn^{2+}. Balzer (1982) has shown that Mn^{2+} concentrations in bottom waters of the Baltic rise above the level predicted by equilibrium with oxides as the underlying sediment becomes anoxic, and stabilize at about the level predicted for $MnCO_3$ saturation.

The high capacity of manganese oxides for the sorption of cations leads to an enrichment of manganese-rich sediments in a number of economically valuable transition metals, particularly copper, nickel, and cobalt. However, the exact mechanism of the incorporation of these metals remains controversial. For example, R. G. Burns and M. Burns (1977a) maintained that these elements substitute within the lattice of the manganese minerals, whereas Glasby (1974)

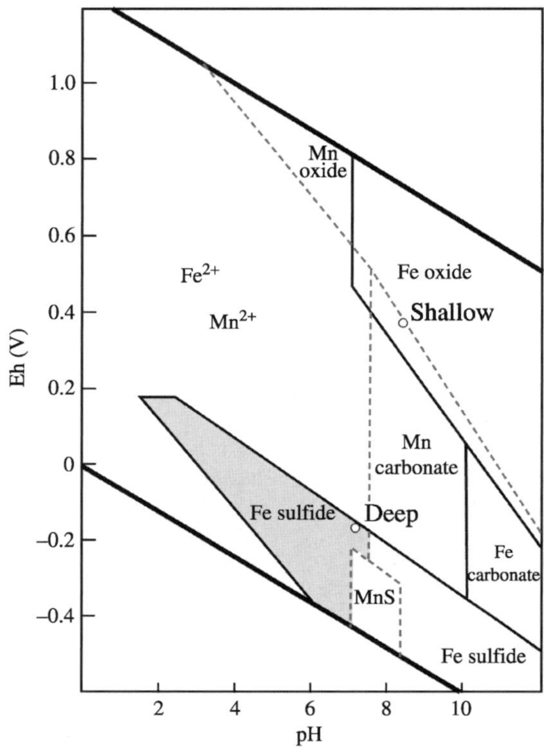

Figure 3 Eh–pH diagram with sulfides included. Note the large field of stability for Fe sulfide compared to Mn sulfide. The points labeled shallow and deep refer to typical compositions of waters from the Black Sea (reproduced by permission of Society of Economic Geologists from *Sedimentary and Diagenetic Mineral Deposits: A Basin Analysis Approach to Extraporation* (eds. E. R. Force, J. J. Eidel, and J. B. Maynard), **1991**, pp. 147–159).

argued that this cannot be the case and proposed instead that the metals are loosely held on exchange sites.

The influence of microorganisms is another problem in the geochemistry of manganese. Bacteria catalyze the oxidation of Fe^{2+} (Nealson, 1997), particularly at low pH where the abiotic reaction is slow. The slow abiotic precipitation rates of manganese oxides also provide an opportunity for bacterial mediation of oxidation (Cowen and Bruland, 1985), and a number of workers have reported bacterial catalysis of manganese oxidation. For example, Mandernack and Tebo (1993) showed that manganese removal from Galapagos vent fluids was strongly inhibited by sodium azide, an agent toxic to bacteria. Juan de Fuca vent fluids did not show this effect, suggesting that the proportion of manganese removed by bacteria catalysis varies greatly between sites.

Manganese removal can be produced both by direct oxidation of Mn^{2+} (Mandernack *et al.*, 1995) and by sorption onto the surface of bacteria (Roitz *et al.*, 2002). Various bacterial groups

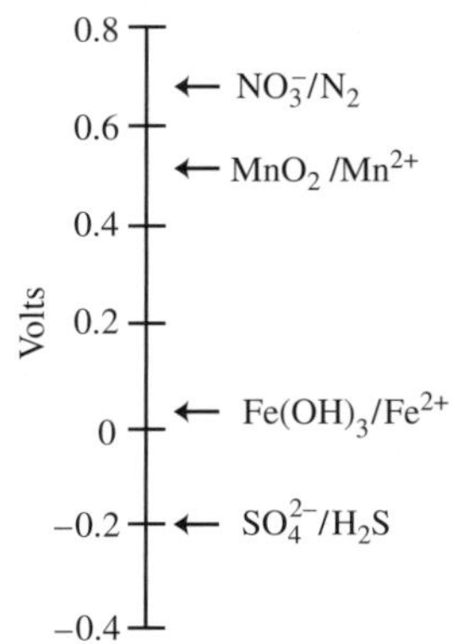

Figure 4 Sequence of redox reactions experienced in modern sediments.

appear to exhibit catalysis, including *Bacillus* and *Leptothrix*, and a variety of manganese oxide minerals can be produced. Zhang *et al.* (2002) have shown that the oxidation process follows standard Michaelis–Menten kinetics:

$$-\frac{d[Mn^{2+}]}{dt} = \frac{X(k[Mn^{2+}])}{(K_s + [Mn^{2+}])} \tag{3}$$

where X is the cell concentration in mg L^{-1}, k is the maximum Mn^{2+} oxidation rate in micromoles of Mn^{2+} (mg cells min)$^{-1}$, and K_s is the so-called "half-velocity constant," which is the concentration of Mn^{2+} when the oxidation rate is half of the maximum. They calculated values for k of 0.0059 and for K_s of 5.7 at pH = 7.5, dissolved oxygen = 8.05 mg L^{-1}, and 25 °C. They also reported that for the abiotic reaction, k_0 is $\sim 10^{-6}$ min^{-1}, so that even a modest cell density would provide faster oxidation than the homogeneous abiotic reaction. Bacteria also seem to be involved in the oxidation of cobalt (Tebo and Nealson, 1984; Moffett and Ho, 1996), which may provide a mechanism whereby these two elements become associated in manganese nodules.

Bacteria also catalyze the reduction of Mn^{4+} to Mn^{2+}. Manganese reduction follows a series of reactions by which bacteria choose a succession of oxidants to metabolize organic matter. This succession produces a stratification of chemical species in sediments (Froelich *et al.*, 1979) or a lateral zonation in aquifers (Chapelle, 1993; Vrobelsky and Chapelle, 1994). This behavior has been referred to as a redox ladder (Langmuir, 1997) and is illustrated in Figure 4. An important characteristic of these successions is that they show little overlap: manganese reduction does not occur in the presence of abundant nitrate (Dollhopf *et al.*, 2000) and iron reduction does not begin until all of the available MnO_2 has been reduced (Myers and Nealson, 1988; Lovley and Phillips, 1988; Wijsman *et al.*, 2002). The kinetics of the process have been investigated by Dollhopf *et al.* (2000) using various strains of *Shewanella putrefaciens*. The appearance of Mn^{2+} as a bacterial product initially follows the expression

$$\ln [Mn^{2+}] = \ln A + B(t) \tag{4}$$

where A and B are constants and t is the time in minutes. B was found to be independent of the strain used and to average 0.040 min^{-1}. A, which reflects the lag time before cell activity begins, varied substantially with the strain used. The results show that bacterial manganese reduction is geologically quite rapid and that the presence of organic matter in sediments or in aquifers should lead to the release of soluble manganese into the overlying water or into the down-flow portion of the aquifer.

7.11.2 DISTRIBUTION OF MANGANESE IN ROCKS AND NATURAL WATERS

An examination of the distribution of manganese among the various reservoirs that make up the Earth reveals much about how the element behaves in geochemical cycles. Table 1 compares manganese and iron in some common rock reservoirs and in some key rock types and types of natural waters. As shown in the previous section, the geochemistry of manganese closely resembles that of iron. Therefore, an understanding of manganese behavior, especially when it comes to the formation of ore deposits, entails an understanding of how manganese and iron differ.

Table 1 Distribution of Mn and Fe.

Rock type	*Mn* (ppm)	*Fe* (ppm)	*Mn/Fe*
Rock reservoirs			
Carbonaceous chondrites	1,900	182,000	0.010
Upper mantle	1,000	64,000	0.016
Oceanic crust	1,300	81,000	0.016
Island-arc andesites	1,100	58,000	0.019
Upper continental crust	600	35,000	0.017
Archean upper crust	1,400	62,000	0.023
Rock types and natural waters			
Basalt	1,550	83,000	0.019
Granite	390	21,100	0.018
Shale	730	50,000	0.015
Black shale (SDO1)	325	65,300	0.005
Sandstone	850	35,000	0.024
Limestone	420	9,500	0.044
Oceanic sediment	2,700	36,000	0.075
River water	0.0082	0.040	0.21
Seawater	72×10^{-6}	250×10^{-6}	0.35

Source: Li (2000).

The rock reservoirs on the modern Earth show a very narrow range of Mn/Fe ratios, ranging only from 0.016 to 0.019, which demonstrates how similar the two elements are in the normal terrestrial rock cycle. Carbonaceous chondrites, which are meteorites that presumably represent the primordial composition of the Earth as a whole, are enriched in iron relative to manganese compared to the Earth's crust and mantle. This difference reflects the concentration of metallic iron, but not of manganese, in the Earth's core. Another variation in composition from the normal crustal value of 0.017 is seen in Archean crust, which averages 0.023, a value that is higher than any common igneous rocks.

When looked at by rock type (Table 1), the Mn/Fe ratio in average basalt, granite, and shale is very close to that of the crustal reservoirs of the modern Earth. Other sedimentary lithologies, however, show pronounced enrichments or depletions in manganese relative to iron, telling us that it is in the Earth's exogenic cycles that we should look for processes that form large accumulations of manganese. In particular, note the large enrichment of manganese in limestones and the strong depletion in black shales. The cause of both is the great insolubility of iron when present in the sulfide pyrite compared to the very limited field of stability of alabandite. This strong contrast between iron and manganese behavior under anoxic conditions is well illustrated by the modern deep versus shallow water of the Black Sea (Figure 5). Dissolved iron is vanishingly low in both the shallow and deep portions of the basins. Manganese, although similarly absent from the shallow water, is much higher than iron in the deep water. There is also a peak in manganese just beneath the redox interface that reflects the redissolution of MnO_2 particles that formed in the shallow water and sank through the interface.

The consequence of this contrast in behavior is that iron tends to be concentrated in deeper-water shales deposited under lower oxygen levels, whereas manganese is concentrated in shallower, more oxygenated environments dominated by limestones. In fact, most large manganese ore deposits are (or were) originally manganese carbonates rather than oxides or sulfides. An ideal factory for the creation of giant manganese accumulations would be a silled basin with low-oxygen bottom waters in the tropics surrounded by fringing carbonate reefs. The giant Molango manganese carbonate deposit, in Mexico, formed as a slope deposit on the margins of such a basin (Okita, 1992).

Another mode of manganese enrichment is suggested by the even greater ratio of manganese to iron in oceanic sediments and by the very high ratio in seawater. Both manganese and iron are leached from oceanic crust at mid-ocean ridge

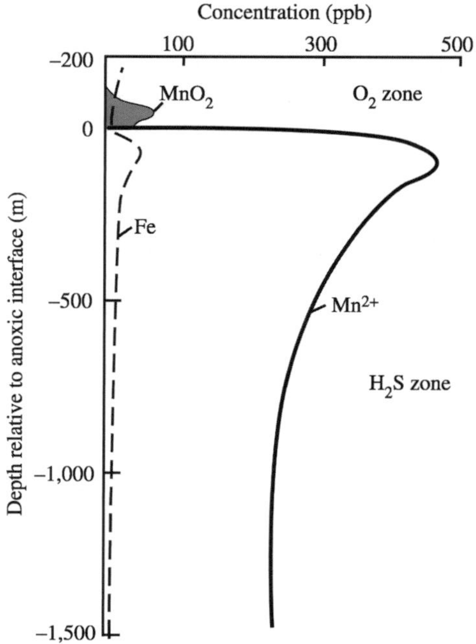

Figure 5 Distribution of dissolved Mn with depth in the Black Sea (reproduced by permission of Society of Economic Geologists from *Sedimentary and Diagenetic Mineral Deposits: A Basin Analysis Approach to Exploration* (eds. E. R. Force, J. J. Eidel, and J. B. Maynard), **1991**, pp. 147–159).

spreading centers and released to the overlying seawater. The iron reacts very rapidly with oxygen to produce iron oxides that fall out close to the vents. Manganese oxide formation is much slower, however, and manganese is dispersed much farther from the vent. Manganese anomalies in seawater are one way of tracking vent plumes which can travel hundreds of kilometers. This process generates an Mn/Fe vector in the sediments that can be used as a prospecting tool to find mineralized vents, as beautifully illustrated for the modern Red Sea by Bäcker *et al.* (1991).

It is interesting to speculate that this process also accounts for the enrichment of Archean crust in manganese. Higher heat flow in the Archean would have resulted in a much higher flux from the ridge-crest hydrothermal systems, and a low-oxygen atmosphere would have prevented the accumulation of manganese oxides in deep-sea sediments, while still retaining much of the iron there as sulfides. Therefore, shallow marine deposits in the Archean should have been strongly enriched in manganese relative to iron. Veizer (1988, figure 5.23) has shown that manganese is strongly enriched in Archean carbonates, with a narrow peak at ~5,000 ppm, compared to Phanerozoic carbonates that show much lower values in a broad range from 10 ppm to 700 ppm. Manganese is so enriched in some older carbonates that their weathering produces lateritic residues with commercial manganese concentrations (Gutzmer and Beukes 1996a, 1997).

Therefore, we have two mechanisms for the addition of manganese to seawater: remobilization from reducing sediments and hydrothermal leaching of ridge-crest basalts. An interesting question emerges from consideration of these two processes: Is there any way to distinguish their products? Furthermore, has the ratio of the processes changed with time?

7.11.3 COMPOSITION OF MANGANESE ACCUMULATIONS

Significant accumulations of manganese, which are defined here as containing >20% manganese, are widely distributed both in time and space. Representative analyses of some of the more prominent occurrences are presented in Table 2. Here a selection of large ore deposits and accumulations from various periods in the Earth's history shows the general chemical character of these accumulations and how they differ with time and with environment. The most obvious differences are between modern deposits of the ocean floor that are remote from hydrothermal influences, the so-called hydrogenous deposits, and all of the other occurrences. The modern hydrothermal deposits are quite similar in many aspects to the ancient ore deposits, leading one to suspect a large hydrothermal component in the source of the manganese in the ancient deposits.

The most prominent geochemical difference, and one that has been exploited in the classification of manganese deposits (see, e.g., Nicholson (1992)), is the extreme concentration of the heavy metals such as cobalt, nickel, and copper in the hydrogenous deposits. A comparison of the analyses in Table 2 for hydrogenous deposits to those for ancient deposits plus modern hydrothermal deposits shows a 10-fold or higher enrichment in the hydrogenous deposits for cobalt, nickel, and copper, but also for lead, thorium, and total rare earth elements (REEs). The ancient deposits and the modern hydrothermal deposits are similar for most elements; the ancient deposits show some enrichment in sulfur, arsenic, and selenium, whereas the modern hydrothermal deposits are relatively enriched in nickel, copper, and molybdenum.

The extreme enrichment of metals in the hydrogenous deposits is usually attributed to their very slow accumulation from oxidizing seawater (Hein *et al.*, 1997) compared to rapid deposition of manganese and iron from suddenly cooled hydrothermal fluids, again under oxidizing conditions. The rate of growth of hydrogenous crusts ranges from 0.5 mm Ma^{-1} to 15 mm Ma^{-1} compared to 20 mm Ma^{-1} to 100 mm Ma^{-1} for hydrothermal crusts. A consideration of the

Table 2 Composition of Mn ores and related accumulations.

Deposit / *Constituent*	*Units*	*Kalahari S. Africa Paleoprot.*	*Tanganshan China Neoprot.*	*Xiangtan China Neoprot.*	*Molango Mexico Jurassic*	*Hydrothermal Pacific Modern*	*Hydrogenous Pacific Modern*	*Hydrogenous Nodules Modern*
Mn	%	36.5	44.9	42.3	28.0	37.0	22.1	18.6
Fe	%	4.22	2.66	1.54	8.53	1.87	15.1	12.5
Na	%	0.01	0.03	0.01	0.00	2.37	1.6	1.7
Mg	%	1.93	1.09	1.23	5.10	1.95	1.26	1.6
Al	%	0.09	0.74	0.44	1.43	1.57	1.01	2.7
Si	%	2.40	1.54	4.75	5.36	7.73	3.69	7.7
P	%	0.01	0.12	0.07	0.06	0.13	1.18	0.25
K	%	0.00	0.01	0.17	0.00	1.01	0.56	0.7
Ca	%	12.2	2.72	1.34	1.73	2.48	4.13	2.3
Ti	%	0.01	0.14	0.10	0.06	0.15	0.77	0.67
C organic	%		2.90	0.60				
C carbonate	%	3.15	8.70	8.70				0.1
$\delta^{13}C_{org}$	per mil				−27.3			
$\delta^{13}C_{org}$	per mil	−9.1			−13.1			
$\delta^{18}O_{carb}$	per mil	−10.4			−3.6			
S total	%	0.04	2.00	0.80	0.20	0.06	0.305	0.47
$\delta^{34}S$	per mil		26.7	52.5	3.2			
Sc	ppm		13	12	2.12	3.8	6.4	10
V	ppm	6	80	53	67	225	515	500
Cr	ppm	16	23	24	15	48	22	35
Co	ppm	50	64	44	132	72	6,400	2,700
Ni	ppm	19	45	24	60	287	5,400	6,600
Cu	ppm	<5	16	11	7	228	1,080	4,500
Zn	ppm	74	19	266	48	238	680	1,200
As	ppm		48	34	31	33	165	140
Se	ppm		2.0	0.5		0.12	0.4	0.6
Rb	ppm	1	1	14	<0.2	<0.2	<0.2	17
Sr	ppm	146	69	101	40	555	1,210	830
Y	ppm	5.3	41	33	8.6	17	166	150
Zr	ppm	2	33	29	15	23	172	560
Nb	ppm	<0.1	25.0	17.5	1.5	9.9	<0.1	50
Mo	ppm	<2	16.7	6.5	2	327	445	400
Sn	ppm	<1	6.5	4.0	<1			2
Sb	ppm		14	2.8	1.1	25	24	40
Ba	ppm	360	50	259	45	1,380	1,700	2,300
Pb	ppm	6	68	8	6	45	1,780	900
Bi	ppm		0.5	<0.1				7.0
Th	ppm	0.4	3.3	1.2	1.4	0.7	33.0	30.0
U	ppm	0.2	0.9	0.7	1.6	2.1	9.6	5.0
La	ppm	3.18	32.0	26	11.8	18.9	202	157
Ce	ppm	3.3	98.8	79.4	18.1	16.3	1,100	530
Pr	ppm	0.5	8.0	6.7	2.4		106	36
Nd	ppm	2.1	32.8	27.8	9.5	7.2	162	158
Sm	ppm	0.33	7.60	5.50	1.70	0.99	42	35
Eu	ppm	0.16	2.35	1.14	0.43	0.28	9.90	9.00
Gd	ppm	0.58	7.52	5.56	1.87		26	32
Tb	ppm	0.09	1.24	0.88	0.29	0.25	7.53	5.40
Dy	ppm	0.55	7.99	5.85	1.49		57.8	31.0
Ho	ppm	0.14	1.55	1.23	0.31		6.60	7.00
Er	ppm	0.42	4.54	3.28	0.85		31.9	18.0
Tm	ppm	0.06	0.58	0.46	0.11		4.30	2.30
Yb	ppm	0.45	3.50	2.75	0.75	0.78	17.7	20.0
Lu	ppm	0.08	0.43	0.38	0.15	0.14	3.34	1.80
Sources		a,b	a	a	a,c,d	e	e	f

[a] Liu *et al.* (in press). [b] Gutzmer (1996). [c] Okita (1987). [d] Liu (1988). [e] Usui and Someya (1997). [f] Li (2000).

distribution of the REEs gives some insight into the relative contribution of hydrothermal fluids and seawater. Figure 6 shows that the hydrogenous accumulations have much higher REE concentrations and a pronounced positive cerium anomaly. The cerium anomaly is thought to result from the oxidation of Ce^{3+} to Ce^{4+} under the oxidizing conditions found in the modern ocean. Tetravalent cerium is much more strongly bound to Fe–Mn oxide surfaces than the trivalent form; cerium is therefore preferentially removed from seawater in the modern ocean; this leaves seawater with a pronounced negative cerium anomaly (Fleet *et al.*, 1983). Oceanic hydrothermal fluids, mostly seawater, have an REE signature close to seawater and so produce deposits with a strong negative cerium anomaly, as seen in Figure 6. For the ancient deposits, both the Molango (Jurassic) and the Kalahari (Paleoproterozoic) deposits have REE plots very close to the hydrothermal signature. The Neoproterozoic deposits (Liu *et al.*, in press), however, have flat REE distributions with a small positive cerium anomaly. This distribution could be obtained by a mixture of hydrothermal and hydrogenous sources. The model curves of Fleet *et al.* (1983, figure 9) suggest that 20–30% hydrogenous manganese mixed with a normal hydrothermal source would produce the cerium anomalies observed in the Neoproterozoic deposits, whereas the Molango and Kalahari deposits received less than 10% hydrogenous manganese.

Europium anomalies have been used to infer a decline with time in the relative contributions of hydrothermal and hydrogenous sources in iron deposits. Many authors (e.g., Derry and Jacobsen, 1990; Bau and Möller, 1993; Klein and Beukes, 1993) have argued that a strong positive anomaly indicates a hydrothermal source, and noted that the anomaly in iron formations declines sharply from the Archean to the Paleoproterozoic to the Neoproterozoic. By contrast, manganese deposits seem to be quite variable in their europium anomalies and to show no trend with time. For example, notice the mixed positive and negative europium anomalies for the Neoproterozoic deposits in Figure 6 and the strong positive anomaly for the young Molango deposit. In addition, modern hydrogenous nodules show a strong positive europium anomaly, inconsistent with their small hydrothermal component. This contrast between the behavior of iron and manganese oxide deposits is an area that needs more study.

The enrichment of the ancient manganese deposits in sulfur-related elements such as arsenic and selenium compared to both the modern hydrothermal and hydrogenous crusts suggests that the ancient deposits formed under lower-oxygen conditions, either in the bottom water of basins or in sediments. Low-oxygen bottom waters can be formed in restricted basins like the present Black Sea or could be caused by an overall low-oxygen ocean, perhaps from total freezing of a "snowball" Earth (Kirschvink *et al.*, 2000). The strong negative cerium anomaly in the Kalahari samples, however, argues against a long period of whole-ocean anoxia, because no manganese nodules and crusts would have been present in deep water to sequester excess Ce^{4+}. Instead, large anoxic basins with restricted access to the open ocean seem to be the most likely source of manganese for large deposits (Okita, 1992; Okita and Shanks, 1992).

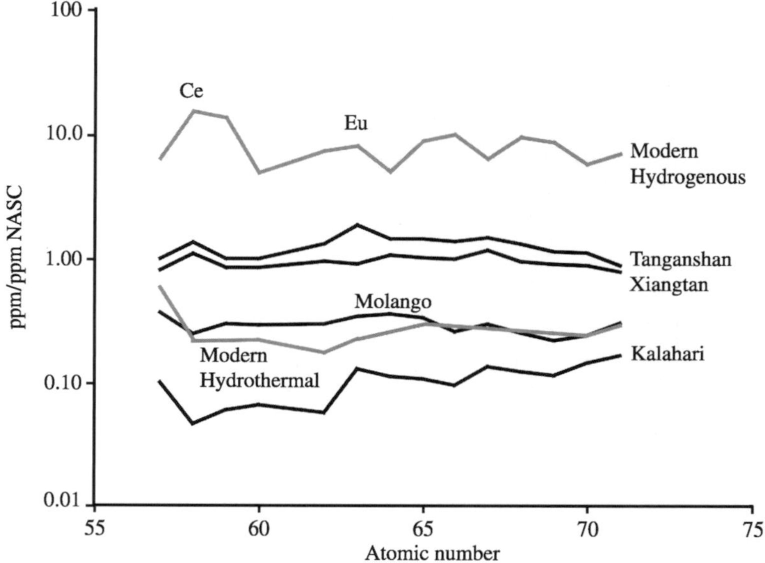

Figure 6 REE distribution for major Mn ore deposits shows variable Eu and Ce anomalies (source Table 2).

7.11.4 COMMON MANGANESE MINERALS

The mineralogy of manganese is complex and analytically difficult. Not only are a large number of phases commonly found, but they are difficult to characterize because of poor crystallinity, fine grain size, intimate intergrowths, and a propensity for alteration during sample handling. Oxides are the most common ore minerals. R. G. Burns and M. Burns (1977a,b) have suggested that manganese oxides consist of subunits of MnO_6^{8-} octahedra. These are arranged in chains or sheets, much as in silicate structures, to give the observed mineral structures. In many cases the chains are arranged in such a way as to form large "tunnels" like those in zeolites. These spaces may be occupied by large cations such as K^+, Ca^{2+}, or Ba^{2+} (Table 3). Complete descriptions of manganese phases, along with X-ray spacings and photomicrographs of polished sections, can be found in Frenzel (1980). See also Turner and Buseck (1983) for a useful summary of oxide structures as seen in transmission electron microscopy.

The mineralogy of the Kalahari manganese field of South Africa has been studied thoroughly. The field is a good type area for examining the behavior of manganese minerals (see especially Kleyenstüber (1986)). Gutzmer and Beukes (1996b) reported 135 mineral phases from the Kalahari deposits, 59 of which are manganese bearing. Of these 59, the oxides (bixbyite, braunite, hausmannite, jacobsite, and manganite) and the carbonates (kutnahorite and rhodochrosite) dominate. The original deposit contained braunite, hematite, and kutnahorite as a finely laminated sediment. The dominant early manganese oxide, braunite, comes in two varieties: type I with the usual structure Mn^{2+} $Mn_6^{3+}SiO_{12}$, and type II, which has less silicon and calcium and divalent manganese to make a structure $Ca_{0.5}$(-Mn^{3+}, $Fe^{3+})_7Si_{0.5}O_{12}$. Much of the original kutnahorite was converted to low-manganese calcite and hausmannite during regional metamorphism. Locally superimposed on these assemblages is a set of fault-related hydrothermal alterations. This hydrothermal activity produced a strong upgrading of the ore in total manganese by removal of CO_2. The final product consists largely of hausmannite and hematite. The type II braunite appears to be a later replacement of the type I structure produced during this metasomatic alteration (Kleyenstüber, 1986). The summary by Nel *et al.* (1986) gives an average manganese content of the unaltered ore as 37%, upgraded to 47% in the hydrothermally altered material, but with a reduction in Mn/Fe ratio from 13.3 to 3.8, suggesting considerable loss of manganese (or addition of iron) during alteration.

Many deposits, including those of the Kalahari, contain manganese carbonates, and most other land-based deposits are thought to have originally had a large proportion of the manganese in carbonates. These phases are usually described as rhodochrosite, but microprobe analyses show the presence of considerable calcium.

Table 3 Common Mn minerals.

Name	*Formula*	*Comments*
Tetravalent oxides		
pyrolusite	β-MnO_2	Single chains of $(MnO_6)^{-8}$ octahedra
Ramsdellite	MnO_2	Double chains
Nsutite	δ-MnO_2	Intergrowths of single and double
Hollandite	$(Ba,K)_{1\text{-}2}Mn_8O_{16}\cdot x\,H_2O$	Open tunnels permit large cation incorporation
Cryptomelane	$K_{1-2}Mn_8O_{16}\cdot x\,H_2O$	
Psilomelane	$(Ba,K,Mn,Co)_3(O,OH)_6Mn_8O_{16}$	= romanechite
Todorokite	$(Na,Ca,K,Ba,Mn,)Mn_3O_7\cdot x\,H_2O$	= 10 Å manganite
Birnessite	$(Ca,Na)(Mn^{2+}Mn^{4+})_7O_{14}\cdot x\,H_2O$	= 7 Å manganite
Lithiophorite	$(Al,Li)(OH)_2MnO_2$	Common in supergene zone
Trivalent oxides		
Bixbyite	α-$(Mn,Fe)_2O_3$	>30% Fe_2O_3
Spinel structures		
Hausmannite	Mn_3O_4	Up to 7 mol.% Fe_3O_4
Jacobsite	Mn_3O_4	>45 mol.% Fe_3O_4
Hydroxides		
Manganite	γ-MnOOH	
Pyrochroite	$Mn(OH)_2$	Brucite structure
Silicates		
Braunite	Mn_7SiO_{12}	
Carbonates		
Kutnahorite	$CaMn(CO_3)_2$	
Rhodochrosite	Mn CO_3	

Source: Maynard (1983).

Experimental work (Goldsmith and Graf, 1957) has revealed that there is a solubility gap in the Mn–Ca carbonates between 50 mol.% Mn (the mineral kutnahorite) and 80 mol.% Mn (calcian rhodochrosite), but many analyses of manganese carbonates from modern sediments show manganese contents that fall within this gap, suggesting that these phases are metastable with respect to kutnahorite and rhodochrosite.

Changes in mineralogy during metamorphism have been described from the Kalahari manganese field of South Africa. Bixbyite, manganite, hausmannite, and a silica-deficient braunite are characteristic of the metamorphosed zone. Heubner (1976) suggested that the persistence of such minerals indicates the presence of locally high oxygen fugacities during metamorphism; manganosite is the only stable phase under conditions prevailing during the metamorphism of common ferruginous rocks. The Kalahari deposits were later subjected to supergene alteration, which resulted in the formation of a distinctive assemblage, including nsutite, todorokite, and lithiophorite.

Like other transition elements, manganese is subject to crystal field effects, but only Mn^{4+}, and to a lesser extent the rare Mn^{3+}, have a crystal field stabilization energy. Accordingly, Mn^{4+} in octahedral positions in minerals should be strongly favored over Mn^{2+} in solution (Crerar *et al.*, 1980, p. 296). Minerals that do contain Mn^{2+}, such as rhodochrosite, tend to be light colored, compared with the dark Mn^{4+} minerals, again because of the splitting of *d* orbitals of Mn^{4+} in the imposed crystal field, which gives rise to excited states with the same spin multiplicity.

7.11.5 BEHAVIOR OF MANGANESE IN IGNEOUS SETTINGS, ESPECIALLY MID-OCEAN RIDGE VENTS

Igneous processes by themselves do not produce manganese enrichments because there is no discrimination between manganese and iron. More strongly oxidizing conditions are needed. A common place to find such conditions is where magmas reach the surface via volcanic activity. The high heat flow around volcanic centers induces hydrothermal circulation of the water in the surrounding rocks. Porphyry Cu–Mo deposits are a product of such circulation in terrestrial settings, where meteoric water is the circulating fluid, whereas Cyprus- or Kuroko-type deposits form in submarine settings with seawater as the fluid (Sillitoe, 1980). The seawater hydrothermal systems can precipitate sulfide minerals either within the volcanic rocks, within overlying sediments, or sometimes even at the surface where the fluids vent into a brine pool. In many cases, however, the superficial sulfides are lost and the only permanent reminder of the presence of seafloor vents is an accumulation of Mn–Fe oxides.

Beginning in the mid-1960s, geologists became aware of the deposition of metal-rich sediments along mid-ocean ridges (e.g., Bonatti and Joensuu, 1966; Boström and Peterson, 1966; see also Mills and Elderfield, 1995, figure 1). The prediction that the high heat flow in these areas should induce hydrothermal circulation of seawater through the oceanic crust (Elder, 1965; Deffeyes, 1970) made it logical to assume that these deposits were related to hydrothermal alteration of basalt by seawater. These predictions have been dramatically confirmed by the direct sampling and photography of active hydrothermal vents on a number of mid-ocean ridge vent sites.

The volume of seawater flowing through such systems is large, and may exercise an important control on seawater chemistry (Maynard, 1976; Wolery and Sleep, 1988; Kadko *et al.*, 1995). The dominant reactions seem to be exchange of Mg^{2+} in seawater for Ca^{2+}, Fe^{2+}, Mn^{2+}, and Ba^{2+} in the basalt. Considerable H_4SiO_4 is also released, and SO_4^{2-} is precipitated as a constituent of $CaSO_4$ and is reduced to H_2S. In addition to these ridge crest, or axial systems, which operate at ~300–400 °C, there are also off-axis systems that are cooler, ~100–150 °C. Manganese plumes seem to be confined to the high-temperature vents on the ridge crests (Murton *et al.*, 1999). The insolubility of iron, manganese, and barium phases in seawater, under oxidizing conditions, leads to their precipitation and consequent enrichment in ridge-crest sediment. Because Fe^{2+} oxidizes at a lower Eh than Mn^{2+}, iron precipitates first in an Eh gradient, and manganese is dispersed farther from the vents. Thus, there is a facies sequence of Fe $\rightarrow$ Mn that can be useful in exploration (e.g., Russell, 1975), and is one indicator of a seafloor–hydrothermal source for an orebody. Observations of modern seafloor vents suggest that plume dimensions can be on the order of hundreds of kilometers (e.g., Lupton, 1995) and that high manganese values can cover areas with the diameter on the order of 50 km (Murton *et al.*, 1999, figure 9).

The rapid precipitation of iron oxides close to the vents has a strong effect on the behavior of REE. The freshly precipitated iron oxides exhibit a very strong absorption of the REE, so that, despite the fact that vent fluids have perhaps 10 times the REE content of seawater, these are all removed close to the vents and the hydrothermal emission actually produces a net removal of REE from seawater (Mitra *et al.*, 1994). Thus, at least in today's high-O_2 oceans, the REE signature of the seawater-hydrothermal systems is not transferred to the bulk seawater. The hydrothermal signature is characterized by a strong positive europium anomaly, produced by europium released from plagioclase during hydrothermal alteration. Modern iron deposits lack a positive europium anomaly unless they are deposited very close to the ridge crest (Olivarez and Owen, 1991), but manganese deposits show a wide variety of positive and negative anomalies (Figure 6; see also Graf *et al.*, 1994; Fan *et al.*, 1999). This variability indicates a strong local control of basin configuration on the proportion of hydrothermal and hydrogenous manganese.

Experimental work on basalt–seawater reactions sheds further light on how hydrothermal Fe–Mn deposits form. For the mobilization of heavy metals from basalt, two factors seem to be important: temperature and the water/rock ratio (e.g., Seyfried and Bischoff, 1981; Seyfried and Ding, 1995). Either high temperatures (>400 °C) or high water/rock ratios (>10 at 300 °C) are required. At the higher temperatures, complexing of the metals by the chloride in seawater becomes significant, and a large proportion of the metals in the basalt can be leached. At lower temperatures, however, the chloride content of seawater is too small for significant complexing, and no leaching takes place unless large amounts of Mg^{2+} pass through the rock. Because most of these systems are thought to operate at 300–400 °C (e.g., Hannington *et al.*, 1995, figure 10), chloride complexing is probably not the important mechanism in leaching and transporting the metals; rather, Mg^{2+} reactions under high water/rock ratios dominate.

There are several ancient analogues of mid-ocean ridge deposits, the best known being the ophiolite-associated copper ore bodies of Cyprus (e.g., Robertson, 1975). Ophiolites, which are interpreted to be former oceanic ridges now exposed on land, have a characteristic stratigraphy. Metal enrichment occurs as epigenetic and syngenetic pyrite–chalcopyrite at the top of the sheeted dike complex and in the basal pillow lavas, and as syngenetic iron-rich or manganese-rich accumulations either directly overlying the sulfides or higher in the pillow-lava sequence. The oxide portions of the Cyprus deposits occur in two forms: iron-rich deposits (ochres) directly overlying the massive sulfides and containing some intermixed sulfides, and manganese-rich deposits (umbers) higher in the section, within or at the top of the upper pillow lavas. The ochres are thought to be products of submarine alteration of the massive sulfides, and are similar in many respects to the hydrothermal oxide deposits of the East Pacific Rise. The umbers, in contrast, are richer in manganese, which imparts a chocolate-brown color, and are interbedded with pelagic sediment. Chemically, they are similar to the ochres, but with a higher Mn/Fe ratio. Because the umbers are found higher in the stratigraphic sequence, separated from the ores by the upper pillow lavas, they are probably products of a separate event, but one related genetically to ore deposition. Robertson (1975, p. 528) proposed that they formed from hydrothermal solutions released into oxidizing seawater on the elevated flanks of the ridge, whereas the massive sulfides formed from solutions released into small anoxic basins within the axial rift. Subsequently, these became superficially oxidized to ochres.

Fluid inclusions, strontium isotopes, and sulfur isotopes indicate that seawater was the hydrothermal fluid. Spooner and Bray (1977) showed that fluid inclusions in quartz co-precipitated with the ore have salinities indistinguishable from that of seawater. $^{87}Sr/^{86}Sr$ ratios for the Cyprus deposits, which in some samples are close to that of Cretaceous seawater, also indicate a large seawater component in the ore-forming fluids with water/rock ratios exceeding 15/1 (Chapman and Spooner, 1977; Spooner *et al.*, 1977). Sulfur isotopes indicate that at least some sulfur from seawater was incorporated in the sulfides. As seawater SO_4^{2-}, which starts at +21 per mil in the hydrothermal recharge, is reduced to sulfide by the iron in basalt at high temperatures, there is an isotope fractionation that depends on the temperature and the extent of reaction (Ohmoto and Goldhaber, 1997). For example, reduction of 20% of the incurrent SO_4^{2-} at 350 °C produces a final sulfide product with $\delta^{34}S$ of +5 per mil, and a 40% extent of reaction results in +7 per mil. Modern deposits from ridges with minimal sediment cover have $\delta^{34}S$ ranging from +1 per mil to +7 per mil compared to near 0 per mil for basalt-derived sulfur (Shanks *et al.*, 1995). This range of values suggests that significant seawater and basalt-derived sulfur are both involved in the process. However, because the proportion of the incurrent SO_4^{2-} that is reduced is unknown, the relative amounts cannot be quantified. The numbers are similar in ancient deposits. For the ophiolitic copper deposits of Notre Dame Bay, Newfoundland, Bachinski (1977) reported an average $\delta^{34}S$ sulfide value of +9.0 per mil and for the Cyprus deposits of +4.8 per mil.

Perhaps the best-studied deposits associated with oceanic ridges are those in the Red Sea, yet they are somewhat anomalous because of the proximity of continental landmasses, and the apparent involvement of continental material in their genesis. Good descriptions of the deposits and their inferred origin can be found in Shanks and Bischoff (1980) and Scholten *et al.* (2000). Overlying the metal-rich sediments is a dense, hot brine, from which they are believed to have been deposited. This lower brine is, in turn, separated from seawater by a second brine of intermediate composition, which is probably a mixture of the bottom brine and seawater. Chemical analyses of the brines show the depletion in Mg^{2+} and enrichment in Ca^{2+}, compared with seawater, that is typical of basalt–seawater interactions, but Na^+ and Cl^- are unusually high. This enrichment is believed to be caused by dissolution of Miocene evaporites bordering the Red Sea. Of particular note is the greater dispersion of manganese than iron into the upper brine. Mapping of the Fe/Mn ratio in bottom sediments of the Atlantis II deep (Scholten *et al.*, 2000, figure 14.5) shows that iron is highly concentrated near the vents, whereas manganese is much more widely dispersed. Furthermore, the Fe/Mn ratio provides a vector that could be used in locating vent positions in older deposits (Cronan, 1976).

By assuming that the lower brine in the Atlantis II deep is a cooled sample of the hydrothermal fluid, Shanks and Bischoff (1977) have reconstructed the conditions of metal transport and deposition in this system. The results indicate that the brines were derived from normal seawater that acquired a high salinity through reaction with evaporites at low temperatures, then was heated to ~200 °C by interaction with recent intrusive rocks of the rift zone at high water/rock ratios. Heavy metals occur mostly as chloride complexes at higher temperatures, but cooling below 150 °C leads to dissociation and precipitation of the metals except manganese as sulfides. The high density of the brine layer restricts circulation, so that the bottom water becomes anoxic and the sulfides are protected from oxidation. An excess of metals over reduced sulfur should lead to precipitation of virtually all of the sulfur but considerable export of metals into surrounding sediments and seawater.

Metalliferous sediments also occur around the volcanic islands of Stromboli and Santorini, as described by Bonatti *et al.* (1972), Puchelt (1973), and Puchelt *et al.* (1973). At Stromboli, the metalliferous sediment is mostly Fe–Mn oxide. X-ray diffraction shows only birnessite, but the presence of cryptocrystalline goethite and amorphous SiO_2 is inferred. On Santorini, the metalliferous sediment is up to 3 m thick, has a high water content, and is X-ray amorphous. Little manganese is present, and CO_2 is much more abundant than at Stromboli, indicating the presence of $FeCO_3$. Sulfides are conspicuously rare at these two localities, but are common in sediments bordering the island of Vulcano, near Stromboli (Honnorez *et al.*, 1973). Abundant fumaroles and hot springs occur on the island and in the surrounding water; the submarine activity is confined to depths of less than 15 m in an elongate zone ~100 m wide, parallel to the shoreline. Chemically, the fumarolic gases are mostly CO_2 and H_2O with ~0.04% H_2S (Honnorez *et al.*, 1973, table 1). Temperatures range from 100 °C to 600 °C, and mixing of the fumarolic gases with seawater lowers its pH to as low as 2. In the sediment, the result is a sulfidation of Fe–Ti grains, cementation of quartz sand grains by pyrite-marcasite, and silicification of volcanic rock fragments. No base-metal sulfides have been reported. Manganese is slightly enriched in the surrounding sediments, and reaches its highest amount, 1,900 ppm, in the deepest, most distal part of the bay in which the fumaroles occur (Valette, 1973, figure 3), a distribution consistent with the manganese dispersion seen in other deposits. Barium, however, does not follow manganese, but is highest (1,000 ppm) near the shoreline around the fumaroles (Valette, 1973, figure 7).

7.11.6 BEHAVIOR OF MANGANESE IN SEDIMENTATION

Numerous examples are known of manganese enrichment in modern sediments that can be examined for analogues of the processes involved in the formation of manganese deposits in ancient rocks. It is convenient to treat these in two groups: (i) oxides, which form the deep-sea nodules and (ii) crusts, and carbonates, which are not as extensively developed in the modern, but are of interest because of the abundance of carbonate or carbonate-derived ores in the ancient.

7.11.6.1 Manganese Nodules and Crusts in Modern Sediments

The world's largest deposit of manganese is found in modern deep-sea sediments. Pelagic ferromanganese nodules cover the ocean floor over large areas, particularly in the central Pacific, and many volcanic edifices are coated by Fe–Mn crusts built on the rock surfaces. The nodules and crusts form the hydrogenetic end-member in what is really a continuous distribution of Fe–Mn accumulations between hydrogenetic and hydrothermal sources (Hein *et al.*, 1997; Hein and Manheim, 2000). The nodules, in turn, can be divided into those that are supplied with metals from the overlying seawater and those that receive a contribution from diagenetic remobilization of metals from the sediment below. There is also a continuum for these processes and they may be significant for the same nodule (Halbach *et al.*, 1982; Cronan, 1997). Mineralogically, the purely hydrogenetic crusts and nodules are dominated by γ-MnO_2 and amorphous FeOOH, epitaxially intergrown. Diagenetic nodules, alternatively, contain 10 Å manganite ("todorokite") and 7 Å manganite ("birnessite"). The distinction between these two phases may be procedural only; in some cases the 10 Å phase can be collapsed to the 7 Å phase by drying (Hein *et al.*, 1997). The manganese and iron components of the crusts and nodules each contribute in different ways to the overall chemical properties of the bulk accumulation. For example, it is found that crusts from the central Pacific show a strong positive correlation between $\sum$REE and %Fe, but a negative correlation with %Mn. Ce/Ce*, alternatively, has a positive correlation with %Mn, but negative with %Fe.

The nodules show interesting areal variations, particularly in minor-element chemistry (Calvert and Price, 1977; Cronan, 1997). Biologic productivity in the overlying surface waters exerts a strong influence on nodule compositions. Areas with low productivity, 40–10 °N and 20–50 °S, have lower nickel and copper in the nodules, but higher iron and cobalt. The increase in Mn–Ni–Cu with increasing productivity is related to increased organic carbon in the sediments, which leads, in turn, to increased mobility for these elements under suboxic diagenetic conditions. Iron does not experience enhanced mobility because of the development of insoluble iron sulfides that tie up the iron within the sediment. Halbach *et al.* (1982) found that, on a microscale, cobalt is associated with silicate- and iron-rich microlaminae. They suggested that in the high electric field at the surface of Si–Fe colloids, Co^{2+} is oxidized to Co^{3+}, a form that is much more strongly bound in the iron-rich phase but is not able to substitute for manganese in the manganese-rich phase of the nodules. Waters with very high surface productivity such as are found near the equator, tend to show a reversal, and to have fewer nodules and less trace element enrichment. This decline may simply be dilution or it may be that conditions in the sediment and immediately adjacent bottom water are too reducing and the manganese and trace elements escape entirely.

Manganese crusts act as a closed system with respect to the rare earths and a number of radioisotopes, which makes them a good recorder of changes in ocean chemistry (e.g., DeCarlo, 1991). Hein and Manheim (2000, figure 9.10) show how changes in neodymium, lead, and beryllium isotopes in manganese crusts track the circulation of deep water from its generation in the North Atlantic through the Indian Ocean and into the North Pacific.

7.11.6.2 Manganese Carbonates in Modern Sediments

Because of the preponderance of manganese carbonates as ore minerals or as protores for deposits in ancient rocks, consideration needs to be given to modern-day accumulations of carbonates as well as oxides. Such occurrences have been described from deep-water sediments of the Panama Basin (Pedersen and Price, 1982), from shallow, near-shore sediments of Scotland (Calvert and Price, 1970), and from several areas of the Baltic (Suess, 1979; Glasby *et al.*, 1997; Lepland and Stevens, 1998).

In the pore waters of the sediments in the Panama Basin, concentrations of Mn^{+2} reach values as high as 160 μM, and the dry sediments contain as much as 3 wt.% Mn. The high manganese concentrations are associated with manganese oxides at the sediment surface, but in one location a zone of manganese carbonate was also found at a depth of ~150 cm in the sediment. Whitish crusts at this depth proved to be coalesced microspheres of manganese carbonate, ~100 μm in diameter, with compositions close to those expected for kutnahorite (Table 4). No other diagenetic phases were found. Carbon isotopes (Table 4) suggest a negligible contribution of organic carbon to the manganese carbonates. The isotopic composition of carbon in the Panama Basin manganese carbonates is close to that of seawater and is probably derived entirely from pore-water HCO_3^- or the dissolution of shell material.

In the Loch Fyne sediments, manganese occurs both in nodular masses of manganese oxide and as concretions of $MnCO_3$, 1–8 cm in diameter at water depths of 180–200 m. The oxide-rich nodules are commonly cemented and replaced by $MnCO_3$ (Calvert and Price, 1970, figures 4 and 5). Over an area of ~10 km^2, manganese concentrations in the surface sediment exceed 5%, sometimes reaching 10%, although the thickness of this surficial layer is only ~20 cm. Other constituents in the sediment are detrital quartz and clay and shell material. There do not appear to be distinct oxide and carbonate facies. The Loch Fyne sediments have carbon in $MnCO_3$ that is somewhat lighter, isotopically, than carbon from the Panama Basin, indicating some contribution from decaying organic matter.

Table 4 Chemical and carbon isotopic composition of modern Mn carbonates.

Locality	*Mn* (mol.%)	*Ca* (mol.%)	*Mg* (mol.%)	*$\delta^{13}C$, per mil PDB*
Panama Basin	48	47	5	+2.6
Loch Fyne	48	45	7	−5.8
Baltic Sea	85	10	5	−13

Source: Maynard (1983).

The Baltic sediments contain a complex diagenetic assemblage that includes among the reduced phases siderite, MnS, and iron phosphates, in addition to the manganese carbonates. Dissolved Mn^{2+} concentrations in the pore waters reach as much as 80 μM. Manganese carbonate occurs as microspheres 5–25 μm across (Suess, 1979, figure 3), embedded in a matrix of amorphous silica. The MnS phase has an unusual hexagonal form, as seen with the scanning electron microscope. Bottcher and Huckriede (1997) reported that this phase has a sulfur isotopic composition of −13 per mil, indicating derivation from bacterial sulfate reduction. Cubic MnS (alabandite) is found in lesser amounts (Lepland and Stevens, 1998). The unusual presence of these manganese sulfides indicates that the environment was iron limited for pyrite formation, i.e., sulfur remained after all reactive iron was consumed (Sternbeck and Sohlenius, 1997). The carbon isotopic values from the $MnCO_3$ (Table 4) are very negative and indicate that more than half of the carbon is organic derived. The dominance of organic-derived carbon in these accumulations indicates that the $MnCO_3$ formed diagenetically within the sediment.

Oxides of manganese are also present in some isolated areas. Glasby *et al.* (1997) reported that abundant concretions of Mn–Fe oxides are found in three main areas of the Baltic: the Gulf of Bothnia, the Gulf of Finland, and the Gulf of Riga, with Mn/Fe ratios highest in the Gulf of Bothnia concretions. The Baltic is brackish in salinity. There is a strong contrast between the salinity of surface waters at 6.5–7.5 ppt compared to that of inflowing bottom water from the North Sea at 15–20 ppt. This density contrast leads to density stratification and the formation of a halocline that, in turn, promotes lowered oxygen in bottom waters and manganese mobility (Neumann *et al.*, 2002). The three areas of Mn–Fe oxide accumulation are remote from the influence of the more saline bottom waters from the North Sea and tend to lack the stratification of the Baltic proper. The shallowness of the basins and the cold climate favor seasonal overturn of the water column. The higher bottom water oxygenation favors oxide development in these areas compared to the Baltic proper. Localized deeps within the main part of the Baltic tend to develop stable anoxic conditions, a situation that leads to a buildup of dissolved Mn^{2+} in the bottom water. Around the margins of these deeps, there is a zone of enrichment of manganese in the sediment near the halocline, where oxidation in the water column leads to deposition of manganese oxides that then convert to $MnCO_3$ during burial. Oscillation of oxidation state, related to oscillations in the intensity of bottom-water inflows, can lead to laminated sediments with manganese-rich and manganese-poor layers that provide a record of oxygenation events in the basin (Huckriede and Meischner, 1996; Burke and Kemp, 2002).

7.11.7 A GENERAL MODEL OF SEDIMENTARY MANGANESE MINERALIZATION

The behavior of manganese in restricted marine basins such as the Black Sea and the Baltic suggests a general model of manganese behavior in which manganese is solubilized in deep-water sediments in anoxic basins and is reprecipitated around the margins of these basins at the point where the redox interface impinges on the seafloor. This model was first articulated by Force and Cannon (1988) from their observations of the modern and detailed facies analysis of a number of ancient deposits. Subsequently the model has been developed in some detail based on stable isotopic studies of Phanerozoic deposits, particularly Molango in Mexico (Okita, 1987, 1992; Okita *et al.*, 1988; Maynard *et al.*, 1990; Okita and Shanks, 1992). See also reviews by Force and Maynard (1991), who emphasized the ancient record and favored a dominant role for basin geometry, and by Calvert and Pedersen (1996), who emphasized the modern and argued for a dominant role of surface-water productivity in controlling manganese distribution.

The process begins with the precipitation of manganese oxides within the water column at the interface between oxidizing and reducing conditions, usually a halocline. Most of the precipitated manganese simply redissolves as it passes downwards through the water column, unless the seafloor is shallow enough to intercept the redox interface (Figure 7). This phenomenon produces what might be called a "manganese compensation depth." Below this depth, all particulate manganese is dissolved in the water column, so that none reaches the bottom and the sediments are low in manganese. At the depth of the halocline, there is a strong enrichment of the surface sediment in manganese oxide particles. At shallower depths, the sediments are again low in manganese, because the oxygenated surface portion of the water column contains virtually no dissolved manganese. Thus, there is a critical depth for manganese enrichment that produces a "bathtub ring" effect around the margins of the basin. Reaction with organic matter in the sediment then converts this manganese oxide to manganese carbonate. Iron is excluded from this cycling, because it is insoluble in the deeper-water sediments as the sulfide.

A key observation supporting this model is a strong correlation between high manganese contents in the rocks and strongly negative carbon

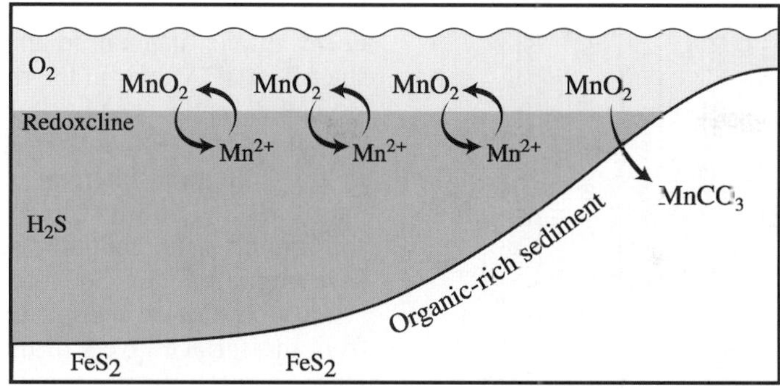

Figure 7 General model for Mn mineralization in euxinic basins.

isotopes (Okita and Shanks, 1992). The production of the $MnCO_3$ mineralization, therefore, required the consumption of large amounts of organic matter and must have occurred during early diagenesis. The process can be represented schematically by the reaction

$$2MnO_2 + CH_2O + HCO_3^- \rightarrow 2MnCO_3 + H_2O + OH^- \quad (5)$$

From Table 2, the organic matter in the sediments has a $\delta^{13}C$ of about -27.3. Combining equal amounts of this carbon with 0 per mil HCO_3^- from seawater would produce a carbonate mineral with $\delta^{13}C$ of about -14 per mil, very close to the observed value. At the same time that the manganese is oxidizing the organic matter, it also attacks any iron sulfide in the sediment (Aller and Rude, 1988; Schippers and Jørgensen, 2001):

$$FeS + 4.5MnO_2 + 4H_2O \rightarrow FeOOH + 4.5\,Mn^{2+} + SO_4^{2-} + 7OH^- \quad (6)$$

From this reaction, the deposit should be very low in sulfur, as is observed, and the pyrite that does form should be very heavy isotopically. This prediction of heavy sulfur is based on the requirement that any pyrite that forms be relatively late, forming after all of the manganese oxide has reacted. Therefore, the degree of contact with the overlying seawater reservoir of sulfate S will be limited, sulfate reduction will go to completion, and the small amount of sulfide that does form will be isotopically close to its parent sulfate compared to normal pyrite in black shale. Okita and Shanks (1992) reported $\delta^{34}S$ values of pyrite from unmineralized black shales of -21 per mil compared to $+3.2$ per mil in the ore zones. Another prediction from this equation is that iron oxides should accompany the $MnCO_3$ mineralization. The ore at Molango is, in fact, distinctly magnetic, and Okita (1992) has reported high concentrations of magnetite and maghemite.

Subsequent work has shown that this model has broad applicability to manganese ore deposits. See, for example, Nyame (1998) on the Nsuta deposit of Ghana, Tang and Liu (1999) for a representative Neoproterozoic deposit of China, and Tsikos (1999) for the Hotazel deposit in the Kalahari manganese field of South Africa. In addition, many deposits that had been thought to be dominated by primary manganese oxide may, in fact, be lateritic weathering products of manganese carbonate protores (Varentsov, 1996).

7.11.8 BEHAVIOR OF MANGANESE IN SOILS AND WEATHERING

Manganese is an important nutrient in soils and can be a limiting factor in plant nutrition. Therefore, it has received considerable attention in the soil science community. Most of this literature focuses on the distribution of available rather than total manganese, which would be of more interest to geochemists trying to reconstruct past soil behavior. An additional limitation is that, because manganese is usually treated as a major element by geochemists but occurs in concentrations less than 1%, most soil analyses of manganese report only one or two significant figures, making judgments about manganese behavior in the soil difficult. The available literature for modern soils (see Maynard, 1992 and references therein) indicates that both iron and manganese have some mobility within the soil but that iron is always retained somewhere within the profile, while manganese may or may not be retained. Table 5 shows that iron and manganese occur in soils at a somewhat higher level than the corresponding parent rocks, but that the Mn/Fe ratio is relatively constant for all parent rock types in the range of 0.013–0.016, similar to the range for crustal reservoirs and rock types given in Table 1.

Sequential extraction of soils shows that the proportion of available to total manganese varies greatly, but averages ~50% (Alloway, 1995, p. 231). As predicted by the Eh–pH diagram for manganese, both lower pH and

Table 5 Distribution of Mn and Fe in soils and paleosols.

Name	*Host rock*	*Age of soil*	*Mn (ppm) soil*	*Fe (ppm) soil*	*Mn/Fe soil*	*Mn (ppm) host rock*	*Fe (ppm) host rock*	*Mn/Fe host rock*	*Source*
Mt. Prinzera, Italy	Ultramafic	Modern	1,600	101,000	0.016	930	58,500	0.016	Venturelli *et al.* (1997)
Belbex, France	Basalt	Modern	1,800	119,000	0.015	1,700	86,200	0.020	Chesworth *et al.* (1981)
San Pedro, Portugal	Granite	Modern	180	13,600	0.013	200	10,400	0.019	Middleburg *et al.* (1988)
Llano, Texas	Granite	0.5 Ga	130	14,300	0.009	220	13,800	0.016	Capo (1984)
Sturgeon, Canada	Basalt	1.1 Ga	2,300	97,800	0.023	2,200	84,900	0.026	Zbinden *et al.* (1988)
Flinflon, Canada	Greenstone	1.8 Ga	840	116,000	0.007	820	32,900	0.025	Holland *et al.* (1989)
Quirke, Canada	Basalt	2.5 Ga	2,000	101,000	0.020	2,300	93,100	0.025	Prasad and Roscoe (1991)
Kalkkloof, South Africa	Ultramafic	2.6 Ga	520	37,900	0.014	800	29,200	0.027	Martini (1994)
Ottosdol, South Africa	Granite	2.8 Ga	290	25,900	0.011	320	23,800	0.013	Grandstaff *et al.* (1986)

lower Eh favor manganese availability to plants. For example, manganese uptake by roots falls off sharply at pH values in the rhizosphere above 5.5. However, manganese availability rises sharply in the summer, presumably reflecting lowered O_2 in the soil atmosphere. These strong shifts in manganese solubility with time and place in the soil provide the potential for considerable manganese migration.

Most profiles show some removal of manganese from the surface layers of the soil. In some cases reprecipitation deeper retains all of this manganese within the profile, while other modern soils show considerable leaching. Middleburg *et al.* (1988) presented a geochemically detailed study of 10 soil profiles on the Iberian peninsula: five show manganese conserved and five show appreciable loss, whereas iron is conserved in all of the profiles. The average manganese loss over the 10 profiles is $\sim$18%.

Thus, modern soils tend to show some leaching of manganese with iron retained, reflecting the greater solubility of manganese in the presence of oxygen. To what extent is this true of ancient soils? Many paleosols have been examined geochemically to try to reconstruct atmospheric conditions of the past, particularly oxygen levels (see Maynard (1992) and Rye and Holland (1998) for reviews). In general, older paleosols, greater than $\sim$2.2 Ga, show considerable iron depletion, whereas younger paleosols show conservation of iron within the profile, which has been interpreted to show that the atmosphere prior to 2.2 Ga held relatively little oxygen (Holland *et al.*, 1989). Manganese, however, tends to show leaching in paleosols of all ages, as can be seen from the Mn/Fe ratios for soils and host rocks given in Table 5, so that manganese behavior in soils is more uniform in time than that of iron.

Manganese ore bodies generally have experienced supergene alteration that has produced a new set of minerals, and, via the removal of carbonate, produced higher grades of ore (Varentsov, 1996). Parc *et al.* (1989) provided a detailed assessment of the progress of weathering in manganese-rich rocks and proposed two sequences of increased oxidation states based on the dominant manganese mineral in the parent rock: rhodochrosite (2) → manganite (3) → cryptomelane (3.94) → nsutite (3.95) → lithiophorite (4) and Mn silicates (2) → birnessite (3.71) → nsutite (3.95) → pyrolusite (4) where the number given in parentheses is the net oxidation state of manganese in the mineral. Cryptomelane is reported as a common phase in most weathering profiles on manganese ores and raises an interesting question of the source of the necessary potassium. Some can come from conversion of potassium silicates to kaolinite in the weathering profile, but even the potassium-free rocks of the Mamatwan ores are converted to a thick layer of cryptomelane (Gutzmer and Beukes, 1996b), which would seem to require some external source of potassium.

A curious aspect of the weathering of manganese ore bodies is the occasional formation of "battery-active" MnO_2, which is a far more valuable product than the ordinary metallurgical grade oxide. Usually the battery-active material is dominated by the mineral nsutite, but much nsutite does not exhibit this property. In fact, manganese exploration crews evaluate the grade of a deposit by making batteries in the field. Why some oxides exhibit this property and others do not has remained something of a mystery. Turner and Buseck (1983) have suggested that it is related to the presence of domains of triple-chain structure in some nsutites.

7.11.9 CONCLUSIONS

Manganese is of great interest in geochemistry, because its minerals are both tracers of redox processes and accumulators of other elements of great geochemical significance. The solubility of manganese compared to iron under reducing and mildly oxidizing conditions leads to its export from low-oxygen environments, be they basalt–hydrothermal systems or euxinic sedimentary basins, and its accumulation in oxidizing environments of the shallow ocean or in low-productivity areas of the deep sea. Therefore, tracking Mn/Fe ratios provides us with a means of reconstructing the oxidation structure of ocean basins or of soils or of groundwater systems.

The strong affinity of manganese oxides for certain transition elements and the REEs provides another way to gain insight into geochemical processes in the oceans. Especially valuable are the records of cerium and europium anomalies preserved in manganese accumulations, which can define the relative proportions of hydrothermal and diagenetic processes in the sedimentary record.

REFERENCES

Aller R. C. and Rude P. D. (1988) Complete oxidation of solid phase sulfide by manganese and bacteria in anoxic marine sediments. *Geochim. Cosmochim. Acta* **52**, 751–765.

Alloway B. J. (1995) *Heavy Metals in Soils*, 2nd edn. Blackie Academic and Professional Publishers, London.

Bachinski D. J. (1977) Sulfur isotopic composition of ophiolitic cupriferous iron sulfide deposits, Notre Dame Bay, Newfoundland. *Econ. Geol.* **72**, 243–257.

Bäcker H., Marchig V., von Stackelberg U., Stoffers P., Puteanus D., and Tufar W. (1991) Hydrothermale Aktivität auf dem Meeresboden. *Geologische Jahrbuch* **D93**, 103–197.

Balzer W. (1982) On the distribution of iron and manganese at the sediment/water interface: thermodynamic versus kinetic control. *Geochim. Cosmochim. Acta* **46**, 1153–1161.

Bau M. and Möller P. (1993) Rare earth element systematics of the chemically precipitated component in Early Precambrian iron formations and the evolution of the terrestrial atmosphere–hydrosphere–lithosphere system. *Geochim. Cosmochim. Acta* **57**, 2239–2249.

Bonatti E. and Joensuu O. (1966) Deep-sea iron deposit from the South Pacific. *Science* **154**, 643–645.

Bonatti E., Honnorez J., Joensuu O., and Rydell H. (1972) Submarine iron deposits from the Mediterranean Sea. In *The Mediterranean Sea* (ed. D. J. Stanley). Hutchinson and Ross, Stroudsburg, PA, pp. 701–710.

Boström K. and Peterson M. N. A. (1966) Precipitates from hydrothermal exhalations on the East Pacific Rise. *Econ. Geol.* **61**, 1258–1265.

Bottcher M. E. and Huckriede H. (1997) First occurrence and stable isotope composition of authigenic γ-MnS in the central Gotland Deep (Baltic Sea). *Mar. Geol.* **137**, 201–205.

Burke I. T. and Kemp A. E. S. (2002) Microfabric analysis of Mn-carbonate laminae deposition and Mn-sulfide formation in the Gotland Deep, Baltic Sea. *Geochim. Cosmochim. Acta* **66**, 1589–1600.

Burns R. G. and Brown B. (1972) Nucleation and mineralogical controls on the composition of manganese nodules. In *Ferromanganese Deposits on the Ocean Floor* (ed. D. R. Horn). Lamont-Doherty Geol. Observatory, New York, pp. 51–61.

Burns R. G. and Burns M. (1977a) The mineralogy and crystal chemistry of deep-sea manganese nodules, a polymetallic resource of the twenty-first century. *Phil. Trans. Roy. Soc. London* **286A**, 283–301.

Burns R. G. and Burns M. (1977b) Mineralogy. In *Marine Manganese Deposits* (ed. G. Glasby). Elsevier, Amsterdam.

Calvert S. E. and Pedersen T. F. (1996) Sedimentary geochemistry of manganese: implications for the environment of formation of manganiferous black shales. *Econ. Geol.* **91**, 36–47.

Calvert S. E. and Price N. B. (1970) Composition of manganese nodules and manganese carbonates from Loch Fyne, Scotland. *Contrib. Mineral. Petrol.* **29**, 215–233.

Calvert S. E. and Price N. B. (1977) Geochemical variation in ferromanganese nodules and associated sediments from the Pacific Ocean. *Mar. Chem.* **5**, 43–74.

Capo R. C. (1984) Petrology and geochemistry of a Cambrian paleosol developed on Precambrian granite, Llano Uplift, Texas. MA Thesis, University of Texas at Austin (unpublished).

Chapelle F. H. (1993) *Ground Water Microbiology and Geochemistry*. Wiley, New York.

Chapman H. J. and Spooner E. T. C. (1977) ^{87}Sr enrichment of ophiolitic sulphide deposits in Cyprus confirms ore formation by circulating seawater. *Earth Planet. Sci. Lett.* **35**, 71–78.

Chesworth W., Dejou J., and Larroque P. (1981) The weathering of basalt and relative mobilities of the major elements at Belbex, France. *Geochim. Cosmochim. Acta* **45**, 1235–1243.

Cowen J. P. and Bruland K. W. (1985) Mineral deposits associated with bacteria. Implications for Fe and Mn marine biogeochemistry. *Deep-Sea Res.* **32**, 253–272.

Crerar D. A., Cormick R. K., and Barnes H. L. (1980a) Geochemistry of manganese, an overview. In *Geology and Geochemistry of Manganese* (eds. I. M. Varentsov and Gy. Grasselly). E. Schweizerbart'sche Verlagsbuchhandlung, Stuttgart, vol. 1, pp. 293–334.

Cronan D. S. (1976) Implications of metal dispersion from hydrothermal systems for mineral exploration on mid-ocean ridges and in island arcs. *Nature* **262**, 567–569.

Cronan D. S. (1997) Some controls on the geochemical variability of manganese nodules with particular reference to the tropical South Pacific. In *Manganese Mineralization: Geochemistry and Mineralogy of Terrestrial and Marine Deposits*, Special Publication Number 119 (eds. K. Nicholson, J. R. Hein, B. Bühn, and S. Dasgupta). Geological Society of London, pp. 139–152.

DeCarlo E. H. (1991) Paleoceanographic implications of rare earth elements variability within a Fe–Mn crust from the central Pacific Ocean. *Mar. Geol.* **98**, 449–467.

Deffeyes K. S. (1970) The axial valley: a steady-state feature of the terrain. In *The Megatectonics of Continents and Oceans* (eds. H. Johnson and B. L. Smith). Rutgers University Press, Rutgers, NJ, pp. 194–222.

Derry L. A. and Jacobsen S. B. (1990) The chemical evolution of Precambrian seawater: evidence from REEs in banded iron formations. *Geochim. Cosmochim. Acta* **54**, 2965–2977.

Dollhopf M. E., Nealson K. H., Simon D. M., and Luther G. W. (2000) Kinetics of Fe(III) and Mn(IV) reduction by the Black Sea strain of *Shewanella putrefaciens* using *in situ* solid state voltammetric Au/Hg electrodes. *Mar. Chem.* **70**, 171–180.

Elder J. W. (1965) Physical processes in geothermal areas. In *Terrestrial Heat Flow*, Monograph No. 8 (ed. W. H. K. Lee). American Geophysical Union, Washington, DC, pp. 221–229.

Fan D.-L., Ye J., Yin L.-M., and Zhang R.-F. (1999) Microbial processes in the formation of the Sinian Gaoyan manganese carbonate ore, Sichuan Province, China. *Ore Deposit Rev.* **15**, 79–93.

Fleet A. J., Boström K., Laubier L., and Smith K. L. (1983) Hydrothermal and hydrogenous ferro-manganese deposits: do they form a continuum? The rare earth element evidence. In *Hydrothermal Processes at Seafloor Spreading Centers* (ed. P. A. Rona). Plenum, New York, pp. 535–555.

Force E. R. and Cannon W. F. (1988) A depositional model for shallow-marine manganese deposits around black-shale basins. *Econ. Geol.* **83**, 83–117.

Force E. R. and Maynard J. B. (1991) Manganese: syngenetic deposits on the margins of anoxic basins. In *Sedimentary and Diagenetic Mineral Deposits: A Basin Analysis Approach to Exploration* (eds. E. R. Force, J. J. Eidel, and J. B. Maynard). Society of Econ. Geologists, El Paso, TX, pp. 147–159.

Frenzel G. (1980) The manganese ore minerals. In *Geology and Geochemistry of Manganese* (eds. I. M. Varentsov and Gy. Grasselly). E. Schweizerbart'sche Verlagsbuchhandlung, Stuttgart, vol. 1, pp. 25–158.

Froelich P. N., Klinkhammer G. P., Bender M. L., Luedtke N. A., Heath G. R., Cullem D., Dauphin P., Hammond D., Hartman B., and Maynard V. (1979) Early oxidation of organic matter in pelagic sediments of the eastern equatorial Atlantic: suboxic diagenesis. *Geochim. Cosmochim. Acta* **43**, 1075–1091.

Glasby G. (1974) Mechanisms of incorporation of manganese and associated trace elements in marine manganese nodules. *Oceanogr. Mar. Biol. Ann. Rev.* **12**, 11–40.

Glasby G. P., Emelyanov E. M., Zhamoida A., Baturin G. N., Leipe T., Bahlo R., and Bonacker P. (1997) Environments of formation of ferromanganese concretions in the Baltic Sea: a critical review. In *Manganese Mineralization: Geochemistry and Mineralogy of Terrestrial and Marine Deposits*, Special Publication No. 119 (eds. K. Nicholson, J. R. Hein, B. Bühn, and S. Dasgupta). Geological Society of London, London, pp. 29–42.

Goldsmith J. R. and Graf D. L. (1957) The system CaO–MnO–CO_2: solid solution and decomposition relations. *Geochim. Cosmochim. Acta* **11**, 310–334.

Graf J. L., O'Connor E. A., and Van Leeuwen P. (1994) Rare earth element evidence of origin and depositional environment of Late Proterozoic ironstone beds and manganese-oxide deposits, SW Brazil and SE Bolivia. *J. South Am. Earth Sci.* **7**, 115–133.

Grandstaff D. E., Edelman M. J., Foster R. W., Zbinden E., and Kimberley M. M. (1986) Chemistry and mineralogy of Precambrian paleosols at the base of the Dominion and Pongola Groups (Transvaal, South Africa). *Precamb. Res.* **32**, 97–131.

Gutzmer J. (1996) Genesis and alteration of the Kalhari and Postmasburg manganese deposits, Griqualand West, South Africa. PhD Dissertation, Rand Afrikaans University (unpublished).

Gutzmer J. and Beukes N. J. (1996a) Karst-hosted fresh-water Paleoproterozoic manganese deposits, Postmasburg, South Africa. *Econ. Geol.* **91**, 1435–1454.

Gutzmer J. and Beukes N. J. (1996b) Mineral paragenesis of the Kalahari manganese field, South Africa. *Ore Geol. Rev.* **11**, 405–428.

Gutzmer J. and Beukes N. J. (1997) Mineralogy and mineral chemistry of oxide-facies manganese ores of the Postmasburg manganese field, South Africa. *Min. Mag.* **61**, 213–231.

Halbach P., Giovanoli R., and von Borstel D. (1982) Geochemical processes controlling the relationship between Co, Mn, and Fe in early diagenetic deep-sea nodules. *Earth Planet. Sci. Lett.* **60**, 226–236.

Hannington M. D., Jonasson I. R., Herzig M., and Petersen S. (1995) Physical and chemical processes of seafloor mineralization at mid-ocean ridges. In *Seafloor Hydrothermal Systems*, Monograph 91 (eds. S. E. Humphris, R. A. Zierenberg, L. S. Mullineaux, and R. E. Thomson). American Geophysical Union, Washington, pp. 115–157.

Hein J. R. and Manheim F. T. (2000) Cobalt-rich ferromanganese crusts in the Pacific. In *Handbook of Marine Mineral Deposits* (ed. D. S. Cronan). CRC Press, Boca Raton, FL, pp. 239–279.

Hein J. R., Koschinsky A., Halbach P., Manheim F. T., Bau M., Kang J.-K., and Lubick N. (1997) Iron and manganese oxide mineralization in the Pacific. In *Manganese Mineralization: Geochemistry and Mineralogy of Terrestrial and Marine Deposits*, Special Publication Number 119 (eds. K. Nicholson, J. R. Hein, B. Bühn, and S. Dasgupta). Geological Society of London, pp. 123–138.

Hem J. D. (1972) Chemical factors that influence the availability of iron and manganese in aqueous systems. *Geol. Soc. Am. Spec. Pap.* **140**, 17–24.

Hem J. D. (1981) Rates of manganese oxidation in aqueous systems. *Geochim. Cosmochim. Acta* **45**, 1369–1374.

Heubner J. S. (1976) The manganese oxides—a bibliographic commentary. In *Oxide Minerals* (ed. D. Rumble). Mineralogical Society of America Short Course Notes, vol. 3, pp. SH1–SH17.

Holland H. D., Feakes C. R., and Zbinden E. A. (1989) The Flin Flon paleosol and the composition of the atmosphere 1.8 BYB. *Am. J. Sci.* **289**, 362–389.

Honnorez J., Honnorez-Guerstein B., Valette J., and Wauschkuhn A. (1973) Present day formation of an exhalative sulfide deposit at Vulcano (Tyrrhenian Sea): Part II. Active crystallization of fumarolic sulfides in the volcanic sediments of the Baia di Levante. In *Ores in Sediments* (eds. G. C. Amstutz and A. J. Bernard). Springer, Heidelberg, pp. 139–166.

Huckriede H. and Meischner D. (1996) Origin and environment of manganese-rich sediments within black-shale basins. *Geochim. Cosmochim. Acta* **60**, 1399–1413.

Kadko D., Baross J., and Alt J. (1995) The magnitude and global implications of hydrothermal flux. In *Seafloor Hydrothermal Systems*, Monograph 91 (eds. S. E. Humphris, R. A. Zierenberg, L. S. Mullineaux, and R. E. Thomson). American Geophysical Union, Washington, DC, pp. 446–466.

Kirschvink J. L., Gaidos E. J., Bertani L. E., Beukes N. J., Gutzmer J., Maepa L. N., and Steinberger R. E. (2000) Paleoproterozoic snowball Earth: extreme climatic and geochemical global change and its biological consequences. *Proc. Natl. Acad. Sci.* **97**, 1400–1405.

Klein C. and Beukes N. J. (1993) Sedimentology and geochemistry of the glaciogenic Late Proterozoic Rapitan iron-formation in Canada. *Econ. Geol.* **88**, 542–565.

Kleyenstüber A. S. E. (1986) The mineralogy of the manganese-bearing Hotazel Formation, of the proterozoic transvaal sequence in Griqualand West, South Africa. In *Mineral Deposits of Southern Africa* (eds. C. R. Anhaeusser and S. Maske). Geological Society of South Africa, Johannesburg, pp. 257–272.

Krauskopf K. B. (1957) Separation of manganese from iron in sedimentary processes. *Geochim. Cosmochim. Acta* **12**, 61–84.

Langmuir D. (1997) *Aqueous Environmental Chemistry*. Prentice-Hall, Upper Saddle River, NJ.

Lepland A. and Stevens R. L. (1998) Manganese authigenesis in the Landsort Deep, Baltic Sea. *Mar. Geol.* **151**, 1–25.

Li Y.-H. (2000) *A Compendium of Geochemistry*. Princeton University Press, Princeton, NJ.

Liu T.-B. (1988) C–S–Fe correlation of shales hosting sedimentary manganese deposits. PhD Dissertation, University of Cincinnati (unpublished).

Liu T.-B., Maynard J. B., and Alten J. Superheavy S isotopes from glacial-associated sediments of the Neoproterozoic of South China: oceanic Anoxia or sulfate limitation? *Geol. Soc. Am. Spec. Pap.* (in press).

Lovley D. R. and Phillips E. J. P. (1988) Manganese inhibition of microbial iron reduction in anaerobic sediments. *Geomicrobiol. J.* **6**, 145–155.

Lupton J. E. (1995) Hydrothermal plumes: near and far field. In *Seafloor Hydrothermal Systems*, Monograph 91 (eds. S. E. Humphris, R. A. Zierenberg, L. S. Mullineaux, and R. E. Thomson). American Geophysical Union, Washington, DC, pp. 317–346.

Mandernack K. W. and Tebo B. M. (1993) Manganese scavenging and oxidation at hydrothermal vents and in vent plumes. *Geochim. Cosmochim. Acta* **57**, 3907–3923.

Mandernack K. W., Post J., and Tebo B. M. (1995) Manganese mineral formation by bacterial spores of the marine *Bacillus*, strain SG-1: evidence for the direct oxidation of Mn(II) to Mn(IV). *Geochim. Cosmochim. Acta* **59**, 4393–4408.

Martini J. E. J. (1994) A Late-Archean–Palaeoproterozoic (2.6 Ga) palaeosol on ultramafics in the Eastern Transvaal, South Africa. *Precamb. Res.* **67**, 159–180.

Maynard J. B. (1976) The long-term buffering of the oceans. *Geochim. Cosmochim. Acta* **40**, 1523–1532.

Maynard J. B. (1983) *Geochemistry of Sedimentary Ore Deposits*. Springer, Heidelberg.

Maynard J. B. (1992) Chemistry of modern soils as a guide to interpreting Precambrian paleosols. *J. Geol.* **100**, 279–289.

Maynard J. B., Okita P. M., May E. D., and Martinez-Vera A. (1990) Palaeogeographic setting of Late Jurassic manganese mineralization in the Molango District, Mexico. In *Sediment Hosted Mineral Deposits*, International Association of Sedimentologists, Special Publication 11 (eds. J. Parnell, L.-J. Ye, and C.-M. Chen). Blackwell, Boston, pp. 17–30.

Middleburg J. J., van der Weijden C. H., and Woitties J. R. W. (1988) Chemical processes affecting the mobility of major, minor and trace elements during weathering of granitic rocks. *Chem. Geol.* **68**, 253–273.

Mills R. A. and Elderfield H. (1995) Hydrothermal activity and the geochemistry of metalliferous sediment. In *Seafloor Hydrothermal Systems,* Monograph 91 (eds. S. E. Humphris, R. A. Zierenberg, L. S. Mullineaux, and R. E. Thomson). American Geophysical Union, Washington, DC, pp. 392–407.

Mitra A., Elderfield H., and Greaves M. J. (1994) Rare earth elements in submarine hydrothermal fluids and plumes form the Mid-Atlantic Ridge. *Mar. Chem.* **46**, 217–235.

Moffett J. W. and Ho J. (1996) Oxidation of cobalt and manganese in seawater via a common microbially catalyzed pathway. *Geochim. Cosmochim. Acta* **60**, 3415–3424.

Murton B. J., Redbourn L. J., German C. R., and Baker E. T. (1999) Sources and fluxes of hydrothermal heat, chemicals and biology within a segment of the Mid-Atlantic Ridge. *Earth Planet. Sci. Lett.* **171**, 301–317.

Myers C. R. and Nealson K. H. (1988) Microbial reduction of manganese oxides: interactions with iron and sulfur. *Geochim. Cosmochim. Acta* **52**, 2727–2732.

Nealson K. H. (1997) Sediment bacteria: Who's there, what are they doing, and what's new? *Ann. Rev. Earth Planet. Sci.* **25**, 403–434.

Nel D. J., Beukes N. J., and De Villiers J. P. R. (1986) The Mamatwan manganese mine of the Kalhari manganese field. In *Mineral Deposits of Southern Africa* (eds. C. R. Anhaeusser and S. Maske) Geological Society of South Africa, Johannesburg, pp. 963–978.

Neumann T., Heiser U., Leosson M. A., and Kersten M. (2002) Early diagenetic processes during Mn carbonate formation: evidence from the isotopic composition of authigenic Ca-rhodochrosite of the Baltic Sea. *Geochim. Cosmochim. Acta* **66**, 867–879.

Nicholson K. (1992) Contrasting mineralogical–geochemical signatures of manganese oxides: guides to metallogenesis. *Econ. Geol.* **87**, 1253–1264

Nyame F. K. (1998) Mineralogy, geochemistry, and genesis of the Nsuta manganese deposit, Ghana. PhD Dissertation, Okayama University (unpublished).

Ohmoto H. and Goldhaber M. B. (1997) Sulfur and carbon isotopes. In *Geochemistry of Hydrothermal Ore Deposits* (ed. H. L. Barnes). Wiley, New York, pp. 517–611.

Okita P. M. (1987) Geochemistry and mineralogy of the Molango manganese orebody, Hidalgo State, Mexico. PhD Dissertation, University of Cincinnati (unpublished).

Okita P. M. (1992) Manganese carbonate mineralization in the Molango District, Mexico. *Econ. Geol.* **87**, 1345–1366.

Okita P. M. and Shanks W. C. (1992) Origin of stratiform sediment-hosted manganese carbonate ore deposits: examples from Molango, Mexico, and TaoJiang, China. *Chem. Geol.* **99**, 139–164.

Okita P. M., Maynard J. B., Spiker E. C., and Force E. R. (1988) Isotopic evidence for organic matter oxidation by manganese reduction in the formation of stratiform manganese carbonate ore. *Geochim. Cosmochim. Acta* **52**, 2679–2685.

Olivarez A. M. and Owen R. M. (1991) The europium anomaly of seawater: implications for fluvial versus hydrothermal REE inputs to the oceans. *Chem. Geol.* **92**, 317–328.

Parc S., Nahon D., Tardy Y., and Vieillard P. (1989) Estimated solubility products and fields of stability for cryptomelane, nsutite, birnessite, and lithiophorite based on natural lateritic weathering sequences. *Am. Mineral.* **74**, 466–475.

Pedersen T. F. and Price N. B. (1982) The geochemistry of manganese carbonate in Panama Basin sediment. *Geochim. Cosmochim. Acta* **46**, 59–68.

Prasad N. and Roscoe S. M. (1991) Profiles of altered zones at ca. 2.45 Ga unconformities beneath Huronian strata, Elliot Lake, Ontario: evidence for Early Aphebian weathering under anoxic conditions. *Geol. Surv. Can., Pap.* **91-1C**, 43–54.

Puchelt H. (1973) Recent iron sediment formation at the Kameni Islands, Santorini (Greece). In *Ores in Sediments* (eds. G. C. Amstutz and A. J. Bernard). Springer, Heidelberg, pp. 227–246.

Puchelt H., Schock H. H., and Schroll E. (1973) Rezente marine Eisenerze auf Santorin, Greichenland: I. Geochemie, Entstehung, Mineralogie. *Geol. Rundsch.* **62**, 786–803.

Robertson A. H. F. (1975) Cyprus umbers: basalt–sediment relationships on a Mesozoic ocean ridge. *J. Geol. Soc. London*. **131**, 511–531.

Roitz J. S., Flegal A. R., and Bruland K. W. (2002) The biogeochemical cycling of manganese in San Francisco Bay:

temporal and spatial variations in surface water concentrations. *Estuar. Coast. Shelf Sci.* **54**, 227–239.

Roy S. (1981) *Manganese Deposits*. Academic Press.

Russell M. J. (1975) Lithogeochemical environment of the Tynagh basemetal deposit, Ireland, and its bearing on ore deposition. *Trans. Inst. Mining Metal.* **84**, B128–B133.

Rye R. and Holland H. D. (1998) Paleosols and the evolution of atmospheric oxygen: a critical review. *Am. J. Sci.* **298**, 621–672.

Schippers A. and Jørgensen B. B. (2001) Oxidation of pyrite and iron sulfide by manganese dioxide in marine sediments. *Geochim. Cosmochim. Acta* **65**, 915–922.

Scholten J. C., Stoffers P., Garbe-Schönberg D., and Moammar M. (2000) Hydrothermal mineralization in the Red Sea. In *Handbook of Marine Mineral Deposits* (ed. D. S. Cronan). CRC Press, Boca Raton, FL, pp. 369–395.

Seyfried W. E. and Bischoff J. L. (1981) Experimental seawater–basalt interaction at 300 °C, 500 bars: chemical exchange, secondary mineral formation and implications for the transport of heavy metals. *Geochim. Cosmochim. Acta* **45**, 135–147.

Seyfried W. E. and Ding K. (1995) Phase equilibria in subseafloor hydrothermal systems: a review of the role of redox, temperature, pH and dissolved Cl on the chemistry of hot spring fluids at mid-ocean ridges. In *Seafloor Hydrothermal Systems*, Monograph 91 (eds. S. E. Humphris, R. A. Zierenberg, L. S. Mullineaux, and R. E. Thomson). American Geophysical Union, Washington, DC, pp. 248–272.

Shanks W. C. and Bischoff J. L. (1977) Ore transport and deposition in the Red Sea geothermal system: a geochemical model. *Geochim. Cosmochim. Acta* **41**, 1507–1519.

Shanks W. C. and Bischoff J. L. (1980) Geochemistry, sulfur isotope composition, and accumulation rates of Red Sea geothermal deposits. *Econ. Geol.* **75**, 445–459.

Shanks W. C., Böhlke J. K., and Seal R. R. (1995) Stable isotopes in mid-ocean ridge hydrothermal systems: interactions between fluids, minerals, and organisms. In *Seafloor Hydrothermal Systems,* Monograph 91 (eds. S. E. Humphris, R. A. Zierenberg, L. S. Mullineaux, and R. E. Thomson). American Geophysical Union, Washington, DC, pp. 194–221.

Sillitoe R. H. (1980) Are porphyry copper and kuroko-type massive sulfide deposits incompatible? *Geology* **8**, 11–14.

Sivaprakash C. (1980) Mineralogy of manganese deposits of Kodura and Garbham, Andhra Pradesh, India. *Econ. Geol.* **75**, 1083–1104.

Spooner E. T. C. and Bray C. J. (1977) Hydrothermal fluids of seawater salinity in ophiolitic sulphide ore deposits in Cyprus. *Nature* **266**, 808–812.

Spooner E. T. C., Chapman H. J., and Smewing J. D. (1977) Strontium isotope contamination and oxidation during ocean floor hydrothermal metamorphism of the ophiolitic rocks of the Troodos Massif, Cyprus. *Geochim. Cosmochim. Acta* **41**, 873–890.

Sternbeck J. and Sohlenius G. (1997) Authigenic sulfide and carbonate mineral formation in Holocene sediments of the Baltic Sea. *Chem. Geol.* **135**, 55–73.

Stumm W. and Morgan J. J. (1996) *Aquatic Chemistry,* 3rd edn. Wiley-Interscience, New York, 1022pp.

Suess E. (1979) Mineral phases formed in anoxic sediments by microbial decomposition of organic matter. *Geochim. Cosmochim. Acta* **43**, 339–352.

Tang S.-Y. and Liu T.-B. (1999) Origin of the early Sinian Minle manganese deposit, Hunan Province, China. *Ore Geol. Rev.* **15**, 71–78.

Tebo B. M. and Nealson K. H. (1984) Microbial mediation of Mn(II) and Co(II) precipitation at the O_2/H_2S interface in two anoxic fjords. *Limnol. Oceanogr.* **29**, 1247–1258.

Tsikos H. (1999) Petrographic and geochemical constraints on the origin and post-depositional history of the Hotazel iron–manganese deposits, Kalahari Manganese Field, South Africa. PhD Dissertation, Rhodes University (unpublished).

Turner S. and Buseck P. R. (1983) Defects in nsutite (γ-MnO_2) and dry-cell battery efficiency. *Nature* **304**, 143–146.

Usui A. and Someya M. (1997) Distribution and composition of marine hydrogenetic and hydrothermal manganese deposits in the northwest Pacific. In *Manganese Mineralization: Geochemistry and Mineralogy of Terrestrial and Marine Deposits,* Special Publication No. 119 (eds. K. Nicholson, J. R. Hein, B. Bühn, and S. Dasgupta). Geological Society of London, London. pp. 177–198.

Valette J. N. (1973) Distribution of certain trace elements in marine sediments surrounding Vulcano Island (Italy). In *Ores in Sediments* (eds. G. C. Amstutz and A. J. Bernard). Springer, Heidelberg, pp. 321–338.

Varentsov I. M. (1996) *Manganese Ores of Supergene Zone: Geochemistry of Formation.* Kluwer Academic, Dordrecht.

Veizer J. (1988) The evolving exogenic cycle. In *Chemical Cycles in the History of the Earth* (eds. C. B. Gregor, R. M. Garrels, F. T. Mackenzie, and J. B. Maynard). Wiley-Interscience, pp. 175–220.

Venturelli G., Contini S., and Bonazzi A. (1997) Weathering of ultramafic rocks and element mobility at Mt. Prinzera, Northern Apennines, Italy. *Min. Mag.* **61**, 765–778.

Vrobelsky D. A. and Chapelle F. H. (1994) Temporal and spatial changes of terminal electron-accepting processes in a petroleum hydrocarbon-contaminated aquifer and the significance for contaminant biodegradation. *Water Resour. Res.* **30**, 1561–1570.

Wijsman J. W. M., Herman P. M. J., Middleburg J. J., and Soetaert K. (2002) A model for early diagenetic processes in sediments of the continental shelf of the Black Sea. *Estuar. Continent. Shelf Sci.* **54**, 403–421.

Wolery T. J. and Sleep N. D. (1988) Interactions of geochemical cycles with the mantle. In *Chemical Cycles in the History of the Earth* (eds. C. B. Gregor, R. M. Garrels, F. T. Mackenzie, and J. B. Maynard). Wiley-Interscience, New York, pp. 77–104.

Zbinden E. A., Holland H. D., and Feakes C. R. (1988) The Sturgeon Falls paleosol and the composition of the atmosphere 1.1 Ga BP. *Precamb. Res.* **42**, 141–163.

Zhang J., Lion L. W., Nelson Y. M., Shuler M. L., and Ghiorse W. C. (2002) Kinetics of Mn(II) oxidation by *Leptothrix discophora* SS1. *Geochim. Cosmochim. Acta* **65**, 773–781.

7.12
Green Clay Minerals

B. Velde

Ecole Normale Supérieure, Paris, France

7.12.1 WHAT ARE WE LOOKING AT?

Color is a problem for scientific study. One aspect is the vocabulary one used to describe color. Mint green, bottle green, and Kelly green are nice names but not of great utility in that people's physical perception of color is not always the same. In some industries, such as colored fabric manufacture, current use is to send a set of standard colors which are matched by the producer. This is similar to the use of the Munsell color charts in geology. None of these processes makes use of physical optical spectral studies. The reason is that they are difficult to obtain and interpret. For a geologist, color is very important but we rarely have the possibility to standardize the method of our color perception. One reason is that color is both a reflective and transmission phenomenon. The thickness of the sample is critical to any transmission characteristics. Hence, a field color determination is different from one made by using a petrographic microscope. Green glauconite in a hand specimen is not the same color in 30 μm thick thin section seen with a microscope using transmitted light.

A second problem is that color in a spectral identification is the result of several absorption emissions, with overlapping signal, forming a complicated spectrum. Interpretation depends very greatly on the spectrum of the light source and the conditions of transmission–reflection of the sample. As a result, for this text, we will not attempt to analyze the physical aspect of green in green clays. In the discussion which follows, reference is made concerning color, to thin section microscopic perception.

Very briefly, green clay minerals are green, because they contain iron. This is perhaps not a great revelation to mineralogists, but it is the key to understanding the origin and stability of green clay minerals. In fact, iron can color minerals either red or green or in various shades of orange and brown. The color most likely depends upon the relative abundance of the iron ion valence in the silicate (clay mineral in our case) structure, the specific bonding of these ions, and other factors. In fact, the reasons for coloration are not known completely, but it is certain that a combination of Fe^{2+} and Fe^{3+} ions is necessary to give a nice green color to clays. In the green clay minerals discussed here, the colors vary greatly as seen under the optical microscope (not always the same as the one seen in hand specimen). Yellow to blue-green hues can be found. However, for the moment, no clear relation between iron content, iron valence ratio, or other factors such as minor transition element concentrations can be found to explain the greenness of green clay minerals. The fact that a clay is green just indicates a combination of the two oxidation states of iron. The color, however, indicates the key to the formation in nature of green clay minerals.

Green clay minerals are in general the product of "mixed valence" conditions of formation, most often in a situation where some iron is reduced from Fe^{3+} and enters into a silicate mineral structure. In general, iron would rather be an oxide when it is in the trivalent state. The moment iron is reduced to a divalent state under surface or near-surface conditions, it looks for a silicate, sulfide, or carbonate to hide in. The reverse is also true, of course. When a silicate is oxidized, Fe^{2+} becoming Fe^{3+}, the iron begins to group together in oxide clumps and eventually exits the silicate structure. This is seen in thin section in altered rocks (weathering or hydrothermal action). The production of trivalent, oxidized iron usually results in a brownish or orange mineral.

If the geology of the formation of green silicate minerals is relatively well defined, especially at near surface or surface conditions, the question remains how much of the iron is in a reduced oxidation state and how? In the case of reduction of iron in surface environments: if most of the iron goes to Fe^{2+}, one mineral is formed; if only part of it is reduced, another is formed. This is the fundamental geochemical aspect of the genesis of green clay minerals; they contain iron in both oxidation states.

Unfortunately modern methods of mineral analysis on a microscopic scale, electron microbeam and others, do not allow the determination of the different oxidation states of iron especially for nonstoichiometric minerals. One can use Mössbauer spectral analysis, but the scales of observations are not the same (Mössbauer needing more material); one method used for observations on a microscale, the other on a macroscale. Given the problems of micro- and macroscale observations, oxidation state information is almost excluded from data gathered since the 1980s or so, and hence information concerning the relations of iron reduction and clay genesis must be taken from older studies. A second, much greater problem is that little X-ray diffraction (XRD) work is done on samples which are analyzed chemically by electron microbeam studies. In the past both types of information, structural and chemical, were available for the same sample. Hence not only do we have no precise chemical data for many samples (oxidation state of iron), but there is a rarity of mineral structural information to go along with the incomplete chemistry. This is critical for the study of clay minerals, because slight chemical changes in a clay mineral are frequently accompanied by changes in its structure, especially when one deals with interstratified clay minerals (mica/smectites for example). In fact, the tendency to obtain more and more precision (analysis of a smaller and smaller sized sample) has led to a total loss of mineralogical data. The Heisenberg principle is unwittingly verified by geologists. We know more about a small part of a sample, but we know less about its whole. As a result, the following discussion is based largely upon old data, those which combine iron oxidation states and XRD information.

7.12.1.1 The Cycle of Transformation

In general, minerals in sedimentary and metamorphic rocks contain ferrous iron (Velde, 1985) which is destined to become iron oxide under conditions of weathering. Oxidation under surface conditions has a tendency to produce iron in the ferric state. Most often the process takes iron out of the silicates and puts it into an oxide phase. In the uppermost layers of mature soils, iron oxide and various silicates, usually non-iron-bearing, are produced. In silicates containing iron, the majority is in the ferric state. The extent of the transformation of iron oxidation state is a rough measure of the maturity of the soil. In the extremely weathered soils one finds only ferric iron and aluminum oxides and hydroxides. These soils are typically red.

Soils become sediments, which eventually find their way to the ocean. In the marine sedimentary environment there is a general tendency towards reducing conditions, through interaction of organic matter with the clays. Hence, reduction of iron can occur in these environments. This does not happen everywhere in sediments, given that red sandstones do exist in large quantities, but when reduction does occur, it has a large impact on the partitioning of iron between oxide and silicate minerals. Lithification (burial diagenesis) tends to increase the reduction of iron in the material and its introduction into silicate (clay mineral) phases (Velde, 1968).

Given this brief background, one should lay down a foundation for the discussion of green clays. First, clay minerals are officially (in mineralogical definitions) composed of crystals less than 2 μm in diameter. This definition was proposed in the nineteenth century when the optical microscope was the major means of mineral investigation. The resolution of the microscopes at the time (and even as of early 2000s) is ~2 μm. Thus, a clay mineral is one that cannot be identified by optical observation (i.e., using different optical properties). Further, a green clay mineral is a mass of unidentifiable grains of green color, which is not very encouraging. However, as XRD techniques became more useful in the 1950s, important new data became available for green and other clay minerals. It might be remarked that one advantage of green clays is that they can be easily recognized both in thin section and hand specimen. Thus, green clays became the subject of investigation as soon as XRD was coupled with microscopic investigations.

7.12.1.2 Geologic Conditions of Green Clay Stability

Green clays belong to the realm of soils, sediments, and sedimentary rocks. A definition of the beginning of metamorphism can, in fact, be based upon the transformation of typical clay minerals into other phases, such as micas and pyrophyllite (Winkler, 1964, Gharrabi *et al.*, 1998; Velde, 1985). The green clays fill the entire space of clay mineral stability and somewhat beyond for several species. For some clay minerals, kinetics are extremely important in their occurrence and stability in geological situations (Velde and Vasseur, 1992). Smectite layers in mixed layer illite/smectite minerals have been eliminated from clay mineral assemblages in Paleozoic rocks at temperatures as low as 80 °C, but they persist to 180 °C in recent sediments. However, it appears that the final transformation of illite, a clay mineral, into mica, a metamorphic mineral, is largely a function of temperature alone, occurring somewhere between 200 °C and 300 °C (Warr *et al.*, 1991), independent of the reaction time (Gharrabi *et al.*, 1998). Chamosites, green clay chlorites, are known to exist in old rocks (Paleozoic) to relatively high temperatures, 200–330 °C (Fernandez and Moro, 1998). Other green clays which form in low-temperature environments (such as berthiérine) persist only to temperatures somewhat below 100 °C but for time spans of hundreds of millions of years (Kim and Lee, 2000).

In general, green clays are stable between surface temperatures (2 °C in the deep ocean) and temperatures somewhat above 300 °C.

7.12.2 DESCRIPTION OF GREEN CLAY MINERALS

7.12.2.1 Potassic Green Clays (Mica-based Structures)

A mica, strictly speaking in our case, is a mineral with one potassium ion or charge on the interlayer site per ten oxygens and two hydroxyl ions, the whole with a negative charge of 22:

$$K_{interlayer}\,(M_{octahedral}M_{tetrahedral})\,O_{10}(OH)_2$$

Potassium is present in the mineral to establish charge balance. Substitution of ions of lower valence in the tetrahedral and octahedral sites leads to an overall charge imbalance which, when high enough, is compensated in low-temperature minerals by potassium. When potassium content or interlayer charge is between about 0.7 and 0.95, the mineral can be called mica like. When the potassium content, i.e., the interlayer charge given in a chemical analysis, is less than 0.7, the mineral which has been analyzed is probably interstratified, i.e., a mixture of mica-like layers (potassic) and lower charge smectite layers (calcium or sodium bearing). Such mixed layering of smectite and mica layers can be detected using XRD. If no XRD information is available for a sample, the distinction between smectite and mica layer is difficult to make. However, this is increasingly the case in modern mineralogical studies and thus one must go on a sort of tradition of compositional range and extrapolate these data from the past to a mineral identification in the present.

The two green clay mica-like minerals (potassic) are glauconite and celadonite. They have very different geological occurrences, but a large overlap in chemical composition (Weaver and Pollard, 1973; Newman and Brown, 1987). The compositional range of glauconites and celadonites can be represented in iron oxide-alumina coordinates (Figure 1). These minerals contain more than 7% K_2O, indicating an approach to a micaceous mineral (i.e., a small number of smectite layers present). There is a tendency for celadonites to have less alumina for equivalent iron contents. The chemical similarity is striking. The color of glauconites differs from that of celadonites in hand specimen, hence the difference in names. Glauconites have a yellow-green hue, while that of celadonite is blue-green. The epitomology of their names does not clarify this distinction.

7.12.2.1.1 Glauconites

Glauconite is almost exclusively found in peloidal forms, i.e., ovoid shapes, in sediments, sandstones, and carbonates. The presence of green oval forms is distinctive and usually a good marker

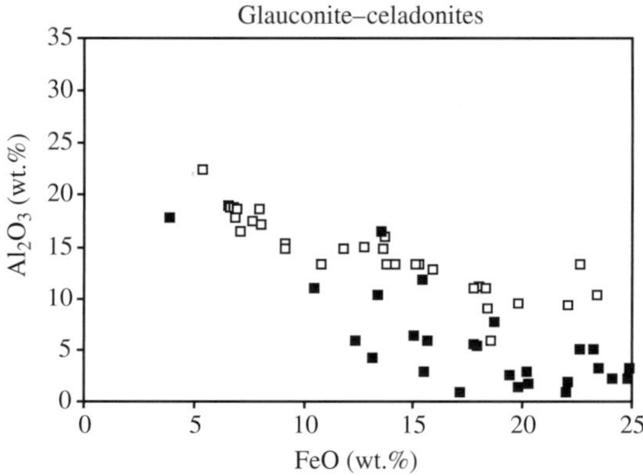

Figure 1 Comparison of glauconite (hollow squares) and celadonites (shaded squares) as a function of their Fe oxide and alumina content (sources Thompson and Hower, 1975; Newmann, 1987; Weaver and Pollard, 1973; Holmes, 1988; Berg-Madsen, 1983).

for field geologists. Consequently, glauconite-bearing rocks are generally well remarked in descriptions of sedimentary rocks and are used to distinguish stratigraphic layers indicated on geologic maps.

Glauconites are part of or form, depending upon where one draws the boundaries, a series of mixed layer minerals. The classic work of Thompson and Hower (1975) showed that the bulk composition of multigrain glauconite samples changes as a function of the mica glauconite (nonexpanding) component determined by XRD. Individual grains in a given sample can show strong differences in chemical composition (Velde and Odin, 1975). Each grain can be inhomogeneous; grain edge compositions differ from compositions at the center (Odin and Lamboy, 1988; Odin and Velde, 1975). There is a correlation between the chemical composition, especially potassium content, and mineral structure, i.e., the mixed layer composition or smectite-mica content. One might use the term glauconite mica for the nonexpanding mineral and glaucony (Odin, 1988) for the general mixed layer mineral, where the proportion of smectitic and micaceous layers is undefined. However, the pure glauconite mica mineral is probably very rarely found in nature. In this text we will use the term glauconite for the mixed layer mineral and glauconite mica when referring to the potassic end-member of the mixed layer series.

(i) Conditions of formation

Physical conditions. The peloidal form common to glauconites is most likely due to an origin of this material as fecal pellets. The survival of such material as independent and individual pellets suggest low depositional rate environments; in fact, one finds present-day glauconites on shallow-water continental shelf depositional sites. Rates of sediment accumulation are low in zones of glauconite occurrence, and the time necessary to form a mature glauconite is estimated to be on the order of a million years (Odin, 1988). The same author details the physical conditions under which glauconites are found forming in present or recent sediments; these are considered to be at a depth near 200 m (Porrenga, 1976). Glauconites are commonly found in the presence of phosphates (Fountain and McClellan, 2000), which indicates that the shelf environment is near a site of deep-ocean upwelling. However, Chafety and Reid (2000) feel that the presence of glauconites in carbonates and in tidal flat sediments in Cambro-Ordovician rocks could indicate different conditions of formation in the past. Föllmi and von Breymann (1992) indicate that a significant portion of glauconite pellets found on the abyssal plain of the Japan Sea is autochthonous. It is possible then that glauconite formation is only restricted to zones of low deposition rate and does not demand a specific water depth.

Chemical conditions. The occurrence of glauconites in sandstones (acid) and carbonates (basic) indicates that pH is not of great importance to their formation. They are found in sedimentary rocks ranging in age from Precambrian to the present day. Hence, their formation seems to be constrained by chemical and physical conditions rather than by specific events. There is no apparent time-dependent reaction which transforms them into new phases.

Glauconites are generally considered to form as a result of the partial reduction of iron in the sedimentary materials of pellets dropped to the bottom of the sea (Hower, 1961; Thompson and Hower, 1975); although a certain portion of iron in glauconites is in the divalent state, there is neither a clear relationship between the abundance of the two valences in glauconites nor is there an evolution toward the glauconite mica end-member (Figure 2).

(ii) Chemical composition of glauconites as mixed layered minerals

Glauconites are iron rich and contain varying proportions of potassium. The iron content varies from one multigrain sample to another, and, in fact, from one grain to another (Velde and Odin, 1975). This property allows one to select different electromagnetic fractions of glauconites from the same geologic sample (Odin, 1988). Electron-microprobe analysis indicates that there can be chemical zoning within single glauconite grains, especially in the outer rim of the grains (Velde and Odin, 1975; Odin and Lamboy, 1988; Parron and Amouric, 1990). If one compares the composition of glauconites to illites and mixed layer illite/smectites nonpeloid forms, there seems to be an almost continuous series of compositions at low-potassium content and only some possible gaps in compositional continuity in the high-potassium compositions. The distinction between glauconite and illite/smectite is classically made on iron content (Velde, 1979, p. 91), but the data reported by Baker (1997) and Berg-Madsen (1983) indicate that potassic micas can range from aluminous to ferric (Figure 3).

For glauconites a more reliable, nonmineralogical, distinction of mica content can be made by using color (green to yellow-green) and peloidal form to distinguish glauconite. Illites themselves are yellow to colorless in thin section, glauconites are green. The darker hues indicate a more micaceous glauconite (less mixed layered smectite). Purified illite tends to be beige or only slightly green in hand specimen. Thus, glauconites are green clays, par excellence.

If one looks at the relations between the total iron and potassium content of some glauconites (Figure 4), it becomes clear that there is not much correlation except perhaps at low-potassium content. Since potassium content is an index of the progress of the mica-forming process, it appears that the amount and oxidation state of iron is not systematically changed during this process. However, there is generally 3 times more Fe_2O_3 than FeO in glauconites, and hence they are ferric minerals. It is possible that the initial steps in glauconite transformation have been overlooked, since these materials are not apt to be very green. The initial transformation of a mixture of sediments containing iron oxide, smectite, and kaolinite to iron smectite could involve the introduction of iron from outside the proto-glauconite pellet. Of course, potassium must be introduced to produce the potassic-micaceous mineral, which is the end-product of the mineral reaction. However, the relation of ferric and ferrous iron as a function of glauconitization (formation of potassic mica) is not at all clear (Figure 5).

Then, one can deduce that there are probably many series of glauconites, whose compositional trend (increase in potassium content) depends on

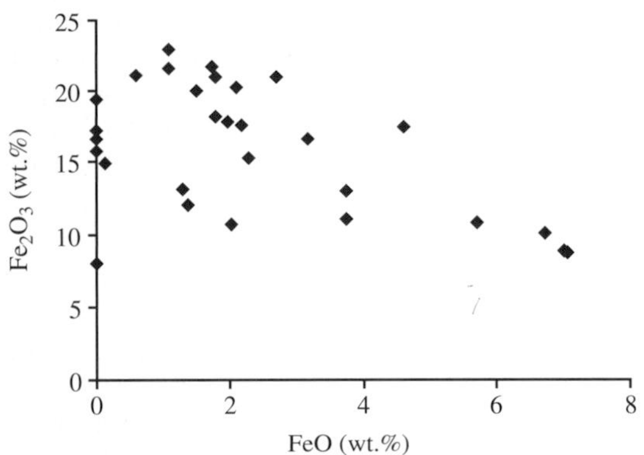

Figure 2 Compositions of glauconite iron contents in weight percent (sources Thompson and Hower, 1975; Velde and Odin, 1975).

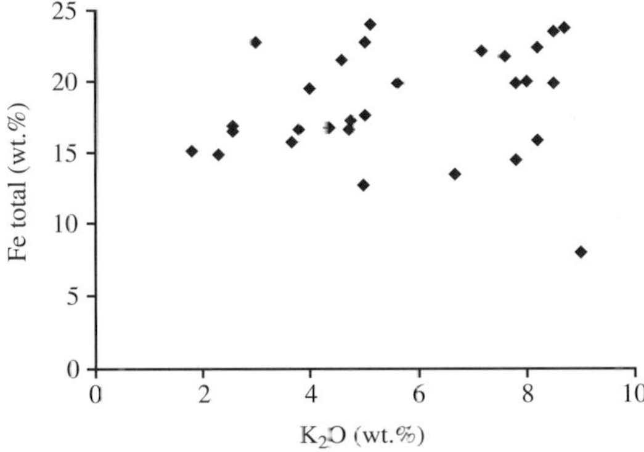

Figure 4 Total iron (oxide weight percent) and K_2O weight percent for glauconites (sources Thompson and Hower, 1975; Velde and Odin, 1975).

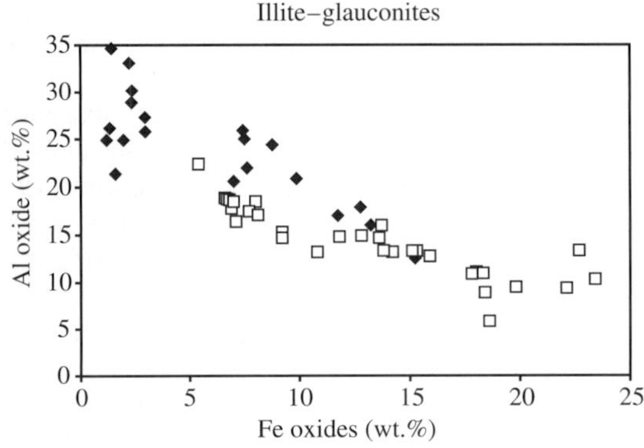

Figure 3 Illite–glauconite compositions as a function of alumina and total iron oxides. Diamonds = illite, hollow squares = glauconite. Data for illite from Weaver and Pollard (1973), Newman and Brown (1978), and Baker (1997). Data for glauconite as in Figure 2.

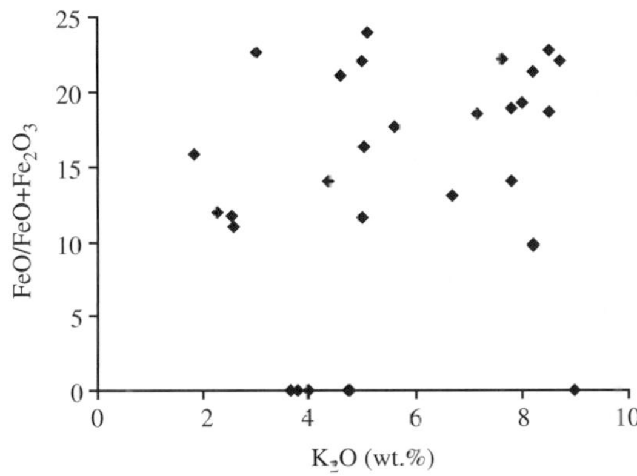

Figure 5 Ratio of ferrous oxide over total iron oxide versus K_2O weight percent for glauconites (sources Thompson and Hower, 1975; Velde and Odin, 1975).

the initial chemistry of the minerals. These evolve to become more glauconite mica-rich with significant amounts of iron in the green clay minerals. In each of the series potassium increases while the iron content and its oxidation state can change.

The maximum potassium content is less than 10 wt.% when the samples show no smectite component by XRD. According to Thompson and Hower (1975) this maximum potassium content represents less than one atom of potassium per unit cell of mica (in fact 0.65 atoms). XRD determinations show that the evolution of glauconite mixed-layer minerals and illite/smectite are similar (Velde and Odin, 1975). However, the geological conditions under which these transformations occur are in general not the same. The glauconite micas show a significant substitution of low-charge ions in both the tetrahedral and octahedral sites, as do illites.

(iii) Limits of stability of glauconite

Weathering. Glauconite is not stable during surface weathering. It tends to oxidize to form ocher in sandstones, which is often used as a pigment for house "paints" in the Mediterranean region. In carbonate rocks it tends to recrystallize into less potassic (micaceous) mixed-layer minerals (Courbe *et al.*, 1981; Loveland, 1981). The destabilization of glauconite in soil or weathering profiles generally occurs in the upper portion, above the zone of water-table fluctuation.

In surface alteration of glauconites in sandstones, these minerals tend to form kaolinite and iron oxide. This mixture is generally considered to represent an assemblage of soil clays indicative of intense weathering conditions. The destabilization of glauconite in weathering environments is more or less the reverse of its formation, one of oxidation of an iron-bearing silicate to form an oxide phase and another silicate. In carbonate rocks the influence of weathering seems to be somewhat mitigated. The glauconite appears to be gradually changed to a mixed-layer mineral with less potassium (glauconite mica) until the oxide-kaolinite stage is reached. The evolution in carbonates would be a sort of "reverse" weathering, which arrives gradually in an alkali-free assemblage, whereas in sandstones the alteration is made, apparently, in one step with the formation of kaolinite and iron oxide.

Metamorphism. Glauconite is stable throughout the range of diagenesis. It finally recrystallizes to biotite or stilpnomelane assemblages under metamorphic conditions (Frey *et al.*, 1973). For the most part it seems that the final recrystallization is abrupt, i.e., there is no indication that the mixed-layer smectite component of the glauconite transforms like illite/smectite interlayer minerals whose smectite components form chlorite. Thus, coexisting glauconite minerals evolve differently from illite/smectites under diagenetic conditions, as postulated by Velde (1977, p56). Glauconite forms a multi-mineral assemblage only at its upper thermal-stability limit. The limit is 320 °C in hydrothermal experiments but is surely lower in natural rocks (Velde and Odin, 1975).

Some glauconites have been identified in hydrothermally altered basalts together with celadonites (Alt *et al.*, 1992; Clayton and Pearce, 2000). This material appears to form a mixed-layer mica-ferric smectite series. The formation of glauconite mixed-layer minerals is therefore not restricted to peloids under shallow-ocean-bottom conditions. However, the identification of glauconite as distinct from ferric illite is difficult. Perhaps these mineral occurrences should be given another name.

(iv) Glauconite as a radiometric indicator

Glauconites have extremely useful properties. Since they form on continental shelves in a relatively short period of time, they can be used to date the sediments in which they form, either via potassium decay or the decay of rubidium, which substitutes for potassium. This type of information is now being used to date the cyclicity of sea-level change. Thus, glauconite dating is useful not only for absolute dating but also to solve problems of sequential stratigraphy (Harris and Whiting, 2000 for example).

7.12.2.1.2 Celadonite

(i) Occurrence

Celadonite occurrence: deep-sea data. Celadonite is almost exclusively found associated with basic-eruptive rocks in deep-marine environments. The Deep Sea Drilling Project (DSDP) has recovered altered basalts from ocean-bottom environments which contain celadonite. Basalts containing celadonite can, of course, also be found on dry land, but these are usually assumed to have experienced a cycle of marine alteration.

Velde (1985) has summarized the relations of celadonite and other clay minerals in altered basaltic rocks based upon the excellent studies in the scientific reports of the DSDP. A brief summary is given here. More recent work has shown this to be approximately correct.

Three major regimes of alteration of basalts occur at or near the ocean floor.

(i) *Interaction of seawater with cooled lava flows in submarine "weathering."* This process forms crusts of basalt-lava surfaces. The reactions concentrate potassium in the weathered products. Clay minerals form at the expense of the glassy parts of the basalt. The new mineral formed is considered to be a nontronite, i.e., a ferric, potassic smectite.

(ii) *Low-temperature hydrothermal activity associated with fracture deposits.* Clays are found in fractures and vugs. Saponites (magnesian smectites) and Fe–Mn oxides are formed. Temperatures are less than 30 °C (LaVerne *et al.*, 1996). This is an oxidation process (Blum *et al.*, 1996).

(iii) *Higher-temperature alteration occurs, which is more closely associated with vein deposits.* Temperatures are thought to be between 50 °C and 70 °C during these events (Blum *et al.*, 1996; Desprairies *et al.*, 1989).

(iv) *Higher-temperature regimes produce green-schist facies minerals.*

The overall tendency in these interactions between fluids at different temperatures and basaltic material is the oxidation of iron and the concentration of potassium. The relatively low concentration of potassium in seawater is surely, in large part, due to these interactions.

Celadonite, then, is a low-temperature mineral, but it is formed above the temperatures where glauconite is assumed to form (near-ocean bottom-water temperature). As far as one can tell from the rather scarce XRD information, no interlayered smectite/celadonite mixed-layer series is usually present. Data from Desprairies *et al.* (1989) and Holmes (1988) suggest that the glauconite is a 10 Å phase, and that the associated nontronite is a potassic smectite (12.8 Å, one hydration layer in the natural, air-dried state). However, information is extremely sparse and this generalization is very tenuous.

Normally weathered (subaerial, surface alteration) basalts can contain nontronite (a ferric smectite) and celadonite which form at the same time but by pseudomorphism of different basalt mineral grains during intermediate stages of alteration (i.e., between rock and soil) as summarized by Righi and Meunier (1995). These observations would lead one to believe that celadonite can be formed in terrestrial environments. If so, celadonite is not entirely restricted to relatively low-temperature hydrothermal formations in marine environments. It does not, apparently, form a significant mixed layer mineral series with smectite minerals as do glauconite and illite.

(ii) Composition

The problem of assessing the chemical data for celadonite is very great. Most analyses have been made using an electron microprobe without any accompanying XRD or other crystallographic determinations. For iron-bearing minerals, which are frequently mixed phases, such methods are difficult to interpret. One does not know if the spot analyzed with the microprobe is single or multiphase, and the oxidation state of the iron is indeterminate. Often, authors attempt to estimate iron oxidation state ratios assuming a stoichiometric formula for the host mineral. However, since we do not know the exact stoichiometric values for ions in the several sites (interlayer, octahedral), such methods are subject to error. Older, more complete data sets can clarify some important issues.

The ideal formula of celadonite mica is $KR^{3+}R^{2+}Si_4O_{10}(OH)_2$ where R^{2+} can be Mg or Fe^{2+}, and R^{3+} can be Fe^{3+} or Al. The MgAl octahedral component is most easily synthesized at high pressure by substitutions forming the muscovite–phengite mineral series (Velde, 1965). However, it is clear that this end-member is not common in natural low-temperature celadonite (Li *et al.*, 1997). The iron end-members show significant solid solution, which can be produced under hydrothermal conditions (Velde, 1972). Classical chemical mineral analyses indicate that the aluminous component is generally less represented in celadonites that in glauconites, but pure $Fe^{2+}Fe^{3+}$ and $Fe^{2+}Al$ end-member octahedral compositions are found in low-grade metamorphic rocks (Weaver and Pollard, 1973; Newman and Brown, 1987). One point concerning celadonite compositions is, however, notable. In general, celadonites have a higher silica content than glauconites; they, therefore, have less substitution of aluminum for silicon in the tetrahedral site of the structure (Table 1). The mica charge imbalance in the 2 : 1 structural layer is found principally in the octahedrally coordinated layer site. Celadonites appear also to have a higher potassium content, i.e., higher total charge imbalance in the 2 : 1 structure than glauconites. Celadonites are, then, on average more "micaceous," with a higher total charge on the structure (potassium content) and they have a charge developed in the octahedral site by substitution of a divalent ion (Fe^{2+} or Mg^{2+}) for a trivalent one (high silica content).

There is overlap in the compositional range of celadonites and glauconite, as mentioned above. Thus, celadonite, glauconite, and illite form an almost continuous series when potassium-rich.

(iii) Distribution

Celadonite-bearing rocks are restricted to zones where basaltic rocks are found, and where these basalts were probably extruded under deep-sea conditions. Their geological occurrence now is, of course, a matter of plate tectonics. One can suspect that initially the altered basalt in which

Table 1 Oxide percent ranges for the potassic mica-like and mixed layer minerals.

	K_{max}	Si	Fe^{3+}	Fe^{2+}
Celadonite	10.3	56.5	6.4–20.6	2.2–5.4
Glauconite	9	53.6	0.0–30.8	0.0–10.2
Illite	11	54.1	0.7–12.3	0–1.9

Source: Weaver and Pollard (1973).

celadonites formed was in an active ocean ridge or perhaps in another zone of tectonic emplacement and hydrothermal activity.

Another, less well-documented occurrence of celadonite is the subaerial alteration of basic and basaltic rocks. Here interaction of surface waters promotes the formation of celadonite, a similar process to that of hydrothermal alteration but at a lower temperature. In both cases there is significant oxidation of the initial iron-bearing rocks. This is reflected in the high ferric iron content of the celadonite minerals.

(iv) Summary

Celadonites are formed during the alteration of basalts. Celadonite is usually found as a replacement of ferromagnesian minerals, frequently pyroxenes. At times hydrothermal alteration yields veins of celadonite in basaltic rocks. The presence of nontronite and celadonite in the same specimens suggests that celadonite is not formed by a series of mineral permutations from smectitic minerals (the ferric iron smectite nontronite). This is in strong contrast to glauconite mica, which forms by the transformation of mixed layer ferric-aluminous smectites. The range of geological conditions of formation (surface or submarine alteration to low-grade metamorphic rocks) suggests that this mineral is stable over the range of conditions generally attributed to clay minerals.

7.12.2.1.3 Summary of potassic green clays

The physical properties of green mica-like clays (color) give two basic groups: blue-green celadonites and yellow-green glauconites. Chemical composition cannot be used to distinguish between these two mineral groups. Further, chemical composition does not readily distinguish illite (a paler micaceous clay mineral) from the glauconite–celadonite minerals. There is a continuum in chemical space for the variables K–Fe^{2+}–Fe^{3+}–Mg–Al–Si. However, these minerals can be distinguished by their geologic conditions of formation. Glauconites are formed in shallow sea environments under slightly oxidizing conditions, whereas celadonites are formed in basalts under conditions of weathering or, more often, as products of hydrothermal alteration. Hence, the mineral names are not mineralogically but geologically based.

Strong overlap at the high potassium end of their compositional ranges (mica-like minerals) and in the lower potassium ranges (interstratified minerals) which also overlap with illite and illite–smectite interlayer minerals suggests that there is a continuum of minerals which change color, depending upon iron content and Fe^{2+}–Fe^{3+} interactions in the structure. Although different names are given to these minerals (Buckley *et al.*, 1978), close inspection of individual green clay minerals in geological samples (HRTEM) indicates significant variations, which show an overlap between the assumed compositional ranges of glauconite–celadonite (Parron and Amouric, 1990; Duplay and Buatier, 1990).

One can generalize these compositional relations in diagrammatic form as shown in Figure 6.

7.12.2.2 Ferrous, Green Clays

Greenish clay, peletal in form, which is not glauconitic, is called berthiérine or a verdine facies mineral (Odin, 1988). This material does not have a tendency to become potassium rich; rather, it is richer in ferrous iron during the course of its mineralogic evolution. The end-member material is nonexpanding and apparently chlorite (7 Å) in structure. This material is not green in thin section. In hand specimen its color varies from black-green to green-black, hence the name Verdine facies, probably derived from French *vert* (green). In thin sections that contain both berthiérine and glauconite, one can easily distinguish the two minerals by their color. Glauconite is yellow-green and berthiérine is yellow-brown. The mineral chamosite, a chlorite, is green in thin section. We will try to clarify these problems of color, the theme of this chapter, in the following sections.

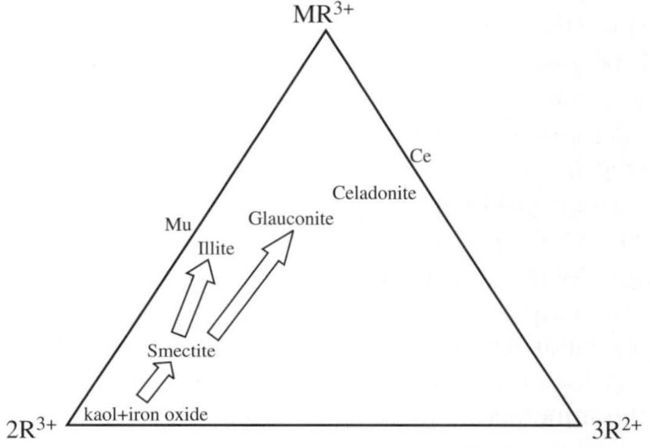

Figure 6 Representation of chemical compositions of potassic, low-temperature micas in MR^{3+}–$2R^{2+}$–$3R^{3+}$ space. The poles represent feldspar, dioctahedral clays, and trioctahedral clays, respectively. M = Na, Ca, and especially K ions, R^{3+} = Al, $Fe^{3+}R^{2+}$ = $Fe^{2+}Mg$. The compositional positions of the minerals Mu (muscovite) kaol (kaolinite), smectite, and mixed layer mica/smectites are indicated. Initial materials are kaolinite (kaol) and iron oxides. A second step is the production of an iron-aluminous smectite and then the formation of either illite via an illite/smectite mixed layer mineral or glauconite via a glauconite mica/iron-smectite mixed layer phase.

7.12.2.2.1 Occurrence of berthiérine and verdine minerals

As far as one can tell from the data presented, verdine facies minerals are found on the surface of shallow marine sediments (Odin, 1988; Thamban and Rao, 2000; Kronen and Glenn, 2000). Berthiérine in sedimentary rocks (Fritz and Toth, 1997; Moore and Hughes, 2000) formed at temperatures below 70 °C (Hornibrook and Longstaffe, 1996). It can apparently form from verdine facies minerals or by various mineral reactions during diagenesis (Fritz and Toth, 1997; Velde, 1989). Verdine facies minerals, mixed layered in nature, are therefore of low-temperature oceanic origin, while berthiérine can possibly form as a pure phase under these conditions or under diagenetic conditions and temperatures up to 70 °C.

(i) Berthiérine

The mineral berthiérine was described by Orcel *et al.* (1949). It was found associated with iron oxide oölites, similar to those of chamosite (see below). For a number of years it lived a peaceful life. Interest in shallow ocean sediments in the 1970s led to more work by Porrenga, Odin, and Giresse (cited in Odin, 1988; Wiewora *et al.*, 1996) on a mineral of similar composition. As more knowledge in the use of more sophisticated XRD machines opened new perspectives, and as more samples became available due to more frequent ocean bottom dredging for scientific purposes, additional mineralogical information became available, and new interpretations were made for the origin of these greenish pellets found in shallow ocean environments. Initial studies indicated that berthiérine is a chlorite group mineral, but as more bulk compositions of these proto-berthiérines (which became the verdine facies) were obtained, it became clear that not all of the pellets were chloritic in composition. Microprobe analyses of berthérines in sedimentary rocks that had experienced low-grade diagenesis, had shown that the minerals were close to chlorite in composition but had a high alumina content (Velde *et al.*, 1974). Odin (1988) has published a full account of this work. We will use these data for the basis of our discussion.

Compositions. Berthiérines are ferrous almost to the total exclusion of Fe^{3+} (Brindley, 1982). Diagenetic and metamorphic chlorites contain small amounts of ferric iron, generally less than 6% (Foster, 1962), and hence there seems to be a slight difference between the chlorite and berthiérine compositions. While chlorites show a range from 3.0 to ~2.5 ions in the octahedral site (Foster, 1962; Hillier and Velde, 1973), berthiérines vary only from 3.0 to 2.75 in octahedral sites. The major characteristics of berthiérines are give in Table 2.

Berthiérines having experienced diagenetic burial conditions contain about one-third as much aluminum as iron; their average alumina content is, therefore, comparable to that of other chlorites (Velde, 1973).

(ii) Verdine minerals and odinite

The definition of the verdine facies is largely due to the work of Odin (1988). Bailey (1988) defined the mineral odinite. Verdines and odinite span much of the range of octahedral site occupation from 2.5 to 2.0 ions, containing much ferric iron and alumina. Assuming that all of the minerals are 7 Å structures, the difference in observed composition of the different phases can be illustrated by their octahedral cation occupancy: Al_2 kaolinite (dioctahedral); $(Mg, Al)_{3-2.5}$ 7 Å trioctahedral chlorite; and $(Mg, Al)_{2-2.5}$ odinite (di-, trioctahedral).

The octahedral occupancy distinguishes these verdine facies or odinite minerals from other sedimentary and in fact all metamorphic phyllosilicate minerals. These relations are clearly demonstrated by Hornibrook and Longstaffe (1996).

(a) XRD and mineral identification

The chemical peculiarities of these minerals are mentioned above. However, the XRD data available at the time of mineral definition also show some very striking features for odinite, which can shed light on the strange chemical features of this verdine mineral.

Odinite is the key mineral in the verdine facies. It was defined by Bailey in 1988, mainly on the basis of chemical analyses provided by Odin, on thermoponderal data, and X-ray powder diagrams. Since clay minerals are fine grained and sheeted, XRD of oriented material is the best method of description. At the time Bailey defined odinite, XRD spectra of good quality for oriented clays were available and were, in fact, quoted by Bailey. Figure 7 presents a modified spectrum given by Odin (1988, pp. 94, 98, 162). Similar diagrams are

Table 2 Ionic occupancy of berthierine, chamosite, and odinite.

	Al tetra	*Vacancy*	*Fe/Fe + Mg*	Fe^{3+}
Berthiérine	0.40–0.90	0–0.25	0.8–1.0	0.01–0.27
Chamosite	0.52–0.73	0.14–0.32	0.52–0.84	0.01–0.60
Odinite	0.15–0.70	0.3–0.54	0.22–0.32	0.72–1.00

Sources: Brindley (1982), Weaver and Pollard (1973), and Bailey (1988).

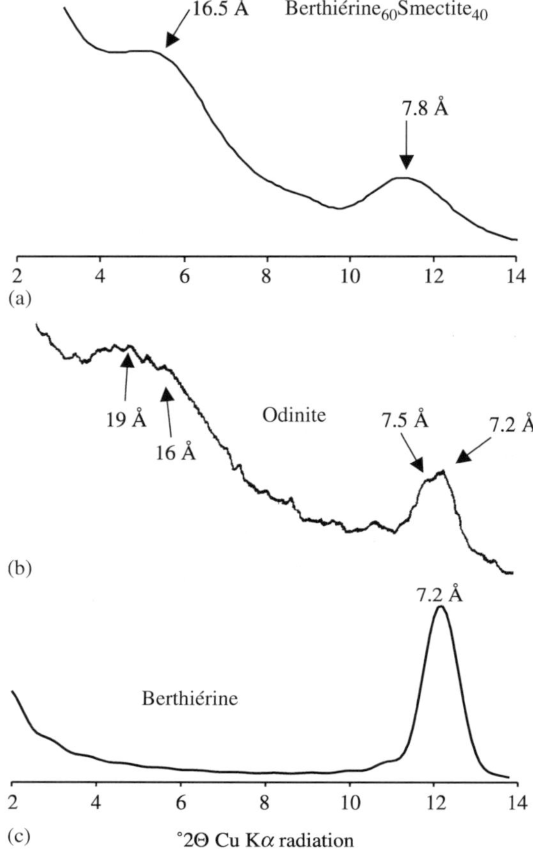

Figure 7 Experimental XRD spectra for odinite and calculated diagrams for hypothetical minerals using the NEWMOD program (Reynolds, 1985) are shown. The berthiérine spectrum was simulated using a ferrous–aluminous 7 Å clay structure with 3–6 coherent diffracting domain structure. The mixed layer 7 Å/smectite mineral berthiérine/nontronite was modeled using 60% berthiérine layers and 3–6 layer coherent diffracting domains in a disordered ($R = 0$) structure (after Odin, 1988, p. 162).

given in Wiewora *et al.* (1996, figure 4) and Hornibrook and Longstaffe (1996, p. 9).

Figure 7(b) shows a typical XRD pattern of oriented clay of odinite and verdine facies minerals. Since the sample was run in the air-dried state, the maximum expected spacing would be 15 Å. The odinite spectra have a high background at low angles, a wide, poorly defined peak above 15 Å in the air-dried state, and a wide, sharper but asymmetric peak near 7 Å. The higher intensity is at 7.2 Å. There is a lower intensity band near 7.5 Å. The intensity of the total 7 Å peak is low compared to the surface area of the wide band above 15 Å. These features suggest that odinite is not simply a ferrous 7 Å mineral based on a serpentine structure as defined in Bailey (1988).

If we model diffractograms of the possible minerals having the proposed odinite structure using the NEWMOD program (Reynolds, 1985), with a ferrous, aluminous serpentine (berthiérine) composition, one obtains a spectrum as in Figure 7(c) with a peak at 7.2 Å. The background to low angles is very low in the 4–6° 2Θ region. This spectrum does not account for the shoulder at 7.5 Å seen in the spectrum of odinite nor the band above 15 Å. If one tries to explain a basal spacing greater than 15 Å in air-dried samples, one has to combine smectite and a 7 Å mineral structure in a mixed layer mineral. The result of such calculations is shown for a 40% ferric dioctahedral smectite–60% ferrous 7 Å chlorite component simulation with a small coherent diffracting domain using the NEWMOD program. A small average diffracting domain (2–6 layer units) gives a wide band above 15 Å and a band near 7.8 Å which is characteristic for the odinite spectra (Table 3).

If we combine the features of the two simulations, we have a composite spectrum with bands at 7.2 Å, 7.5 Å, and >15 Å. We can see by using these simulations that the XRD results for odinite are probably due to a mineral which is a mixture of a ferrous 7 Å chlorite (berthiérine) and a ferric smectite/ferrous 7 Å chlorite mixed layer mineral. Different proportions of each structural element will give different bulk compositions.

A mixture of dioctahedral smectite and berthiérine gives the mineral spacings for the *060* spacings observed by Bailey and Odin, which are between those of clearly aluminous dioctahedral and trioctahedral minerals.

(b) Chemical characteristics and XRD information

As a point of reference for the following discussion, Figure 8 shows the range of solid solution for high-temperature (nonclay mineral) chlorites. This range can serve as a guide to compare the compositions of chlorite and chlorite-like green clays.

The striking mineralogical characteristic of odinite as defined by Bailey (1988) is the octahedral site occupancy which is between 2 and 2.5, a range not found in chlorites formed at higher temperatures (Foster, 1962; Weaver and Pollard, 1973). Since we have both chemical and XRD data for the verdine facies minerals, it should be possible to understand how one can have a bridging composition between di- and trioctahedral phyllosilicate minerals in sediments which is unknown in sedimentary (diagenetic facies) and metamorphic rocks. For comparison, Figure 6 shows the compositional range of high-temperature (diagenesis and higher) chlorite minerals in $Si-R^{2+}-R^{3+}$ coordinates where $R = Fe^{3+}$, Al; and $R^{2+} = Fe^{2+}$, Mg.

The relations of verdine and berthiérine mineral chemistry can be shown diagrammatically as in Figure 9, where the mineral compositional zones

Table 3 General classification of green and associated clay minerals.

1. 7 Å minerals (serpentine–kaolinite groups)
   (1) Trioctahedral (Fe, Mg, Al)$_{3-2.5}$ (Si, Al)$_2$O$_5$(OH)$_8$
      Berthiérine (*Fe*)
      Serpentine (*Mg*)
   (2) Dioctahedral (Al)$_2$ Si$_2$O$_5$(OH)$_8$
      Kaolinite (*Al*)

2. 14 Å minerals (chlorites)
   (1) Trioctahedral (Fe, Mg, Al)$_{2-2.5}$(Si, Al)$_2$O$_5$(OH)$_8$
      Chamosite (*Fe, Al*)
      Chlorite (*Mg, Al*)

3. 10 Å minerals
   (1) Mica and micaceous
      (a) Trioctahedral K(Mg, Fe, Al)$_{3-2.5}$(Si,Al)$_4$ O$_{10}$(OH)$_2$
         Biotite
      (b) Dioctahedral K(Mg, Fe, Al)$_2$ (Si,Al)$_4$ O$_{10}$(OH)$_2$
         Muscovite (*Al*)
         Illite (*K* < 0.9)(*Al*)
         Glauconite (*K* < 0.9)(*Al, Fe*)
         Celadonite (*K* < 0.9)(*Al, Fe, Mg*)
   (2) Talc, pyrophyllite (neutral lattice)
      (a) Trioctahedral (Mg, Fe)$_3$ (Si)$_4$ O$_{10}$ (OH)$_2$
         Talc (*Mg*)
      (b) Dioctahedral (Al)$_2$ (Si)$_4$ O$_{10}$ (OH)$_2$
         Pyrophyllite (*Al*)
   (3) Smectites (low charge, variable spacing)
      (a) Trioctahedral $M^{+}_{(<0.5)}$(Mg, Fe, Al)$_{3-2.5}$(Si,Al)$_4$O$_{10}$(OH)$_2$
         Saponite (*Mg, Al*)
      (b) Dioctahedral $M^{+}_{(<0.5)}$ (Mg, Fe, Al)$_2$ (Si,Al)$_4$ O$_{10}$ (OH)$_2$
         Montmorillonite (*Al, Mg*)
         Beidellite (*Al*)
         Nontronite (Fe^{3+})

4. Interstratified clay minerals
   Illite/smectite
   Glauconite/smectite
   Berthiérine/smectite (odinite)
   Kaolinite/smectite
   Chlorite/mica

(*Al*) = dominantions in octahedral site, glauconite = green clay mineral.

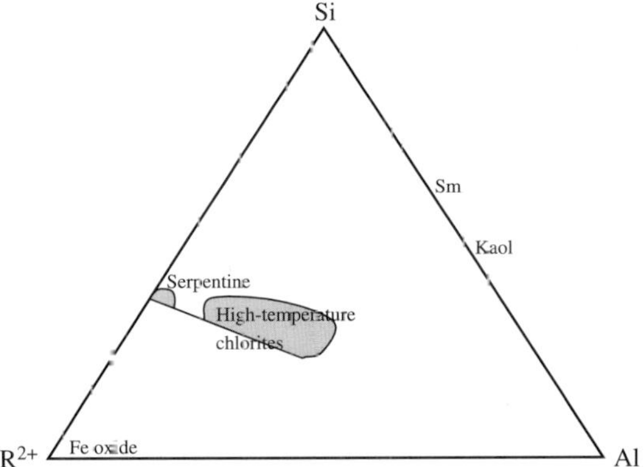

Figure 8 Compositional ranges of high-temperature (burial diagenesis and metamorphic) chlorites and metamorphic serpentines. Si–iR^{2+}–Al coordinates where $R^{2+} = Fe^{2+}$, Mg. Sm = smectite, kaol = kaolinite (source Velde, 1973).

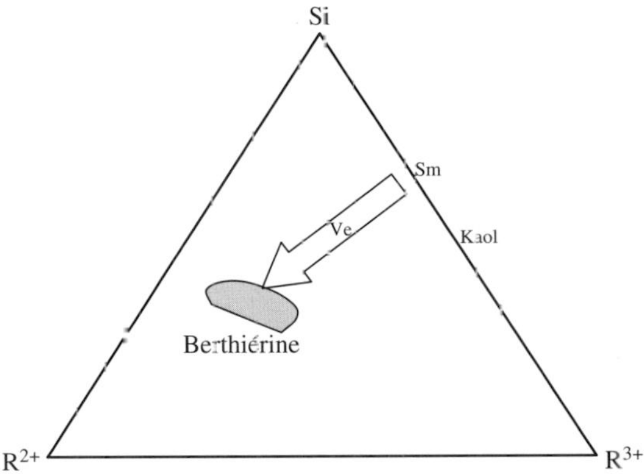

Figure 9 Diagram indicating the chemical evolution of ferric smectite material towards berthiérine (Be). The arrow follows chemical compositions of verdine (Ve) materials. Si–R^{2+}–R^{3+} coordinates where R = Fe^{3+}, Al and $R^{2+} = Fe^{2+}$, Mg. Sm = smectite, kaol = kaolinite.

of smectite and berthiérine are shown, and where the arrow covers the compositions of the verdine minerals (see Velde, 1989). The alignment between smectite (silica rich on the diagram) and the berthiérines reported by Brindley (1988) is striking.

Berthiérine, as shown by Brindley (1982) is essentially a trioctahedral mineral, following the line of trioctahedral chlorites in Figure 7. In our simulations of the XRD spectra of odinite, we use a ferrous serpentine and a ferric dioctahedral smectite component. Translated into constituent ions of a mineral structure, this mineral combination will give a bulk average composition between nontronite (ferric, dioctahedral smectite) and berthiérine (trioctahedral chlorite), a region where the octahedral occupancy average would be between di- and trioctahedral. Varying proportions of berthiérine (shown by the 7.2 Å peak in the XRD spectra) and the mixed layer mineral will change the proportions of di- and trivalent iron This would result in the compositional range shown by Hornibrook and Longstaffe (1996) and Bailey (1988) for verdine facies minerals.

If we look back to the singular chemical features of odinite as having large amounts of ferric iron present and having an overall low occupation of the octahedral site for a trioctahedral mineral, a mixed layered mineral of ferric, dioctahedral smectite (nontronite), and berthiérine (ferrous 7 Å chlorite) would give the overall chemical characteristics of odinite. Thus, one can fit the XRD data by using a ferric smectite/7 Å ferrous chlorite mixed layer mineral which can be fit to the chemical data given by Bailey for odinite.

The author proposes a redefinition of odinite not as a 7 Å chlorite but as a mixed layer dioctahedral ferric smectite/7 Å mineral and a 7 Å ferrous chlorite.

7.12.2.2.2 Chamosite

This mineral was named and identified before berthiérine and before the verdine facies minerals. Hence, its range of composition and mineralogical characteristics have changed somewhat with time. At present, chamosite is a deep green 14 Å mineral (chlorite) with a very high alumina content (see Bailey, 1980). However, the definition of chamosite based on XRD characteristics requires caution. Chamosite is generally considered to be a 14 Å mineral. However, interstratification of 7 Å and 14 Å layers in the chlorite structure can rapidly diminish the distinctive 14 Å peak, making identification difficult (Hillier and Velde, 1992). In fact, 15% of a 7 Å phase interlayered with 14 Å chlorite will entirely eliminate the evidence for the 14 Å mineral present in 85% abundance. Thus, the distinction of berthiérine (7 Å) from chamosite (14 Å) is not easy to obtain. The original berthiérine of Orcel *et al.* (1949) was found in an iron ore deposit, and there has been confusion regarding which mineral became what.

In general, the minerals now identified as chamosite are found in iron ore bodies of sedimentary origin (e.g., Maynard, 1986; Fernandez and Moro, 1998; Wiewora *et al.*, 1998; Kim and Lee, 2000). Chamosite associated with iron oxides appears to follow a compositional trend from iron oxides plus kaolinite to chlorite, as indicated in Figure 8, using the data of Velde (1989). The recombination of iron oxide in the presence of kaolinite gives an aluminous, ferrous mineral, chamosite. This mineral is formed under burial conditions where ferric iron oxide is reduced to ferrous iron which is rapidly incorporated into a 7 Å chlorite mineral. Both chamosite and berthiérine result from the reduction of ferric iron to ferrous iron.

7.12.3 NONCHLORITE, NONMICACEOUS GREEN CLAY MINERALS

This mineral category, slightly negative in its characteristics, represents a small fraction of the green clays encounterd in nature. However, these minerals hold the key to understanding the origin and geochemical position of the other minerals, commonly called green clay minerals.

7.12.3.1 Nontronite

This mineral is a dioctahedral smectite with a large portion of the trivalent sites occupied by ferric iron ions. Some nontronites are orange, some brown, and some green. As modern studies (postelectron microprobe) often fail to give information such as color in their mineralogical descriptions, we will consider very briefly the general context of nontronite occurrence.

Nontronite is essentially a ferric, dioctahedral smectite with significant amounts of aluminum substituting for iron. A very general formula could be written as

$$M^{+}(Mg, Fe^{2+}Fe^{3+}Al^{3+})_2(Si^{4+})_4O_{10}(OH)_2 \cdot nH_2O$$

where M is the exchange ion, and n is the undetermined number of water molecules.

Usually the mineral is silica rich with only limited substitution of aluminum ions in the tetrahedral position. Only minor amounts of Mg and Fe^{2+} are found in the octahedral position.

Nontronites are found generally in altered basic rocks, often basalts, either as products of subaerial weathering or as products of the interaction of hydrothermal waters with basalts on the ocean bottom (see Velde, 1985; Righi and Meunier, 1996) or as mineral transformations of pelletal iron oxides and kaolinite (Pedro *et al.*, 1978) in freshwater environments. The surface weathering occurrence seems to form almost pure smectite minerals (i.e., without interlayering of a micaceous component). Weathered basalts, serpentines, and gabbros all form nontronite in the intermediate stages of surface weathering (between rock an soil). The co-formation of nontronite, saponite, talc, chlorite, and beidellite in these rocks suggests that there is no complete solid solution between other smectites (aluminous, dioctahedral, or trioctahedral) under these chemical conditions. In these contexts nontronite contains mainly calcium as the exchange ion.

Formation of nontronite from pelletal freshwater sedimentary material recalls the formation of berthiérine or glauconite, but in a freshwater context (Pedro *et al.*, 1978). Again there seems to be little tendency to form a mixed layer mineral.

Deep-sea occurrences (see Velde (1985) for an overview) are varied, but the material is generally associated with basalts. Nontronite seems to be formed directly from basaltic glass weathering at a very low rate in deep-sea environments. It is, however, not directly associated with celadonite. XRD data are sparse, and hence there is always the possibility that there is a tendency to form a mixed layer nontronite/celadonite mineral. Nontronite can be found as a product of precipitation from solution around hydrothermal vents where there seems to be no silicate precursor and no interlayering with celadonite (Bischoff, 1972). In some cases, however (Schrader *et al*., 1980), it appears to be interlayered with a nonexpanding layer. Thus, it is possible that at temperatures above surface conditions (7–20 °C), interlayering of ferric smectite and celadonite can occur.

In general, it appears that nontronite is not associated with the formation of potassic green micas (celadonite or glauconite mica) but appears to be associated with the formation of berthiérine, the ferrous chlorite (see Section 7.12.2.2.2).

7.12.3.2 Talc

This mineral, $R_3^{2+}Si_4O_{10}(OH)_2$, is generally considered to be of metamorphic or hydrothermal origin, found in high magnesian, low silica rocks such as serpentines. However, some occurrences of talc have been reported in weathered basic rocks (see Righi and Meunier, 1996). Most often alteration talcs are only slightly green in thin section. Talc appears in the early stages of weathering and does not seem to persist into the soil clay mineralogy. Talc of rather high iron content ($Fe_{1.49}$, $Mg_{4.5}$) was identified in deep-sea hyrothermal deposits (Lonsdale *et al.*, 1980). A special talc, minnesotaite, is a ferrous talc found in low-grade metamorphic iron ores. This mineral is green, and of much higher-temperature origin than those of weathering formation.

It appears that talc can have a complete range of Fe–Mg content in different low-temperature environment.

7.12.4 GEOCHEMICAL ORIGIN OF GREEN CLAYS

The above sections describe the occurrence of green clay minerals giving compositions, XRD characteristics, and general situations of formation. This information can be used to establish a general geochemical framework for the formation of green (iron-bearing) clay minerals.

7.12.4.1 Shallow Ocean Bottom, Marine Green Clays—Glauconite and Berthiérine

Shallow sediments under conditions of low sedimentation rate can produce either glauconites or berthiérine–verdine minerals. Given the mineralogical information presented above, one can think of the glauconite/ferric-aluminous smectite mixed layered mineral series culminating in the formation of the micaeous (nonexpanding) mineral glauconite and the berthiérine/smectite series (verdine) of mixed layered minerals culminating in the formation of berthiérine as two similar sequences of mineral change, whose initial starting point is the mineral material in fecal pellets. Since the starting points are similar, a dioctahedral iron-rich smectite plus probably kaolinite and ferric iron oxide, there must be some fundamental geochemical difference in the conditions of formation of the two mineral series.

Porrenga (1976) and Odin (1988) have described the occurrence of glauconite and berthiérine facies minerals in present-day and recent sediments. From their summaries it appears that the berthiérine mineral sequences are present in zones near river deltas, whereas glauconites are more often found in shelf areas of lower sedimentation rate. The major difference in the chemical evolution of the two clay types is the change from ferric iron to ferrous in the berthiérine. There is a much more moderate change in iron oxidation state in the glauconites and an addition of potassium. The evolution of glauconite is towards a potassic, ferric clay mineral, whereas that of berthiérine is towards an alkali-free, ferrous mineral. Hence, one can deduce that the reducing power of the environment on the sedimentary peloid materials is very important. The difference between "fresh" sediments and those that are more reworked could reflect a difference in the organic matter content of the materials in the two environments of sedimentation. This could be the motor for the development of the differences in ferrous or ferric iron in the end-product minerals. Reduction of iron in river delta, organic-rich sediments would produce berthiérine. In the absence of a strong reducing agent, the iron in the peletal material remains essentially ferric and accumulates potassium to form glauconite mica. These two clay

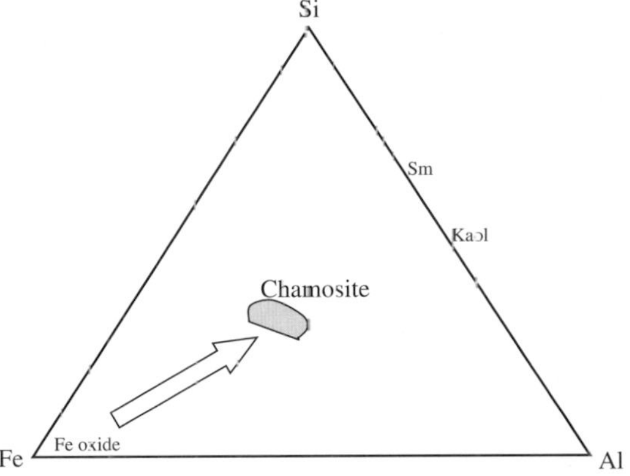

Figure 10 Representation of chemical evolution of iron oxide and kaolinitic material towards the chamosite mineral composition. Arrow follows compositions of oolitic material. Si–Fe–Al coordinates. Sm = smectite, kaol = kaolinite (source Velde, 1989).

minerals are thus formed in different geochemical sedimentary environments. Temperature of formation is always low, near that of the ocean shelf.

Chemically, there is not a significant difference between the early stages of glauconite and berthiérine formation. During glauconite formation the system is largely oxidizing, and during berthiérine formation the system is reducing. Glauconite tends to become more alkali rich and berthiérine silica poor. When berthiérine is formed, silica leaves the local pelletal system (Figure 9); during glauconite formation potassium enters the system as seen in Figure 11.

This figure indicates the change in composition as a function of alkali ions combined with a charge inducing substitution (K^+R^{3+}), divalent ions in octahedal ($3R^{2+}$), and trivalent ions in octahedral occupancy ($2R^{3+}$) as they occur in clay mineral proportions (see Velde, 1985). The starting point, ferrric, aluminous smectite and ferric iron evolves towards potassic, ferric glauconite mica in one instance, and to ferrous berthiérine containing no alkali in the other.

7.12.4.2 Alteration Environments

Some green clays reflect other environments. Celadonite is formed in or from basalts or gabbros at low or medium (low-grade metamorphism, probably below 250 °C) temperatures. Here the key to mineral origin seems to be the presence of basic rocks, which are silica poor and low in potassium. Temperature in the range of clay mineral stability is not a limiting factor during formation. Hydrothermal action and weathering both produce celadonite minerals. In these chemical environments there is a high magnesium and a smaller iron content in the presence of low silica activity. The clay mineral most likely to be produced is ferric smectite which is a common mineral in basalt alteration. However, when sufficient potassium is available in this environment, this potassium engenders the formation of the sole mica stable at low temperatures, a potassic one.

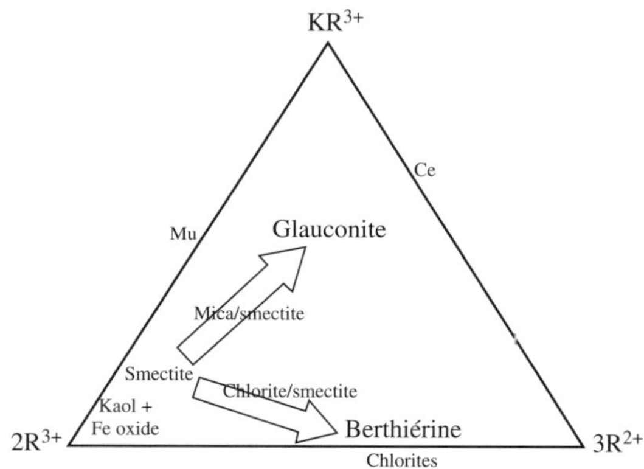

Figure 11 Representation of the evolution of clay pellets in shallow shelf sediment areas according to the oxido-reduction conditions locally present. Lower arrow shows berthiérine formation through reduction of iron, shifting the pellet composition from the ferric ($R^{3+} = Fe^{3+}$) pole to the ferrous pole ($R^{2+} = Fe^{2+}$). This reaction passes through a chemical evolution by the formation of a berthiérine/smectite mixed layer mineral (chl in the figure). The arrow towards glauconite indicates the change in composition with increase in potassium and some reduction of ferric iron. The diagram represents feldspar, dioctahedral clays, and trioctahedral clays, respectively. $R^{2+} = Fe^{2+}$, $R^{3+} = Al$, Fe^{3+}. The compositional positions of the minerals Mu (muscovite) kaol (kaolinite) and end-member celadonite (Ce) are indicated.

7.12.4.3 Diagenesis Reactions

Chamosite, a 14 Å chlorite, is due to the transformation of ferric iron to a ferrous state in the presence of a high aluminum mineral, kaolinite. The oxidation state change occurs during burial diagenesis, perhaps through a combination of temperature and probably breakdown of some organic matter. The combination of abundant iron and kaolinite produces this mineral, again in a special geochemical environment. Some authors report that berthiérine (7 Å chlorite) transforms isochemically into a 14 Å mineral, chamosite (Hornibrook and Longstaffe, 1996). Here one finds a mineral change which is apparently related to the structure of the mineral without significant change in composition. Other green 14 Å chlorites are found in sandstones with roughly the chamosite, high-iron and high-aluminum, composition. They appear to be due to mineral precipitation in pore spaces. These occurrences are numerous and well reported in the literature dealing with sandstone diagenesis and petroleum accumulation. In these instances the name chamosite is rarely used however, and the term chlorite is preferred. During the

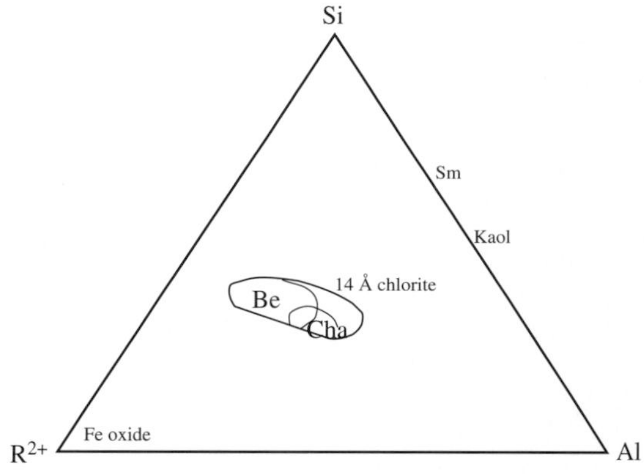

Figure 12 Chlorite compositional ranges.

formation of chlorite in sandstone diagenesis, it is likely that mobile iron oxides become ferrous through changes in the reducing potential of pore fluids and are the source of mineral genesis. In the case of chamosite, the structure of the mineral distinguishes it from lower-temperature chlorites, but its chemical composition overlaps with that of berthiérines. This is a type of polymorphic relation, as is the case for kyanite—sillimanite–andalusite where mineral names distinguish different structures of minerals with the same composition.

Berthiérine is used to designate an essentially 7 Å iron-rich mineral, chamosite, a 14 Å chlorite with a wide range of Fe–Mg ratio. This is indicated by the combined data from Figures 8–10 shown in Figure 12.

7.12.5 GENERAL REFLECTIONS

In the above descriptions it seems clear that green clay mineral names can reflect either compositional differences, mineral structure differences, or differences in geological occurrence. The potassic, mica-like minerals, illite–glauconite–celadonite, are examples of the latter. Chamosite–berthiérine is an example of structural differences giving rise to different mineral names. At times mineral names can signify a mixture of structural components (odinite); in other examples, the mixed component minerals have no specific mineral names and are simply called mixed layer minerals. All of this is not of great consequence to the initiated, but it lends confusion to the learning process of young people who wish to understand the significance of this geological material. Geologist-mineralogists can claim a work-in-progress excuse, because we find new relations continually due to more exhaustive research. However, it would be nice to clean up the mess from time to time. The problems outlined above are based on work done in the 1980s or even earlier.

REFERENCES

Alt J. C., France-Lanord C., Floyd P. A., Castillo P., and Galy A. (1992) Low-temperature hydrothermal alteration of Jurassic ocean floor crust. Site 801. *Proc. ODP Sci. Results* **129**, 415–431.

Bailey S. W. (1988) Odinite: a new dioctahedral–trioctahedral Fe^{3+}-rich 1/1 clay mineral. *Clay Min.* **23**, 237–247.

Baker J. C. (1997) Green ferric clay in non-marine sandstones of the rewan group, southern bowen basin, Eastern Australia. *Clay Min.* **32**, 499–506.

Berg-Madsen V. (1983) High-alumina glaucony from middle Cambrian of Öland and Bornholm, southern Baltocandia. *J. Sedim. Petrol.* **53**, 875–893.

Bischoff J. L. (1972) A ferroan nontronite from the Red Sea geothermal system. *Clays Clay Min.* **20**, 217–223.

Brindley G. W. (1982) Chemical composition of berthiérine: a review. *Clays Clay Min.* **30** 152–155.

Buckley H. A., Bevan J. C., Brown K. M., Johnson L. R., and Farmer V. C. (1978) Glauconite and celadonite: two separate mineral species. *Min. Mag.* **42**, 373–378.

Chafety H. S. and Reid A. (2000) Syndepositional shallow-water precipitation of glauconite minerals. *Sed. Geol.* **136**, 29–42.

Clayton T. and Pearce R. B. (2000) Alteration mineralogy of cretaceous basalt from ODP site 1001, Leg 165 Caribbean Sea. *Clay Min.* **35**, 719–733.

Courbe C., Velde B., and Meunier A. (1981) Weathering of glauconites: a reversal of the glauconitization process in a soil profile in western France. *Clay Min.* **16**, 231–243.

Desprairies A., Tremblay P., and Laloy C. (1989) Secondary mineral assemblages in a volcanic sequence drilled during ODP leg 104 in the Norwegian Sea. In *Proceedings of Ocean Drilling Program Scientific Reports* (eds. O. Eldholm and E. Taylor). vol. 104, pp. 397–409.

Duplay J. and Buatier M. (1990) The problem of differentiation of glauconite and celadonite. *Chem. Geol.* **84**, 264–266.

Fernandez A. and Moro M. C. (1998) Origin and depositional environment of Ordovocian stratiform iron mineralization from Zamora (NW Iberian Peninsula). *Mineralium Deposita* **33**, 606–619.

Föllmi K. and von Breymann M. (1992) Phosphates and glauconites of sites 789 and 799. In *Proceedings of Ocean Drilling Program*, Scientific Results (eds. N. J. Stewart and W. R. Winkler, *et al.*), vol. 127, pp. 63–72.

Foster M. D. (1962) Interpretation of the composition and a classification of chlorites. *US Geol. Surv. Prof. Pap.* **414A**, 33.

Fountain K. B. and McClellan G. H. (2000) Mineralogical and geochemical evidence or the origin of phosphorite nodules of the upper West Florida slope. *Soc. Econ. Pal. Min.* **66**, 201–220.

Frey M., Hunziker J. C., Roggwiller P., and Schindler C. (1973) Progressiv niedriggradige Metamorphose glaukonitführender Horisonte in den Helvetischen Alpen. *Contrib. Mineral Petrol.* **39**, 185–218.

Fritz S. T. and Toth T. A. (1997) An Fe-berthiérine from Cretaceous laterite: Part I. Characterisation and Part II. Estimation of Eh, pH, and pCO_2 conditions of formation. *Clays Clay Min.* **45**, 54–586.

Gharrabi M., Velde B., and Sagon J.-P. (1998) The transformation of illite to muscovite in pelitic rocks: constraints from X-ray diffraction. *Clays Clay Min.* **46**, 79–88.

Giresse P. and Odin G. (1973) Nature minéralogique et origine des glauconites du plateau continental du Gabon et du Congo. *Sedimentology* **20**, 457–488.

Harris L. C. and Whiting B. M. (2000) Sequence-stratigraphic signification of miocene to pliocene glauconite-rich layers on and offshore of the US mid-Atlantic margin. *Sedim. Geol.* **134**, 129–147.

Hillier S. and Velde B. (1991) Octahedral occupancy and the chemical composition of diagenetic (low temperature) chlorites. *Clay Min.* **26**, 149–168.

Hillier S. and Velde B. (1992) Chlorite interstratified with 7A mineral: an example from offshore Norway and possible implications for the interpretation of the composition of diagenetic chlorites. *Clay Min.* **27**, 475–486.

Holmes M. A. (1988) Evidence for continuous and discontinuous alteration in DSDP hole 418A basalts and its significance to natural gamma-ray log readings. In: *Proceedings of Ocean Drilling Project Scientific Results* (eds. M. H. Salisbury and J. H. Scott, *et al.*), vol. 102, pp. 135–149.

Hornibrook E. R. C. and Longstaffe F. J. (1996) Berthiérine from lower cretaceous clearwater formation, Alberta, Canada. *Clays Clay Min.* **44** 1–21.

Hower J. (1961) Some factors concerning the nature and origin of glauconites. *Am. Mineral.* **46**, 313–334.

Kim Y. and Lee Y. I. (2000) Ironstones and green marine clays in the dongjeom formation (Early Ordovician) of Korea. *Sedim. Geol.* **130**, 65–80.

Kronen J. D. and Glenn C. R. (2000) Pristine to reworked verdine: keys to sequence stratigraphy in mixed carbonate-siliclastic for reef sediments (Great Barrier Reef). SEPM Special Publication 66, Marine Authigenesis from Global to Microbial, pp. 387–403 (536pp.).

LaVerne C., Belarouchi A., and Honnorez J. (1996) Alteration mineralogy and chemistry of the upper oceanic crust from hole 896A, Costa Rica rift. In *Proceedings of Ocean Drilling Program Scientific Reports* (eds. J. C. Alt, H. Kinoshita, L. B. Stokking, and P. J. Michael), vol. 148, pp. 151–170.

Li G., Peacor D. R., Coombs D. S., and Kawachi Y. (1997) Solid solution in the celadonite family: the new minerals ferroceladonite, and ferroaluminum celadonite. *Am. Mineral.* **82**, 503–511.

Lonsdale P. F., Bischoff J. L., Burns V. M., Kastner M., and Sweeney R. E. (1980) A high temperature hydrothermal deposit on the seabed at a Gulf of California spreading center. *Earth Planet. Sci. Lett.* 420–498.

Loveland P. J. (1981) Weathering of a soil glauconite in Southern England. *Geoderma* **40**, 40–42.

Moore D. M. and Hughes R. E. (2000) Ordovician and Pennsylvanian berthiérine-bearing flint clays. *Clays Clay Min.* **48**, 145–149.

Newman A. C. D. (ed.) and Brown G. (1987) The chemical composition of clays. In *Chemistry of Clays and Clay Minerals*, Mineral. Soc. Monogr. 6, pp. 2–116 (480pp.).

Odin G. S. (ed.) (1988) Green marine clays, developments in sedimentology Elsevier, **45**, 444pp.

Odin G. and Lamboy M. (1988) Glaucony from the margin off northwestern Spain. In *Green Marine Clays*, Developments in Sedimentology 45 (ed. G. Odin). Elsevier, pp. 249–276 (444pp.).

Orcel J., Hénin S., and Caillère S. (1949) Sur les silicates phylliteux des minerais de fer oolithiques. *C. R. Acad. Sci. Paris* **229**, 134–135.

Parron D. and Amouric M. (1990) Crystallochemical heterogeneity of glauconites and the related problem of glauconite–celadonite distinction. *Chem. Geol.* **84**, 286–289.

Porrenga D. H. (1976) Glauconite and chamosite as depth indicators in marine environments. *Mar. Geol.* **5**, 495–501.

Righi D. and Meunier A. (1995) Origin of clays by rock weathering. In *Origin and Mineralogy of Clays* (ed. B. Velde). Springer, pp. 43–161 (334pp.).

Schrader E. L., Rosenthal B. S., Furbish W. J., and Mattey D. P. (1980) Mineralogy and geochemistry of hydrothermal and pelagic sediments from mounds hydrothermal field, galapagos spreading center, DSDP Leg 54. *J. Sedim. Petrol.* **50**, 917–928.

Thamban M. and Rao V. P. (2000) Distribution and composition of glaucony facies from sediments of the western continental margin of India. SEPM Special Publication 66, Marine Authigenesis from Global to Microbial, pp. 233–244(536pp.).

Thompson G. R. and Hower J. (1975) The mineralogy of glauconite. *Clays Clay Min.* **23**, 289–300.

Velde B. (1965) Phengitic micas: synthesis, stability and natural occurrence. *Am. J. Sci.* **262**, 886–913.

Velde B. (1968) The effect of chemical reduction on the stability of pyrophyllite and kaolinite in pelitic rocks. *J. Sedim. Petrol.* **39**, 13–16.

Velde B. (1972) Celadonite micas: solid solutions and stability. *Contrib. Mineral. Petrol.* **37**, 235–247.

Velde B. (1973) Phase equilibria studies in the system MgO–Al_2O_3–SiO_2–H_2O chlorites and associated minerals. *Min. Mag.* **39**, 297–312.

Velde B. (1977) *Clays and Clay Minerals in Natural and Synthetic Systems*. Elsevier, 218pp.

Velde B. (1985) *Clay Minerals: A Physio-chemical Explanation of their Occurrence*. Elsevier, 427pp.

Velde B. (1989) Phyllosilicate formation in berthiérine peloids in Phanerozoic ironstones. In *Phanerozoic Ironsones*. Special Publication 46 (eds. T. P. Young and W. E. G. Taylor). Geol. Soc. G B, pp. 3–8.

Velde B. and Odin G. (1975) Further information related to the origin of glauconite. *Clays Clay Min.* **23**, 376–381.

Velde B. and Vasseur G. (1992) Estimation of the diagenetic smectite to illite transformation in time–temperature space. *Am. Mineral.* **77**, 967–976.

Velde B., Raoult J. F., and Leikine M. (1974) Metamorphosed berthiérine pellets in Mid-Cretaceous rocks from north-eastern Algeria. *J. Sedim. Petrol* **44**, 1275–1280.

Velde B., Pedro G., and Carmouze J.-P. (1978) Pelletal nontronite formation in recent sediments of Lake Chad. *Chem. Geol.* **23**, 139–149.

Warr L. N., Primmer T. J., and Robinson D. (1991) Variscan very low-grade metamorphism in southwest England: a distathermal and thrust-related origin. *J. Metamorph. Geol.* **9**, 751–764.

Weaver C. E. and Pollard L. D. (1973) *The Chemistry of Clay Minerals*. Elsevier, Amsterdam.

Wiewora A., Lacka B., and Giresse P. (1996) Characterization and origin of 1 : 1 phyllosilicates within peloid of recent, Holocene and Miocene deposits of the Congo Basin. *Clays Clay Min.* **44**, 587–598.

Winkler H. G. F. (1964) Das P–T Feld der Diagenese und Niedrigtemperierten Metamorphose auf Grund von Mineralreaktionen. *Beiträge für Mineral. Petrol.* **10**, 70–93.

7.13
Chronometry of Sediments and Sedimentary Rocks

W. B. N. Berry

University of California, Berkeley, CA, USA

7.13.1 INTRODUCTION

Sedimentary rocks are commonly organized into discrete strata. The strata are composed of materials, diverse particles of inorganic and/or organic origin, that reflect aspects of the environmental conditions under which they got accumulated. Sequences of sedimentary-rock layers were seen and studied initially in cliffs, man-made exposures, and sites where the vegetation was not thick enough to obscure the rock layers. It was in mines, however, that sequences of strata came to be examined closely. Miner's observations of the succession of sedimentary rock layers they saw and quarried below the Earth's surface gave birth to a domain, viz. stratigraphy, and an understanding of sedimentary rocks. Berry (1968, 1987) discusses—from a historical perspective, the geological timescale—how an economic imperative became a significant force in the development of chronometry of sedimentary rocks.

7.13.2 CHRONOMETRY BASED ON THE FOSSIL RECORD—FIRST STEPS

Coal miners came to regard many layers of sedimentary rock as "friends," commonly giving them interesting names (Stinking Vein, the Dungy Drift, Kingswood Toad, etc.), or in some cases, simply numbers. Miners working in the British coal mines of the late 1700s understood the orderly succession of coal-bearing layers as well as the layers above and below them so well, that even in the dim candlelight of a mine shaft, they could discern where they were in a sequence of layers. Bits of clam or snail shell, wood, or fossils commonly helped them distinguish one layer from another. William Smith—in his professional work as a surveyor—has successfully applied his knowledge of the orderly succession of strata in many mines, and understanding that fossils were useful in distinguishing one stratum from another.

Winchester (2001, p. 71), in his wonderfully incisive discussion on William Smith's life and contribution to geology, pointed out: "The stratigraphical order in which the different types of rock were arranged in the coalfield, as the local miners knew and as William Smith learned from them all too rapidly, had an utterly predictable regularity to it." Winchester (2001, pp. 71–72) deftly portrayed development of Smith's understanding of the consistent orderliness of strata as follows:

> Smith would see and come to know the strata intimately as he saw them one by one, again and again, as the great winding chain (of the bucket in he stood to be lowered down into a mine) lowered him still further down through the measures.
> Smith learned both from records of the arrangement of strata in coal mining areas, such as that documentated by John Strachey in 1719 (see Fig. 1) and from talks with miners that each coal bed was considered unique. Because miners could distinguish one coal bed from others, they had a basic understanding of the succession of strata in a coal mine, and they could recite the sequence of coal beds from mine entry downwards. Smith became interested in and made observations of rocks intercalated between the coals. For example, Smith saw that certain sandstones bore features similar to those in modern dunes or beaches. He could see that these sands gave way to rocks made of muds similar to those found in the banks or beds of rivers. He noted, as Strachey (1719) had before him, (see Fig. 1), that some rock layers were crowded with fossils. Certain of these fossils were similar to those found in rivers, others were similar to creatures living in modern nearshore ocean environments (cockles in Fig. 1). and others resembled modern ferns. Smith realized that, in most mines, a sequence of sandstones–mudstones–coals was repeated over and over, as Strachey depicted (see Fig. 1) and described (1725).

Smith visited a number of coal mines, finding in each the same stratal sequence. The numerous mine tours led him to realize that the miners could recognize individual coal seams by certain unique characteristics, which included the fossil content.

An opportunity to explore further on his observations came when he was employed to conduct surveys needed to plan a route for a canal that would carry coal from the mines to markets. Smith realized that cutting a new lengthy canal would mean slicing open sequences of rock layers he had not studied previously. By close scrutinizing the rock layers in the pathway of the canal, Smith observed rocks, called Lias, lying above red rocks that he had seen above the coal in many mines.

Fossil collecting was an attractive pastime in Europe in the 1700s. Generally, collectors displayed their treasures extracted from diverse rock layers in a glass-fronted cabinet so that visitors could admire these treasures. Seldom were the specific rock layers from which the fossils were obtained noted. The desired goal was to have a collection of visually interesting objects. In some households, the most treasured specimens were given the same esthetic value as fine china. Indeed, china and fossils were displayed side by side. Smith viewed many such collections in the course of duties as a land surveyor. Among the collections he examined in the town of Bath were those of the Reverend Joseph Townsend and the Reverend Benjamin Richardson. Townsend was educated as a doctor and in the training had acquired a considerable interest in science. Those interests led him to purchase an interest in certain Irish mines which, in turn, led to mingling religious activities with observations of natural processes. He acquired a considerable amount of knowledge on rocks and their associated minerals and fossils. Richardson was an avid fossil collector, who met Smith at a meeting of the Agricultural Society. Richardson had assembled an unusually large fossil collection of which he was quite proud. He invited Smith to his home in Bath to see this collection. Smith saw that Richardson's fossil collection was organized on the basis of the types of the organisms. For example, ammonites were in one drawer and corals were in another. Richardson told Smith that he had no knowledge of the rock layers from which his treasured fossils had come. Smith responded by saying that he could arrange the fossils according to the sedimentary-rock layers from which each had been obtained. That organization, Smith suggested, would enable Richardson to see the progression of life forms from primitive and less advanced to the modern and relatively more sophisticated and advanced. Richardson agreed to Smith's suggested reorganization of his collection. Within about a day's time, Smith had completed it. Richardson was amazed at how rapidly the task had been accomplished. He questioned Smith on how he did it. Smith said that he had several years' experience observing the orderly succession of sedimentary-rock layers and their contained fossils. These observations have been made both beneath the ground in mines and above the ground in various exposures, including those made in cutting through rock layers for canals. Smith pointed out that as he went to different parts of Britain, the same rock layers occurred in the same superpositional order and each stratum bore essentially the same fossils in each area studied. Because the fossil content of each layer was consistent from area to area, Smith recognized that fossils could be used to identify the presence of any given rock layer. Smith was well aware that miners had been doing essentially that for many years in their quest for economically valued resources.

Richardson realized that Smith's observations were revolutionary. He also understood that they needed confirmation. Richardson contacted his long-time friend Townsend. Together they devised a test of Smith's idea. They pointed to a nearby hill on which a church had been built. They asked Smith to predict the rock layers and the fossil content of each rock layer that lay beneath the church. Smith told them what fossils should be found on the hill slopes and the superpositional order of each of the fossil-bearing rock layers they should encounter as they climbed up the hill. Excitedly, the three men went to the hill, climbed it, and examined the rock layers and collected fossils from many layers. At every step, they confirmed the predictions Smith had made.

The three men dined at Townsend's home in Bath on the evening of June 11, 1799. After dinner, Richardson and Townsend spread a large piece of paper on the dining table and asked Smith to tell them what he knew of the order of strata in the vicinity of Bath and the fossils found in each stratum. Smith recited the orderly superposition of rock layers that he knew so well, from chalk at the top as the youngest to coal at the base as the oldest. He cited 23 rock layers and he indicated the fossil content, the basic rock characteristics and the thickness of each layer. The tabulation that Smith dictated was designated "The Order of Strata and their Embedded Organic Remains, in the vicinity of Bath; examined and proved prior to 1799." One copy of that table of strata is preserved in the files of the Geological Society of London. This table is a milestone in development of an understanding of Earth history and in man's ability to reckon the passage of time. Winchester (2001, p. 134) commented on the significance of the table: "For the first time the earth had a provable history, a written record that paid no heed or obeisance to religious teaching and dogma, that declared its independence from the kind of faith that is no more than blind acceptance of absurdity. A science—an elemental basic science that would in due course allow mankind to exploit the almost limitless treasures of the underworld—had at last broken free from the age-old constraints of doctrine and canonical instruction."

The basic principle embodied in the table of strata that Smith dictated to his ecclesiastic associates in June 1799 has come to be known as the principle of faunal (and floral) succession (see discussions in Berry, 1968 and 1987 and in Kleinpell, 1979). Smith used that principle to predict what rock layers and the fossils they contained lay unseen underground in his everyday work. Smith realized that as searches were carried out for more coal and other natural resources—e.g, ores, building stones, sand and gravel, etc.—the principle would be invaluable in finding what lies underneath the surface of the earth. Predictions on the underneath strata at any site could be based on analyses of rocks and faunas .

The principle of faunal succession proved economically valuable in the quest and recovery of many natural resources used by mankind. The principle also led to the understanding that certain fossil aggregates are unique in the time of their occurrence and, therefore, could be used as the basis for a time unit. Two elements were required for such a use: (i) uniqueness of the fossil aggregate and (ii) its position in the overall succession of rock layers. The time units founded upon such fossil aggregates in the years after recognition of the principle of faunal succession were defined relatively broadly and even somewhat vaguely. However, they were fit together to form the basic elements of a geological time scale still in use. Darwin used that timescale in his discussion of organic evolution through natural selection.

Although Smith was encouraged by his friends to publish his table of strata and the information on which it was based, he was not able to do so. He had to continue his active professional work as a surveyor to provide a living for his family. As he pursued his surveying duties, however, he added to his knowledge the rock and fossil successions in previously unexplored areas. His ever-expanding knowledge of sedimentary rock sequences and the fossils found in many rock layers led him to set the goal of using his knowledge to make a geological map of England and Wales. That map would be based upon faunal succession and the active tracing of such rock layers. Smith envisioned a colored map that would depict the rock units as they are spread throughout England and Wales. Inasmuch as most of the rocks, he wished to depict, lay unseen under forests, plowed fields, ponds, and streams, he had to base his map upon his principle of faunal succession. The basic observations were time-consuming and, commonly, relatively tedious. Nevertheless, Smith pressed onward with his project on producing a map (see Simon Winchester's book *The Map that Changed the World* for a comprehensive and insightful account of Smith's career leading to production of the map). Smith hand-colored initial copies of the map and delivered a copy to the Board of Agriculture on May 23, 1815. Copies became available to the public by August of that year. The map clearly demonstrated the applicability and validity of the principle of faunal (and floral) succession. Not long after the map was published, the fundamental units of a timescale based upon faunal aggregates began to be developed. Berry (1968, 1987) wrote a brief history of that timescale. This history indicates that many of the units were initially relatively broadly defined, being founded upon accumulations of large groups of fossils obtained from relatively large aggregates of strata. Each unit was characterized by its superpositional relationship to subjacent and superjacent units and its unique fossil aggregate.

Initially, Smith's map demonstrated that the fossil content of each unit was essential to its recognition over a wide geographic area. Subsequently, the identity of the contained fossils came on focus for study. Miners knew most of the fossil plants or animal remains that typified the various layers, and they could predict where to dig to find more coal or other resources. Many private citizens interested in collecting some of "nature's wonders" also had a basic understanding of the fossils that were characteristic of certain rock layers. That knowledge was gradually integrated by a number of fossil collectors into a general understanding of the spectrum of fossils that characterized each major rock aggregate, the superpositional relationships of which were known. Fundamentally, most of the units of the timescale came to be recognized by bulk faunal- and/or floral-fossil aggregates obtained from large-scale clusters of sedimentary rocks.

Some of the units, now considered Periods in what is known as the geological timescale, were recognized because their rock or mineral contents were economically valuable. The need for building materials and coal led to numerous detailed studies of sedimentary rocks in many areas. Fossils were obtained from a number of layers of these rocks and were used to identify the positions of the most economically valuable layers, as they were traced from productive areas to prospective sites for more raw materials. The Carboniferous Period as a unit in the geologist's timescale is an example. Coal-bearing strata had been examined in many western European countries for almost a century before Smith clambered down into the mines in Somerset County, England to see the rock layer successions that the local miners knew so well (Figure 1). Miners in other western European countries, notably Belgium, France, and Germany knew the rock-layer superpositional order in mines in which they worked. The Belgian J. J. D'Omalius d'Halloy, described coal seams and subjacent and superjacent strata to the coal seen in Belgian mines. He designated the coal-bearing sequence as the Terrain Bituminifere. A few years later Conybeare and Phillips (1822) wrote a remarkable summary of British geology, *Outlines of the Geology of England and Wales.* Smith's map, available less than seven years earlier than publication of this remarkable summary, provided the essential chronometric understanding for the work. In the Conybeare–Phillips analysis, coal-bearing strata were grouped with superjacent lime rocks and subjacent red sandstones within a Carboniferous Order. Fossils from Carboniferous Order rocks in Britain and Ireland were described, respectively, by John Phillips and Frederick McCoy. Laurent G. de Koninck devoted much of his professional life in recording the fossils from Belgian Carboniferous rocks. Miners were very familiar with the remains of plants they found in the coals. Some of them collected fossil plants they found interesting or decorative. Some coal-measure plants were obtained by collectors for display in their own cabinets of nature's interesting objects. The French geologist Brogniart published a precise account of certain of these fossil floras in 1821. The many studies of floras and faunas from Carboniferous order rocks led to a fundamental understanding of the bulk aggregate of fossils by which the Carboniferous Period could be recognized as a unique interval in the history of the earth. The superpositional relationships of the group of rocks bearing those fossils had been documented in numerous mines throughout western Europe. Therefore, the fossil content could be acknowledged as indicative of a certain, unique, interval of deep or geologic time.

Giovanni Arduino described rocks that formed certain mountains in Italy as Tertiary because not only did they overlie older rocks but also they contained shell and rock particles that came from older, subjacent rocks. Because the subjacent rocks were termed "Secondary," the rocks above them were called "Tertiary." The Tertiary rocks seen in Italy commonly contained numbers of relatively modern-looking shells of marine clams, snails and sea urchins. Fossils bearing close resemblance have had been reported to be found at many sites in western Europe. Tertiary rocks were recognized widely throughout western Europe. The area around Paris includes one of the most extensively examined successions of Tertiary rocks. Cuvier and Brongniart provided a detailed account of these strata and their contained fossils early in the nineteenth century. Similar faunas were found and recorded in areas near London. The presence of Tertiary fossils and the allure of important scientific "work" in and around Paris as well as in sedimentary rock sequences in French resorts drew attention from Charles Lyell. Although he studied law, earthly processes fascinated Lyell. He became friendly with Roderick Murchison at Geological Society of London meetings, an association that led to an invitation from Murchison to join him and his wife on a geological tour of the continent for several months in 1828. Lyell accepted the invitation. He and the Murchisons traveled across France studying rocks and collecting fossils. Lyell parted from the Murchisons in the Fall of 1828 to tour Italy before returning to Paris, and, eventually, London by the end of February 1829. Lyell's observations of his collections of Tertiary fossils from many sites in France and Italy led him to realize that they could be divided. The divisions could be established

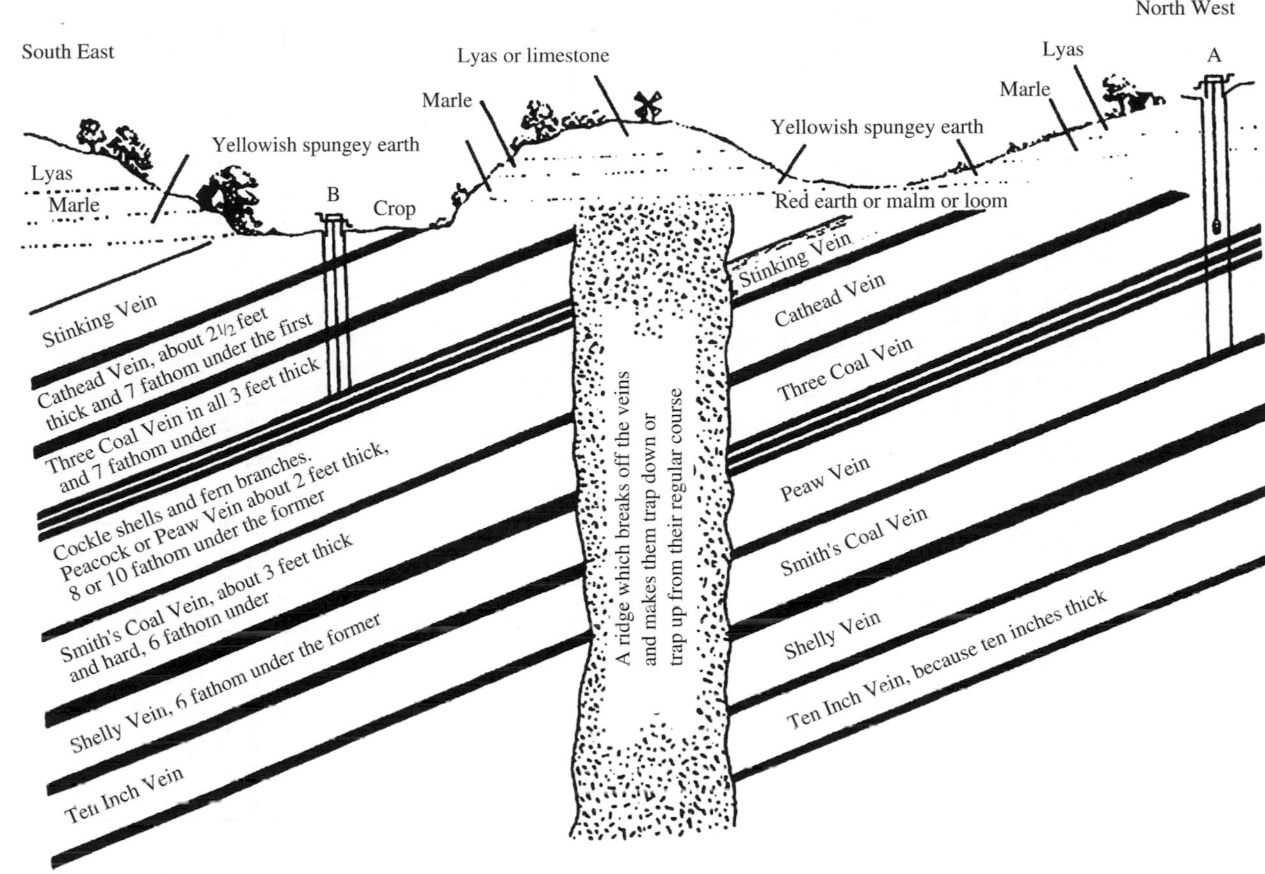

Figure 1 John Strachey's sketch of the occurrence of coal beds and adjacent strata in Somerset near Bristol, England. Note that each coal bed is named and that the occurrence of fossils ("cockle shells and fern branches") is indicated (source Strachey, 1719).

upon the degree of modernity of fossil clam and snail shells. Lyell's idea for division was founded upon analyses of a large number of marine-mollusc shells obtained from many sedimentary-rock successions from a number of areas. Lyell did not consider possible boundaries between the divisions he proposed.

Lyell (1833, pp. xii–xiii) stated in his introduction to the third volume of his *Principles of Geology* that by January 1829 he:

> had fully decided on attempting to establish four subdivisions of the great Tertiary epoch, the same which are fully illustrated in the present work. I considered the basin of Paris and London to be the type of the first division; the beds of the Superga, of the second; the Subapennine strata of northern Italy, of the third; and Ischia and Val di Noto, of the fourth.

Lyell found that most Tertiary sedimentary rocks bore abundant fossils, enabling him to collect masses of fossils from virtually every site visited. When he arrived in Paris, Lyell told Jules Desnoyers of his thoughts on Tertiary divisions. Desnoyers commented to Lyell that Gerard Deshayes had reached similar conclusions based upon his studies of fossil clam and snail shells housed in museum collections. Lyell contacted Deshayes and persuaded him to examine more than 4×10^4 specimens of Tertiary and modern clam and snail shells that represented more than 8,000 species. Deshayes compiled tables of the species using the stratigraphic position of each as the guide. Initially, Lyell and Deshayes agreed to three Tertiary divisions, which were, from oldest to youngest, Eocene, Miocene, and Pliocene. Subsequently, Lyell divided the Pliocene to an older and a newer Pliocene. The divisions were based upon the percentage of still-living marine-mollusc species. The percentages had relatively broad ranges and the faunas were so numerous that the percentage of still-living marine-mollusc species method for dividing the Tertiary came to be widely used across Europe. Darwin found that he could use it as well in his studies of Tertiary sedimentary rocks and their fossil content in South America.

Lyell's newer Pliocene contained ~90% still-living species. Lyell (1833, p. 54) noted that in his Older Pliocene "the proportion of recent species varies from upwards of a third to somewhat more than half the entire number." The Miocene was typified by "rather less than eighteen to one hundred" still extant species (Lyell, 1833, p. 54), and the Eocene had ~3% and 0.5% extant mollusc species. In 1846, Edward Forbes studied floral and faunal changes that took place during the last glaciation and used the designation Pleistocene for that glacial interval. Pleistocene quickly thereafter became the designation used by most students of the last major glaciation. Lyell concluded that the newer Pliocene was synonymous with Pleistocene and made that change in print in 1873. Lyell (1833) commented that his units might be divided through more precise floral and faunal studies. Divisions of certain of Lyell's Tertiary divisions were recognized, as predicted. The lower part of Lyell's Eocene was split off to form the Paleocene by Schimper in a study of fossil floras in 1874. The sedimentary rocks and fauna of Lyell's upper part of the Eocene were designated as Oligocene by von Beyrich in 1854.

7.13.3 REFINEMENTS IN CHRONOMETRY USING FOSSILS

Lyell (1833, p. 59) analyzed the stratigraphic occurrences of species from the lower to upper part of the Tertiary stratigraphic succession and commented:

> This increase of existing species, and gradual disappearance of the extinct, as we trace the series of formations from the older to the newer, is strictly analogous, as we before observed, to the fluctuations of a population such as might be recorded at successive periods, from the time when the oldest of the individuals now living was born to the present moment. The disappearance of persons who never were contemporaries of the greater part of the present generation, would be seen to have kept pace with the birth of those who now rank amongst the oldest men living, just as the Eocene and Miocene species are observed to have given place to those Pliocene testacea which are now contemporary with man.

Lyell's remarks foreshadowed the next steps in developing a timescale using fossils.

The French paleontologist Alcide d'Orbigny proposed stages as chronometric divisions that were more precise than Lyell's Tertiary divisions. D'Orbigny was engaged in preparing a comprehensive analysis of fossils and rocks in which they were found in France during the 1830s and 1840s. In the course of that work, he became dismayed at the plethora of local rock-unit names and successions of fossiliferous strata that had been described in exquisite detail. What concerned d'Orbigny was that few of the successions as well as local rock units had been related or correlated with each other. In his monumental compendium on Jurassic strata, *Terrains Jurassiques*, d'Orbigny (1842, p. 9) commented:

> Geologists, in their classifications permit themselves to be influenced by mineralogic composition of beds, whereas I take for my point of study ... the annihilation of an assemblage of organisms and replacement by another. I proceed solely on the identity of faunal composition ...

D'Orbigny recognized that the similarity of faunal assemblages was the key to correlating the many sedimentary-rock units and their contained faunas. Furthermore, he was well aware that certain groups of strata were characterized by aggregates of fossils that were unique to them. A fossil aggregate might occur in a single layer or in several layers of rock. The essential ingredient to d'Orbigny was that each aggregate could be recognized over a broad area. Furthermore, d'Orbigny asserted that in his studies of Jurassic rocks throughout France, the same aggregates of fossils always occurred in the same superpositional relationships. D'Orbigny gave the designation *stage* to each group of strata that bore the same fossil aggregate. He named most of the stages after geographic sites at which rocks bearing many of the fossils that typified each stage could be found.

D'Orbigny consulted geologists/paleontologists who had collected fossil aggregates closely similar to those he found in France in other parts of Europe as well as in Russia, India, and central America. Their comments suggested to him that the fossil aggregates characteristic of most stages could be widely found outside France. That idea led d'Orbigny to the conclusion that the faunas that characterized each stage were divisions that "nature has delineated with bold strokes across the whole earth." Comprehensive studies of Jurassic faunas in countries outside France demonstrated that d'Orbigny was overenthusiastic in his assertion that his stage faunas were global in their distribution. Nonetheless, most elements in his faunal aggregates could be recognized relatively widely across some of western Europe and, therefore, many of d'Orbigny's stage names were used by practicing geologists in many countries. Ultimately, because at least a few fossils in aggregates that characterize each stage were found in many different areas across the globe, stages came to be accepted as the basic unit for international correlations and chronometric discussions.

Hancock (1977, p. 11) analyzed d'Orbigny's concept of stage, and concluded that each of d'Orbigny's stages "is the major body of strata less than a system (anywhere in the world) that contains at least some of a long list of fossils which are peculiar to that piece of the total stratigraphic column." Hancock's (1977) analyses of d'Orbigny's stages makes clear that each stage is based on its fossil content and that the faunas that characterize each stage occur in a stratigraphic superpositional context.

The basic ingredient for making as precise, short-duration chronometric time units as is possible using fossils in superpositional order was elucidated by the German geologist F. A. Quenstedt. He advocated careful, detailed measurements of sequences of rock layers coupled with precise collecting of fossils layer by layer (see discussion in Berry, 1968). When Quenstedt examined occurrences of fauna in Jurassic strata in German sites, he found that he had difficulty using d'Orbigny's stages because certain fossil taxa indicated by d'Orbigny to be found only in one stage could be found occurring with fauna indicative of the subjacent and/or superjacent stage fauna. Quenstedt criticized d'Orbigny's work because it did not involve precise measurements of strata and careful positioning of each fossil collected from each stratal layer. Quenstedt found that many of the associations of fossils said by d'Orbigny to characterize a stage were actually faunal aggregates from a cluster of strata. Quenstedt maintained that geologists interested in precise chronometry should make a very detailed study of all the layers through which any fossil species could be found. Indeed, Quenstedt suggested, rock layer and faunal studies should be conducted at the scale of centimeters. Such detailed observations should be made in as many rock layer sequences as possible in any given area. After such observations had been made, Quenstedt believed that such analyses would lead, through comparison of the ranges of many species in many stratigraphic sections at several localities over a somewhat broad area, to a relatively precise understanding of the succession of individual species. That understanding, Quenstedt indicated, would result in relatively precise chronometry.

Precise measurements of rock-layer successions and collection of faunas, as Quensedt advocated, were being carried out by many geologists working in the 1830s and 1840s in a number of areas in Europe. The majority of these studies involved collecting ammonites from Jurassic strata (see Arkell, 1933).

Although Quenstedt asserted that detailed measurements of sequences of rock layers and their contained fossils could lead to relatively precise chronometries, it was one of his students, Albert Oppel, who demonstrated the validity and applicability of his mentor's assertion. Oppel initially studied the Jurassic rocks and faunas in the Schwabian area of the Jura Mountains in Germany, following the precise rock-layer measuring and fossil collecting proclaimed necessary by his mentor. Then, he traveled to many exposures of Jurassic strata in Europe, obtaining and studying fossils, and carefully measuring successions of rock layers. After extensive studies of a number of fossil species and the layers in which each occurred throughout a broad area in western Europe, Oppel formulated the principle which permits establishment of the most-precise, short-duration chronometric units possible using fossil occurrences in layered rocks.

Oppel called these units zones. He grouped the zones to broader units, which were essentially stages.

In his studies, Oppel noted as carefully as possible, the ranges of every species he found through all strata in which it occurred in every sequence of rock layers he examined. Oppel (1856–1858, p. 3) pointedly stated that he took care to "investigate the vertical distribution of each individual species at many different places ignoring the mineralogic character of the beds." From analyses of the plots of the vertical ranges of every species in all succession of rock layers examined, he realized that there were groups of strata characterized by closely similar aggregates of fossil species. He believed such bodies were time synchronous over the area in which they occurred. Oppel saw from his plots of vertical occurrences or stratigraphic ranges that some species occurred only in a short vertical distance in a few rock layers. Other taxa occurred through many layers. Still other species were found to have intermediate stratigraphic ranges. The differences among species in their occurrences in rock-layer sequences led to recognition of a pattern of overlapping ranges. Oppel discovered that, using certain appearances of new species, he could quite clearly point out boundaries between rock layers bearing distinctive and unique fossil aggregates. His studies resulted in recognition of a succession of 33 aggregates. He gave the designation *zone* to such bodies of rock bearing a unique aggregate of species. In Oppel's study, the base of each zone was chosen at the initial occurrence of one or a few species and the top of each zone was the base of the superjacent zone.

Hancock (1977, p. 12) pointed out that although the word "zone" had been used by geologists for many years with a number of different meanings, it was Oppel who gave the term a chronometric identity. Hancock (1977, p. 12) went on to state:

> Even today a brief perusal of Oppel's book impresses with its spread of detail.In eight separate districts of western Europe, the Jurassic rocks are subdivided into 33 zones correlated on the basis of their fossil content. Oppel's contemporaries outside Germany were completely bowled over; even the French admitted that it was pertinent to France and published a Tabular Summary (Laugel, 1858).

Arkell (1933) wrote a masterful review of the British Jurassic rocks and faunas. In that work, he considered Oppel's zones so vital in understanding British Jurassic history that he provided the following translation of the passage he considered as Oppel's most concise statement of his method of recognizing zones.

> Comparison has often been made between whole groups of beds, but it has not been shown that each horizon, identifiable in any place by a number of peculiar and constant species, is to be recognised with the same degree of certainty in distant regions. This task is admittedly a hard one, but it is only by carrying it out that an accurate correlation of a whole system can be assured. It necessarily involves exploring the vertical range of each separate species of the beds; by this means will be brought into prominence those zones which, through the constant and exclusive occurrence of certain species, mark themselves off from their neighbors as distinct horizons. In this way is obtained an ideal profile, of which the component parts of the same age in the various districts are characterised always by the same species (Oppel, 1856, p. 3; translated in Arkell, 1933, p. 16).

Hancock's (1977, p. 12) discussion of chronometry drew attention to Oppel's careful, precise rock-layer measurements and fossil collecting in establishing a method by which the "record of irreversible evolution of life on earth" could be documented. Hancock (1977, p. 12) pointed out that Oppel's complete work was published in the same year that "Alfred Russell Wallace and Charles Robert Darwin read their joint paper to the Linnean Society of London, 'On the Tendency of Species to Form Varieties, and on the Perpetuation of Varieties and Species by Natural Means of Selection.'" Hancock (1977, p. 12) commented that "Oppel himself remarked that the more accurately the fossils are examined and species defined, the greater the number of zonal divisions that could be recognized."

Oppel grouped his zones into stages. He pointed out (Oppel, 1856–1858, pp. 814–815) that "the stages which have been introduced by d'Orbigny were first supposed to represent stages or zones, but only later did it become clear that the majority of his stages could be further subdivided into more zones." Oppel clustered his zones to form groups of zones or stages, finding that, in general, he could recognize and use d'Orbigny's Jurassic Stages. Oppel (1856–1858) did not simply divide d'Orbigny's stages, rather he grouped his zones to form stages. When Oppel determined that certain zones did not fit into one of d'Orbigny's stages the stage definition was modified.

Hancock (1977) pointed out that the methods d'Orbigny and Oppel used were accepted slowly and reluctantly by many geologists. In the United States, e.g., the entities stages and zones were ignored until the 1930s. Many of the fundamental European studies of Jurassic rocks and fossils involved use of zones, founded primarily upon studies of ammonites. Charles Lapworth introduced Oppel's zone methodology to the Paleozoic through his work with graptolites.

Lapworth and his students, E. M. R. Wood and Gertrude Elles, carried out the precise stratigraphic measuring advocated by Quenstedt and Oppel in collecting graptolites and examining their stratigraphic ranges in rock-layer sequences throughout the British Isles. Using Oppel's methodology, Elles and Wood delineated nineteen graptolite zones within the Ordovician and Silurian in Britain. Lapworth (1879, p. 3), in a masterful analyses of Lower Paleozoic faunas, commented that "... we have no reliable chronological scale in geology but such as is afforded by the relative magnitude of zoological change." Graptolite zones, documented using Oppel's methodology, proved to be extraordinarily valuable tools in working out geologic structures in the Victoria (Australia) goldfields. Recognition of graptolite zones guided miners to noses of anticlines on which gold was concentrated. Similarly, graptolite zones have been used successfully in the search for gold in Nevada, USA. Both the Australian and Nevada gold occurrences proved to be among the world's most valuable at the time of their exploitation.

Use of zones based upon Oppel's methodology has proven to be of significant economic value in the 1920s and 1930s. On both the North American Gulf Coast and West Coast, oil finding was enhanced through the study of fossil benthic foraminiferans obtained through precise determinations of rock-layer positions of each species. Whole cores of rock layers were obtained in drilling for oil. These cores through numbers of rock layers gave precise rock–layer position of fossils encountered underground. Complete cores from any well proved difficult and expensive to obtain. Development of well side-wall cores reduced the need to obtain complete cores yet they could result in sampling most levels of the strata through which the driller's bit penetrated.

7.13.4 OIL RECOVERY IN CALIFORNIA USING FOSSIL-BASED CHRONOMETRY

R. M. Kleinpell (1938) employed Oppel's techniques in a study of the occurrences of fossil benthic foraminiferans in Miocene oil-bearing succession in California. Kleinpell entered the domain of oil exploration in California at a time when geologists were somewhat perplexed at how most effectively to ascertain geologic ages of the complexly folded and faulted Tertiary strata that contained California's oil resources (see Berry, 2000). Geologists concerned with finding oil soon realized that fossils were the only tools the oil-well driller had to determine rock unit and rock-layer position information when drilling beneath the ground surface. Nearly all of the rocks encountered in the initial exploration of California's oil-producing areas were fine-grained and didn't have obvious fossils. The problem was, simply, if oil was found in one well at some depth beneath the ground within a relatively thick, homogenous-appearing succession of rocks, how could one identify this oil-bearing position in other wells, given the folded and faulted aspects of California's oil-bearing sedimentary rocks? Fossils simply had to be obtained from the rock layers lying unseen underground to answer this question. In contrast to initial observations of strata encountered in drilling for oil in California, R. M. Kleinpell did see that the tiny shells of benthic foraminiferans are relatively plentiful in these rocks. He examined many rocks in cores from numerous wells drilled for oil. A well-read California geologist, Ralph Reed, was concerned greatly with expanding California's petroleum industry. When he learned that Kleinpell had found multitudes of fossil foraminiferans in rocks in wells already drilled for oil, he suggested to Kleinpell that the German Jurassic ammonite specialist, Albert Oppel, had published a methodology for analyzing fossil occurrences in precise stratigraphic positions that might "work" in California's oil exploration. Kleinpell studied Oppel's methods as a graduate student at Stanford and, working under the guidance of Hubert Schenck, Kleinpell used Oppel's methodology in his doctoral dissertation study of California's Miocene strata. Those strata were the most highly oil-bearing known at the time.

Kleinpell (1938) followed Oppel's procedures by tabulating the precise stratigraphic level at which every foraminiferan species he identified was found in every well core and every surface section he examined. That information permitted him to ascertain the stratigraphic ranges of benthic foraminiferan species found in mid-Tertiary rocks studied throughout much of central California. In all, Kleinpell analyzed the stratigraphic ranges of ~200 benthic foraminiferan species in ~200 stratigraphic sequences. His study resulted in delineation of six stages in strata considered to be approximately of Miocene age. Two or three zones were recognized within most of the stages. Each zone is recognized by a unique association of species. Every zone bears the name of one species in the unique association.

Documentation of the California Miocene zones and stages comprised Kleinpell's doctoral dissertation at Stanford which he completed in 1933. Upon fulfilling the doctoral degree requirements, Kleinpell went to work as a consulting oil geologist using his zones and stages to find oil successfully (see discussion in Berry (2000)). The book, *Miocene Stratigraphy of California*, which is based on Kleinpell's doctoral dissertation, was published by the American Association of Petroleum Geologists in 1938.

Kleinpell's zones and stages have been used with great success in California oil exploration. As wells were drilled and surface samples taken, foraminiferans were extracted from rock matrix. The associations were studied and species in each identified. Then the species associations found were compared with those unique associations of taxa that characterized each zone. When a match was recognized, the presence of the zone was documented. Certain zones were divided into subzones using the same (Oppel's methodology) procedures used to recognize the zones. Many geologists were employed to sit beside the well being drilled to obtain and prepare samples for foraminiferan study. Careful matching of species associations found at ever-deeper stratigraphic positions led to precise pinpointing of oil-bearing strata. Kleinpell and others involved in oil exploration simply matched associations of species found with those unique associations of taxa that Kleinpell indicated were characteristic of his zones. Use of Kleinpell's zones and stages resulted in recovery of many millions of barrels of oil from California's oil-bearing strata. A great deal of that oil was used to fuel American and Allies' efforts in World War II. At one time, California's oil, most of which was found using Kleinpell's zones and stages, amounted to about one-fourth of the world's production.

Because it seemed to Kleinpell that, among Americans and certainly among Californians, perhaps only Ralph Reed was aware of Oppel's work, Kleinpell included a number of lengthy quotes in German from Oppel in his 1938 book on the California Miocene stratigraphy. Over the years, based on comments from friends and associates, Kleinpell realized that few, if any, actually read these German passages. Therefore, he (Kleinpell, 1979) translated certain passages he thought were especially pertinent from Oppel's work (1856–1858). Certain portions of the passages Kleinpell (1979) translated are included herein. They include the same passage that Arkell (1933) cited in his work on the British Jurassic. Both Arkell and Kleinpell considered page 3 of Oppel's compendium as Oppel's most significant statement on zones. Kleinpell's (1979) translations are:

Page 3 of Oppel (1856–1858): "... it becomes necessary to explore, without regard to the mineralogic nature of the beds, the vertical distribution of every single species in the various localities, and then to erect those zones which may be distinguished from the adjacent one by the constant and unique occurrences of certain species. Thereby one obtains an ideal profile of the contemporaneous subdivisions which are always again characterized, in different areas, by the same species."

Page 4 "Since individual horizons may often be distinguished one from another more clearly than can one entire stage be distinguished from another stage, I have still retained the groupings of Jurassic strata into stages, since thereby the piecing together of the occurrences of the less well known fossils is facilitated. At the conclusion of every stage, I list those species upon which the classification and correlation of the beds is preeminently based."

Page 13 "... when I erect the zone of Am. raricostatus as the uppermost division of the lower Liassic. Above it begins the first Paxillose (Bel. Elongatus), as well as other species, which characterize the lowest zone of the middle Liassic. Although the paleontologic distinctions between adjacent (juxtaposed) bounding beds of the two stages is seldom pronounced, much as is the case of the distinctions between two neighboring zones within the same stage, so even here these distinctions in most cases can be readily accomplished."

Pages 814–815 "So, first of all we had to distinguish between 3 Jurassic zones. Many of them showed a remarkable uniformity through the entire terrain under consideration here, others are, on the contrary, only recognized with difficulty, partly because of the change in prevailing facies in some localities, ...the stages which have been introduced by D'Orbigny were first supposed to represent stages or zones, but only later did it become clear that the majority of his stages could be further subdivided into more zones. D'Orbigny has almost exclusively selected only locality names for the designation of his Jurassic stages."

In post-World War II America, need for oil grew tremendously. To help meet that need, Kleinpell not only used but also, perhaps more significantly, taught a number of students and professional associates how to use his California version of Oppel's methodology. In time, certain California oil geologists realized that Kleinpell's application of Oppel's methods, which had proven so significant in California petroleum exploration, were being overlooked by many involved in oil exploration. Accordingly, they persuaded Kleinpell to review his ideas and thoughts on Oppel's methods in a work entitled *Criteria in Correlation: Relevant Principles of Science* which was published in 1979 by the Pacific Division of the American Association of Petroleum Geologists. In that work, Kleinpell (1979, p. 12) pointed out:

> Oppel was the first to fully recognize and use the overlapping ranges of a few taxa that remain

diagnostic regardless of the particular facies in which they occur. This provides the finest level of refinement in biochronology and is made possible by recognizing the fractions of taxonomic ranges where consistent joint occurrences of such faithful taxa are found. Such units of fractional ranges are called Zones and constitute the bases of the smallest chronologic units that can be designated using sound scientific principle.

Kleinpell (1979, p. 12) stated that zones characterized according to Oppel's method may be recognized throughout a faunal province, and that zones established are not applicable throughout the world "except fortuitously or during the occasional flourishing of cosmopolitan faunas." Kleinpell (1979, p. 12) went on to point out that Davies (1934, p. 56) had described Oppel's zone method in a textbook for oil-field geologists as follows:

> Molluscan species are, as a rule, of small value as zone-fossils; but associations of species may be of value. A long-lived species has as contemporaries in its youth other species that preceded it in extinction, and new species arose to be its contemporaries in old age. Consequently, if the time-range of a great number of species were tabulated, it might be found that even small divisions in time were characterized by a particular association or overlap of species, some of which lived no later, others no earlier. Even then, it is not to be supposed that such an overlap was contemporaneous over very wide areas.

Kleinpell (1979, pp. 15–16) concluded his discussion of Oppel's zonation methodology as follows:

> The chronologic dimension of an Oppelian Zone, then, is diagnosed by the unique congregation of fossil species, the vertical stratigraphic ranges of which consistently overlap. The boundaries of such a Zone are two horizons; in paleontological correlations two critical horizons, not simply one, are involved: a *lower* horizon than which the age-diagnostic congregation cannot have occurred any earlier, and an *upper* horizon than which the age-diagnostic congregation cannot have occurred any later. Thus, Oppelian Zonation provides the greatest possible refinement of the general prehistoric time scale that has been made available to us on the basis of paleontological evidence.
>
> The strengths of Opplian Zonation lie in: (1) emphasizing careful induction from many stratigraphic sequences throughout the province so that a disciplined generalization is achieved; (2) use of the refined time-stratigraphic phenomenon of overlapping ranges of selected taxons (congregations); (3) acknowledging the limitations placed on any faunal chronology by the biogeographic facies problem; and (4) minimizing dependence of a Zone's units on ecological factors by considering all biofacies within a province.

Zones and stages based upon Oppel's procedures have been used by students for studying Mesozoic rocks and faunas for many years. Many of these studies involve very detailed analyses of the occurrences and evolution of ammonites. Minutely detailed layer-by-layer ammonite collecting at many European localities at which Jurassic strata are exposed has led to recognition of subzones within certain zones that are based upon ammonite faunas. Oppel predicted that such subzones might be recognized as divisions of his zones using the same basic methodology that he had used to distinguish zones. Callomon (1995) pointed out that certain Jurassic subzones based on ammonite faunas may have had durations of less than a million years (Figure 2).

Harland *et al.* (1990) comprehensively reviewed the development and use of many of the stages recognized in the Paleozoic, Mesozoic, and Cenozoic. They cited certain zonal successions based upon different organisms for each of the systems (Figure 3).

7.13.5 PRINCIPLES OF CHOROLOGY: THE SCIENCE OF THE DISTRIBUTION OF ORGANISMS

As Kleinpell (1979) pointed out, Oppel made an attempt to consider fossil taxa from a number of different facies which represented different environmental conditions in ascertaining the association or congregation of taxa that are used to denote a zone. Kleinpell (1979, p. 18) stated specifically: "one of the most important elements in Oppelian stages and zones is that ecologic facies are selected before a reliable chronology is attained." Kleinpell (1979) indicated that species from rocks that accumulated under different environmental conditions should be included within the faunas used to characterize a zone. In his discussion of the science of chorology, Kleinpell (1979) attempted to draw attention to the fact that ecologic and biogeographic factors constrain the distribution of organisms and that these constraints must be factored into any attempt to use zones characterized using Oppel's methodology.

Oppel's faunas, which were primarily ammonites, came from a large portion of western Europe. Oppel realized, however, that certain zones could not be recognized throughout the entire area and that, in such instances, only a stage or group of zones, could be recognized. This realization led, ultimately, to an understanding that zones delineated using Oppel's methodology are confined to a biogeographic province.

When d'Orbigny discussed his concept of stage, he thought the faunas that typified each were "global" in distribution. That idea was not consistent with the discoveries made in studies of distribution patterns of modern organisms by

Period	Epoch	JURASSIC Stage			PLANKTONIC AMMONITE ZONES (from Cope. *et al* 1980 a and b)	DINOFLAGRLLATE ZONES (from Woollam and Riding 1983)		ZONATIONS NANNOFOSSIL ZONES (from van Hinte, 1978b)
Jurrassic	Malm	Tithonian	Volgian (Vol)	Portlandian (Por)	*Subcraspedites lamplughi*	*Gochteodinia villosa*		*Nannoconus colomi*
					Subcraspedites preplicomphalus			
					Subcraspedites primitivus			
					?Titanites (Paracraspedites) oppressus			
					Titanites anquilormis	*Ctenidodinium cumulum/Ctenidodinium panneum*		
					Galbanites (Kerberites) kerbeus			
					Galbanites okusensis			
					Glaucolithites glaucolithus			
					Progalbanifes albani			
				Kimmeridgian (Kim) Late	*Virgatopavlovia fittoni*	*Glossodinium dimorphum/Dingodinium tuberosum*		*Parhabdolithus embergeri*
					Pavlovia rotunda			
					Pavlovia pallasioides			
					Pectinatites (Pectinatites) pectinatus			
					Pectinafites Arkellites hudlestoni			
					Pect (Virggatosphinctoides) wheatleyensis			
					Pectinatites (Virgato.) scitulus			*Watznaueria communis*
		Tth			*Pectinatites (Virgato.) elegans*			
		Kimmeridgian		Early	*Aulacostephanus autissiodorensis*			
					Aulacostephanus eudoxus	*Scnniodinium luridum*		
					Aulacostephaodies mutabilis			
					Rasenia cymodoce			
		Kim			*Pictonia baylei*	*Gonyaulacysta jurassica Scriniodinium crystallinium*		
		Oxfordian		Late	*Amoebocews rosenkrantzi*			*Vekshinella stradneri*
					Amoeboceras regulare			
					Amoeboceras serratum			
					Amoeboceras glosense			
				Mid	*Cardioceras tenuiserratum*			
					Cardioceras densiplicatum	*Acanthaulax senta*		
				Early	*Cardioceras cordatum*			*Actinozygus geometricus*
	M/m	Oxf			*Ouenstedtoceras mariae*	*Wanaea fimbriata*		*Diadozgus dorsetense*
	Dogger	Callovian			*Ouenstedtoceras (Lamberticeras) lamberti*	*Wanaea thysanota*		*Discorhabdus jungi*
					Peltoceras athleta			
					Erymnoceras coronatum	*Ctenidodinium ornatum/Ctenidodinium continum*		*Podorhabdus rahla*
					Kosmoceras (Gulielmites) jason			*Podorhabdus escaigi*
					Sigaloceras calloviense			*Stephanolithion bigoti*
				Clv	*Macrocephalites (M.) macrocephalus*	*Ctenidodinium combaziril Ctenidodium sellwoodii*		*Stephanolithion hexum*
		Bathonian		Late	*Clydoniceras (Clydoniceras) discus*			*Stephanolithion speciosum var. cotum*
					Oppelia (Oxycerites) aspidoides			
					Procerites hodsoni			*Diazomatolithus lehmani*
				Mid	*Morrisiceras (morrisiceras)morrisi*			
					Tulites (Tulites) subcontractus			
					Procerites progracilis			
				Early	*Asphinctites tenuiplicatus*			
		Bth			*Zigiagiceras (Zigiagiceras) zigzag*			
		Bajocian		Late	*Parkinsonia parkinsoni*	*Nannoceratopais gracilis*	*Acanthaulax crispa*	*Stephanolithion speciosum s.s.*
					Strenoceras (Garantiena)garantiana			
					Strenoceras sublurcatum		*Nannoceratopsis gracilis*	
				Early	*Stephanoceras humphriesianum*			
					Emileia (Otoites) souzei			
					Witcheiia laeviuscula			
		Baj			*Hyperlioceras discites*			
		Aalenian			*Graphoceras concavum*		*Mancodinium semitabulatum*	
					Ludwigia ,nurchisonae			
	Dog			Aal	*Leioceras opalinum*			*Discorhabdus tubus*
	Lias	Toarcian			*Dumortieria levesquei*			*Podorhabdus cylindratus*
					Grammoceras thouarsense			
					Haugia variabilis			
					Hildoceras bifrons			
					Harpoceras falciferum			
				Toa	*Dactylioceras tenuicostaum*			
		Pliensbachian			*Pleuroceras spinatum*		*Luehndea spinosa*	
					Amaltheus margaritatus			
					Prodactylioceras davoei	*Liasidium variabile*		
					Tragophylloceras ibex			*Crepidolithus crassus*
				Plb	*Uptonia jamesoni*			
		Sinemurian		Late	*Echioceras raricostatum*			
					Oxynoticeras oxynotum			*Palaeopontosphaera dubia*
					Asteroceras obtusum			
				Early	*Caenisiles turnei*	*Dapcodinium priscum*		*Parhabdolithus liasicus*
					Arnioceras semicostamum			*Parhabdolithus marthae*
		Sin			*Arietites bucklandi*			*Crucirhabdus primulus*
		Hettangian			*Schlotheimia angula*			*Annulithus arkelli*
					Alsatites liasicus			
J	Lia			Het	*Psilocerus planorbis*			

Figure 2 Jurassic stages, zones, and subzones recognized using Oppel's methods using stratigraphic ranges of ammonite species. Synchroneity between ammonite-bearing rocks that accumulated in shelf environments and sediments that accumulated in deep-ocean settings is also shown. Zones are based on dinoflagellates and coccolithophores (source Calloman, 1995).

Alfred Russell Wallace and many others since the mid 1850s which demonstrated that most organisms are limited in their distribution to certain geographic areas. Modern land and marine organisms are members of faunal or floral aggregates that collectively comprise the fauna or flora of a biogeographic province. Analyses of the distribution of fossil floras and faunas have demonstrated that biogeographic provinces essentially similar to those in the modern world have existed throughout the Phanerozoic. Biogeographic provinces are separated by climatic and/or physical

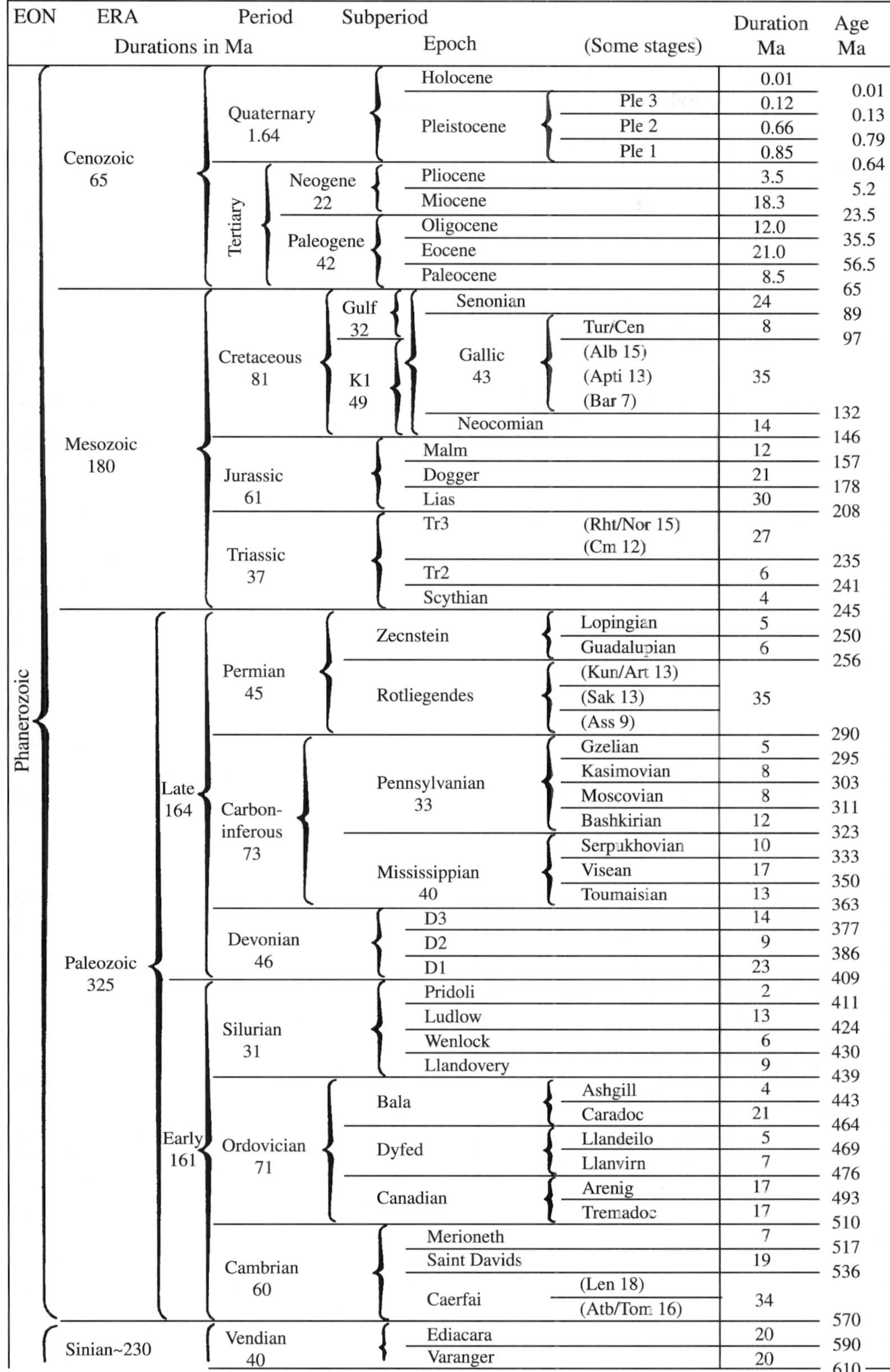

EON	ERA (Durations in Ma)	Period	Subperiod	Epoch	(Some stages)	Duration Ma	Age Ma
Phanerozoic	Cenozoic 65	Quaternary 1.64		Holocene		0.01	0.01
				Pleistocene	Ple 3	0.12	0.13
					Ple 2	0.66	0.79
					Ple 1	0.85	0.64
		Tertiary	Neogene 22	Pliocene		3.5	5.2
				Miocene		18.3	23.5
			Paleogene 42	Oligocene		12.0	35.5
				Eocene		21.0	56.5
				Paleocene		8.5	65
	Mesozoic 180	Cretaceous 81	Gulf 32	Senonian		24	89
				Gallic 43	Tur/Cen	8	97
			K1 49		(Alb 15) (Apti 13) (Bar 7)	35	132
				Neocomian		14	146
		Jurassic 61		Malm		12	157
				Dogger		21	178
				Lias		30	208
		Triassic 37		Tr3	(Rht/Nor 15) (Cm 12)	27	235
				Tr2		6	241
				Scythian		4	245
	Paleozoic 325 (Late 164)	Permian 45		Zecnstein	Lopingian	5	250
					Guadalupian	6	256
				Rotliegendes	(Kun/Art 13) (Sak 13) (Ass 9)	35	290
		Carbon-inferous 73	Pennsylvanian 33		Gzelian	5	295
					Kasimovian	8	303
					Moscovian	8	311
					Bashkirian	12	323
			Mississippian 40		Serpukhovian	10	333
					Visean	17	350
					Tournaisian	13	363
		Devonian 46		D3		14	377
				D2		9	386
				D1		23	409
	(Early 161)	Silurian 31		Pridoli		2	411
				Ludlow		13	424
				Wenlock		6	430
				Llandovery		9	439
		Ordovician 71		Bala	Ashgill	4	443
					Caradoc	21	464
				Dyfed	Llandeilo	5	469
					Llanvirn	7	476
				Canadian	Arenig	17	493
					Tremadoc	17	510
		Cambrian 60		Merioneth		7	517
				Saint Davids		19	536
				Caerfai	(Len 18) (Atb/Tom 16)	34	570
	Sinian~230	Vendian 40		Ediacara		20	590
				Varanger		20	610

Figure 3 Phanerozoic timescale (source Harland *et al.*, 1990).

barriers. Obviously, e.g., land masses are barriers to distributions of marine organisms. Among land-dwelling organisms, physical barriers such as mountains and deserts are barriers to distribution. Water temperatures commonly are barriers to distribution of whole marine provincial faunas. Availability of certain food resources or nutrient supplies may serve as barriers to distribution of many organisms.

Biogeographic provinces limit the distributions of those species that characterize zones delineated following Oppel's methodology. Stages, which are characterized by faunas that are more inclusive than the faunas that characterize zones, may not be limited to a single biogeographic province. Stage faunas may occur more widely. Some of them may be recognized throughout a biogeographic region or realm.

Many oil company-sponsored studies conducted in the quest for oil resources demonstrated the great value inherent in assessing environmental and ecologic relationships within the context of stages and zones delineated using Oppel's methodology. One such study is cited herein to document the economic value of using chronometric units delineated following Oppel's methods. Bandy and Arnal had access to a large volume of detailed stratigraphic, sedimentologic, and fossil-occurrence data from rocks encountered in numbers of wells drilled into oil-bearing Middle Tertiary strata in the San Joaquin Basin in California. They (Bandy and Arnal, 1969) used Kleinpell's zones and stages as chronometric units. Their study was part of an extensive program sponsored by the Gulf Oil Corporation to reconstruct the environmental history of the San Joaquin Basin with the goal of enhancing oil exploration and, ultimately, recovery. The basin lies in the southern part of California's Central Valley. The basin is bounded on the south by the Tehachapi Mountains, on the west by the California Coast Ranges, on the east by the Sierras, and it passes laterally northward into the Sacramento valley.

Bandy and Arnal (1969) examined more 5,000 mid-Tertiary foraminifer-bearing samples taken from 109 wells drilled for petroleum. The foraminifers were identified to species and a chronometric determination was made for every association recovered using Kleinpell's (1938) zones and stages as the chronometric scale (see Figure 4). Each sample was assigned to a zone and stage in the Kleinpell scheme. Most of the zone and stage assignments in the Bandy and Arnal study (see Figures 5–7) made by a consulting firm were headed by Stanley Beck. Then, using studies of the sedimentary constituents of each sample as well as ecologic analyses of the benthic foraminiferan species, Bandy and Arnal (1969) assigned each sample to a biofacies which they suggested was indicative of certain environmental conditions. They (Bandy and Arnal, 1969, p. 787) reported that "recent studies have indicated that many important bathyal species have rather similar upper depth limits in contrasting oceanic areas, regardless of differences in temperature, oxygen, and salinity." They compared the fossil foraminiferan associations indicative of a biofacies with those from modern environments and concluded that the biofacies they recognized could be deduced to be indicative of the depth of the seafloor. They (Bandy and Arnal, 1969, p. 787) noted that "examination of the various biofacies in the Tertiary strata of the San Joaquin Basin illustrates the entire gamut from the marsh habitat with a totally arenaceous fauna of *Ammobaculites* and *Ammobaculites* to the other extreme, an abyssal fauna." Because they (Bandy and Arnal, 1969) used biofacies as indicative of seafloor depth, they could plot a depth for each association within each stage. Those plots were used to construct paleobathymetric maps for each stage (see Figure 5). They evaluated fossil aggregates obtained from debris flows to indicate sites at which such flows had carried faunas down from shallower environments. Changes in bathymetry during each stage, as well as the areal distribution pattern for each depth interval were discussed. For example, Bandy and Arnal (1969, pp. 797–798) commented in their discussion on the Saucesian stage that "changes in bottom depth during the Saucesian include one prominent area of shoaling along the entire northern side of the basin and a few areas of subsidence generally within the deeper areas. There were 1000 to 2000 feet of shoaling in the northern area and as much as 2000 feet of deepening in a few places." Inasmuch as the study was of oil-bearing sites, Bandy and Arnal (1969, p. 798) pointed out that "location of oil fields with respect to basin topography suggests that major producing areas are in deeper waters, especially near the steeper slopes of the sea floor and in areas of rapid tectonism." Bandy and Arnal (1969) discussed basin subsidence for each stage, which they termed "paleotectonism." For the Saucesian Stage (see Figures 6 and 7), they inferred that the greatest thicknesses of Saucesian sediments were the sites of greatest subsidence, and that they were in the deepest waters at the south end of the basin. They (Bandy and Arnal, 1969) assessed changes in bathymetry in their analyses of sediment accumulations for each stage. For the Saucesian, e.g., they (Bandy and Arnal, 1969, p. 799) noted that the relevant data showed that "uplift was prominent in at least three areas within the basin." They (Bandy and Arnal, 1969, p. 816) concluded this study by pointing out that the oil-producing strata were "laid down on or near slopes of the basin where there are rather rapid changes in facies."

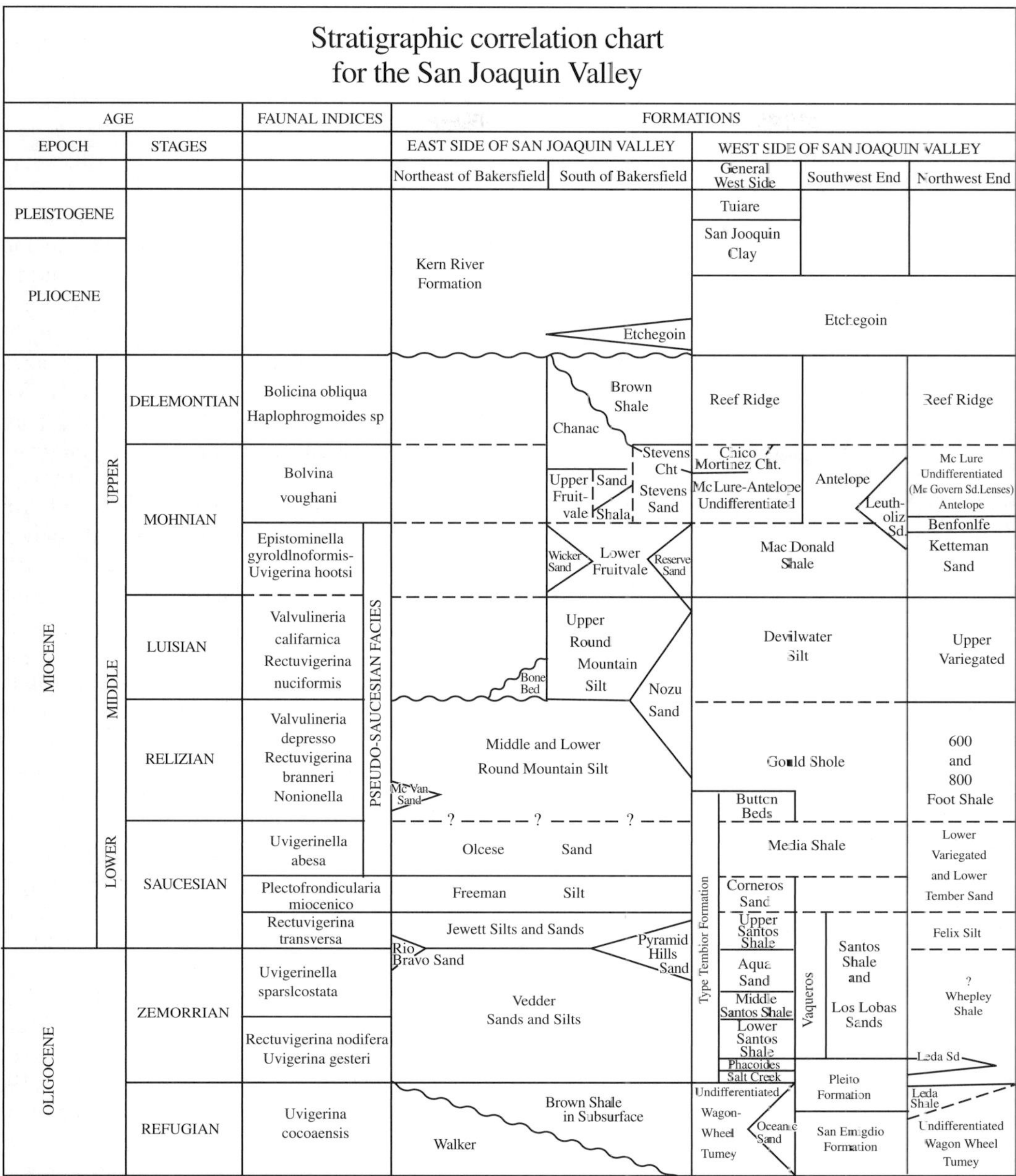

Figure 4 Kleinpell's mid-Tertiary stages and zones and rock unit correlations using these chronometric units for the southern San Joaquin Valley, California. This table was developed by Stanley Beck for the Gulf Oil Company for use by Bandy and Arnal (1969) in their discussion of the mid-Tertiary sedimentary history and petroleum potential of the area.

7.13.6 CONSTRAINTS ON CHRONOMETRY IMPOSED BY CHOROLOGY

As analyses of the stratigraphic distributions of organisms developed, the need to understand the constraints imposed by chorology on the extent to which zones defined using Oppel's methodology could be recognized and used became more and more apparent. In the case of graptolites, e.g., not only were graptolite zones found to be limited to a single faunal province, but also, within a province, they were found to be most recognizable in strata that formed on the margins of shelves or platforms. Rarely are graptolites found in inner shelf settings. Conodonts, an extinct fossil group, are abundant in carbonate rocks that formed in a spectrum of shelf environments, primarily those that lay within the tropics. Conodont zones are limited to a biogeographic province. Within a province, relatively inner- or shallow-shelf fauna differ from deeper-shelf faunas.

Studies of benthic foramineferans revealed certain clusters of taxa which were primarily limited to certain depths on the seafloor (e.g., the faunas that typify the pseudo-Saucesian of Bandy and Arnal (1969) which reflects a deep

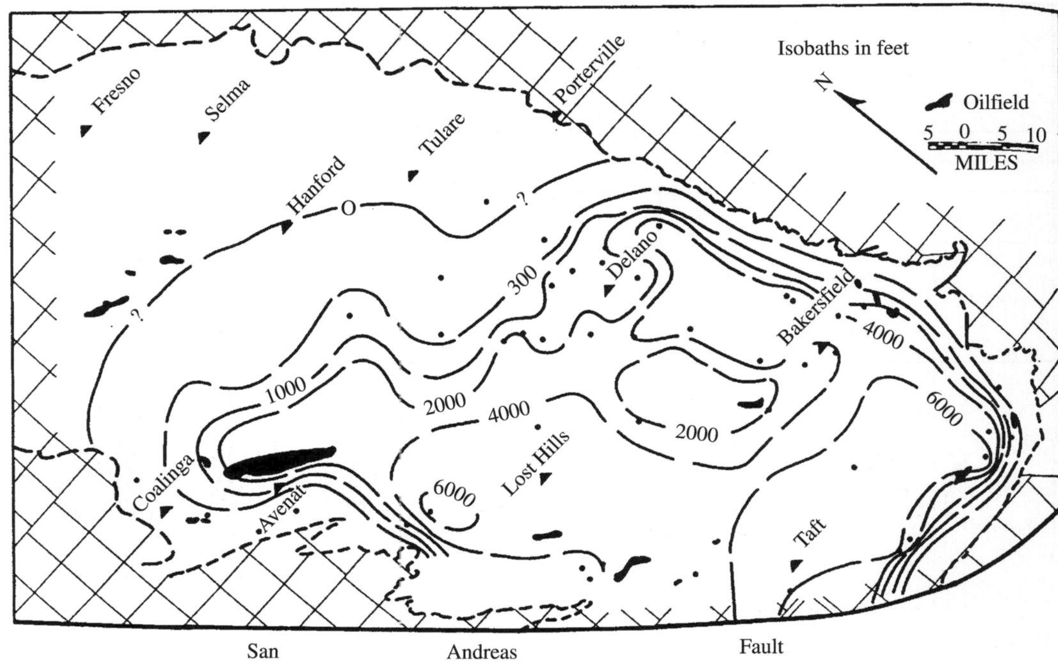

Figure 5 Map showing paleobathymetry of the area at the end of the Relizian Stage (source Bandy and Arnal, 1969, figure 10).

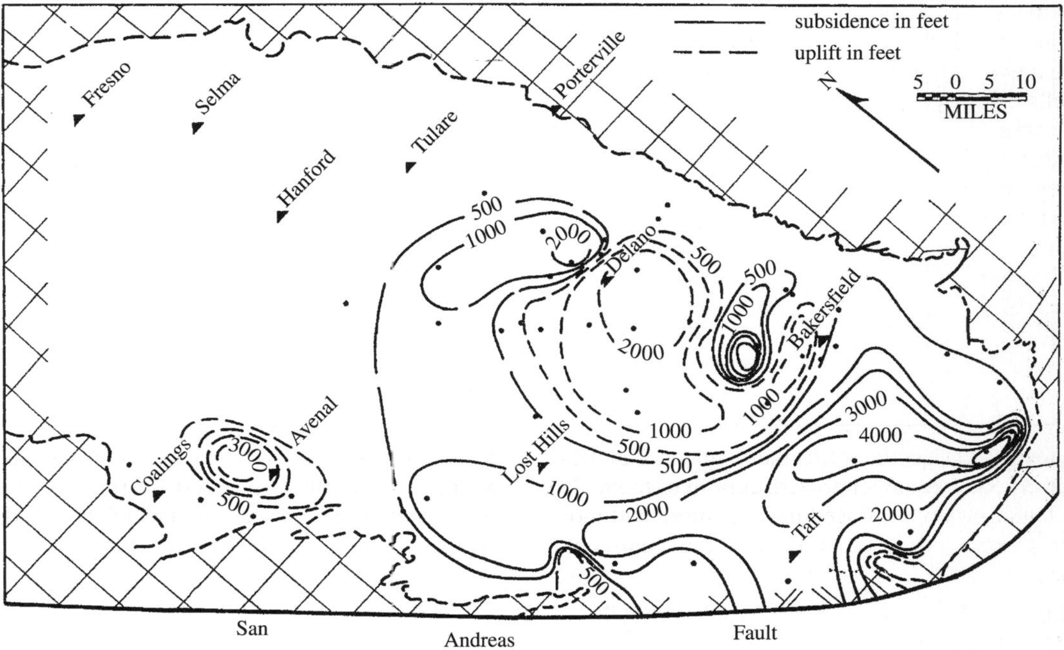

Figure 6 Paleotectonic map of the Saucesian stage indicating areas of uplift and of subsidence (source Bandy and Arnal, 1969, figure 9).

bathyal seafloor environment). Remains of planktic foraminferans were primarily recovered from strata that accumulated on the slopes of platforms as well as on the floors of the open ocean. Bandy and Arnal's (1969) study indicated that planktic foraminiferans are relatively rare in most basinal strata. When cores of sediments beneath the ocean floors were taken from many areas, the sediments were found to be composed primarily of shells of organisms that were planktic in life. Shells of planktic foraminiferans were found to be plentiful in many deep ocean-floor sites. Plates of the planktic algae, coccolithoporids, are recovered in large numbers at many deep ocean-floor sites,

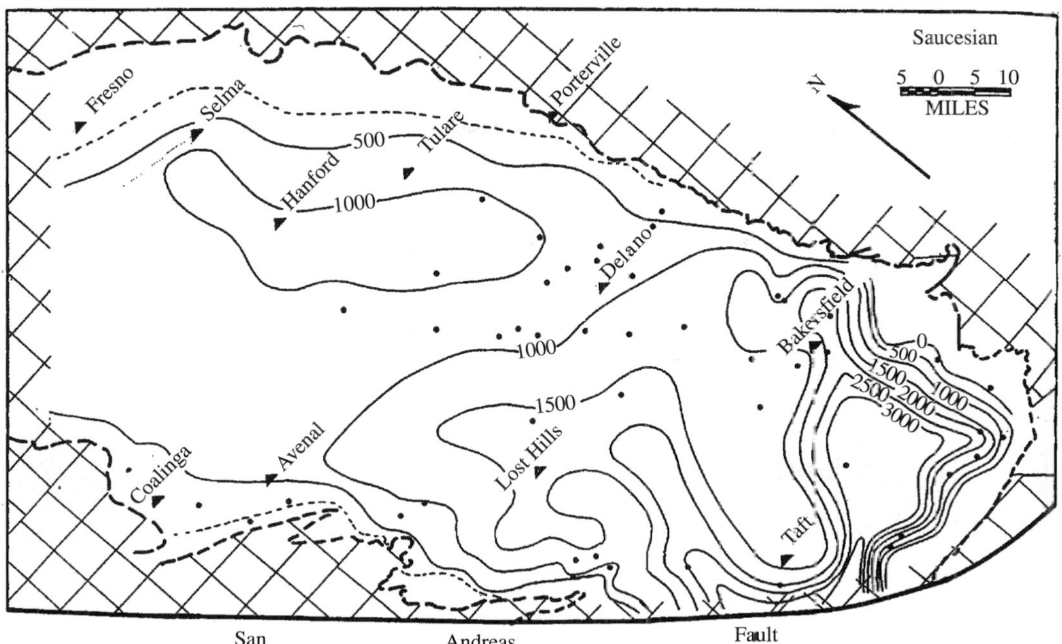

Figure 7 Isopach map of Saucesian strata (source Bandy and Arnal, 1969, figure 8).

especially those under tropical oceans. Locally, especially in sediments near the equatorial region of the Pacific, radiolarian shells occur in large numbers. Shells of these three types of planktic organisms occur in stratal succession in cores of wells drilled into deep ocean-floor sediments that are similar to stratal sequences in cores recovered in drilling for oil. The sedimentary record in many deep ocean-floor cores may extend back in time from the present day into the early Tertiary and, in some sites, into the Cretaceous and Jurassic. Shells are so plentiful in most of the deep ocean-floor cores and recovery may be made at such closely spaced intervals that evolutionary development of certain lineages may be traced closely. In initial studies of the deep-ocean sediment record, zones were delineated using essentially Oppel's methodology. As more cores were taken, subzones of these initial zones were recognized, certain of them also following Oppel's procedures. As more and more cores were taken from deep-ocean sediments, specific evolutionary developments in certain taxonomic lineages, e.g., among globorotalid (a type of planktic foraminiferan) species, have been recognized and used to divide the original zones. Additionally, studies of the occurrences of planktic taxa in numerous cores have led to the recognition that the first and last appearances of many species occur at approximately the same stratigraphic level within a province. That realization has led to use of first and last appearance datums as a means of dividing zones. These datums are termed first appearance datums (FADs) and last appearance datums (LADs). The FADs and LADs are considered to be essentially time synchronous within a province. Time of migration of an organism from place to place within a province is thought to have been relatively slight when viewed in the context of geologic time. Datums are used for precise chronometry in studies of Cenozoic sediments (Berggren *et al.*, 1995; Aubry, 1995).

Complete recovery of sediments in cores taken from deep-ocean drilling sites proved difficult to achieve in certain instances, thus care must be taken in examining cores from the deep ocean to ascertain if core recovery has been complete. If not, those levels at which materials are missing are noted and gaps or missing portions of the sediment and fossil record are considered when chronometric analyses are made.

The cores do provide such vast numbers of fossil shells the stratigraphic occurrences of which may be so precisely noted that many divisions of planktic foraminiferan and coccolithophorid zones seem to be only a few tens of thousands of years in duration (see Figure 8). The core record is somewhat similar to stratigraphic sections studied on land and to similar sequences seen in mines, but the age duration of most cores is significantly longer and the fossils in them far more numerous. The deep-ocean sediments bearing these fossiliferous successions lie on top of volcanic rocks the radiometric dates of which may be obtained and the magnetic polarity of which can be determined. The combined record of fossil remains, magnetic polarities, and links to radiometric chronologies provided by drilling into the seafloor opened the way for developing more precise chronometries than seems possible to achieve in most land-based successions.

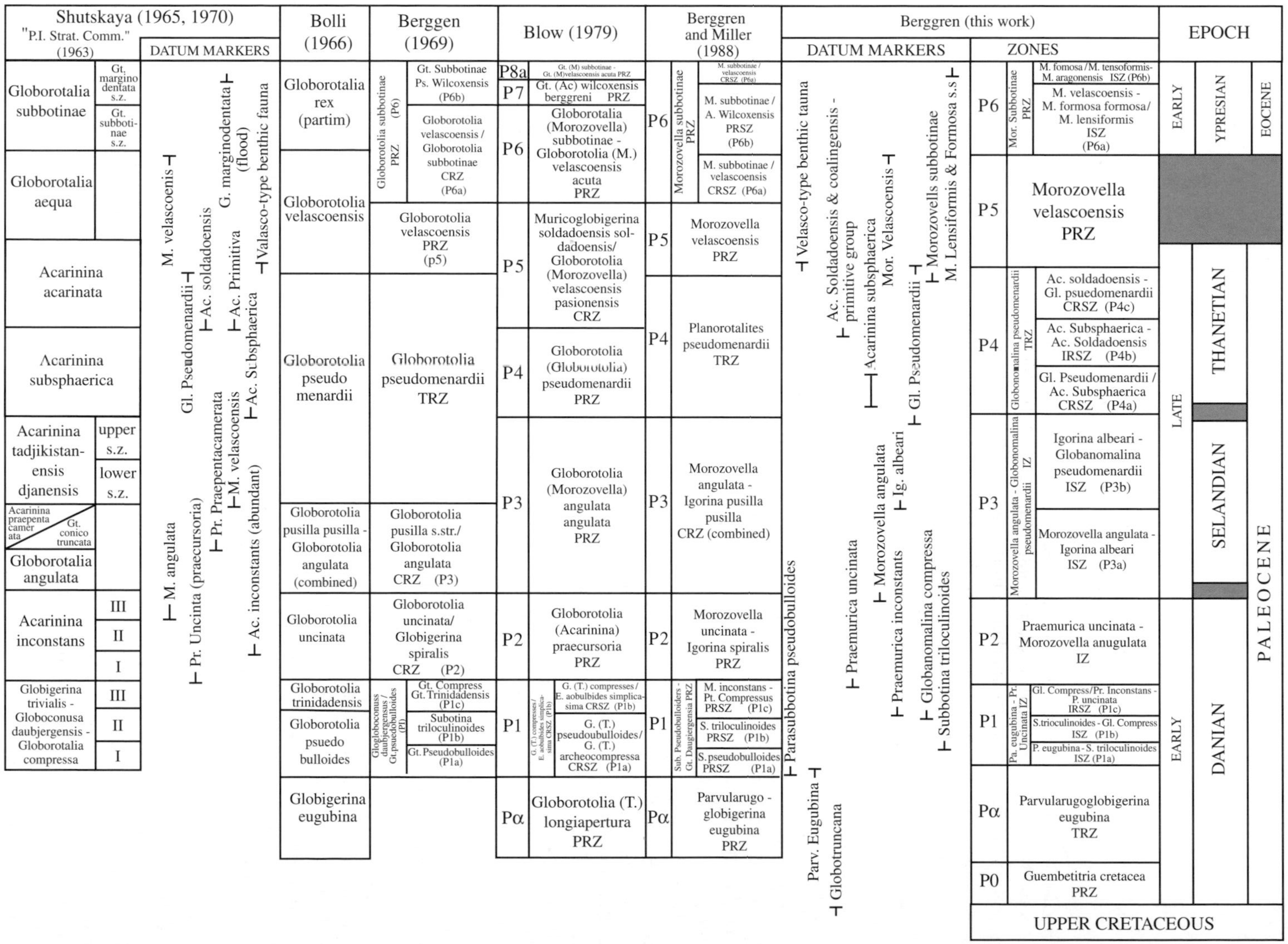

Shutskaya (1965, 1970) "P.I. Strat. Comm." (1963)		Bolli (1966)	Berggen (1969)	Blow (1979)		Berggren and Miller (1988)		Berggren (this work) ZONES		EPOCH		
Globorotalia subbotinae	Gt. margino dentata s.z.; Gt. subbotinae s.z.	Globorotalia rex (partim)	Globorotolia subbotinae PRZ (P6): Gt. Subbotinae Ps. Wilcoxensis (P6b); Globorotolia velascoensis / Globorotolia subbotinae CRZ (P6a)	P8a; P7; P6	Gt. (M) subbotinae - Gt. (M)velascoensis acuta PRZ; Gt. (Ac) wilcoxensis berggreni PRZ; Globorotalia (Morozovella) subbotinae - Globorotolia (M.) velascoensis acuta PRZ	P6	Morozovella subbotinae PRZ: M. subbotinae / velascoensis CRSZ (P6a); M. subbotinae / A. Wilcoxensis PRSZ (P6b); M. subbotinae / velascoensis CRSZ (P6a)	P6	Mor. Subbotinae PRZ: M. fomosa/M. tensoformis-M. aragonensis ISZ (P6b); M. velascoensis - M. formosa formosa/ M. lensiformis ISZ (P6a)	EARLY	YPRESIAN	EOCENE
Globorotalia aequa		Globorotolia velascoensis	Globorotolia velascoensis PRZ (p5)	P5	Muricoglobigerina soldadoensis soldadoensis/ Globorotolia (Morozovella) velascoensis pasionensis CRZ	P5	Morozovella velascoensis PRZ	P5	Morozovella velascoensis PRZ			
Acarinina acarinata												
Acarinina subsphaerica		Globorotolia pseudo menardii	Globorotolia pseudomenardii TRZ	P4	Globorotolia (Globorotolia) pseudomenardii PRZ	P4	Planorotalites pseudomenardii TRZ	P4	Globonomalina pseudomenardii TRZ: Ac. soldadoensis - Gl. psuedomenardii CRSZ (P4c); Ac. Subsphaerica - Ac. Soldadoensis IRSZ (P4b); Gl. Pseudomenardii / Ac. Subsphaerica CRSZ (P4a)	LATE	THANETIAN	PALEOCENE
Acarinina tadjikistanensis djanensis	upper s.z.; lower s.z.											
Acarinina praepenta camerata / Gt. conico truncata												
Globorotalia angulata		Globorotolia pusilla pusilla - Globorotolia angulata (combined)	Globorotolia pusilla s.str./ Globorotolia angulata CRZ (P3)	P3	Globorotolia (Morozovella) angulata angulata PRZ	P3	Morozovella angulata - Igorina pusilla pusilla CRZ (combined)	P3	Morozovella angulata - Globonomalina pseudomenardii IZ: Igorina albeari - Globanomalina pseudomenardii ISZ (P3b); Morozovella angulata - Igorina albeari ISZ (P3a)		SELANDIAN	
Acarinina inconstans	III; II; I	Globorotolia uncinata	Globorotolia uncinata/ Globigerina spiralis CRZ (P2)	P2	Globorotolia (Acarinina) praecursoria PRZ	P2	Morozovella uncinata - Igorina spiralis PRZ	P2	Praemurica uncinata - Morozovella anugulata IZ	EARLY	DANIAN	
Globigerina trivialis - Globoconusa daubjergensis - Globorotalia compressa	III; II; I	Globorotolia trinidadensis; Globorotolia psuedo bulloides	Glogloboconuss daubjergensus / Gt.psuedobulloides (P1): Gt. Compress Gt. Trinidadensis (P1c); Subotina triloculinoides (P1b); Gt. Pseudobulloides (P1a)	P1	G. (T.) compresses / E. aobulbides simplicasima CRSZ (P1b): G. (T.) compresses / E. aobulbides simplicasima CRSZ (P1b); G. (T.) pseudoubulloides/ G. (T.) archeocompressa CRSZ (P1a)	P1	Sub. Pseudobulloiders - Gt. Daugiergensia PRZ: M. inconstans - Pt. Compressus PRSZ (P1c); S. triloculinoides PRSZ (P1b); S. pseudobulloides PRSZ (P1a)	P1	Pa. eugubina - Pr. Uncinata IZ: Gl. Compress/Pr. Inconstans - P. uncinata IRSZ (P1c); S.trioculinoides - Gl. Compress ISZ (P1b); P. eugubina - S. triloculinoides ISZ (P1a)			
		Globigerina eugubina		Pα	Globorotolia (T.) longiapertura PRZ	Pα	Parvularugo - globigerina eugubina PRZ	Pα	Parvularugoglobigerina eugubina TRZ			
								P0	Guembetitria cretacea PRZ			
								UPPER CRETACEOUS				

Shutskaya (1965, 1970) – DATUM MARKERS:
- M. velascoenis
- Gl. Pseudomenardii
- Ac. soldadoensis
- G. marginodentata (flood)
- Ac. Primitiva
- Valasco-type benthic fauna
- Pr. Praepentacamerata
- M. velascoensis
- Ac. Subsphaerica
- M. angulata
- Pr. Uncinta (praecursoria)
- Ac. inconstants (abundant)

Berggren (this work) – DATUM MARKERS:
- Velasco-type benthic tauna
- Ac. Soldadoensis & coalingensis - primitive group
- Acarinina subsphaerica
- Mor. Velascoensis
- Gl. Pseudomenardii
- Morozovells subbotinae
- M. Lensiformis & Formosa s.s
- Praemurica uncinata
- Morozovella angulata
- Ig. albeari
- Praemurica inconstants
- Globanomalina compressa
- Subbotina triloculinoides
- Parasubbotina pseudobulloides
- Parv. Eugubina
- Globotruncana

Figure 8 Comparative Paleocene planktic foraminiferan zonations illustrating zones from the work of several authors based on fossil aggregates and datums (source Berggren *et al.*, 1995, figure 7).

Studies of fossils recovered from ocean cores indicate that zones based on planktic organisms are limited to biogeographic provinces. These provinces appear to be reflective of latitudinal distributions of water temperatures, and, to some extent, water masses.

7.13.7 RADIOCHRONOMETRY

Chronometry based on the fossil record gave geologists, especially those involved in the search for natural resources, a valuable tool with which they could direct recovery of natural resources. William Smith demonstrated that the use of chronometric units were used to guide routes for shipping those resources to markets. Being as valuable as the chronometry, it cannot indicate absolute ages of rocks. That development had to wait until about the turn of the 1800s to the 1900s. The French physicist Antoine Henri Becquerel discovered in 1896 that uranium spontaneously emits rays that can cloud photographic plates in the dark. He called this process radioactivity. The American chemist, B. B. Boltwood noted that radioactive decay of uranium leads ultimately to lead. Armed with that observation, Boltwood analyzed uranium minerals obtained from many rocks found at numerous sites around the globe. He found that uranium minerals from older rocks contained more lead than uranium minerals in younger rocks. Using an estimated rate of decay of uranium to lead, Boltwood generated some preliminary ages for certain Paleozoic and older rocks in 1907. Boltwood's assessments were followed by those of Arthur Homes who was able to refine a uranium to lead decay rate and to propose absolute ages for many parts of the timescale based not only upon the fossil record but also older rocks, those that comprise the pre-Phanerozoic or Precambrian. Improvements in technology and in ascertaining naturally occurring daughter elements produced by radiometric decay from a parent element occurring in rocks followed the initial studies. These methods are discussed in significant depth by Harland *et al.* (1990, pp. 73–103). Radiochronometry is discussed in many textbooks. Of them, Boggs (2001, pp. 592–601) is as inclusive as any.

Potassium, which occurs in many volcanic rocks, has an isotope, potassium-40, which decays to argon-39. This parent–daughter ratio proved widely useful in studies of a number of volcanic rocks and became perhaps the most widely used radiochronometric method. It has been used to date many parts of the sedimentary-rock record through analyses of the relationships of volcanic rocks with sedimentary rocks. In an effort to achieve greater precision in radiochronometry, specialists in the study of decay of potassium-40 to argon-39 found that if they converted potassium-39 to argon-39 by irradiation with neutrons in a nuclear reactor they could let argon-39 proxy for potassium-40. To accomplish this, they had to know the ratio of potassium-39 to potassium-40. The process involving irradiation permitted analysis just for argon isotopes, viz., argon-40 and argon-39. Age of the containing mineral and, presumably, rock may then be determined from the argon-40 to argon-39 ratio. The assumption in this procedure is that that the potassium-39 to potassium-40 ratio in nature has been essentially constant and that the amount produced by irradiation depends on the duration of the irradiation and the total amount of potassium present. Inasmuch as the amount of potassium-39 that is converted to argon-39 during irradiation is not known, samples are irradiated with what is called a "fluence monitor mineral" or, in essence, a standard mineral the age of which is known. This procedure allows calculation of an irradiation coefficient which may then be used in calculations of the age of the unknown sample. Clearly, calibration of the "fluence monitor mineral" or standard influences the accuracy of argon-40 to argon-39 ages. In discussing the argon–argon method, Swisher (in Berggren *et al.* (1995, p. 133)) commented that the precision of argon-40/argon-39 ages "far exceeds the accuracy of the age of any currently available monitor mineral." Swisher (in Berggren *et al.* (1995, p. 133)) discussed details of the argon/argon methodology and noted that, for purposes of intercalibration among laboratories, one of two monitor minerals are commonly used. The argon/argon procedure has proven to be valuable because small samplescan be analyzed and corrections may be made for argon leaks from the sample.

Another widely used radiochronometric method involves study of the isotopes of carbon. Studies on these isotopes demonstrated that carbon-14 is taken in consistently by living organisms but that when the organism dies, carbon-14 starts to decay. Thus, a ratio of carbon-14 to the stable carbon isotopes, one of which is carbon-12, can provide an age in years since death of the organism. The ages of organic materials from the present as far back in time to $\sim 5 \times 10^4$ yr of age can be determined, with appropriate corrections for materials somewhat greater than 10^4 yr old.

Sediments have been dated by examination of the potassium/argon ratio in certain glauconites. Glauconite systems take time to close off the loss of argon, thus dates derived from glauconites in sediments are significantly younger than the age of the enclosing sediments.

Another radiochronometric method used with sedimentary rocks is the decay of uranium-238, which decays through several daughters one of which is thorium-230. Uranium-238 is relatively soluble in seawater and so is detectable in seawater analyses. Thorium-230 becomes included in certain minerals that are incorporated into sea-bottom sediment. Thorium-230 decays, with a half-life of 7.5×10^4 yr, to radium-226. Cores of seafloor sediment exhibit a decrease in throrium-230. By comparing the amount of thorium at any given depth in a sediment core with the amount in the surface sediment, an age for different sediment levels in the core can be determined.

Radiochronologic methods are especially valuable for many volcanic and plutonic rocks and organic substances, but, in general, they have limited applicability in determining the absolute ages of sedimentary rocks. Organic materials that occur as components of relatively young sediments may be studied to indicate a minimum age for the rock or sediment. For most sediments and sedimentary rocks, especially those older than a few tens of thousands of years, the absolute age must be ascertained by examination of the relationships with volcanic rocks. For much of the fossil-based timescale, especially the Paleozoic and early Mesozoic, absolute ages have to be worked out by looking at nearby volcanic rocks or, perhaps, obtaining a minimum age from a plutonic rock that cuts through the sediments. Because of experimental errors involved in radiochronologic methodology, zones and some stages, notably those in the Paleozoic, provide more precise estimates of age than radiochronology in many studies of sedimentary rocks.

Dating of the sedimentary and, indeed, the record of other types of rocks for the pre-Phanerozoic or Precambrian essentially requires use of radiochronologic methods. In portions of the Precambrian, notably the latter part, fossil remains of algae and bacteria have been useful in certain cases. For most of the Precambrian, however, radiochronology is essentially the only useful method. Harland *et al.* (1990) reviewed Precambrian radiochronology and ages of Phanerozoic rocks.

7.13.8 MAGNETIC FIELD POLARITY AND CHRONOMETRY

Remanent magnetism studies developed during the 1960s. The study was led primarily by two groups of geophysicists, one in California and the other in Australia. These scientists studied the remanent magnetism of a number of volcanic rocks on land and discovered that the magnetic field in certain ones of them was the reverse of the modern magnetic field. As they studied volcanic rocks of different ages that had formed over the last few thousand years, they realized that the Earth's magnetic field did reverse itself at intervals throughout the past. From that initial recognition, a sequence of normal and reversed magnetic field polarities was documented. The next step taken was to use potassium–argon radiometric dating methodology to determine the ages of the volcanic rocks the remanent magnetism of which had been determined. Following that step, the absolute or age in years of each reversal could be established. Initially, the polarity reversal scale was developed for only about the last 5 Myr. At the time the dating was carried out, the error percent in the potassium–argon dates was greater than a number of the shorter polarity reversals. Enhanced techniques in the potassium–argon dating method, notably development of the argon–argon method, allowed the polarity reversal scale to be extended back in time into the Cretaceous and even the latter part of the Jurassic.

Although the polarity reversal time scale was based initially on studies of volcanic rocks on land, geophysicists realized that the ocean floors seemingly provided a marvelously continuous record of volcanic activity that extended back in time through the Middle Jurassic. Fluctuations in magnetic field intensity they saw in records of seafloor volcanic rocks came to be interpreted as being indicative of the polarity reversals seen in the volcanic rocks on land. The records obtained in the initial studies started with newly formed volcanic rocks near the crests of submarine oceanic ridges and went from the ridges into older and older rock. The ages of the seafloor volcanic rocks could, to some degree, be determined from the paleontologic ages of the oceanic sediments superjacent to the volcanic-rock basement.

The seafloor magnetic field intensity records began with towing a magnetometer from a ship back and forth over a large segment of the northeastern part of the Pacific Ocean. Analysis of the results from that study revealed that the earth's magnetic field intensities were aligned in relatively long, linear "stripes" that lay parallel with a deep ocean ridge crest from which new crustal material was emanating. This discovery provided the means by which the polarity reversal scale documented on land could be extended back in time and, where possible, calibrated with both radiochronology and fossil-based chronometry.

Originally, the ages of oceanic magnetic field fluctuations were determined by extrapolating their ages based on an assumption of the rate of movement of the volcanic material away from the place it formed, the oceanic ridge crest. Initially, the rate of spreading since the Late Cretaceous was thought to be constant at $\sim 1.9 \text{ cm yr}^{-1}$. Subsequent studies indicated that the rate has not been constant but rather that it has decreased from 70 mm yr^{-1} late in the Cretaceous to ~32 mm yr^{-1} at present. Furthermore, even that rate of decrease has not been constant.

With the use of sensitive detectors of magnetism, the oceanic-sediment record has come into focus in polarity reversal analyses. Although the magnetism signal is relatively weak in sediment, it can be detected, notably in iron-bearing oceanic sediments. Polarity determinations from oceanic sediment have an advantage because the sediments commonly are fossiliferous. In instances where they are, a paleontologic age may be determined for the same sediment yielding a magnetic field determination. Cores taken from deep-ocean sediments have provided age and magnetic polarity information back in time into about the mid-Jurassic.

Harland *et al.*, 1990, p. 142) commented that the magnetic field reversals seen in the ocean-floor volcanic rock record is the "richest single source of information about magnetic reversals." They went on to point out that "gaps and duplications are commonly present on the profiles" of seafloor volcanic rock magnetic field changes "because of the jumping of ridges to new positions." They also pointed out (Harland *et al.*, 1990, p. 246) that the fossil record in oceanic sediments and/or radiochronologic methods have to be used to calibrate the seafloor magnetic field fluctuations.

Geophysicists number the magnetic polarity reversal scale starting at 1 for the present-day ridge-center volcanism. The polarity time intervals are called chrons. The Cenozoic chrons are designated as C and a number. The Mesozoic chrons are designated M followed by a number. Where extensive studies of remanent magnetism of oceanic materials reveal reversals within a time of normal or reversed polarity, the number of a chron is followed by a letter, e.g., in C5a, C5b, C5c. Because many of the chrons may be dated paleontologically, study of such chrons has resulted in recognition of an Integrated Magnetobiostratigraphic Scale (IMBS). Berggren *et al.* (1995, pp. 130–132) discussed terminology of chrons and chron divisions (see Figure 9) as components of "The Integrated Magnetobiostratigraphic Scale."

Aubry (1995) drew attention to certain problems that could arise in using only magnetic chronology units. Two normal polarity units could be separated by an unconformity but the unconformity might not be detected. For example, an unconformity could eliminate detection of a normal magnetic unit if two reversed units lay on each side of that unconformity. Aubry (1995, p. 215) pointed out that only by use of the fossil evidence for dating in conjunction with the magnetic reversal data could such uncertainties be eliminated. Aubry stated cogently (1995, p. 215) that "biostratigraphy remains an essential component in the science of stratigraphy. In most instances, magnetozones and stable isotope events can only be confidently identified with the support of biostratigraphy." Both the comments made by Aubry (1995) and by Harland *et al.* (1990, pp. 144–147) concerning uncertainties in calibrating the seafloor magnetic field fluctuations indicate the value of and, in fact, necessity of using the fossil record in sediments to have confidence in magnetic chronometry.

7.13.9 ORBITAL CHRONOMETRY

Herbert *et al.* (1995) pointed out from a review of Cretaceous to Pleistocene hemipelagites and pelagites that these deposits display cyclical variations that "ripple in an almost unbroken wave train from the Pleistocene Ice Age world into the warm Cretaceous Period." The cyclical deposits commonly display alternations in color, from light to dark, or show up as subtle changes in grain size and sorting. The forces that lead to this cyclical variation appear to be "variations in insolation as a function of latitude and season, cause by quasi-cyclical changes in the Earth's orbit" (Herbert *et al.*, 1995, p. 81). These changes in the Earth's orbit, commonly called the "Milankovitch cycles," "can be partitioned into the precessional index (modern mean period circa 21 ky), obliquity, (modern mean period circa 41 ky), and eccentricity (modern mean periods of 95, 123, and 413 ky)" (Herbert *et al.*, 1995, p. 81). Herbert *et al.* (1995, p. 81) describe it, using orbital chronometry as a means of providing enhanced refinement of other chronometric methods, as follows:

Just as tick marks of a yardstick measure distance, sedimentary cycles of orbital origin count time at a high precision as compared with most chronological methods. Furthermore, because cyclic marine sediments generally contain stratigraphically useful microfossils and often retain polarity reversal information, the "Milankovitch" Clock could be widely applicable to time scale problems. When anchored accurately in a magnetochronologic and biochronologic framework (which must ultimately be calibrated by radiometric methods), orbital cycles in sediments provide a chronometer that functions at one to two orders of magnitude better resolution than standard techniques."

Having made that general statement of the potential inherent studies of cylicity in hemielagites and pelagites, Herbert *et al.* (1995, p. 82) caution: "The orbital rhythms imprinted on sediments do not constitute perfect chronometers. Repeat times are only quasi-periodic, due to the complicated modulation patterns of the basic cycles." They (Herbert *et al.*, 1995, p. 82)

Middle–late miocene timescale

TIME (Ma): 5, 6, 7, 8, 9, 10, 11, 12, 13, 14, 15

CHRONS: C2Ar; C3n (1, 2, 3, 4n); C3r; C3An (1, 2n); C3Ar; C3Bn; C3Br (1, 2, 3r); C4n (1, 2n); C4r (1, 2r); C4Ar (1, 2, 3); C5n (1, 2n); C5r (1, 2, 3r); C5An (1, 2n); C5Ar (1, 2r/n, 3r); C5AAn; C5AAr; C5ABn; C5ABr; C5ACn; C5ADn; C5ADr; C5Bn (1, 2n); C5Br

POLARITY

EPOCH: PLIOCENE, EARLY; MIOCENE, LATE; MIOCENE, MIDDLE

AGE: ZANCLIAN; MESSINIAN; TORTONIAN; SERRAVALLIAN; LANGHIAN

PLANKTONIC FORAMINIFERA

(SUB)TROPICAL — Berggren (this work): PL1 b Gt. cibaoensis-Glb. nepenthes ISZ; PL1 a Gt. tumida-Gt. cibaoensis IRZ; G. tumida-Glb. nepenthes IZ; M14 Gt. lenguaensis - Gt. tumida IZ; M13 b *Gd. extremus/ Gt. plesiotumida- Gt. lenguaensis ISZ*; M13 a *N. acostaensis- Gd. extremus/ Gt. plesiotumida- ISZ*; M12 *N. mayen-N. acostaensis IZ*; M11 *Glb. nepenthes/N. mayeri Conc. RZ*; M10 *Gt. f. robusta - Glb. nepenthes IZ*; M9 b *Gt. f. robusta Tot. RZ*; M9 a *Gt. f. lobala Lin. Z*; M8 *Gt. fohsi s.s Lin. Z*; *Gt. f. lobata- Gt. f. robusta IZ*; M7 *Gl. peripheroacuta Lin. Z*; M6 *O. sutur. - Gl. peripher. IZ*; M5 b *Pr. glomerosa- O. sutualis ISZ*; *Pr. sicana O. suturalis IZ*

Blow (1969): N 19; N18; N 17; N 16; N 15; N 14; N 13; N 12; N 11; N 10; N9; N8

TRANSITIONAL — Berggren and others (1983a); this work: Gt. puncticulata IZ; *Gt. sphericomiozea IZ*; Mt10 Gt. conomiozea/ Gt. mediterranea- Gt. sphericomiozea IZ; Mt9 *N. mayeri- Gt. conomiozea IZ*; Mt8 *Glb. nepenthes/M. mayeri Conc. RZ*; Mt7 *Gt. peripheroronda- Glb. nepenthes IZ*; Mt6 *Orb. suturalis/ Gl. peripheroronda Conc. RZ*; Mt5 b *Pr. glomerosa- O.suturalis ISZ*; *Pr. sicana O. suturalis IZ*

(SUB)ANTARTIC — Berggren (1992): AN 7 *Neogloboquadrina pachyderma TRZ*; AN6 *Gt. scitula PRZ*; AN 5 *N.nympha TRZ*; AN 4 *Gt. miozea PRZ*

CALCAREOUS NANNOPLANKTON

Martini (1971): NN14; NN13; NN12; NN11 (a, b, c, d); NN10; NN 9b; NN9 a-b; NN9a; NN8; NN7; NN6; NN5; NN4

Bukry (1973, 1975): CN10 (a, b, c); CN9 (a, b, c, d); CN8 (a, b); CN 7b; CN7 a-b; CN7a; CN6; CN5b; CN5 (a); CN4; CN3

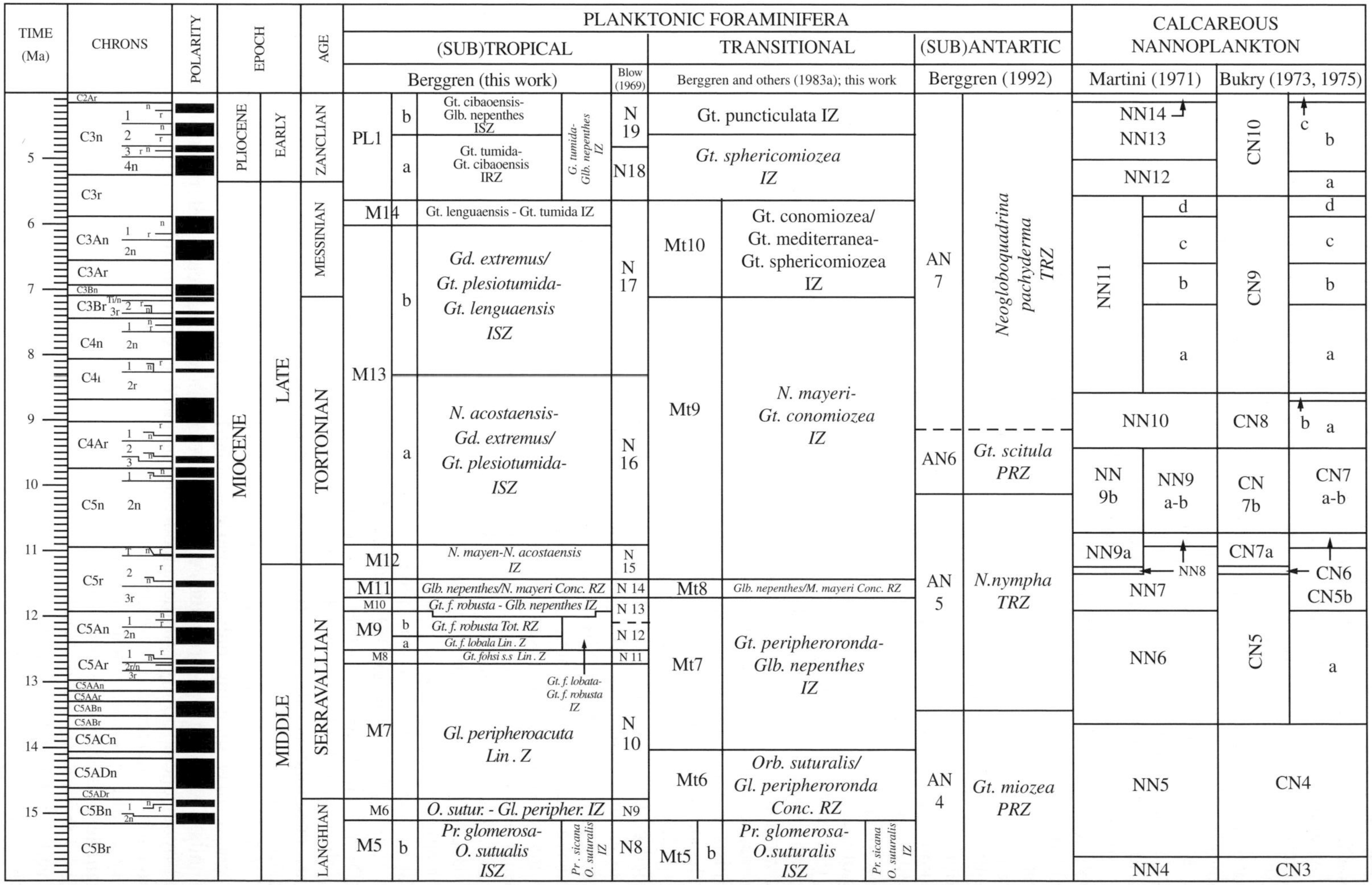

Figure 9 Middle-Late Miocene timescale indicating relationships among fossil-based, radiometric and magnetic field reversal chronometries. Note that three zone successions based on developments among planktic foraminiferans are indicated. These three different zonal sequences reflect biogeographic provincialism in these organisms. That provincialism resulted from ocean surface-water temperature differences. The calcareous nannoplankton or coccolithophore zonal scales of two different specialists are shown. These reflect different perceptions of the fossil aggregates that characterize each zone (source Berggren *et al.*, 1995, figure 5).

suggested that, at present, orbital cycles may be most useful for dating sediments of late Miocene–Pleistocene age. Herbert *et al.* (1995, p. 82) commented calculations used in celestial mechanics suggest that eccentricity periods "should stay constant over time" and that the "mean periods of obliquity and precessional obliquity should increase gradually over Earth's history due to tidal friction."

Precise times of duration of certain Cretaceous stages (defined by their diagnostic faunas) seem to be possible, based on integration of orbital chronometry with radiometric techniques. Orbital chronometry is a significant component in relatively precise Pliocene–Pleistocene age analyses (Berggren *et al.*, 1995, pp. 131–132).

7.13.9.1 Aurichorology—The Golden Spikes and Global Statotype Section and Points

One development aimed at creating greater precision in defining boundaries of chronometric units, especially those based on near-shore marine and land dwelling organisms, is the process called "aurichorology" by Walsh (2000). In that process a so-called "golden spike" is driven into the stratigraphic level of the lowest occurrence of a selected taxon of a fauna that characterizes a zone or stage. The stratigraphic section at which this particular occurrence is seen commonly has been selected after considerable careful study. This procedure was initiated in the process of selecting the boundary between the Silurian and Devonian. Intensive study of graptolite zone faunas within that boundary interval focused ultimately on the evolutionary development of a single taxon, *Monograptus uniformis uniformis*, on which to found the base of a graptolite zone as well as the boundary between two systems. That taxon is one of the significant components of the graptolite fauna that characterizes the basal zone of the Devonian System. The next step, following that decision, was to identify a stratigraphic section at which that faunal occurrence could be found relatively easily. One other consideration in that selection was co-occurrence of other taxa in close stratigraphic proximity that could be used to enhance correlation. The procedure approved by the International Stratigraphic Commission involved driving a marker (called, perhaps euphemistically, the "golden spike") into the specific rock layer which bears the stratigraphically lowest *M. uniformis uniformis* in one thoughtfully selected section. That stratigraphic section, chosen after considerable debate among international specialists in the study of graptolites and the rocks and other faunas of both the Silurian and the Devonian, is at Klonk near Prague, Czech Republic. The base of the Devonian defined by this process became the top of the Silurian by definition. Geologists have adopted the procedures used in selecting the Silurian–Devonian boundary as the fundamental steps to be followed by which boundaries between other systems and divisions of systems, series, stages, and zones (characterized in the Oppel methodology) are identified.

The many discussions involving the procedures to define the base of the Devonian and, therefore, the Silurian–Devonian boundary ultimately led to widespread acceptance of the procedures in defining boundaries between units in the chronometric scale. Accordingly, studies of sedimentary rocks and their contained fauna and flora were focused on those taxa whose lowest stratigraphic occurrences can be documented as precisely as possible to be used as the base of zones, stages, and systems as well as the stratigraphic sections. That stratigraphic position, once chosen by vote of a group of stratigraphic paleontologists who are specialists in the study of the fossil organisms and time interval under analysis, is considered to be unmovable for a number of years. This procedure is thought, by many, to be the most appropriate for defining chronometric units because it specifies a particular evolutionary event seen in a specific stratigraphic and geographic place irreversible in time. The stratigraphic section and point in it chosen for the "golden spike" is referred to as a global statotype section and point (GSSP). Although the process has been formally codified in international stratigraphic practice (Remaine, 2000), its usefulness and even validity have been challenged (Fortey, 1993; Aubry *et al.*, 2000). Remaine (2000, p. 213) stated that chronometric units are "*characterized*" by their fossil content and are "*defined* by their lower boundary which is in the same time the upper boundary of the underlying" unit. Fortey (1993) commented on the procedure involved in selecting GSSPs, both from the viewpoint of understanding the realities of the stratigraphic and fossil record and the fragility of human nature. One development arising from the quest for the "best" GSSPs, and the evolutionary development upon which a definition of a chronometric unit was laid is, as Fortey (1993) pointed out:

> ... the spread of an idea of the ideal biostratigraphic section, a kind of platonic rock section equipped with perfect properties for international correlation: continuous, confacial and conformable, fossiliferous throughout, yet with cryptic breaks minimized, replete with fossils of several groups, which are arranged in evolutionary series. Furthermore, such sections have an horizon suitable for hammering in a 'golden spike' for the base of a chronostratigraphic

interval to immediate international satisfaction. Such sections rarely, if ever, exist in nature. Yet their pursuit has been one of the motivating forces behind various Working Groups of the International Geological Correlation Programme. It is curious to find such an idealistic concept holding sway in the geological sciences, which are so generally pragmatic. The importance of sound criteria for international correlation is not to be gainsaid, but this is, perhaps, a different matter from the relentless pursuit of a perfect section which is likely to prove a chimera.

The search for GSSPs has also led many to overlook the realities of how organisms are constrained in their distribution by faunal/floral provinces, and by the environments in which they live. Chorology is a significant component of chronometry using the evolution of organisms regardless of the—essentially imagined—precision inherent in GSSPs.

7.13.10 TERMINOLOGIES

An almost overwhelming plethora of terms—accompanied by a multitude of misunderstandings—accompanies the subject of the chronometry of sedimentary rocks. Many terms have been codified (see Hedberg, 1965, 1976, 1978; Murphy and Salvador, 1999; Remaine, 2000; Remaine *et al.*, 1996; Berggren *et al.*, 1995; Walsh, 2000, 2001). The terms and their meanings have been so hotly contested and debated that understanding of the real value inherent in chronometry of sedimentary rocks has essentially been lost. As the recovery of petroleum in California based on studies such as that of Bandy and Arnal, discussed herein, using the chronometric units elucidated by Kleinpell, sedimentary-rock chronometry has been and continues to be of great value to the global economy. Those concerned with recovery of natural resources kept their attention directed toward that goal. Those not so directly concerned with recovery of natural resources entered into debates on terminology.

7.13.11 SUMMARY

It was the need to tap increasingly larger volumes of nature's resources and get them to treatment facilities and, ultimately, to consumers that motivated William Smith to lead the way in documenting the need for a potential chronometry. As the practical geologic work carried out by William Smith, Albert Oppel, Charles Lapworth, Robert Kleinpell, and others concerned with recovery of natural resources has revealed that the procedures in chronometry using fossils are basically simple. Many fossiliferous rock sequences in a relatively broad area must be studied and measured carefully and precisely. Numerous fossils must be extracted, from these stratigraphic sections, from as many layers as possible. These fossils must be identified and analyzed critically, seeking unique associations that characterize certain intervals in the stratigraphic record. Such associations, if selected after careful analyses of range of species, are unique in time and so may be used in chronometry. This is the basic procedure used by Oppel, who followed in the footsteps of his mentor, August Quenstedt, in documenting the need for detailed measurements of strata and precision in locating stratigraphic positions of species. As Kleinpell (1979) pointed out, and as work with Jurassic ammonite zones and subzones, especially, demonstrated, Oppel's methodology provides the students of the Phanerozoic sedimentary rock record with the most effective means of chronometry. Units in that chronometry are limited in the areas in which they may be used by distribution patterns of organisms that are in the domain of chorology.

Fossil-based chronometry has been enhanced through studies of radiochronometry and magnetobiochronometry. Today, explorations of deep ocean-floor sediments provides data that enables precise chronometry of the Mesozoic to modern geologic history of the oceans. Orbital chronometry seems to promise enhanced precision in chronometry of the Late Mesozoic and younger sedimentary record.

The need for sedimentary-rock chronometry and its application remains as valuable a tool in finding and recovering nature's resources today as it was in William Smith's day. Deep ocean-floor explorations have added an interesting, exciting, and still-expanding dimension to sedimentary-rock chronometry.

REFERENCES

Arkell W. J. (1933) *The Jurassic System in Great Britain.* Clarendon Press.

Aubry M.-P. (1995) From chronology to stratigraphy: interpreting the Lower and Middle Eocene stratigraphic record in the Atlantic Ocean. In *Geochronology, Time Scales and Global Stratigraphic Correlation*, SEPM Special Publication 54 (eds. W. A. Berggren, D. V. Kent, M.-P. Aubry, and J. Hardenbol), pp. 213–274.

Aubry M.-P., Berggren W. A., Van Couvering J. A., and Steininger F. (2000) Should the golden spike glitter? *Episodes* **23**, 203–210.

Bandy O. L. and Arnal R. E. (1969) Middle Tertiary basin development in California. *Geol. Soc. Am. Bull.* **80**, 783–820.

Berggren W. A., Kent D. V., Swisher C. C., III, and Aubry M.-P. (1995) A revised Cenozoic geochronology and chronostratigraphy. In *Geochronology, Time Scales and Global Stratigraphic Correlation*, SEPM Special Publication 54 (eds. W. A. Berggren, D. V. Kent, M.-P. Aubry, and J. Hardenbol), pp. 129–212.

Berry W. B. N. (1968) *Growth of a Prehistoric Time Scale.* W. H. Freeman.

Berry W. B. N. (1987) *Growth of a Prehistoric Time Scale*, revised edn. Blackwell.

Berry W. B. N. (2000) R. M. Kleinpell's zones and stages: an oppelian biostratigraphic solution to a challenge in the search for petroleum in California. *Earth Sci. Hist.* **19**, 161–174.

Boggs S., Jr. (2001) *Principles of Sedimentology and Stratigraphy*, 3rd edn. Prentice Hall.

Callomon J. H. (1995) Time from Fossils: S.S. Buckman and Jurassic high-resolution geochronology. In *Milestones in Geology*, Geol. Soc. London Mem. (ed. M. J. LeBas), vol. 16, pp. 127–150.

Conybeare W. D. and Phillips W. (1822) *Outlines of the Geology of England and Wales.* William Phillips, George Yard.

Davies A. M. (1934) *Tertiary Faunas, a Text-book for Oilfield Paleontologists and Students of Geology*. Thomas Murby and Co.

d'Orbigny A. D. (1842) *Paleontologie Francaise, Terraines Jurassiques: Part 1*. Cephalopodes. Masson.

Fortey R. A. (1993) Charles Lapworth and the biostratigraphic paradigm. *Geol. Soc. London J.* **150**, 209–218.

Hancock J. M. (1977) The historic development of concepts of biostratigraphic correlation. In *Concepts and Methods of Biostratigraphy* (eds. E. G. Kauffman and J. E. Hazel). Hutchinson and Ross, pp. 3–22.

Harland W. B., Armstrong R. L., Cox A. V., Craig L. E., Smith A. G., and Smith D. G. (1990) *A Geologic Time Scale 1989*, 2nd edn. Cambridge University Press.

Hedberg H. D. (1965) Chronostratigraphy and biostratigraphy. *Geol. Mag.* **102**, 451–461.

Hedberg H. D. (ed.) (1976) *International Stratigraphic Guide* Wiley.

Hedberg H. D. (1978) Stratotypes and a geochronologic scale. In *Contributions to the Geologic Time Scale*, Am. Assoc. Petrol. Geol. Studies in Geology (ed. G. V. Cohee), vol. 6, pp. 33–38.

Herbert T. D., Premoli-Silva I., Erba E., and Fischer A. G. (1995) Orbital chronology of Cretaceous–Paleocene marine sediments. In *Geochronology, Time Scales and Stratigraphic Correlation*, SEPM Spec. Publ. (eds. W. A. Berggren, D. V. Kent, M.-P. Aubry, and J. Hardenbol), vol. 54, pp. 81–93.

Kleinpell R. M. (1938) *Miocene Stratigraphy of California*. Am. Assoc. Petrol. Geol.

Kleinpell R. M. (1979) *Criteria in Correlation: Relevant Principles of Science*. Pacific Section Am. Assoc. Petrol. Geol.

Lapworth C. (1879) On the tripartite classification of the Lower Paleozoic rocks. *Geol. Mag.* **6**, 1–15.

Laugel F. (1859) Un tablaeu resume de la classification du terrain jurassique etablier par le docteur Albert Oppel. *Soc. Geol. France 2* **15**, 657–664.

Lyell C. (1833) *Principles of Geology*. vol. 3, J. Murray.

Murphy M. A. and Salvador A. (19991) International stratigraphic guide—an abridged version. *Episodes* **22**, 255–271.

Oppel A. (1856–1858) *Die Juraformation Englands, Frankreichs und des sudwestlichen Deutschlands*. Wurttemb. Naturwiss. Verein.

Remane J. (2000) Comments on the paper of Aubry *et al.* (2000). *Episodes* **23**, 211–213.

Remane J., Bassett M. G., Cowie J. W., Gohrbandt K. H., Lane R., Michelsen O., and Naiwen W. (1996) Revised guidelines for the establishment of global chronostratigraphic standards by the international commission on stratigraphy. *Episodes* **19**, 77–81.

Strachey J. (1719) A curious description of the strata observed in the coal mines of mendip in Somersetshire, being a letter of John Strachey Esq. To Dr. Robert Welsted, M.D. and R.S.Soc. and by him communicated to the society. *Roy. Soc. (London) Phil. Trans.* **30**(360), 968–973.

Strachey J. (1725) An account of the strata in coal mines, etc. *Royal Society (London) Philos. Trans.* **33**(391), 395–398.

Walsh S. L. (2000) Eubiostratigraphic units, quasibiostratigraphic units, and "assemblage zones." *J. Vert. Paleo.* **20**, 761–775.

Walsh S. L. (2001) Notes on geochronologic and chronostratigraphic units. *Geol. Soc. Am. Bull.* **113**, 704–713.

Winchester S. (2001) *The Map that Changed the World*. Harper Collins.

7.14 The Geochemistry of Mass Extinction

L. R. Kump
The Pennsylvania State University, PA, USA

7.14.1 INTRODUCTION

The course of biological evolution is inextricably linked to that of the environment through an intricate network of feedbacks that span all scales of space and time. Disruptions to the environment have biological consequences, and vice versa. Fossils provide the *prima facie* evidence for biotic disruptions: catastrophic losses of global biodiversity at various times in the Phanerozoic. However, the forensic evidence for the causes and environmental consequences of these mass extinctions resides primarily in the geochemical composition of sedimentary rocks deposited during the extinction intervals. Thus, advancement in our understanding of mass extinctions requires detailed knowledge obtained from both paleontological and geochemical records.

This chapter reviews the state of knowledge concerning the geochemistry of the "big five" extinctions of the Phanerozoic (e.g., Sepkoski, 1993): the Late Ordovician (Hirnantian; 440 Ma), the Late Devonian (an extended or multiple event with its apex at the Frasnian–Famennian (F–F) boundary; 367 Ma), the Permian–Triassic (P–Tr; 251 Ma), the Triassic–Jurassic (Tr–J; 200 Ma), and the Cretaceous–Tertiary (K–T; 65 Ma). The focus on the big five is a matter of convenience, as there is a continuum in extinction rates from "background" to "mass extinction." Although much of the literature on extinctions centers on the causes and extents of biodiversity loss, in recent years paleontologists have begun to focus on recoveries (see, e.g., Hart, 1996; Kirchner and Weil, 2000; Erwin, 2001 and references therein).

To the extent that the duration of the recovery interval may reflect a slow relaxation of the environment from perturbation, analysis of the geochemical record of recovery is an integral part of this effort. In interpreting the geochemical and biological records of recovery, we need to maintain a clear distinction among the characteristics of the global biota: their biodiversity (affected by differences in origination and extinction rates) and ecosystem function (guild structure, complexity of interactions, productivity). Geochemical records reflect attributes of ecosystem function, not biodiversity; low-diversity recovery faunas and floras may support pre-event productivities. Thus, geochemical and biodiversity recovery intervals are interdependent but not equivalent, and may not be of equal duration.

From the biological point of view, there is an inevitable lag between peak extinction rates and peak origination rates, and the durations and underlying causes of the lags are topics of debate. Both intrinsic (e.g., the fact that ecospace is created as biodiversity increases producing positive feedback) and external (environmental) constraints are possible. Kirchner and Weil (2000) performed a time-series analysis of extinction and origination-rate data, and concluded that the lag is ~10 Myr and independent of the magnitude of the event. Erwin (2001) raised the possibility that the 10 Myr lag may be an artifact of the coarseness of the timescales utilized, and discussed possible environmental and ecological limits on rate of recovery from mass extinction.

The comparison of the geochemical records of the five major mass extinctions of the Phanerozoic reveals few commonalities. Most, but not all, exhibit sharp drops in the carbon isotopic composition (δ^{13}C) of the surface ocean, indicating substantial disruptions to the global carbon cycle. The P–Tr and F–F events are associated with indicators of widespread anoxia and enhanced pyrite burial (positive δ^{34}S excursions), whereas the Late Ordovician extinction occurred during a brief interlude of oxic conditions from general anoxia. Some are associated with sea-level transgressions from previous lowstands (P–Tr, Tr–J, K–T), but the Late Ordovician and F–F occurred during sea-level falls. Long-term climates change across all events, but span major coolings (Late Ordovician, F–F) to prominent warmings (P–Tr, Tr–J, K–T).

Evidence for extraterrestrial influence is strong for the K–T, suggestive for the Tr–J and Late Permian, and missing for the F–F and Late Ordovician. What these times have in common is that all were times of biotic and environmental change. Long-term trends toward extreme environmental conditions presaged the Late Ordovician, F–F, and P–Tr events, whereas the Tr–J and K–T seem to have been abrupt shocks to the Earth system, perhaps belying their extraterrestrial cause. However, even for the K–T extinction there is indication of environmental and biotic change before the known impact event and mass extinction (e.g., Keller *et al.*, 1993; Barrera, 1994; Abramovich and Keller, 2002).

7.14.2 ISOTOPE RECORDS OF THE MAJOR MASS EXTINCTIONS

7.14.2.1 Carbon Isotope Record

The carbon isotopic composition of marine limestones is generally interpreted to reflect the rate of organic carbon burial (F_{org}, mol yr^{-1}), although a number of other factors are actually involved and may be as significant for particular events or trends. The time rate of change of the δ^{13}C of the ocean, δ_o, can be expressed as (e.g., Kump and Arthur, 1999)

$$\frac{d\delta_o}{dt} = \frac{F_{in}(\delta_{in} - \delta_o) - F_{org}\Delta}{M} \tag{1}$$

where F_{in} is the combined inorganic carbon input from weathering and volcanism (mol yr^{-1}), δ_{in} is its isotopic composition, Δ is the isotopic difference between contemporaneous sedimented organic carbon (δ_{org}) and carbonate carbon (δ_{carb}, $\approx\delta_o$), approximately −25‰, and M is the combined oceanic and atmospheric inorganic carbon reservoir size (mol). Because both $(\delta_{in} - \delta_o)$ and Δ are negative numbers, an increase in δ_o with time indicates either a reduction in the rate of carbon input, an increase in the isotopic composition of the input, an increase in organic carbon burial, a decrease in Δ (Δ becomes more negative), or some combination of these. The burial of carbonate carbon does not enter into this equation explicitly, because its isotopic composition is close to, and usually simplified to be equivalent to, δ_o.

When considering values representative of long (million years) intervals of Earth history, steady state can be assumed, and δ_o can then be expressed as

$$\delta_o = \delta_{in} - \frac{F_{org}}{F_{in}}\Delta \tag{2}$$

Because at steady state F_{in} equals the sum of F_{org} and the carbonate burial rate (F_{carb}), F_{org}/F_{in} reflects the fraction of the burial flux which is organic (f_{org}). From this equation, then, one sees that an interpretation of elevated δ_o as high rates of F_{org} assumes that F_{in}, Δ, and δ_{in} are all unchanging in time. As argued below, during particular intervals of Earth history that assumption may be seriously in error.

With these caveats in mind, the overall Phanerozoic pattern in δ_o suggests that f_{org} generally increased through the Paleozoic, collapsed at the P–Tr boundary, and then oscillated without net trend through the Mesozoic–Cenozoic (Figure 1(a)). The Late Ordovician and Late Devonian were times of stepwise increases in δ_o and perhaps f_{org}, whereas the Tr–J and K–T events occurred at times when δ_o was near its

Mesozoic–Cenozoic mean. Closer inspection of the extinction intervals reveals positive excursions during the Late Ordovician (Figure 1(b)) and F–F (Figure 1(c)) extinctions, and negative excursions at the P–Tr, Tr–J, and K–T boundaries (Figures 1(c)–(f)). Note that in some cases, the δ_{org} values are displayed in Figure 1. In such cases, well-studied boundary sections are shales, but, where available, limestone $\delta^{13}C$ values substantiate the trends.

7.14.2.2 Sulfur Isotope Record

The basis of interpretation of the sulfur isotope record is essentially identical to that for carbon; pyrite sulfur simply substitutes for organic carbon in the burial term of Equation (1). However, the long residence time of sulfur (sulfate) in the ocean (~50 Myr; Holser *et al.*, 1988) means that the time rate of change term on the left-hand side of Equation (1) is important, and cannot be neglected, and steady state cannot be assumed (i.e., Equation (2) cannot be used for the sulfur cycle).

The early Paleozoic exhibits heavy $\delta^{34}S$ values, and $\delta^{34}S$ declines ~20‰ through the Paleozoic reaching a minimum of ~10‰ just before the P–Tr boundary (Figure 2(a)). A sharp rise in $\delta^{34}S$ in the earliest Triassic foreshadows the general trend through the rest of the Phanerozoic, with $\delta^{34}S$ rising by ~10‰. Apparently, pyrite sulfur burial rates were high in the early Paleozoic but declined through the Paleozoic, perhaps in response to an increasing proportion of organic carbon burial on land. A general decline in terrestrial coal basins from the Carboniferous to recent may be the explanation for the increase in $\delta^{34}S$ from the late Paleozoic to now.

Mass extinctions occurred during times of both heavy (Late Ordovician, Late Devonian) and light (P–Tr, Tr–J, K–T) $\delta^{34}S$. Large positive excursions in $\delta^{34}S$ followed the P–Tr, Tr–J, and perhaps the Late Ordovician (Goodfellow *et al.*, 1992) events, while a dramatic decline in $\delta^{34}S$ was initiated in the Late Devonian. The K–T event had no notable effect on the sulfur isotopic composition of the ocean.

To the extent that $\delta^{34}S$ and $\delta^{13}C$ can be interpreted in terms of changes in the burial proportions of pyrite sulfur and organic carbon, consideration of both records together may shed some light on causal factors and/or environmental responses to these extinction events. The Late Ordovician and Late Devonian events were followed by positive excursions in $\delta^{13}C$ and either no or negative excursions in $\delta^{34}S$. A positive excursion in $\delta^{13}C$ should be accompanied by a positive excursion in $\delta^{34}S$ if enhanced burial of marine organic carbon was the cause, because carbon and sulfur contents of marine sediments tend to covary (Berner and Raiswell, 1983). The sulfur isotope records, instead, suggest that terrestrial organic carbon burial was enhanced during the Silurian and Carboniferous (Kump, 1992).

That the Carboniferous was a period of enhanced terrestrial organic carbon burial is well established in the literature (e.g., Berner and Canfield, 1989). Perhaps the initial establishment of terrestrial ecosystems in the Silurian was responsible for the isotope signature of terrestrial burial during this period. In contrast, the positive $\delta^{34}S$ excursions after the P–Tr and Tr–J extinction events are best interpreted in terms of enhanced marine pyrite sulfur (and organic carbon) burial (e.g., Isozaki, 1997) but reduced global organic carbon burial. These excursions significantly postdate the extinction events, but because of the long response time of $\delta^{34}S$, one should focus on the slope of its curve rather than the absolute value to get a sense of fluxes (e.g., Richter and Turekian, 1993).

7.14.2.3 Strontium Isotope Record

The strontium isotope record of seawater reflects the relative magnitude of radiogenic (crystalline rock weathering) and nonradiogenic (basaltic weathering and high- and low-temperature seafloor reactions) sources to the ocean (e.g., Palmer and Elderfield, 1985). Periods of low $^{87}Sr/^{86}Sr$ are generally interpreted to reflect periods when seafloor hydrothermal activity was high, whereas periods of high ratios are generally interpreted to reflect times of elevated rates of continental weathering. In reality, the strontium cycle has multiple influences that are poorly constrained by the isotope record alone (e.g., Kump, 1989).

For the strontium isotope record, the present is the key to the Cambrian; in between, the $^{87}Sr/^{86}Sr$ ratio is seen to fall in roller-coaster fashion through the Paleozoic and early Mesozoic to a minimum in the Late Jurassic, and then to rise in steps to the present (Figure 3(a)). The Late Ordovician, Late Devonian (Figure 3(b)), and P–Tr events occurred just after local minima in $^{87}Sr/^{86}Sr$ and thus mark times of significant change in the strontium cycle. In contrast, the Tr–J event follows a local maximum value in the ratio, and marks a time of declining $^{87}Sr/^{86}Sr$. The K–T event occurred contemporaneously with a local maximum in the ratio, on both the multimillion and 10^5-year timescales (MacLeod *et al.*, 2001; Figures 3(a) and (c)).

7.14.2.4 Oxygen Isotope Record

The most complete oxygen isotope record for the Phanerozoic is that of Veizer *et al.* (1999).

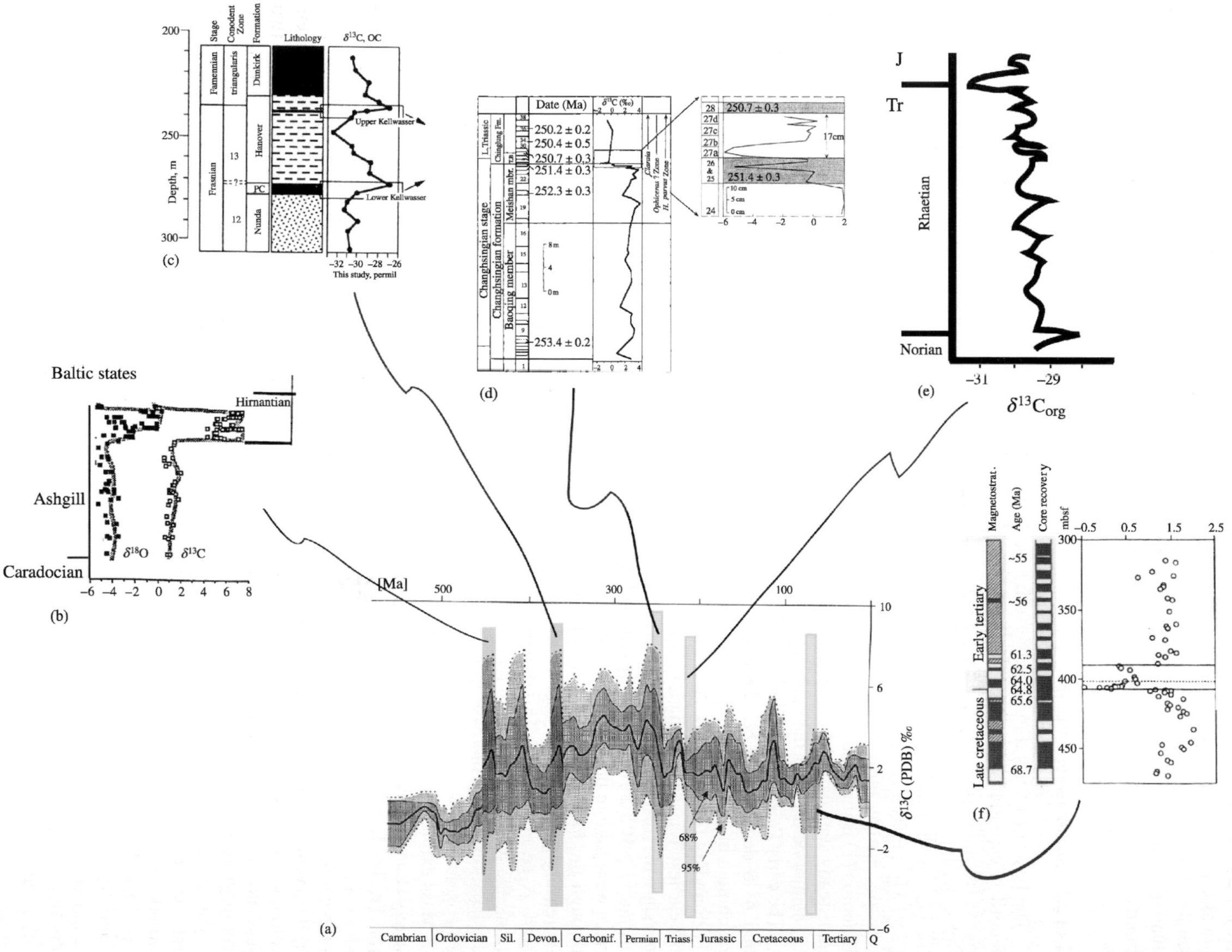

(a)
[Ma]
500
300
100
δ13C (PDB) ‰
68%
95%
Cambrian
Ordovician
Sil.
Devon.
Carbonif.
Permian
Triass
Jurassic
Cretaceous
Tertiary
Q
(b)
Baltic states
Hirnantian
Ashgill
Caradocian
δ18O
δ13C
(c)
Stage
Conodent Zone
Formation
Lithology
δ13C, OC
Depth, m
Famennian
Frasnian
triangularis
Dunkirk
Hanover
Nunda
PC
Upper Kellwasser
Lower Kellwasser
This study, permil
(d)
Date (Ma)
δ13C (‰)
Changhsingian stage
Changhsingian formation
Baoqing member
Meishan mbr.
L.Triassic
250.2 ± 0.2
250.4 ± 0.5
250.7 ± 0.3
251.4 ± 0.3
252.3 ± 0.3
253.4 ± 0.2
Clarkina
Ophiceras ? Zone
H. parvus Zone
17cm
(e)
J
Tr
Rhaetian
Norian
δ13Corg
(f)
Magnetostrat.
Age (Ma)
Core recovery
mbsf
Early tertiary
Late cretaceous

Although there is considerable controversy regarding the extent to which the data have been compromised by diagenesis (e.g., Mii *et al.*, 1997), the general correspondence between other climatic indicators and $\delta^{18}O$ is encouraging (Veizer *et al.*, 2000; Figure 4). Toward that end, it is interesting to note that with the exception of the Late Ordovician, four of the big five Phanerozoic mass extinctions occurred during times with relatively depleted $\delta^{18}O$ values (warm intervals).

In detail, we find that the Late Ordovician was a local peak in $\delta^{18}O$, presumably reflecting the cool conditions of the time and, for the Hirnantian, the growth of continental ice sheets. The Late Devonian ushered in a cooling trend. A similar trend in $\delta^{18}O$ was initiated in the Late Permian and continued through the Triassic and Jurassic, but interpreting this in terms of gradual cooling is inconsistent with other climate indicators (e.g., Frakes *et al.*, 1992).

7.14.3 INTERPRETING THE GEOCHEMICAL RECORDS OF MASS EXTINCTION

7.14.3.1 Late Ordovician

The first of the mass extinctions of the Phanerozoic occurred during the last stage of the Ordovician Period, the Hirnantian, at ca. 440 Ma. The extinction appears to have occurred in two pulses (see Sheehan, 2001 for a thorough review of this event). The first was associated with a glacio-eustatic sea-level fall of 70–100 m as ice sheets developed on Gondwana, which at the time was situated at the South Pole. Large expanses of tropical to subtropical epicontinental seas replete with carbonate platforms, and diverse benthic and planktonic faunas became subaerially exposed. Marginal anoxic zones became fully oxygenated as oceanic mixing rates apparently intensified. After this first wave of extinction, a more cosmopolitan "Hirnantian" fauna evolved, only to suffer considerable losses at the end of the glaciation as sea-level rose and shelfal anoxia was re-established.

The marine carbon isotope record provides some clues to the environmental causes and consequences of this event, but diverse interpretations have been published (e.g., Wilde *et al.*, 1990; Brenchley *et al.*, 1994; Kump *et al.*, 1999). Isotopic data from the western US (Kump *et al.*, 1999; Finney *et al.*, 1999) confirm that a global positive excursion of 5–7‰ brackets the Hirnantian glaciation and extinction events (Middleton *et al.*, 1991; Long, 1993; Brenchley *et al.*, 1994; Wang *et al.*, 1997; Figure 1(b)). The glaciation apparently was confined largely to the Hirnantian, and thus of short duration (~0.5 Myr; Brenchley *et al.*, 1994).

In contrast, other Phanerozoic extinctions are associated with negative excursions, and have generally been interpreted to represent the loss of surface water productivity (photosynthesis discriminates in favor of ^{12}C, creating ^{13}C-enriched surface waters; see below). Brenchley *et al.* (1994) proposed that the positive $\delta^{13}C$ excursion reflected increased marine biological productivity promoted by more vigorous mixing of the ocean during glaciation. The resultant drawdown of atmospheric CO_2 furthered the glaciation through positive climatic feedback. Alternatively, CO_2 drawdown occurred before the glaciation in response to Taconic orogeny and associated enhanced weatherability of the continents (Kump *et al.*, 1999). A climatic threshold was reached that allowed for the establishment and autocatalytic growth of Gondwanan ice sheets through positive ice-albedo feedback. The spread of ice reduced the exposure area and thus weathering rate of silicate rocks (the long-term sink for atmospheric CO_2), so CO_2 levels rose.

The limestone-dominated rivers provided isotopically heavy carbon to the oceans (compared to pre-excursion rivers that had a higher proportion of carbon obtained from shale-derived fossil-carbon weathering at higher latitudes), driving the positive excursion, and perhaps masking the effects of the loss of surface water productivity on $\delta^{13}C$. Ultimately, high CO_2 created a sufficiently strong greenhouse effect to overcome the cooling effects of high albedo, and the ice sheets collapsed. As Sheehan (2001) points out, the two hypotheses are not mutually exclusive, although they present contrasting predictions for the time course of atmospheric CO_2 through the event. Weak proxy evidence of atmospheric CO_2 rise through the event, in the form of a reduction in the

---

Figure 1 Phanerozoic carbon isotope record. Mass extinction intervals are shaded in gray (widths do not correspond to durations of inserts): (a) global marine carbonate record (after Veizer *et al.*, 1999); (b) marine carbonate record from the Late Ordovician of the Baltic States (after Brenchley *et al.*, 1994); (c) Late Devonian marine organic carbon record from New York State (after Murphy *et al.*, 2000); (d) Late Permian marine carbonate record from China (after Bowring *et al.*, 1998); (e) Late Triassic marine organic carbon record from Canada (after Ward *et al.*, 2001); (f) Late Cretaceous–early Tertiary record of the carbon isotopic difference between fine fraction and benthic carbonate (left panel), between shallow dwelling planktonic and benthic foraminifera (open symbols, right panel) and between more deeper dwelling planktonic and benthic foraminifera (filled symbols, right panel) from the south Atlantic (DSDP Site 528; after D'Hondt *et al.*, 1998).

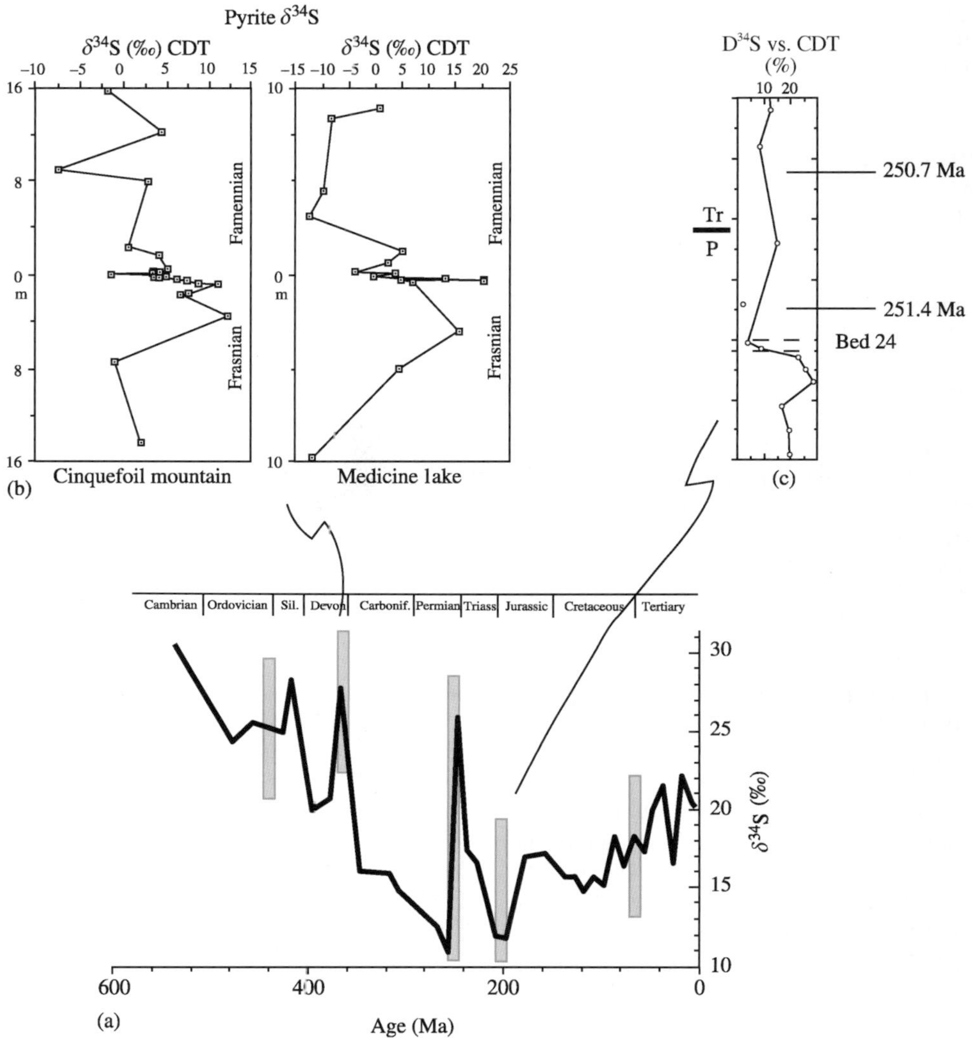

Figure 2 Phanerozoic sulfur isotope record. Mass extinction intervals are shaded in gray (widths do not correspond to durations of inserts): (a) global marine sulfate record (after Strauss, 1999); (b) pyrite sulfur isotope composition from the Late Devonian of Canada (after Wang *et al.*, 1996); (c) Permian–Triassic sulfate–sulfur isotope record from China (after Kaiho *et al.*, 2001; note in Figure 2(a) that a large positive excursion follows in the Early Triassic).

isotopic difference between limestone and kerogen δ^{13}C, is supportive of the weathering hypothesis (Kump *et al.*, 1999).

Lack of a deep-water (benthic) carbon isotope record hinders an assessment of the extent to which pelagic ecosystem function (biological pumping of organic matter, nutrients, and trace metals to the deep sea) was disrupted during the event. In contrast, as discussed below, there is clear evidence in the form of a collapsed δ^{13}C gradient from surface to deep sea for a shutdown of the biological pump during the K–T event (Figure 1(f)). The fossil record indicates that although the pelagic biota was certainly not immune to Late Ordovician extinction, recovery of abundance and diversity was rapid (Sheehan, 2001).

Fossil soil carbonate δ^{13}C (e.g., Cerling, 1991; Mora *et al.*, 1991) and/or the concentration and δ^{13}C of carbonate substituted in goethites in ancient soils (Yapp and Poths, 1992) have been proposed as p_{CO_2} barometers. In the case of the Late Ordovician (for which there are very few paleosols), Yapp and Poths (1992) derived a very high estimate of paleo-p_{CO_2} ($\sim$16 times present atmospheric level). Presuming that the paleosol was deposited at the time of glaciation, these high CO_2 levels seem paradoxical (Kasting, 1992). Climate models show that glaciation can occur at high p_{CO_2} (Crowley and Baum, 1991) under certain paleogeographical conditions (e.g., a large continent with its coastline at a pole), but Gibbs *et al.* (1997) argued for a maximum p_{CO_2} of $\sim$8–10 × present under Ordovician conditions of

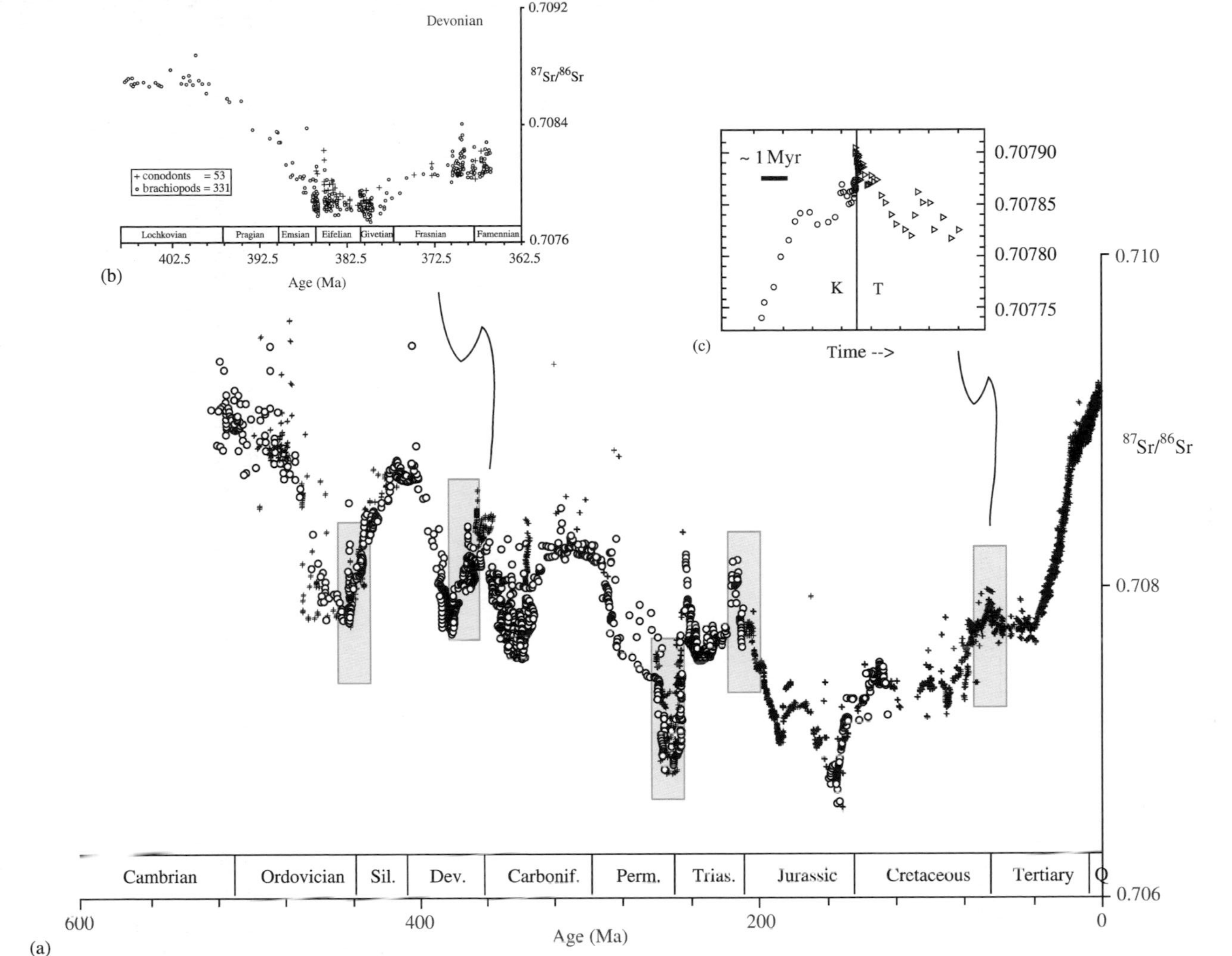

Figure 3 Phanerozoic strontium isotope record from marine carbonates. Mass extinction intervals are shaded in gray (widths do not correspond to durations of inserts): (a) global trends in strontium isotopic composition of the oceans (after Veizer *et al.*, 1999); (b) expanded record from the Late Devonian (after Veizer *et al.*, 1999); (c) Cretaceous–Tertiary record, displaying a pronounced excursion at the K–T boundary (after Martin and Macdougall, 1991).

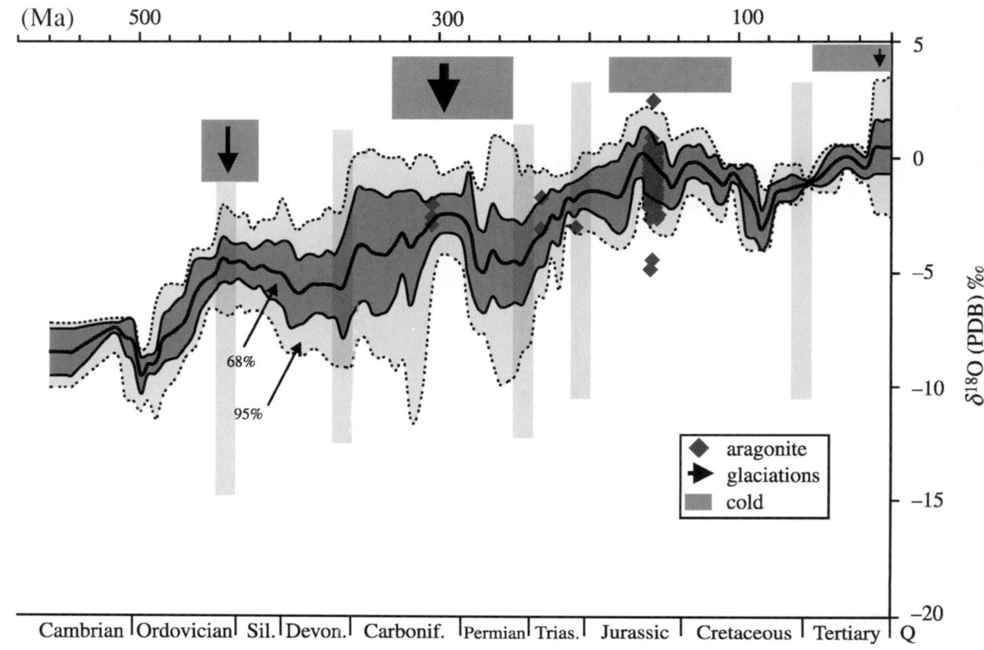

Figure 4 Phanerozoic variations in the oxygen isotopic composition of marine carbonates. Mass extinction intervals are shaded in gray (after Veizer *et al.*, 1999).

paleogeography and (reduced) solar luminosity. If the paleosols record CO_2 levels at the end of the glacial period, the paradox of high atmospheric p_{CO_2} during glaciation is reconciled; under the weathering hypothesis, p_{CO_2} rises during the glaciation to levels sufficient to overcome the ice-albedo effect.

Further support for the weathering hypothesis for Late Ordovician glaciation comes from the strontium isotope record. A multimillion-year decline in the marine record of $^{87}Sr/^{86}Sr$ in limestones is reversed in the early Late Ordovician (Veizer *et al.*, 1999), a likely consequence of collisional tectonics that initiated at this time and continued (as did the rise in $^{87}Sr/^{86}Sr$) through the Silurian (Richter *et al.*, 1992). The increase (from 0.7078 to 0.7088) rivals the Cenozoic rise, which has been attributed to the Himalayan uplift and associated with the progressive cooling leading to Quaternary glaciation (Raymo *et al.*, 1988).

The large shifts in strontium isotopic composition of the ocean probably reflect a substantial increase in the supply of radiogenic strontium facilitated by the unroofing of older continental crust during orogeny (Richter *et al.*, 1992). Climatic cooling compensated for tectonically increased weatherability of the continents via feedbacks between climate, CO_2, and global weathering (Kump and Arthur, 1997).

The known sulfate–sulfur isotope record (Figure 2(a)) is too crude to allow detailed interpretations of the Late Ordovician event. A pyrite sulfur isotope record from northwestern Canada exhibits a trend toward heavy values in the Early Silurian (Goodfellow *et al.*, 1992). Determining whether this is simply a basinal effect or reflective of a global trend must await further isotopic work; a focus on extracting trace sulfate from limestones promises a much higher resolution record for this event in the near future.

Efforts to find trace-metal evidence of extraterrestrial impact at the Ordovician–Silurian boundary have been unsuccessful (e.g., Orth *et al.*, 1986; Wang *et al.*, 1992). Peaks in iridium abundance at the boundary have been linked to reductions in sedimentation rate; the persistent cosmic source of iridium is otherwise diluted by high terrigenous or biogenic sedimentation.

Overall, the Late Ordovician extinction appears to be the result of purely terrestrial phenomena. High sea-level stands of the early Paleozoic allowed for animal diversification on shallow-water, epicontinental carbonate platforms that proved, however, to be highly sensitive to glacio-eustatic effects on ecospace availability and lateral shifts in the oxic–anoxic interface. Tectonic activity facilitated the establishment of Gondwanan ice sheets which robbed the shallow seas of water, leading to extinction, establishment of recovery ecosystems, and then the destruction of these as the ice sheets melted, perhaps catastrophically.

7.14.3.2 Late Devonian

The Late Devonian was a time of widespread, shallow epicontinental seas that supported abundant and diverse warm-water metazoan communities. This biotic bliss was terminated in a series of extinctions extending over perhaps 3 Myr (McGhee, 1996) that had lasting effects on reef-building stromatoporoids, corals, brachiopods, and fish (e.g., McLaren, 1982). Reefs recovered in the Famennian but, at least in Western Australia, were dominated by cyanobacteria (Playford *et al.*, 1984).

Global cooling may have played a role in the preferential extinction of warm-water faunas (including coral reefs; McGhee, 1996). Widespread anoxia has also been invoked to explain this interval of elevated extinction (e.g., Joachimski and Buggisch, 1993). Associated eutrophication is invoked by Murphy *et al.* (2000) to explain the demise of Devonian carbonate platforms. Their hypothesis is supported by substantial increases in the C/N and C/P ratios of organic matter preserved in the Kellwasser deposits. Preferential release of nutrients is argued to occur during early diagenesis when overlying waters are anoxic, supporting high levels of productivity in the water column (Van Cappellen and Ingall, 1994, 1996).

Anoxia was certainly a characteristic of the Late Devonian, but in the western US, based on geochemical proxies, anoxia ended 6 m below (~100 kyr before) the major F–F extinction. Bratton *et al.* (1999) discuss the possibility that this was a local phenomenon and that anoxia persisted through the F–F boundary elsewhere, but favor the alternative hypothesis that other sections suffered depositional hiatus or erosion of the latest Frasnian sediments. A positive $\delta^{13}C$ excursion (in both carbonate and organic carbon) began in the Frasnian but continued well into the Famennian (Wang *et al.*, 1996). A positive pyrite sulfur isotope excursion also occurred at this time. If these excursions indicate enhanced organic carbon and pyrite sulfur burial under widespread anoxic conditions, then it would seem that such conditions persisted well beyond the F–F boundary extinction.

Wang *et al.* (1996) identified a brief negative $\delta^{13}C$ excursion at the F–F boundary in Alberta, Canada that may reflect the temporary loss of the biological pump. They argue that productivity collapse was the result of an asteroid/comet impact at the F–F boundary, a time otherwise under biotic stress as the result of widespread warmth and anoxia. Murphy *et al.* (2000) detail a similar carbon isotope stratigraphy, in this case in organic matter, with two peaks from a baseline of $\delta^{13}C = -31‰$ to $-27‰$, representing the lower Kellwasser (Frasnian) and upper Kellwasser (F–F boundary) episodes of black shale deposition in Europe (Figure 1(c)).

Between these two events, $\delta^{13}C$ drops to $-32‰$ just prior to the F–F boundary. Perhaps this is the brief negative excursion of Wang *et al.* (1996). Correlation between organic carbon and carbonate carbon $\delta^{13}C$ records is imprecise because of the additional possibility of productivity or atmospheric p_{CO_2}-induced variations in isotopic fractionation that could generate phase offsets between the two records (Kump and Arthur, 1999). Nevertheless, in this case, based on paired analyses from the same section, the inorganic and organic records appear to be in phase, suggesting that atmospheric CO_2 levels were sufficiently high that any changes did not significantly affect the isotopic fractionation that occurs during photosynthesis (Joachimski *et al.*, 2002).

The impact hypothesis for this extinction dates back to McLaren (1970). There is some support for extraterrestrial impact at this time. Iridium anomalies have been identified at or near the F–F (Playford *et al.*, 1984; Wang *et al.*, 1991) and Devonian–Carboniferous (Wang *et al.*, 1993) boundaries. However, the Devonian–Carboniferous anomaly has no supporting evidence for impact, and elemental ratios are not chondritic. Redox changes have been invoked to explain this anomaly (Wang *et al.*, 1993). Stronger evidence for impact is present for the F–F boundary (as summarized in Wang *et al.*, 1996), including microtektites, meteoritic elemental ratios, known impact craters, and high-energy (tsunami?) deposits. However, reported iridium enrichments at the stratotype area for the F–F boundary in southern France were not subsequently substantiated (Girard *et al.*, 1997), and no F–F iridium anomaly was found in New York State (McGhee *et al.*, 1984).

7.14.3.3 Permian–Triassic

The largest extinction event of the Phanerozoic occurred in the latest Permian, a time when both shallow and deep marine environments appear to have experienced widespread anoxia. As a result, anoxia has figured prominently into proposed extinction mechanisms for this time, although models for extinction that invoke multiple causality are currently in favor (e.g., Erwin, 1993, 1995; Kozur, 1998).

The latest Permian isotope record displays abrupt negative excursion in carbonate and organic carbon isotopes (e.g., Baud *et al.*, 1989; Magaritz *et al.*, 1991; Holser *et al.*, 1991; Wang *et al.*, 1994; Bowring *et al.*, 1998; de Wit *et al.*, 2002; Figure 1(d)), and a substantial increase in the sulfur isotopic composition of evaporite sulfate minerals in the Early Triassic following a minimum value in the Late Permian (e.g., Kajiwara *et al.*, 1994; Scholle, 1995; Strauss, 1997, 1999; Figure 2(c)).

These proxies may represent the widespread development of oceanic anoxia and the establishment of strong chemical stratification of the ocean (Gruszczynski *et al.*, 1989). Unusual chemistry (e.g., anoxia and widespread carbonate supersaturation) extended into the Early Triassic and may have been a contributor to the long recovery interval (e.g., Hallam, 1991; Woods *et al.*, 1999).

The coincidence of Siberian flood basalt emplacement (Renne *et al.*, 1995) with the anoxia and extinction event suggests a causative role for volcanism (see also Kozur, 1998). However, as more firmly established for the K–T event, asteroid or cometary impact remains a viable explanation for potentially catastrophic change in the latest Permian, and may have been the trigger for the Siberian traps and the catastrophic ocean overturn. Tentative evidence for a cometary impact exists in the form of noble gases in fullerenes (Becker *et al.*, 2001), unusual Ni-rich grains (Kaiho *et al.*, 2001), and an impact crater whose broad range of allowable ages includes the Late Permian (Mory *et al.*, 2000a). However, the fullerene results have been challenged (Farley and Mukhopadhyay, 2001; Becker and Poreda, 2001; Isozaki, 2001), and the crater may be considerably older than P–Tr (Reimold and Koeberl 2000; Mory *et al.*, 2000b).

Numerical modeling indicates that Permian deep-water anoxia required either low atmospheric p_{O_2} or warm bottom-water source regions together with elevated oceanic nutrient (phosphate) concentrations (Hotinski *et al.*, 2001). Warm source regions acquire lower oxygen concentrations at equilibrium with the atmosphere before sinking, providing less oxygen to deep waters. When upwelled to the surface, higher phosphate concentrations in deep waters intensify the biological pump and thus increase O_2 demand in deep waters. These factors prove to be much more important than sluggish circulation itself, which does reduce O_2 supply to deep waters but also reduces the strength of biological pumping and thus the O_2 demand in deep waters (e.g., Sarmiento *et al.*, 1988; Hotinski *et al.*, 2000, 2001). In fact, another model failed to generate anoxia under reasonable Permian conditions (Zhang *et al.*, 2001). Subsequent analysis of the Zhang *et al.* (2001) results (Hotinski *et al.*, 2002) indicated that surface forcings in their model generated a more vigorous circulation of cooler water than in the model of Hotinski *et al.* (2001).

The transgression of anoxic deep waters with sea-level rise may have been the direct kill mechanism (Wignall, 1990; Hallam, 1991; Wignall and Hallam, 1992, 1993; Wignall and Twitchett, 1996; Cirilli *et al.*, 1998). Alternatively, anoxic deep waters may have undergone periodic catastrophic upwelling, induced by cooling or other surface forcings, causing significant transients in CO_2 and perhaps H_2S surface-water concentrations (Knoll *et al.*, 1996). These concentrations would have induced CO_2 toxicity (hypercapnia), especially in calcifying organisms, and could have been lethal. Selectivity of marine extinctions for those organisms, especially sensitive to high aqueous CO_2 concentration, supports the hypercapnia explanation for the extinction (Knoll *et al.*, 1996). Terrestrial extinctions could have been produced by the climate changes attendant upon the equilibration of the atmosphere with surface waters enriched in CO_2. This notion is supported by evidence that widespread deep-water anoxia developed considerably before the main extinction event (Isozaki, 1997).

Research on the climatic and carbon isotopic effect of destabilization of methane clathrates (e.g., Dickens *et al.*, 1995, 1997; Dickens, 2001) during a (possibly) similar event in the Paleocene suggests yet another hypothesis for the end-Permian disaster: Siberian trap volcanism led to warming and rearrangement of ocean circulation patterns, bringing warm intermediate waters into contact with continental shelf sediments, leading to the catastrophic release of methane to the ocean, and generating anoxia, negative $\delta^{13}C$ excursions (Krull and Retallack, 2000; de Wit *et al.*, 2002), and enhanced global warming (Wignall, 2001).

7.14.3.4 Triassic–Jurassic

Of the Phanerozoic "big five," the mass extinction that occurred at the Tr–J boundary has received the least attention from geoscientists. The event affected the diversity of both terrestrial and marine ecosystems, and may have created the opportunity for the rise to dominance of dinosaurs by selective extinction of the nondinosauran competitors (Olsen *et al.*, 2002). Its timing coincides with the emplacement of a large igneous province (the Central American Magmatic Province) and associated volcanic activity (Marzoli *et al.*, 1999; Hesselbo *et al.*, 2002). Interestingly, the event appears to have initiated on land some several hundred kiloyears before it did in the ocean, suggesting a trigger to which terrestrial ecosystems were most sensitive, followed by prolonged environmental change that adversely affected marine ecosystems (Palfy *et al.*, 2000). Others have challenged this interpretation, arguing that the events on land and sea were synchronous (Hesselbo *et al.*, 2002). Paleobotanical data (stomatal density) suggest that atmospheric CO_2 levels increased fourfold across the boundary, perhaps increasing leaf temperatures to lethal limits (McElwain *et al.*, 1999), but soil carbonate isotope proxies suggest that CO_2 levels were constant over this interval (Tanner *et al.*, 2001). As of late 2002, the issue is unresolved (Beerling, 2002; Tanner, 2002).

The geochemical record of the Tr–J extinction is rather limited, but recent additions have been made that are producing a clearer picture of the event. Carbon isotope records from organic matter in boundary sections from Canada (Ward *et al.*, 2001), England, and Greenland (Hesselbo *et al.*, 2002), and carbonates in Hungary (Palfy *et al.*, 2001) document a negative excursion of 2–4‰ centered on the boundary. Ward *et al.* (2001) ascribed this to a collapse of marine productivity, and likened the response to the P–Tr and K–T extinction records. There is no apparent shift in the strontium isotopic composition of the ocean across the event (Hallam, 1994). A modest iridium anomaly together with a substantial increase in fern spore abundance occurs precisely at the Tr–J boundary (Olsen *et al.*, 2002), indicating that a meteorite impact may have triggered the mass extinction.

7.14.3.5 Cretaceous–Tertiary

The fossil record shows that the species composition of terrestrial and marine ecosystems suffered a nearly complete turnover at the K–T boundary. Marked iridium anomalies (e.g., Alvarez *et al.*, 1980; Kyte *et al.*, 1985) and other indicators of asteroid or comet impact (several papers in Sharpton and Ward, 1990) including the Chicxulub impact crater (Hildebrand *et al.*, 1991) have provided compelling evidence for an extraterrestrial cause of this extinction. However, the carbon isotope record of this event has shed the most light on how this mass extinction affected the operation of the Earth system. Through preferential incorporation of ^{12}C into organic matter during photosynthesis, the biological pump establishes an isotopic gradient of 1–3‰ between the surface and deep (e.g., Kroopnick, 1974).

The isotope record of the latest Cretaceous to earliest Tertiary, however, shows an essentially complete collapse of this gradient, which for some sites is interpreted to have lasted from hundreds of thousands (Zachos *et al.*, 1989) to 3 Myr (D'Hondt *et al.*, 1998). Hsü and McKenzie (1985) referred to this state as the "Strangelove Ocean," because it was their conclusion that it represented a near cessation of primary production in the surface ocean. This interval is now recognized in other regions of the world (e.g., Stott and Kennett, 1989; Keller and Lindinger, 1989; Barrera and Keller, 1990; D'Hondt *et al.*, 1998), and amongst other groups of organisms, including mollusks that lived in offshore shelf environments (Hansen *et al.*, 1993).

A number of suggestions for the cause of the $\delta^{13}C$ gradient collapse have been made, including cessation of biological productivity or enhanced oceanic overturn (Zachos *et al.*, 1989), but Kump (1991) has shown that the loss of the biological pump is the explanation most consistent with mass balance constraints on the carbon isotopic system. Interestingly, it is conceivable that productivity in the surface ocean remained high, perhaps aided by an explosion in the abundance of stress-adapted plankton (Hollander *et al.*, 1993; Percival and Fischer, 1977).

If so, the loss of the biological pump would have to be explained by a loss in the ability of the ocean system to aggregate fine-grained organic matter into large particles, or provide the ballast (usually dense biogenic mineral material, especially $CaCO_3$) to facilitate sinking (Armstrong *et al.*, 2002). Any zooplankton that appeared soon after the event may not have created sufficiently large fecal pellets to facilitate sinking (D'Hondt *et al.*, 1998). A dearth of coarser-grained calcareous material in their fecal pellets may also have contributed to a weak biological pump.

There are other geochemical tracers of biosphere response to mass extinction besides carbon isotopes. For example, trace-metal contents of K–T sediments may provide additional evidence about the extinction itself and the ensuing "Strangelove Ocean" recovery interval (Vogt, 1972; Hsü *et al.*, 1982; Officer and Drake, 1985). Erickson and Dickson (1987) calculated that the metal burden of a 10 km meteorite would increase the metal content of the ocean several fold. They used the present-day residence times of the elements to argue for rather fast removal of the short residence-time elements (e.g., iron) and longer times, thousands of years, for others (e.g., nickel and copper). However, the major removal mechanism for many metals is in association with the biological pump. Thus, the enrichment of metals as a direct result of meteorite vaporization and solubilization (and perhaps as the result of an interval of intense continental weathering due to nitric acid rain; Macdougall, 1988) is likely to persist much longer in the absence of the pump.

Over even longer intervals of time, the metal content of the oceans would continue to rise as rivers, aerosols, and hydrothermal activity provided metals to the ocean at a rate much faster than their rate of removal in the absence of the biological pump. Many short-lived elements today might have become long-lived ones during the "Strangelove Ocean" interval, with ocean mixing homogenizing their distributions. Thus, the persistence of the "Strangelove Ocean" in the face of high rates of diversification and expansion of stress-related biota may have been due to the persistence of toxic concentrations of metals (Leary and Rampino, 1990).

Direct proxies of water-column metal contents have only been established in one case for the K–T boundary (Stott and Delaney, 1988). These investigators found a positive excursion in the Cd/Ca ratio in benthic foraminifera at site 690C (Weddell Sea, Antarctica), but just before the K–T boundary. There was no apparent change in the ratio across the boundary, which Stott and Delaney (1988) interpreted to indicate that there was little change in productivity at this site as a result of the extinction.

7.14.4 SUMMARY WITH EXTENSIONS

It can safely be said that there are no universal geochemical precursors or responses to extinction events in the Phanerozoic (Table 1). The Paleozoic extinctions are associated with indicators of anoxia and the Mesozoic extinctions have indications of asteroid or comet impact. Even within these groupings there are significant differences. The Paleozoic events exhibit both positive and negative carbon and sulfur isotope excursions; if anoxia is important, shouldn't the perturbations of the carbon and sulfur cycle be similar in all cases and leave a consistent isotopic signature? Oddly enough, only the strontium isotopic signatures of the events share commonalities. In each of the Paleozoic extinction events, $^{87}Sr/^{86}Sr$ reached a minimum just prior to the event. Is it this minimum that is significant (suggesting that weathering rates were low, and thus that the oceans were starved of nutrients prior to the event; Berner, 1989) or is it the upswing in the ratio (suggesting a significant increase in continental weathering) that is important? The answer to these questions is the subject of ongoing research by the author and a host of others.

Nevertheless, it may be useful to consider the Paleozoic and Mesozoic extinctions to be the result of two phenomena: oceanic anoxia in the Paleozoic and asteroid–comet impact in the Mesozoic.

7.14.4.1 Paleozoic Suffocation

The development of widespread deep-ocean anoxia is a diabolical state that certainly increases the likelihood of mass extinction. Its association with all three Paleozoic mass extinctions and with other minor extinctions of the Phanerozoic is compelling circumstantial evidence for its role in these biotic turnovers (e.g., Arthur *et al.*, 1987; Arthur and Sageman, 1994; Wilde *et al.*, 1990; Hallam, 1998; Harries and Little, 1999). Loss of oxygen from deep waters is certain to kill any aerobic organisms inhabiting abyssal plain or mid-ocean ridge environments, but the fossil record of these ecosystems is sparse. An expanded oxygen-minimum zone that transgresses onto a continental shelf creates inimical environments only to the extent that wind-driven air–sea exchange of

Table 1 Geochemical and environmental phenomena associated with mass extinction.

Extinction	*$\delta^{13}C$ excursion*	*$\delta^{34}S$ excursion*	*$^{87}Sr/^{86}Sr$ excursion*	*Extra-terrestrial Ir?*	*Anoxia?*	*Global temperature change*	*Sea-level change*
Cretaceous–Tertiary	(−)	(0)	(+) (short term)	Y	N	↑	↑ after Maastr. lowstand[a]
Triassic–Jurassic	(−)	(0)	(0)	Y[b]	N	↑	↑ after Late Triassic ↓
Permian–Triassic	(−)	(+)	(+) (from Paleozoic minimum)	N, but Fe–Si–Ni grains and fullerenes	Y	↑	↑ from Late Permian lowstand
Late Devonian (F–F)	Generally (+) but (−) right at F–F boundary[c]	(+)	(+) (from Devonian minimum)	N (but Y above and below F–F boundary)	Y (but may have ended just before F–F)[d]	↑[e]	↓
Late Ordovician	(+) (1st wave) (−) (2nd wave)	(0)	(+) (from Ordovician minimum)	N	Y, before and after glaciation	↓ then ↑	↓ then ↑

[a] Keller *et al.* (1993). [b] Olsen *et al.* (2002). [c] Wang *et al.* (1996). [d] Copper (1998). [e] Bratton *et al.* (1999).

oxygen cannot overcome the flux of reductants from the open ocean.

Numerical modeling of the Cretaceous Western Interior Seaway indicates that this exchange is rapid and suggests that instead it is the flux of nutrients (e.g., phosphate) that stimulates epeiric sea productivity and locally reduces the oxygen content of seaway waters (cf. Jewell, 1993; Slingerland *et al.*, 1996). Rapid overturn of an anoxic ocean can produce elevated p_{CO_2} in surface waters (Knoll *et al.*, 1996) that may be toxic, especially to calcifying organisms, and could effect a global warming episode that persisted for millennia. The trigger for the overturn could be renewed thermohaline circulation as a result of cooling, as is associated with the Late Ordovician and F–F events. Asteroid or comet impact would also mix the ocean instantly.

Relatively less investigated consequence of mixing an anoxic ocean would be the development of high partial pressures of H_2S in surface waters and a substantial flux of the gas to the atmosphere. H_2S is highly toxic to aerobic organisms (Bagarinao, 1992). It is also an efficient scavenger of nutrient metals such as iron and molybdenum, creating waters that are unsupportive of primary production. For these reasons, vigorous mixing of sulfidic waters to the surface presents a kill mechanism that is perhaps even more effective than that of CO_2. However, the surface-water H_2S concentration and thus the air–sea flux would be reduced to an unknown amount by sulfide oxidation in surface waters. In fact, this suggests a test of the hypothesis: well-preserved P–Tr boundary sediments may contain biomarkers of sulfide-oxidizing bacteria (e.g., the photoautotrophic green-sulfur bacteria biomarker *isorenieratene*; e.g., Simons and Kenig, 2001).

7.14.4.2 Mesozoic Menace

While anoxia emerges as the common theme of Paleozoic mass extinction, the Mesozoic mass extinctions (Tr–J and K–T) occurred during well-oxygenated intervals. For these events, asteroid or cometary impact emerge as the most likely proximal causes of the extreme type of environmental change required for mass extinction. The perplexing question is why there isn't better evidence for extraterrestrial impact and extinction in the Paleozoic. The well-known impactor size–frequency relationship (e.g., Alvarez, 1987) argues for a 100 Myr average waiting time between 10 km asteroid impacts of the sort that wreaked havoc at the K–T boundary. The odds that there were no K–T sized impactors during the Paleozoic (three recurrence intervals) are e^{-3} or 0.05.

Relatively recently work on the K–T impact has emphasized the importance of target rock composition in determining the magnitude of the environmental disruption following the impact (e.g., O'Keefe and Ahrens, 1989; Sharpton *et al.*, 1996; Pope *et al.*, 1997), and this factor has increased in importance subsequently because of indications that the dust generated by the impact may not serve to reduce sunlight effectively (Pope, 2002). Chicxulub target rocks are limestones with gypsum evaporites. Vaporization of these materials would have instantaneously transferred tremendous quantities of CO_2 and SO_2 to the atmosphere. The short-term cooling from sulfuric acid aerosols would be followed by a substantial warming from the greenhouse effects of CO_2. Perhaps Paleozoic impacts were into seafloor basalt or other volatile-poor rocks, and thus their climatic effects were less.

A Paleozoic oceanic impact at a time of widespread anoxia would instantaneously replace the surface ocean with anoxic, sulfidic, high p_{CO_2} deep waters. CO_2 and H_2S would degass to the atmosphere; CO_2 would equilibrate, and H_2S would oxidize. However, an H_2S-rich plume could last days to weeks, and spread across the land surface (A. Pavlov, personal communication). Such a scenario may explain the F–F and P–Tr extinction events; if the impactor was a comet, a smaller iridium anomaly would result and may have escaped detection. If so, only the Late Ordovician extinction remains as a likely candidate for a purely terrestrial extinction mechanism (glacio-eustatic sea-level fall causing shallow-marine habitat loss).

ACKNOWLEDGMENTS

The author acknowledges support from the NASA Astrobiology Institute, and the NSF Geology and Paleontology and Biocomplexity Programs.

REFERENCES

Abramovich S. and Keller G. (2002) High stress late Maastrichtian paleoenvironment, inference from planktonic foraminifera in Tunisia. *Palaeogeogr. Palaeoclimat. Palaeoecol.* **178**, 145–164.

Alvarez L. W. (1987) Mass extinctions caused by large bolide impacts. *Phys. Today* **40**, 24–33.

Alvarez L. W., Alvarez W., Asaro F., and Michel H. V. (1980) Extraterrestrial cause for the Cretaceous–Tertiary extinction. *Science* **208**, 1095–1108.

Armstrong R. A., Lee C., Hedges J. I., Honjo S., and Wakeham S. G. (2002) A new, mechanistic model for organic carbon fluxes in the ocean based on the quantitative association of POC with ballast minerals. *Deep-Sea Res. II* **49**, 219–236.

Arthur M. A. and Sageman B. B. (1994) Marine black shales, depositional mechanisms and environments of ancient deposits. *Ann. Rev. Earth Planet. Sci.* **22**, 499–551.

Arthur M. A., Schlanger S. O., and Jenkyns H. C. (1987) The Cenomanian–Turonian oceanic anoxic event: II.

Paleoceaonographic controls on organic matter production and preservation. *Geol. Soc. London Spec. Publ.* **26**, 401–420.

Bagarinao T. (1992) Sulfide as an environmental factor and toxicant, tolerance and adaptations in aquatic organisms. *Aquat. Toxicol.* **24**, 21–62.

Barrera E. (1994) Global environmental changes preceding the Cretaceous–Tertiary boundary, Early–Late Maastrichtian transition. *Geology* **22**, 877–880.

Barrera E. and Keller G. (1990) Stable isotope evidence for gradual environmental changes and species survivorship across the Cretaceous/Tertiary boundary. *Paleoceanography* **5**, 867–890.

Baud A., Magaritz M., and Holser W. T. (1989) Permian–Triassic of the Tethys, carbon isotope studies. *Geol. Rundsch.* **78**, 649–677.

Becker L. and Poreda R. J. (2001) An extraterrestrial impact at the Permian–Triassic boundary? (reply). *Science* **293**, 2343.

Becker L., Poreda R. J., Hunt A. G., Bunch T. E., and Rampino M. (2001) Impact event at the Permian–Triassic boundary: evidence from extraterrestrial noble gases in fullerenes. *Science* **291**, 1530–1533.

Beeling D. (2002) CO_2 and the end-Triassic mass extinction. *Nature* **415**, 386–387.

Berner R. A. (1989) Drying, O_2 and mass extinction: discussion. *Nature* **340**, 603–604.

Berner R. A. and Canfield D. E. (1989) A new model for atmospheric oxygen over Phanerozoic time. *Am. J. Sci.* **289**, 333–361.

Berner R. A. and Raiswell R. (1983) Burial of organic carbon and pyrite sulfur in sediments over Phanerozoic time, a new theory. *Geochim. Cosmochim. Acta.* **47**, 855–862.

Bowring S. A., Erwin D. H., Jin Y. G., Martin M. W., Davidek K., and Wang W. (1998) U/Pb zircon geochronology and tempo of the end-Permian mass extinction. *Science* **280**, 1035–1045.

Bratton J. F., Berry W. B. N., and Morrow J. R. (1999) Anoxia pre-dates Frasnian–Famennian boundary mass extinction horizon in the Great Basin, USA. *Palaeogeogr. Palaeoclimat. Palaeoecol.* **154**, 275–292.

Brenchley P. J., Marshall J. D., Carden G. A. F., Robertson D. B. R., Long D. G. F., Meidla T., Hints L., and Anderson T. F. (1994) Bathymetric and isotopic evidence for a short-lived Late Ordovician glaciation in a greenhouse period. *Geology* **22**, 295–298.

Cerling T. E. (1991) Carbon dioxide in the atmosphere: evidence from Cenozoic and Mesozoic Paleosols. *Am. J. Sci.* **291**, 377–400.

Cirilli S., Radrizzani C. P., Ponton M., and Radrizzani S. (1998) Stratigraphical and palaeoenvironmental analysis of the Permian Triassic transition in the Badia Valley (southern Alps Italy). *Palaeogeogr. Palaeoclimat. Palaeoecol.* **138**, 85–113.

Copper P. (1998) Evaluating the Frasnian–Famennian mass extinction, comparing brachiopod faunas. *Acta Palaeontologica Polonica* **43**, 137–154.

Crowley T. J. and Baum S. K. (1991) Towards reconciliation of Late Ordovician (~440 Ma) glaciation with very high CO_2 levels. *J. Geophys. Res.* **96**, 22597–22610.

de Wit M. J., Ghosh J. G., de Villiers S., Rakotosolofo N., Alexander J., Tripathi A., and Looly C. (2002) Multiple organic carbon isotope reversals across the Permo–Triassic boundary of terrestrial Gondwana sequences, clues to extinction patterns and delayed ecosystem recovery. *J. Geol.* **110**, 227–240.

D'Hondt S., Donaghay P., Zachos J. C., Luttenberg D., and Lindinger M. (1998) Organic carbon fluxes and ecological recovery from the Cretaceous–Tertiary mass extinction. *Science* **282**, 276–279.

Dickens G. R. (2001) On the fate of past gas, what happens to methane released from a bacterially mediated gas hydrate capacitor? *Geochem. Geophys. Geosys.* **2**, U1–U5.

Dickens G. R., O'Neil J. R., Rea D. K., and Owen R. M. (1995) Dissociation of oceanic methane hydrate as a cause of the carbon isotope excursion at the end of the Paleocene. *Paleoceanography* **10**, 965–971.

Dickens G. R., Castillo M. M., and Walker J. C. G. (1997) A blast of gas in the latest Paleocene: simulating first-order effects of massive dissociation of oceanic methane hydrate. *Geology* **25**, 259–262.

Erickson D. J., III. and Dickson S. M. (1987) Global trace-element biogeochemistry at the K/T boundary, oceanic and biotic response to a hypothetical meteorite impact. *Geology* **15**, 1014–1017.

Erwin D. H. (1993) *The Great Paleozoic Crisis, Life and Death in the Permian.* Columbia University Press, New York.

Erwin D. H. (1995) The end-Permian mass extinction. In *The Permian of Northern Pangea* (eds. P. A. Scholle, T. M. Peryt, and D. S. Ulmer-Scholle). Springer, New York, vol. 1, pp. 20–34.

Erwin D. H. (2001) Lessons from the past, biotic recoveries from mass extinctions. *Proc. Natl. Acad. Sci.* **98**, 5399–5403.

Farley K. A. and Mukhopadhyay S. (2001) An extraterrestrial impact at the Permian–Triassic boundary? (critical comment) *Science* **293**, 2343.

Finney S. C., Berry W. B. N., Cooper J. D., Ripperdan R. L., Sweet W. C., Jacobson S. R., Soufiane A., Achab A., and Noble P. J. (1999) Late Ordovician mass extinction: a new perspective from stratigraphic sections in central Nevada. *Geology* **27**, 215–218.

Frakes L. A., Jane E. F., and Jozef I. S. (1992) *Climate Modes of the Phanerozoic: The History of the Earth's Climate over the Past 600 Million Years.* Cambridge University Press, Cambridge.

Gibbs M., Barron E. J., and Kump L. R. (1997) An atmospheric pCO_2 threshold for glaciation in the Late Ordovician. *Geology* **25**, 447–450.

Girard C., Robin E., Rocchia R., Froget L., and Feist R. (1997) Search for impact remains at the Frasnian–Famennian boundary in the stratotype area, southern France. *Palaeogeogr. Palaeoclimat. Palaeoecol.* **132**, 391–397.

Goodfellow W. D., Nowlan G. S., McCracken A. D., Lanz A. C., and Gregoire D. C. (1992) Geochemical anomalies near the Ordovician–Silurian boundary, Northern Yukon Territory, Canada. *Hist. Biol.* **6**, 1–23.

Gruszczynski M., Halas S., Hoffman A., and Malkowski K. (1989) A brachiopod calcite record of the oceanic carbon and oxygen isotope shifts at the Permian/Triassic transition. *Nature* **337**, 64–68.

Hallam A. (1991) Why was there a delayed radiation after the end-Palaeozoic extinctions? *Hist. Biol.* **5**, 257–262.

Hallam A. (1994) Strontium isotope profiles of Triassic–Jurassic boundary sections in England and Austria. *Geology* **22**, 1079–1082.

Hallam A. (1998) Mass extinctions in Phanerozoic time. *Geol. Soc. Spec. Publ.* **140**, 259–274.

Hansen T. A., Upshaw B., III, Kauffman E. G., and Gose W. (1993) Patterns of molluscan extinction and recovery across the Cretaceous–Tertiary boundary in East Texas: report on new outcrops. *Cretaceous Res.* **14**, 685–706.

Hart M. B. (1996) *Biotic Recovery from Mass Extinction Events*. Geol. Soc. London, Spec. Publ. 102, pp. 265–277.

Harries P. J. and Little C. T. S. (1999) The early Toarcian (Early Jurassic) and the Cenomanian–Turonian (Late Cretaceous) mass extinctions: similarities and contrasts. *Palaeogeogr. Palaeoclimat. Palaeoecol.* **154**, 39–66.

Hesselbo S. P., Robinson S. A., Surlyk F., and Piasecki S. (2002) Terrestrial and marine extinction at the Triassic–Jurassic boundary synchronized with major carbon-cycle perturbation, a link to initiation of massive volcanism? *Geology* **30**, 251–254.

Hildebrand A. R., Penfield G. T., Kring D. A., Pilkington D., Camargo A., Jacobsen S. B., and Boynton W. V. (1991) Chicxulub crater, a possible Cretaceous–Tertiary boundary impact crater on the Yucatan peninsula. *Geology* **19**, 867–871.

Hollander D. J., McKenzie J. A., and Hsü K. J. (1993) Carbon isotope evidence for unusual plankton blooms and fluctuations of surface water CO_2 in "Strangelove Ocean" after terminal Cretaceous event. *Palaeogeogr. Palaeoclimat. Palaeoecol.* **104**, 229–237.

Holser W. T., Schidlowski M., Mackenzie F. T., and Maynard J. B. (1988) Biogeochemical cycles of carbon and sulfur. In *Chemical Cycles in the Evolution of the Earth* (eds. C. B. Gregor, R. M. Garrels, F. T. Mackenzie, and J. B. Maynard). Wiley-Interscience, New York, pp. 105–174.

Holser W. T., Schönlaub H. P., Boeckelmann K., and Magaritz M. (1991) The Permian–Triassic of the Gartnerkofel-1 core (Carnic Alps, Austria), synthesis and conclusions. *Abhandlungen der Geologischen Bundesanstalt* **45**, 5–16.

Hotinski R. M., Kump L. R., and Najjar R. G. (2000) Opening Pandora's box: the impact of open system modeling on interpretations of anoxia. *Paleoceanography* **15**, 267–279.

Hotinski R. M., Bice K. L., Kump L. R., Najjar R. G., and Arthur M. A. (2001) Ocean stagnation and end-Permian anoxia. *Geology* **29**, 7–10.

Hotinski R. M., Kump L. R., and Bice K. L. (2002) Comment on Zhang *et al.* (2001) *Paleoceanography* **17**, 1052, doi: 10.1029/2001PA000680.

Hsü K. J. and McKenzie J. A. (1985) A "Strangelove" ocean in the earliest Tertiary. In *The Carbon Cycle and Atmospheric CO_2: Natural Variations Archean to Present* (eds. E. T. Sundquist and W. S. Broecker). Am. Geophys. Union, Washington, DC, pp. 487–492.

Hsü K. J., He Q., McKenzie J. A., Weissert H., Perch-Nielsen K., Oberhaensli H., Kelts K., LaBrecque J., Tauxe L., Kraehenbuehl U., Percival S. F., Jr., Wright R., Karpoff A. M., Petersen N., Tucker P., Poore R. Z., Gombos A. M., Pisciotto K. A., Carman M. F., Jr., and Schreiber E. (1982) Mass mortality and its environmental and evolutionary consequences. *Science* **216**, 249–256.

Isozaki Y. (1997) Permo–Triassic boundary: superanoxia and stratified superocean, records from lost deep-sea. *Science* **276**, 235–238.

Isozaki Y. (2001) An extraterrestrial impact at the Permian–Triassic boundary? (critical comment). *Science* **293**, 2343a.

Jewell P. W. (1993) Water-column stability, residence times, and anoxia in the Cretaceous North American seaway. *Geology* **21**, 579–582.

Joachimski M. M. and Buggisch W. (1993) Anoxic events in the late Frasnian—causes of the Frasnian–Famennian faunal crisis. *Geology* **21**, 675–678.

Joachimski M. M., Pancost R. D., Freeman K. H., Ostertag-Henning C., and Buggisch W. (2002) Carbon isotope geochemistry of the Frasnian–Famennian transition. *Palaeogeogr. Palaeoclimat. Palaeoecol.* **181**, 91–109.

Kaiho K., Kajiwara Y., Nakano T., Miura Y., Kawahata H., Tazaki K., Ueshima M., Chen Z., and Shi G. R. (2001) End-Permian catastrophe by a bolide impact, evidence of a gigantic release of sulfur from the mantle. *Geology* **29**, 815–818.

Kajiwara Y., Yamakita S., Ishida K., Ishiga H., and Imai A. (1994) Development of a largely anoxic stratified ocean and its temporary massive mixing at the Permian/Triassic boundary supported by the sulfur isotopic record. *Palaeogeogr. Palaeoclimat. Palaeoecol.* **111**, 367–379.

Kasting J. F. (1992) Paradox lost and paradox found. *Nature* **355**, 676–677.

Keller G. and Lindinger M. (1989) Stable isotopic, TOC and $CaCO_3$ record across the Cretaceous/Tertiary boundary at El Kef, Tunisia. *Palaeogeogr. Palaeoclimat. Palaeoecol.* **73**, 243–266.

Keller G. R., Barrera E., Schmitz B., and Mattson E. (1993) Gradual mass extinction, species survivorship, and long-term environmental changes across the Cretaceous–Tertiary boundary in high latitudes. *Geol. Soc. Am. Bull.* **105**, 979–997.

Kirchner J. W. and Weil A. (2000) Delayed biological recovery from extinctions throughout the fossil record. *Nature* **404**, 177–180.

Knoll A. H., Bambach R. K., Canfield D. E., and Grotzinger J. P. (1996) Comparative Earth history and Late Permian mass extinction. *Science* **273**, 452–457.

Kozur H. W. (1998) Some aspects of the Permian–Triassic boundary and possible causes for the biotic crisis around this boundary. *Palaeogeogr. Palaeoclimat. Palaeoecol.* **143**, 227–272.

Kroopnick P. M. (1974) The dissolved O_2–CO_2–^{13}C system in the eastern equatorial Pacific. *Deep-Sea Res.* **21**, 211–227.

Krull E. S. and Retallack G. J. (2000) $\delta^{13}C$ depth profiles from paleosols across the Permian–Triassic boundary, evidence for methane release. *Geol. Soc. Am. Bull.* **112**, 1459–1472.

Kump L. R. (1989) Alternative modeling approaches to the geochemical cycles of carbon, sulfur, and strontium isotopes. *Am. J. Sci.* **289**, 390–410.

Kump L. R. (1991) Interpreting carbon-isotope excursions: strangelove oceans. *Geology* **19**, 299–302.

Kump L. R. (1992) Coupling of the carbon and sulfur biogeochemical cycles over Phanerozoic time. In *Interactions of C, N, P and S Biogeochemical Cycles*, NATO ASI Series (eds. R. Wollast, F. T. Mackenzie, and L. Chou). Springer, Berlin, pp. 475–490.

Kump L. R. and Arthur M. A. (1997) Global chemical erosion during the Cenozoic, Weatherability balances the budget. In *Tectonic Uplift and Climate Change* (ed. W. Ruddiman). Plenum, New York, pp. 399–426.

Kump L. R. and Arthur M. A. (1999) Interpreting carbon-isotope excursions, carbonates and organic matter. *Chem. Geol.* **161**, 181–198.

Kump L. R., Arthur M., Patzkowsky M., Gibbs M., Pinkus D. S., and Sheehan P. (1999) A weathering hypothesis for glaciation at high atmospheric pCO_2 in the Late Ordovician. *Palaeogeogr. Palaeoclimat. Palaeoecol.* **152**, 173–187.

Kyte F. T., Smit J., and Wasson J. T. (1985) Siderophile interelement variations in the Cretaceous–Tertiary boundary sediments from Caravaca, Spain. *Earth Planet. Sci. Lett.* **73**, 183–195.

Leary P. N. and Rampino M. R. (1990) A multi-causal model of mass extinctions, increase in trace metals in the oceans. In *Extinction Events in Earth History* (eds. E. G. Kauffman and O. H. Walliser). Springer, New York, pp. 45–55.

Long D. G. F. (1993) Oxygen and carbon isotopes and event stratigraphy near the Ordovician–Silurian boundary, Anticosti Island, Quebec. *Palaeogeogr. Palaeoclimat. Palaeoecol.* **104**, 49–59.

Macdougall J. D. (1988) Seawater strontium isotopes, acid rain, and the Cretaceous–Tertiary boundary. *Science* **239**, 485–487.

MacLeod K. G., Huber B. T., and Fullagar P. D. (2001) Evidence for a small (approximately 0.000030) but resolvable increase in seawater $^{87}Sr/^{86}Sr$ ratios across the Cretaceous–Tertiary boundary. *Geology* **29**, 303–306.

Magaritz M., Krishnamurthy R. V., and Holser W. T. (1991) Parallel trends in organic and inorganic carbon isotopes across the Permian/Triassic boundary. *Am. J. Sci.* **291**, 727–739.

Martin E. E. and Macdougall J. D. (1991) Seawater Sr isotopes at the Cretaceous/Tertiary boundary. *Earth Planet. Sci. Lett.* **104**, 166–180.

Marzoli A., Renne P. R., Piccirillo E. M., Ernesto M., Bellieni G., and de Min A. (1999) Extensive 200-million-year-old continental flood basalts of the Central Atlantic Magmatic Province. *Science* **284**, 616–618.

McElwain J. C., Beerling D. J., and Woodward F. I. (1999) Fossil plants and global warming at the Triassic–Jurassic boundary. *Science* **285**, 1386–1390.

McGhee G. R., Jr. (1996) *The Late Devonian Mass Extinction: The Frasnian/Famennian Crisis.* Columbia University Press, New York.

McGhee G. R., Jr., Gilmore J. S., Orth C. J., and Olsen E. (1984) No geochemical evidence for an asteroidal impact at

Late Devonian mass extinction horizon. *Nature* **398**, 629–631.

McLaren D. J. (1970) Presidential address, time, life and boundaries. *J. Paleontol.* **44**, 801–815.

McLaren D. J. (1982) Frasnian–Famennian extinctions. *Geol. Soc. Am. Spec. Pap.* **190**, 477–484.

Middleton P. D., Marshall J. D., and Brenchley P. J. (1991) Evidence for isotopic change associated with Late Ordovician glaciation, from brachiopods and marine cements of central Sweden. In *Advances in Ordovician Geology* (eds. C. R. Barnes and S. H. Williams). Geol. Surv. Canada, Ottawa, pp. 313–323.

Mii H.-S., Grossman E. L., and Yancey T. E. (1997) Stable carbon and oxygen isotope shifts in Permian seas of West Spitsbergen: global change or diagenetic artifact? *Geology* **25**, 227–230.

Mora C. I., Driese S. G., and Seager P. G. (1991) Carbon dioxide in the Paleozoic atmosphere: evidence from carbon-isotope compositions of pedogenic carbonate. *Geology* **19**, 1017–1020.

Mory A. J., Iasky R. P., Glikson A. Y., and Pirajno F. (2000a) Woodleigh, Carnarvon Basin, Western Australia, a new 120 km diameter impact structure. *Earth Planet. Sci. Lett.* **177**, 119–228.

Mory A. J., Iasky R. P., Glikson A. Y., and Pirajno F. (2000b) Response to "Critical comment on A. J. Mory *et al.* (2000a)" by W. U. Reimold and C. Koeberl. *Earth Planet. Sci. Lett.* **184**, 359–365.

Murphy A. E., Sageman B. B., and Hollander D. J. (2000) Eutrophication by decoupling of the marine biogeochemical cycles of C, N, and P: a mechanism for the Late Devonian mass extinction. *Geology* **28**, 427–430.

Officer C. B. and Drake C. L. (1985) Terminal Cretaceous environmental events. *Science* **227**, 1161–1167.

O'Keefe J. D. and Ahrens T. J. (1989) Impact production of CO_2 by the Cretaceous/Tertiary extinction bolide and the resultant heating of the Earth. *Nature* **338**, 247–249.

Olsen P. E., Kent D. V., Sues H.-D., Koeberl C., Huber H., Montanari A., Rainforth E. C., Fowell S. J., Szajna M. J., and Hartline B. W. (2002) Ascent of dinosaurs linked to an iridium anomaly at the Triassic–Jurassic boundary. *Science* **296**, 1305–1307.

Orth C. J., Gilmore J. S., Quintana L. R., and Sheehan P. M. (1986) Terminal Ordovician extinction, geochemical analysis of the Ordovician/Silurian boundary, Anticosti Island, Quebec. *Geology* **14**, 433–436.

Palfy J., Mortensen J. K., Carter E. S., Smith P. L., Friedman R. M., and Tipper H. W. (2000) Timing the end-Triassic mass extinction: first on land, then in the sea? *Geology* **28**, 39–42.

Palfy J., Demeny A., Haas J., Hetenyi M., Orchard M. J., and Veto I. (2001) Carbon isotope anomaly and other geochemical changes at the Triassic–Jurassic boundary from marine section in Hungary. *Geology* **29**, 1047–1050.

Palmer M. R. and Elderfield H. (1985) Sr isotope composition of seawater over the past 75 m.y. *Nature* **314**, 526–528.

Percival S. F., Jr. and Fischer A. G. (1977) Changes in calcareous nanoplankton in the Cretaceous–Tertiary biotic crisis at Zumaya, Spain. *Evol. Theor.* **2**, 1–35.

Playford P. E., McLaren D. J., Orth C. J., Gilmore J. S., and Goodfellow W. D. (1984) Iridium anomaly in the Upper Devonian of the Canning Basin, Western Australia. *Science* **226**, 437–439.

Pope K. O. (2002) Impact dust not the cause of the Cretaceous–Tertiary mass extinction. *Geology* **30**, 99–102.

Pope K. O., Baines K. H., Ocampo A. C., and Ivanov B. A. (1997) Energy, volatile production, and climatic effects of the Chicxulub Cretaceous/Tertiary impact. *J. Geophys. Res.* **102**, 21645–21664.

Raymo M. E., Ruddiman W. F., and Froelich P. N. (1988) Influence of late Cenozoic mountain building on ocean geochemical cycles. *Geology* **16**, 649–653.

Reimold W. U. and Koeberl C. (2000) Critical comment on A. J. Mory *et al.* (2000a)—Discussion. *Earth Planet. Sci. Lett.* **184**, 353–357.

Renne P. R., Zichao Z., Richards M. A., Black M. T., and Basu A. R. (1995) Synchrony and causal relations between Permian–Triassic boundary crises and Siberian flood volcanism. *Science* **269**, 1413–1416.

Richter F. M. and Turekian K. K. (1993) Simple models for the geochemical response of the ocean to climatic and tectonic forcing. *Earth Planet. Sci. Lett.* **119**, 121–131.

Richter F. M., Rowley D. B., and DePaolo D. J. (1992) Sr isotope evolution of seawater, the role of tectonics. *Earth Planet. Sci. Lett.* **109**, 11–23.

Sarmiento J. L., Herbert T., and Toggweiler J. R. (1988) Causes of anoxia in the world ocean. *Global Biogeochem. Cycles* **2**, 115–128.

Scholle P. A. (1995) Carbon and sulfur isotope stratigraphy of the Permian and adjacent intervals. In *The Permian of Northern Pangea* (eds. P. A. Scholle, T. M. Peryt, and D. S. Ulmer-Scholle). Springer, New York, vol. 1, pp. 133–152.

Sepkoski J. J., Jr. (1993) Ten years in the library: new data confirm paleontological patterns. *Paleobiology* **19**, 43–51.

Sharpton V. L. and Ward P. D. (eds.) (1990) *Global Catastrophes in Earth History*. Geol. Soc. Am. Spec. Paper 147, Boulder, 631pp.

Sharpton V. L., Marin L. E., Carney J. L., Lee S., Ryder G., Schuraytz B. C., Sikora P., and Spudis P. D. (1996) A model of the Chicxulub impact basin based on evaluation of geophysical data, well logs, and drill core samples. *Spec. Pap. Geol. Soc. Am.* **307**, 55–74.

Sheehan P. M. (2001) The Late Ordovician mass extinction. *Ann. Rev. Earth Planet. Sci.* **29**, 331–364.

Simons D.-J. H. and Kenig F. (2001) Molecular fossil constraints on the water column structure of the Cenomanian–Turonian Western Interior Seaway, USA. *Palaeogeogr. Palaeoclimat. Palaeoecol.* **169**, 129–152.

Slingerland R., Kump L. R., Arthur M., Fawcett P., Sageman B., and Barron E. (1996) Estuarine circulation in the Turonian Western Interior Seaway of North America. *Bull. Geol. Soc. Am.* **108**, 941–952.

Stott L. D. and Delaney M. L. (1988) Cd/Ca in benthic foraminifera and stable isotopes across the Cretaceous–Tertiary boundary at Site 690C (Leg 113), Weddell Sea, Antarctic. *EOS* **69**, 1243.

Stott L. D. and Kennett J. P. (1989) New constraints on early Tertiary palaeoproductivity from carbon isotopes in foraminifera. *Nature* **342**, 526–529.

Strauss H. (1997) The isotopic composition of sedimentary sulfur through time. *Palaeogeogr. Palaeoclimat. Palaeoecol.* **132**, 97–118.

Strauss H. (1999) Geological evolution from isotope proxy signals—sulfur. *Chem. Geol.* **161**, 89–101.

Tanner L. H. (2002) Triassic–Jurassic atmospheric CO_2 spike. *Nature* **415**, 387–388.

Tanner L. H., Hubert J. F., Coffey B. P., and McInerney D. P. (2001) Stability of atmospheric CO_2 levels across the Triasic/Jurassic boundary. *Nature* **411**, 675–677.

Van Cappellen P. and Ingall E. D. (1994) Benthic phosphorus regeneration, net primary production, and ocean anoxia: a model of the coupled marine biogeochemical cycles of carbon and phosphorus. *Paleoceanography* **9**, 677–682.

Van Cappellen P. and Ingall E. D. (1996) Redox stabilization of the atmosphere and oceans by phosphorus-limited marine productivity. *Science* **271**, 493–496.

Veizer J., Ala D., Azmy K., Bruckschen P., Buhl D., Bruhn F., Carden G. A. F., Diener A., Ebneth S., Godderis Y., Jasper T., Korte C., Pawellek F., Podlaha O. G., and Strauss H. (1999) $^{87}Sr/^{86}Sr$, $\delta^{13}C$ and $\delta^{18}O$ evolution of Phanerozoic seawater. *Chem. Geol.* **161**, 59–89.

Veizer J., Godderis Y., and Francois L. M. (2000) Evidence for decoupling of atmospheric CO_2 and global climate during the Phanerozoic eon. *Nature* **408**, 698–701.

Vogt P. R. (1972) Evidence for global synchronism in mantle plume convection, and possible significance for geology. *Nature* **240**, 338–342.

Wang K., Orth C. J., Attrep M., Jr., Chatterton B. D. E., Hou H., and Geldsetzer H. H. J. (1991) Geochemical evidence for a catastrophic biotic event at the Frasnian/Famennian boundary in south China. *Geology* **19**, 776–779.

Wang K., Chatterton B. D. E., Attrep M., Jr., and Orth C. J. (1992) Iridium abundance maxima at the latest Ordovician mass extinction horizon, Yangtze Basin, China, terrestrial or extraterrestrial? *Geology* **20**, 39–42.

Wang K., Attrep M., Jr., and Orth C. J. (1993) Global iridium anomaly, mass extinction, and redox change at the Devonian–Carboniferous boundary. *Geology* **21**, 1071–1074.

Wang K., Geldsetzer H. H. J., and Krouse H. R. (1994) Permian–Triassic extinction. Organic delta-C-13 evidence from British-Columbia, Canada. *Geology* **22**, 580–584.

Wang K., Geldsetzer H. H. J., Goodfellow W. D., and Krouse H. R. (1996) Carbon and sulfur isotope anomalies across the Frasnian–Famennian extinction boundary, Alberta, Canada. *Geology* **24**, 187–191.

Wang K., Chatterton B. D. E., and Wang Y. (1997) An organic carbon isotope record of Late Ordovician to Early Silurian marine sedimentary rocks, Yangtze Sea, South China, implications for CO_2 changes during the Hirnantian glaciation. *Palaeogeogr. Palaeoclimat. Palaeoecol.* **132**, 147–158.

Ward P. D., Haggart J. W., Carter E. S., Wilbur D., Tipper H. W., and Evans T. (2001) Sudden productivity collapse associated with the Triassic–Jurassic boundary mass extinction. *Science* **292**, 1148–1151.

Wignall P. B. (1990) Observations on the evolution and classification of dysaerobic communities. *Paleontol. Soc. Spec. Publ.* **5**, 99–111.

Wignall P. B. (2001) Large igneous provinces and mass extinctions. *Earth Sci. Rev.* **53**, 1–33.

Wignall P. B. and Hallam A. (1992) Anoxia as a cause of the Permian/Triassic mass extinction, facies evidence from Northern Italy and the western United States. *Palaeogeogr. Palaeoclimat. Palaeoecol.* **93**, 21–46.

Wignall P. B. and Hallam A. (1993) Griesbachian (Early Triassic) palaeoenvironmental changes in the Salt Range, Pakistan and southeast China and their bearing on the Permo–Triassic extinction. *Palaeogeogr. Palaeoclimat. Palaeoecol.* **102**, 215–237.

Wignall P. B. and Twitchett R. J. (1996) Ocean anoxia and the end-Permian mass extinction. *Science* **272**, 1155–1158.

Wilde P., Quinby-Hunt M. S., and Berry W. B. N. (1990) Vertical advection from oxic or anoxic water from the main pycnocline as a cause of rapid extinction or rapid radiations. In *Extinction Events in Earth History* (eds. E. G. Kauffman and O. H. Walliser). Springer, New York, pp. 85–98.

Woods A. D., Bottjer D. J., Mutti M., and Morrison J. (1999) Lower Triassic large sea-floor carbonate cements: their origin and a mechanism for the prolonged biotic recovery from the end-Permian mass extinction. *Geology* **27**, 645–648.

Yapp C. J. and Poths H. (1992) Ancient atmospheric CO_2 pressures inferred from natural goethites. *Nature* **355**, 342–344.

Zachos J. C., Arthur M. A., and Dean W. E. (1989) Geochemical evidence for suppression of pelagic marine productivity at the Cretaceous/Tertiary boundary. *Nature* **337**, 61–64.

Zhang R., Follows M. J., Grotzinger J. P., and Marshall J. (2001) Could the Late Permian deep ocean have been anoxic? *Paleoceanography* **16**, 317–329.

7.15
Evolution of Sedimentary Rocks

J. Veizer

Ruhr University, Bochum, Germany and University of Ottawa, ON, Canada

and

F. T. Mackenzie

University of Hawaii, Honolulu, HI, USA

7.15.1 INTRODUCTION

For almost a century, it has been recognized that the present-day thickness and areal extent of Phanerozoic sedimentary strata increase progressively with decreasing geologic age. This pattern has been interpreted either as reflecting an increase in the rate of sedimentation toward the present (Barrell, 1917; Schuchert, 1931; Ronov, 1976) or as resulting from better preservation of the younger part of the geologic record (Gilluly, 1949; Gregor, 1968; Garrels and Mackenzie, 1971a; Veizer and Jansen, 1979, 1985).

Study of the rocks themselves led to similarly opposing conclusions. The observed secular (=age) variations in relative proportions of lithological types and in chemistry of sedimentary rocks (Daly, 1909; Vinogradov *et al*., 1952; Nanz, 1953; Engel, 1963; Strakhov, 1964, 1969; Ronov, 1964, 1982) were mostly given an evolutionary interpretation. An opposing, uniformitarian, approach was proposed by Garrels and Mackenzie (1971a). For most isotopes, the consensus favors deviations from the present-day steady state as the likely cause of secular trends.

This chapter attempts to show that recycling and evolution are not opposing, but complementary, concepts. It will concentrate on the lithological and chemical attributes of sediments, but not deal with the evolution of sedimentary mineral deposits (Veizer *et al*., 1989) and of life (Sepkoski, 1989), both well amenable to the outlined conceptual treatment. The chapter relies heavily on Veizer (1988a) for the sections dealing with general recycling concepts, on Veizer (2003) for the discussion of isotopic evolution of seawater, and on Morse and Mackenzie (1990) and Mackenzie and Morse (1992) for discussion of carbonate rock recycling and environmental attributes.

7.15.2 THE EARTH SYSTEM

The lithosphere, hydrosphere, atmosphere, and biosphere, or rocks, water, air, and life are all part of the terrestrial exogenic system that is definable by the rules and approaches of general system science theory, with its subsets, such as population dynamics and hierarchical structures.

7.15.2.1 Population Dynamics

The fundamental parameters essential for quantitative treatment of population dynamics are the population size (A_0) and its recycling rate. A_0 is normalized in the subsequent discussion to one population (or 100%) and the rates of recycling relate to this normalized size. Absolute rates can be established by multiplying this relative recycling rate (parameter b below) by population size.

A steady state natural population, characterized by a continuous generation/destruction (birth/death) cycle, is usually typified by an age structure similar to that in Figure 1, the cumulative curve defining all necessary parameters of a given population. These are its *half-life* τ_{50}, *mean age* τ_{mean}, and *oblivion age* or life expectancy τ_{max}.

For steady state first-order (=single population) systems, the survival rate of constituent units can be expressed as

$$A_{t^*} = A_0 \, e^{-kt^*} \quad (1)$$

where A_{t^*} is the cumulative fraction of the surviving population older than t^*, $A_0 = 1$ (one population), t^* is age (not time), and k is the rate constant for the recycling process. In the subsequent discussion, the recycling rate is often considered in the form of a *recycling proportionality constant b*, which is related to the above

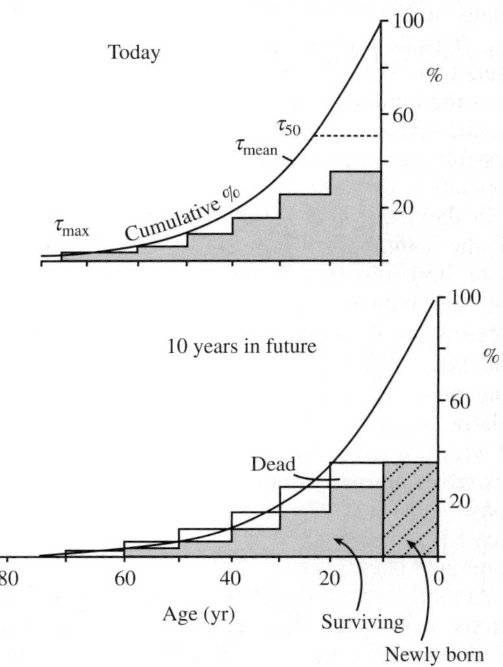

Figure 1 Simplified age distribution pattern for a steady state extant population. In this case, the natality/mortality rate is 35% of the total population for a 10-yr interval ($b = 35 \times 10^{-3}\ \text{yr}^{-1}$). τ_{max} is defined as the fifth percentile. In practice τ_{max} is the age at which the resolution of the database becomes indistinguishable from the background. The b value for the same population is inversely proportional to the available time resolution T (Equation (2)). Today's "instantaneous" rates of deposition and erosion of sediments exceed those calculated from the geological record based on time resolution of 10^6–10^7 yr (Sadler, 1981). It is therefore essential to stipulate the resolution T in consideration of rates (after Veizer and Jansen, 1985).

equation through formalism:

$$b = 1 - e^{-kT} \quad (2)$$

where T is the time resolution or duration of recycling (cf. Veizer and Jansen, 1979, 1985). In general, the larger the b value—i.e., the faster the rate of recycling—the steeper the slope of the cumulative curve and the shorter the τ_{50} and τ_{max} of the population. For a steady state extant population, generation per unit time must equal combined destruction for all age groups during the same time interval. Consequently, the cumulative slope remains the same but propagates into the future (Figure 1).

The above terminology is applicable to internally (cannibalistically) recycling populations. In an external type of recycling, the influx and efflux cause a similar age structure, but the terminology differs. In this case, the average duration an individual unit resides within the population is termed the *residence time* τ_{res}. Mathematically, τ_{res} is similar to the cannibalistic τ_{mean} and it relates to the above parameters as $\tau_{max} \geq \tau_{res} \geq \tau_{50}$. It is this alternative—populations interconnected by external fluxes—that is usually referred to as the familiar box model by natural scientists. Frequently, box models are nothing more than one possible arrangement for propagation of cyclic populations.

In the subsequent discussion, the terminology of the cannibalistic populations is employed. Note, however, that the age distribution patterns and the recycling rates calculated from these patterns are a consequence of both cannibalistic and external recycling. At this stage, we lack the data and the criteria for quantification of their relative significance. Nevertheless, from the point of view of preservation probability, it may be desirable, but not essential, to know whether the constituent units (geologic entities) have been created and destroyed by internal, external, or combined phenomena.

Among natural populations, two major deviations from the ideal pattern are ubiquitous (Pielou, 1977; Lerman, 1979). The first deviation consists of populations with excessive proportions of young units (e.g., planktonic larval stages) because their destruction rate is very high, but chances for survival improve considerably with maturation (type II in Figure 2). The mathematical formalism for such populations (e.g., Lerman, 1979) is a power-law function,

$$A_{t^*} = A_0(1 + kt^*)^{-z} \quad (3)$$

where the exponent z increases for populations with progressively larger destruction rates of young units.

The other common exception consists of populations with suppressed destruction of young units

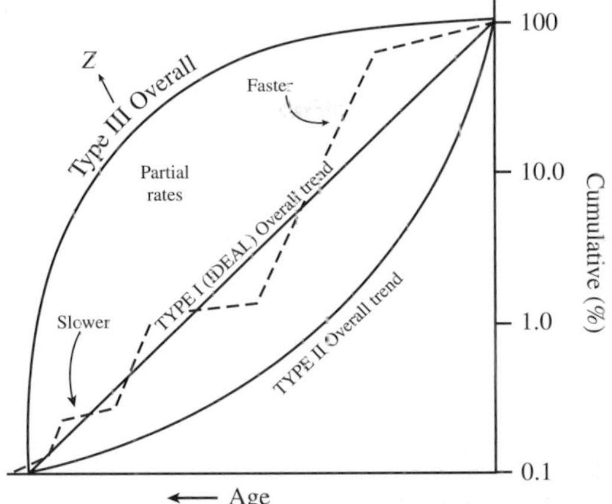

Figure 2 Cumulative age distribution functions for various populations plotted on a semilogarithmic scale (after Veizer, 1988a).

(type III, Figure 2). In these instances, the destruction rates increase rapidly as the life span τ_{max} is approached. Mathematical expression for this relation is

$$A_{t^*} = A_0[1 - (kt^*)^z] \quad (4)$$

where the exponent z increases for populations with progressively larger mortality rates of old units.

These relationships are valid for populations of constant size. For non-steady-state populations with stable age structures—i.e., those with overall rates of growth or decline much slower than the rates of recycling of their constituents—the age distributions approach the pattern of the constant-size populations. The calculated recycling rates are therefore identical. For populations where the overall growth (decline) approaches the rate of recycling of constituent units, independent criteria are required to differentiate recycling from the growth (decline) component.

The above discussion assumed a quasi-continuous generation/destruction process for a first-order system, but natural variability causes oscillations, at a hierarchy of frequencies and amplitudes, around the smooth overall patterns. Furthermore, geological processes are usually discrete and episodic phenomena. Because of all these factors, partial intervals have generation/destruction rates that deviate from the smooth average rates (Figure 2), with the connecting tangents having either shallower or steeper slopes. Note again that a given partial slope may reflect deviations in generation, in destruction, or in their combined effect. Usually, the problem is not resolvable, but the combined effect is the most likely alternative. As the population ages, the magnitude of this higher-order scatter diminishes to the level of uncertainties in the database (Figure 2). Quantitative interpretation of such higher-order features from the preserved record is therefore possible only for a length of time roughly comparable to the life-span τ_{max} of a given population. It would be pointless to attempt quantification of oceanic spreading rates from the fragmentary record of pre-Jurassic ophiolites. Any such quantification must rely on some derivative signal, such as isotopic composition of seawater, which may be preserved in coeval sediments. In contrast to the fast cycling oceanic crust, the higher-order scatter for slowly cycling populations (e.g., continental crust) remains considerable, because it has not yet been smoothed out by the superimposed recycling. Such populations still retain vestiges of ancient episodic events.

For geologic entities (e.g., crustal segments, mineral deposits, tectonic domains, and fossils), the age distribution patterns can be extracted from their stratigraphic and geochronologic assignments. At present, only major features of the record can be quantitatively interpreted, because the database is usually not of the desired reliability.

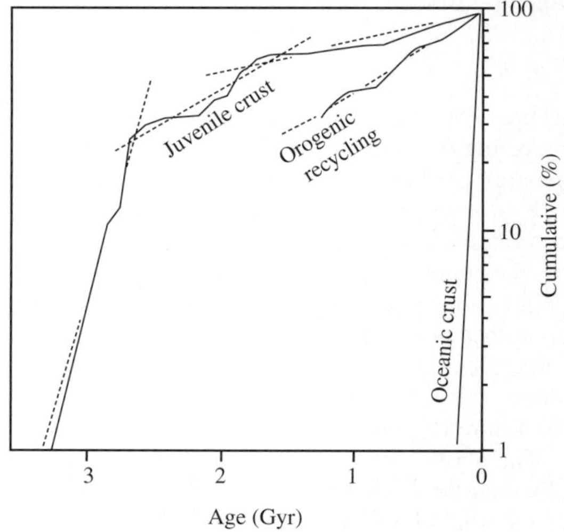

Figure 3 Cumulative age distribution of oceanic and continental crusts. Continental crust based on Condie (2001), oceanic crust after Sclater *et al.* (1981), and Rowley (2002).

7.15.3 GENERATION AND RECYCLING OF THE OCEANIC AND CONTINENTAL CRUST

The concept of global tectonics (Dietz, 1961; Hess, 1962; Morgan, 1968; Le Pichon *et al.*, 1973) combined the earlier proposals of continental drift and seafloor spreading into a unified theory of terrestrial dynamics. It introduced the notion of continual generation and destruction of oceanic crust and implied similar consequences for other tectonic realms.

The present-day age distribution pattern of the *oceanic crust* is well known (Sprague and Pollack, 1980; Sclater *et al.*, 1981; Rowley, 2002), and the plate tectonic concept of ocean floor generation/subduction well established. The age distribution pattern (Figure 3) conforms to systematics with a half-life (τ_{50}) of ~60 Myr, translating to generation/destruction of ~3.5 km^2 (or ~20 km^3) of oceanic crust per year. The maximal life-span τ_{max} (oblivion age) for its tectonic settings and their associated sediments is therefore less than 200 Myr.

The situation for continental crust is more complex and still dogged by the controversy (e.g., Sylvester, 2000) that pits the proponents of its near-complete generation in the early planetary history, followed only by crust/mantle recycling (Armstrong, 1981 and the adherents) against those advocating an incremental growth in the course of geologic evolution (see Taylor and McLennan, 1985). The latest summary of the volume/age estimates for the continental crust (Condie, 2001) provides a more definitive constraint for the discussion of the issue. The total volume of the continental crust is estimated at 7.177×10^9 km^3 (Cogley, 1984), and the age distribution pattern of its juvenile component (Figure 3) suggests a tripartite evolution, with ~25% ~4.0–2.6 Gyr old, another 35% added between ~2.6 Gyr and 1.7 Gyr, and the remaining 40% subsequently. The observed growth pattern is of a sinusoidal (logistic) type (Veizer and Jansen, 1979), with commencement of large-scale crustal generation at ~4 Gyr, accelerating growth rate that culminated in major phases of crustal generation and cratonisation during the ~2.6–1.7 Gyr time span, and a declining rate subsequently.

The above crustal generation pattern is only a minimal estimate, based on the assumption that no continental crust was recycled into the mantle. The other limiting alternative can be based on the proposition that the continental crust attained its present-day steady state ~1.75 Gyr ago. If this were the case, today's preserved post 1.75 Gyr crust is only about one-half of that generated originally, with the equivalent amount recycled into the mantle. This recycling may go hand in hand with orogenic activity that has a τ_{50} of ~800 Myr (Figure 3), a value in good agreement with the previous estimates for the low- and high-grade metamorphic reworking rates (τ_{50} of 673 Myr and 987 Myr, respectively) by Veizer and Jansen (1985). Furthermore, in the post-1.6 Ga record, the orogenic segments are composed, on average, of about equal amounts of juvenile (2.7×10^9 km^3) and recycled (2.6×10^9 km^3) crust (figure 4 in Condie, 2001). This would suggest that each orogenic episode results in incorporation of about one-half of the juvenile crust into the reworked crustal segment, with the other half being subducted into the mantle.

Considering the above scenarios as limiting alternatives, the long-term average rate of continental crust generation would be $\sim$1.7 $\pm$ 0.1 $km^3 yr^{-1}$ (1.1 $\pm$ 0.5 $km^3 yr^{-1}$ by Reymer and Schubert, 1984) for no mantle recycling and about twice that much for the alternative where about one-half of the juvenile crust contributes to the orogenic buildup of the continents, while the other half is recycled into the mantle.

7.15.4 GLOBAL TECTONIC REALMS AND THEIR RECYCLING RATES

Tectonic setting is the principal controlling factor of lithology, chemistry, and preservation of sediment accumulations in their depocenters, the sedimentary basins. The latest classification of Ingersoll and Busby (1995) assigns sedimentary basins into five major groups based on their relationship to plate boundaries (Figure 4). It groups together the basins that are associated with the divergent, interplate, convergent, transform, and hybrid settings, and recognizes 26 basin types. This classification, while it cannot take into account the entire complexity of natural systems, implies that basins associated with transform and convergent boundaries are more prone to destruction than basins associated with divergent and intraplate settings, particularly those developed on continental crust (Figure 4).

This qualitative observation is consistent with the prediction of oblivion ages for specific tectonic realms based on the concept of population dynamics (Veizer and Jansen, 1985). Theoretically, if τ_{max} is taken as the fifth percentile, the oblivion ages for specific tectonic realms should be a factor of $\sim$4.5 times the respective half-lives, but empirically, due to deviations from the ideal type I age pattern, the τ_{max} is usually some $\sim$3.0–3.5 times τ_{50}. This qualification notwithstanding, the short-lived basins are erased faster from the geologic record and the degree of tectonic diversity must be a function of time. The diversity diminishes as the given segment of the solid earth ages, and the rate of memory loss is inversely proportional to recycling rates of the constituent tectonic realms. For a steady state system, the calculated theoretical preservational probabilities are depicted in Figure 5. This reasoning shows that the realms of the oceanic domain (basins of active margins to immature orogenic belts) should have only $\leq$5% chance of survival in crustal segments older than $\sim$100–300 Myr, while the platformal and intracratonic basins can survive for billions of years.

Due to rather poor inventories, the proposed systematics should be viewed as nothing more than a conceptual framework, but it nevertheless helps to visualize the probability of preservation of sedimentary packages in the geologic record.

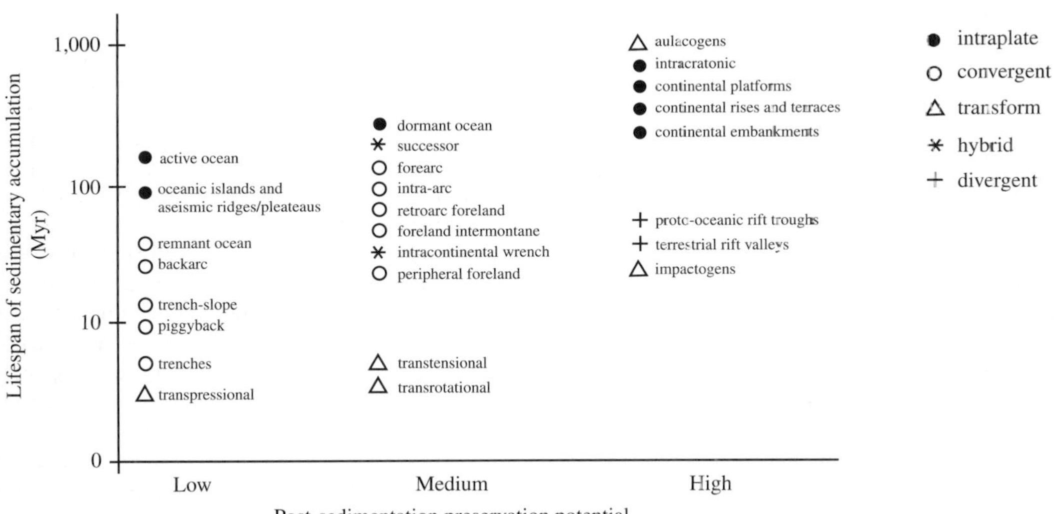

Figure 4 Typical life spans for sedimentary basins versus their post-sedimentation preservation potential. "Preservation potential" refers to average amount of time during which basins will not be uplifted and eroded, or be tectonically destroyed during and following sedimentation. Sedimentary or volcanic fill may be preserved as accretionary complexes during and after basin destruction (true of all strata deposited on oceanic crust). Basins with full circles, particularly intraplate continental margins, are "preserved" in the sense of retaining their basement, but they are likely to be subcreted beneath or within suture belts, and are difficult to recognize in the ancient record in such settings (after Ingersoll and Busby, 1995).

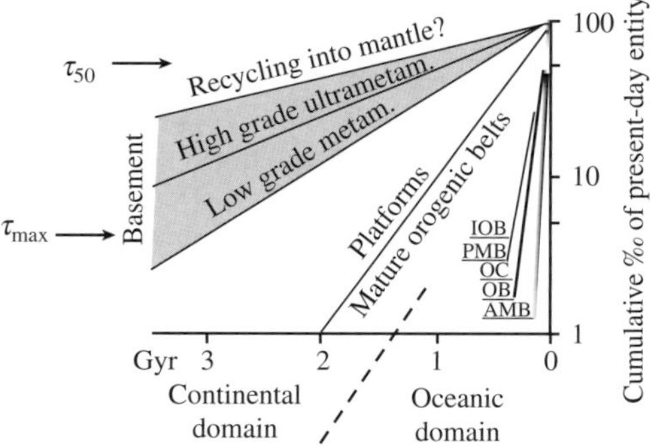

Figure 5 Preservation probabilities for major global tectonic realms. AMB = active margin basins, OB = oceanic intraplate basins, OC = oceanic crust, PMB = passive margin basins, IOB = immature orogenic belts. Rates were derived on the assumption that all deviations from the ideal pattern of the type I age distributions were a consequence of the poor quality of the available database (after Veizer, 1988b).

7.15.5 PRESENT-DAY SEDIMENTARY SHELL

The present-day mass of global sediments is $\sim 2.7 \times 10^{24}$ g (Ronov, 1982; Hay *et al.*, 2001). Of these, ~72.6% are situated within the confines of the present-day continents (orogenic belts 51.9%, platforms 20.7%), 12.9% at passive margin basins, 5.5% at active margin basins, and the sediments covering the ocean floor account for ~8.3% of the total (Ronov, 1982; Gregor, 1985; Veizer and Jansen, 1985).

The apparent decline of sedimentary thicknesses in progressively older sections (Barrell, 1917; Gilluly, 1949) is reflected also in the latest inventory of mass/age distribution of the global sedimentary mass, which declines exponentially with age (Figure 6). This exponential decline is not clearly discernible from the mass/age distribution of sediments within the confines of the continental crust (Ronov, 1993), but adding the mass of sediments presently associated with the passive margin tectonic settings (Gregor, 1985) and the sediments on the ocean floor (Hay *et al.*, 1988), the pattern clearly emerges. Hence, the preservation of the sedimentary record is a function of tectonic setting, with sediments on continental crust surviving well into the Precambrian, while the continuous record of passive margin sediments ends at ~250 Myr ago and that of the ocean floor sediments at ~100 Myr ago (Figure 6).

The original concept of recycling, as developed by Garrels and Mackenzie (1971a), was based solely on the "continental" database assembled by the group at the Vernadsky Institute of Geochemistry in Moscow (Ronov, 1949, 1964, 1968, 1976, 1982, 1993). The former authors proposed that the present-day mass/age distribution of global sedimentary mass is consistent with a half-mass age of ~600 Myr, resulting in deposition and destruction of about five sedimentary masses over the entire geologic history. Furthermore, the observed temporal relationship of clastics/carbonates/evaporites led Garrels and Mackenzie (1971a) to propose a concept of differential recycling rates for different lithologics based on their susceptibility to chemical weathering, with clastics having half-mass age of ~600 Myr, carbonates ~200 Myr, and evaporites ~100 Myr.

Subsequently, Veizer (1988a) and Hay *et al.* (1988) pointed out that the concept of differential recycling, although valid, can only partially be based on the susceptibility to chemical weathering. In a layer cake stratigraphy, the removal of carbonate strata, for instance, would necessarily result in a collapse of all the overlying strata, regardless of their lithology. On a macroscale, therefore, the sediments are removed en-masse, with chemical weathering rates coupled to the physical ones as (Millot *et al.*, 2002)

$$\text{Chem} = 0.39(\text{Phy})^{0.66} \tag{5}$$

This point of view is supported also by the fact that the particulate load accounts for ~3/4 of the present-day fluvial sediment flux (Garrels and Mackenzie, 1971a).

Indeed, the detailed consideration of temporal lithological trends does not follow the pattern anticipated from recycling based on their susceptibility to chemical weathering. For example, during the Phanerozoic, the relative proportion of carbonates increases, and of clastics decreases, with age (Figure 7). Similarly, the lithological trends of Ronov (1964) that span the entire geological time span, if recast into the present concept (Figure 8), show that it is particularly the most ancient sequences that contain the highly labile immature clastics (arkoses and graywackes). Furthermore, limestones, evaporites, and phosphorites appear to have a similar age distribution pattern, intermediate between that of the passive margin basins and platforms, while dolostones plot on a platformal trend. This suggests that the Ronov (1964) type of secular distribution of lithologies is a reflection of preservation probabilities of different tectonic settings, each having its own type of sediment assemblages.

The above view is clearly supported by the mass/age distribution of lithologies within the same tectonic domain. For example, carbonates, chert, red clay, and terrigeneous sediments on the ocean floor (Hay *et al.*, 1988) all have the same type of age distribution pattern that is controlled by a single variable, the rate of spreading and subduction of the ocean floor. This sedimentary mass also differs lithologically from its "continental" counterpart, because it is comprised of

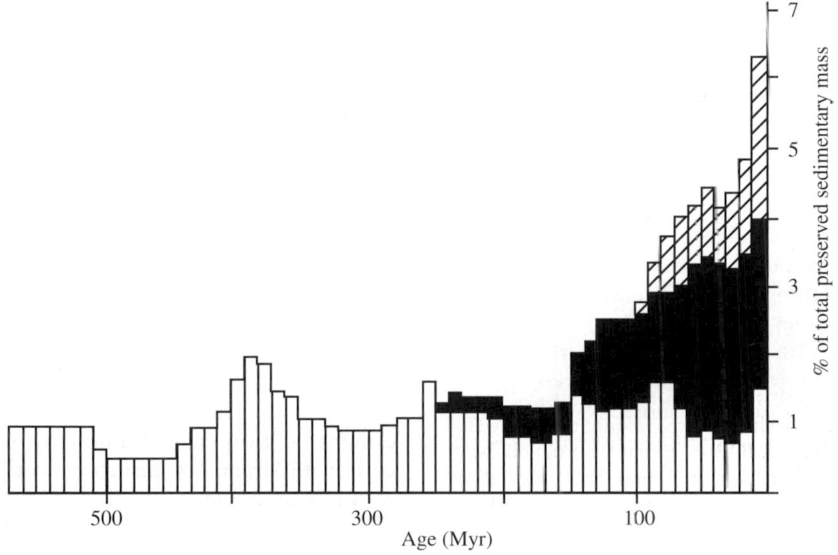

Figure 6 The mass/age distribution of preserved sedimentary mass deposited on continental crust (vertical lines), basin of passive margins (black), and on the oceanic floor (cross-hatched) (courtesy of W. W. Hay).

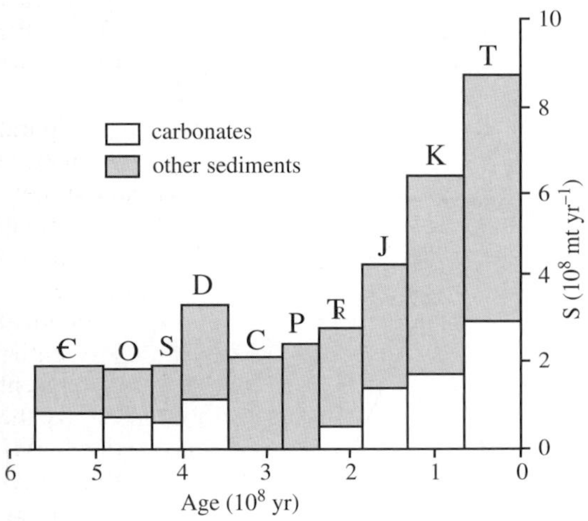

Figure 7 Surviving amounts of epicontinental terrigenous-clastic and marine-carbonates during the Phanerozoic (after Morse and Mackenzie, 1990).

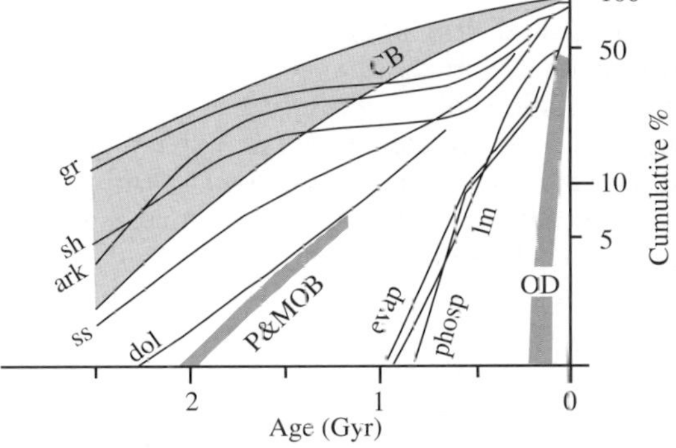

Figure 8 Observed cumulative mass/age distributions of major sedimentary lithological types. Explanation of abbreviations: gr denotes graywackes; sh, shales; ark, arkoses; ss, sandstones; dol, dolostones; evap, evaporites; lm, limestones; phosp, phosphorites; CB, continental basement; P, platforms; MOB, mature orogenic belts; OD, oceanic domain (after Veizer, 1988c).

76 wt. % terrigeneous material, 7% calcium carbonate, 10% opal, and 7% mineral bound water (Plank and Langmuir, 1998). Based on its age distribution pattern (Figure 6), the average rate of pelagic sediment subduction is $\sim 1 \times 10^{21}$ g Myr^{-1} (Hay *et al.*, 1988), an estimate in good agreement with an upper limit of $1.1 \pm 0.5 \times 10^{21}$ g Myr^{-1} for sediment subduction based on Sm/Nd isotopic constraints (Veizer and Jansen, 1985) discussed later in the text. The estimates based on direct measurements, however, suggest a sediment subduction rate of ~ 1.4 to 1.8×10^{21} g Myr^{-1} (Rea and Ruff, 1996; Plank and Langmuir, 1998), the difference likely accounted for by material entering the trenches from the adjacent accretionary wedge. This material can be scraped off the subducting slab, uplifted, eroded, and rapidly recycled (Hay *et al.*, 2001).

7.15.6 TECTONIC SETTINGS AND THEIR SEDIMENTARY PACKAGES

Is the proposition that differential recycling of sediments is controlled by tectonic settings supported by the observational data? Could this explain, for example, the paradox of increasing carbonate/clastic ratio with age during the Phanerozoic (Figure 7)?

The inventories of lithologies for basin types are sparse. Veizer and Ernst (1996) attempted to quantify the sedimentary fill of the North American Phanerozoic basins, and the results are presented in Figure 9. Due to inherent limitations of the database and its interpretation, and particularly due to inconsistencies in lithological descriptions, the sedimentary facies were grouped into three categories only: coarse clastics (sandstones, siltstones, conglomerates, and arkoses), fine clastics (shales, graywackes), and "chemical" sediments (carbonates, evaporites, and cherts). These limitations notwithstanding, the compilation shows that basins associated with immature tectonic settings, such as arc-trench systems, are filled chiefly by clastic sediments, mostly immature fine grained graywackes and shales (forearc). Foreland basins, situated usually on the continental side of the continental-margin/arch-trench system, contain a higher proportion of coarse, often mature, clastics and some chemical sediments. This is even more the case for the passive margin (continental rise, terrace, and embankment) settings. Finally, carbonate sedimentation predominates in the intracratonic settings. Preservation probability of tectonic settings can therefore explain the tendency for the average lithology shifting from clastics towards carbonates with increasing age during the Phanerozoic.

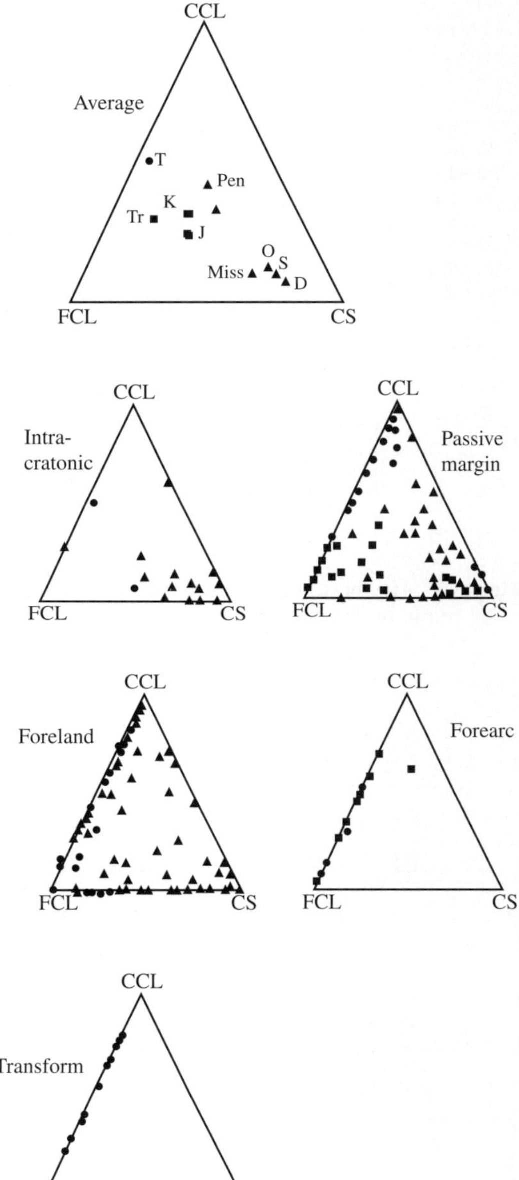

Figure 9 Relative proportions of lithological types within intracratonic to transform basins and the averages for geological periods. Based on the North American cross-sections in Cook and Bally (1975). An independent compilation by Berry and Wilkinson (1994) from the same source, but based on different criteria, yielded comparable temporal patterns. CCL coarse clastics; FCL fine clastics; CS "chemical" sediments. A global review based on the Ronov (1982) database also yields comparable patterns (after Veizer and Ernst, 1996).

7.15.7 PETROLOGY, MINERALOGY, AND MAJOR ELEMENT COMPOSITION OF CLASTIC SEDIMENTS

7.15.7.1 Provenance

The petrology of *coarse clastic* (conglomerate-size) *sediments* is controlled in the first instance by their provenance that, in turn, is a function of tectonic setting. Advancing tectonic stability is accompanied by an increasingly mature composition of the clasts (Figure 10). For first cycle sediments (Cox and Lowe, 1995), the early arc stage of tectonic evolution is dominated by volcanic clasts, from the growing and accreting volcanic pile. Subsequently, plutonic and metamorphic lithologies dominate the orogenic and uplift stages. The post-tectonic conglomerates contain a high proportion of recycled sedimentary clasts. Mineralogically, the evolution is from clasts with abundant plagioclase, to K-feldspar rich, and finally to quartz (chert) dominated clasts.

A similar provenance control relates also the *sandstone* petrology to tectonic setting, as expressed in the Q (quartz)—F (feldspar)—L (unstable lithic fragments) ternary diagrams (Figure 11) of Dickinson *et al.* (1983). Again, increasing tectonic maturity shifts the mode of sandstone petrology from L towards the F/Q tangent, terminating in the recycled Q mode.

Fine clastic sediments, mostly *mudrocks*, in contrast to their coarser counterparts, are either derived by first cycle weathering of silicate minerals or glass, or from recycling of older mudrocks. Physical comminution plays only a secondary role. The average shale is composed of ~40–60% clay minerals, 20–30% quartz, 5–10% feldspar and minor iron oxide, carbonate, organic matter, and other components (Yaalon, 1962; Shaw and Weaver, 1965). Granitic source rocks produce shales richer in kaolinite and illite, the mafic ones richer in smectites (Cox and Lowe, 1995).

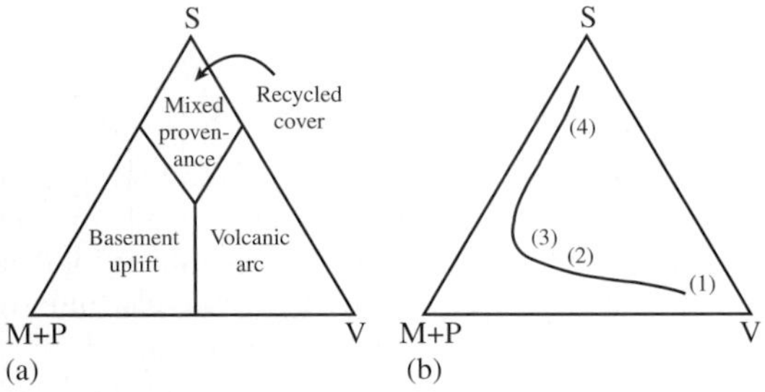

Figure 10 Idealized evolution of conglomerate clast composition. S = sedimentary clasts; M + P = metamorphic + plutonic clasts; V = volcanic clasts: (a) Approximate compositional fields for conglomerate clast populations of differing provenance. (b) Generalized stages in the evolution of conglomerate clast compositions on a crustal block: (1) volcanic arc stage; (2) dissected arc/accretion orogen stage; (3) post-tectonic granite/basement uplift stage; and (4) sediment-recycling-dominated stage (after Cox and Lowe, 1995).

Geochemical processes associated with weathering and soil formation are dominated by alteration of feldspars (and volcanic glass), feldspars accounting for 70% of the upper crust, if the relatively inert quartz is discounted (Taylor and McLennan, 1985). Advancing weathering leads to a shift towards an aluminum rich composition that can be approximated by the chemical index of alteration (CIA) of Nesbitt and Young (1984)

$$\mathrm{CIA}=100(\mathrm{Al_2O_3}/(\mathrm{Al_2O_3}+\mathrm{CaO}^*+\mathrm{Na_2O}+\mathrm{K_2O})) \quad (6)$$

The suspended sediments of major rivers clearly reflect this alteration trend and plot on a tangent between the source (upper continental crust (UCC)) and the clays, the end-products of weathering (Figure 12).

The overall outcome is a depletion of the labile Ca–Na plagioclases in the sediments of progressively more stable tectonic settings, the trend being more pronounced in the fine-grained muddy sediments than in their coarser counterparts (Figure 13).

7.15.7.2 Transport Sorting

Transport processes, involving first cycle as well as recycled components, result in further

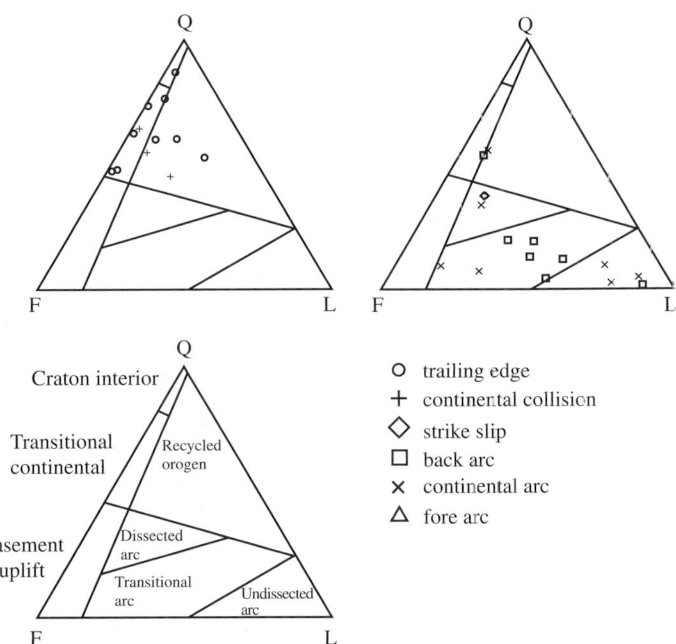

Figure 11 Ternary diagrams of framework quartz (Q)–feldspar (F)–unstable lithic fragments (L) for sands. Provenance fields from Dickinson *et al.* (1983) (after McLennan *et al.*, 1990).

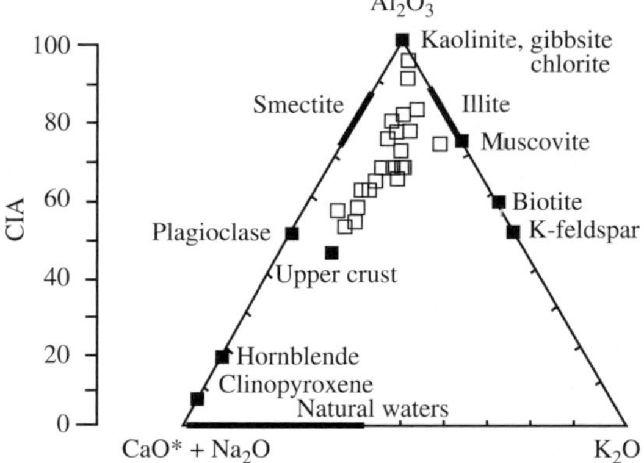

Figure 12 Ternary plots of molecular proportions of Al_2O_3–(Na_2O + CaO^*)–K_2O with the Chemical Index of Alteration (CIA) scale shown on the left. Also plotted are selected idealized igneous and sedimentary minerals and the range of typical natural waters. Squares are suspended sediments from major rivers throughout the world representing a variety of climatic regimes. CaO^* is the silicate bound concentration only (after McLennan *et al.*, 2003).

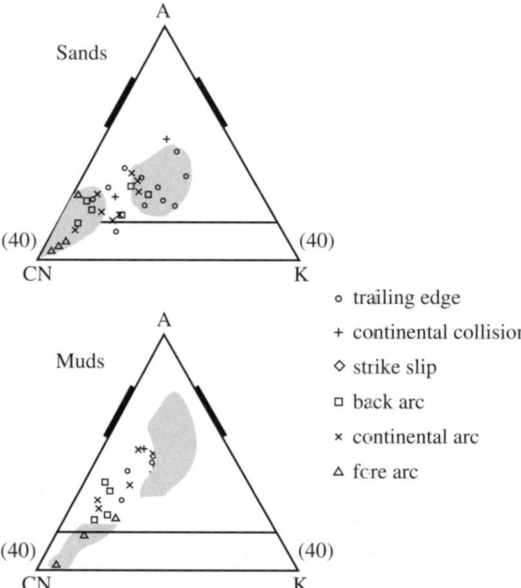

Figure 13 Ternary plots of mole fraction Al_2O_3(A)–$CaO + Na_2O$ in silicates (CN)–K_2O(K) (note that the lower part of the ternary diagrams, $A < 40$, is not shown). The plagioclase-K-feldspar join, at $A = 50$, and illite and smectite compositions (thick bars) as in Figure 12. Fields encompassing data from Fore Arc and Trailing Edge tectonic settings are shown in stippled patterns (after Mc Lennan *et al.*, 1990).

sorting by grain size and density. For *sandstone* components, the higher stability of quartz, compared to feldspar and lithic grains, results in an increasing SiO_2/Al_2O_3 ratio and a decrease in concentration of trace elements that were associated chiefly with the labile aluminosilicate minerals (McLennan *et al.*, 2003). Simultaneously, the labile nature of plagioclase relative to K-feldspar leads to a rise in the K_2O/Na_2O ratio. As for provenance, the overall shift in major element composition is towards the A–K tangent (Figure 13). More importantly, transport processes are the main factor that separates the sand- and the mud-size fractions. As for sandstones, *mudstones* also evolve towards the A/K tangent, but with increasing maturity they shift more towards the A apex of the ternary diagram (Figure 13).

7.15.7.3 Sedimentary Recycling

The processes of mechanical weathering and dissolution in recycling systems lead to the diminution of grain size of all mineralogical constituents, as well as to progressively mono-mineralic quartz-rich sediments (Cox and Lowe, 1995). Sedimentary recycling is particularly effective in redistributing the trace elements, a topic discussed in the next section.

7.15.8 TRACE ELEMENT AND ISOTOPIC COMPOSITION OF CLASTIC SEDIMENTS

As already pointed out, the suspended load of rivers falls on the tangent connecting the UCC and its clay rich weathering products (Figure 12). The chemical composition of clastic sediments reflects therefore that of the UCC, albeit depleted for those elements that are leached out, and transported as dissolved load, to be eventually concentrated in seawater and precipitated as (bio)chemical sediments. The variable residence times (τ) of these elements in seawater are a reflection of their relative mobility, with the logarithm of τ directly proportional to the logarithm of the ratio of a given element in seawater to its upper crustal abundance (Figure 14). The elements in the upper right corner (sodium, calcium, magnesium, strontium) are rapidly mobilized during sedimentary processes, while those in the lower left corner, such as titanium, zirconium, hafnium, niobium, tantalum, thorium, zirconium, nickel, cobalt, rare-earth elements (REEs) are mostly transferred from the UCC into the clastic sedimentary mass (McLennan *et al.*, 2003). It is the latter assemblage of elements, and even more so their ratios, that are useful for provenance studies. Because of their coherent behavior in geochemical processes, the REE and the elements such as thorium, zirconium, scandium, titanium... (Taylor and McLennan, 1985) are particularly suitable for tracing the ultimate provenence of clastic

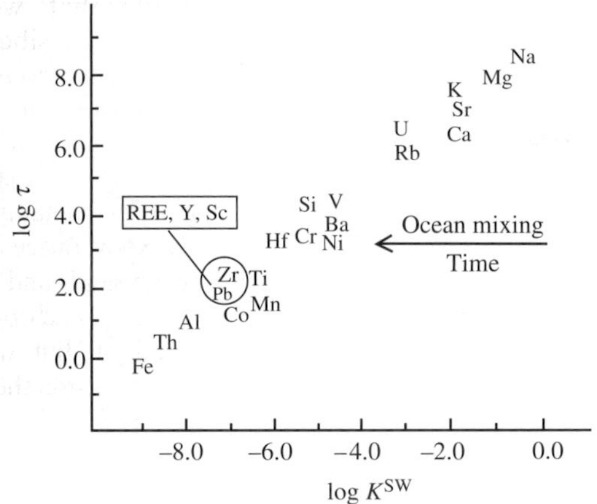

Figure 14 Plot of log τ (residence time in years) versus log K^{SW} (concentration in seawater/concentration in upper continental crust) for selected elements (after McLennan *et al.*, 2003).

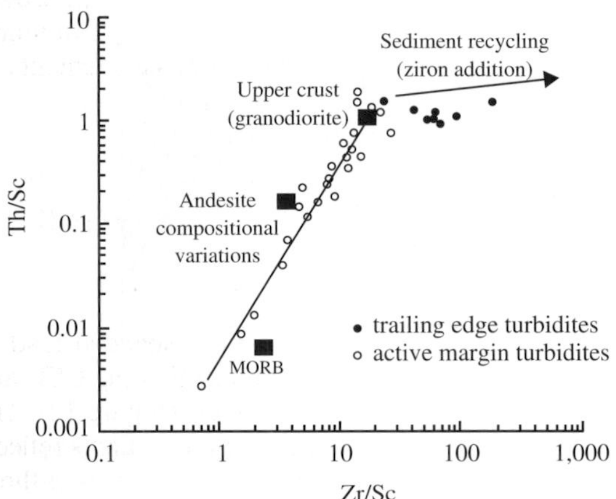

Figure 15 Th/Sc versus Zr/Sc for modern turbidites from active and passive margin setting (after McLennan *et al.*, 2003).

sediments (McLennan *et al.*, 1990). An example is given in Figure 15, where the modern active margin turbidites reflect directly the composition of the source, with Th/Sc and Zr/Sc ratios increasing in tandem with increasing igneous differentiation of the source rocks. The coherent trends break down in the trailing margin turbidites because these contain mostly recycled components of the older sediments. Repeated sedimentary recycling tends to enrich the sand-size fraction in heavy minerals. Heavy minerals, such as zircon (uranium, hafnium), monazite (thorium), chromite (chromium), titanium-minerals (ilmenite, titanite, rutile), or cassiterite (tin) are the dominant carriers of trace elements in sandstones. In trailing edge turbidites, the recycling tends to concentrate the much more abundant zircon (carrier phase of zirconium) than monazite (thorium), thus increasing the Zr/Sc but not so much the Th/Sc ratio.

Since a number of isotope systematics (U, Th/Pb, Lu/Hf, Sm/Nd) in clastic sediments is essentially controlled by the heavy mineral fraction, such considerations are of considerable importance for any geological interpretations.

Again, modern turbidites provide a classic example (Taylor and McLennan, 1985). Their Th/Sc ratio in active arc settings straddles the mafic to felsic join, with the bulk of samples reflecting the dominant andesitic component (Figure 16). Their ε_{Nd} of $\sim+5$ is that of modern oceanic crust. The turbidites in progressively more evolved tectonic settings contain increasing proportions of recycled sedimentary components, become more quartzose, have more negative ε_{Nd} and higher Th/Sc ratios. The observation that the more "evolved" tectonic settings incorporate recycled components from progressively older sources is confirmed by the neodymium model ages of these clastic sediments that increase from ~250 Myr from fore arc settings to some 1.8 Gyr for the trailing age settings (Figure 17).

7.15.9 SECULAR EVOLUTION OF CLASTIC SEDIMENTS

7.15.9.1 Tectonic Settings and Lithology

Based on the tectonic concept of differential preservational probabilities, the progressively older segments of the continental crust should retain only the remnants of the most stable tectonic settings, that is they should be increasingly composed of basement and its platformal to intracratonic sedimentary cover. This is the case throughout the Phanerozoic and Proterozoic, but not for the oldest segment, the Archean. Compared to the Proterozoic, the Archean contains a disproportionate abundance of the perishable greenstones (e.g., Windley, 1984; Condie, 1989, 2000) with mid-ocean ridge basalt (MORB) and oceanic plateau basalt (OPB) affinities and immature clastic lithologies, particularly graywackes (Figure 8). Regardless of their precise present-day analogue, these greenstones are an expression of the ephemeral oceanic tectonic domain in the sense of Veizer and Jansen (1985) (Figure 5). The temporal distribution of greenstones is, therefore, entirely opposite to that expected from continuous recycling, regardless of its actual rate. The fact that so many of them survived to this day, despite this recycling, argues for their excessive original abundance and entrainment into the growing and stabilizing continents. How is this tectonic and lithological evolution, from an oceanic to continental domain

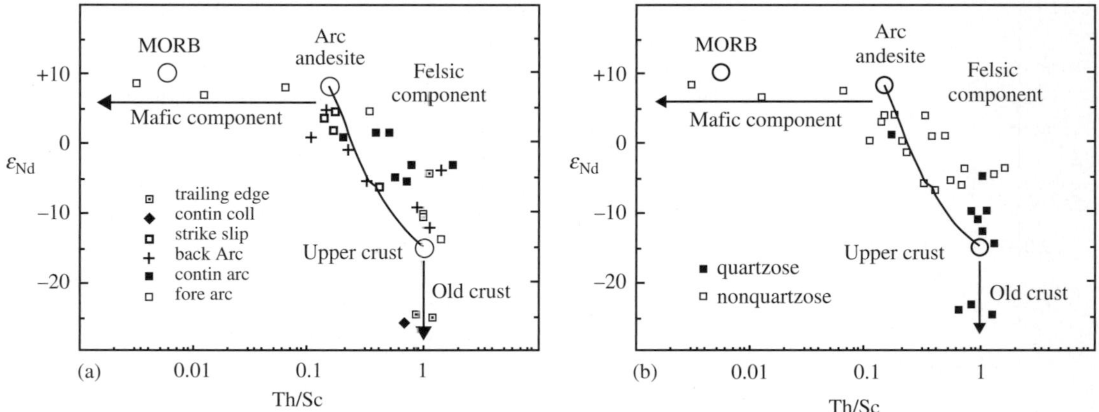

Figure 16 Plot of ε_{Nd} versus Th/Sc ratio for deep-sea turbidites according to tectonic setting of deposition (a) and according to quartz content (b). Also shown are compositions of various geochemical reservoirs (from Taylor and McLennan, 1985) and mixing relationships between average island arc andesite and upper continental crust (after McLennan *et al.*, 1990).

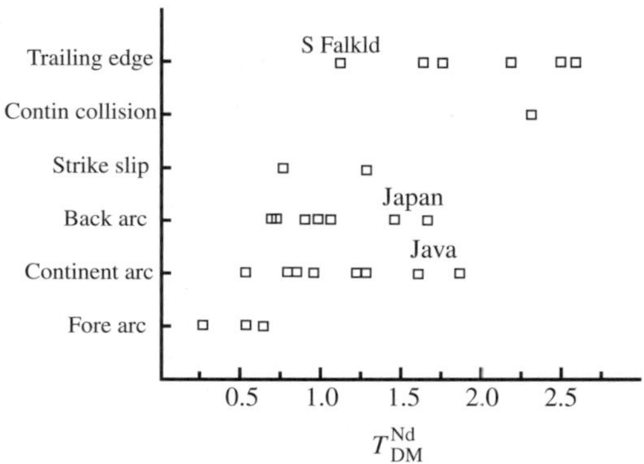

Figure 17 Neodymium model ages (relative to depleted mantle) for recent turbidites from continental margins. The large range in model ages indicates that the rate of addition of new mantle-derived material is highly variable during sedimentary recycling (after McLennan *et al.*, 1990).

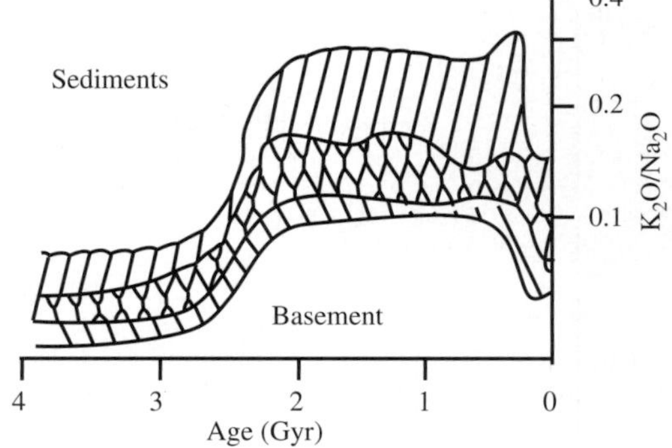

Figure 18 The K_2O/Na_2O ratio for undifferentiated "continental" sediments and their basement, based on the data of Engel *et al.* (1974) (after Veizer, 1988a).

around the Archean/Proterozoic transition, reflected in the chemistry of clastic sediments?

7.15.9.2 Chemistry

The K_2O/Na_2O ratio of sediments and continental basement increased considerably at about the Archean/Proterozoic transition (Figure 18), as already pointed out by Engel *et al.* (1974). Subsequently, Taylor (1979) and Taylor and McLennan (1985) emphasized that the REE showed similar trends in their overall concentrations, in LREE/HREE and Yb/La ratios, and in the decline of the size of the europium anomaly. The general interpretation of these data was based on the proposition that the compositions of the continental crust, and of "continental" sediments, reflect a major cratonisation event that spanned the Archean/Proterozoic transition and resulted in an UCC of more felsic nature.

This interpretation was questioned by Gibbs *et al.* (1988) and Condie (1993), who argued that the global averages for clastic rocks reflect only the variable proportions of facies associated with the predominant tectonic settings at any given time and not the change in the composition of the continental crust. This is undoubtedly the case, but—as discussed in the previous section—the types of the Archean tectonic settings are exactly the opposite to that expected from preservation probabilities based on the recycling concept. This feature must therefore reflect the fact that immature tectonic settings and lithologies were the norm in the Archean and they still dominate the preserved record, despite the high rate of recycling. Furthermore, the secular trends are present regardless whether the tectonic facies assemblages are considered together or separately (Figure 19). Combined, or separately, the Th/Sc

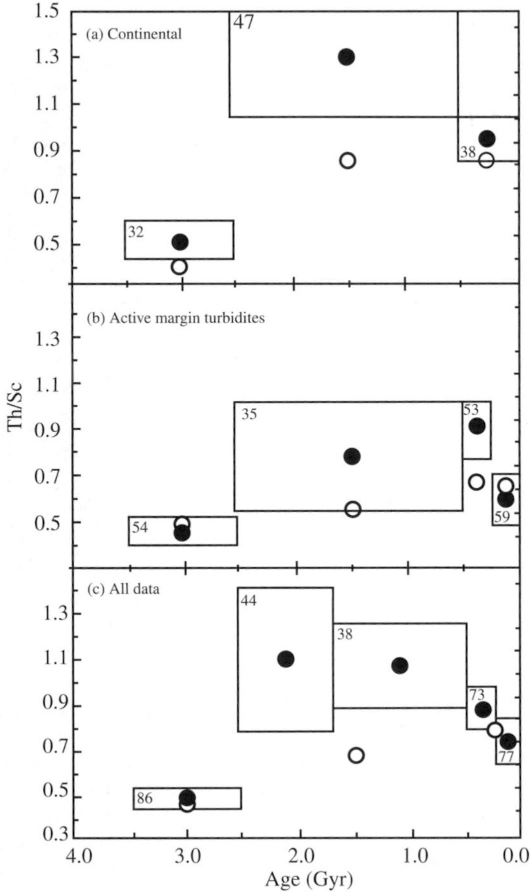

Figure 19 Secular variation in Th/Sc ratio for sedimentary rocks deposited in stable continental regions (shales) and tectonically active regions (shales/graywackes). Sample numbers are shown in or near boxes. Solid symbols and boxes are arithmetic means and 95% confidence intervals. Open circles are sediment averages reported by Condie (1993) (after McLennan *et al.*, in press).

secular trends show the same features as the K_2O/Na_2O parameter, with a rise in the ratio at the Archean/Proterozoic boundary and a reversal of the trend for the youngest portion of the record. Trends similar to that of Th/Sc one were observed also for Th, U, Th/U ratio (McLennan *et al.*, 2003) and La/Sc ratio (Condie, 1993). This is to a degree true also for the Eu/Eu^* anomaly, although the magnitude is smaller than previously believed.

The Archean clastic rocks are also often enriched in chromium, nickel, and cobalt (Danchin, 1967; Condie, 1993), but this feature may only be of regional significance, mostly for the Kaapval Craton and the Pilbara block, where it may reflect the ubiquity of komatiites in the source regions of the sediments. Nevertheless, considering the frequent discrepancies in the anticipated Ni/Cr ratios, in lower than expected MgO content and, at times, high abundance of incompatible elements (Condie, 1993; McLennan *et al.*, 1990), the ultramafic provenance is not an unequivocal explanation and some secondary processes, such as weathering, may have played a role in repartitioning of these elements.

The apparent decline of the K_2O/Na_2O and Th/Sc ratios in the youngest segment of the secular trend is due to the fact that the youngest segments contain mostly the transient immature tectonic settings with their immature clastic assemblages. These are prone to destruction with advancing tectonic maturation (Figure 9) and their preservation into the Paleozoic and Proterozoic is therefore limited.

7.15.9.3 Isotopes

The discussion of the chemistry of clastic sediments suggested an overall mafic to felsic evolution of global sediments, and presumably of UCC, their ultimate source, in the course of geologic history, with a major evolutionary step across the Archean/Proterozoic transition. The response of isotopes to this evolutionary scenario can best be gauged by consideration of the REE isotope systematics, such as the Sm/Nd and Lu/Hf.

The major fractionation of REE is accomplished during igneous differentiation of rocks from the mantle, resulting in lower parent to daughter ratios of the crustal products for both, the $^{147}Sm/^{143}Nd$ (McCulloch and Wasserburg, 1978) and the $^{176}Lu/^{177}Hf$ (White and Patchett, 1982) systematics. Although exceptions do exist (e.g., McLennan *et al.*, 2003; Patchett, 2003), the subsequent igneous, metamorphic, and sedimentary history of the rocks usually does not affect their inherited parent/daughter ratios. For crustal rocks, including clastic sediments, it is therefore possible to calculate the time when the original material "departed" from the mantle, the latter approximated by the chondritic uniform reservoir (CHUR) evolutionary trend (Figure 20). The intercept with the CHUR is the model age. A similar reasoning applies also to the Lu/Hf systematics, with one exception. In clastic sediments, the usual carrier phase for hafnium is the heavy mineral zircon that tends to be fractionated into the sand fraction by the processes of sedimentary recycling. Mature sands therefore may contain an "excess" of hafnium and low Lu/Hf ratios (Patchett, 2003).

When model ages are plotted against stratigraphic ages of clastic sediments it becomes clear that the former are at best similar to, but mostly higher than their stratigraphic ages (Figure 21). Furthermore, the discrepancy increases from the Archean to today, from an average "excess" of ~250 Myr to ~1.8 Gyr. This is true for both isotope systematics and for muds as well as

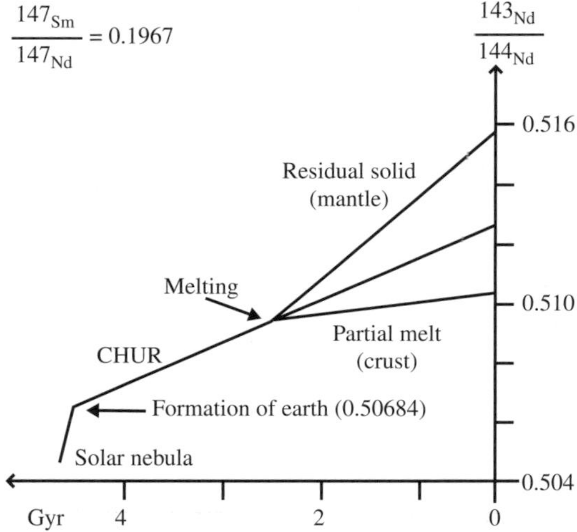

Figure 20 Theoretical evolution of Sm/Nd system in the course of planetary evolution.

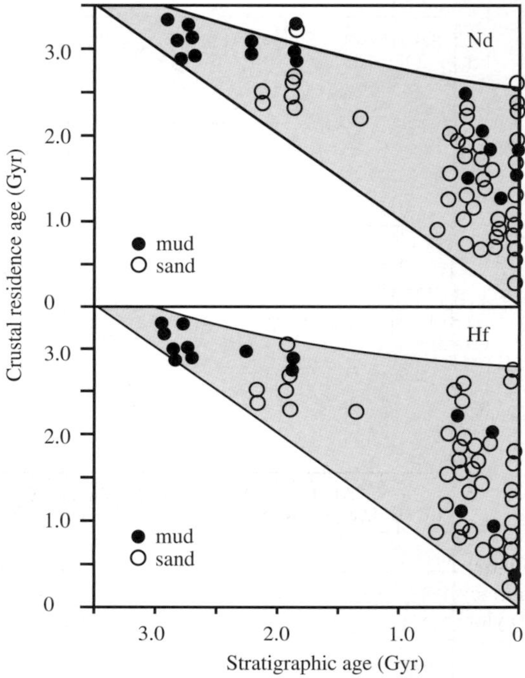

Figure 21 Stratigraphic age versus Nd- and Hf-crustal residence ages. Model ages were calculated using linear ε evolution from 0 to +10 for Nd and 0 to +16 for Hf, from 4.56 Gyr to present. The similarity of the model age systematics underscores the overall coherent behavior of the Sm–Nd and Lu–Hf isotopic systems in the sedimentary environment (after Vervoort *et al.*, 1999).

sands. These excess ages are usually interpreted as "crustal residence ages" (O'Nions *et al.*, 1983; Allègre and Rousseau, 1984), with an implication that they reflect the evolution of the ultimate crystalline crustal source. The average present-day modal excess of ~1.8 Gyr (Figure 21) is indeed a measure of the average ultimate provenance and thus of the mean age of the continental crust. In reality, however, sedimentary recycling is much faster than the metamorphism/erosion of the crystalline basement (Figure 5) and the excess model ages are therefore a consequence of the cannibalistic recycling of the ancient sedimentary mass (Veizer and Jansen, 1979, 1985). These relationships can be utilized for evaluation of the degree of cannibalistic recycling.

7.15.10 SEDIMENTARY RECYCLING

Sedimentary accumulations were ultimately derived from disintegration of the UCC and the global sedimentary mass should therefore have a chemical composition comparable to this part of the crust. This indeed is mostly the case (Figure 22). Compared with the UCC, the composition of present-day average global sediments (AS) for most elements does not deviate more than ±50% from that of the UCC. Exceptions are the enrichments in boron, calcium, vanadium, chromium, iron, cobalt, and nickel, and the depletion in sodium. In addition, the sediments (Goldschmidt, 1933; Rubey, 1951; Vinogradov, 1967; Ronov, 1968) are strongly enriched in excess volatiles and have a higher oxidation state.

The anomalous enrichment in calcium and depletion in sodium are a consequence of hydrothermal exchange between the ocean floor and seawater, processes discussed later in the text. Hydrothermal processes can also account for most of the other elements enriched in sediments. Vanadium, chromium, cobalt, nickel plus tin are even more enriched in the sediments of the ocean floor (global subducting sediment (GLOSS)) than the AS (Figure 22) due to stronger impact of the hydrothermal systems (Plank and Langmuir, 1998). Nevertheless, the general overall absence of large anomalies in the normalized average composition of sediments suggests that the exogenic ± endogenic inputs and sinks for most major elements—except possibly calcium and sodium—were balanced throughout most of geologic history, a proposition supported by the fact that the average composition of the subducting sediments (GLOSS; Plank and Langmuir, 1998) is also approaching that of the continental crust (Figure 22).

Because the continental sedimentary mass accounts for the bulk of the present-day global sediments, the growth of the global sedimentary mass should be a function of the growth of the

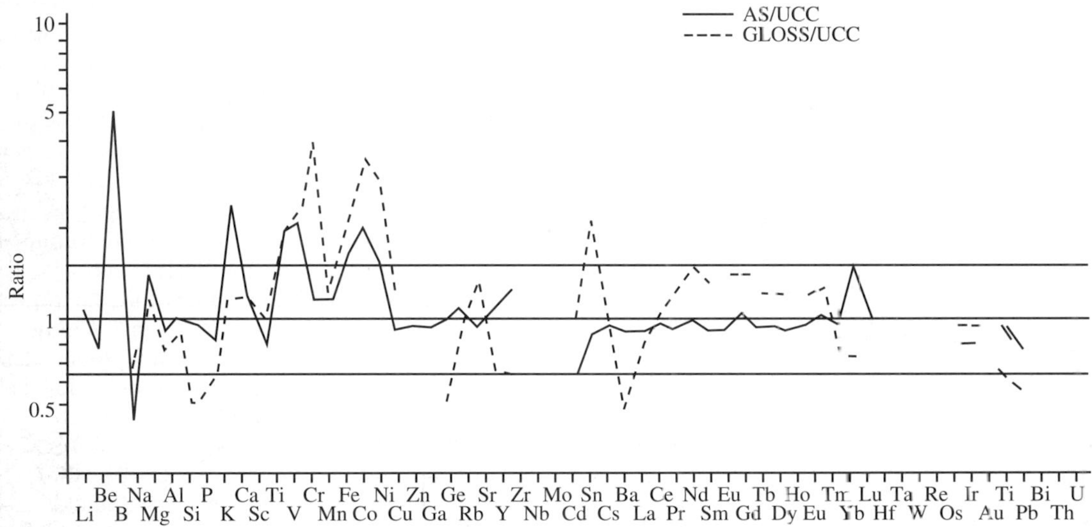

Figure 22 The elemental composition of average global sediment (AS) and global subducting sediment (GLOSS) normalized to upper continental crust (UCC). AS and UCC from McLennan and Murray (1999) and McLennan and Taylor (1999), respectively. Ti, Nb, Cs, and Ta are corrected as proposed in Plank and Langmuir (1998). GLOSS from Plank and Langmuir (1998).

continents. It is feasible, therefore, that at least some of this mass, particularly during the early stages of the earth's history and at the beginning of each tectonomagmatic cycle, evolved on oceanic or intermediate type crust. If so, it would have been derived from a source more mafic than the present-day UCC. In an entirely cannibalistic (closed) recycling system, this composition would have been perpetuated indefinitely regardless of the nature of the later continents.

In reality we must be dealing with a partially open system, because some sediments are being subducted while others are being formed at the expense of primary igneous and metamorphic rocks. Estimates based on Sm/Nd systematics (Figure 23) indicate that the sedimentary cycle is ~90 ± 5% cannibalistic, attaining its near present-day steady state around the Archean/Proterozoic transition. The *first-order* features of secular trends, such as the K_2O/Na_2O, Th/Sc (Figures 18 and 19), REE, U, Th, and Th/U (Collerson and Kamber, 1999; McLennan *et al.*, 2003) can be explained, provided the Archean was the time when the sedimentary mass was mostly growing by addition of the first-cycle sediments from erosion of contemporaneous young (≤250 Myr old) igneous precursors. Subsequent to the large-scale cratonization events, and subsequent to establishment of a substantial global sedimentary mass at ~2.5 ± 0.5 Gyr, cannibalistic sediment—sediment recycling became the dominant feature of sedimentary evolution. The general absence of neodymium model ages much in excess of their stratigraphic ages in most Archean sediments (Figure 23) is consistent with the absence of the inherited old detrital components. This observation strongly argues against the presence of large continental landmasses prior to ~3 Gyr.

7.15.11 OCEAN/ATMOSPHERE SYSTEM

The previous discussion dealt with the solid earth component of the exogenic cycle, a system that is recycled on 10^6-10^9 yr timescales. In contrast, the ocean, atmosphere, and life are recycled at much faster rates and the continuity of the past record is lost rapidly. For quantitative evaluations we have to rely therefore on proxy signals embedded in marine (bio)chemical sediments.

Other contributions in this Treatise (Volumes 6–8) deal with the lithological and chemical aspects of evolution of specific types of (bio)-chemical sediments, such as cherts, phosphorites, hydrocarbons, and evaporites, and we will therefore concentrate on the most ubiquitous category, the carbonate rocks (see also Section 7.15.12). This carrier phase also contains the largest number of chemical and isotopic tracers.

As already pointed out in Figure 7, the relative proportion of carbonate rocks within the continental realm generally decreases in the course of the Phanerozoic, and the Mesozoic and Cenozoic "deficiency" of carbonates was attributed to a tectonic cause, the ubiquity of transient immature tectonic settings. Another reason is the general northward drift of continents, which resulted in a progressive decline in the shelf areas that fell within the confines of the tropical climatic belt

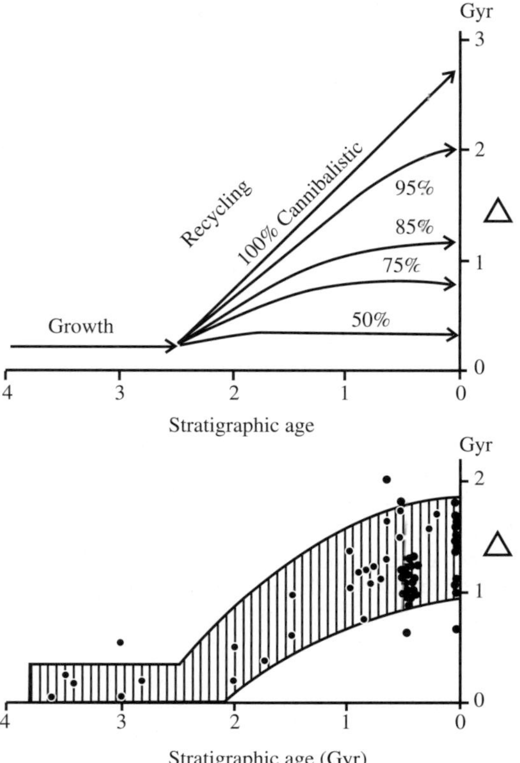

Figure 23 Models of Sm/Nd excess ages for sedimentary rocks. The Sm/Nd systematics dates the time of fractionation from the mantle. Regardless of whether most or only some of the sediments were generated during early terrestrial history, they would inherit Sm/Nd systematics from their igneous precursors. In a cannibalistic sedimentary recycling, these ancient systematics will be perpetuated and, as a consequence, Sm/Nd of all second-stage sediments should exceed their stratigraphic ages (Figure 21) and the Δ (Sm/Nd model age minus stratigraphic age) should increase toward the present, with a 45° slope being an upper limit for a completely closed system. In order to generate the observed smaller Δ, it is necessary to add sediments formed from a young source. The upper figure represents model calculations based on the assumption that prior to 2.5 Ga the sedimentary mass was growing through addition of first-cycle sediments. The post-Archean evolution assumes cannibalistic recycling of the steady state mass, and the slopes represent the degree of cannibalism for this recycling. The bottom part is a collation of experimental data (after Veizer and Jansen, 1985).

(Walker *et al.*, 2002; Bluth and Kump, 1991; Kiessling, 2002). Finally, in the course of the Mesozoic and Cenozoic, the locus of carbonate sedimentation migrated from the shelves to the pelagic realm, mirroring the role that calcareous shells of foraminiferans, pteropods, and coccolithophorids commenced to play in the carbonate budget (Kuenen, 1950). This environmental shift may have been accompanied by the deepening of the carbonate compensation depth (Ross and Wilkinson, 1991), which may have enlarged the oceanic areas that were sufficiently shallow for preservation of pelagic carbonates. Another feature of carbonate sedimentation is the relative scarcity of dolostones in Cenozoic and Mesozoic sequences, compared to their Paleozoic and particularly Proterozoic counterparts (Chilingar, 1956; Veizer, 1985). Carbonate sediments as such are mostly associated with low latitude sedimentary environments, but for the Phanerozoic the dolostones/total carbonate ratio within the tropical belt increases polewards (Berry and Wilkinson, 1994), that is towards arid climatic zones. This suggests that the process of dolomitization is a near-surface phenomenon. The more or less consistent offset of the modes of $\delta^{18}O$ between dolostones and limestones, of ~2–3‰, throughout the entire geological history (Shields and Veizer, 2002) is also consistent with such an interpretation, because the observed $\Delta^{18}O_{dol\text{-}cal}$ is a near-equilibrium value (Land, 1980). These carbonates are therefore either primary marine precipitates, or more likely they are early diagenetic products of stabilization and dolomitization of carbonate precursors. Pore waters at this stage were still in contact with the overlying seawater and/or contained an appreciable seawater component. The high frequency of dolostones in ancient sequences may again be principally a reflection of the ubiquity of shelf, epicontinental, and platformal tectonic settings preserved from the Paleozoic and Proterozoic times. Changes in seawater chemistry, such as Mg/Ca ratio, saturation state, or SO_4 content, may have been a complementary factor.

7.15.11.1 The Chemical Composition of Ancient Ocean

Earlier studies (e.g., Holland, 1978, 1984) assumed that the chemical composition of seawater during the Phanerozoic was comparable to the present-day one. Subsequently, experimental data on fluid inclusions in halite (Lowenstein *et al.*, 2001; Horita *et al.*, 1991, 2002) and on carbonate cements (Cicero and Lohmann, 2001) suggested that at least the magnesium, calcium, and strontium, and their ratios, in Phanerozoic seawater may have been variable (see Chapter 6.21). Steuber and Veizer (2002) assembled a continuous record of Sr/Ca variations for the Phanerozoic oceans (Figure 24) that covaries positively with the "accretion rate of the oceanic crust" (Gaffin, 1987) and negatively with the less well-known Mg/Ca ratio. Such a covariance would suggest that we are dealing with coupled phenomena and they proposed that all these variables are ultimately driven by tectonics, specifically seafloor spreading rates that, in turn, control the associated hydrothermal and

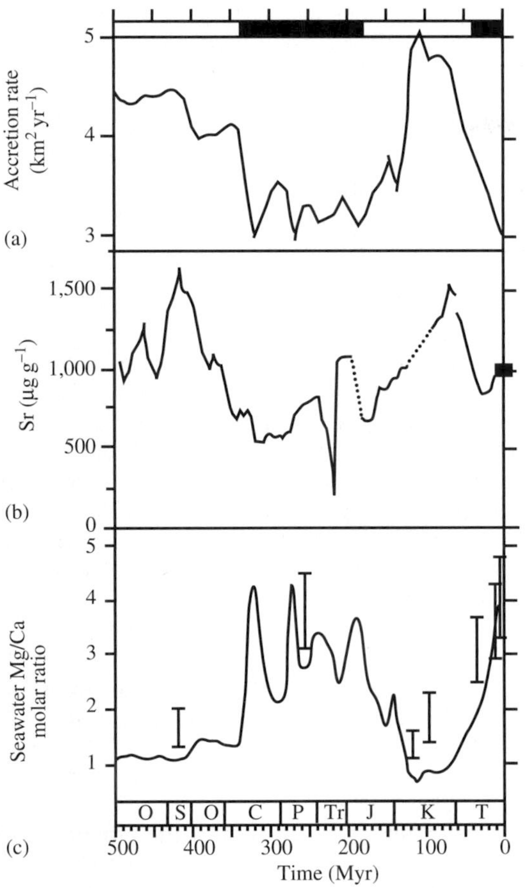

Figure 24 (a) "Accretion rate of oceanic crust" (Gaffin, 1987); (b) trends in Sr concentrations in biological LMC (Steuber and Veizer, 2002; Lear *et al.*, 2003); and (c) Mg/Ca ratio of seawater. Vertical bars are based on experimental data from fluid inclusions in halite (Lowenstein *et al.*, 2001), while the trend line results from model calculations (Stanley and Hardie, 1998). The bars at the top represent "calcitic" (blank) and "aragonitic" (black) seas (Sandberg, 1983) (after Steuber and Veizer, 2002).

low-temperature alteration processes. Since the hydrothermal alteration of young oceanic crust effectively exchanges magnesium for calcium (see Chapter 6.07), the accretion of the oceanic crust would modulate the Mg/Ca ratio of seawater. At high accretion rates, the low Mg/Ca ratio favors precipitation of calcite and, as a result, higher retention of strontium in seawater. At slow accretion rates, the high Mg/Ca ratio favors aragonite precipitation and a high rate of strontium removal from seawater.

The above scenario is consistent with the observation that calcite was the dominant mineralogy of carbonate skeletal components in the early to mid-Paleozoic and the mid-Jurassic to mid-Tertiary (Figure 25), the times of high Sr/Ca and low Mg/Ca ratio (Figure 24). Aragonite mineralogy, however, dominated the mid-Carboniferous to Jurassic and the Tertiary to Quaternary intervals (Sandberg, 1983; Kiessling, 2002) with opposite chemical attributes. The changing Mg/Ca ratio of seawater can also be at least in part responsible for the general scarcity of $MgSO_4$-bearing potash minerals (Hardie, 1996) in the Paleozoic and Mesozoic marine evaporites.

While all the above trends and their correlations are likely real, the proposed causative mechanism is being questioned lately due to the proposition that seafloor spreading rates have been about constant since at least 180 Myr (Rowley, 2002). If so, the sea-level stands cannot be inverted into "accretion rates of oceanic crust," as done by Gaffin (1987). A causative mechanism for all these covariant phenomena remains therefore enigmatic.

The reconnaissance studies of fluid inclusions (Horita *et al.*, 1991, 2002) suggest also that early Paleozoic seawater was ~2.5 × depleted in SO_4, compared to its present-day counterpart. From model considerations, based on the mineralogy and volume of evaporites, claims have been made also for changes in the potassium concentration of Phanerozoic oceans (Hardie, 1996), and for an increase in the total salinity, from the modern 35 ppt to ~50 ppt in the Cambrian (Hay *et al.*, 2001). The experimental confirmation for all these theoretical assertions is presently not available (see Chapter 6.21).

On timescales of billion of years, ancient Precambrian carbonates appear to have been enriched in Fe^{2+} and Mn^{2+}, if compared to their Phanerozoic counterparts (Veizer, 1985). In part, this may be a reflection of diagenetic alteration processes that tend to raise the iron and manganese contents of successor phases (Brand and Veizer, 1980; Veizer, 1983). However, the Archean manganese concentrations, in the 10^3–10^4 ppm range, are likely not explained by diagenetic processes alone. Accepting that the redox state of the Archean and early Proterozoic oceans may have been lower than that of their Phanerozoic counterparts (Cloud, 1976), the high Fe^{2+} and Mn^{2+} content of contemporaneous carbonates may reflect higher concentrations of these elements in the ancient oceans.

7.15.11.2 Isotopic Evolution of Ancient Oceans

In contrast to chemistry, where the secular trends are still mostly obscured by the natural scatter in the database, the isotope evolution of seawater is better resolved.

7.15.11.2.1 Strontium isotopes

In modern oceans the concentration of strontium is ~8 ppm and its residence time is ~4–8 Myr

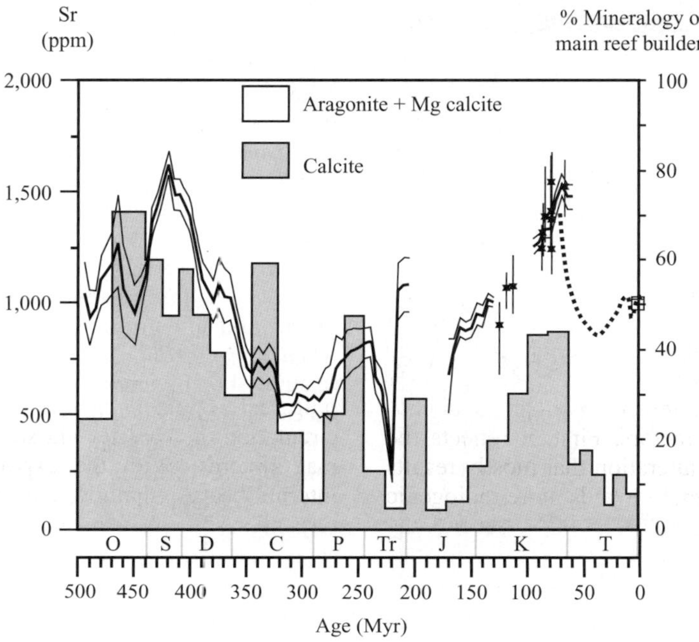

Figure 25 Predominant skeletal mineralogy, calcite (shaded) versus aragonite + high-Mg calcite (blank), in Phanerozoic reefs. Mineralogy from Kiessling (2002), Sr data from Steuber and Veizer (2002) and Lear *et al.* (2003) (see also Figure 24) (courtesy of T. Steuber).

(Holland, 1984). The present isotopic ratio $^{87}Sr/^{86}Sr$ is 0.7092 (McArthur, 1994), controlled essentially by two fluxes, the "mantle" and the "river" flux. The former represents strontium exchanged between seawater and oceanic crust ($^{87}Sr/^{86}Sr$ ~0.703) in hydrothermal systems on the ocean floor. The latter, reflecting the more fractionated composition of the continental crust, feeds into the oceans more radiogenic strontium, with an average isotope ratio for rivers of ~0.711 (Wadleigh *et al.*, 1985; Goldstein and Jacobsen, 1987; Palmer and Edmond, 1989). Note, nevertheless, that the latter may vary from 0.703 to 0.730 or more, depending on whether the river is draining a young volcanic terrane or an old granitic shield. The third input is the flux of strontium from diagenesis of carbonates, which results in expulsion of some strontium from the solid phase during precursor to product (usually aragonite to low-magnesium calcite) recrystallization, but this flux is not large enough to influence the isotopic composition of seawater. A simple balance calculation based on isotopes, therefore, shows that strontium in seawater originates ~3/4 from the "river" flux and ~1/4 from the "mantle" flux, generating the modern value of 0.7092. This value is uniform with depth and into marginal seas. Even water bodies such as Hudson Bay, with a salinity ~1/2 of the open ocean due to large riverine influx, have this same isotope ratio. This is because the rivers are very dilute relative to seawater, with strontium concentrations usually 1,000 times less, and their impact is felt only if the proportion of seawater in the mixtures is less than 10%.

The above considerations show that the strontium isotopic composition of seawater is controlled essentially by tectonic evolution, that is, by relative contributions from weathering processes on continents and from the intensity of submarine hydrothermal systems. Over geological time, however, the isotopic compositions of these two fluxes have evolved, because ^{87}Sr is a decay product of ^{87}Rb:

$$\left(\frac{^{87}\mathrm{Sr}}{^{86}\mathrm{Sr}}\right)_\mathrm{p} = \left(\frac{^{87}\mathrm{Sr}}{^{86}\mathrm{Sr}}\right)_\mathrm{o} + \left(\frac{^{87}\mathrm{Rb}}{^{86}\mathrm{Sr}}\right)(e^{\lambda t} - 1) \quad (7)$$

where p = present, o = initial ratio at the formation of the Earth 4.5 Gyr ago (0.699), λ = decay constant ($1.42 \times 10^{-11}\ \mathrm{yr}^{-1}$), and t = time since the beginning (e.g., formation of the Earth 4.5 Gyr ago).

From Equation (7) it is evident that the term $(^{87}Sr/^{86}Sr)_p$ for coeval rocks originating from the same source $(^{87}Sr/^{86}Sr)_o$ depends only on their Rb/Sr ratios. Since this ratio is ~6 times larger for the more fractionated continental rocks than for the basalts (~0.15 to 0.027; Faure, 1986), the $^{87}Sr/^{86}Sr$ of the continental crust at any given time considerably exceeds that of the mantle and oceanic crust, increasing to the present-day values of ~0.730 for the average continental crust, as opposed to ~0.703 for the oceanic crust. The rivers draining the continents are less radiogenic

(~0.711) than the crust itself due to the fact that most riverine strontium originates from the weathering of carbonate rocks rather than from their silicate counterparts. The former, as marine sediments, inherited their strontium from seawater, which—as discussed above—contains also the less radiogenic strontium from hydrothermal sources.

The lower envelope of the strontium isotopic trend during the *Precambrian* (Figure 26) straddles the mantle values until about the Archean/Proterozoic transition, afterwards deviating towards more radiogenic values, reflecting the input from the continental crust. The large spread of values above the lower envelope is a consequence of several factors. First, it reflects the impact of secondary alteration that mostly results in resetting of the signal towards more radiogenic values (Veizer and Compston, 1974). Second, the scatter also includes higher order oscillations in the strontium isotope ratio of seawater which cannot be as yet resolved for the Precambrian due to poor stratigraphic resolution and inadequate geochronological control.

From the above discussion, it is clear that the primary control on strontium isotopic composition of the first order (billion years trend) for seawater will be exercised by the growth pattern of the continental crust. Two competing hypotheses dominate this debate:

(i) The generation of the entire continental crust was an early event and the present-day scarcity of older remnants is a consequence of their destruction (recycling) by subsequent tectonic processes (Armstrong, 1981; Sylvester *et al.*, 1997).

(ii) The continents were generated episodically over geologic history, with major phases of continent formation in the late Archean and early Proterozoic, and attainment of a near modern extent by ~1.8 Gyr (Veizer and Jansen, 1979; Taylor and McLennan, 1985; McCulloch and Bennett, 1994).

Model calculations by Goddéris and Veizer (2000) of seawater $^{87}Sr/^{86}Sr$ evolution for the two alternatives, and for the coeval mantle, show that the measured experimental data fit much better with the second pattern of continental growth. Note that this scenario is also fully compatible with the evolution of sediments and their chemistry discussed in the preceding sections of this chapter. The Archean oceans were "mantle buffered" (Veizer *et al.*, 1982) by vigorous circulation of seawater via submarine hydrothermal systems. With the exponential decline of internal heat dissipation, the vigor of the hydrothermal system also declined and at the same time the flux of radiogenic strontium from growing continents brought in by rivers started to assert itself. This tectonically controlled transition from "mantle" to "river buffered" oceans across the Archean/Proterozoic transition is a first order feature of terrestrial evolution, with consequences for other isotopic systematics, for redox state of the ocean/atmosphere system and for other related phenomena (see Chapter 6.21).

The resolution of the database is considerably better for the *Phanerozoic* than for the Precambrian due to higher quality of samples and to much better biostratigraphic resolution, with duration of biozones from ~1 Myr in the Cenozoic to ~5 Myr in the early Paleozoic. The first data documenting the strontium isotopic variations in Phanerozoic seawater were published by Peterman *et al.* (1970), with subsequent advances by Veizer and Compston (1974) and Burke *et al.* (1982).

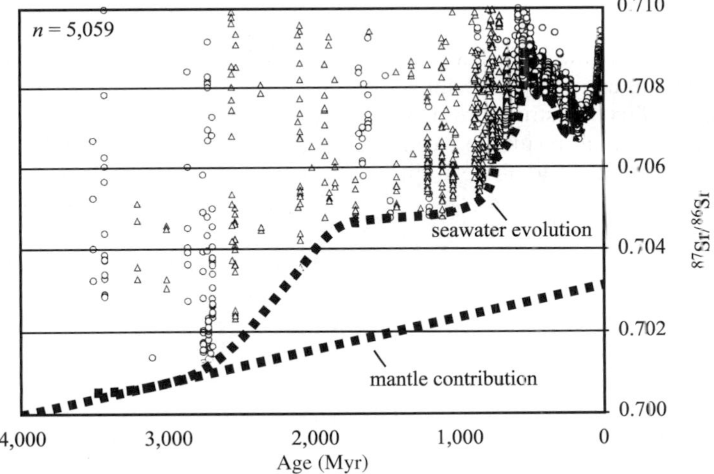

Figure 26 Strontium isotopic composition of sedimentary carbonate rocks during geologic history. Reproduced from Shields and Veizer (2002). Circles and triangles represent samples with good and poor age assignment, respectively.

The latest version, by Veizer *et al.* (1999), is reproduced in Figure 27. Overall, this second order Phanerozoic trend shows a decline in $^{87}Sr/^{86}Sr$ values from the Cambrian to the Jurassic, followed by a steep rise to today's values, with superimposed third order oscillations at 10^7 yr frequency. Because of the better quality of samples, the experimental data indicate an existence of still higher order $^{87}Sr/^{86}Sr$ oscillations within biozones. However, since these samples often do not originate from the same profile, their relative ages within a biozone are difficult to discern. They have to be treated therefore as coeval and the secular trend thus becomes a band (Figure 28).

In general, it is again tectonics that is the cause of the observed Phanerozoic trend, with the "mantle" input of greater relative importance at times of the troughs and the "river" flux dominating in the Tertiary and early Paleozoic. Nevertheless, it is difficult to correlate the overall trend or the superimposed oscillations with specific tectonic events. The problem arises from the fact that model solutions do not produce unique answers. The "river" flux is likely the major reason for the observed $^{87}Sr/^{86}Sr$ oscillations, because the changes in seafloor spreading rates, apart from being disputed (Rowley, 2002), are relatively sluggish and the strontium isotope ratio of the "mantle" flux is relatively constant at ~0.703. The "river" flux, however, may vary widely in both strontium elemental flux and its isotope ratio. For example, the rapid Tertiary rise in $^{87}Sr/^{86}Sr$ (Figure 27) is commonly interpreted as reflecting the uplift of the Himalayas. Accepting this to be the case, it still remains an open question whether the rise is due to higher flux of "river" strontium (increased weathering rate), its more radiogenic nature (unroofing of older core complexes), or both (cf. the contributions in Ruddiman, 1997). For these reasons, it is difficult to utilize the strontium isotopic curve of seawater as a direct proxy for continental weathering rates in model considerations. Nevertheless, it is intriguing that the Phanerozoic seawater strontium isotope curve correlates surprisingly well with the estimated past sediment fluxes (Hay *et al.*, 2001) that were reconstructed by the "population dynamics" approach discussed in the introductory section of this chapter.

7.15.11.2.2 Osmium isotopes

The isotope ^{187}Os is generated by β decay of ^{187}Re. In many ways the systematics and the presently known secular evolution of $^{187}Os/^{186}Os$ in seawater is similar to that of $^{87}Sr/^{86}Sr$ (Peucker-Ehrenbring and Ravizza, 2000). The present-day $^{187}Os/^{186}Os$ of UCC is ~1.2–1.3, runoff ~1.4 (due likely to preferential weathering of radiogenic black shales), seawater, ~1.06 and that of meteorites and mantle, ~0.13. In contrast to strontium with a seawater residence time of 4–8 Myr, the residence time of osmium is on the order of 10^4 yr due to the effective scavenging of osmium by Fe/Mn crusts and organic-rich sediments. This enables tracing of short-term fluctuations in seawater composition, such as the Quaternary glacial/interglacial cycles, a feat difficult to replicate by the buffered strontium system.

Metalliferous sediments usually have low Re/Os ratios and reflect well the isotopic

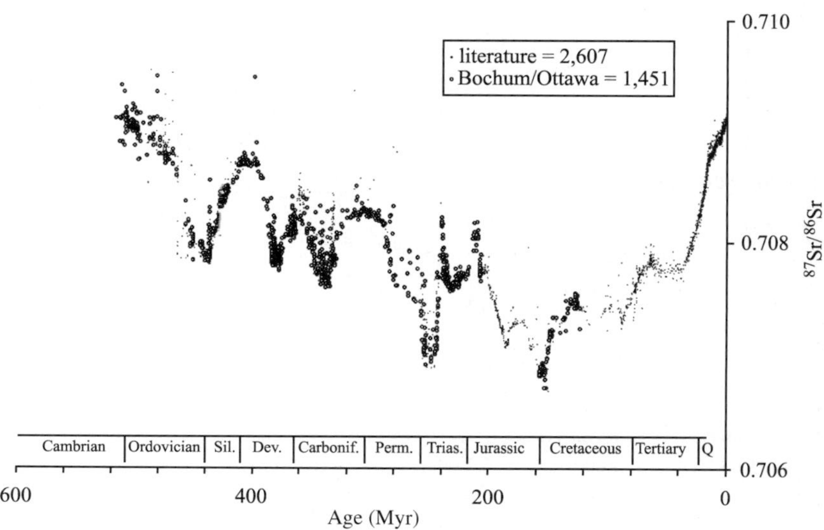

Figure 27 $^{87}Sr/^{86}Sr$ variations for the Phanerozoic based on 4,055 samples of brachiopods, belemnites, and conodonts. Normalized to NBS 987 of 0.710240 (after Veizer *et al.*, 1999).

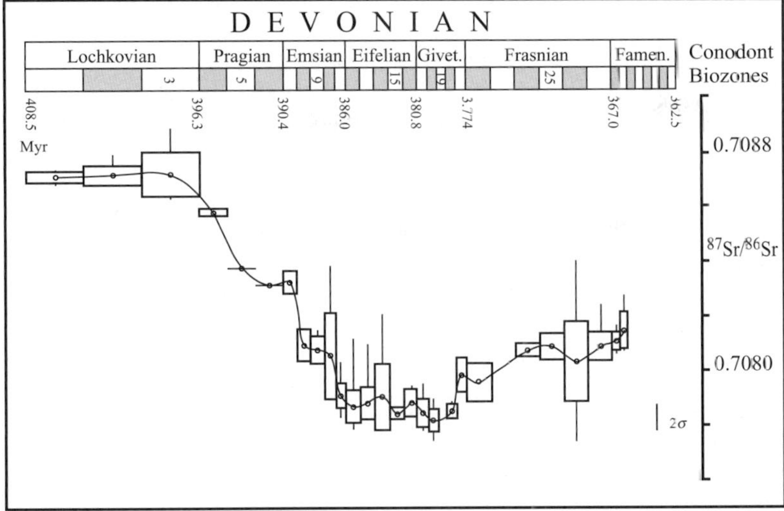

Figure 28 $^{87}Sr/^{86}Sr$ variations during the Devonian based on conodont biozones. Explanations: circle = mean; box = $\pm 1\sigma$; vertical line = minimum and maximum. The 2 σ in the lower right corner is an average 2 σ for the NBS 987 standard. Note that only brachiopods are included in this. Figures 1–37 are conodont biozones (after Veizer *et al.*, 1999).

composition of seawater. Their disadvantage is the slow accumulation rate that limits temporal resolution. Organic rich shales, alternatively, have high Re/Os ratios, requiring age correction by the isochron technique. Corals and carbonate sediments do not appear to preserve the hydrogenous (seawater) $^{187}Os/^{186}Os$ record.

The presently known data for seawater $^{187}Os/^{186}Os$ cover mostly the Cenozoic, with fragmentary results for the Cretaceous and the Jurassic. As for $^{87}Sr/^{86}Sr$, the $^{187}Os/^{186}Os$ declines from present-day value of ~1.06 to 0.15 at the K/T boundary. The sudden drop, from ~0.8 to 0.15 at the K/T boundary likely reflects the cosmic input from the meteoric impact. The $^{187}Os/^{186}Os$ ratio for the Mesozoic oscillates between 0.8 and 0.15, but the details are not yet resolved. For further discussion of Re/Os systematics, see Chapter 6.20.

7.15.11.2.3 Sulfur isotopes

In contrast to strontium and osmium isotopes, the isotopes of sulfur are strongly fractionated by biological processes, particularly during the dissimilatory bacterial reduction of sulfate to sulfide. The laboratory results for this step are anywhere from +4 to −46‰ (CDT), but even larger fractionations have been observed in natural systems (Harrison and Thode, 1958; Chambers and Trudinger, 1979; Habicht and Canfield, 1996).

The geologic record is characterized by a dearth of *Precambrian* evaporitic sequences, including their sulfate facies. Stratiform barites do exist, but they may be, at least in part, of hydrothermal origin. The scarcity of evaporites is partly due to their poor survival rates in the face of tectonic processes, but another reason may be low pO_2 levels in the contemporaneous ocean/atmosphere system, particularly during the Archean (Veizer, 1988a). In addition, most Archean sulfides, such as pyrites, contain $\delta^{34}S$ close to 0‰ CDT, the value typical of the mantle (Figure 29), rather than the expected highly negative ones characteristic of bacterial dissimilatory sulfate reduction. These observations were usually interpreted (e.g., Schidlowski *et al.*, 1983; Hayes *et al.*, 1992) as being due to biological evolution, where it is assumed that the invention of oxygen generating photosynthesis and of bacterial sulfate reduction were only later developments. In that case, most of the sulfur in the Archean host phases would have originated from mantle sources and carried its isotopic signature. Only with the onset of these two biological processes, in about the late Archean or early Proterozoic, was enough oxygen generated to stabilize sulfate in seawater and to initiate its bacterial reduction to H_2S, the latter eventually forming sulfide minerals, such as pyrite. This development resulted in the burial of large quantities of sulfides depleted in ^{34}S in the sediments, causing the residual sulfate in the ocean to shift towards heavier values. The result is the bifurcation of $\delta^{34}S$ sulfate/sulfide values at the time of "invention" of bacterial dissimilatory sulfate reduction (Figure 29).

The above scenario is appealing, but not mandatory. As shown by Godderis and Veizer (2000), the same "logistic" scenario of continental growth that generated the strontium isotope trend can also generate the observed $\delta^{34}S$ pattern

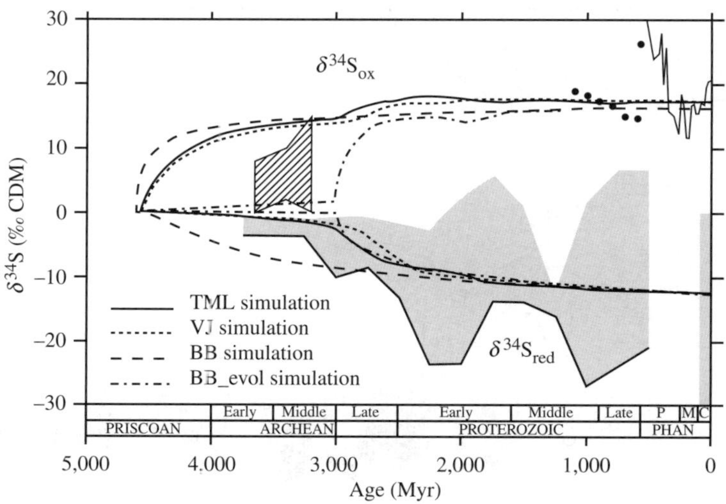

Figure 29 Model sulfur isotopic evolution in the course of geologic history. The lower trend (labeled $\delta^{34}S_{red}$) represents the $\delta^{34}S$ for sulfides. The upper curves (labeled $\delta^{34}S_{ox}$) are the $\delta^{34}S$ of marine sulfates. TML and VJ simulations assume a logistic type of continental growth as proposed by Taylor and McLennan (1985) and Veizer and Jansen (1979), respectively. BB simulation assumes an instantaneous generation of continental crust, BB-evol simulation assumes instantaneous continental generation, but with delayed "invention" of oxygen generating photosynthesis. The Phanerozoic trend as in Figure 30. Dots represent measurements of Precambrian sulfates (Claypool *et al.*, 1980) and the hatched field represents sulfates from Holser *et al.* (1988) (after Goddéris and Veizer, 2000).

(Figure 29) and the growth of sulfate in the oceans. This explanation has the advantage that a single scenario generates all these (and other) evolutionary patterns. In short, the early "mantle" buffered oceans (Veizer *et al.*, 1982) had a large consumption of oxygen in the submarine hydrothermal systems (Wolery and Sleep, 1988; see Chapter 6.21) because they operated at considerably higher rates than today. The capacity of this sink declined exponentially in the course of geologic history, reflecting the decay in the dissipation of the heat from the core and mantle. As a result, the buffering of the ocean was taken over by the continental "river" flux. In summary, the bacterial dissimilatory sulfate reduction must have been extant at the time of bifurcation of the $\delta^{34}S$ record, but the isotope data do not provide a definitive answer as to the timing of this invention. Tectonic evolution would override its impact even if established much earlier.

The Precambrian $\delta^{34}S$ record is spotty for sulfide-S and almost nonexistent for sulfate-S (Canfield and Teske, 1996; Strauss, 1993). A fragmentary record for the latter exists only for the latest Neoproterozoic (Strauss, 1993), suggesting a large shift from ~20 to 33 ± 2‰ at the transition into the Phanerozoic.

The $\delta^{34}S_{sulfate}$ variations in *Phanerozoic* oceans, based on evaporites (Holser and Kaplan, 1966; Claypool *et al.*, 1980), form an overall trough-like trend similar to strontium isotopes (Figure 30). Note, however, the large age uncertainties for, and the large gaps between, the studied evaporitic sequences. This is due to their episodic occurrence and uncertain chronology and is part of the reason for the large spread in the coeval $\delta^{34}S$ values despite the fact that the $\delta^{34}S_{sulfate}$ in seawater is spatially homogeneous (Longinelli, 1989). Another reason for this large spread in the $\delta^{34}S$ values is the evolution of sulfur isotopes in the course of the evaporative process, from sulfate to chloride to late salt facies. A recent development of the technique that enabled measurement of $\delta^{34}S$ in structurally bound sulfate in carbonates (Kampschulte and Strauss, 1998; Kampschulte, 2001) yielded a Phanerozoic secular curve with much greater temporal resolution (Figure 30).

The $\delta^{34}S_{sulfate}$ and $\delta^{13}C_{carbonate}$ secular curves correlate negatively (Veizer *et al.*, 1980), suggesting that it is the redox balance (peddling of oxygen between the carbon and sulfur cycles) that controls the $\delta^{34}S$ variations in Phanerozoic oceans. Note, nevertheless, that a physical geological scenario for this coupling is as yet not clarified. If redox balance is indeed a major control mechanism, it would suggest that the withdrawal of ^{32}S due to pyrite burial in sediments was twice as large as today in the early Paleozoic versus about one-half in the late Paleozoic (Kump, 1989).

In addition to the long-term 10^8 yr trends, shorter spikes, on 10^5–10^6 yr scales (e.g., Permian/Triassic transition; Holser, 1977), do exist, but their "catastrophic" geological causes are not as yet resolved.

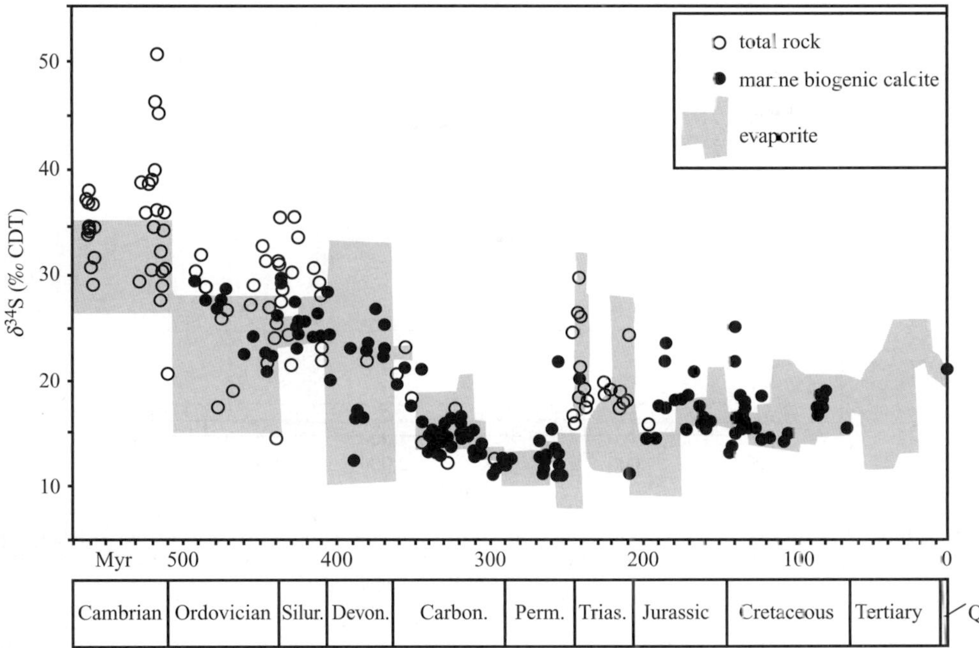

Figure 30 Sulfur isotopic composition of Phanerozoic seawater based on measurements of sulfur structurally bound in calcitic shells as well as evaporites. Note that the shell samples are mostly the same as those of the Sr, O, and C isotopes in Veizer *et al.* (1999) (after Kampschulte, 2001 by courtesy of the author).

7.15.11.2.4 Carbon isotopes

The two dominant exogenic reservoirs of carbon are carbonate rocks and organic matter in sediments. They are linked in the carbon cycle via atmospheric CO_2 and the carbon species dissolved in the hydrosphere. The $\delta^{13}C$ for the total dissolved carbon (TDC) in seawater is ~+1 ± 0.5‰ (PDB), with surficial waters generally heavier and deep waters lighter than this average (Kroopnick, 1980; Tan, 1988). Atmospheric CO_2 in equilibrium with TDC of marine surface water has a $\delta^{13}C$ of ~−7‰. CO_2 is preferentially utilized by photosynthetic plants for production of organic carbon causing further depletion in ^{13}C (Equation 8):

$$6CO_2 + 6H_2O \rightarrow C_6H_{12}O_6 + 6O_2 \qquad (8)$$

Most land plants utilize the so-called C_3, or Calvin pathway (O'Leary, 1988), that results in tissue with a $\delta^{13}C_{org}$ of ~−25‰ to −30‰. The situation for aquatic plants is somewhat different because they utilize dissolved and not gaseous CO_2. Tropical grasses, however, utilize the C_4 (Hatch-Slack or Kranz) pathway and have a $\delta^{13}C$ of some −10‰ to −15‰. A third group that combines these two pathways, the CAM plants (algae and lichens), has intermediate $\delta^{13}C$ values. In detail, the nature of the discussed variations is far more complex (Deines, 1980; Sackett, 1989) and depends on the type of organic compounds involved. For our purposes, however, it is only essential to realize that C_{org} is strongly depleted in ^{13}C. This organic matter, which is very labile, is easily oxidized into CO_2 that inherits the ^{13}C-depleted signal.

The $\delta^{13}C$ of mantle carbon is ~−5‰ PDB (Schidlowski *et al.*, 1983; Hayes *et al.*, 1992) and in the absence of life and its photosynthetic capabilities, this would also be the isotopic composition of seawater. Yet, as far back as 3.5 Gyr ago, and possibly as far as ~4 Gyr ago (Schidlowski *et al.*, 1983), the carbonate rocks (~seawater) had $\delta^{13}C$ at ~0‰ PDB (Figure 31). This suggests that a reservoir of reduced organic carbon that accounted for ~1/5 of the entire exogenic carbon existed already some 4 Gyr ago, "pushing up" the residual 4/5 of carbon, present in the oxidized form in the ocean/atmosphere system, from −5‰ to 0‰ PDB. This is an oxidized/reduced partitioning similar to that we have today. Stated in a simplified manner, life with its photosynthetic capabilities, and possibly of present-day magnitude, can be traced almost as far back as we have a rock record. This photosynthesis may or may not have been generating oxygen as its byproduct, but was essential in order to "lift" the seawater $\delta^{13}C$ to values similar to the present-day ones. In order to sustain seawater $\delta^{13}C$ at this level during the entire geologic history, it is necessary that the input and output in the carbon cycle have the same $\delta^{13}C$. Since the input from the mantle, via volcanism and hydrothermal systems, has a $\delta^{13}C$ of −5‰ and the subducted carbonates are 0‰, the subduction process must

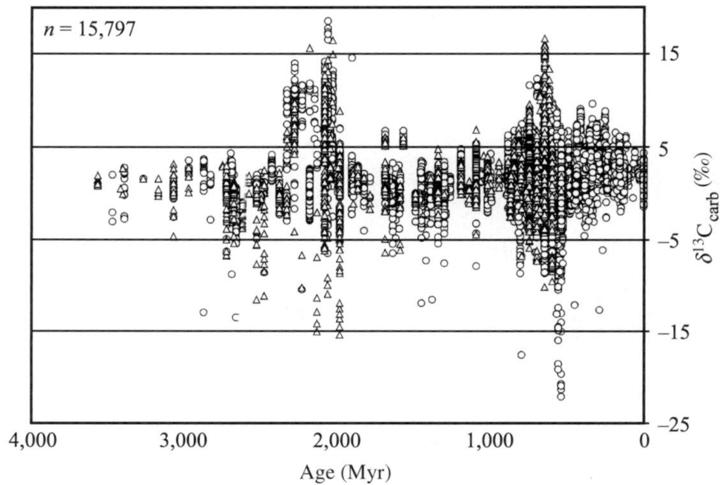

Figure 31 The $\delta^{13}C$ composition of carbonate rocks during geologic history. Reproduced from Shields and Veizer (2002). Triangles—dolostones, circles—limestones and fossil shells. For possible explanation of the large Paleo- and Neoproterozoic spreads, see Rothman *et al.* (2003).

involve also a complementary ^{13}C-depleted component, organic matter. This is possible to contemplate as long as oceanic waters were not fully oxygenated, such as may have been the case in the Archean. This is either because oxygen generating photosynthesis was "invented" as late as the late Archean or early Proterozoic (Cloud, 1976; Holland, 1984), or because tectonic evolution led to a progressive oxygenation of the ocean/atmosphere system due to a switchover from a "mantle"—to a "river"—buffered ocean system (Goddéris and Veizer, 2000). For the latter alternative, and in analogy to sulfur, it is possible to argue that oxygen generating photosynthesis (photosystem 2) may have been extant as far back as we have the geologic record, without necessarily inducing oxygenation of the early ocean/atmosphere system (but see Lasaga and Ohmoto, 2002). Whatever the cause, the oxygenation of the system in the early Proterozoic would have resulted in oxidation of organic matter that was settling down through the water column. Today only ~1% of organic productivity reaches the ocean floor and ~0.1% survives into sedimentary rocks. As a result, the addition of mantle carbon, coupled with the subduction loss of the ^{13}C-enriched limestone carbon, would slowly force the $\delta^{13}C$ of seawater back to mantle values. In order to sustain the near 0‰ PDB of seawater during the entire geologic history, it is necessary to lower the input of mantle carbon into the ocean/atmosphere system by progressively diminishing the impact of hydrothermal and volcanic activity over geologic time.

Superimposed on this invariant *Precambrian* $\delta^{13}C$ seawater trend are two intervals with very heavy (and very light) values, at ~2.2 Gyr and in the Neoproterozoic (Figure 31). The former has been interpreted as a result of the invention of oxygen generating photosynthesis that resulted in the sequestration of huge quantities of organic matter (Karhu and Holland, 1996) into coeval sediments and the Neoproterozoic interval was the time of the proposed "snowball earth" (Hoffman *et al.*, 1998). At this stage, the reasons for the high frequency of the anomalous $\delta^{13}C$ values during these two intervals are not well understood, but it is interesting that both were associated with large glaciations, as was the later discussed ^{13}C-enriched Permo/Carboniferous interval.

The sampling density and time resolution in the *Phanerozoic* enabled the delineation of a much better constrained secular curve (Figure 32), with a maximum in the late Permian, but even in this case we are dealing with a band of data, reflecting the fact that the $\delta^{13}C_{DIC}$ of seawater is not uniform in time and space, that organisms can incorporate metabolic carbon into their shells (vital effect), and that some samples may also contain a diagenetic overprint. Superimposed on the overall trend are higher oscillations, at 10^7 yr and shorter timescales, but their meaning is not yet understood.

Frakes *et al.* (1992) proposed that the $\delta^{13}C_{carbonate}$ (seawater) becomes particularly heavy at times of glaciations, and that such times are also characterized by low CO_2 levels. The coincidences of the $\delta^{13}C$ peaks with the late Ordovician and Permocarboniferous glacial episodes appear to support this proposition, but the Mesozoic/Cenozoic record is divergent. Accepting the validity of the present $\delta^{13}C$ trend, it is possible to calculate the model p_{CO_2} levels of ancient atmospheres. Three Phanerozoic p_{CO_2} reconstructions exist (Berner and Kothavala, 2001; Berner and Streif, 2001; Rothman, 2002)

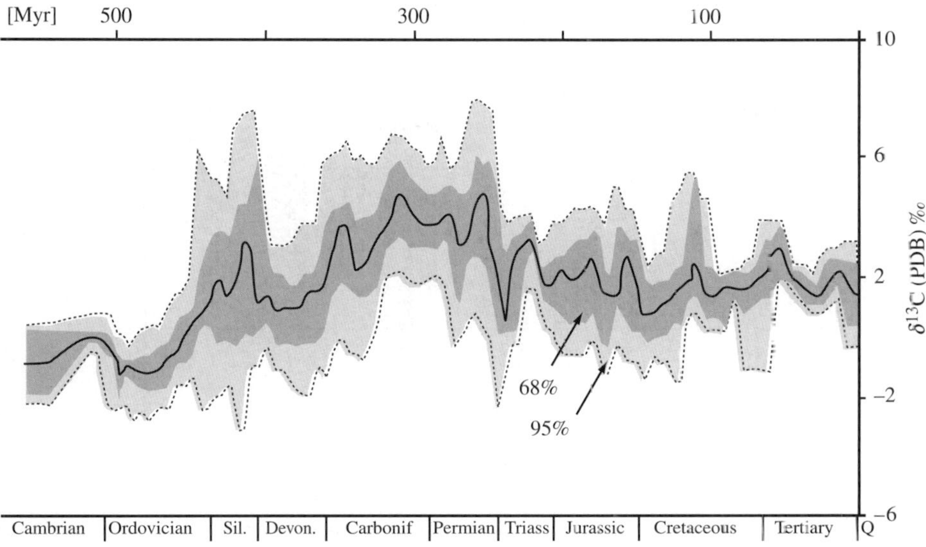

Figure 32 The Phanerozoic $\delta^{13}C$ trend for LMC shells. The running mean for ~4,500 samples is based on a 20 Ma window and 5 Ma forward step. The shaded areas around the running mean include the 68% ($\pm 1\sigma$) and 95% ($\pm 2\sigma$) of all data (after Veizer *et al.*, 1999).

that are internally inconsistent and not one of them shows any correlation with the paleoclimate deduced from sedimentological criteria (Veizer *et al.*, 2000; Boucot and Gray, 2001; Veizer, 2003; Figure 35). This led Veizer *et al.* (2000) to conclude that either the estimates of paleo-CO_2 were unreliable or there was no direct relationship between p_{CO_2} levels and climate for most of the Phanerozoic.

Higher order peaks, at a 10^6 yr resolution, have been observed in the geologic record, particularly in deep sea borehole sections and are discussed by Ravizza and Zachos (see Chapter 6.20).

7.15.11.2.5 Oxygen isotopes

The oxygen isotope record of some 10,000 limestones and low-magnesium calcitic fossils (Shields and Veizer, 2002) shows a clear trend of ^{18}O depletion with age of the rocks (Figure 33). This isotope record in ancient marine carbonates (but also cherts and phosphates) is one of the most controversial topics of isotope geochemistry. It centers on the issue of the primary versus post-depositional origin of the secular trend (e.g., Land, 1995 versus Veizer, 1995). Undoubtedly, diagenesis, and other post-depositional phenomena reset the $\delta^{18}O$, usually to more negative values, during stabilization of original metastable phases (e.g., aragonite, high-magnesium calcite), into the more stable phase, diagenetic low-magnesium calcite. Every carbonate rock is subjected to this stabilization stage and most, if not all, of its internal components are reset. The only exception can be the original low-magnesium calcitic shells of some organisms, such as brachiopods, belemnites, and foraminifera. Yet, the overall bulk rock depletions, relative to these stable phases, are ~2–3‰ (Veizer *et al.*, 1999) and not some 7‰ or more as is the case for the Precambrian limestones (Figure 33). The rocks, once diagenetically stabilized become relatively "inert" to further resetting. The retention of $\Delta^{18}O_{dolomite-calcite}$ of ~3‰ during the entire geologic history (Shields and Veizer, 2002) is also consistent with such an interpretation.

The observed *Precambrian* $\delta^{18}O$ secular trend is therefore real, albeit shifted by 2–3‰ to lighter values, and likely reflects the changing $\delta^{18}O$ of seawater. The exchange of oxygen at $T > 350\,°C$ between percolating seawater and oceanic crust results in ^{18}O enrichment of the water and ultimately oceans. The opposite happens at $T < 350\,°C$ (Muehlenbachs, 1998; Gregory, 1991; Wallmann, 2001). One interpretation could be that over geologic history this "isotopically neutral" crossover point migrated to shallower depths, thus reducing the profile of the low-T alteration relative to the deeper one, perhaps due to blanketing of the ocean floor by pelagic biogenic sediments during the Phanerozoic that sealed the off-ridge oceanic crust from seawater percolation.

The *Phanerozoic* trend (Figure 34) is based on ~4,500 samples of low-magnesium calcitic fossils from about 100 localities worldwide. The reasons for believing that it is essentially a primary trend were discussed in detail by Veizer *et al.* (1999).

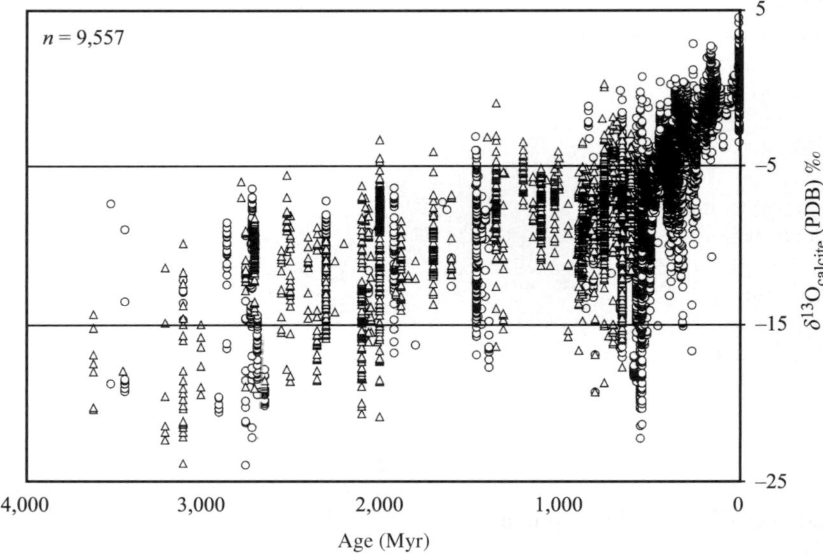

Figure 33 Oxygen isotopic composition of limestones and calcitic shells during geologic history. Triangles and circles represent samples with good and poor age assignment, respectively (after Shields and Veizer, 2002).

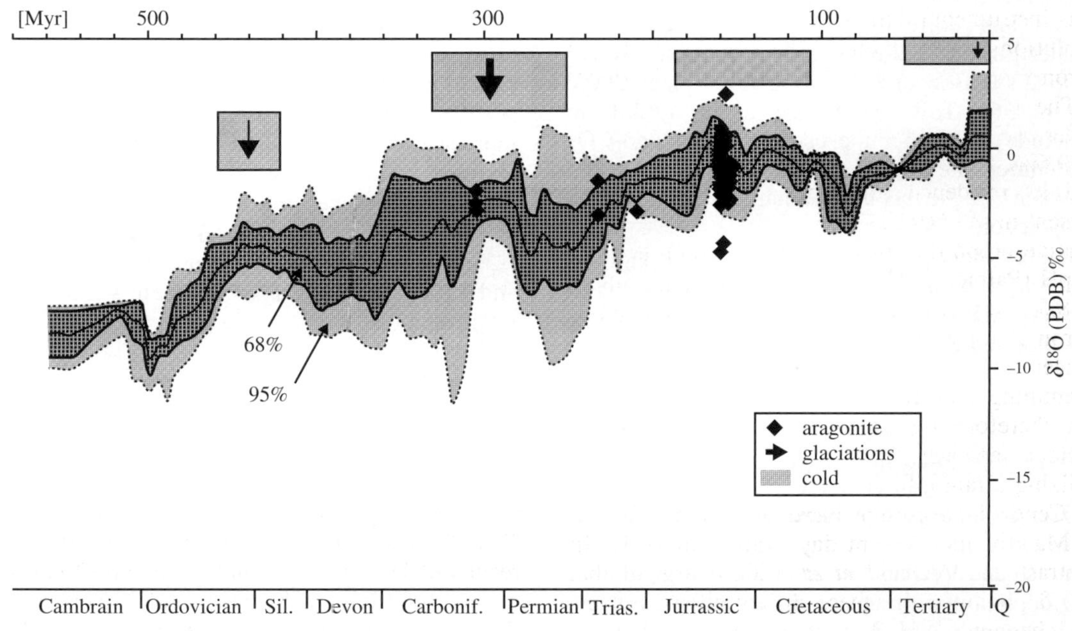

Figure 34 Phanerozoic $\delta^{18}O$ trend for low-Mg calcitic and aragonitic shells. Note that this is a trend for $CaCO_3$, offset by the fractionation factor α from that of seawater. See Figure 32 for further explanations (after Veizer *et al.*, 1999).

The pattern in Figure 34 has considerable implications for our understanding of past climate, but is still subject to debate. The models of Muehlenbachs (1998) and Gregory (1991) claimed that due to a balance of high and low temperature reactions during interaction of the water cycle with the lithosphere, the $\delta^{18}O$ of the oceans should have been buffered near its present-day value. If seawater always had $\delta^{18}O$ of ~0‰ SMOW (standard mean ocean water), the primary nature of the $\delta^{18}O$ record (Figure 34) would demand cooling oceans in the course of the Phanerozoic. With such an assumption, however, the early- to mid-Paleozoic ocean temperatures would have to have been in excess of 40 °C, even at times of glaciations. This is an unpalatable proposition, not only climatologically, but also in view of the similarity in faunal assemblages, in our case brachiopods, during this entire time span. Accepting that the $\delta^{18}O$ of past seawater was evolving towards ^{18}O-enriched values (Wallmann, 2001) and detrending the data accordingly (Figure 35), the superimposed second order structure of the curve correlates well with the Phanerozoic paleoclimatic record (cf. also Boucot and Gray, 2001). The observed structure, therefore, likely reflects paleotemperatures. If so, this would indicate that global climate swings were not confined to higher latitudes, but involved equatorial regions as well. As already pointed out, neither the $\delta^{18}O$ nor the paleoclimate record correlate with the model p_{CO_2} estimates for the ancient atmosphere.

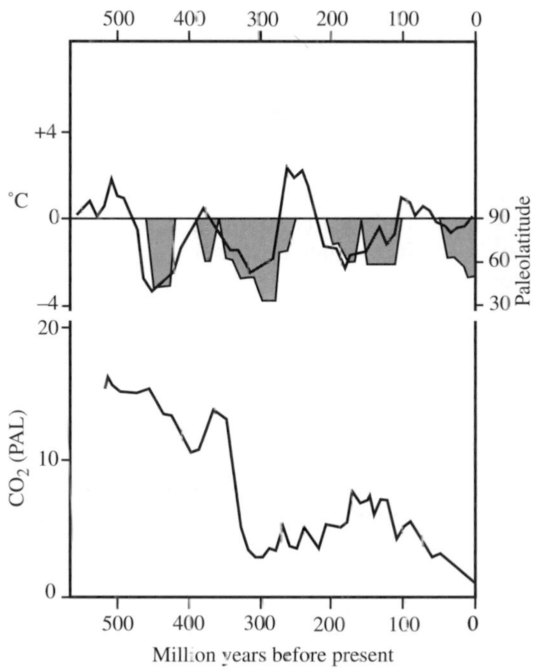

Figure 35 Reconstructed variations in mean temperature of shallow low-latitude seawater during the Phanerozoic based on the data in Figure 34. Note the good agreement of the cooling episodes with the extended latitudinal dispersion of ice rafted debris (shaded histograms). In the subsequent publication, Shaviv and Veizer (2003) showed that the proposed temperature variations correlated with the intensity of cosmic-ray flux reaching the Earth. The p_{CO2} (PAL—present-day atmospheric level) is that for the Geocarb model of Berner (1994).

7.15.11.2.6 Isotope tracers in developmental stages

The advances in instrumentation and particularly the arrival of the multicollector inductively coupled plasma mass spectrometers (MC-ICPMS) opened a window for a number of new tracers that were difficult to tackle with the old instrumentation. A pattern for seawater evolution is thus emerging for isotopes of boron and calcium.

The present-day boron concentration in seawater and its $\delta^{11}B$ are uniform at 4.5 ppm and −39.6‰, respectively (Lemarchand *et al.*, 2000) and its residence time is ~14 Myr. Boron is present in seawater as $B(OH)_4^-$ and $B(OH)_3$. The relative proportion of these species is a function of pH (Palmer and Swihart, 1996), with $B(OH)_3$ 19.8‰ enriched in ^{11}B relative to $B(OH)_4^-$. Boron incorporation into carbonate skeletons, at concentrations of ~10–60 ppm, is from $B(OH)_4^-$ (Hemming and Hanson, 1992) and their $\delta^{11}B$ can therefore be used for tracing the pH of ancient seawater. Pearson and Palmer (2000), utilizing foraminiferal calcite, argued that the pH of Cenozoic seawater increased from ~7.4 at 60 Ma to its present-day value of 8.1. In contrast, Lemarchand *et al.* (2000) argued that the $\delta^{11}B$ trend in these foraminifera reflects the changing $^{11}B/^{10}B$ composition of seawater at constant pH, a development largely due to scavenging of boron by an increasing flux of clastic sediments.

The residence time of *calcium*, ~1 Myr versus the mixing rates of ocean of ~1,000 yr, means that its isotopes are distributed homogeneously in seawater (Skulan *et al.*, 1997), with $^{40}Ca/^{44}Ca$ equal to 45.143 (Schmitt *et al.*, 2001). Modern carbonate shells show a variation of ~4‰. These shells are ~1‰ enriched relative to magmatic rocks, but compared to seawater they are depleted by 1–3‰, depending on the trophic level of the organism (Skulan *et al.*, 1997). Zhu and MacDougall (1998) showed that calcium isotope composition of the shells was both species and temperature dependent and that river water is depleted by 2‰ relative to seawater. The temperature dependency of calcium isotope fractionation enables this tracer to be a potential paleotemperature proxy (Nägler *et al.*, 2000). The seawater $^{44}Ca/^{40}Ca$ secular variations are indicated by the results of De La Rocha and De Paolo (2000) for the last 160 Myr, and by the data of Zhu (1999) for the entire 3.4 Gyr of earth history. The latter indicate that, in analogy to strontium isotopes (Veizer and Compston, 1976), the Archean samples have $^{44}Ca/^{40}Ca$ ratios similar to the earth mantle, with the crustal-like values first appearing at the Archean/Proterozoic transition. However, in contrast to the strontium isotope trend, the calcium isotope ratios appear to dip towards mantle values also at ~1.6 Gyr ago.

In the near future, the isotopes of *silicon, iron, and magnesium* also will likely develop into useful paleoceanographic tracers, but at this stage their utility for pre-Quaternary studies is limited.

7.15.12 MAJOR TRENDS IN THE EVOLUTION OF SEDIMENTS DURING GEOLOGIC HISTORY

7.15.12.1 Overall Pattern of Lithologic Types

The lithologic composition and the relative percentages of sedimentary and volcanic rocks preserved within the confines of present-day continents in crustal segments of various ages (Ronov, 1964; Budyko *et al.*, 1985) are shown in Figure 36. It should be remembered that with increasing geologic age, the total sedimentary rock mass diminishes (Figure 6) and a given volume percentage of rock 3 Gyr ago represents much less mass than an equal percentage of rocks 200 Myr old. Despite this limitation, some general trends in lithologic rock types agreed on by most investigators are evident in this summary.

The outcrops of very old Archean rocks are few and thus may not be representative of the original sediment compositions deposited. Nevertheless, it appears that carbonate rocks are relatively rare in the Archean. Based on data from the limited outcrops, Veizer (1973) concluded that Archean carbonate rocks are predominantly limestones. During the early Proterozoic, the abundance of carbonates increases markedly, and for most of this Era the preserved carbonate rock mass is typified by the ubiquity of early diagenetic, and perhaps primary, dolostones (Veizer, 1973; Grotzinger and James, 2000). In the Phanerozoic, carbonates constitute ~30% of the total sedimentary mass, with sandstones and shales accounting for the rest. The Phanerozoic record of carbonates will be elaborated upon in the subsequent text.

Other highlights of the lithology-age distribution of Figure 36 are: (i) a marked increase in the abundance of submarine volcanogenic rocks and immature sandstones (graywackes) with increasing age; (ii) a significant percentage of arkoses in the early and middle Proterozoic, and an increase in the importance of mature sandstones (quartz-rich) with decreasing age; red beds are significant rock types of Proterozoic and younger age deposits; (iii) a significant "bulge" in the relative abundance of banded iron-ore formations, chert-iron associations, in the early Proterozoic; and (iv) a lack of evaporitic sulfate (gypsum, anhydrite) and salt (halite, sylvite) deposits in sedimentary rocks older than ~800 Myr. Marine evaporites owe their existence to a unique combination of tectonic, paleogeographic, and sea-level conditions. Seawater bodies must be restricted to some degree, but also must exchange with the open ocean to permit large volumes of seawater to enter these restricted basins and evaporate. Environmental settings of evaporite deposition may occur on cratons or in rifted basins. Figure 37 illustrates

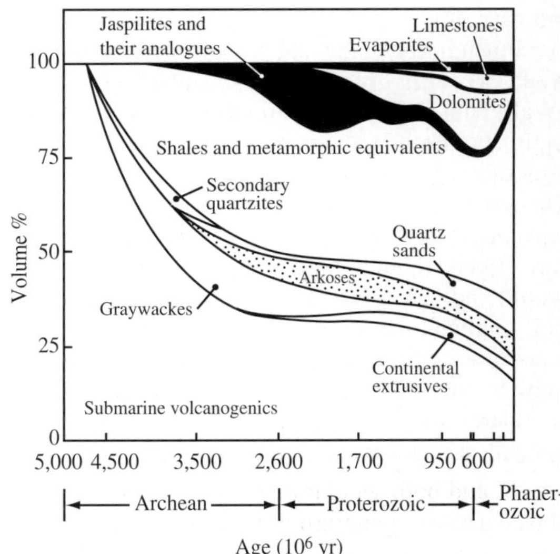

Figure 36 Volume percent of sedimentary rocks as a function of age. Extrapolation beyond ~3 Ga is hypothetical (after Ronov, 1964).

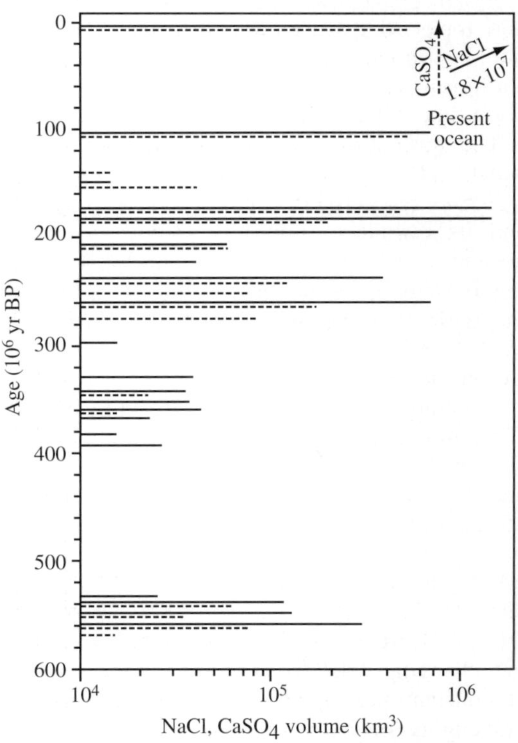

Figure 37 Phanerozoic distribution of NaCl (dark line) and $CaSO_4$ (dashed line) in marine evaporites (after Holser, 1984).

that because evaporite deposition requires an unusual combination of circumstances, a "geological accident" (Holser, 1984), the intensity of deposition has varied significantly during geologic time. This conclusion implies that the volume preserved of NaCl and $CaSO_4$ per unit of time in the Phanerozoic reasonably reflects the volume deposited, and because of the lack of any secular trend in evaporite mass per unit time, there has been only minor differential recycling of these rocks relative to other lithologies. Because the oceans are important reservoirs for these components, such large variations in the rate of NaCl and $CaSO_4$ output from the ocean to evaporites imply changes in the salinity and chemistry of seawater (Hay *et al.*, 2001).

These trends in lithologic features of the sedimentary rock mass are a consequence of evolution of the surface environment of the planet as well as recycling and post-depositional processes, and both secular and cyclic processes have played a role in generating the lithology-age distribution we see today (Veizer, 1973, 1988a; Mackenzie, 1975). For the past 1.5–2.0 Gyr, the Earth has been in a near present-day steady state, and the temporal distribution of rock types since then has been controlled primarily by recycling in response to plate tectonic processes.

Because sedimentary carbonates are important rock types in terms of providing mineralogical, chemical, biological, and isotopic data useful in interpretation of the history of Earth's surface environment, the following sections discuss these rock types in some detail. The discussion is mainly limited to the Phanerozoic because of the more complete database for this Eon than for the Precambrian.

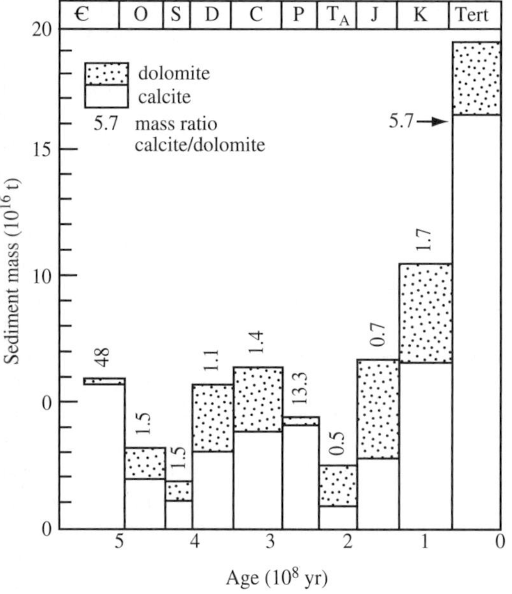

Figure 38 The Phanerozoic sedimentary carbonate mass distribution as a function of geologic age. Period masses of calcite and dolomite and the period mass ratios of calcite/dolomite are also shown (after Mackenzie and Morse, 1992).

7.15.12.2 Phanerozoic Carbonate Rocks

7.15.12.2.1 Mass-age distribution and recycling rates

As stated above, carbonate rocks comprise ~30% of the mass of Phanerozoic sediments. Given and Wilkinson (1987) reevaluated all the existing data on Phanerozoic carbonate rocks, their masses, and their relative calcite and dolomite contents (Figure 38). It can be seen that, as with the total sedimentary mass (Garrels and Mackenzie, 1971a,b), the mass of carbonate rock preserved is pushed toward the front of geologic time. The Tertiary, Carboniferous, and Cambrian periods are times of significant carbonate preservation, whereas the preservation of Silurian and Triassic carbonates is minimal.

The survival rates of the carbonate masses for different Phanerozoic systems are shown in Figure 39, together with the Gregor (1985) plot for the total sedimentary mass. The difference between the survival rate of the total carbonate mass and that of dolomite is the mass of limestone surviving per interval of time. The half-life of all the post-Devonian sedimentary mass is 130 Myr, and for a constant mass with a constant probability of destruction, the mean sedimentation rate is $\sim 100 \times 10^{14}$ g yr^{-1}. The modern global erosional flux is $\sim 200 \times 10^{14}$ g yr^{-1}, of which ~15% is particulate and dissolved carbonate. Although the data are less reliable for the survival rate of Phanerozoic carbonate sediments than for the total sedimentary mass, the half-life of the post-Permian carbonate mass is ~86 Myr. This gives a mean sedimentation rate of $\sim 35 \times 10^{14}$ g carbonate per year, compared to the present-day carbonate flux of 30×10^{14} g yr^{-1} (Morse and Mackenzie, 1990). The difference in half-lives between the total sedimentary mass, which is principally sandstone and shale, and the carbonate mass probably is a consequence of the more rapid recycling of the carbonate mass at a rate ~1.5 times that of the total mass.

This is not an unlikely situation. With the advent of abundant carbonate-secreting, planktonic organisms in the Jurassic, the site of carbonate deposition shifted significantly from shallow-water areas to the deep sea. This gradual shift will increase still further the rate of destruction (by eventual subduction) of the global carbonate mass relative to the total sedimentary mass because the recycling rate of oceanic crust (the "b" values of Veizer and Jansen, 1985; Veizer, 1988a) exceeds that of the "continental" sediments by a factor of 6. Also Southam and Hay (1981), using a half-life of 100 Myr for pelagic sediment, estimated that as much as 50% of all sedimentary rock formed by weathering of igneous rock may have been lost by subduction during the past 4.5 Gyr.

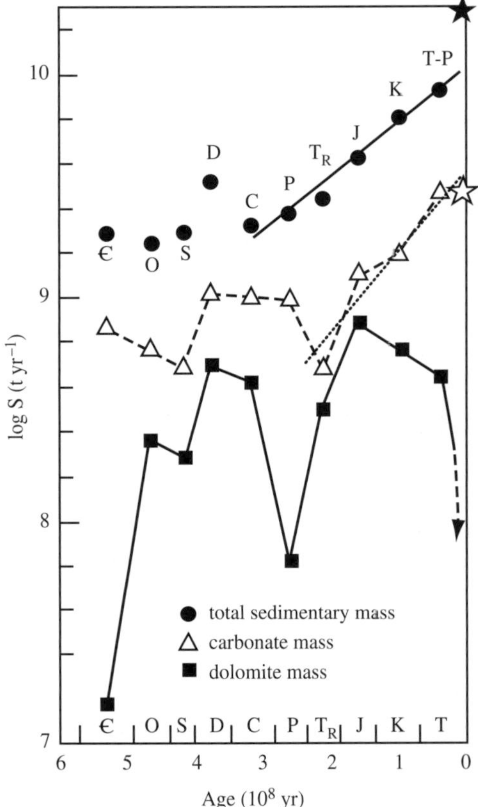

Figure 39 Phanerozoic sedimentary rock mass–age relationships expressed as the logarithm of the survival rate in tons per year versus time. The straight lines are best fits to the total mass data (solid line) and to the carbonate mass data (dotted line) for particular intervals of Phanerozoic time. The difference between the logarithm of S for the carbonate mass and that of the dolomite mass is the survival rate of the calcite mass. Filled star is present total riverine flux to the oceans, whereas open star is carbonate flux (after Mackenzie and Morse, 1992).

Thus, it appears, as originally suggested by Garrels and Mackenzie (1969, 1971a, 1972), that the carbonate component of the sedimentary rock mass may have a cycling rate slightly different from that of the total sedimentary mass. These authors argued that the differential recycling rates for different lithologies were related to their resistance to chemical weathering and transport. Evaporites are the most easily soluble, limestones are next, followed by dolostones, and shales and sandstones are the most inert. Although resistance to weathering may play a small role in the selective destruction of sedimentary rocks, it is likely, as argued previously, that differences in the recycling rates of different tectonic regimes in which sediments are deposited are more important.

7.15.12.2.2 Dolomite/calcite ratios

For several decades it has been assumed that the Mg/Ca ratio of carbonate rocks increases with increasing rock age (see Daly, 1909; Vinogradov and Ronov, 1956a,b; Chilingar, 1956; Figure 40). In these summaries, the magnesium content of North American and Russian Platform carbonates is relatively constant for the latest 100 Myr, and then increases gradually, very close to, if not the same as, the commencement of the general increase in the magnesium content of pelagic limestones (Renard, 1986). The dolomite content in deep sea sediments also increases erratically with increasing age back to ~125 Myr before present (Lumsden, 1985). Thus, the increase in magnesium content of carbonate rocks with increasing age into at least the early Cretaceous appears to be a global phenomenon, and to a first approximation, is not lithofacies related. In the 1980s, the accepted truism that dolomite abundance increases relative to limestone with increasing age has been challenged by Given and Wilkinson (1987). They reevaluated all the existing data and concluded that dolomite abundances do vary significantly throughout the Phanerozoic but may not increase systematically with age (Figure 40). Yet the meaning of these abundance curves, and indeed their actual validity, is still controversial (Zenger, 1989).

Voluminous research on the "dolomite problem" (see Hardie, 1987, for discussion) has shown that the reasons for the high magnesium content of carbonates are diverse and complex. Some dolomitic rocks are primary precipitates; others were deposited as $CaCO_3$ and then converted entirely or partially to dolomite before deposition of a succeeding layer; still others were dolomitized by migrating underground waters tens or hundreds of millions of years after deposition. It is therefore exceedingly important to know the distribution of the calcite/dolomite ratios of carbonate rocks through geologic time. This information has a bearing on the origin of dolomite, as well as on the properties of the coeval atmosphere–hydrosphere system (Given and Wilkinson, 1987; Wilkinson and Algeo, 1989; Berner, 1990; Morse and Mackenzie, 1990; Mackenzie and Morse, 1992; Arvidson and Mackenzie, 1999; Arvidson *et al.*, 2000; Holland and Zimmermann, 2000). For example, it could be argued that if the dolomite/calcite ratio progressively increases with age of the rock units (Figure 40), the trend principally reflects enhanced susceptibility of older rock units to processes of dolomitization. Such a trend would then be only a secondary feature of the sedimentary carbonate rock mass due to progressive diagenesis and seawater driven dolomitization (Garrels and Mackenzie, 1971a; Mackenzie, 1975;

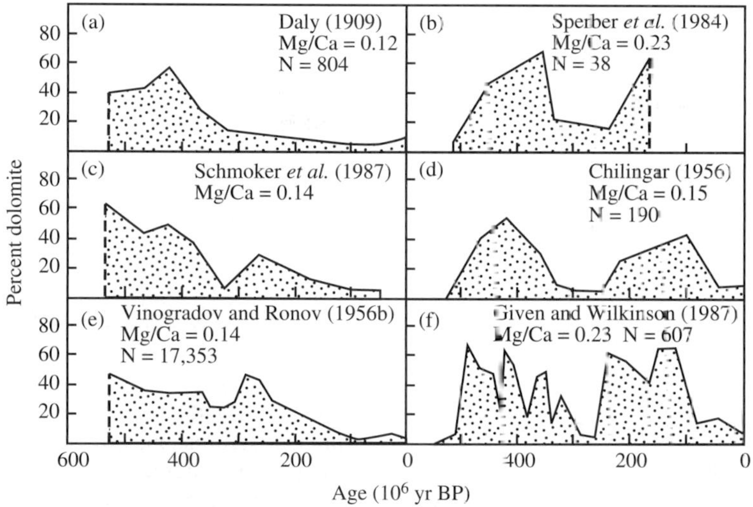

Figure 40 Estimates ((a) Daly (1909), (b) Sperber *et al.* (1984), (c) Schmoker *et al.* (1987), (d) Chilingar (1956), (e) Vinogradov and Ronov (1956b), and (f) Given and Wilkinson (1987)) of percent dolomite in Phanerozoic cratonic carbonate rocks as a function of age. Mg/Ca = average ratio (after Wilkinson and Algeo, 1989).

Holland and Zimmermann, 2000). Alternatively, if the trend in the calcite/dolomite ratio is cyclic in nature, this cyclicity could be interpreted as representing environmental change in the ocean-atmosphere system (Wilkinson and Algeo, 1989). For the discussion here, we accepted the data of Given and Wilkinson (1987) on the Ca/Mg ratio of Phanerozoic sedimentary carbonates (Figure 38) to calculate the mass ratios of these carbonate components as a function of age, but do realize that such data are still a matter of controversy (Zenger, 1989; Holland and Zimmermann, 2000). In reality, it is most likely that both cyclical and secular compositional changes in the ocean-atmosphere-sediment system, as well as diagenesis, contribute to dolomite abundance during geologic time.

The period-averaged mass ratio of calcite to dolomite is anomalously high for the Cambrian and Permian System rocks (Figure 38). For the remainder of the Phanerozoic, it appears to oscillate within the 1.1 ± 0.6 range, except for the limestone-rich Tertiary. Comparison with the generalized Phanerozoic sea-level curves of Vail *et al.* (1977) and Hallam (1984) (Figure 42) hints that dolomites are more abundant at times of higher sea levels. Mackenzie and Agegian (1986, 1989) and Given and Wilkinson (1987) were the first to suggest this possible cyclicity in the calcite/dolomite ratio during the Phanerozoic, and Lumsden (1985) observed a secular decrease in dolomite abundance in deep marine sediments from the Cretaceous to Recent, corresponding to the general fall of sea level during this time interval. These cycles in calcite/dolomite ratios correspond crudely to the Fischer (1984) two Phanerozoic super cycles and to the Mackenzie and Pigott (1981) oscillatory and submergent tectonic modes.

7.15.12.2.3 *Ooids and ironstones*

Although still somewhat controversial (e.g., Bates and Brand, 1990), the textures of ooids appear to vary during Phanerozoic time. Sorby (1879) first pointed out the petrographic differences between ancient and modern ooids: ancient ooids commonly exhibit relict textures of a calcitic origin, whereas modern ooids are dominantly made of aragonite. Sandberg (1975) reinforced these observations by study of the textures of some Phanerozoic ooids and a survey of the literature. His approach, and that of others who followed, was to employ the petrographic criteria, among others, of Sorby: i.e., if the microtexture of the ooid is preserved, then the ooid originally had a calcite mineralogy; if the ooid exhibits textural disruption, its original mineralogy was aragonite. The textures of originally aragonitic fossils are usually used as checks to deduce the original mineralogy of the ooids. Sandberg (1975) observed that ooids of inferred calcitic composition are dominant in rocks older than Jurassic.

Following this classical work, Sandberg (1983, 1985) and several other investigators (Mackenzie and Pigott, 1981; Wilkinson *et al.*, 1985; Bates and Brand, 1990) attempted to quantify further this relationship. Figure 41 is a schematic diagram representing a synthesis of the inferred mineralogy of ooids during the Phanerozoic. This diagram is highly tentative, and more data are needed to document the trends. However, it appears that while originally aragonitic ooids are found

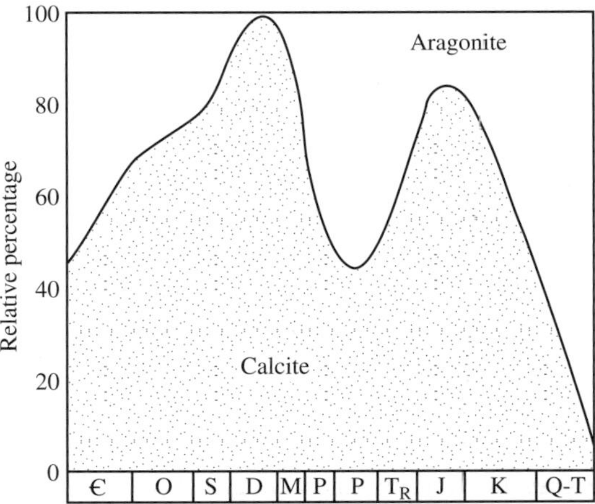

Figure 41 Inferred mineralogy of Phanerozoic ooids (after Morse and Mackenzie, 1990).

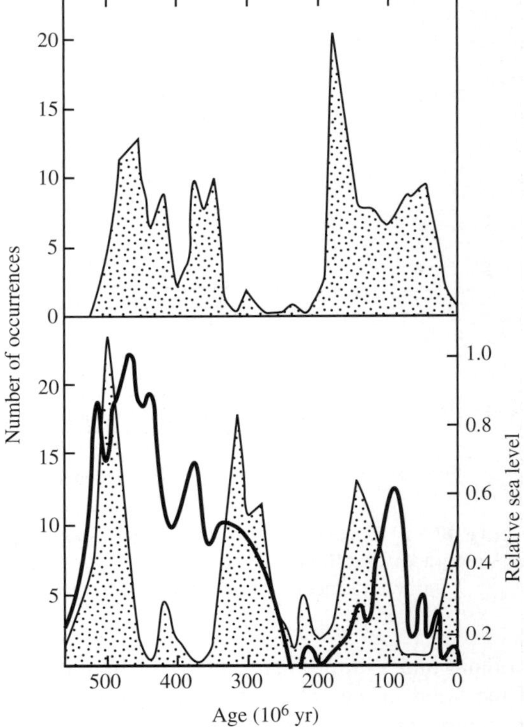

Figure 42 Number of occurrences of Phanerozoic ironstones (upper diagram, source Van Houten and Bhattacharyya, 1982) and oolitic limestones (lower diagram, source Wilkinson *et al.*, 1985) as a function of geologic age. The relative sea-level curve is that of Hallam (1984). Minima in occurrences appear to correlate with times of sea level withdrawal from the continents (after Morse and Mackenzie, 1990).

throughout the Phanerozoic, an oscillatory trend in the relative percentage of calcite versus aragonite ooids may be superimposed on a long-term evolutionary decrease in ooids with an inferred original calcitic mineralogy. Although the correlation is not strong, the two major maxima in the sea-level curves of Vail *et al.* (1977) and Hallam (1984) appear to coincide with times when calcite ooids were important seawater precipitates (Figure 42). Wilkinson *et al.* (1985) found that the best correlation between various data sets representing global eustasy and ooid mineralogy is that of inferred mineralogy with percentage of continental freeboard. Sandberg (1983) further concluded that the cyclic trend in ooid mineralogy correlates with cyclic trends observed for the inferred mineralogy of carbonate cements. Van Houten and Bhattacharyya (1982; and later Wilkinson *et al.*, 1985) showed that the distribution of Phanerozoic ironstones (hematite and chamosite oolitic deposits) also exhibits a definite cyclicity (Figure 42) that too appears to covary with the generalized sea-level curve. Minima appear to coincide with times of sea-level withdrawal from the continents.

7.15.12.2.4 Calcareous shelly fossils

In some similarity to the trends observed for the inorganic precipitates of ooids and ironstones, the mineralogy of calcareous fossils during the Phanerozoic also shows a cyclic pattern (Figure 25) with calcite being particularly abundant during high sea levels of the early to mid-Paleozoic and the Cretaceous (Stanley and Hardie, 1998; Kiessling, 2002). Overall (Figure 43) there is a general increase in the diversity of major groups of calcareous organisms such as coccolithophorids, pteropods, hermatypic corals, and coralline algae. It is noteworthy that the major groups of pelagic and benthic organisms contributing to carbonate sediments in today's ocean first appeared in the fossil record during the middle Mesozoic and progressively became more abundant. What is most evident in Figure 43 is a long-term increase in the production of biogenic carbonates dominated by aragonite and magnesian calcite mineralogies. Because organo-detrital carbonates are such an important part of the Phanerozoic carbonate rock record, this increase in metastable mineralogies played an important role in the pathway of diagenesis of carbonate sediments. The ubiquity of low magnesian calcite skeletal organisms for much of the Paleozoic led to production of calcitic organo-detrital sediments whose original bulk chemical and mineralogical composition was closer to that of their altered and lithified counterparts of Cenozoic age.

7.15.12.2.5 The carbonate cycle in the ocean

The partitioning of carbonate burial between shoal-water and deep sea realms has varied in a cyclic pattern through Phanerozoic time

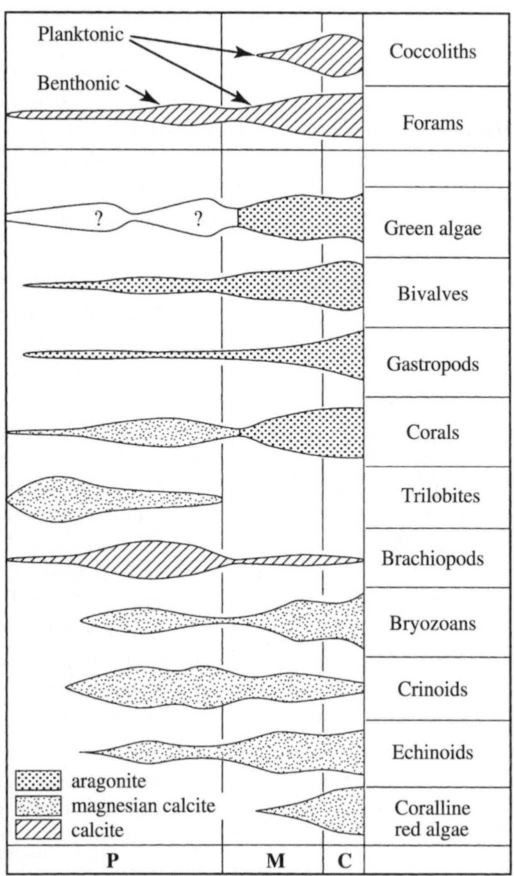

Figure 43 Mineralogical evolution of benthic and planktonic organism diversity during the Phanerozoic based on summaries of Milliken and Pigott (1977) and Wilkinson (1979). P is Paleozoic, M is Mesozoic, and C is Cenozoic.

(Morse and Mackenzie, 1990). The variation in the magnitudes of the fluxes of $(Ca,Mg)CO_3$ to the two environments through time is difficult to assess; even today's fluxes are probably not known within a factor of 2. To gain some impression of the fluxes involved, a tentative model of the carbonate carbon cycle in the world's oceans is shown in Figure 44. This is a representation of the mean state of the cycle during the most recent glacial to interglacial transition. About 18×10^{12} moles of calcium and magnesium (equivalent to 216×10^6 metric tons of carbon) accumulate yearly as carbonate minerals (Morse and Mackenzie, 1990), mainly as biological precipitates. Of this flux $\sim 6 \times 10^{12}$ moles are deposited as calcium and magnesium carbonates in shoal-water areas (Milliman, 1974; Smith, 1978; Wollast, 1993; Milliman *et al.*, 1999), and the remainder accumulates as calcareous oozes in the pelagic realm. The 12×10^{12} moles of carbonate accumulated annually in the deep sea are only ~17% of the annual carbonate production rate of 72×10^{12} mol of

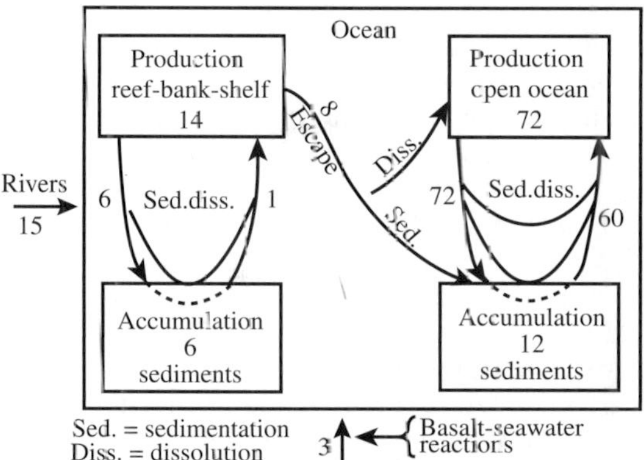

Figure 44 Tentative model of global ocean carbonate cycle. Fluxes are in units of 10^{12} mol C yr^{-1} as $(Ca,Mg)CO_3$ and represent estimates of fluxes averaged over the most recent glacial–interglacial transition (after Mackenzie and Morse, 1992).

the open ocean photic zone. This efficient recycling of carbonate carbon in the open ocean water column and at the sediment-water interface is a well-known feature of the marine carbon cycle (Broecker and Peng, 1982). It is important to note that much shoal-water carbonate production ends up in sediments of reefs, banks, etc., so that, in contrast to the pelagic realm, production rate more closely approximates sedimentation rate. However there is an escape of carbonate sediment from shoal-water areas to the deep sea, where it is deposited or dissolved (Land, 1979; Kiessling, 2002). The magnitude of this flux is poorly known but may affect the chemistry of open-ocean regions owing to dissolution of the carbonate debris (Droxler *et al.*, 1988; Agegian *et al.*, 1988; Sabine and Mackenzie, 1975). Furthermore, its accumulation on the slopes of banks may act as a record of paleoenvironmental change (Droxler *et al.*, 1983).

7.15.12.3 Geochemical Implications of the Phanerozoic Carbonate Record

The reasons for the mass-age relationships discussed previously are not totally clear. A number of investigators (Mackenzie and Pigott, 1981; Sandberg, 1985; Wilkinson *et al.*, 1985; Wilkinson and Given, 1986; Wilkinson and Algeo, 1989; Mackenzie and Agegian, 1989; Stanley and Hardie, 1998; Arvidson *et al.*, 2000) concluded that these observations are the result of changing atmosphere–hydrosphere environmental conditions through the Phanerozoic. Others argued that, for example, the ooid observations are not statistically significant (Bates and Brand, 1990) or that the Given and Wilkinson (1987) mass-age database is not valid

(Holland and Zimmermann, 2000). The latter authors do accept, however, that there is generally lower dolomite abundance in carbonate sediments deposited during the past 200 Myr. These qualifications notwithstanding, a number of previously described parameters (Sr/Ca, Mg/Ca, aragonite/calcite, possibly dolomite/calcite, and frequency of ooids and iron ores) appear to be related in some degree to sea-level stands, the latter at least in part a reflection of plate tectonic activity during the Phanerozoic.

Rowley (2002) in his recent publication argues that the rate of oceanic plate production may not have varied significantly for the latest 180 Myr. If so, this may have major implications on our understanding of the model linkage of tectonics to sea-level change, atmospheric CO_2, seawater chemistry and related phenomena. Nevertheless, while the 30–40% variations in seafloor spreading rates during the latest 100 Myr (Delaney and Boyle, 1986) are probably not justifiable, we cannot at present dismiss entirely the proposition that hydrothermal exchange between seafloor and ocean, and presumably the rate of plate generation, might still have varied somewhat. It may be possible therefore that the first-order changes in sea level can still be driven by the accretion of the ridges, but, at the same time, we should also search for alternative linkages of sea-level stands to the mineralogical and chemical properties of the Phanerozoic sedimentary cycle.

In the standard reasoning, the first-order changes in sea level are driven by the accretion of ridges: high accretion rate, high sea level; low accretion rate, low sea level. Regardless of the tie between oceanic plate production and sea level, extended times of global high sea level may have been times of enhanced atmospheric CO_2 levels (Berner *et al.*, 1983; Lasaga *et al.*, 1985; Berner, 2000), higher temperatures (not necessarily related solely to atmospheric CO_2 concentrations), probably lower seawater Mg/Ca ratios (Lowenstein *et al.*, 2001; Dickson, 2002), different saturation states of seawater (Arvidson *et al.*, 2000), and perhaps different seawater sulfate concentrations than at present. The converse is true for first-order global sea-level low stands. It appears that the environmental conditions for early dolomitization, formation of calcitic ooids and cements, and preponderance of calcitic reef-building organisms are best met during extended times of global high sea levels, with ubiquitous shallow-water and sabkha-like environments (calcite seas; Sandberg, 1983). Dolomitization of precursor calcite and aragonite phases, either in marine waters or in mixed continental-marine waters, would be enhanced under these conditions. Furthermore, the potentially lowered pH of marine waters during times of enhanced atmospheric carbon dioxide would favor syndepositional or later dolomitization in mixed marine-meteoric waters, because the range of seawater-meteoric compositional mixtures over which calcite could be dissolved and dolomite precipitated is expanded (Plummer, 1975). Perhaps superimposed on the hypothesized Phanerozoic cyclic dolomite/calcite ratio is a longer term trend in which dolomite abundance increases with increasing age, particularly in rocks older than 200 Myr, due to favorable environmental conditions as well as to advancing late diagenetic and burial dolomitization. During the past 150 Myr, this magnesium has been transferred out of the dolomite reservoir ("bank" of Holland and Zimmermann, 2000) into the magnesium silicate reservoir by precipitation of silicates and dissolution of dolomite (Garrels and Mackenzie, 1972; Garrels and Perry, 1974; Holland and Zimmermann, 2000) and to a lesser extent into the ocean reservoir, this accounting in part for the increasing Mg/Ca ratio of seawater during the past 150 Myr (Lowenstein *et al.*, 2001).

Thus, it appears that the apparent trends in Phanerozoic carbonate mineralogy are related to changes in atmosphere–hydrosphere conditions that are driven in part by plate tectonic mechanisms. However, we are aware that this tentative proposition requires collection of more data on the detailed chemistry and mineralogy of Phanerozoic carbonate sequences worldwide as well as resolution of the problem related to the past production rates of the oceanic lithosphere.

ACKNOWLEDGMENT

We would like to dedicate this chapter to Robert M. Garrels.

REFERENCES

Agegian C. R., Mackenzie F. T., Tribble J. S., and Sabine C. (1988) Carbonate production and flux from a mid-depth bank ecosystem. *Natl. Undersea Res. Prog. Res. Report* **88**(1), 5–32.

Allègre C. J. and Rousseau D. (1984) The growth of the continents through geological time studied by Nd isotope analysis of shales. *Earth Planet. Sci. Lett.* **67**, 19–34.

Armstrong R. L. (1981) Radiogenic isotopes: the case for crustal recycling on a near-steady-state no-continental-growth Earth. *Phil. Trans. Roy. Soc. London A Ser.* **301**, 443–472.

Arvidson R. S. and Mackenzie F. T. (1999) The dolomite problem: control of precipitation kinetics by temperature and saturation state. *Am. J. Sci.* **299**, 257–288.

Arvidson R. S., Mackenzie F. T., and Guidry M. W. (2000) Ocean/atmosphere history and carbonate precipitation rates: a solution to the dolomite problem. In *Marine Authigenesis: from Global to Microbial*, SEPM Spec. Publ. No. 65 (eds. C. R. Glenn, L. Prévôt-Lucas, and J. Lucas), SEPM, Tulsa, pp. 1–5.

Bates N. R. and Brand U. (1990) Secular variation of calcium carbonate mineralogy: an evaluation of oöid and micrite chemistries. *Geol. Rundsch.* **79**, 27–46.

Barrell J. (1917) Rhythms and the measurement of geological time. *Geol. Soc. Am. Bull.* **28**, 745–904.

Berner R. A. (1990) Atmospheric carbon dioxide levels over Phanerozoic time. *Science* **249**, 1382–1386.

Berner R. A. (1994) GEOCARB: II. A revised model of atmospheric CO_2 over Phanerozoic time. *Am. J. Sci.* **294**, 56–91.

Berner R. A. (2000) The effect of the rise of land plants on atmospheric CO_2 during the paleozoic. In *Plants Invade the Land: Evolutionary and Environmental Approaches* (eds. P. G. Gensel and D. Edwards). Columbia University Press, New York, pp. 173–178.

Berner R. A. and Kothavala Z. (2001) GEOCARB: III. A revised model of atmospheric CO_2 over phanerozoic time. *Am. J. Sci.* **301**, 182–204.

Berner U. and Streif H. (2001) *Klimafakten, 2001, Der Rückblick—Ein Schlüssel für die Zukunft*. Schweizerbart'sche Verlagsbuchhandlung, Science Publishers, Stuttgart.

Berner R. A., Lasaga A. C., and Garrels R. M. (1983) The carbonate silicate geochemical cycle and its effect on atmospheric carbon dioxide over the past 100 million years. *Am. J. Sci.* **283**, 641–683.

Berry J. P. and Wilkinson B. H. (1994) Paleoclimatic and tectonic control on the accumulation of North American cratonic sediment. *Geol. Soc. Am. Bull.* **106**, 855–865.

Bluth G. J. and Kump L. R. (1991) Phanerozoic paleogeology. *Am. J. Sci.* **291**, 284–308.

Boucot A. J. and Gray J. (2001) A critique of Phanerozoic climatic models involving changes in the CO_2 content of the atmosphere. *Earth Sci. Rev.* **56**, 1–159.

Brand U. and Veizer J. (1980) Chemical diagenesis of a multicomponent carbonate system: I. Trace elements. *J. Sedim. Petrol.* **50**, 1219–1236.

Broecker W. S. and Peng T. H. (1982) *Tracers in the Sea*. Eldigio Press, Palisades, NY.

Budyko M. I., Ronov A. B., and Yanshin A. L. (1985) *History of the Earth's Atmosphere*. Springer, Heidelberg.

Burke W. H., Denison R. F., Hetherington E. A., Koepnick R. F., Nelson H. F., and Otto J. B. (1982) Variation of seawater $^{87}Sr/^{86}Sr$ throughout Phanerozoic time. *Geology* **10**, 516–519.

Canfield D. E. and Teske A. (1996) Late Proterozoic rise in atmospheric oxygen concentration inferred from phylogenetic and sulphur-isotope studies. *Nature* **382**, 127–132.

Chambers L. A. and Trudinger P. A. (1979) Microbiological fractionation of stable sulfur isotopes: a review and critique. *Geomicrobiol. J.* **1**, 249–293.

Chilingar G. V. (1956) Relationship between Ca/Mg ratio and geologic age. *Am. Assoc. Petrol. Geol. Bull.* **40**, 2256–2266.

Cicero A. D. and Lohmann K. C. (2001) Sr/Mg variation during rock-water interaction: implications for secular changes in the elemental chemistry of ancient seawater. *Geochim. Cosmochim. Acta* **65**, 741–761.

Claypool G. E., Holser W. T., Kaplan I. R., Sakai H., and Zak I. (1980) The age curves of sulfur and oxygen isotopes in marine sulfate and their mutual interpretation. *Chem. Geol.* **28**, 199–260.

Cloud P. E. (1976) Major features of crustal evolution. *Trans. Geol. Soc. S. Afr.* **79**, 1–32.

Cogley J. G. (1984) Continental margins and the extent and number of continents. *Rev. Geophys. Space Phys.* **22**, 101–122.

Collerson K. D. and Kamber B. S. (1999) Evolution of the continents and the atmosphere inferred from the Th–U–Nb systematics of the depleted mantle. *Science* **283**, 1519–1522.

Condie K. C. (1989) *Plate Tectonics and Crustal Evolution*. Pergamon, London.

Condie K. C. (1993) Chemical composition and evolution of the upper continental crust: contrasting results from surface samples and shales. *Chem. Geol.* **104**, 1–37.

Condie K. C. (2000) Episodic continental growth models: afterthoughts and extensions. *Tectonophysics* **322**, 153–162.

Condie K. C. (2001) Continental growth during formation of Rodinia at 1.35–0.9 Ga. *Gondwana Res.* **4**, 5–16.

Cook T. D. and Bally A. W. (1975) *Shell Atlas: Stratigraphic Atlas of North and Central America*. Princeton University Press, Princeton, NJ.

Cox R. and Lowe D. R. (1995) A conceptual review of regional-scale controls on the composition of clastic sediment and the co-evolution of continental blocks and their sedimentary cover. *J. Sedim. Res.* **A65**, 1–12.

Daly R. A. (1909) First calcareous fossils and evolution of limestones. *Geol. Soc. Am. Bull.* **20**, 153–170.

Danchin R. V. (1967) Chromium and nickel in the Fig Tree shale from South Africa. *Science* **158**, 261–262.

Deines P. (1980) The isotopic composition of reduced organic carbon. In *Handbook of Environment Isotope Gechemistry: Vol. 1. The Terrestrial Environment* (eds. P. Fritz and J. C. Fontes). Elsevier, Amsterdam, pp. 329–406.

Delaney M. L. and Boyle E. A. (1986) Lithium in foraminiferal shells: implications for high-temperature hydrothermal circulation fluxes and oceanic generation rates. *Earth Planet. Sci. Lett.* **80**, 91–105.

De La Rocha C. L. and De Paolo J. (2000) Isotopic evidence for variations in the marine calcium cycle over the Cenozoic. *Science* **289**, 1176–1178.

Dickinson W. R., Beard L. S., Brakenridge G. R., Erjavec J. L., Ferguson R. C., Inman K. F., Knepp R. A., Lindberg F. A., and Ryberg P. T. (1983) Provenance of North American Phanerozoic sandstones in relation to tectonic setting. *Geol. Soc. Am. Bull.* **94**, 222–235.

Dickson J. A. D. (2002) Fossil echinoderms as monitor of the Mg/Ca ratio of Phanerozoic oceans. *Science* **298**, 1222–1224.

Dietz R. S. (1961) Continent and ocean basin evolution by spreading of the seafloor. *Nature* **190**, 854–857.

Droxler A. W., Schlager W., and Wallon C. C. (1983) Quaternary aragonite cycles and oxygen-isotopic records in Bahamian carbonate ooze. *Geology* **11**, 235–239.

Droxler A. W., Morse J. W., and Kornicker W. A. (1988) Controls on carbonate mineral accumulation in Bahamian basins and adjacent Atlantic ocean sediments. *J. Sedim. Petrol.* **58**, 120–130.

Engel A. E. J. (1963) Geologic evolution of North America. *Science* **140**, 143–152.

Engel A. E. J., Itson S. P., Engel C. G., Stickney D. M., and Cray E. J. (1974) Crustal evolution and global tectonics, a petrogenic view. *Bull. Geol. Soc. Am.* **85**, 843–858.

Faure G. (1986) *Principles of Isotope Geology*. Wiley, New York.

Fischer A. G. (1984) The two Phanerozoic super cycles. In *Catastrophes in Earth History* (eds. W. A. Berggren and J. A. Vancouvering). Princeton University Press, NJ, pp. 129–148.

Frakes L. A., Francis J. E., and Syktus J. I. (1992) *Climate Mode of the Phanerozoic: The History of the Earth's Climate over the Past 600 Million Years*. Cambridge University Press, Cambridge, UK.

Gaffin S. (1987) Ridge volume dependence of seafloor generation rate and inversion using long term sea level change. *Am. J. Sci.* **287**, 596–611.

Garrels R. M. and Mackenzie F. T. (1969) Sedimentary rock types: relative proportions as a function of geological time. *Science* **163**, 570–571.

Garrels R. M. and Mackenzie F. T. (1971a) *Evolution of Sedimentary Rocks*. Norton, New York.

Garrels R. M. and Mackenzie F. T. (1971b) Gregor's denudation of the continents. *Nature* **231**, 382–383.

Garrels R. M. and Mackenzie F. T. (1972) A quantitative model for the sedimentary rock cycle. *Mar. Chem.* **1**, 22–41.

Garrels R. M. and Perry E. A., Jr. (1974) Chemical history of the oceans deduced from postdepositional changes in sedimentary rocks. In *Studies in Paleo-Oceanography Special Publication Society Economic Paleontologists and Mineralogists*, **20**, (ed. W. W. Hay), SPEM, Tulsa, OK, pp. 193–204.

Gibbs A. K., Montgomery C. W., O'Day P. A., and Erslev E. A. (1988) Crustal evolution revisited: reply to comments by S. M. McLennan *et al.*, on "The Archean–Proterozoic transition: evidence from the geochemistry of metasedimentary rocks from Guyana and Montana". *Geochim. Cosmochim. Acta* **52**, 793–795.

Gilluly J. (1949) Distribution of mountain building in geologic time. *Geol. Soc. Am. Bull.* **60**, 561–590: **120**, 135–139.

Given R. K. and Wilkinson B. H. (1987) Dolomite abundance and stratigraphic age: constraints on rates and mechanisms of Phanerozoic dolostone formation. *J. Sedim. Petrol.* **57**, 1068–1079.

Goddéris Y. and Veizer J. (2000) Tectonic control of chemical and isotopic composition of ancient oceans: the impact of continental growth. *Am. J. Sci.* **300**, 434–461.

Goldschmidt V. M. (1933) Grundlagen der quantitativen Geochemie. *Fortschr. Mineral. Kristallog. Petrogr.* **17**, 1–112.

Goldstein S. J. and Jacobsen S. B. (1987) The Nd and Sr isotope systematics of river water dissolve material: implications for the source of Nd and Sr in seawater. *Chem. Geol.* **66**, 245–272.

Gregor C. B. (1968) The rate of denudation in post-Algonkian time. *Proc. Koninkl. Ned. Akad. Wetenschap. Ser. B: Phys. Sci.* **71**, 22–30.

Gregor C. B. (1985) The mass-age distribution of Phanerozoic sediments. In *Geochronology and the Geologic Record*, Mem. No. 10 (ed. N. J. Snelling). Geol. Soc. London, London, UK, pp. 284–289.

Gregory R. T. (1991) Oxygen isotope history of seawater revisited: composition of seawater. In *Stable Isotope Geochemistry: a Tribute to Samuel Epstein*, Geochem. Soc. Spec. Publ. 3 (eds. H. P. Tailor, Jr., J. R. O'Neil, and I. R. Kaplan), Min. Soc. America, Washington, DC, pp. 65–76.

Grotzinger J. P. and James N. P. (2000) *Precambrian Carbonates, Evolution and Understanding: Carbonate Sedimentation and Diagenesis in the Evolving Precambrian World.* SEPM Special Publication # 67.

Habicht K. A. and Canfield D. E. (1996) Sulphur isotope fractionation in modern microbial mats and the evolution of the sulphur cycle. *Nature* **382**, 342–343.

Hallam A. (1984) Pre-Quaternary sea-level changes. *Ann. Rev. Earth Planet. Sci.*, **12**, 205–243.

Hardie L. A. (1987) Perspectives on dolomitization: a critical view of some current views. *J. Sedim. Petrol.* **57**, 166–183.

Hardie L. A. (1996) Secular variation in seawater chemistry: an explanation of the coupled secular variations in the mineralogies of marine limestones and potash evaporites over the past 600 Myr. *Geology* **24**, 279–283.

Harrison A. G. and Thode H. G. (1958) Mechanism of the bacterial reduction of sulfate from isotope fractionation studies. *Faraday Soc. Trans.* **54**, 84–92.

Hay W. W., Sloan J. L. II, and Wold C. N. (1988) Mass/age distribution and composition of sediments on the ocean floor and the global rate of sediment subduction. *J. Geophys. Res.* **93**, 14933–14940.

Hay W. W., Wold C. N., Söding E., and Flügel S. (2001) Evolution of sediment fluxes and ocean salinity. In *Geologic Modelling and Simulations: Sedimentary Systems* (eds. D. F. Merriam and J. C. Davis). Kluwer Academic/Plenum, Dordrecht, pp. 153–167.

Hayes J. M., Des Marais D. J., Lambert J. B., Strauss H., and Summons R. E. (1992) Proterozoic biogeochemistry. In *The Proterozoic Biosphere: A Multidisciplinary Study* (eds. J. W. Schopf and C. Klein). Cambridge University Press, Cambridge, UK, pp. 81–134.

Hemming N. G. and Hanson G. N. (1992) Boron isotope composition and concentration in modern marine carbonates. *Geochim. Cosmochim. Acta* **56**. 537–543.

Hess H. H. (1962) History of ocean basins. In *Petrologic Studies* (eds. A. E. J. Engel, H. L. James, and B. F. Leonard). Geological Society of America, Bouldev, co., pp. 599–620.

Hoffman P. F., Kaufman A. J., Halverson G. P., and Schrag D. P. (1998) A Neoproterozoic snowball Earth. *Science* **281**, 1342–1346.

Holland H. D. (1978) *The Chemistry of the Atmosphere and Oceans*. Wiley, New York.

Holland H. D. (1984) *The Chemical Evolution of the Atmosphere and Oceans*. Princeton University Press, Princeton, NJ.

Holland H. D. and Zimmerman H. (2000) The dolomite problem revisited. *Int. Geol. Rev.* **2**, 481.

Holser W. T. (1977) Catastrophic chemical events in the history of the ocean. *Nature* **267**, 403–408.

Holser W. T. (1984) Gradual and abrupt shifts in ocean chemistry during Phanerozoic time. In *Patterns of Change in Earth Evolution* (eds. H. D. Holland and A. F. Trendall). Springer, Heidelberg, pp. 123–143.

Holser W. T. and Kaplan I. R. (1966) Isotope geochemistry of sedimentary sulfates. *Chem. Geol.* **1**, 93–135.

Holser W. T., Schidlowski M., McKenzie F. T., and Maynard J. B. (1988) Geochemical cycles of carbon and sulfur. In *Chemical Cycles in the Evolution of the Earth* (eds. C. B. Gregor, R. M. Garrels, F. T. Mackenzie, and J. B. Maynard). Wiley, New York, pp. 105–173.

Horita J., Friedman T. J., Lazar B., and Holland H. D. (1991) The composition of Permian seawater. *Geochim. Cosmochim. Acta* **55**, 417–432.

Horita J., Zimmermann H., and Holland H. D. (2002) The chemical evolution of seawater during the Phanerozoic: implications from the record of marine evaporites. *Geochim. Cosmochim. Acta* **66**, 3733–3756.

Ingersoll R. V. and Busby C. J. (1995) Tectonics of sedimentary basins. In *Tectonics of Sedimentary Basins* (eds. C. J. Busby and R. V. Ingersoll). Blackwell, Oxford, 1–51.

Kampschulte A. (2001) Schwefelisotopenuntersuchungen an strukturell substituierten Sulfaten in marinen Karbonaten des Phanerozoikums—Implikationen für die geochemische Evolution des Meerwassers und Korrelation verschiedener Stoffkreisläufe. PhD Thesis, Ruhr Universität, Bochum.

Kampschulte A. and Strauss H. (1998) The isotopic composition of trace sulphates in Paleozoic biogenic carbonates: implications for coeval seawater and geochemical cycles. *Min. Mag.* **62A**, 744–745.

Karhu J. A. and Holland H. D. (1996) Carbon isotopes and the rise of atmospheric oxygen. *Geology* **24**, 867–870.

Kiessling W. (2002) Secular variations in the Phanerozoic reef ecosystem. *SEPM Spec. Publ.* **72**, 625–690.

Kroopnick P. (1980) The distribution of ^{13}C in the Atlantic ocean. *Earth Planet. Sci. Lett.* **49**, 469–484.

Kuenen Ph. H. (1950) *Marine Geology*. Wiley, New York.

Kump L. R. (1989) Alternative modeling approaches to the geochemical cycles of carbon, sulfur, and strontium isotopes. *Am. J. Sci.* **289**, 390–410.

Land L. S. (1979) The fate of reef-derived sediment on the north Jamaica island slope. *Mar. Geol.* **29**, 55–71.

Land L. S. (1980) The isotopic and trace element geochemistry of dolomite: the state of the art. *SEPM Spec. Publ.* **28**, 87–110.

Land L. S. (1995) Comment on "Oxygen and carbon isotopic composition of Ordovician brachiopods: implications for coeval seawater" by H. Qing and J. Veizer. *Geochim. Cosmochim. Acta* **59**, 2843–2844.

Lasaga A. C. and Ohmoto H. (2002) The oxygen geochemical cycle: dynamics and stability. *Geochem. Cosmochim. Acta* **66**, 361–381.

Lasaga A. C., Berner R. A., and Garrels R. M. (1985) An improved geochemical model of atmospheric CO_2 fluctuations over the past 100 million years. In *The Carbon Cycle and Atmospheric CO_2: Natural Variations Archean to Present*, Geophysical Monograph 32 (eds. E. T. Sundquist and W. S. Broecker). American Geophysical Union, Washington, DC, pp. 397–411.

Lear C. H., Elderfield H., and Wilson P. A. (2003) A Cenozoic seawater Sr/Ca record from benthic foraminiferal calcite and its application in determining global weathering fluxes. *Earth Planet. Sci. Lett.* **208**, 69–84.

Lemarchand D., Gaillardet J., Lewin E., and Allègre J. C. (2000) The influence of rivers on marine boron isotopes and implications for reconstructing past ocean pH. *Nature* **408**, 951–954.

Le Pichon X., Francheteau J., and Bonnin J. (1973) *Plate Tectonics*. Elsevier, Amsterdam.

Lerman A. (1979) *Geochemical Processes: Water and Sediment Environments*. Wiley, New York.

Longinelli A. (1989) Oxygen-18 and sulphur-34 in dissolved oceanic sulphate and phosphate. In *Handbook of Environmental Isotope Geochemistry* (eds. P. Fritz and J. C. Fontes). Elsevier, Amsterdam, pp. 219–255.

Lowenstein T. K., Timofeeft M. N., Brennan S. T., Hardie L. A., and Demicco R. V. (2001) Oscillations in Phanerozoic seawater chemistry: evidence from fluid inclusions. *Science* **294**, 1086–1088.

Lumsden D. N. (1985) Secular variations in dolomite abundance in deep marine sediments. *Geology* **13**, 766–769.

Mackenzie F. T. (1975) Sedimentary cycling and the evolution of seawater. In *Chemical Oceanography*, 2nd edn. (eds. J. P. Riley and G. Skirrow). Academic Press, London, vol. 1, pp. 309–364.

Mackenzie F. T. and Agegian C. (1986) Biomineralization, atmospheric CO_2 and the history of ocean chemistry. In *Proc. 5th Int. Conf. Biomineral. Department of Geology, University Texas, Arlington, 2.*

Mackenzie F. T. and Agegian C. (1989) Biomineralization and tentative links to plate tectonics. In *Origin, Evolution, and Modern Aspects of Biomineralization in Plants and Animals* (ed. R. E. Crick). Plenum, New York, pp. 11–28.

Mackenzie F. T. and Morse J. W. (1992) Sedimentary carbonates through Phanerozoic time. *Geochim. Cosmochim. Acta* **56**, 3281–3295.

Mackenzie F. T. and Pigott J. P. (1981) Tectonic controls of Phanerozoic sedimentary rock cycling. *J. Geol. Soc. London* **138**, 183–196.

McArthur J. M. (1994) Recent trends in strontium isotope stratigraphy. *Terra Nova* **6**, 331–358.

McCulloch M. T. and Bennett V. C. (1994) Progressive growth of the Earth's continental crust and depleted mantle: geochemical constraints. *Geochim. Cosmochim. Acta* **58**, 4717–4738.

McCulloch M. T. and Wasserburg G. J. (1978) Sm–Nd and Rb–Sr chronology of continental crust formation. *Science* **200**, 1003–1011.

McLennan S. M. and Murray R. W. (1999) Geochemistry of sediments. In *Encyclopedia of Geochemistry* (eds. C. P. Marshall and R. W. Fairbridge). Kluwer, Dordrecht, pp. 282–292.

McLennan S. M. and Taylor S. R. (1999) Earth's continental crust. In *Encyclopedia of Geochemistry* (eds. C. P. Marshall and R. W. Fairbridge). Kluwer, Dordrecht, pp. 145–151.

McLennan S. M., Taylor S. R., McCulloch M. T., and Maynard J. B. (1990) Geochemical and Nd–Sr isotopic composition of deep-sea turbidites: crustal evolution and plate tectonic associations. *Geochim. Cosmochim. Acta* **54**, 2015–2050.

McLennan S. M., Bock B., Hemming S. R., Hurowitz J. A., Lev S. M., and McDaniel D. K. (2003) The roles of provenance and sedimentary processes in the geochemistry of sedimentary rocks. In *Geochemistry of Sediments and Sedimentary Rocks: Evolutionary Considerations to Mineral Deposit-Forming Environments*. (ed. D. R. Lentz), Geol. Assoc. Canada GEOtext. St. John's, Nfld, vol. 5, pp. 1–31.

McLennan S. M., Taylor S. R., and Hemming S. R. Composition, differentiation and evolution of continental crust: constraints from sedimentary rocks and heat flow. In *Evolution and Differentiation of Continental Crust* (eds. M. Brown and T. Rushmer). Cambridge University Press (in press).

Milliken K. L. and Pigott J. D. (1977) Variation of oceanic Mg/Ca ratio through time-implications for the calcite sea. *Geol Soc. Am. South-Central Meet. (abstr.)*, 64–65.

Milliman J. D. (1974) *Recent Sedimentary Carbonates: 1. Marine Carbonates*. Springer, Heidelberg.

Milliman J. D., Troy P. J., Balch W. M., Adams A. K., Li Y.-H., and Mackenzie F. T. (1999) Biologically mediated dissolution of calcium carbonate above the chemical lysocline? *Deep Res. I* **46**, 1653–1669.

Millot R., Gaillardet J., Dupre B., and Allègre J. C. (2002) The global control of silicate weathering rates and the coupling with physical erosion: new insights from rivers on the Canadian shield. *Earth Planet. Sci. Lett.* **196**, 83–98.

Morgan W. J. (1968) Rises, trenches, great faults, and crustal blocks. *J. Geophys. Res.* **73**, 1959–1982.

Morse J. W. and Mackenzie F. T. (1990) *Geochemistry of Sedimentary Carbonates*. Elsevier, Amsterdam.

Muehlenbachs K. (1998) The oxygen isotopic composition of the oceans, sediments and the seafloor. *Chem. Geol.* **145**, 263–273.

Nägler T. F., Eisenhauer A., Müller A., Hemleben C., and Kramers J. (2000) The δ^{44}Ca-temperature calibration on fossil and cultured *Globigerinoides sacculifer*: new tool for reconstruction of past sea surface temperatures. *Geochem. Geophys. Geosys.* **1** 2000900091.

Nanz R. H., Jr. (1953) Chemical composition of Precambrian slates with notes on the geochemical evolution of lutites. *J. Geol.* **61**, 51–64.

Nesbitt H. W. and Young G. M. (1984) Predictions of some weathering trends of plutonic and volcanic rocks based on thermodynamic and kinetic considerations. *Geochim. Cosmochim. Acta* **48**, 1523–1534.

O'Leary M. H. (1988) Carbon isotopes in photosynthesis. *Bioscience* **38**, 328–336.

O'Nions R. K., Hamilton P. J., and Hooker P. J. (1983) A Nd isotope investigation of sediments related to crustal development in the British Isles. *Earth Planet. Sci. Lett.* **63**, 329–338.

Palmer M. R. and Edmond J. M. (1989) Strontium isotope budget of the modern ocean. *Earth Planet. Sci. Lett.* **92**, 11–26.

Palmer M. R. and Swihart G. H. (1996) Boron isotope geochemistry: an overview. *Rev. Mineral.* **33**, 709–744.

Patchett P. J. (2003) Provenance and crust-mantle evolution studies based on radiogenic isotopes in sedimentary rocks. In *Geochemistry of Sediments and Sedimentary Rocks: Evolutionary Considerations to Mineral Deposit—Forming Environments*, Geological Association of Canada GEOtext (ed. D. R. Lentz). St. John's, Nfld, vol. 5, pp. 89–97.

Pearson P. N. and Palmer M. R. (2000) Atmospheric carbon dioxide concentrations over the past 60 million years. *Nature* **406**, 695–699.

Peterman Z. E., Hedge C. E., and Tourtelot H. A. (1970) Isotopic composition of strontium in seawater throughout Phanerozoic time. *Geochim. Cosmochim. Acta* **34**, 105–120.

Peucker-Ehrenbring B. and Ravizza G. (2000) The marine osmium isotope record. *Terra Nova* **12**, 205–219.

Pielou E. C. (1977) *Mathematical Ecology*. Wiley, New York.

Plank T. and Langmuir C. H. (1998) The chemical composition of subducting sediment and its consequences for the crust and mantle. *Chem. Geol.* **145**, 325–394.

Plummer L. N. (1975) Mixing of seawater with calcium carbonate ground water. *Geol. Soc. Am. Mem.* **142**, 219–236.

Rea D. K. and Ruff L. J. (1996) Composition and mass flux of sediment entering the world's subduction zones: implications for global sedimentary budgets, great earthquakes, and volcanism. *Earth Planet. Sci. Lett.* **140**, 1–12.

Renard M. (1986) Pelagic carbonate chemostratigraphy (Sr, Mg, ^{18}O, ^{13}C). *Mar. Micropaleontol.* **10**, 117–164.

Reymer A. and Schubert G. (1984) Phanerozoic additions to the continental crust and crustal growth. *Tectonics* **3**, 63–77.

Ronov A. B. (1949) A history of the sedimentation and epeirogenic movements of the European part of the USSR (based on the volumetric method). *AN SSSR Geofiz. Inst. Trudy* **3**, 1–390. (in Russian).

Ronov A. B. (1964) Common tendencies in the chemical evolution of the Earth's crust, ocean and atmosphere. *Geochem. Int.* **1**, 713–737.

Ronov A. B. (1968) Probable changes in the composition of seawater during the course of geological time. *Sedimentology* **10**, 25–43.

Ronov A. B. (1976) Global carbon geochemistry, volcanism, carbonate accumulation, and life. *Geokhimiya* **8**, 1252–1257; *Geochem. Int.* **13**, 175–196.

Ronov A. B. (1982) The Earth's sedimentary shell (quantitative patterns of its structure, compositions, and evolution). *Int. Geol. Rev.* **24**, 1313–1388.

Ronov A. B. (1993) *Stratisphere or Sedimentary Layer of the Earth*. Nauka, Russian.

Ross S. K. and Wilkinson B. M. (1991) Planktogenic/eustatic control on cratonic/oceanic carbonate accumulation. *J. Geol.* **99**, 497–513.

Rothman D. H. (2002) Atmospheric carbon dioxide levels for the last 500 million years. *Proc. Natl. Acad. Sci.* **99**, 4167–4171.

Rothman D. H., Hayes J. M., and Summons R. (2003) Dynamics of the Neoproterozoic carbon cycle. *Proc. Natl. Acad. Sci* **100**, 8124–8129.

Rowley D. B. (2002) Rate of plate creation and destruction: 180 Ma to present. *Geol. Soc. Am. Bull.* **114**, 927–933.

Rubey W. W. (1951) Geologic history of seawater: an attempt to state the problem. *Geol. Soc. Am. Bull.* **62**, 1111–1148.

Ruddiman W. F. (1997) *Tectonic Uplift and Climate Change*. Plenum, New York.

Sabine C. and Mackenzie F. T. (1975) Bank-derived carbonate sediment transport and dissolution in the Hawaiian Archipelago. *Aquat. Geochem.* **1**, 189–230.

Sackett W. M. (1989) Stable carbon isotope studies on organic matter in the marine environment. In *Handbook of Environmental Isotope Geochemistry* (eds. P. Fritz and J. C. Fontes). Elsevier, Heidelberg, vol. 3, pp. 139–169.

Sadler P. M. (1981) Sediment accumulation rates and the completeness of stratigraphic sections. *J. Geol.* **89**, 569–584.

Sandberg P. A. (1975) New interpretation of Great Salt Lake oöids and of ancient non-skeletal carbonate mineralogy. *Sedimentology* **22**, 497–538.

Sandberg P. A. (1983) An oscillating trend in non-skeletal carbonate mineralogy. *Nature* **305**, 19–22.

Sandberg P. A. (1985) Nonskeletal aragonite and pCO_2 in the Phanerozoic and Proterozoic. In *The Carbon Cycle and Atmospheric CO_2: Natural Variations Archean to Present*, Geophys. Monogr. Ser. 32 (eds. E. T. Sundquist and W. S. Broecker). American Geophysical Union, Washington, DC, pp. 585–594.

Schidlowski M., Hayes J. M., and Kaplan I. R. (1983) Isotopic inferences of ancient biochemistries: carbon, sulfur, hydrogen, and nitrogen. In *Earth's Earliest Biosphere* (ed. J. W. Schopf). Princeton University Press, Princeton, NJ, pp. 149–186.

Schmitt A.-D., Bracke G., Stille P., and Kiefel B. (2001) The calcium isotope composition of modern seawater determined by thermal ionisation mass spectrometry. *Geostand. Newslett.* **25**, 267–275.

Schuchert C. (1931) Geochronology of the age of the Earth on the basis of sediments and life. *Natl. Res. Council Bull.* **80**, 10–64.

Sclater J. G., Parsons B., and Jaupart C. (1981) Oceans and continents: similarities and differences in the mechanism of heat loss. *J. Geophys. Res.* **86**, 11535–11552.

Sepkoski J. J., Jr. (1989) Periodicity in extinction and the problem of catastrophism in the history of life. *J. Geol. Soc. London* **146**, 7–19.

Shaviv N. J. and Veizer J. (2003) Celestial driver of Phanerozoic climate?. *GSA Today* **13**(7), 4–10.

Shaw D. B. and Weaver C. E. (1965) The mineralogical composition of shales. *J. Sedim. Petrol.* **35**, 213–222.

Shields G. and Veizer J. (2002) The Precambrian marine carbonate isotope database: version 1. *Geochem. Geophys. Geosys.* **3**(6), June 6, 2002, p. 12 (http://g-cubed.org/gc2002/2001GC000266).

Skulan J., De Paolo D. J., and Owens T. L. (1997) Biological control of calcium isotopic abundences in the global calcium cycle. *Geochim. Cosmochim. Acta* **61**, 2505–2510.

Smith S. V. (1978) Coral reef area and contributions of reefs to processes and resources of the world's oceans. *Nature* **273**, 225–226.

Sorby H. C. (1879) The structure and origin of limestones. *Proc. Geol. Soc. London* **35**, 56–95.

Southam J. R. and Hay W. W. (1981) Global sedimentary mass balance and sea level changes. In *The Oceanic Lithosphere, The Sea* (ed. C. Emiliani). Wiley, New York, vol. 7, pp. 1617–1684.

Sperber C. M., Wilkinson B. H., and Peacor D. R. (1984) Rock composition, dolomite stoichiometry, and rock/water reactions in dolomitic carbonate rocks. *J. Geol.* **92**, 609–622.

Sprague D. and Pollack H. N. (1980) Heat flow in the Mesozoic and Cenozoic. *Nature* **285**, 393–395.

Stanley S. M. and Hardie L. A. (1998) Secular oscillations in the carbonate mineralogy of reef-building and sediment-producing organisms driven by tectonically forced shifts in seawater chemistry. *Palaeogeogr. Palaeoclimat. Palaeoecol.* **144**, 3–19.

Steuber T. and Veizer J. (2002) A Phanerozoic record of plate tectonic control of seawater chemistry and carbonate sedimentation. *Geology* **30**, 1123–1126.

Strakhov N. M. (1964) States and development of the external geosphere and formation of sedimentary rocks in the history of the Earth. *Int. Geol. Rev.* **6**, 1466–1482.

Strakhov N. M. (1969) *Principles of Lithogenesis*. Oliver and Boyd, Edinburgh, Vol. 2.

Strauss H. (1993) The sulfur isotopic record of Precambrian sulfates: new data and a critical evaluation of the existing record. *Precamb. Res.* **63**, 225–246.

Sylvester P. J. (2000) Continental formation, growth and recycling. *Tectonophysics* **322**, 163–190.

Sylvester P. J., Campbell I. H., and Bowyer D. A. (1997) Niobium/uranium evidence for early formation of the continental crust. *Science* **275**, 521–523.

Tan F. C. (1988) Stable carbon isotopes in dissolved inorganic carbon in marine and estuarine environments. In *Handbook of Environmental Isotope Geochemistry* (eds. P. Fritz and J. C. Fontes). Elsevier, Amsterdam, vol. 3, pp. 171–190.

Taylor S. R. (1979) Chemical composition and evolution of continental crust: the rare earth element evidence. In *The Earth: Its Origin, Structure and Evolution* (ed. M. W. McElhinny). Academic Press, New York, pp. 353–376.

Taylor S. R. and McLennan S. M. (1985) *The Continental Crust: its Composition and Evolution*. Blackwell, Oxford, UK.

Vail P. R., Mitchum R. W., and Thompson S. (1977) Seismic stratigraphy and global changes of sea level. 4, Global cycles of relative changes of sea level. *AAPG Mem.* **26**, 83–97.

Van Houten F. B. and Bhattacharyya D. P. (1982) Phanerozoic oölitic ironstones-geologic record and facies. *Ann. Rev. Earth Planet. Sci.* **10**, 441–458.

Veizer J. (1973) Sedimentation in geologic history: recycling versus evolution or recycling with evolution. *Contrib. Mineral. Petrol.* **38**, 261–278.

Veizer J. (1983) Trace element and isotopes in sedimentary carbonates. *Rev. Mineral.* **11**, 265–300.

Veizer J. (1985) Carbonates and ancient oceans: isotopic and chemical record on timescales of 10^7–10^9 years. *Geophys. Monogr. Am. Geophys. Union* **32**, 595–601.

Veizer J. (1988a) The evolving exogenic cycle. In *Chemical Cycles in the Evolution of the Earth* (eds. C. B. Gregor, R. M. Garrels, F. T. Mackenzie, and J. B. Maynard). Wiley, New York, pp. 175–220.

Veizer J. (1988b) Continental growth: comments on "The Archean–Proterozoic transition: evidence from Guyana and Montana" by A. K. Gibbs C. W. Montgomery P. A. O'Day and E. A. Erslev. *Geochim. Cosmochim. Acta* **52**, 789–792.

Veizer J. (1988c) Solid Earth as a recycling system: temporal dimensions of global tectonics. In *Physical and Chemical Weathering in Geochemical Cycle* (eds. A. Lerman and M. Meyback). Reidel, Dordrecht, pp. 357–372.

Veizer J. (1995) Reply to the comment by L. S. Land on "Oxygen and carbon isotopic composition of Ordovician brachiopods: implications for coeval seawater: discussion". *Geochim. Cosmochim. Acta* **59**, 2845–2846.

Veizer J. (2003) Isotopic evolution of seawater on geological timescales: sedimentological perspective. In *Geochemistry of Sedimentary Rocks: Secular Evolutionary Considerations to Mineral Deposit-forming Environments*, Geological Association of Canda, GEOtext (ed. D. R. Lentz). St. John's, Nfld, vol. 5, pp. 99–114.

Veizer J. and Compston W. (1974) $^{87}Sr/^{86}Sr$ composition of seawater during the Phanerozoic. *Geochim. Cosmochim. Acta* **38**, 1461–1484.

Veizer J. and Compston W. (1976) $^{87}Sr/^{86}Sr$ in Precambrian carbonates as an index of crustal evolution. *Geochim. Cosmochim. Acta* **40**, 905–914.

Veizer J. and Ernst R. E. (1996) Temporal pattern of sedimentation: Phanerozoic of North America. *Geochem. Int.* **33**, 64–76.

Veizer J. and Jansen S. L. (1979) Basement and sedimentary recycling and continental evolution. *J. Geol.* **87**, 341–370.

Veizer J. and Jansen S. L. (1985) Basement and sedimentary recycling: 2. Time dimension to global tectonics. *J. Geol.* **93**, 625–643.

Veizer J., Holser W. T., and Wilgus C. K. (1980) Correlation of $^{13}C/^{12}C$ and $^{34}S/^{32}S$ secular variations. *Geochim. Cosmochim. Acta* **44**, 579–587.

Veizer J., Compston W., Hoefs J., and Nielsen H. (1982) Mantle buffering of the early oceans. *Naturwissenschaften* **69**, 173–180.

Veizer J., Laznicka P., and Jansen S. L. (1989) Mineralization through geologic time: recycling perspective. *Am. J. Sci.* **289**, 484–524.

Veizer J., Ala D., Azmy K., Bruckschen P., Buhl D., Bruhn F., Carden G. A. F., Diener A., Ebneth S., Goddéris Y., Jasper T., Korte C., Pawellek F., Podlaha O. G., and Strauss H. (1999) $^{87}Sr/^{86}Sr$, $\delta^{13}C$ and $\delta^{18}O$ evolution of Phanerozoic seawater. *Chem. Geol.* **161**, 59–88.

Veizer J., Goddéris Y., and François L. M. (2000) Evidence for decoupling of atmospheric CO_2 and global climate during the Phanerozoic eon. *Nature* **408**, 698–701.

Vervoort J. D., Patchett P. J., Blichert-Toft J., and Albarede F. (1999) Relationship between Lu–Hf and Sm–Nd isotopic systems in the global sedimentary system. *Earth Planet. Sci. Lett.* **168**, 79–99.

Vinogradov A. P. (1967) The formation of the oceans. *Izv. Akad. Nauk SSSR, Ser. Geol.* **4**, 3–9.

Vinogradov A. P. and Ronov A. B. (1956a) Composition of the sedimentary rocks of the Russian platform in relation to the history of its tectonic movements. *Geochemistry* **6**, 533–559.

Vinogradov A. P. and Ronov A. B. (1956b) Evolution of the chemical composition of clays of the Russian platform. *Geochemistry* **2**, 123–129.

Vinogradov A. P., Ronov A. B., and Ratynskii V. Y. (1952) Evolution of the chemical composition of carbonate rocks. *Izv. Akad. Nauk SSSR Ser. Geol.* **1**, 33–60.

Wadleigh M. A., Veizer J., and Brooks C. (1985) Strontium and its isotopes in Canadian rivers: fluxes and global implications. *Geochim. Cosmochim. Acta* **49**, 1727–1736.

Walker L. J., Wilkinson B. H., and Ivany L. C. (2002) Continental drift and Phanerozoic carbonate accumulation in shallow-shelf and deep-marine settings. *J. Geol.* **110**, 75–87.

Wallmann K. (2001) The geological water cycle and the evolution of marine $\delta^{18}O$. *Geochim. Cosmochim. Acta* **65**, 2469–2485.

White W. M. and Patchett P. J. (1982) Hf–Nd–Sr and incompatible-element abundances in island arcs: implications for magma origin and crust-mantle evolution. *Earth Planet. Sci. Lett.* **67**, 167–185.

Wilkinson B. H. (1979) Biomineralization, paleoceanography, and the evolution of calcareous marine organisms. *Geology* **7**, 524–527.

Wilkinson B. H. and Algeo T. J. (1989) Sedimentary carbonate record of calcium–magnesium cycling. *Am. J. Sci.* **289**, 1158–1194.

Wilkinson B. H. and Given R. K. (1986) Secular variation in abiotic marine carbonates: constraints on Phanerozoic atmospheric carbon dioxide contents and oceanic Mg/Ca ratios. *J. Geol.* **94**, 321–334.

Wilkinson B. H., Owen R. M., and Carroll A. R. (1985) Submarine hydrothermal weathering, global eustasy and carbonate polymorphism in Phanerozoic marine oolites. *J. Sedim. Petrol.* **55**, 171–183.

Windley B. F. (1984) *The Evolving Continents*, 2nd edn. Wiley, New York.

Wolery T. J. and Sleep N. H. (1988) Interaction of the geochemical cycles with the mantle. In *Chemical Cycles in the Evolution of the Earth* (eds. C. B. Gregor, R. M. Garrels, F. T. Mackenzie, and J. B. Maynard). Wiley, New York, pp. 77–104.

Wollast R. (1993) The relative importance of biomineralization and dissolution of $CaCO_3$ in the global carbon cycle. In *Past and Present Biomineralization Processes*. Bulletin de l'Institut Oceanographie, *Monaco* No. 13 (ed. Francoise Doumenge), pp. 13–35.

Yaalon D. H. (1962) Mineral composition of average shale. *Clay Mineral Bull.* **5**, 31–36.

Zenger D. H. (1989) Dolomite abundance and stratigraphic age: constraints on rates and mechanisms of Phanerozoic dolostone formation. *J. Sedim. Petrol.* **59**, 162–164.

Zhu P. (1999) *Calcium Isotopes in the Marine Environment.* PhD Thesis, University of California, San Diego.

Zhu P. and MacDougall J. D. (1998) Calcium isotopes in the marine environment and the oceanic calcium cycle. *Geochim. Cosmochim. Acta* **62**, 1691–1698.

Volume Subject Index

The index is in letter-by-letter order, whereby hyphens and spaces within index headings are ignored in the alphabetization (e.g. Arabian–Nubian Shield precedes Arabian Sea). Terms in parentheses are excluded from the initial alphabetization. In line with normal materials science practice, compound names are not inverted but are filed under substituent prefixes.

The index is arranged in set-out style, with a maximum of three levels of heading. Location references refer to the page number. Major discussion of a subject is indicated by bold page numbers. Page numbers suffixed by *f* or *t* refer to figures or tables.